Table of Atomic Masses*

Element	Symbol	Atomic Number	Atomic Mass
Actinium	Ac	89	[227]§
Aluminum	Al	13	26.98
Americium	Am	95	[243]
Antimony	Sb	51	121.8
Argon	Ar	18	39.95
Arsenic	As	33	74.92
Astatine	At	85	[210]
Barium	Ba	56	137.3
Berkelium	Bk	97	[247]
Beryllium	Be	4	9.012
Bismuth	Bi	83	209.0
Bohrium	Bh	107	[264]
Boron	B	5	10.81
Bromine	Br	35	79.90
Cadmium	Cd	48	112.4
Calcium	Ca	20	40.08
Californium	Cf	98	[251]
Carbon	C	6	12.01
Cerium	Ce	58	140.1
Cesium	Cs	55	132.90
Chlorine	Cl	17	35.45
Chromium	Cr	24	52.00
Cobalt	Co	27	58.93
Copernicium	Cn	112	[285]
Copper	Cu	29	63.55
Curium	Cm	96	[247]
Darmstadtium	Ds	110	[271]
Dubnium	Db	105	[262]
Dysprosium	Dy	66	162.5
Einsteinium	Es	99	[252]
Erbium	Er	68	167.3
Europium	Eu	63	152.0
Fermium	Fm	100	[257]
Flerovium	Fl	114	[289]
Fluorine	F	9	19.00
Francium	Fr	87	[223]
Gadolinium	Gd	64	157.3
Gallium	Ga	31	69.72
Germanium	Ge	32	72.59
Gold	Au	79	197.0
Hafnium	Hf	72	178.5
Hassium	Hs	108	[265]
Helium	He	2	4.003
Holmium	Ho	67	164.9
Hydrogen	H	1	1.008
Indium	In	49	114.8
Iodine	I	53	126.9
Iridium	Ir	77	192.2
Iron	Fe	26	55.85
Krypton	Kr	36	83.80
Lanthanum	La	57	138.9
Lawrencium	Lr	103	[260]
Lead	Pb	82	207.2
Livermorium	Lv	116	[293]
Lithium	Li	3	6.9419
Lutetium	Lu	71	175.0
Magnesium	Mg	12	24.31
Manganese	Mn	25	54.94
Meitnerium	Mt	109	[268]
Mendelevium	Md	101	[258]
Mercury	Hg	80	200.6
Molybdenum	Mo	42	95.94
Neodymium	Nd	60	144.2
Neon	Ne	10	20.18
Neptunium	Np	93	[237]
Nickel	Ni	28	58.69
Niobium	Nb	41	92.91
Nitrogen	N	7	14.01
Nobelium	No	102	[259]
Osmium	Os	76	190.2
Oxygen	O	8	16.00
Palladium	Pd	46	106.4
Phosphorus	P	15	30.97
Platinum	Pt	78	195.1
Plutonium	Pu	94	[244]
Polonium	Po	84	[209]
Potassium	K	19	39.10
Praseodymium	Pr	59	140.9
Promethium	Pm	61	[145]
Protactinium	Pa	91	[231]
Radium	Ra	88	226
Radon	Rn	86	[222]
Rhenium	Re	75	186.2
Rhodium	Rh	45	102.9
Roentgenium	Rg	111	[272]
Rubidium	Rb	37	85.47
Ruthenium	Ru	44	101.1
Rutherfordium	Rf	104	[261]
Samarium	Sm	62	150.4
Scandium	Sc	21	44.96
Seaborgium	Sg	106	[263]
Selenium	Se	34	78.96
Silicon	Si	14	28.09
Silver	Ag	47	107.9
Sodium	Na	11	22.99
Strontium	Sr	38	87.62
Sulfur	S	16	32.07
Tantalum	Ta	73	180.9
Technetium	Tc	43	[98]
Tellurium	Te	52	127.6
Terbium	Tb	65	158.9
Thallium	Tl	81	204.4
Thorium	Th	90	232.0
Thulium	Tm	69	168.9
Tin	Sn	50	118.7
Titanium	Ti	22	47.88
Tungsten	W	74	183.9
Uranium	U	92	238.0
Vanadium	V	23	50.94
Xenon	Xe	54	131.3
Ytterbium	Yb	70	173.0
Yttrium	Y	39	88.91
Zinc	Zn	30	65.38
Zirconium	Zr	40	91.22

*The values given here are to four significant figures where possible. §A value given in parentheses denotes the mass of the longest-lived isotope.

Chemistry

Volume I

Ninth Edition

Steven S. Zumdahl | Susan A. Zumdahl

CENGAGE
Learning·

Australia • Brazil • Japan • Korea • Mexico • Singapore • Spain • United Kingdom • United States

CENGAGE Learning·

Chemistry: Volume I, Ninth Edition

Chemistry, 9th Edition
Steven S. Zumdahl | Susan A. Zumdahl

© 2014, 2010 Cengage Learning. All Rights Reserved.

Senior Manager, Student Engagement:

Linda deStefano

Janey Moeller

Manager, Student Engagement:

Julie Dierig

Marketing Manager:

Rachael Kloos

Manager, Production Editorial:

Kim Fry

Manager, Intellectual Property Project Manager:

Brian Methe

Senior Manager, Production and Manufacturing:

Donna M. Brown

Manager, Production:

Terri Daley

For product information and technology assistance, contact us at
Cengage Learning Customer & Sales Support, 1-800-354-9706

For permission to use material from this text or product,
submit all requests online at **cengage.com/permissions**
Further permissions questions can be emailed to
permissionrequest@cengage.com

This book contains select works from existing Cengage Learning resources and was produced by Cengage Learning Custom Solutions for collegiate use. As such, those adopting and/or contributing to this work are responsible for editorial content accuracy, continuity and completeness.

Compilation © 2014 Cengage Learning

ISBN-13: 978-1-305-31858-8
ISBN-10: 1-305-31858-7

WCN: 01-100-101

Cengage Learning
5191 Natorp Boulevard
Mason, Ohio 45040
USA

Cengage Learning is a leading provider of customized learning solutions with office locations around the globe, including Singapore, the United Kingdom, Australia, Mexico, Brazil, and Japan. Locate your local office at:
international.cengage.com/region.

Cengage Learning products are represented in Canada by Nelson Education, Ltd.
For your lifelong learning solutions, visit **www.cengage.com/custom.**
Visit our corporate website at **www.cengage.com.**

Printed in the United States of America

Brief Contents

Chapter 1

Chemical Foundations

A high-performance race car uses chemistry for its structure, tires, and fuel. (© Maria Green/Alamy)

When you start your car, do you think about chemistry? Probably not, but you should. The power to start your car is furnished by a lead storage battery. How does this battery work, and what does it contain? When a battery goes dead, what does that mean? If you use a friend's car to "jump-start" your car, did you know that your battery could explode? How can you avoid such an unpleasant possibility? What is in the gasoline that you put in your tank, and how does it furnish energy to your car so that you can drive it to school? What is the vapor that comes out of the exhaust pipe, and why does it cause air pollution? Your car's air conditioner might have a substance in it that is leading to the destruction of the ozone layer in the upper atmosphere. What are we doing about that? And why is the ozone layer important anyway?

All of these questions can be answered by understanding some chemistry. In fact, we'll consider the answers to all of these questions in this text.

Chemistry is around you all the time. You are able to read and understand this sentence because chemical reactions are occurring in your brain. The food you ate for breakfast or lunch is now furnishing energy through chemical reactions. Trees and grass grow because of chemical changes.

Chemistry also crops up in some unexpected places. When archaeologist Luis Alvarez was studying in college, he probably didn't realize that the chemical elements iridium and niobium would make him very famous when they helped him solve the problem of the disappearing dinosaurs. For decades scientists had wrestled with the mystery of why the dinosaurs, after ruling the earth for millions of years, suddenly became extinct 65 million years ago. In studying core samples of rocks dating back to that period, Alvarez and his coworkers recognized unusual levels of iridium and niobium in these samples—levels much more characteristic of extraterrestrial bodies than of the earth. Based on these observations, Alvarez hypothesized that a large meteor hit the earth 65 million years ago, changing atmospheric conditions so much that the dinosaurs' food couldn't grow, and they died—almost instantly in the geologic timeframe.

Chemistry is also important to historians. Did you realize that lead poisoning probably was a significant contributing factor to the decline of the Roman Empire? The Romans had high exposure to lead from lead-glazed pottery, lead water pipes, and a sweetening syrup called *sapa* that was prepared by boiling down grape juice in lead-lined vessels. It turns out that one reason for sapa's sweetness was lead acetate ("sugar of lead"), which formed as the juice was cooked down. Lead poisoning, with its symptoms of lethargy and mental malfunctions, certainly could have contributed to the demise of the Roman society.

Chemistry is also apparently very important in determining a person's behavior. Various studies have shown that many personality disorders can be linked directly to imbalances of trace elements in the body. For example, studies on the inmates at Stateville Prison in Illinois have linked low cobalt levels with violent behavior. Lithium salts have been shown to be very effective in controlling the effects of manic-depressive disease, and you've probably at some time in your life felt a special "chemistry" for another person. Studies suggest there is literally chemistry going on between two people who are attracted to each other. "Falling in love" apparently causes changes in the chemistry of the brain; chemicals are produced that give that "high" associated with a new relationship. Unfortunately, these chemical effects seem to wear off over time, even if the relationship persists and grows.

The importance of chemistry in the interactions of people should not really surprise us. We know that insects communicate by emitting and receiving chemical signals via molecules called *pheromones*. For example, ants have a very complicated set of chemical signals to signify food sources, danger, and so forth. Also, various female sex attractants have been isolated and used to lure males into traps to control insect populations. It would not be surprising if humans also emitted chemical signals that we were not aware of on a conscious level. Thus chemistry is pretty interesting and pretty important. The main goal of this text is to help you understand the concepts of chemistry so that you can better appreciate the world around you and can be more effective in whatever career you choose.

1.1 | Chemistry: An Overview

Since the time of the ancient Greeks, people have wondered about the answer to the question: What is matter made of? For a long time, humans have believed that matter is composed of atoms, and in the previous three centuries, we have collected much indirect evidence to support this belief. Very recently, something exciting has happened— for the first time we can "see" individual atoms. Of course, we cannot see atoms with the naked eye; we must use a special microscope called a *scanning tunneling microscope* (STM). Although we will not consider the details of its operation here, the STM uses an electron current from a tiny needle to probe the surface of a substance. The STM pictures of several substances are shown in Fig. 1.1. Notice how the atoms are connected to one another by "bridges," which, as we will see, represent the electrons that interconnect atoms.

So, at this point, we are fairly sure that matter consists of individual atoms. The nature of these atoms is quite complex, and the components of atoms don't behave much like the objects we see in the world of our experience. We call this world the *macroscopic world*—the world of cars, tables, baseballs, rocks, oceans, and so forth. One of the main jobs of a scientist is to delve into the macroscopic world and discover its "parts." For example, when you view a beach from a distance, it looks like a continuous solid substance. As you get closer, you see that the beach is really made up of individual grains of sand. As we examine these grains of sand, we find that they are composed of silicon and oxygen atoms connected to each other to form intricate shapes (Fig. 1.2). One of the main challenges of chemistry is to understand the connection between the macroscopic world that we experience and the *microscopic world* of atoms and molecules. To truly understand chemistry, you must learn to think on the atomic level. We will spend much time in this text helping you learn to do that.

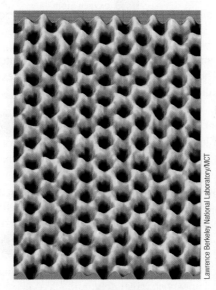

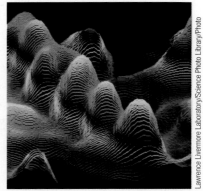

Figure 1.1 | Scanning tunneling microscope images.

An image showing the individual carbon atoms in a sheet of graphene.

Scanning tunneling microscope image of DNA.

Figure 1.2 | Sand on a beach looks uniform from a distance, but up close the irregular sand grains are visible, and each grain is composed of tiny atoms.

O
Si

Chuck Place. Inset: Jeremy Burgess/SPL/Photo Researchers, Inc.

Critical Thinking

The scanning tunneling microscope allows us to "see" atoms. What if you were sent back in time before the invention of the scanning tunneling microscope? What evidence could you give to support the theory that all matter is made of atoms and molecules?

One of the amazing things about our universe is that the tremendous variety of substances we find there results from only about 100 different kinds of atoms. You can think of these approximately 100 atoms as the letters in an alphabet from which all the "words" in the universe are made. It is the way the atoms are organized in a given substance that determines the properties of that substance. For example, water, one of the most common and important substances on the earth, is composed of two types of atoms: hydrogen and oxygen. Two hydrogen atoms and one oxygen atom are bound together to form the water molecule:

oxygen atom

hydrogen atom

water molecule

When an electric current passes through it, water is decomposed to hydrogen and oxygen. These *chemical elements* themselves exist naturally as diatomic (two-atom) molecules:

oxygen molecule   written O_2

hydrogen molecule   written H_2

We can represent the decomposition of water to its component elements, hydrogen and oxygen, as follows:

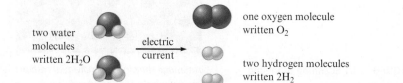

two water
molecules
written $2H_2O$

electric
current

one oxygen molecule
written O_2

two hydrogen molecules
written $2H_2$

Notice that it takes two molecules of water to furnish the right number of oxygen and hydrogen atoms to allow for the formation of the two-atom molecules. This reaction explains why the battery in your car can explode if you jump-start it improperly. When you hook up the jumper cables, current flows through the dead battery, which contains water (and other things), and causes hydrogen and oxygen to form by decomposition of some of the water. A spark can cause this accumulated hydrogen and oxygen to explode, forming water again.

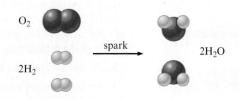

This example illustrates two of the fundamental concepts of chemistry: (1) Matter is composed of various types of atoms, and (2) one substance changes to another by reorganizing the way the atoms are attached to each other.

These are core ideas of chemistry, and we will have much more to say about them.

Science: A Process for Understanding Nature and Its Changes

How do you tackle the problems that confront you in real life? Think about your trip to school. If you live in a city, traffic is undoubtedly a problem you confront daily. How do you decide the best way to drive to school? If you are new in town, you first get a map and look at the possible ways to make the trip. Then you might collect information about the advantages and disadvantages of various routes from people who know the area. Based on this information, you probably try to predict the best route. However, you can find the best route only by trying several of them and comparing the results. After a few experiments with the various possibilities, you probably will be able to select the best way. What you are doing in solving this everyday problem is applying the same process that scientists use to study nature. The first thing you did was collect relevant data. Then you made a prediction, and then you tested it by trying it out. This process contains the fundamental elements of science.

1. Making observations (collecting data)
2. Suggesting a possible explanation (formulating a hypothesis)
3. Doing experiments to test the possible explanation (testing the hypothesis)

Scientists call this process the *scientific method*. We will discuss it in more detail in the next section. One of life's most important activities is solving problems—not "plug and chug" exercises, but real problems—problems that have new facets to them, that involve things you may have never confronted before. The more creative you are at solving these problems, the more effective you will be in your career and your personal life. Part of the reason for learning chemistry, therefore, is to become a better problem solver. Chemists are usually excellent problem solvers because to master chemistry, you have to master the scientific approach. Chemical problems are frequently very complicated—there is usually no neat and tidy solution. Often it is difficult to know where to begin.

1.2 | The Scientific Method

Science is a framework for gaining and organizing knowledge. Science is not simply a set of facts but also a plan of action—a *procedure* for processing and understanding certain types of information. Scientific thinking is useful in all aspects of life, but in this text we will use it to understand how the chemical world operates. As we said in our previous discussion, the process that lies at the center of scientific inquiry is called

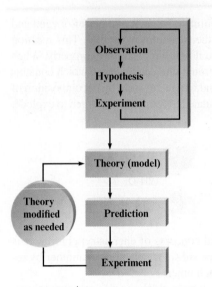

Figure 1.3 | The fundamental steps of the scientific method.

the **scientific method**. There are actually many scientific methods, depending on the nature of the specific problem under study and the particular investigator involved. However, it is useful to consider the following general framework for a generic scientific method (Fig. 1.3):

Steps in the Scientific Method

1. *Making observations.* Observations may be *qualitative* (the sky is blue; water is a liquid) or *quantitative* (water boils at 100°C; a certain chemistry book weighs 2 kg). A qualitative observation does not involve a number. A quantitative observation (called a **measurement**) involves both a number and a unit.

2. *Formulating hypotheses.* A **hypothesis** is a *possible* explanation for an observation.

3. *Performing experiments.* An experiment is carried out to test a hypothesis. This involves gathering new information that enables a scientist to decide whether the hypothesis is valid—that is, whether it is supported by the new information learned from the experiment. Experiments always produce new observations, and this brings the process back to the beginning again.

To understand a given phenomenon, these steps are repeated many times, gradually accumulating the knowledge necessary to provide a possible explanation of the phenomenon.

Scientific Models

Once a set of hypotheses that agrees with the various observations is obtained, the hypotheses are assembled into a theory. A **theory**, which is often called a **model**, is a set of tested hypotheses that gives an overall explanation of some natural phenomenon.

It is very important to distinguish between observations and theories. An observation is something that is witnessed and can be recorded. A theory is an *interpretation*— a possible explanation of why nature behaves in a particular way. Theories inevitably change as more information becomes available. For example, the motions of the sun and stars have remained virtually the same over the thousands of years during which humans have been observing them, but our explanations—our theories—for these motions have changed greatly since ancient times.

The point is that scientists do not stop asking questions just because a given theory seems to account satisfactorily for some aspect of natural behavior. They continue doing experiments to refine or replace the existing theories. This is generally done by using the currently accepted theory to make a prediction and then performing an experiment (making a new observation) to see whether the results bear out this prediction.

Always remember that theories (models) are human inventions. They represent attempts to explain observed natural behavior in terms of human experiences. A theory is actually an educated guess. We must continue to do experiments and to refine our theories (making them consistent with new knowledge) if we hope to approach a more complete understanding of nature.

As scientists observe nature, they often see that the same observation applies to many different systems. For example, studies of innumerable chemical changes have shown that the total observed mass of the materials involved is the same before and after the change. Such generally observed behavior is formulated into a statement called a **natural law**. For example, the observation that the total mass of materials is not affected by a chemical change in those materials is called the **law of conservation of mass**.

Note the difference between a natural law and a theory. A natural law is a summary of observed (measurable) behavior, whereas a theory is an explanation of behavior. A law summarizes what happens; a theory (model) is an attempt to explain why it happens.

In this section we have described the scientific method as it might ideally be applied (Fig. 1.4). However, it is important to remember that science does not always progress

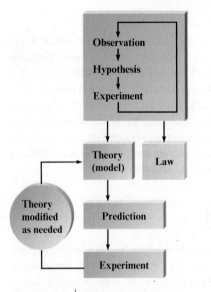

Figure 1.4 | The various parts of the scientific method.

Chemical connections
A Note-able Achievement

Post-it Notes, a product of the 3M Corporation, revolutionized casual written communications and personal reminders. Introduced in the United States in 1980, these sticky-but-not-too-sticky notes have now found countless uses in offices, cars, and homes throughout the world.

The invention of sticky notes occurred over a period of about 10 years and involved a great deal of serendipity. The adhesive for Post-it Notes was discovered by Dr. Spencer F. Silver of 3M in 1968. Silver found that when an acrylate polymer material was made in a particular way, it formed cross-linked microspheres. When suspended in a solvent and sprayed on a sheet of paper, this substance formed a "sparse mono-layer" of adhesive after the solvent evaporated. Scanning electron microscope images of the adhesive show that it has an irregular surface, a little like the surface of a gravel road. In contrast, the adhesive on cello-phane tape looks smooth and uniform, like a superhighway. The bumpy surface of Silver's adhesive caused it to be sticky but not so sticky to produce permanent adhesion, because the number of contact points between the binding surfaces was limited.

When he invented this adhesive, Silver had no specific ideas for its use, so he spread the word of his discovery to his fellow employees at 3M to see if anyone had an application for it. In addition, over the next several years development was carried out to improve the adhesive's properties. It was not until 1974 that the idea for

Post-it Notes popped up. One Sunday Art Fry, a chemical engineer for 3M, was singing in his church choir when he became annoyed that the book-mark in his hymnal kept falling out. He thought to himself that it would be nice if the bookmark were sticky enough to stay in place but not so sticky that it couldn't be moved. Luckily, he remembered Silver's glue—and the Post-it Note was born.

For the next three years, Fry worked to overcome the manufac-turing obstacles associated with the product. By 1977 enough Post-it Notes were being produced to supply 3M's corporate headquarters, where the employees quickly became addicted to their many uses. Post-it Notes are now available in 62 colors and 25 shapes.

In the years since the introduction of Post-it Notes, 3M has heard some

remarkable stories connected to the use of these notes. For example, a Post-it Note was applied to the nose of a corporate jet, where it was intended to be read by the plane's Las Vegas ground crew. Someone forgot to remove it, however. The note was still on the nose of the plane when it landed in Minneapolis, having survived a takeoff, a landing, and speeds of 500 miles per hour at temperatures as low as −56°F. Stories on the 3M Web site describe how a Post-it Note on the front door of a home survived the 140-mile-per-hour winds of Hurricane Hugo and how a foreign official accepted Post-it Notes in lieu of cash when a small bribe was needed to cut through bureaucratic hassles.

Post-it Notes have definitely changed the way we communicate and remember things.

Robert Boyle (1627–1691) was born in Ireland. He became especially interested in experiments involving air and developed an air pump with which he produced evacuated cylinders. He used these cylinders to show that a feather and a lump of lead fall at the same rate in the absence of air resistance and that sound cannot be produced in a vacuum. His most famous experiments involved careful measurements of the volume of a gas as a function of pressure. In his book Boyle urged that the ancient view of elements as mystical substances should be abandoned and that an element should instead be defined as anything that cannot be broken down into simpler substances. This concept was an important step in the development of modern chemistry.

smoothly and efficiently. For one thing, hypotheses and observations are not totally independent of each other, as we have assumed in the description of the idealized scientific method. The coupling of observations and hypotheses occurs because once we begin to proceed down a given theoretical path, our hypotheses are unavoidably couched in the language of that theory. In other words, we tend to see what we expect to see and often fail to notice things that we do not expect. Thus the theory we are testing helps us because it focuses our questions. However, at the same time, this focusing process may limit our ability to see other possible explanations.

It is also important to keep in mind that scientists are human. They have prejudices; they misinterpret data; they become emotionally attached to their theories and thus lose objectivity; and they play politics. Science is affected by profit motives, budgets, fads, wars, and religious beliefs. Galileo, for example, was forced to recant his astronomical observations in the face of strong religious resistance. Lavoisier, the father of modern chemistry, was beheaded because of his political affiliations. Great progress in the chemistry of nitrogen fertilizers resulted from the desire to produce explosives to fight wars. The progress of science is often affected more by the frailties of humans and their institutions than by the limitations of scientific measuring devices. The scientific methods are only as effective as the humans using them. They do not automatically lead to progress.

Critical Thinking

What if everyone in the government used the scientific method to analyze and solve society's problems, and politics were never involved in the solutions? How would this be different from the present situation, and would it be better or worse?

1.3 | Units of Measurement

Making observations is fundamental to all science. A quantitative observation, or *measurement,* always consists of two parts: a *number* and a scale (called a *unit*). Both parts must be present for the measurement to be meaningful.

In this textbook we will use measurements of mass, length, time, temperature, electric current, and the amount of a substance, among others. Scientists recognized long ago that standard systems of units had to be adopted if measurements were to be useful. If every scientist had a different set of units, complete chaos would result. Unfortunately, different standards were adopted in different parts of the world. The two major systems are the *English system* used in the United States and the *metric system* used by most of the rest of the industrialized world. This duality causes a good deal of trouble; for example, parts as simple as bolts are not interchangeable between machines built using the two systems. As a result, the United States has begun to adopt the metric system.

Most scientists in all countries have used the metric system for many years. In 1960, an international agreement set up a system of units called the *International System* (*le Système International* in French), or the **SI system**. This system is based on the metric system and units derived from the metric system. The fundamental SI units are listed in Table 1.1. We will discuss how to manipulate these units later in this chapter.

Chemical connections
Critical Units!

How important are conversions from one unit to another? If you ask the National Aeronautics and Space Administration (NASA), very important! In 1999, NASA lost a $125 million Mars Climate Orbiter because of a failure to convert from English to metric units.

The problem arose because two teams working on the Mars mission were using different sets of units. NASA's scientists at the Jet Propulsion Laboratory in Pasadena, California, assumed that the thrust data for the rockets on the Orbiter they received from Lockheed Martin Astronautics in Denver, which built the spacecraft, were in metric units. In reality, the units were English. As a result, the Orbiter dipped 100 km lower into the Mars atmosphere than planned, and the friction from the atmosphere caused the craft to burn up.

NASA's mistake refueled the controversy over whether Congress should require the United States to

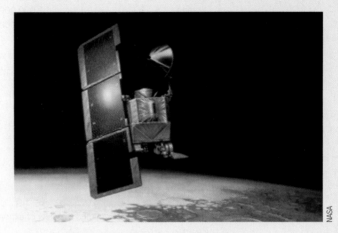
Artist's conception of the lost Mars Climate Orbiter.

switch to the metric system. About 95% of the world now uses the metric system, and the United States is slowly switching from English to metric. For example, the automobile industry has adopted metric fasteners, and we buy our soda in 2-L bottles.

Units can be very important. In fact, they can mean the difference

between life and death on some occasions. In 1983, for example, a Canadian jetliner almost ran out of fuel when someone pumped 22,300 lb of fuel into the aircraft instead of 22,300 kg. Remember to watch your units!

Because the fundamental units are not always convenient (expressing the mass of a pin in kilograms is awkward), prefixes are used to change the size of the unit. These are listed in Table 1.2. Some common objects and their measurements in SI units are listed in Table 1.3.

One physical quantity that is very important in chemistry is *volume*, which is not a fundamental SI unit but is derived from length. A cube that measures 1 meter (m) on

Soda is commonly sold in 2-L bottles—an example of the use of SI units in everyday life.

Table 1.1 | Fundamental SI Units

Physical Quantity	Name of Unit	Abbreviation
Mass	kilogram	kg
Length	meter	m
Time	second	s
Temperature	kelvin	K
Electric current	ampere	A
Amount of substance	mole	mol
Luminous intensity	candela	cd

Table 1.2 | Prefixes Used in the SI System (The most commonly encountered are shown in blue.)

Prefix	Symbol	Meaning	Exponential Notation*
exa	E	1,000,000,000,000,000,000	10^{18}
peta	P	1,000,000,000,000,000	10^{15}
tera	T	1,000,000,000,000	10^{12}
giga	G	1,000,000,000	10^{9}
mega	M	1,000,000	10^{6}
kilo	k	1,000	10^{3}
hecto	h	100	10^{2}
deka	da	10	10^{1}
—	—	1	10^{0}
deci	d	0.1	10^{-1}
centi	c	0.01	10^{-2}
milli	m	0.001	10^{-3}
micro	μ	0.000001	10^{-6}
nano	n	0.000000001	10^{-9}
pico	p	0.000000000001	10^{-12}
femto	f	0.000000000000001	10^{-15}
atto	a	0.000000000000000001	10^{-18}

*See Appendix 1.1 if you need a review of exponential notation.

each edge is represented in Fig. 1.5. This cube has a volume of $(1 \text{ m})^3 = 1 \text{ m}^3$. Recognizing that there are 10 decimeters (dm) in a meter, the volume of this cube is $(1 \text{ m})^3 = (10 \text{ dm})^3 = 1000 \text{ dm}^3$. A cubic decimeter, that is, $(1 \text{ dm})^3$, is commonly called a *liter (L),* which is a unit of volume slightly larger than a quart. As shown in Fig. 1.5, 1000 L is contained in a cube with a volume of 1 cubic meter. Similarly, since 1 decimeter equals 10 centimeters (cm), the liter can be divided into 1000 cubes, each with a volume of 1 cubic centimeter:

$$1 \text{ L} = (1 \text{ dm})^3 = (10 \text{ cm})^3 = 1000 \text{ cm}^3$$

Table 1.3 | Some Examples of Commonly Used Units

Length	A dime is 1 mm thick. A quarter is 2.5 cm in diameter. The average height of an adult man is 1.8 m.
Mass	A nickel has a mass of about 5 g. A 120-lb person has a mass of about 55 kg.
Volume	A 12-oz can of soda has a volume of about

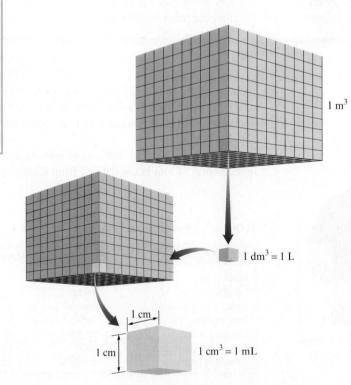

Figure 1.5 | The largest cube has sides 1 m in length and a volume of 1 m³. The middle-sized cube has sides 1 dm in length and a volume of 1 dm³, or 1 L. The smallest cube has sides 1 cm in length and a volume of 1 cm³, or 1 mL.

Figure 1.6 | Common types of laboratory equipment used to measure liquid volume.

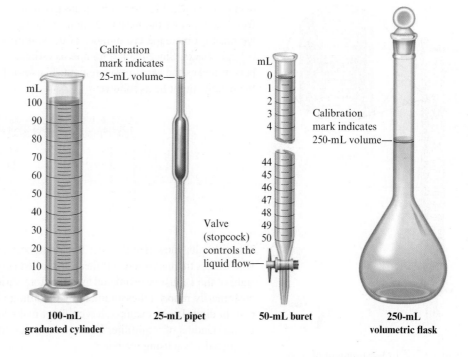

Calibration mark indicates 25-mL volume—

Calibration mark indicates 250-mL volume—

Valve (stopcock) controls the liquid flow—

100-mL graduated cylinder   **25-mL pipet**   **50-mL buret**   **250-mL volumetric flask**

Also, since 1 cm³ = 1 milliliter (mL),

$$1 \text{ L} = 1000 \text{ cm}^3 = 1000 \text{ mL}$$

Thus 1 liter contains 1000 cubic centimeters, or 1000 milliliters.

Chemical laboratory work frequently requires measurement of the volumes of liquids. Several devices for the accurate determination of liquid volume are shown in Fig. 1.6.

An important point concerning measurements is the relationship between mass and weight. Although these terms are sometimes used interchangeably, they are *not* the same. ***Mass** is a measure of the resistance of an object to a change in its state of motion.* Mass is measured by the force necessary to give an object a certain acceleration. On the earth we use the force that gravity exerts on an object to measure its mass. We call this force the object's **weight**. Since weight is the response of mass to gravity, it varies with the strength of the gravitational field. Therefore, your body mass is the same on the earth and on the moon, but your weight would be much less on the moon than on the earth because of the moon's smaller gravitational field.

Because weighing something on a chemical balance involves comparing the mass of that object to a standard mass, the terms *weight* and *mass* are sometimes used interchangeably, although this is incorrect.

Critical Thinking

What if you were not allowed to use units for one day? How would this affect your life for that day?

1.4 | Uncertainty in Measurement

The number associated with a measurement is obtained using some measuring device. For example, consider the measurement of the volume of a liquid using a buret (shown in Fig. 1.7 with the scale greatly magnified). Notice that the meniscus of the liquid

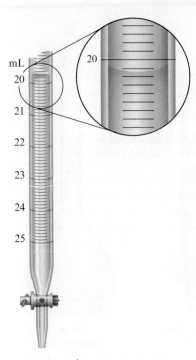

Figure 1.7 | Measurement of volume using a buret. The volume is read at the bottom of the liquid curve (called the meniscus).

A measurement always has some degree of uncertainty.

Uncertainty in measurement is discussed in more detail in Appendix 1.5.

occurs at about 20.15 mL. This means that about 20.15 mL of liquid has been delivered from the buret (if the initial position of the liquid meniscus was 0.00 mL). Note that we must estimate the last number of the volume reading by interpolating between the 0.1-mL marks. Since the last number is estimated, its value may be different if another person makes the same measurement. If five different people read the same volume, the results might be as follows:

Person	Results of Measurement
1	20.15 mL
2	20.14 mL
3	20.16 mL
4	20.17 mL
5	20.16 mL

These results show that the first three numbers (20.1) remain the same regardless of who makes the measurement; these are called *certain* digits. However, the digit to the right of the 1 must be estimated and therefore varies; it is called an *uncertain* digit. We customarily report a measurement by recording all the certain digits plus the *first* uncertain digit. In our example it would not make any sense to try to record the volume to thousandths of a milliliter because the value for hundredths of a milliliter must be estimated when using the buret.

It is very important to realize that a *measurement always has some degree of* **uncertainty**. The uncertainty of a measurement depends on the precision of the measuring device. For example, using a bathroom scale, you might estimate the mass of a grapefruit to be approximately 1.5 lb. Weighing the same grapefruit on a highly precise balance might produce a result of 1.476 lb. In the first case, the uncertainty occurs in the tenths of a pound place; in the second case, the uncertainty occurs in the thousandths of a pound place. Suppose we weigh two similar grapefruits on the two devices and obtain the following results:

	Bathroom Scale	Balance
Grapefruit 1	1.5 lb	1.476 lb
Grapefruit 2	1.5 lb	1.518 lb

Do the two grapefruits have the same mass? The answer depends on which set of results you consider. Thus a conclusion based on a series of measurements depends on the certainty of those measurements. For this reason, it is important to indicate the uncertainty in any measurement. This is done by always recording the certain digits and the first uncertain digit (the estimated number). These numbers are called the **significant figures** of a measurement.

The convention of significant figures automatically indicates something about the uncertainty in a measurement. The uncertainty in the last number (the estimated number) is usually assumed to be ±1 unless otherwise indicated. For example, the measurement 1.86 kg can be taken to mean 1.86 ± 0.01 kg.

Example 1.1

Uncertainty in Measurement

In analyzing a sample of polluted water, a chemist measured out a 25.00-mL water sample with a pipet (see Fig. 1.6). At another point in the analysis, the chemist used a graduated cylinder (see Fig. 1.6) to measure 25 mL of a solution. What is the difference between the measurements 25.00 mL and 25 mL?

Solution

Even though the two volume measurements appear to be equal, they really convey different information. The quantity 25 mL means that the volume is between 24 mL and 26 mL, whereas the quantity 25.00 mL means that the volume is between 24.99 mL and 25.01 mL. The pipet measures volume with much greater precision than does the graduated cylinder.

See Exercise 1.33

When making a measurement, it is important to record the results to the appropriate number of significant figures. For example, if a certain buret can be read to ±0.01 mL, you should record a reading of twenty-five milliliters as 25.00 mL, not 25 mL. This way at some later time when you are using your results to do calculations, the uncertainty in the measurement will be known to you.

Precision and Accuracy

Two terms often used to describe the reliability of measurements are *precision* and *accuracy*. Although these words are frequently used interchangeably in everyday life, they have different meanings in the scientific context. **Accuracy** refers to the agreement of a particular value with the true value. **Precision** refers to the degree of agreement among several measurements of the same quantity. Precision reflects the *reproducibility* of a given type of measurement. The difference between these terms is illustrated by the results of three different dart throws shown in Fig. 1.8.

Two different types of errors are illustrated in Fig. 1.8. A **random error** (also called an *indeterminate error*) means that a measurement has an equal probability of being high or low. This type of error occurs in estimating the value of the last digit of a measurement. The second type of error is called **systematic error** (or *determinate error*). This type of error occurs in the same direction each time; it is either always high or always low. Fig. 1.8(a) indicates large random errors (poor technique). Fig. 1.8(b) indicates small random errors but a large systematic error, and Fig. 1.8(c) indicates small random errors and no systematic error.

In quantitative work, precision is often used as an indication of accuracy; we assume that the *average* of a series of precise measurements (which should "average out" the random errors because of their equal probability of being high or low) is accurate, or close to the "true" value. However, this assumption is valid only if

a

Neither accurate nor precise.

c

Accurate and precise.

b

Precise but not accurate.

Figure 1.8 | The results of several dart throws show the difference between precise and accurate.

systematic errors are absent. Suppose we weigh a piece of brass five times on a very precise balance and obtain the following results:

Weighing	Result
1	2.486 g
2	2.487 g
3	2.485 g
4	2.484 g
5	2.488 g

Normally, we would assume that the true mass of the piece of brass is very close to 2.486 g, which is the average of the five results:

$$\frac{2.486 \text{ g} + 2.487 \text{ g} + 2.485 \text{ g} + 2.484 \text{ g} + 2.488 \text{ g}}{5} = 2.486 \text{ g}$$

However, if the balance has a defect causing it to give a result that is consistently 1.000 g too high (a systematic error of +1.000 g), then the measured value of 2.486 g would be seriously in error. The point here is that high precision among several measurements is an indication of accuracy *only* if systematic errors are absent.

Example 1.2

Precision and Accuracy

To check the accuracy of a graduated cylinder, a student filled the cylinder to the 25-mL mark using water delivered from a buret (see Fig. 1.6) and then read the volume delivered. Following are the results of five trials:

Trial	Volume Shown by Graduated Cylinder	Volume Shown by the Buret
1	25 mL	26.54 mL
2	25 mL	26.51 mL
3	25 mL	26.60 mL
4	25 mL	26.49 mL
5	25 mL	26.57 mL
Average	25 mL	26.54 mL

Is the graduated cylinder accurate?

Solution

Precision is an indication of accuracy only if there are no systematic errors.

The results of the trials show very good precision (for a graduated cylinder). The student has good technique. However, note that the average value measured using the buret is significantly different from 25 mL. Thus this graduated cylinder is not very accurate. It produces a systematic error (in this case, the indicated result is low for each measurement).

See Question 1.11

1.5 | Significant Figures and Calculations

Calculating the final result for an experiment usually involves adding, subtracting, multiplying, or dividing the results of various types of measurements. Since it is very important that the uncertainty in the final result is known correctly, we have developed rules for counting the significant figures in each number and for determining the correct number of significant figures in the final result.

Leading zeros are never significant figures.

Captive zeros are always significant figures.

Trailing zeros are sometimes significant figures.

Exact numbers never limit the number of significant figures in a calculation.

Rules for Counting Significant Figures

1. *Nonzero integers.* Nonzero integers always count as significant figures.
2. *Zeros.* There are three classes of zeros:
   a. *Leading zeros* are zeros that *precede* all the nonzero digits. These do not count as significant figures. In the number 0.0025, the three zeros simply indicate the position of the decimal point. This number has only two significant figures.
   b. *Captive zeros* are zeros *between* nonzero digits. These always count as significant figures. The number 1.008 has four significant figures.
   c. *Trailing zeros* are zeros at the *right end* of the number. They are significant only if the number contains a decimal point. The number 100 has only one significant figure, whereas the number 1.00×10^2 has three significant figures. The number one hundred written as 100. also has three significant figures.
3. *Exact numbers.* Many times calculations involve numbers that were not obtained using measuring devices but were determined by counting: 10 experiments, 3 apples, 8 molecules. Such numbers are called *exact numbers*. They can be assumed to have an infinite number of significant figures. Other examples of exact numbers are the 2 in $2\pi r$ (the circumference of a circle) and the 4 and the 3 in $\frac{4}{3}\pi r^3$ (the volume of a sphere). Exact numbers also can arise from definitions. For example, 1 inch is defined as *exactly* 2.54 centimeters. Thus, in the statement 1 in = 2.54 cm, neither the 2.54 nor the 1 limits the number of significant figures when used in a calculation.

Exponential notation is reviewed in Appendix 1.1.

Note that the number 1.00×10^2 above is written in **exponential notation**. This type of notation has at least two advantages: The number of significant figures can be easily indicated, and fewer zeros are needed to write a very large or very small number. For example, the number 0.000060 is much more conveniently represented as 6.0×10^{-5} (the number has two significant figures).

Interactive Example 1.3

Sign in at http://login.cengagebrain.com to try this Interactive Example in **OWL**.

Significant Figures

Give the number of significant figures for each of the following results.

a. A student's extraction procedure on tea yields 0.0105 g of caffeine.

b. A chemist records a mass of 0.050080 g in an analysis.

c. In an experiment a span of time is determined to be 8.050×10^{-3} s.

Solution

a. The number contains three significant figures. The zeros to the left of the 1 are leading zeros and are not significant, but the remaining zero (a captive zero) is significant.

b. The number contains five significant figures. The leading zeros (to the left of the 5) are not significant. The captive zeros between the 5 and the 8 are significant, and the trailing zero to the right of the 8 is significant because the number contains a decimal point.

c. This number has four significant figures. Both zeros are significant.

See Exercises 1.27 through 1.30

To this point we have learned to count the significant figures in a given number. Next, we must consider how uncertainty accumulates as calculations are carried out. The detailed analysis of the accumulation of uncertainties depends on the type of calculation involved and can be complex. However, in this textbook we will use the following simple rules that have been developed for determining the appropriate number of significant figures in the result of a calculation.

Rules for Significant Figures in Mathematical Operations

1. *For multiplication or division,* the number of significant figures in the result is the same as the number in the least precise measurement used in the calculation. For example, consider the calculation

$$4.56 \times 1.4 = 6.38 \xrightarrow{\text{Corrected}} 6.4$$

          ↑   ↑
    Limiting term has   Two significant
    two significant   figures
    figures

   The product should have only two significant figures, since 1.4 has two significant figures.

2. *For addition or subtraction,* the result has the same number of decimal places as the least precise measurement used in the calculation. For example, consider the sum

$$
\begin{array}{r}
12.11 \\
18.0 \quad \leftarrow \text{Limiting term has one decimal place} \\
\underline{1.013} \\
31.123 \xrightarrow{\text{Corrected}} 31.1 \\
\end{array}
$$

                          ↑
                 One decimal place

   The correct result is 31.1, since 18.0 has only one decimal place.

Note that for multiplication and division, significant figures are counted. For addition and subtraction, the decimal places are counted.

In most calculations you will need to round numbers to obtain the correct number of significant figures. The following rules should be applied when rounding.

Rules for Rounding

1. In a series of calculations, carry the extra digits through to the final result, *then* round.

2. If the digit to be removed

   a. is less than 5, the preceding digit stays the same. For example, 1.33 rounds to 1.3.

   b. is equal to or greater than 5, the preceding digit is increased by 1. For example, 1.36 rounds to 1.4.

Rule 2 is consistent with the operation of electronic calculators.

Although rounding is generally straightforward, one point requires special emphasis. As an illustration, suppose that the number 4.348 needs to be rounded to two significant figures. In doing this, we look *only* at the *first number* to the right of the 3:

4.348
   ↑
Look at this number to
round to two significant figures.

Do not round sequentially. The number 6.8347 rounded to three significant figures is 6.83, not 6.84.

The number is rounded to 4.3 because 4 is less than 5. It is incorrect to round sequentially. For example, do *not* round the 4 to 5 to give 4.35 and then round the 3 to 4 to give 4.4.

When rounding, *use only the first number to the right of the last significant figure.*

It is important to note that Rule 1 above usually will not be followed in the examples in this text because we want to show the correct number of significant figures in each step of a problem. This same practice is followed for the detailed solutions given in the *Solutions Guide*. However, when you are doing problems, you should carry extra digits throughout a series of calculations and round to the correct number of significant figures only at the end. This is the practice you should follow. The fact that your rounding procedures are different from those used in this text must be taken into account when you check your answer with the one given at the end of the book or in the *Solutions Guide*. Your answer (based on rounding only at the end of a calculation) may differ in the last place from that given here as the "correct" answer because we have rounded after each step. To help you understand the difference between these rounding procedures, we will consider them further in Example 1.4.

Significant Figures in Mathematical Operations

Carry out the following mathematical operations, and give each result with the correct number of significant figures.

a. $1.05 \times 10^{-3} \div 6.135$

b. $21 - 13.8$

c. As part of a lab assignment to determine the value of the gas constant (R), a student measured the pressure (P), volume (V), and temperature (T) for a sample of gas, where

$$R = \frac{PV}{T}$$

The following values were obtained: $P = 2.560$, $T = 275.15$, and $V = 8.8$. (Gases will be discussed in detail in Chapter 5; we will not be concerned at this time about the units for these quantities.) Calculate R to the correct number of significant figures.

Solution

a. The result is 1.71×10^{-4}, which has three significant figures because the term with the least precision (1.05×10^{-3}) has three significant figures.

b. The result is 7 with no decimal point because the number with the least number of decimal places (21) has none.

c. $R = \dfrac{PV}{T} = \dfrac{(2.560)(8.8)}{275.15}$

The correct procedure for obtaining the final result can be represented as follows:

$$\frac{(2.560)(8.8)}{275.15} = \frac{22.528}{275.15} = 0.0818753$$

$$= 0.082 = 8.2 \times 10^{-2} = R$$

The final result must be rounded to two significant figures because 8.8 (the least precise measurement) has two significant figures. To show the effects of rounding at intermediate steps, we will carry out the calculation as follows:

Rounded to two
significant figures
↓

$$\frac{(2.560)(8.8)}{275.15} = \frac{22.528}{275.15} = \frac{23}{275.15}$$

Now we proceed with the next calculation:

$$\frac{23}{275.15} = 0.0835908$$

Rounded to two significant figures, this result is

$$0.084 = 8.4 \times 10^{-2}$$

This number must be rounded to two significant figures.

Note that intermediate rounding gives a significantly different result than that obtained by rounding *only at the end*. Again, we must reemphasize that in *your* calculations you should round only at the end. However, because rounding is carried out at intermediate steps in this text (to always show the correct number of significant figures), the final answer given in the text may differ slightly from the one you obtain (rounding only at the end).

See Exercises 1.35 through 1.38

There is a useful lesson to be learned from Part c of Example 1.4. The student measured the pressure and temperature to greater precision than the volume. A more precise value of R (one with more significant figures) could have been obtained if a more precise measurement of V had been made. As it is, the efforts expended to measure P and T very precisely were wasted. Remember that a series of measurements to obtain some final result should all be done to about the same precision.

1.6 | Learning to Solve Problems Systematically

One of the main activities in learning chemistry is solving various types of problems. The best way to approach a problem, whether it is a chemistry problem or one from your daily life, is to ask questions such as the following:

1. What is my goal? Or you might phrase it as: Where am I going?
2. Where am I starting? Or you might phrase it as: What do I know?
3. How do I proceed from where I start to where I want to go? Or you might say: How do I get there?

We will use these ideas as we consider unit conversions in this chapter. Then we will have much more to say about problem solving in Chapter 3, where we will start to consider more complex problems.

1.7 | Dimensional Analysis

It is often necessary to convert a given result from one system of units to another. The best way to do this is by a method called the **unit factor method** or, more commonly, **dimensional analysis**. To illustrate the use of this method, we will consider several unit conversions. Some equivalents in the English and metric systems are listed in Table 1.4. A more complete list of conversion factors given to more significant figures appears in Appendix 6.

Consider a pin measuring 2.85 cm in length. What is its length in inches? To accomplish this conversion, we must use the equivalence statement

$$2.54 \text{ cm} = 1 \text{ in}$$

If we divide both sides of this equation by 2.54 cm, we get

$$1 = \frac{1 \text{ in}}{2.54 \text{ cm}}$$

This expression is called a *unit factor.* Since 1 inch and 2.54 cm are exactly equivalent, multiplying any expression by this unit factor will not change its *value.*

Table 1.4 | English–Metric Equivalents

Length	1 m = 1.094 yd
	2.54 cm = 1 in
Mass	1 kg = 2.205 lb
	453.6 g = 1 lb
Volume	1 L = 1.06 qt
	1 ft³ = 28.32 L

The pin has a length of 2.85 cm. Multiplying this length by the appropriate unit factor gives

$$2.85 \text{ cm} \times \frac{1 \text{ in}}{2.54 \text{ cm}} = \frac{2.85}{2.54} \text{ in} = 1.12 \text{ in}$$

Note that the centimeter units cancel to give inches for the result. This is exactly what we wanted to accomplish. Note also that the result has three significant figures, as required by the number 2.85. Recall that the 1 and 2.54 in the conversion factor are exact numbers by definition.

Unit Conversions I

A pencil is 7.00 in long. What is its length in centimeters?

Solution

Where are we going?

To convert the length of the pencil from inches to centimeters

What do we know?

> The pencil is 7.00 in long.

How do we get there?

Since we want to convert from inches to centimeters, we need the equivalence statement 2.54 cm = 1 in. The correct unit factor in this case is $\dfrac{2.54 \text{ cm}}{1 \text{ in}}$:

$$7.00 \text{ in} \times \frac{2.54 \text{ cm}}{1 \text{ in}} = (7.00)(2.54) \text{ cm} = 17.8 \text{ cm}$$

Here the inch units cancel, leaving centimeters, as requested.

See Exercises 1.41 and 1.42

Note that two unit factors can be derived from each equivalence statement. For example, from the equivalence statement 2.54 cm = 1 in, the two unit factors are

$$\frac{2.54 \text{ cm}}{1 \text{ in}} \quad \text{and} \quad \frac{1 \text{ in}}{2.54 \text{ cm}}$$

Consider the direction of the required change to select the correct unit factor.

How do you choose which one to use in a given situation? Simply look at the *direction* of the required change. To change from inches to centimeters, the inches must cancel. Thus the factor 2.54 cm/1 in is used. To change from centimeters to inches, centimeters must cancel, and the factor 1 in/2.54 cm is appropriate.

Problem-Solving Strategy

Converting from One Unit to Another
> To convert from one unit to another, use the equivalence statement that relates the two units.
> Derive the appropriate unit factor by looking at the direction of the required change (to cancel the unwanted units).
> Multiply the quantity to be converted by the unit factor to give the quantity with the desired units.

Unit Conversions II

You want to order a bicycle with a 25.5-in frame, but the sizes in the catalog are given only in centimeters. What size should you order?

Solution

Where are we going?
To convert from inches to centimeters

What do we know?
> The size needed is 25.5 in.

How do we get there?
Since we want to convert from inches to centimeters, we need the equivalence statement 2.54 cm = 1 in. The correct unit factor in this case is $\dfrac{2.54 \text{ cm}}{1 \text{ in}}$:

$$25.5 \text{ in} \times \frac{2.54 \text{ cm}}{1 \text{ in}} = 64.8 \text{ cm}$$

See Exercises 1.41 and 1.42

To ensure that the conversion procedure is clear, a multistep problem is considered in Example 1.7.

Unit Conversions III

A student has entered a 10.0-km run. How long is the run in miles?

Solution

Where are we going?
To convert from kilometers to miles

What do we know?
> The run is 10.00 km long.

How do we get there?
This conversion can be accomplished in several different ways. Since we have the equivalence statement 1 m = 1.094 yd, we will proceed by a path that uses this fact. Before we start any calculations, let us consider our strategy. We have kilometers, which we want to change to miles. We can do this by the following route:

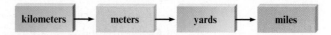

To proceed in this way, we need the following equivalence statements:

$$1 \text{ km} = 1000 \text{ m}$$
$$1 \text{ m} = 1.094 \text{ yd}$$
$$1760 \text{ yd} = 1 \text{ mi}$$

To make sure the process is clear, we will proceed step by step:

Kilometers to Meters

$$10.0 \text{ km} \times \frac{1000 \text{ m}}{1 \text{ km}} = 1.00 \times 10^4 \text{ m}$$

Meters to Yards

$$1.00 \times 10^4 \, \cancel{m} \times \frac{1.094 \, yd}{1 \, \cancel{m}} = 1.094 \times 10^4 \, yd$$

Note that we should have only three significant figures in the result. However, since this is an intermediate result, we will carry the extra digit. Remember, round off only the final result.

Yards to Miles

$$1.094 \times 10^4 \, \cancel{yd} \times \frac{1 \, mi}{1760 \, \cancel{yd}} = 6.216 \, mi$$

Note in this case that 1 mi equals exactly 1760 yd *by designation*. Thus 1760 is an exact number.

Since the distance was originally given as 10.0 km, the result can have only three significant figures and should be rounded to 6.22 mi. Thus

$$10.0 \, km = 6.22 \, mi$$

Alternatively, we can combine the steps:

$$10.0 \, \cancel{km} \times \frac{1000 \, \cancel{m}}{1 \, \cancel{km}} \times \frac{1.094 \, \cancel{yd}}{1 \, \cancel{m}} \times \frac{1 \, mi}{1760 \, \cancel{yd}} = 6.22 \, mi$$

See Exercises 1.41 and 1.42

In the text we round to the correct number of significant figures after each step to show the correct significant figures for each calculation. However, since you use a calculator and combine steps on it, you should round only at the end.

In using dimensional analysis, your verification that everything has been done correctly is that you end up with the correct units. In doing chemistry problems, you should always include the units for the quantities used. Always check to see that the units cancel to give the correct units for the final result. This provides a very valuable check, especially for complicated problems.

Study the procedures for unit conversions in the following examples.

Interactive Example 1.8

Sign in at http://login.cengagebrain.com to try this Interactive Example in OWL.

Unit Conversions IV

The speed limit on many highways in the United States is 55 mi/h. What number would be posted in kilometers per hour?

Solution

Where are we going?

To convert the speed limit from 55 miles per hour to kilometers per hour

What do we know?

> The speed limit is 55 mi/h.

How do we get there?

We use the following unit factors to make the required conversion:

Result obtained by rounding only at the end of the calculation

$$\frac{55 \, \cancel{mi}}{h} \times \frac{1760 \, \cancel{yd}}{1 \, \cancel{mi}} \times \frac{1 \, \cancel{m}}{1.094 \, \cancel{yd}} \times \frac{1 \, km}{1000 \, \cancel{m}} = 88 \, km/h$$

Note that all units cancel except the desired kilometers per hour.

See Exercises 1.49 through 1.51

Unit Conversions V

A Japanese car is advertised as having a gas mileage of 15 km/L. Convert this rating to miles per gallon.

Solution

Where are we going?

To convert gas mileage from 15 kilometers per liter to miles per gallon

What do we know?

> The gas mileage is 15 km/L.

How do we get there?

We use the following unit factors to make the required conversion:

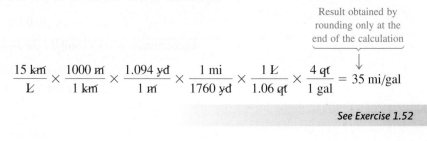

Result obtained by rounding only at the end of the calculation

$$\frac{15 \text{ km}}{\text{L}} \times \frac{1000 \text{ m}}{1 \text{ km}} \times \frac{1.094 \text{ yd}}{1 \text{ m}} \times \frac{1 \text{ mi}}{1760 \text{ yd}} \times \frac{1 \text{ L}}{1.06 \text{ qt}} \times \frac{4 \text{ qt}}{1 \text{ gal}} = 35 \text{ mi/gal}$$

See Exercise 1.52

Unit Conversions VI

The latest model Corvette has an engine with a displacement of 6.20 L. What is the displacement in units of cubic inches?

Solution

Where are we going?

To convert the engine displacement from liters to cubic inches

What do we know?

> The displacement is 6.20 L.

How do we get there?

We use the following unit factors to make the required conversion:

$$6.20 \text{ L} \times \frac{1 \text{ ft}^3}{28.32 \text{ L}} \times \frac{(12 \text{ in})^3}{(1 \text{ ft})^3} = 378 \text{ in}^3$$

Note that the unit factor for conversion of feet to inches must be cubed to accommodate the conversion of ft^3 to in^3.

See Exercise 1.56

1.8 | Temperature

Three systems for measuring temperature are widely used: the Celsius scale, the Kelvin scale, and the Fahrenheit scale. The first two temperature systems are used in the physical sciences, and the third is used in many of the engineering sciences. Our purpose here is to define the three temperature scales and show how conversions from one scale to another can be performed. Although these conversions can be carried out routinely on most calculators, we will consider the process in some detail here to illustrate methods of problem solving.

The three temperature scales are defined and compared in Fig. 1.9. Note that the size of the temperature unit (the *degree*) is the same for the Kelvin and Celsius scales. The fundamental difference between these two temperature scales is their zero points. Conversion between these two scales simply requires an adjustment for the different zero points.

$T_K = T_C + 273.15$

$$\text{Temperature (Kelvin)} = \text{temperature (Celsius)} + 273.15$$

or

$T_C = T_K - 273.15$

$$\text{Temperature (Celsius)} = \text{temperature (Kelvin)} - 273.15$$

For example, to convert 300.00 K to the Celsius scale, we do the following calculation:

$$300.00 - 273.15 = 26.85°C$$

Note that in expressing temperature in Celsius units, the designation °C is used. The degree symbol is not used when writing temperature in terms of the Kelvin scale. The unit of temperature on this scale is called a *kelvin* and is symbolized by the letter K.

Converting between the Fahrenheit and Celsius scales is somewhat more complicated because both the degree sizes and the zero points are different. Thus we need to consider two adjustments: one for degree size and one for the zero point. First, we must account for the difference in degree size. This can be done by reconsidering Fig. 1.9. Notice that since 212°F = 100°C and 32°F = 0°C,

$$212 - 32 = 180 \text{ Fahrenheit degrees} = 100 - 0 = 100 \text{ Celsius degrees}$$

Thus 180° on the Fahrenheit scale is equivalent to 100° on the Celsius scale, and the unit factor is

$$\frac{180°F}{100°C} \quad \text{or} \quad \frac{9°F}{5°C}$$

or the reciprocal, depending on the direction in which we need to go.

Next, we must consider the different zero points. Since 32°F = 0°C, we obtain the corresponding Celsius temperature by first subtracting 32 from the Fahrenheit

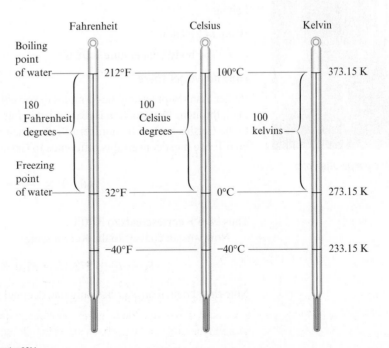

Figure 1.9 | The three major temperature scales.

temperature to account for the different zero points. Then the unit factor is applied to adjust for the difference in the degree size. This process is summarized by the equation

$$(T_F - 32°F)\frac{5°C}{9°F} = T_C \tag{1.1}$$

where T_F and T_C represent a given temperature on the Fahrenheit and Celsius scales, respectively. In the opposite conversion, we first correct for degree size and then correct for the different zero point. This process can be summarized in the following general equation:

$$T_F = T_C \times \frac{9°F}{5°C} + 32°F \tag{1.2}$$

Equations (1.1) and (1.2) are really the same equation in different forms. See if you can obtain Equation (1.2) by starting with Equation (1.1) and rearranging.

At this point it is worthwhile to weigh the two alternatives for learning to do temperature conversions: You can simply memorize the equations, or you can take the time to learn the differences between the temperature scales and to understand the processes involved in converting from one scale to another. The latter approach may take a little more effort, but the understanding you gain will stick with you much longer than the memorized formulas. This choice also will apply to many of the other chemical concepts. Try to think things through!

Understand the process of converting from one temperature scale to another; do not simply memorize the equations.

Interactive Example 1.11

Sign in at http://login.cengagebrain.com to try this Interactive Example in **OWL**.

A nurse taking the temperature of a patient.

Thinkstock/Getty Images

Temperature Conversions I

Normal body temperature is 98.6°F. Convert this temperature to the Celsius and Kelvin scales.

Solution

Where are we going?

To convert the body temperature from degrees Fahrenheit to degrees Celsius and to kelvins.

What do we know?

> The body temperature is 98.6°F.

How do we get there?

Rather than simply using the formulas to solve this problem, we will proceed by thinking it through. The situation is diagramed in Fig. 1.10. First, we want to convert 98.6°F to the Celsius scale. The number of Fahrenheit degrees between 32.0°F and 98.6°F is 66.6°F. We must convert this difference to Celsius degrees:

$$66.6°F \times \frac{5°C}{9°F} = 37.0°C$$

Thus 98.6°F corresponds to 37.0°C.

Now we can convert to the Kelvin scale:

$$T_K = T_C + 273.15 = 37.0 + 273.15 = 310.2 \text{ K}$$

Note that the final answer has only one decimal place (37.0 is limiting).

See Exercises 1.57, 1.59, and 1.60

Figure 1.10 | Normal body temperature on the Fahrenheit, Celsius, and Kelvin scales.

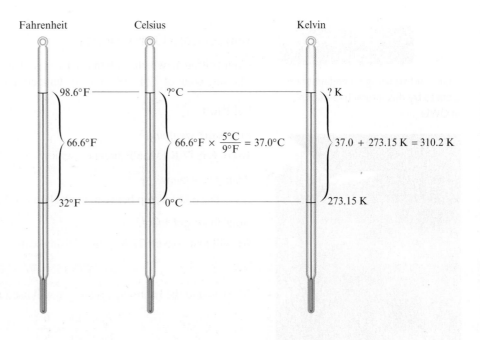

Example 1.12

Temperature Conversions II

One interesting feature of the Celsius and Fahrenheit scales is that $-40°C$ and $-40°F$ represent the same temperature, as shown in Fig. 1.9. Verify that this is true.

Solution

Where are we going?
To show that $-40°C = -40°F$

What do we know?

> The relationship between the Celsius and Fahrenheit scales

How do we get there?
The difference between 32°F and $-40°F$ is 72°F. The difference between 0°C and $-40°C$ is 40°C. The ratio of these is

$$\frac{72°F}{40°C} = \frac{8 \times 9°F}{8 \times 5°C} = \frac{9°F}{5°C}$$

as required. Thus $-40°C$ is equivalent to $-40°F$.

See Exercise 1.61

Since, as shown in Example 1.12, $-40°$ on both the Fahrenheit and Celsius scales represents the same temperature, this point can be used as a reference point (like 0°C and 32°F) for a relationship between the two scales:

$$\frac{\text{Number of Fahrenheit degrees}}{\text{Number of Celsius degrees}} = \frac{T_F - (-40)}{T_C - (-40)} = \frac{9°F}{5°C}$$

$$\frac{T_F + 40}{T_C + 40} = \frac{9°F}{5°C} \qquad (1.3)$$

where T_F and T_C represent the same temperature (but not the same number). This equation can be used to convert Fahrenheit temperatures to Celsius, and vice versa, and may be easier to remember than Equations (1.1) and (1.2).

Liquid nitrogen is so cold that water condenses out of the surrounding air, forming a cloud as the nitrogen is poured.

Richard Magna/Fundamental Photographs © Cengage Learning

Temperature Conversions III

Liquid nitrogen, which is often used as a coolant for low-temperature experiments, has a boiling point of 77 K. What is this temperature on the Fahrenheit scale?

Solution

Where are we going?

To convert 77 K to the Fahrenheit scale

What do we know?

> The relationship between the Kelvin and Fahrenheit scales

How do we get there?

We will first convert 77 K to the Celsius scale:

$$T_C = T_K - 273.15 = 77 - 273.15 = -196°C$$

To convert to the Fahrenheit scale, we will use Equation (1.3):

$$\frac{T_F + 40}{T_C + 40} = \frac{9°F}{5°C}$$

$$\frac{T_F + 40}{-196°C + 40} = \frac{T_F + 40}{-156°C} = \frac{9°F}{5°C}$$

$$T_F + 40 = \frac{9°F}{5°C}(-156°C) = -281°F$$

$$T_F = -281°F - 40 = -321°F$$

See Exercises 1.57, 1.59, and 1.60

1.9 | Density

A property of matter that is often used by chemists as an "identification tag" for a substance is **density**, the mass of substance per unit volume of the substance:

$$\text{Density} = \frac{\text{mass}}{\text{volume}}$$

The density of a liquid can be determined easily by weighing an accurately known volume of liquid. This procedure is illustrated in Example 1.14.

Determining Density

A chemist, trying to identify an unknown liquid, finds that 25.00 cm³ of the substance has a mass of 19.625 g at 20°C. The following are the names and densities of the compounds that might be the liquid:

Compound	Density in g/cm³ at 20°C
Chloroform	1.492
Diethyl ether	0.714
Ethanol	0.789
Isopropyl alcohol	0.785
Toluene	0.867

Which of these compounds is the most likely to be the unknown liquid?

Solution

Where are we going?

To calculate the density of the unknown liquid

What do we know?

> The mass of a given volume of the liquid.

How do we get there?

To identify the unknown substance, we must determine its density. This can be done by using the definition of density:

$$\text{Density} = \frac{\text{mass}}{\text{volume}} = \frac{19.625 \text{ g}}{25.00 \text{ cm}^3} = 0.7850 \text{ g/cm}^3$$

This density corresponds exactly to that of isopropyl alcohol, which therefore most likely is the unknown liquid. However, note that the density of ethanol is also very close. To be sure that the compound is isopropyl alcohol, we should run several more density experiments. (In the modern laboratory, many other types of tests could be done to distinguish between these two liquids.)

See Exercises 1.67 and 1.68

There are two ways of indicating units that occur in the denominator. For example, we can write g/cm³ or g cm⁻³. Although we will use the former system here, the other system is widely used.

Besides being a tool for the identification of substances, density has many other uses. For example, the liquid in your car's lead storage battery (a solution of sulfuric acid) changes density because the sulfuric acid is consumed as the battery discharges. In a fully charged battery, the density of the solution is about 1.30 g/cm³. If the density falls below 1.20 g/cm³, the battery will have to be recharged. Density measurement is also used to determine the amount of antifreeze, and thus the level of protection against freezing, in the cooling system of a car.

The densities of various common substances are given in Table 1.5.

Table 1.5 | Densities of Various Common Substances* at 20°C

Substance	Physical State	Density (g/cm³)
Oxygen	Gas	0.00133
Hydrogen	Gas	0.000084
Ethanol	Liquid	0.789
Benzene	Liquid	0.880
Water	Liquid	0.9982
Magnesium	Solid	1.74
Salt (sodium chloride)	Solid	2.16
Aluminum	Solid	2.70
Iron	Solid	7.87
Copper	Solid	8.96
Silver	Solid	10.5
Lead	Solid	11.34
Mercury	Liquid	13.6
Gold	Solid	19.32

*At 1 atmosphere pressure.

1.10 | Classification of Matter

Before we can hope to understand the changes we see going on around us—the growth of plants, the rusting of steel, the aging of people, the acidification of rain—we must find out how matter is organized. **Matter**, best defined as anything occupying space and having mass, is the material of the universe. Matter is complex and has many levels of organization. In this section we will introduce basic ideas about the structure of matter and its behavior.

We will start by considering the definitions of the fundamental properties of matter. Matter exists in three **states**: solid, liquid, and gas. A *solid* is rigid; it has a fixed volume and shape. A *liquid* has a definite volume but no specific shape; it assumes the shape of its container. A *gas* has no fixed volume or shape; it takes on the shape and volume of its container. In contrast to liquids and solids, which are only slightly compressible, gases are highly compressible; it is relatively easy to decrease the volume of a gas. Molecular-level pictures of the three states of water are given in Fig. 1.11. The different properties of ice, liquid water, and steam are determined by the different arrangements of the molecules in these substances. Table 1.5 gives the states of some common substances at 20°C and 1 atmosphere pressure.

Most of the matter around us consists of **mixtures** of pure substances. Wood, gasoline, wine, soil, and air all are mixtures. The main characteristic of a mixture is that it has *variable composition.* For example, wood is a mixture of many substances, the proportions of which vary depending on the type of wood and where it grows. Mixtures can be classified as **homogeneous** (having visibly indistinguishable parts) or **heterogeneous** (having visibly distinguishable parts).

A homogeneous mixture is called a **solution**. Air is a solution consisting of a mixture of gases. Wine is a complex liquid solution. Brass is a solid solution of copper and zinc. Sand in water and iced tea with ice cubes are examples of heterogeneous mixtures. Heterogeneous mixtures usually can be separated into two or more homogeneous mixtures or pure substances (for example, the ice cubes can be separated from the tea).

Mixtures can be separated into pure substances by physical methods. A **pure substance** is one with constant composition. Water is a good illustration of these ideas. As we will discuss in detail later, pure water is composed solely of H_2O molecules, but the water found in nature (groundwater or the water in a lake or ocean) is really a mixture.

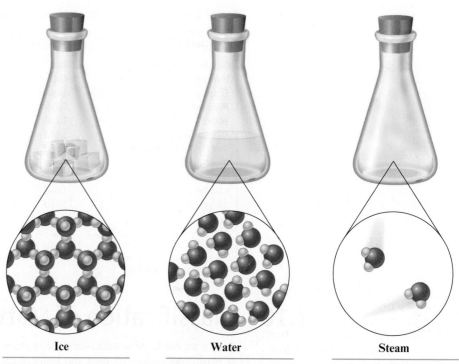

Ice	Water	Steam
Solid: The water molecules are locked into rigid positions and are close together.	**Liquid:** The water molecules are still close together but can move around to some extent.	**Gas:** The water molecules are far apart and move randomly.

Figure 1.11 | The three states of water (where red spheres represent oxygen atoms and blue spheres represent hydrogen atoms).

Seawater, for example, contains large amounts of dissolved minerals. Boiling seawater produces steam, which can be condensed to pure water, leaving the minerals behind as solids. The dissolved minerals in seawater also can be separated out by freezing the mixture, since pure water freezes out. The processes of boiling and freezing are **physical changes**. When water freezes or boils, it changes its state but remains water; it is still composed of H_2O molecules. A physical change is a change in the form of a substance, not in its chemical composition. A physical change can be used to separate a mixture into pure compounds, but it will not break compounds into elements.

One of the most important methods for separating the components of a mixture is **distillation**, a process that depends on differences in the volatility (how readily substances become gases) of the components. In simple distillation, a mixture is heated in a device such as that shown in Fig. 1.12. The most volatile component vaporizes at the lowest temperature, and the vapor passes through a cooled tube (a condenser), where it condenses back into its liquid state.

The simple, one-stage distillation apparatus shown in Fig. 1.12 works very well when only one component of the mixture is volatile. For example, a mixture of water and sand is easily separated by boiling off the water. Water containing dissolved minerals behaves in much the same way. As the water is boiled off, the minerals remain behind as nonvolatile solids. Simple distillation of seawater using the sun as the heat source is an excellent way to desalinate (remove the minerals from) seawater.

However, when a mixture contains several volatile components, the one-step distillation does not give a pure substance in the receiving flask, and more elaborate methods are required.

Another method of separation is simple **filtration**, which is used when a mixture consists of a solid and a liquid. The mixture is poured onto a mesh, such as filter paper, which passes the liquid and leaves the solid behind.

A third method of separation is **chromatography**. Chromatography is the general name applied to a series of methods that use a system with two *phases* (states) of matter: a mobile phase and a stationary phase. The *stationary phase* is a solid, and the *mobile phase* is either a liquid or a gas. The separation process occurs because the

The term *volatile* refers to the ease with which a substance can be changed to its vapor.

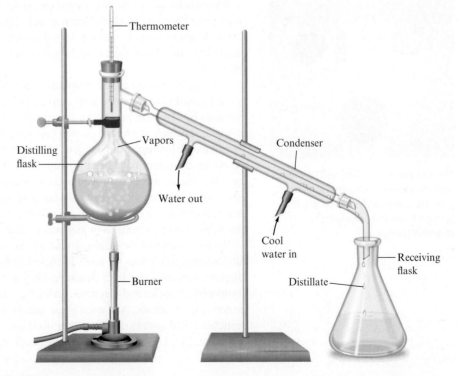

Figure 1.12 | Simple laboratory distillation apparatus. Cool water circulates through the outer portion of the condenser, causing vapors from the distilling flask to condense into a liquid. The nonvolatile component of the mixture remains in the distilling flask.

Figure 1.13 | Paper chromatography of ink. (a) A dot of the mixture to be separated is placed at one end of a sheet of porous paper. (b) The paper acts as a wick to draw up the liquid.

Photos © Charles D. Winters

Kristen Brochmann/Fundamental Photographs

The element mercury (top left) combines with the element iodine (top right) to form the compound mercuric iodide (bottom). This is an example of a chemical change.

components of the mixture have different affinities for the two phases and thus move through the system at different rates. A component with a high affinity for the mobile phase moves relatively quickly through the chromatographic system, whereas one with a high affinity for the solid phase moves more slowly.

One simple type of chromatography, **paper chromatography**, uses a strip of porous paper, such as filter paper, for the stationary phase. A drop of the mixture to be separated is placed on the paper, which is then dipped into a liquid (the mobile phase) that travels up the paper as though it were a wick (Fig. 1.13). This method of separating a mixture is often used by biochemists, who study the chemistry of living systems.

It should be noted that when a mixture is separated, the absolute purity of the separated components is an ideal. Because water, for example, inevitably comes into contact with other materials when it is synthesized or separated from a mixture, it is never absolutely pure. With great care, however, substances can be obtained in very nearly pure form.

Pure substances are either compounds (combinations of elements) or free elements. A **compound** is a substance with *constant composition* that can be broken down into elements by chemical processes. An example of a chemical process is the electrolysis of water, in which an electric current is passed through water to break it down into the free elements hydrogen and oxygen. This process produces a chemical change because the water molecules have been broken down. The water is gone, and in its place we have the free elements hydrogen and oxygen. A **chemical change** is one in which a given substance becomes a new substance or substances with different properties and different composition. **Elements** are substances that cannot be decomposed into simpler substances by chemical or physical means.

We have seen that the matter around us has various levels of organization. The most fundamental substances we have discussed so far are elements. As we will see in later chapters, elements also have structure: They are composed of atoms, which in turn are composed of nuclei and electrons. Even the nucleus has structure: It is composed of protons and neutrons. And even these can be broken down further, into elementary particles called *quarks*. However, we need not concern ourselves with such details at this point. Fig. 1.14 summarizes our discussion of the organization of matter.

Figure 1.14 | The organization of matter.

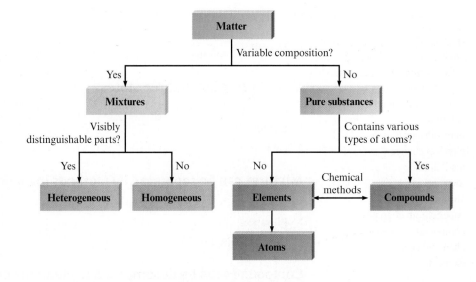

For review

Key terms

Section 1.2
scientific method
measurement
hypothesis
theory
model
natural law
law of conservation of mass

Section 1.3
SI system
mass
weight

Section 1.4
uncertainty
significant figures
accuracy
precision
random error
systematic error

Section 1.5
exponential notation

Section 1.7
unit factor method
dimensional analysis

Section 1.9
density

Scientific method
› Make observations
› Formulate hypotheses
› Perform experiments

Models (theories) are explanations of why nature behaves in a particular way.
› They are subject to modification over time and sometimes fail.

Quantitative observations are called measurements.
› Measurements consist of a number and a unit.
› Measurements involve some uncertainty.
› Uncertainty is indicated by the use of significant figures.
    › Rules to determine significant figures
    › Calculations using significant figures
› Preferred system is the SI system.

Temperature conversions
› $T_K = T_C + 273.15$
› $T_C = (T_F - 32°F)\left(\dfrac{5°C}{9°F}\right)$
› $T_F = T_C\left(\dfrac{9°F}{5°C}\right) + 32°F$

Key terms

Section 1.10

matter
states (of matter)
homogeneous mixture
heterogeneous mixture
solution
pure substance
physical change
distillation
filtration
chromatography
paper chromatography
compound
chemical change
element

Density

> Density $= \dfrac{\text{mass}}{\text{volume}}$

Matter can exist in three states:

> Solid
> Liquid
> Gas

Mixtures can be separated by methods involving only physical changes:

> Distillation
> Filtration
> Chromatography

Compounds can be decomposed to elements only through chemical changes.

Review questions   *Answers to the Review Questions can be found on the Student website (accessible from* **www.cengagebrain.com**).

1. Define and explain the differences between the following terms.
   a. law and theory
   b. theory and experiment
   c. qualitative and quantitative
   d. hypothesis and theory

2. Is the scientific method suitable for solving problems only in the sciences? Explain.

3. Which of the following statements could be tested by quantitative measurement?
   a. Ty Cobb was a better hitter than Pete Rose.
   b. Ivory soap is 99.44% pure.
   c. Rolaids consumes 47 times its weight in excess stomach acid.

4. For each of the following pieces of glassware, provide a sample measurement and discuss the number of significant figures and uncertainty.

5. A student performed an analysis of a sample for its calcium content and got the following results:

   14.92%   14.91%   14.88%   14.91%

   The actual amount of calcium in the sample is 15.70%. What conclusions can you draw about the accuracy and precision of these results?

6. Compare and contrast the multiplication/division significant figure rule to the significant figure rule applied for addition/subtraction in mathematical operations.

7. Explain how density can be used as a conversion factor to convert the volume of an object to the mass of the object, and vice versa.

8. On which temperature scale (°F, °C, or K) does 1 degree represent the smallest change in temperature?

9. Distinguish between physical changes and chemical changes.

10. Why is the separation of mixtures into pure or relatively pure substances so important when performing a chemical analysis?

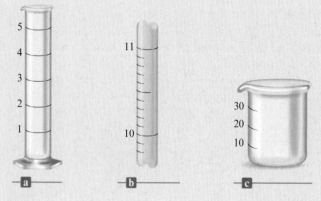

Active Learning Questions

These questions are designed to be used by groups of students in class.

1. **a.** There are 365 days per year, 24 hours per day, 12 months per year, and 60 minutes per hour. Use these data to determine how many minutes are in a month.

   b. Now use the following data to calculate the number of minutes in a month: 24 hours per day, 60 minutes per hour, 7 days per week, and 4 weeks per month.

   c. Why are these answers different? Which (if any) is more correct? Why?

2. You go to a convenience store to buy candy and find the owner to be rather odd. He allows you to buy pieces in multiples of four, and to buy four, you need $0.23. He only allows you to do this by using 3 pennies and 2 dimes. You have a bunch of pennies and dimes, and instead of counting them, you decide to weigh them. You have 636.3 g of pennies, and each penny weighs 3.03 g. Each dime weighs 2.29 g. Each piece of candy weighs 10.23 g.

   a. How many pennies do you have?

   b. How many dimes do you need to buy as much candy as possible?

   c. How much should all these dimes weigh?

   d. How many pieces of candy could you buy? (number of dimes from part b)

   e. How much would this candy weigh?

   f. How many pieces of candy could you buy with twice as many dimes?

3. When a marble is dropped into a beaker of water, it sinks to the bottom. Which of the following is the best explanation?

   a. The surface area of the marble is not large enough to be held up by the surface tension of the water.

   b. The mass of the marble is greater than that of the water.

   c. The marble weighs more than an equivalent volume of the water.

   d. The force from dropping the marble breaks the surface tension of the water.

   e. The marble has greater mass and volume than the water.

   Justify your choice, and for choices you did not pick, explain what is wrong about them.

4. You have two beakers, one filled to the 100-mL mark with sugar (the sugar has a mass of 180.0 g) and the other filled to the 100-mL mark with water (the water has a mass of 100.0 g). You pour all the sugar and all the water together in a bigger beaker and stir until the sugar is completely dissolved.

   a. Which of the following is true about the mass of the solution? Explain.

      i. It is much greater than 280.0 g.
      ii. It is somewhat greater than 280.0 g.
      iii. It is exactly 280.0 g.
      iv. It is somewhat less than 280.0 g.
      v. It is much less than 280.0 g.

   b. Which of the following is true about the volume of the solution? Explain.

      i. It is much greater than 200.0 mL.
      ii. It is somewhat greater than 200.0 mL.
      iii. It is exactly 200.0 mL.
      iv. It is somewhat less than 200.0 mL.
      v. It is much less than 200.0 mL.

5. You may have noticed that when water boils, you can see bubbles that rise to the surface of the water.

   a. What is inside these bubbles?

      i. air
      ii. hydrogen and oxygen gas
      iii. oxygen gas
      iv. water vapor
      v. carbon dioxide gas

   b. Is the boiling of water a chemical or physical change? Explain.

6. If you place a glass rod over a burning candle, the glass appears to turn black. What is happening to each of the following (physical change, chemical change, both, or neither) as the candle burns? Explain each answer.

   a. the wax   **b.** the wick   **c.** the glass rod

7. Which characteristics of a solid, a liquid, and a gas are exhibited by each of the following substances? How would you classify each substance?

   a. a bowl of pudding   **b.** a bucketful of sand

8. Sketch a magnified view (showing atoms/molecules) of each of the following and explain:

   a. a heterogeneous mixture of two different compounds

   b. a homogeneous mixture of an element and a compound

9. Paracelsus, a sixteenth-century alchemist and healer, adopted as his slogan: "The patients are your textbook, the sickbed is your study." Is this view consistent with using the scientific method?

10. What is wrong with the following statement?

   "The results of the experiment do not agree with the theory. Something must be wrong with the experiment."

11. Why is it incorrect to say that the results of a measurement were accurate but not precise?

12. What data would you need to estimate the money you would spend on gasoline to drive your car from New York to Chicago? Provide estimates of values and a sample calculation.

13. Sketch two pieces of glassware: one that can measure volume to the thousandths place and one that can measure volume only to the ones place.

14. You have a 1.0-cm³ sample of lead and a 1.0-cm³ sample of glass. You drop each in separate beakers of water. How do the volumes of water displaced by each sample compare? Explain.

15. Consider the addition of 15.4 to 28. What would a mathematician say the answer is? What would a scientist say? Justify the scientist's answer, not merely citing the rule, but explaining it.

16. Consider multiplying 26.2 by 16.43. What would a mathematician say the answer is? What would a scientist say? Justify the scientist's answer, not merely citing the rule, but explaining it.

A blue question or exercise number indicates that the answer to that question or exercise appears at the back of this book and a solution appears in the *Solutions Guide*, as found on PowerLecture.

Questions

17. The difference between a *law* and a *theory* is the difference between *what* and *why*. Explain.

18. The scientific method is a dynamic process. What does this mean?

19. Explain the fundamental steps of the scientific method.

20. What is the difference between random error and systematic error?

21. A measurement is a quantitative observation involving both a number and a unit. What is a qualitative observation? What are the SI units for mass, length, and volume? What is the assumed uncertainty in a number (unless stated otherwise)? The uncertainty of a measurement depends on the precision of the measuring device. Explain.

22. To determine the volume of a cube, a student measured one of the dimensions of the cube several times. If the true dimension of the cube is 10.62 cm, give an example of four sets of measurements that would illustrate the following.

    a. imprecise and inaccurate data

    b. precise but inaccurate data

    c. precise and accurate data

    Give a possible explanation as to why data can be imprecise or inaccurate. What is wrong with saying a set of measurements is imprecise but accurate?

23. What are significant figures? Show how to indicate the number one thousand to 1 significant figure, 2 significant figures, 3 significant figures, and 4 significant figures. Why is the answer, to the correct number of significant figures, not 1.0 for the following calculation?

$$\frac{1.5 - 1.0}{0.50} =$$

24. A cold front moves through and the temperature drops by 20 degrees. In which temperature scale would this 20 degree change represent the largest change in temperature?

25. When the temperature in degrees Fahrenheit (T_F) is plotted vs. the temperature in degrees Celsius (T_C), a straight-line plot results. A straight-line plot also results when T_C is plotted vs. T_K (the temperature in kelvins). Reference Appendix A1.3 and determine the slope and y-intercept of each of these two plots.

26. Give four examples illustrating each of the following terms.

    a. homogeneous mixture   d. element

    b. heterogeneous mixture   e. physical change

    c. compound   f. chemical change

Exercises

In this section similar exercises are paired.

Significant Figures and Unit Conversions

27. Which of the following are exact numbers?

    a. There are *100* cm in 1 m.

    b. One meter equals *1.094* yards.

c. We can use the equation

$$°F = \frac{9}{5}°C + 32$$

to convert from Celsius to Fahrenheit temperature. Are the numbers $\frac{9}{5}$ and *32* exact or inexact?

d. $\pi = 3.1415927$.

28. Indicate the number of significant figures in each of the following:

    a. This book contains more than 1000 pages.

    b. A mile is about 5300 ft.

    c. A liter is equivalent to 1.059 qt.

    d. The population of the United States is approaching 3.1×10^2 million.

    e. A kilogram is 1000 g.

    f. The Boeing 747 cruises at around 600 mi/h.

29. How many significant figures are there in each of the following values?

    a. 6.07×10^{-15} e. 463.8052

    b. 0.003840   f. 300

    c. 17.00   g. 301

    d. 8×10^8 h. 300.

30. How many significant figures are in each of the following?

    a. 100   e. 0.0048

    b. 1.0×10^2 f. 0.00480

    c. 1.00×10^3 g. 4.80×10^{-3}

    d. 100.   h. 4.800×10^{-3}

31. Round off each of the following numbers to the indicated number of significant digits, and write the answer in standard scientific notation.

    a. 0.00034159 to three digits

    b. 103.351×10^2 to four digits

    c. 17.9915 to five digits

    d. 3.365×10^5 to three digits

32. Use exponential notation to express the number 385,500 to

    a. one significant figure.

    b. two significant figures.

    c. three significant figures.

    d. five significant figures.

33. You have liquid in each graduated cylinder shown:

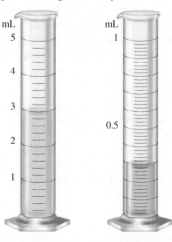

You then add both samples to a beaker. How would you write the number describing the total volume? What limits the precision of this number?

34. The beakers shown below have different precisions.

a. Label the amount of water in each of the three beakers to the correct number of significant figures.

b. Is it possible for each of the three beakers to contain the exact same amount of water? If no, why not? If yes, did you report the volumes as the same in part a? Explain.

c. Suppose you pour the water from these three beakers into one container. What should be the volume in the container reported to the correct number of significant figures?

35. Evaluate each of the following, and write the answer to the appropriate number of significant figures.

a. $212.2 + 26.7 + 402.09$

b. $1.0028 + 0.221 + 0.10337$

c. $52.331 + 26.01 - 0.9981$

d. $2.01 \times 10^2 + 3.014 \times 10^3$

e. $7.255 - 6.8350$

36. Perform the following mathematical operations, and express each result to the correct number of significant figures.

a. $\dfrac{0.102 \times 0.0821 \times 273}{1.01}$

b. $0.14 \times 6.022 \times 10^{23}$

c. $4.0 \times 10^4 \times 5.021 \times 10^{-3} \times 7.34993 \times 10^2$

d. $\dfrac{2.00 \times 10^6}{3.00 \times 10^{-7}}$

37. Perform the following mathematical operations, and express the result to the correct number of significant figures.

a. $\dfrac{2.526}{3.1} + \dfrac{0.470}{0.623} + \dfrac{80.705}{0.4326}$

b. $(6.404 \times 2.91)/(18.7 - 17.1)$

c. $6.071 \times 10^{-5} - 8.2 \times 10^{-6} - 0.521 \times 10^{-4}$

d. $(3.8 \times 10^{-12} + 4.0 \times 10^{-13})/(4 \times 10^{12} + 6.3 \times 10^{13})$

e. $\dfrac{9.5 + 4.1 + 2.8 + 3.175}{4}$

(Assume that this operation is taking the average of four numbers. Thus 4 in the denominator is exact.)

f. $\dfrac{8.925 - 8.905}{8.925} \times 100$

(This type of calculation is done many times in calculating a percentage error. Assume that this example is such a calculation; thus 100 can be considered to be an exact number.)

38. Perform the following mathematical operations, and express the result to the correct number of significant figures.

a. $6.022 \times 10^{23} \times 1.05 \times 10^2$

b. $\dfrac{6.6262 \times 10^{-34} \times 2.998 \times 10^8}{2.54 \times 10^{-9}}$

c. $1.285 \times 10^{-2} + 1.24 \times 10^{-3} + 1.879 \times 10^{-1}$

d. $\dfrac{(1.00866 - 1.00728)}{6.02205 \times 10^{23}}$

e. $\dfrac{9.875 \times 10^2 - 9.795 \times 10^2}{9.875 \times 10^2} \times 100$ (100 is exact)

f. $\dfrac{9.42 \times 10^2 + 8.234 \times 10^2 + 1.625 \times 10^3}{3}$ (3 is exact)

39. Perform each of the following conversions.

a. 8.43 cm to millimeters

b. 2.41×10^2 cm to meters

c. 294.5 nm to centimeters

d. 1.445×10^4 m to kilometers

e. 235.3 m to millimeters

f. 903.3 nm to micrometers

40. a. How many kilograms are in 1 teragram?

b. How many nanometers are in 6.50×10^2 terameters?

c. How many kilograms are in 25 femtograms?

d. How many liters are in 8.0 cubic decimeters?

e. How many microliters are in 1 milliliter?

f. How many picograms are in 1 microgram?

41. Perform the following unit conversions.

a. Congratulations! You and your spouse are the proud parents of a new baby, born while you are studying in a country that uses the metric system. The nurse has informed you that the baby weighs 3.91 kg and measures 51.4 cm. Convert your baby's weight to pounds and ounces and her length to inches (rounded to the nearest quarter inch).

b. The circumference of the earth is 25,000 mi at the equator. What is the circumference in kilometers? in meters?

c. A rectangular solid measures 1.0 m by 5.6 cm by 2.1 dm. Express its volume in cubic meters, liters, cubic inches, and cubic feet.

42. Perform the following unit conversions.

a. 908 oz to kilograms

b. 12.8 L to gallons

c. 125 mL to quarts

d. 2.89 gal to milliliters

e. 4.48 lb to grams

f. 550 mL to quarts

43. Use the following exact conversion factors to perform the stated calculations:

$$5\tfrac{1}{2} \text{ yd} = 1 \text{ rod}$$
$$40 \text{ rods} = 1 \text{ furlong}$$
$$8 \text{ furlongs} = 1 \text{ mile}$$

a. The Kentucky Derby race is 1.25 miles. How long is the race in rods, furlongs, meters, and kilometers?

b. A marathon race is 26 miles, 385 yards. What is this distance in rods, furlongs, meters, and kilometers?

44. Although the preferred SI unit of area is the square meter, land is often measured in the metric system in hectares (ha). One hectare is equal to 10,000 m². In the English system, land is often measured in acres (1 acre = 160 rod^2). Use the exact conversions and those given in Exercise 43 to calculate the following.

a. 1 ha = _____ km^2

b. The area of a 5.5-acre plot of land in hectares, square meters, and square kilometers

c. A lot with dimensions 120 ft by 75 ft is to be sold for $6500. What is the price per acre? What is the price per hectare?

45. Precious metals and gems are measured in troy weights in the English system:

$$24 \text{ grains} = 1 \text{ pennyweight (exact)}$$
$$20 \text{ pennyweight} = 1 \text{ troy ounce (exact)}$$
$$12 \text{ troy ounces} = 1 \text{ troy pound (exact)}$$
$$1 \text{ grain} = 0.0648 \text{ g}$$
$$1 \text{ carat} = 0.200 \text{ g}$$

a. The most common English unit of mass is the pound avoirdupois. What is 1 troy pound in kilograms and in pounds?

b. What is the mass of a troy ounce of gold in grams and in carats?

c. The density of gold is 19.3 g/cm^3. What is the volume of a troy pound of gold?

46. Apothecaries (druggists) use the following set of measures in the English system:

$$20 \text{ grains ap} = 1 \text{ scruple (exact)}$$
$$3 \text{ scruples} = 1 \text{ dram ap (exact)}$$
$$8 \text{ dram ap} = 1 \text{ oz ap (exact)}$$
$$1 \text{ dram ap} = 3.888 \text{ g}$$

a. Is an apothecary grain the same as a troy grain? (See Exercise 45.)

b. 1 oz ap = _____ oz troy.

c. An aspirin tablet contains 5.00×10^2 mg of active ingredient. What mass in grains ap of active ingredient does it contain? What mass in scruples?

d. What is the mass of 1 scruple in grams?

47. For a pharmacist dispensing pills or capsules, it is often easier to weigh the medication to be dispensed than to count the individual pills. If a single antibiotic capsule weighs 0.65 g, and a pharmacist weighs out 15.6 g of capsules, how many capsules have been dispensed?

48. A children's pain relief elixir contains 80. mg acetaminophen per 0.50 teaspoon. The dosage recommended for a child who weighs between 24 and 35 lb is 1.5 teaspoons. What is the range of acetaminophen dosages, expressed in mg acetaminophen/kg body weight, for children who weigh between 24 and 35 lb?

49. Science fiction often uses nautical analogies to describe space travel. If the starship *U.S.S. Enterprise* is traveling at warp factor 1.71, what is its speed in knots and in miles per hour? (Warp 1.71 = 5.00 times the speed of light; speed of light = 3.00×10^8 m/s; 1 knot = 2030 yd/h.)

50. The world record for the hundred meter dash is 9.58 s. What is the corresponding average speed in units of m/s, km/h, ft/s, and mi/h? At this speed, how long would it take to run 1.00×10^2 yards?

51. Would a car traveling at a constant speed of 65 km/h violate a 40 mi/h speed limit?

52. You pass a road sign saying "New York 112 km." If you drive at a constant speed of 65 mi/h, how long should it take you to reach New York? If your car gets 28 miles to the gallon, how many liters of gasoline are necessary to travel 112 km?

53. You are in Paris, and you want to buy some peaches for lunch. The sign in the fruit stand indicates that peaches cost 2.45 euros per kilogram. Given that 1 euro is equivalent to approximately $1.32, calculate what a pound of peaches will cost in dollars.

54. In recent years, there has been a large push for an increase in the use of renewable resources to produce the energy we need to power our vehicles. One of the newer fuels that has become more widely available is E85, a mixture of 85% ethanol and 15% gasoline. Despite being more environmentally friendly, one of the potential drawbacks of E85 fuel is that it produces less energy than conventional gasoline. Assume a car gets 28.0 mi/gal using gasoline at $3.50/gal and 22.5 mi/gal using E85 at $2.85/gal. How much will it cost to drive 500. miles using each fuel?

55. Mercury poisoning is a debilitating disease that is often fatal. In the human body, mercury reacts with essential enzymes leading to irreversible inactivity of these enzymes. If the amount of mercury in a polluted lake is 0.4 μg Hg/mL, what is the total mass in kilograms of mercury in the lake? (The lake has a surface area of 100 mi^2 and an average depth of 20 ft.)

56. Carbon monoxide (CO) detectors sound an alarm when peak levels of carbon monoxide reach 100 parts per million (ppm). This level roughly corresponds to a composition of air that contains 400,000 μg carbon monoxide per cubic meter of air (400,000 μg/m^3). Assuming the dimensions of a room are 18 ft $\times$ 12 ft $\times$ 8 ft, estimate the mass of carbon monoxide in the room that would register 100 ppm on a carbon monoxide detector.

Temperature

57. Convert the following Fahrenheit temperatures to the Celsius and Kelvin scales.

a. $-459°F$, an extremely low temperature

b. $-40.°F$, the answer to a trivia question

c. $68°F$, room temperature

d. 7×10^7 °F, temperature required to initiate fusion reactions in the sun

58. A thermometer gives a reading of 96.1°F $\pm$ 0.2°F. What is the temperature in °C? What is the uncertainty?

59. Convert the following Celsius temperatures to Kelvin and to Fahrenheit degrees.

a. the temperature of someone with a fever, 39.2°C

b. a cold wintery day, $-25°C$

c. the lowest possible temperature, $-273°C$

d. the melting-point temperature of sodium chloride, 801°C

60. Convert the following Kelvin temperatures to Celsius and Fahrenheit degrees.

a. the temperature that registers the same value on both the Fahrenheit and Celsius scales, 233 K

b. the boiling point of helium, 4 K

c. the temperature at which many chemical quantities are determined, 298 K

d. the melting point of tungsten, 3680 K

61. At what temperature is the temperature in degrees Fahrenheit equal to twice the temperature in degrees Celsius?

62. The average daytime temperatures on the earth and Jupiter are 72°F and 313 K, respectively. Calculate the difference in temperature, in °C, between these two planets.

63. Use the figure below to answer the following questions.

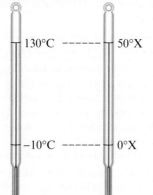

a. Derive the relationship between °C and °X.

b. If the temperature outside is 22.0°C, what is the temperature in units of °X?

c. Convert 58.0°X to units of °C, K, and °F.

64. Ethylene glycol is the main component in automobile antifreeze. To monitor the temperature of an auto cooling system, you intend to use a meter that reads from 0 to 100. You devise a new temperature scale based on the approximate melting and boiling points of a typical antifreeze solution (−45°C and 115°C). You wish these points to correspond to 0°A and 100°A, respectively.

a. Derive an expression for converting between °A and °C.

b. Derive an expression for converting between °F and °A.

c. At what temperature would your thermometer and a Celsius thermometer give the same numerical reading?

d. Your thermometer reads 86°A. What is the temperature in °C and in °F?

e. What is a temperature of 45°C in °A?

Density

65. A material will float on the surface of a liquid if the material has a density less than that of the liquid. Given that the density of water is approximately 1.0 g/mL, will a block of material having a volume of 1.2×10^4 in^3 and weighing 350 lb float or sink when placed in a reservoir of water?

66. For a material to float on the surface of water, the material must have a density less than that of water (1.0 g/mL) and must not react with the water or dissolve in it. A spherical ball has a radius of 0.50 cm and weighs 2.0 g. Will this ball float or sink when placed in water? (*Note:* Volume of a sphere = $\frac{4}{3}\pi r^3$.)

67. A star is estimated to have a mass of 2×10^{36} kg. Assuming it to be a sphere of average radius 7.0×10^5 km, calculate the average density of the star in units of grams per cubic centimeter.

68. A rectangular block has dimensions 2.9 cm × 3.5 cm × 10.0 cm. The mass of the block is 615.0 g. What are the volume and density of the block?

69. Diamonds are measured in carats, and 1 carat = 0.200 g. The density of diamond is 3.51 g/cm^3.

a. What is the volume of a 5.0-carat diamond?

b. What is the mass in carats of a diamond measuring 2.8 mL?

70. Ethanol and benzene dissolve in each other. When 100. mL of ethanol is dissolved in 1.00 L of benzene, what is the mass of the mixture? (See Table 1.5.)

71. A sample containing 33.42 g of metal pellets is poured into a graduated cylinder initially containing 12.7 mL of water, causing the water level in the cylinder to rise to 21.6 mL. Calculate the density of the metal.

72. The density of pure silver is 10.5 g/cm^3 at 20°C. If 5.25 g of pure silver pellets is added to a graduated cylinder containing 11.2 mL of water, to what volume level will the water in the cylinder rise?

73. In each of the following pairs, which has the greater mass? (See Table 1.5.)

a. 1.0 kg of feathers or 1.0 kg of lead

b. 1.0 mL of mercury or 1.0 mL of water

c. 19.3 mL of water or 1.00 mL of gold

d. 75 mL of copper or 1.0 L of benzene

74. **a.** Calculate the mass of ethanol in 1.50 qt of ethanol. (See Table 1.5.)

b. Calculate the mass of mercury in 3.5 in^3 of mercury. (See Table 1.5.)

75. In each of the following pairs, which has the greater volume?

a. 1.0 kg of feathers or 1.0 kg of lead

b. 100 g of gold or 100 g of water

c. 1.0 L of copper or 1.0 L of mercury

76. Using Table 1.5, calculate the volume of 25.0 g of each of the following substances at 1 atm.

a. hydrogen gas

b. water

c. iron

Chapter 5 discusses the properties of gases. One property unique to gases is that they contain mostly empty space. Explain using the results of your calculations.

77. The density of osmium (the densest metal) is 22.57 g/cm^3. If a 1.00-kg rectangular block of osmium has two dimensions of 4.00 cm × 4.00 cm, calculate the third dimension of the block.

78. A copper wire (density = 8.96 g/cm^3) has a diameter of 0.25 mm. If a sample of this copper wire has a mass of 22 g, how long is the wire?

Classification and Separation of Matter

79. Match each description below with the following microscopic pictures. More than one picture may fit each description. A picture may be used more than once or not used at all.

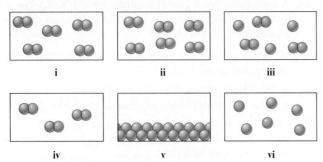

i   ii   iii

iv   v   vi

    a. a gaseous compound
    b. a mixture of two gaseous elements
    c. a solid element
    d. a mixture of a gaseous element and a gaseous compound

80. Define the following terms: solid, liquid, gas, pure substance, element, compound, homogeneous mixture, heterogeneous mixture, solution, chemical change, physical change.

81. What is the difference between homogeneous and heterogeneous matter? Classify each of the following as homogeneous or heterogeneous.
    a. a door
    b. the air you breathe
    c. a cup of coffee (black)
    d. the water you drink
    e. salsa
    f. your lab partner

82. Classify the following mixtures as homogeneous or heterogeneous.
    a. potting soil
    b. white wine
    c. your sock drawer
    d. window glass
    e. granite

83. Classify each of the following as a mixture or a pure substance.
    a. water
    b. blood
    c. the oceans
    d. iron
    e. brass
    f. uranium
    g. wine
    h. leather
    i. table salt
    Of the pure substances, which are elements and which are compounds?

84. Suppose a teaspoon of magnesium filings and a teaspoon of powdered sulfur are placed together in a metal beaker. Would this constitute a mixture or a pure substance? Suppose the magnesium filings and sulfur are heated so that they react with each other, forming magnesium sulfide. Would this still be a "mixture"? Why or why not?

85. If a piece of hard, white blackboard chalk is heated strongly in a flame, the mass of the piece of chalk will decrease, and eventually the chalk will crumble into a fine white dust. Does this change suggest that the chalk is composed of an element or a compound?

86. During a very cold winter, the temperature may remain below freezing for extended periods. However, fallen snow can still disappear, even though it cannot melt. This is possible because a solid can vaporize directly, without passing through the liquid state. Is this process (sublimation) a physical or a chemical change?

87. Classify the following as physical or chemical changes.
    a. Moth balls gradually vaporize in a closet.
    b. Hydrofluoric acid attacks glass and is used to etch calibration marks on glass laboratory utensils.
    c. A French chef making a sauce with brandy is able to boil off the alcohol from the brandy, leaving just the brandy flavoring.
    d. Chemistry majors sometimes get holes in the cotton jeans they wear to lab because of acid spills.

88. The properties of a mixture are typically averages of the properties of its components. The properties of a compound may differ dramatically from the properties of the elements that combine to produce the compound. For each process described below, state whether the material being discussed is most likely a mixture or a compound, and state whether the process is a chemical change or a physical change.
    a. An orange liquid is distilled, resulting in the collection of a yellow liquid and a red solid.
    b. A colorless, crystalline solid is decomposed, yielding a pale yellow-green gas and a soft, shiny metal.
    c. A cup of tea becomes sweeter as sugar is added to it.

Additional Exercises

89. Lipitor, a pharmaceutical drug that has been shown to lower "bad" cholesterol levels while raising "good" cholesterol levels in patients taking the drug, had over $11 billion in sales in 2006. Assuming one 2.5-g pill contains 4.0% of the active ingredient by mass, what mass in kg of active ingredient is present in one bottle of 100 pills?

90. In Shakespeare's *Richard III,* the First Murderer says:
    "Take that, and that! [*Stabs Clarence*]
    If that is not enough, I'll drown you in a malmsey butt within!"
    Given that 1 butt = 126 gal, in how many liters of malmsey (a foul brew similar to mead) was the unfortunate Clarence about to be drowned?

91. The contents of one 40. lb bag of topsoil will cover 10. square feet of ground to a depth of 1.0 inch. What number of bags is needed to cover a plot that measures 200. by 300. m to a depth of 4.0 cm?

92. In the opening scenes of the movie *Raiders of the Lost Ark,* Indiana Jones tries to remove a gold idol from a booby-trapped pedestal. He replaces the idol with a bag of sand of approximately equal volume. (Density of gold = 19.32 g/cm³; density of sand ≈ 2 g/cm³.)
    a. Did he have a reasonable chance of not activating the mass-sensitive booby trap?
    b. In a later scene, he and an unscrupulous guide play catch with the idol. Assume that the volume of the idol is about 1.0 L. If it were solid gold, what mass would the idol have? Is playing catch with it plausible?

93. A parsec is an astronomical unit of distance where 1 parsec = 3.26 light years (1 light year equals the distance traveled by light in one year). If the speed of light is 186,000 mi/s, calculate the distance in meters of an object that travels 9.6 parsecs.

94. You are driving 65 mi/h and take your eyes off the road for "just a second." What distance (in feet) do you travel in this time?

95. This year, like many past years, you begin to feel very sleepy after eating a large helping of Thanksgiving turkey. Some people attribute this sleepiness to the presence of the amino acid tryptophan in turkey. Tryptophan can be used by the body to produce serotonin, which can calm the brain's activity and help to bring on sleep.

    a. What mass in grams of tryptophan is in a 0.25-lb serving of turkey? (Assume tryptophan accounts for 1.0% of the turkey mass.)

    b. What mass in grams of tryptophan is in 0.25 quart of milk? (Assume tryptophan accounts for 2.0% of milk by mass and that the density of milk is 1.04 kg/L.)

96. Which of the following are chemical changes? Which are physical changes?

    a. the cutting of food

    b. interaction of food with saliva and digestive enzymes

    c. proteins being broken down into amino acids

    d. complex sugars being broken down into simple sugars

    e. making maple syrup by heating maple sap to remove water through evaporation

    f. DNA unwinding

97. A column of liquid is found to expand linearly on heating. Assume the column rises 5.25 cm for a 10.0°F rise in temperature. If the initial temperature of the liquid is 98.6°F, what will the final temperature be in °C if the liquid has expanded by 18.5 cm?

98. A 25.00-g sample of a solid is placed in a graduated cylinder, and then the cylinder is filled to the 50.0-mL mark with benzene. The mass of benzene and solid together is 58.80 g. Assuming that the solid is insoluble in benzene and that the density of benzene is 0.880 g/cm³, calculate the density of the solid.

99. For each of the following, decide which block is more dense: the orange block, the blue block, or it cannot be determined. Explain your answers.

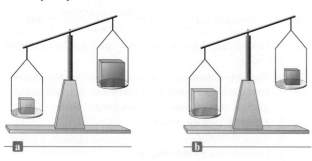

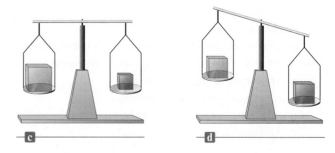

100. According to the *Official Rules of Baseball*, a baseball must have a circumference not more than 9.25 in or less than 9.00 in and a mass not more than 5.25 oz or less than 5.00 oz. What range of densities can a baseball be expected to have? Express this range as a single number with an accompanying uncertainty limit.

101. The density of an irregularly shaped object was determined as follows. The mass of the object was found to be 28.90 g ± 0.03 g. A graduated cylinder was partially filled with water. The reading of the level of the water was 6.4 cm³ ± 0.1 cm³. The object was dropped in the cylinder, and the level of the water rose to 9.8 cm³ ± 0.1 cm³. What is the density of the object with appropriate error limits? (See Appendix 1.5.)

102. The chemist in Example 1.14 did some further experiments. She found that the pipet used to measure the volume of the liquid is accurate to ±0.03 cm³. The mass measurement is accurate to ±0.002 g. Are these measurements sufficiently precise for the chemist to distinguish between isopropyl alcohol and ethanol?

ChemWork Problems

These multiconcept problems (and additional ones) are found interactively online with the same type of assistance a student would get from an instructor.

103. The longest river in the world is the Nile River with a length of 4,145 mi. How long is the Nile in cable lengths, meters, and nautical miles?

    Use these exact conversions to help solve the problem:

    $$6 \text{ ft} = 1 \text{ fathom}$$
    $$100 \text{ fathoms} = 1 \text{ cable length}$$
    $$10 \text{ cable lengths} = 1 \text{ nautical mile}$$
    $$3 \text{ nautical miles} = 1 \text{ league}$$

104. Secretariat is known as the horse with the fastest run in the Kentucky Derby. If Secretariat's record 1.25-mi run lasted 1 minute 59.2 seconds, what was his average speed in m/s?

105. The hottest temperature recorded in the United States is 134°F in Greenland Ranch, CA. The melting point of phosphorus is 44°C. At this temperature, would phosphorus be a liquid or a solid?

106. The radius of a neon atom is 69 pm, and its mass is 3.35×10^{-23} g. What is the density of the atom in grams per cubic centimeter (g/cm³)? Assume the nucleus is a sphere with volume = $\frac{4}{3}\pi r^3$.

107. Which of the following statements is(are) *true*?

    a. A spoonful of sugar is a mixture.

    b. Only elements are pure substances.

c. Air is a mixture of gases.

d. Gasoline is a pure substance.

e. Compounds can be broken down only by chemical means.

108. Which of the following describes a chemical property?

a. The density of iron is 7.87 g/cm^3.

b. A platinum wire glows red when heated.

c. An iron bar rusts.

d. Aluminum is a silver-colored metal.

Challenge Problems

109. A rule of thumb in designing experiments is to avoid using a result that is the small difference between two large measured quantities. In terms of uncertainties in measurement, why is this good advice?

110. Draw a picture showing the markings (graduations) on glassware that would allow you to make each of the following volume measurements of water, and explain your answers (the numbers given are as precise as possible).

a. 128.7 mL   b. 18 mL   c. 23.45 mL

If you made these measurements for three samples of water and then poured all of the water together in one container, what total volume of water should you report? Support your answer.

111. Many times errors are expressed in terms of percentage. The percent error is the absolute value of the difference of the true value and the experimental value, divided by the true value, and multiplied by 100.

$$\text{Percent error} = \frac{|\text{true value} - \text{experimental value}|}{\text{true value}} \times 100$$

Calculate the percent error for the following measurements.

a. The density of an aluminum block determined in an experiment was 2.64 g/cm^3. (True value 2.70 g/cm^3.)

b. The experimental determination of iron in iron ore was 16.48%. (True value 16.12%.)

c. A balance measured the mass of a 1.000-g standard as 0.9981 g.

112. A person weighed 15 pennies on a balance and recorded the following masses:

3.112 g	3.109 g	3.059 g
2.467 g	3.079 g	2.518 g
3.129 g	2.545 g	3.050 g
3.053 g	3.054 g	3.072 g
3.081 g	3.131 g	3.064 g

Curious about the results, he looked at the dates on each penny. Two of the light pennies were minted in 1983 and one in 1982. The dates on the 12 heavier pennies ranged from 1970 to 1982. Two of the 12 heavier pennies were minted in 1982.

a. Do you think the Bureau of the Mint changed the way it made pennies? Explain.

b. The person calculated the average mass of the 12 heavy pennies. He expressed this average as 3.0828 g ± 0.0482 g. What is wrong with the numbers in this result, and how should the value be expressed?

113. On October 21, 1982, the Bureau of the Mint changed the composition of pennies (see Exercise 112). Instead of an alloy of 95% Cu and 5% Zn by mass, a core of 99.2% Zn and 0.8% Cu with a thin shell of copper was adopted. The overall composition of the new penny was 97.6% Zn and 2.4% Cu by mass. Does this account for the difference in mass among the pennies in Exercise 112? Assume the volume of the individual metals that make up each penny can be added together to give the overall volume of the penny, and assume each penny is the same size. (Density of Cu = 8.96 g/cm^3; density of Zn = 7.14 g/cm^3.)

114. As part of a science project, you study traffic patterns in your city at an intersection in the middle of downtown. You set up a device that counts the cars passing through this intersection for a 24-hr period during a weekday. The graph of hourly traffic looks like this.

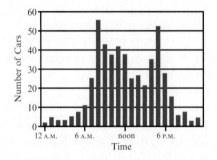

a. At what time(s) does the highest number of cars pass through the intersection?

b. At what time(s) does the lowest number of cars pass through the intersection?

c. Briefly describe the trend in numbers of cars over the course of the day.

d. Provide a hypothesis explaining the trend in numbers of cars over the course of the day.

e. Provide a possible experiment that could test your hypothesis.

115. Sterling silver is a solid solution of silver and copper. If a piece of a sterling silver necklace has a mass of 105.0 g and a volume of 10.12 mL, calculate the mass percent of copper in the piece of necklace. Assume that the volume of silver present plus the volume of copper present equals the total volume. Refer to Table 1.5.

$$\text{Mass percent of copper} = \frac{\text{mass of copper}}{\text{total mass}} \times 100$$

116. Make molecular-level (microscopic) drawings for each of the following.

a. Show the differences between a gaseous mixture that is a homogeneous mixture of two different compounds, and a gaseous mixture that is a homogeneous mixture of a compound and an element.

b. Show the differences among a gaseous element, a liquid element, and a solid element.

117. Confronted with the box shown in the diagram, you wish to discover something about its internal workings. You have no tools and cannot open the box. You pull on rope B, and it moves rather freely. When you pull on rope A, rope C appears to be pulled slightly into the box. When you pull on rope C, rope A almost disappears into the box.*

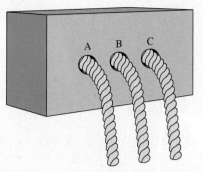

a. Based on these observations, construct a model for the interior mechanism of the box.

b. What further experiments could you do to refine your model?

118. An experiment was performed in which an empty 100-mL graduated cylinder was weighed. It was weighed once again after it had been filled to the 10.0-mL mark with dry sand. A 10-mL pipet was used to transfer 10.00 mL of methanol to the cylinder. The sand–methanol mixture was stirred until bubbles no longer emerged from the mixture and the sand looked uniformly wet. The cylinder was then weighed again. Use the data obtained from this experiment (and displayed at the end of this problem) to find the density of the dry sand, the density of methanol, and the density of sand particles. Does the bubbling that occurs when the methanol is added to the dry sand indicate that the sand and methanol are reacting?

---

*From Yoder, Suydam, and Snavely, *Chemistry* (New York: Harcourt Brace Jovanovich, 1975), pp. 9–11.

Mass of cylinder plus wet sand	45.2613 g
Mass of cylinder plus dry sand	37.3488 g
Mass of empty cylinder	22.8317 g
Volume of dry sand	10.0 mL
Volume of sand plus methanol	17.6 mL
Volume of methanol	10.00 mL

Integrative Problems

These problems require the integration of multiple concepts to find the solutions.

119. The U.S. trade deficit at the beginning of 2005 was $475,000,000. If the wealthiest 1.00% of the U.S. population (297,000,000) contributed an equal amount of money to bring the trade deficit to $0, how many dollars would each person contribute? If one of these people were to pay his or her share in nickels only, how many nickels are needed? Another person living abroad at the time decides to pay in pounds sterling (£). How many pounds sterling does this person contribute (assume a conversion rate of 1 £ = $1.869)?

120. The density of osmium is reported by one source to be 22610 kg/m^3. What is this density in g/cm^3? What is the mass of a block of osmium measuring 10.0 cm × 8.0 cm × 9.0 cm?

121. At the Amundsen-Scott South Pole base station in Antarctica, when the temperature is −100.0°F, researchers who live there can join the "300 Club" by stepping into a sauna heated to 200.0°F then quickly running outside and around the pole that marks the South Pole. What are these temperatures in °C? What are these temperatures in K? If you measured the temperatures only in °C and K, can you become a member of the "300 Club" (that is, is there a 300.-degree difference between the temperature extremes when measured in °C and K)?

Atoms, Molecules, and Ions

Polarized light micrograph of crystals of tartaric acid. (Sinclair Stammers/Photo Researchers, Inc.)

Where does one start in learning chemistry? Clearly we must consider some essential vocabulary and something about the origins of the science before we can proceed very far. Thus, while Chapter 1 provided background on the fundamental ideas and procedures of science in general, Chapter 2 covers the specific chemical background necessary for understanding the material in the next few chapters. The coverage of these topics is necessarily brief at this point. We will develop these ideas more fully as it becomes appropriate to do so. A major goal of this chapter is to present the systems for naming chemical compounds to provide you with the vocabulary necessary to understand this book and to pursue your laboratory studies.

Because chemistry is concerned first and foremost with chemical changes, we will proceed as quickly as possible to a study of chemical reactions (Chapters 3 and 4). However, before we can discuss reactions, we must consider some fundamental ideas about atoms and how they combine.

2.1 | The Early History of Chemistry

Chemistry has been important since ancient times. The processing of natural ores to produce metals for ornaments and weapons and the use of embalming fluids are just two applications of chemical phenomena that were utilized prior to 1000 B.C.

The Greeks were the first to try to explain why chemical changes occur. By about 400 B.C. they had proposed that all matter was composed of four fundamental substances: fire, earth, water, and air. The Greeks also considered the question of whether matter is continuous, and thus infinitely divisible into smaller pieces, or composed of small, indivisible particles. Supporters of the latter position were Demokritos* of Abdera (*c*. 460–*c*. 370 B.C.) and Leucippos, who used the term *atomos* (which later became *atoms*) to describe these ultimate particles. However, because the Greeks had no experiments to test their ideas, no definitive conclusion could be reached about the divisibility of matter.

The next 2000 years of chemical history were dominated by a pseudoscience called alchemy. Some alchemists were mystics and fakes who were obsessed with the idea of turning cheap metals into gold. However, many alchemists were serious scientists, and this period saw important advances: The alchemists discovered several elements and learned to prepare the mineral acids.

The foundations of modern chemistry were laid in the sixteenth century with the development of systematic metallurgy (extraction of metals from ores) by a German, Georg Bauer (1494–1555), and the medicinal application of minerals by a Swiss alchemist/physician known as Paracelsus (full name: Philippus Theophrastus Bombastus von Hohenheim [1493–1541]).

The first "chemist" to perform truly quantitative experiments was Robert Boyle (1627–1691), who carefully measured the relationship between the pressure and volume of air. When Boyle published his book *The Skeptical Chymist* in 1661, the quantitative sciences of physics and chemistry were born. In addition to his results on the quantitative behavior of gases, Boyle's other major contribution to chemistry consisted of his ideas about the chemical elements. Boyle held no preconceived notion about the number of elements. In his view, a substance was an element unless it could be broken down into two or more simpler substances. As Boyle's experimental definition of an element became generally accepted, the list of known elements began to grow, and the Greek system of four elements finally died. Although Boyle was an excellent scientist,

*Democritus is an alternate spelling.

Roald Hoffmann

Figure 2.1 | The Priestley Medal is the highest honor given by the American Chemical Society. It is named for Joseph Priestley, who was born in England on March 13, 1733. He performed many important scientific experiments, among them the discovery that a gas later identified as carbon dioxide could be dissolved in water to produce *seltzer.* Also, as a result of meeting Benjamin Franklin in London in 1766, Priestley became interested in electricity and was the first to observe that graphite was an electrical conductor. However, his greatest discovery occurred in 1774 when he isolated oxygen by heating mercuric oxide.

Because of his nonconformist political views, Priestley was forced to leave England. He died in the United States in 1804.

he was not always right. For example, he clung to the alchemists' views that metals were not true elements and that a way would eventually be found to change one metal into another.

The phenomenon of combustion evoked intense interest in the seventeenth and eighteenth centuries. The German chemist Georg Stahl (1660–1734) suggested that a substance he called "phlogiston" flowed out of the burning material. Stahl postulated that a substance burning in a closed container eventually stopped burning because the air in the container became saturated with phlogiston. Oxygen gas, discovered by Joseph Priestley (1733–1804),* an English clergyman and scientist (Fig. 2.1), was found to support vigorous combustion and was thus supposed to be low in phlogiston. In fact, oxygen was originally called "dephlogisticated air."

2.2 | Fundamental Chemical Laws

By the late eighteenth century, combustion had been studied extensively; the gases carbon dioxide, nitrogen, hydrogen, and oxygen had been discovered; and the list of elements continued to grow. However, it was Antoine Lavoisier (1743–1794), a French chemist (Fig. 2.2), who finally explained the true nature of combustion, thus clearing the way for the tremendous progress that was made near the end of the eighteenth century. Lavoisier, like Boyle, regarded measurement as the essential operation of chemistry. His experiments, in which he carefully weighed the reactants and products of various reactions, suggested that *mass is neither created nor destroyed.* Lavoisier's verification of this **law of conservation of mass** was the basis for the developments in chemistry in the nineteenth century. Mass is neither created nor destroyed in a chemical reaction.

Oxygen is from the French *oxygène,* meaning "generator of acid," because it was initially considered to be an integral part of all acids.

Lavoisier's quantitative experiments showed that combustion involved oxygen (which Lavoisier named), not phlogiston. He also discovered that life was supported by a process that also involved oxygen and was similar in many ways to combustion. In 1789 Lavoisier published the first modern chemistry textbook, *Elementary Treatise on Chemistry,* in which he presented a unified picture of the chemical knowledge assembled up to that time. Unfortunately, in the same year the text was published, the French Revolution broke out. Lavoisier, who had been associated with collecting taxes for the government, was executed on the guillotine as an enemy of the people in 1794.

After 1800, chemistry was dominated by scientists who, following Lavoisier's lead, performed careful weighing experiments to study the course of chemical reactions and to determine the composition of various chemical compounds. One of these chemists, a Frenchman, Joseph Proust (1754–1826), showed that *a given compound always contains exactly the same proportion of elements by mass.* For example, Proust found that the substance copper carbonate is always 5.3 parts copper to 4 parts oxygen to 1 part

*Oxygen gas was actually first observed by the Swedish chemist Karl W. Scheele (1742–1786), but because his results were published after Priestley's, the latter is commonly credited with the discovery of oxygen.

Figure 2.2 | Antoine Lavoisier was born in Paris on August 26, 1743. Although Lavoisier's father wanted his son to follow him into the legal profession, young Lavoisier was fascinated by science. From the beginning of his scientific career, Lavoisier recognized the importance of accurate measurements. His careful weighings showed that mass is conserved in chemical reactions and that combustion involves reaction with oxygen. Also, he wrote the first modern chemistry textbook. It is not surprising that Lavoisier is often called the father of modern chemistry.

To help support his scientific work, Lavoisier invested in a private tax-collecting firm and married the daughter of one of the company executives. His connection to the tax collectors proved fatal, for radical French revolutionaries demanded his execution, which occurred on the guillotine on May 8, 1794.

carbon (by mass). The principle of the constant composition of compounds, originally called "Proust's law," is now known as the **law of definite proportion**. A given compound always contains exactly the same proportion of elements by mass.

Proust's discovery stimulated John Dalton (1766–1844), an English schoolteacher (Fig. 2.3), to think about atoms as the particles that might compose elements. Dalton reasoned that if elements were composed of tiny individual particles, a given compound should always contain the same combination of these atoms. This concept explained why the same relative masses of elements were always found in a given compound.

But Dalton discovered another principle that convinced him even more of the existence of atoms. He noted, for example, that carbon and oxygen form two different compounds that contain different relative amounts of carbon and oxygen, as shown by the following data:

	Mass of Oxygen That Combines with 1 g of Carbon
Compound I	1.33 g
Compound II	2.66 g

Dalton noted that compound II contains twice as much oxygen per gram of carbon as compound I, a fact that could easily be explained in terms of atoms. Compound I might

Figure 2.3 | John Dalton (1766–1844), an Englishman, began teaching at a Quaker school when he was 12. His fascination with science included an intense interest in meteorology, which led to an interest in the gases of the air and their ultimate components, atoms. Dalton is best known for his atomic theory, in which he postulated that the fundamental differences among atoms are their masses. He was the first to prepare a table of relative atomic weights.

Dalton was a humble man with several apparent handicaps: He was not articulate and he was color-blind, a terrible problem for a chemist. Despite these disadvantages, he helped to revolutionize the science of chemistry.

be CO, and compound II might be CO_2.* This principle, which was found to apply to compounds of other elements as well, became known as the **law of multiple proportions**: When two elements form a series of compounds, the ratios of the masses of the second element that combine with 1 g of the first element can always be reduced to small whole numbers.

To make sure the significance of this observation is clear, in Example 2.1 we will consider data for a series of compounds consisting of nitrogen and oxygen.

| Example 2.1 | Illustrating the Law of Multiple Proportions |

The following data were collected for several compounds of nitrogen and oxygen:

	Mass of Nitrogen That Combines with 1 g of Oxygen
Compound A	1.750 g
Compound B	0.8750 g
Compound C	0.4375 g

Show how these data illustrate the law of multiple proportions.

Solution

For the law of multiple proportions to hold, the ratios of the masses of nitrogen combining with 1 g of oxygen in each pair of compounds should be small whole numbers. We therefore compute the ratios as follows:

$$\frac{A}{B} = \frac{1.750}{0.8750} = \frac{2}{1}$$

$$\frac{B}{C} = \frac{0.8750}{0.4375} = \frac{2}{1}$$

$$\frac{A}{C} = \frac{1.750}{0.4375} = \frac{4}{1}$$

These results support the law of multiple proportions.

See Exercises 2.37 and 2.38

The significance of the data in Example 2.1 is that compound A contains twice as much nitrogen (N) per gram of oxygen (O) as does compound B and that compound B contains twice as much nitrogen per gram of oxygen as does compound C.

These data can be explained readily if the substances are composed of molecules made up of nitrogen atoms and oxygen atoms. For example, one set of possibilities for compounds A, B, and C is

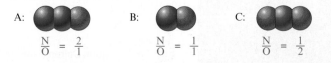

Now we can see that compound A contains two atoms of N for every atom of O, whereas compound B contains one atom of N per atom of O. That is, compound A contains twice as much nitrogen per given amount of oxygen as does compound B. Similarly, since compound B contains one N per O and compound C contains one N

*Subscripts are used to show the numbers of atoms present. The number 1 is understood (not written). The symbols for the elements and the writing of chemical formulas will be illustrated further in Sections 2.6 and 2.7.

per *two* Os, the nitrogen content of compound C per given amount of oxygen is half that of compound B.

Another set of compounds that fits the data in Example 2.1 is

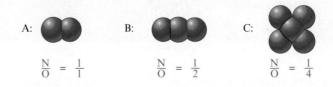

$$\frac{N}{O} = \frac{1}{1} \qquad \frac{N}{O} = \frac{1}{2} \qquad \frac{N}{O} = \frac{1}{4}$$

Verify for yourself that these compounds satisfy the requirements.

Still another set that works is

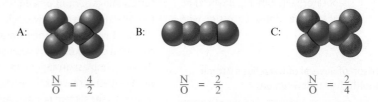

$$\frac{N}{O} = \frac{4}{2} \qquad \frac{N}{O} = \frac{2}{2} \qquad \frac{N}{O} = \frac{2}{4}$$

See if you can come up with still another set of compounds that satisfies the data in Example 2.1. How many more possibilities are there?

In fact, an infinite number of other possibilities exists. Dalton could not deduce absolute formulas from the available data on relative masses. However, the data on the composition of compounds in terms of the relative masses of the elements supported his hypothesis that each element consisted of a certain type of atom and that compounds were formed from specific combinations of atoms.

2.3 | Dalton's Atomic Theory

In 1808 Dalton published *A New System of Chemical Philosophy,* in which he presented his theory of atoms:

These statements are a modern paraphrase of Dalton's ideas.

> ### Dalton's Atomic Theory
>
> **1.** Each element is made up of tiny particles called atoms.
> **2.** The atoms of a given element are identical; the atoms of different elements are different in some fundamental way or ways.
> **3.** Chemical compounds are formed when atoms of different elements combine with each other. A given compound always has the same relative numbers and types of atoms.
> **4.** Chemical reactions involve reorganization of the atoms—changes in the way they are bound together. The atoms themselves are not changed in a chemical reaction.

It is instructive to consider Dalton's reasoning on the relative masses of the atoms of the various elements. In Dalton's time water was known to be composed of the elements hydrogen and oxygen, with 8 g of oxygen present for every 1 g of hydrogen. If the formula for water were OH, an oxygen atom would have to have 8 times the mass of a hydrogen atom. However, if the formula for water were H_2O (two atoms of hydrogen for every oxygen atom), this would mean that each atom of oxygen is 16 times as massive as *each* atom of hydrogen (since the ratio of the mass of one oxygen to that of *two* hydrogens is 8 to 1). Because the formula for water was not then known, Dalton could not specify the relative masses of oxygen and hydrogen unambiguously. To solve the problem, Dalton made a fundamental assumption: He decided that nature would be as simple as possible. This assumption led him to conclude that the formula for water should be OH. He thus assigned hydrogen a mass of 1 and oxygen a mass of 8.

Chemical connections
Berzelius, Selenium, and Silicon

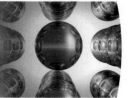

Jöns Jakob Berzelius was probably the best experimental chemist of his generation and, given the crudeness of his laboratory equipment, maybe the best of all time. Unlike Lavoisier, who

Comparison of Several of Berzelius's Atomic Masses with the Modern Values

Element	Atomic Mass	
	Berzelius's Value	Current Value
Chlorine	35.41	35.45
Copper	63.00	63.55
Hydrogen	1.00	1.01
Lead	207.12	207.2
Nitrogen	14.05	14.01
Oxygen	16.00	16.00
Potassium	39.19	39.10
Silver	108.12	107.87
Sulfur	32.18	32.07

could afford to buy the best laboratory equipment available, Berzelius worked with minimal equipment in very plain surroundings. One of Berzelius's students described the Swedish chemist's workplace: "The laboratory consisted of two ordinary rooms with the very simplest arrangements; there were neither furnaces nor hoods, neither water system nor gas. Against the walls stood some closets with the chemicals, in the middle the mercury trough and the blast lamp table. Beside this was the sink consisting of a stone water holder with a stopcock and a pot standing under it. [Next door in the kitchen] stood a small heating furnace."

In these simple facilities, Berzelius performed more than 2000 experiments over a 10-year period to determine accurate atomic masses for the 50 elements then known. His success can be seen from the data in the table at left. These remarkably accurate values attest to his experimental skills and patience.

Besides his table of atomic masses, Berzelius made many other major contributions to chemistry. The most important of these was the invention of a simple set of symbols for the elements along with a system for writing the formulas of compounds to replace the awkward symbolic representations of the alchemists. Although some chemists, including Dalton, objected to the new system, it was gradually adopted and forms the basis of the system we use today.

In addition to these accomplishments, Berzelius discovered the elements cerium, thorium, selenium, and silicon. Of these elements,

Using similar reasoning for other compounds, Dalton prepared the first table of **atomic masses** (sometimes called **atomic weights** by chemists, since mass is often determined by comparison to a standard mass—a process called *weighing*). Many of the masses were later proved to be wrong because of Dalton's incorrect assumptions about the formulas of certain compounds, but the construction of a table of masses was an important step forward.

Although not recognized as such for many years, the keys to determining absolute formulas for compounds were provided in the experimental work of the French chemist Joseph Gay-Lussac (1778–1850) and by the hypothesis of an Italian chemist named Amadeo Avogadro (1776–1856). In 1809 Gay-Lussac performed experiments in which he measured (under the same conditions of temperature and pressure) the volumes of gases that reacted with each other. For example, Gay-Lussac found that

Joseph Louis Gay-Lussac, a French physicist and chemist, was remarkably versatile. Although he is now known primarily for his studies on the combining of volumes of gases, Gay-Lussac was instrumental in the studies of many of the other properties of gases. Some of Gay-Lussac's motivation to learn about gases arose from his passion for ballooning. In fact, he made ascents to heights of over 4 miles to collect air samples, setting altitude records that stood for about 50 years. Gay-Lussac also was the codiscoverer of boron and the developer of a process for manufacturing sulfuric acid. As chief assayer of the French mint, Gay-Lussac developed many techniques for chemical analysis and invented many types of glassware now used routinely in labs. Gay-Lussac spent his last 20 years as a lawmaker in the French government.

The Alchemists' Symbols for Some Common Elements and Compounds

Substance	Alchemists' Symbol
Silver	
Lead	
Tin	
Platinum	
Sulfuric acid	
Alcohol	
Sea salt	

selenium and silicon are particularly important in today's world. Berzelius discovered selenium in 1817 in connection with his studies of sulfuric acid. For years selenium's toxicity has been known, but only recently have we become aware that it may have a positive effect on human health. Studies have shown that trace amounts of selenium in the diet may protect people from heart disease and cancer. One study based on data from 27 countries showed an inverse relationship between the cancer death rate and the selenium content of soil in a particular region (low cancer death rate in areas with high selenium content). Another research paper reported an inverse relationship between the selenium content of the blood and the incidence of breast cancer in women. A study reported in 1998 used the toenail clippings of 33,737 men to show that selenium seems to protect against prostate cancer. Selenium is also found in the heart muscle and may play an important role in proper heart function. Because of these and other studies, selenium's reputation has improved, and many scientists are now studying its function in the human body.

Silicon is the second most abundant element in the earth's crust, exceeded only by oxygen. As we will see in Chapter 10, compounds involving silicon bonded to oxygen make up most of the earth's sand, rock, and soil. Berzelius prepared silicon in its pure form in 1824 by heating silicon tetrafluoride (SiF_4) with potassium metal. Today, silicon forms the basis for the modern microelectronics industry centered near San Francisco in a place that has come to be known as "Silicon Valley." The technology of the silicon chip (see figure) with its printed circuits has transformed computers from room-sized monsters with thousands of unreliable vacuum tubes to desktop and notebook-sized units with trouble-free "solid-state" circuitry.

A chip capable of transmitting 4,000,000 simultaneous phone conversations.

2 volumes of hydrogen react with 1 volume of oxygen to form 2 volumes of gaseous water and that 1 volume of hydrogen reacts with 1 volume of chlorine to form 2 volumes of hydrogen chloride. These results are represented schematically in Fig. 2.4.

In 1811 Avogadro interpreted these results by proposing that *at the same temperature and pressure, equal volumes of different gases contain the same number of particles.* This assumption (called **Avogadro's hypothesis**) makes sense if the distances between the particles in a gas are very great compared with the sizes of the particles. Under these conditions, the volume of a gas is determined by the number of molecules present, not by the size of the individual particles.

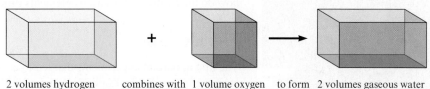

2 volumes hydrogen   combines with   1 volume oxygen   to form   2 volumes gaseous water

1 volume hydrogen   combines with   1 volume chlorine   to form   2 volumes hydrogen chloride

Figure 2.4 | A representation of some of Gay-Lussac's experimental results on combining gas volumes.

Figure 2.5 | A representation of combining gases at the molecular level. The spheres represent atoms in the molecules.

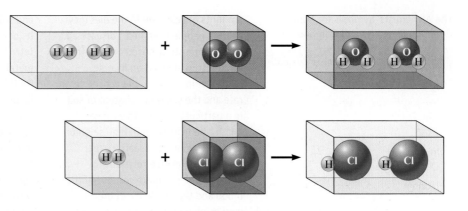

If Avogadro's hypothesis is correct, Gay-Lussac's result,

2 volumes of hydrogen react with 1 volume of oxygen ⟶ 2 volumes of water vapor

can be expressed as follows:

2 molecules* of hydrogen react with 1 molecule of oxygen ⟶ 2 molecules of water

These observations can best be explained by assuming that gaseous hydrogen, oxygen, and chlorine are all composed of diatomic (two-atom) molecules: H_2, O_2, and Cl_2, respectively. Gay-Lussac's results can then be represented as shown in Fig. 2.5. (Note that this reasoning suggests that the formula for water is H_2O, not OH as Dalton believed.)

Unfortunately, Avogadro's interpretations were not accepted by most chemists, and a half-century of confusion followed, in which many different assumptions were made about formulas and atomic masses.

During the nineteenth century, painstaking measurements were made of the masses of various elements that combined to form compounds. From these experiments a list of relative atomic masses could be determined. One of the chemists involved in contributing to this list was a Swede named Jöns Jakob Berzelius (1779–1848), who discovered the elements cerium, selenium, silicon, and thorium and developed the modern symbols for the elements used in writing the formulas of compounds.

There are seven elements that occur as diatomic molecules:

H_2, N_2, O_2, F_2, Cl_2, Br_2, I_2

2.4 | Early Experiments to Characterize the Atom

On the basis of the work of Dalton, Gay-Lussac, Avogadro, and others, chemistry was beginning to make sense. The concept of atoms was clearly a good idea. Inevitably, scientists began to wonder about the nature of the atom. What is an atom made of, and how do the atoms of the various elements differ?

The Electron

The first important experiments that led to an understanding of the composition of the atom were done by the English physicist J. J. Thomson (Fig. 2.6), who studied electrical discharges in partially evacuated tubes called **cathode-ray tubes** (Fig. 2.7) during

---

*A *molecule* is a collection of atoms (see Section 2.6).

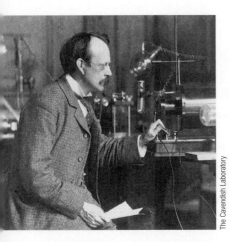

The Cavendish Laboratory

Figure 2.6 | J. J. Thomson (1856–1940) was an English physicist at Cambridge University. He received the Nobel Prize in physics in 1906.

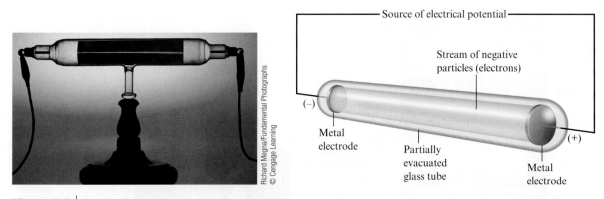

Figure 2.7 │ A cathode-ray tube. The fast-moving electrons excite the gas in the tube, causing a glow between the electrodes. The green color in the photo is due to the response of the screen (coated with zinc sulfide) to the electron beam.

the period from 1898 to 1903. Thomson found that when high voltage was applied to the tube, a "ray" he called a *cathode ray* (because it emanated from the negative electrode, or cathode) was produced. Because this ray was produced at the negative electrode and was repelled by the negative pole of an applied electric field (Fig. 2.8), Thomson postulated that the ray was a stream of negatively charged particles, now called **electrons**. From experiments in which he measured the deflection of the beam of electrons in a magnetic field, Thomson determined the *charge-to-mass ratio* of an electron:

$$\frac{e}{m} = -1.76 \times 10^8 \text{ C/g}$$

where e represents the charge on the electron in coulombs (C) and m represents the electron mass in grams.

One of Thomson's primary goals in his cathode-ray tube experiments was to gain an understanding of the structure of the atom. He reasoned that since electrons could be produced from electrodes made of various types of metals, *all* atoms must contain electrons. Since atoms were known to be electrically neutral, Thomson further assumed that atoms also must contain some positive charge. Thomson postulated that an atom consisted of a diffuse cloud of positive charge with the negative electrons embedded randomly in it. This model, shown in Fig. 2.9, is often called the *plum pudding model* because the electrons are like raisins dispersed in a pudding (the positive charge cloud), as in plum pudding, a favorite English dessert.

A classic English plum pudding in which the raisins represent the distribution of electrons in the atom.

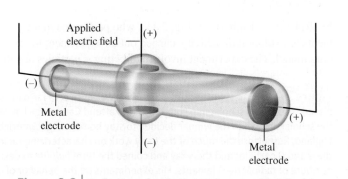

Figure 2.8 │ Deflection of cathode rays by an applied electric field.

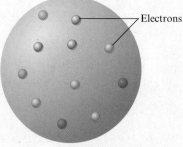

Figure 2.9 │ The plum pudding model of the atom.

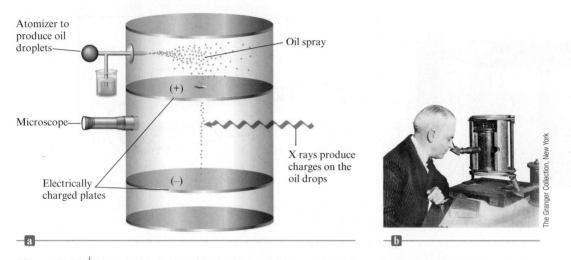

Figure 2.10 | (a) A schematic representation of the apparatus Millikan used to determine the charge on the electron. The fall of charged oil droplets due to gravity can be halted by adjusting the voltage across the two plates. This voltage and the mass of the oil drop can then be used to calculate the charge on the oil drop. Millikan's experiments showed that the charge on an oil drop is always a whole-number multiple of the electron charge. (b) Robert Millikan using his apparatus.

In 1909 Robert Millikan (1868–1953), working at the University of Chicago, performed very clever experiments involving charged oil drops. These experiments allowed him to determine the magnitude of the electron charge (Fig. 2.10). With this value and the charge-to-mass ratio determined by Thomson, Millikan was able to calculate the mass of the electron as 9.11×10^{-31} kg.

Radioactivity

In the late nineteenth century, scientists discovered that certain elements produce high-energy radiation. For example, in 1896 the French scientist Henri Becquerel found accidentally that a piece of a mineral containing uranium could produce its image on a photographic plate in the absence of light. He attributed this phenomenon to a spontaneous emission of radiation by the uranium, which he called **radioactivity**. Studies in the early twentieth century demonstrated three types of radioactive emission: gamma (γ) rays, beta (β) particles, and alpha (α) particles. A γ ray is high-energy "light"; a β particle is a high-speed electron; and an α particle has a 2+ charge, that is, a charge twice that of the electron and with the opposite sign. The mass of an α particle is 7300 times that of the electron. More modes of radioactivity are now known, and we will discuss them in Chapter 19. Here we will consider only α particles because they were used in some crucial early experiments.

The Nuclear Atom

In 1911 Ernest Rutherford (Fig. 2.11), who performed many of the pioneering experiments to explore radioactivity, carried out an experiment to test Thomson's plum pudding model. The experiment involved directing α particles at a thin sheet of metal foil,

Figure 2.11 | Ernest Rutherford (1871–1937) was born on a farm in New Zealand. In 1895 he placed second in a scholarship competition to attend Cambridge University but was awarded the scholarship when the winner decided to stay home and get married. As a scientist in England, Rutherford did much of the early work on characterizing radioactivity. He named the α and β particles and the γ ray and coined the term *half-life* to describe an important attribute of radioactive elements. His experiments on the behavior of α particles striking thin metal foils led him to postulate the nuclear atom. He also invented the name *proton* for the nucleus of the hydrogen atom. He received the Nobel Prize in chemistry in 1908.

Figure 2.12 | Rutherford's experiment on α-particle bombardment of metal foil.

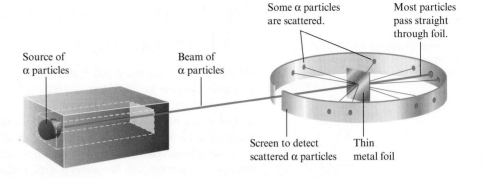

Source of
α particles

Beam of
α particles

Some α particles
are scattered.

Most particles
pass straight
through foil.

Screen to detect
scattered α particles

Thin
metal foil

as illustrated in Fig. 2.12. Rutherford reasoned that if Thomson's model were accurate, the massive α particles should crash through the thin foil like cannonballs through gauze, as shown in Fig. 2.13(a). He expected the α particles to travel through the foil with, at the most, very minor deflections in their paths. The results of the experiment were very different from those Rutherford anticipated. Although most of the α particles passed straight through, many of the particles were deflected at large angles, as shown in Fig. 2.13(b), and some were reflected, never hitting the detector. This outcome was a great surprise to Rutherford. (He wrote that this result was comparable with shooting a howitzer at a piece of paper and having the shell reflected back.)

Rutherford knew from these results that the plum pudding model for the atom could not be correct. The large deflections of the α particles could be caused only by a center of concentrated positive charge that contains most of the atom's mass, as illustrated in Fig. 2.13(b). Most of the α particles pass directly through the foil because the atom is mostly open space. The deflected α particles are those that had a "close encounter" with the massive positive center of the atom, and the few reflected α particles are those that made a "direct hit" on the much more massive positive center.

In Rutherford's mind these results could be explained only in terms of a **nuclear atom**—an atom with a dense center of positive charge (the **nucleus**) with electrons moving around the nucleus at a distance that is large relative to the nuclear radius.

Critical Thinking

You have learned about three different models of the atom: Dalton's model, Thomson's model, and Rutherford's model. What if Dalton was correct? What would Rutherford have expected from his experiments with gold foil? What if Thomson was correct? What would Rutherford have expected from his experiments with gold foil?

Figure 2.13 | Rutherford's experiment.

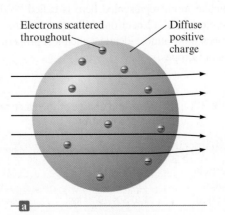

Electrons scattered
throughout

Diffuse
positive
charge

The expected results of the metal foil experiment if Thomson's model were correct.

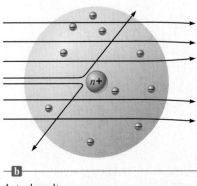

Actual results.

2.5 | The Modern View of Atomic Structure: An Introduction

The forces that bind the positively charged protons in the nucleus will be discussed in Chapter 19.

In the years since Thomson and Rutherford, a great deal has been learned about atomic structure. Because much of this material will be covered in detail in later chapters, only an introduction will be given here. The simplest view of the atom is that it consists of a tiny nucleus (with a diameter of about 10^{-13} cm) and electrons that move about the nucleus at an average distance of about 10^{-8} cm from it (Fig. 2.14).

As we will see later, the chemistry of an atom mainly results from its electrons. For this reason, chemists can be satisfied with a relatively crude nuclear model. The nucleus is assumed to contain **protons**, which have a positive charge equal in magnitude to the electron's negative charge, and **neutrons**, which have virtually the same mass as a proton but no charge. The masses and charges of the electron, proton, and neutron are shown in Table 2.1.

Two striking things about the nucleus are its small size compared with the overall size of the atom and its extremely high density. The tiny nucleus accounts for almost all the atom's mass. Its great density is dramatically demonstrated by the fact that a piece of nuclear material about the size of a pea would have a mass of 250 million tons!

The *chemistry* of an atom arises from its electrons.

An important question to consider at this point is, *"If all atoms are composed of these same components, why do different atoms have different chemical properties?"* The answer to this question lies in the number and the arrangement of the electrons. The electrons constitute most of the atomic volume and thus are the parts that "intermingle" when atoms combine to form molecules. Therefore, the number of electrons possessed by a given atom greatly affects its ability to interact with other atoms. As a result, the atoms of different elements, which have different numbers of protons and electrons, show different chemical behavior.

A sodium atom has 11 protons in its nucleus. Since atoms have no net charge, the number of electrons must equal the number of protons. Therefore, a sodium atom has 11 electrons moving around its nucleus. It is *always* true that a sodium atom has 11 protons and 11 electrons. However, each sodium atom also has neutrons in its nucleus, and different types of sodium atoms exist that have different numbers of neutrons. For example, consider the sodium atoms represented in Fig. 2.15. These two atoms are **isotopes**, or *atoms with the same number of protons but different numbers of neutrons.* Note that the symbol for one particular type of sodium atom is written

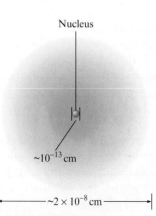

If the atomic nucleus were the size of this ball bearing, a typical atom would be the size of this stadium.

Mass number ⟶ ^A_ZX ⟵ Element symbol
Atomic number ⟶

Mass number ⟶ $^{23}_{11}$Na ⟵ Element symbol
Atomic number ⟶

where the **atomic number** Z (number of protons) is written as a subscript, and the **mass number** A (the total number of protons and neutrons) is written as a superscript. (The particular atom represented here is called "sodium twenty-three." It has 11 electrons, 11 protons, and 12 neutrons.) Because the chemistry of an atom is due to its electrons, isotopes show almost identical chemical properties. In nature most elements contain mixtures of isotopes.

Nucleus

$\sim10^{-13}$ cm

$\sim2 \times 10^{-8}$ cm

Figure 2.14 | A nuclear atom viewed in cross section. Note that this drawing is not to scale.

Table 2.1 | The Mass and Charge of the Electron, Proton, and Neutron

Particle	Mass	Charge*
Electron	9.109×10^{-31} kg	1−
Proton	1.673×10^{-27} kg	1+
Neutron	1.675×10^{-27} kg	None

*The magnitude of the charge of the electron and the proton is 1.60×10^{-19} C.

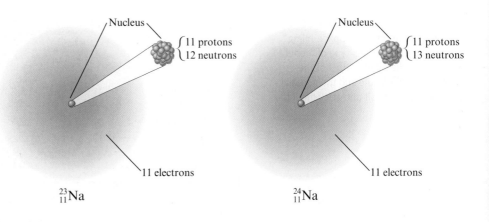

Figure 2.15 | Two isotopes of sodium. Both have 11 protons and 11 electrons, but they differ in the number of neutrons in their nuclei.

$^{23}_{11}\text{Na}$ $^{24}_{11}\text{Na}$

Critical Thinking

The average diameter of an atom is 2×10^{-10} m. What if the average diameter of an atom were 1 cm? How tall would you be?

Interactive Example 2.2

Sign in at http://login.cengagebrain.com to try this Interactive Example in OWL.

Writing the Symbols for Atoms

Write the symbol for the atom that has an atomic number of 9 and a mass number of 19. How many electrons and how many neutrons does this atom have?

Solution

The atomic number 9 means the atom has 9 protons. This element is called *fluorine*, symbolized by F. The atom is represented as

$$^{19}_{9}\text{F}$$

and is called *fluorine nineteen*. Since the atom has 9 protons, it also must have 9 electrons to achieve electrical neutrality. The mass number gives the total number of protons and neutrons, which means that this atom has 10 neutrons.

See Exercises 2.59 through 2.62

2.6 | Molecules and Ions

From a chemist's viewpoint, the most interesting characteristic of an atom is its ability to combine with other atoms to form compounds. It was John Dalton who first recognized that chemical compounds are collections of atoms, but he could not determine the structure of atoms or their means for binding to each other. During the twentieth century, we learned that atoms have electrons and that these electrons participate in bonding one atom to another. We will discuss bonding thoroughly in Chapters 8 and 9; here, we will introduce some simple bonding ideas that will be useful in the next few chapters.

The forces that hold atoms together in compounds are called **chemical bonds**. One way that atoms can form bonds is by *sharing electrons*. These bonds are called **covalent bonds**, and the resulting collection of atoms is called a **molecule**. Molecules can be represented in several different ways. The simplest method is the **chemical formula**, in which the symbols for the elements are used to indicate the types of atoms present and subscripts are used to indicate the relative numbers of atoms. For example, the formula for carbon dioxide is CO_2, meaning that each molecule contains 1 atom of carbon and 2 atoms of oxygen.

Figure 2.16 | (a) The structural formula for methane. (b) Space-filling model of methane. This type of model shows both the relative sizes of the atoms in the molecule and their spatial relationships. (c) Ball-and-stick model of methane.

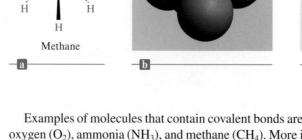

Methane

Ammonia

Examples of molecules that contain covalent bonds are hydrogen (H_2), water (H_2O), oxygen (O_2), ammonia (NH_3), and methane (CH_4). More information about a molecule is given by its **structural formula**, in which the individual bonds are shown (indicated by lines). Structural formulas may or may not indicate the actual shape of the molecule. For example, water might be represented as

$$H{-}O{-}H \quad \text{or} \quad \overset{O}{\underset{H \quad H}{\diagup\diagdown}}$$

The structure on the right shows the actual shape of the water molecule. Scientists know from experimental evidence that the molecule looks like this. (We will study the shapes of molecules further in Chapter 8.)

The structural formula for ammonia is shown in the margin at left. Note that atoms connected to the central atom by dashed lines are behind the plane of the paper, and atoms connected to the central atom by wedges are in front of the plane of the paper.

In a compound composed of molecules, the individual molecules move around as independent units. For example, a molecule of methane gas can be represented in several ways. The structural formula for methane (CH_4) is shown in Fig. 2.16(a). The **space-filling model** of methane, which shows the relative sizes of the atoms as well as their relative orientation in the molecule, is given in Fig. 2.16(b). **Ball-and-stick models** are also used to represent molecules. The ball-and-stick structure of methane is shown in Fig. 2.16(c).

A second type of chemical bond results from attractions among ions. An **ion** is an atom or group of atoms that has a net positive or negative charge. The best-known ionic compound is common table salt, or sodium chloride, which forms when neutral chlorine and sodium react.

To see how the ions are formed, consider what happens when an electron is transferred from a sodium atom to a chlorine atom (the neutrons in the nuclei will be ignored):

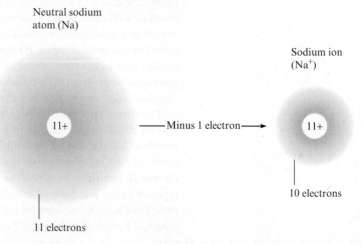

Neutral sodium
atom (Na)

11+

——Minus 1 electron——>

Sodium ion
(Na^+)

11+

10 electrons

11 electrons

Na$^+$ is usually called the *sodium ion* rather than the sodium cation. Also Cl$^-$ is called the *chloride ion* rather than the chloride anion. In general, when a specific ion is referred to, the word *ion* rather than *cation* or *anion* is used.

With one electron stripped off, the sodium, with its 11 protons and only 10 electrons, now has a net 1+ charge—it has become a *positive ion*. A positive ion is called a **cation**. The sodium ion is written as Na$^+$, and the process can be represented in short-hand form as

$$Na \longrightarrow Na^+ + e^-$$

If an electron is added to chlorine,

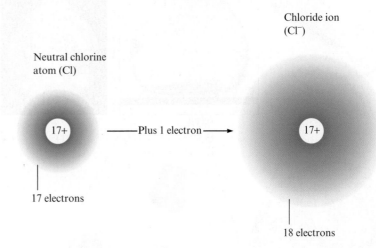

Chloride ion (Cl$^-$)

Neutral chlorine atom (Cl)

17+ ——— Plus 1 electron ——→ 17+

17 electrons

18 electrons

the 18 electrons produce a net 1− charge; the chlorine has become an *ion with a negative charge*—an **anion**. The chloride ion is written as Cl$^-$, and the process is represented as

$$Cl + e^- \longrightarrow Cl^-$$

Because anions and cations have opposite charges, they attract each other. This *force of attraction between oppositely charged ions* is called **ionic bonding**. As illustrated in Fig. 2.17, sodium metal and chlorine gas (a green gas composed of Cl$_2$ molecules) react to form solid sodium chloride, which contains many Na$^+$ and Cl$^-$ ions packed together and forms the beautiful colorless cubic crystals.

A solid consisting of oppositely charged ions is called an **ionic solid**. Ionic solids can consist of simple ions, as in sodium chloride, or of **polyatomic** (many atom) **ions**, as in ammonium nitrate (NH$_4$NO$_3$), which contains ammonium ions (NH$_4$$^+$) and nitrate ions (NO$_3$$^-$). The ball-and-stick models of these ions are shown in Fig. 2.18.

2.7 | An Introduction to the Periodic Table

In a room where chemistry is taught or practiced, a chart called the **periodic table** is almost certain to be found hanging on the wall. This chart shows all the known elements and gives a good deal of information about each. As our study of chemistry progresses, the usefulness of the periodic table will become more obvious. This section will simply introduce it to you.

A simplified version of the periodic table is shown in Fig. 2.19. The letters in the boxes are the symbols for the elements; these abbreviations are based on the current element names or the original names (Table 2.2). The number shown above each symbol is the *atomic number* (number of protons) for that element. For example, carbon (C) has atomic number 6, and lead (Pb) has atomic number 82. Most of the elements are **metals**. Metals have characteristic physical properties such as efficient conduction of heat and electricity, malleability (they can be hammered into thin sheets), ductility (they can be pulled into wires), and (often) a lustrous appearance. Chemically, metals tend to *lose* electrons to form positive ions. For example, copper is a typical metal. It

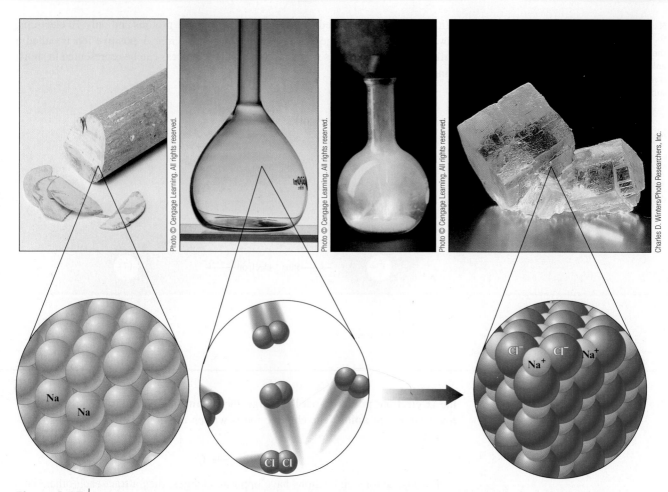

Figure 2.17 │ Sodium metal (which is so soft it can be cut with a knife and which consists of individual sodium atoms) reacts with chlorine gas (which contains Cl_2 molecules) to form solid sodium chloride (which contains Na^+ and Cl^- ions packed together).

Figure 2.18 │ Ball-and-stick models of the ammonium ion (NH_4^+) and the nitrate ion (NO_3^-). These ions are each held together by covalent bonds.

is lustrous (although it tarnishes readily); it is an excellent conductor of electricity (it is widely used in electrical wires); and it is readily formed into various shapes, such as pipes for water systems. Copper is also found in many salts, such as the beautiful blue copper sulfate, in which copper is present as Cu^{2+} ions. Copper is a member of the transition metals—the metals shown in the center of the periodic table.

The relatively few **nonmetals** appear in the upper-right corner of the table (to the right of the heavy line in Fig. 2.19), except hydrogen, a nonmetal that resides in the

Table 2.2 │ The Symbols for the Elements That Are Based on the Original Names

Current Name	Original Name	Symbol
Antimony	Stibium	Sb
Copper	Cuprum	Cu
Iron	Ferrum	Fe
Lead	Plumbum	Pb
Mercury	Hydrargyrum	Hg
Potassium	Kalium	K
Silver	Argentum	Ag
Sodium	Natrium	Na
Tin	Stannum	Sn
Tungsten	Wolfram	W

The periodic table

1 1A																		18 8A
1 H	2 2A												13 3A	14 4A	15 5A	16 6A	17 7A	2 He
3 Li	4 Be												5 B	6 C	7 N	8 O	9 F	10 Ne
11 Na	12 Mg	3	4	5	6	7	8	9	10	11	12		13 Al	14 Si	15 P	16 S	17 Cl	18 Ar
19 K	20 Ca	21 Sc	22 Ti	23 V	24 Cr	25 Mn	26 Fe	27 Co	28 Ni	29 Cu	30 Zn		31 Ga	32 Ge	33 As	34 Se	35 Br	36 Kr
37 Rb	38 Sr	39 Y	40 Zr	41 Nb	42 Mo	43 Tc	44 Ru	45 Rh	46 Pd	47 Ag	48 Cd		49 In	50 Sn	51 Sb	52 Te	53 I	54 Xe
55 Cs	56 Ba	57 La*	72 Hf	73 Ta	74 W	75 Re	76 Os	77 Ir	78 Pt	79 Au	80 Hg		81 Tl	82 Pb	83 Bi	84 Po	85 At	86 Rn
87 Fr	88 Ra	89 Ac†	104 Rf	105 Db	106 Sg	107 Bh	108 Hs	109 Mt	110 Ds	111 Rg	112 Cn		113 Uut	114 Fl	115 Uup	116 Lv	117 Uus	118 Uuo

Alkali metals — (left column, Group 1A below H)

Alkaline earth metals (Group 2A)

Transition metals (Groups 3–12)

Halogens (Group 17)

Noble gases (Group 18)

*Lanthanides	58 Ce	59 Pr	60 Nd	61 Pm	62 Sm	63 Eu	64 Gd	65 Tb	66 Dy	67 Ho	68 Er	69 Tm	70 Yb	71 Lu
†Actinides	90 Th	91 Pa	92 U	93 Np	94 Pu	95 Am	96 Cm	97 Bk	98 Cf	99 Es	100 Fm	101 Md	102 No	103 Lr

Figure 2.19 | The periodic table.

Metals tend to form positive ions; nonmetals tend to form negative ions.

Elements in the same vertical column in the periodic table form a *group* (or *family*) and generally have similar properties.

Samples of chlorine gas, liquid bromine, and solid iodine.

upper-left corner. The nonmetals lack the physical properties that characterize the metals. Chemically, they tend to *gain* electrons in reactions with metals to form negative ions. Nonmetals often bond to each other by forming covalent bonds. For example, chlorine is a typical nonmetal. Under normal conditions it exists as Cl_2 molecules; it reacts with metals to form salts containing Cl^- ions (NaCl, for example); and it forms covalent bonds with nonmetals (for example, hydrogen chloride gas, HCl).

The periodic table is arranged so that elements in the same vertical columns (called **groups** or **families**) have *similar chemical properties*. For example, all of the **alkali metals**, members of Group 1A—lithium (Li), sodium (Na), potassium (K), rubidium (Rb), cesium (Cs), and francium (Fr)—are very active elements that readily form ions with a 1+ charge when they react with nonmetals. The members of Group 2A—beryllium (Be), magnesium (Mg), calcium (Ca), strontium (Sr), barium (Ba), and radium (Ra)—are called the **alkaline earth metals**. They all form ions with a 2+ charge when they react with nonmetals. The **halogens**, the members of Group 7A—fluorine (F), chlorine (Cl), bromine (Br), iodine (I), and astatine (At)—all form diatomic molecules. Fluorine, chlorine, bromine, and iodine all react with metals to form salts containing ions with a 1− charge (F^-, Cl^-, Br^-, and I^-). The members of Group 8A—helium (He), neon (Ne), argon (Ar), krypton (Kr), xenon (Xe), and radon (Rn)—are known as the **noble gases**. They all exist under normal conditions as monatomic (single-atom) gases and have little chemical reactivity.

Chemical connections
Hassium Fits Right In

Hassium, element 108, does not exist in nature but must be made in a particle accelerator. It was first created in 1984 and can be made by shooting magnesium-26 ($^{26}_{12}Mg$) atoms at curium-248 ($^{248}_{96}Cm$) atoms. The collisions between these atoms produce some hassium-265 ($^{265}_{108}Hs$) atoms. The position of hassium in the periodic table (see Fig. 2.19) in the vertical column containing iron, ruthenium, and osmium suggests that hassium should have chemical properties similar to these metals.

However, it is not easy to test this prediction—only a few atoms of hassium can be made at a given time and they last for only about 9 seconds. Imagine having to get your next lab experiment done in 9 seconds!

Amazingly, a team of chemists from the Lawrence Berkeley National Laboratory in California, the Paul Scherrer Institute and the University of Bern in Switzerland, and the Institute of Nuclear Chemistry in Germany have done experiments to characterize the chemical behavior of hassium. For

example, they have observed that hassium atoms react with oxygen to form a hassium oxide compound of the type expected from its position on the periodic table. The team has also measured other properties of hassium, including the energy released as it undergoes nuclear decay to another atom.

This work would have surely pleased Dmitri Mendeleev (see Fig. 7.24), who originally developed the periodic table and showed its power to predict chemical properties.

Note from Fig. 2.19 that alternate sets of symbols are used to denote the groups. The symbols 1A through 8A are the traditional designations, whereas the numbers 1 to 18 have been suggested recently. In this text the 1A to 8A designations will be used.

The horizontal rows of elements in the periodic table are called **periods**. Horizontal row 1 is called the *first period* (it contains H and He); row 2 is called the *second period* (elements Li through Ne); and so on.

Another format of the periodic table will be discussed in Section 7.11.

We will learn much more about the periodic table as we continue with our study of chemistry. Meanwhile, when an element is introduced in this text, you should always note its position on the periodic table.

2.8 | Naming Simple Compounds

When chemistry was an infant science, there was no system for naming compounds. Names such as sugar of lead, blue vitrol, quicklime, Epsom salts, milk of magnesia, gypsum, and laughing gas were coined by early chemists. Such names are called *common names*. As chemistry grew, it became clear that using common names for compounds would lead to unacceptable chaos. Nearly 5 million chemical compounds are currently known. Memorizing common names for these compounds would be an impossible task.

The solution, of course, is to adopt a *system* for naming compounds in which the name tells something about the composition of the compound. After learning the system, a chemist given a formula should be able to name the compound or, given a name, should be able to construct the compound's formula. In this section we will specify the most important rules for naming compounds other than organic compounds (those based on chains of carbon atoms).

We will begin with the systems for naming inorganic **binary compounds**—compounds composed of two elements—which we classify into various types for easier recognition. We will consider both ionic and covalent compounds.

Binary Ionic Compounds (Type I)

Binary ionic compounds contain a positive ion (cation) always written first in the formula and a negative ion (anion). In naming these compounds, the following rules apply:

A monatomic cation has the same name as its parent element.

Naming Type I Binary Compounds

1. The cation is always named first and the anion second.
2. A monatomic (meaning "one-atom") cation takes its name from the name of the element. For example, Na^+ is called sodium in the names of compounds containing this ion.
3. A monatomic anion is named by taking the root of the element name and adding *-ide*. Thus the Cl^- ion is called chloride.

In formulas of ionic compounds, simple ions are represented by the element symbol: Cl means Cl^-, Na means Na^+, and so on.

Some common monatomic cations and anions and their names are given in Table 2.3.

The rules for naming binary ionic compounds are illustrated by the following examples:

Compound	Ions Present	Name
NaCl	Na^+, Cl^-	Sodium chloride
KI	K^+, I^-	Potassium iodide
CaS	Ca^{2+}, S^{2-}	Calcium sulfide
Li_3N	Li^+, N^{3-}	Lithium nitride
CsBr	Cs^+, Br^-	Cesium bromide
MgO	Mg^{2+}, O^{2-}	Magnesium oxide

Interactive Example 2.3

Sign in at http://login.cengagebrain .com to try this Interactive Example in **OWL**.

Naming Type I Binary Compounds

Name each binary compound.

a. CsF   **b.** $AlCl_3$ **c.** LiH

Solution

a. CsF is cesium fluoride.

b. $AlCl_3$ is aluminum chloride.

c. LiH is lithium hydride.

Notice that, in each case, the cation is named first and then the anion is named.

See Exercise 2.71

Table 2.3 | Common Monatomic Cations and Anions

Cation	Name	Anion	Name
H^+	Hydrogen	H^-	Hydride
Li^+	Lithium	F^-	Fluoride
Na^+	Sodium	Cl^-	Chloride
K^+	Potassium	Br^-	Bromide
Cs^+	Cesium	I^-	Iodide
Be^{2+}	Beryllium	O^{2-}	Oxide
Mg^{2+}	Magnesium	S^{2-}	Sulfide
Ca^{2+}	Calcium	N^{3-}	Nitride
Ba^{2+}	Barium	P^{3-}	Phosphide
Al^{3+}	Aluminum		

Formulas from Names

So far we have started with the chemical formula of a compound and decided on its systematic name. The reverse process is also important. For example, given the name calcium chloride, we can write the formula as $CaCl_2$ because we know that calcium forms only Ca^{2+} ions and that, since chloride is Cl^-, two of these anions will be required to give a neutral compound.

Binary Ionic Compounds (Type II)

In the binary ionic compounds considered earlier (Type I), the metal present forms only a single type of cation. That is, sodium forms only Na^+, calcium forms only Ca^{2+}, and so on. However, as we will see in more detail later in the text, there are many metals that form more than one type of positive ion and thus form more than one type of ionic compound with a given anion. For example, the compound $FeCl_2$ contains Fe^{2+} ions, and the compound $FeCl_3$ contains Fe^{3+} ions. In a case such as this, the *charge on the metal ion must be specified.* The systematic names for these two iron compounds are iron(II) chloride and iron(III) chloride, respectively, where the *Roman numeral indicates the charge of the cation.*

Another system for naming these ionic compounds that is seen in the older literature was used for metals that form only two ions. *The ion with the higher charge has a name ending in* -ic, *and the one with the lower charge has a name ending in* -ous. In this system, for example, Fe^{3+} is called the ferric ion, and Fe^{2+} is called the ferrous ion. The names for $FeCl_3$ and $FeCl_2$ are then ferric chloride and ferrous chloride, respectively. In this text we will use the system that employs Roman numerals. Table 2.4 lists the systematic names for many common type II cations.

Table 2.4 | Common Type II Cations

Ion	Systematic Name
Fe^{3+}	Iron(III)
Fe^{2+}	Iron(II)
Cu^{2+}	Copper(II)
Cu^+	Copper(I)
Co^{3+}	Cobalt(III)
Co^{2+}	Cobalt(II)
Sn^{4+}	Tin(IV)
Sn^{2+}	Tin(II)
Pb^{4+}	Lead(IV)
Pb^{2+}	Lead(II)
Hg^{2+}	Mercury(II)
Hg_2^{2+}*	Mercury(I)
Ag^+	Silver[†]
Zn^{2+}	Zinc[†]
Cd^{2+}	Cadmium[†]

*Note that mercury(I) ions always occur bound together to form Hg_2^{2+} ions.
[†]Although these are transition metals, they form only one type of ion, and a Roman numeral is not used.

Interactive Example 2.4

Sign in at http://login.cengagebrain .com to try this Interactive Example in **OWL**.

Formulas from Names for Type I Binary Compounds

Given the following systematic names, write the formula for each compound:

a. potassium iodide

b. calcium oxide

c. gallium bromide

Solution

Name	Formula	Comments
a. potassium iodide	KI	Contains K^+ and I^-
b. calcium oxide	CaO	Contains Ca^{2+} and O^{2-}
c. gallium bromide	$GaBr_3$	Contains Ga^{3+} and Br^-
		Must have $3Br^-$ to balance charge of Ga^{3+}

See Exercise 2.71

Interactive Example 2.5

Sign in at http://login.cengagebrain .com to try this Interactive Example in **OWL**.

Naming Type II Binary Compounds

1. Give the systematic name for each of the following compounds:
   a. CuCl   **b.** HgO   **c.** Fe_2O_3

2. Given the following systematic names, write the formula for each compound:
   a. Manganese(IV) oxide
   b. Lead(II) chloride

Solution

All of these compounds include a metal that can form more than one type of cation. Thus we must first determine the charge on each cation. This can be done by recognizing that a compound must be electrically neutral; that is, the positive and negative charges must exactly balance.

1.

Formula	Name	Comments
a. CuCl	Copper(I) chloride	Because the anion is Cl^-, the cation must be Cu^+ (for charge balance), which requires a Roman numeral I.
b. HgO	Mercury(II) oxide	Because the anion is O^{2-}, the cation must be Hg^{2+} [mercury(II)].
c. Fe_2O_3	Iron(III) oxide	The three O^{2-} ions carry a total charge of 6−, so two Fe^{3+} ions [iron(III)] are needed to give a 6+ charge.

2.

Name	Formula	Comments
a. Manganese(IV) oxide	MnO_2	Two O^{2-} ions (total charge 4−) are required by the Mn^{4+} ion [manganese(IV)].
b. Lead(II) chloride	$PbCl_2$	Two Cl^- ions are required by the Pb^{2+} ion [lead(II)] for charge balance.

See Exercise 2.72

Mercury(II) oxide.

A compound containing a transition metal usually requires a Roman numeral in its name.

Note that the use of a Roman numeral in a systematic name is required only in cases where more than one ionic compound forms between a given pair of elements. This case most commonly occurs for compounds containing transition metals, which often form more than one cation. Elements that form only one cation do not need to be identified by a Roman numeral. Common metals that do not require Roman numerals are the Group 1A elements, which form only 1+ ions; the Group 2A elements, which form only 2+ ions; and aluminum, which forms only Al^{3+}. The element silver deserves special mention at this point. In virtually all its compounds, silver is found as the Ag^+ ion. Therefore, although silver is a transition metal (and can potentially form ions other than Ag^+), silver compounds are usually named without a Roman numeral. Thus AgCl is typically called silver chloride rather than silver(I) chloride, although the latter name is technically correct. Also, a Roman numeral is not used for zinc compounds, since zinc forms only the Zn^{2+} ion.

As shown in Example 2.5, when a metal ion is present that forms more than one type of cation, the charge on the metal ion must be determined by balancing the positive and negative charges of the compound. To do this you must be able to recognize the common cations and anions and know their charges (see Tables 2.3 and 2.5). The procedure for naming binary ionic compounds is summarized in Fig. 2.20.

Critical Thinking

We can use the periodic table to tell us something about the stable ions formed by many atoms. For example, the atoms in column 1 always form 1+ ions. The transition metals, however, can form more than one type of stable ion. What if each transition metal ion had only one possible charge? How would the naming of compounds be different?

Figure 2.20 | Flowchart for naming binary ionic compounds.

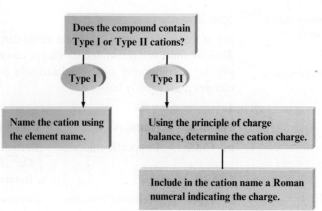

Does the compound contain Type I or Type II cations?

Type I

Type II

Name the cation using the element name.

Using the principle of charge balance, determine the cation charge.

Include in the cation name a Roman numeral indicating the charge.

Various chromium compounds dissolved in water. From left to right: $CrCl_2$, $K_2Cr_2O_7$, $Cr(NO_3)_3$, $CrCl_3$, K_2CrO_4.

Richard Megna/Fundamental Photographs © Cengage Learning

Interactive Example 2.6

Sign in at http://login.cengagebrain.com to try this Interactive Example in **OWL**.

Naming Binary Compounds

1. Give the systematic name for each of the following compounds:

   a. $CoBr_2$ **b.** $CaCl_2$ **c.** Al_2O_3

2. Given the following systematic names, write the formula for each compound:

   a. Chromium(III) chloride

   b. Gallium iodide

Solution

1.

Formula	Name	Comments
a. $CoBr_2$	Cobalt(II) bromide	Cobalt is a transition metal; the compound name must have a Roman numeral. The two Br^- ions must be balanced by a Co^{2+} ion.
b. $CaCl_2$	Calcium chloride	Calcium, an alkaline earth metal, forms only the Ca^{2+} ion. A Roman numeral is not necessary.
c. Al_2O_3	Aluminum oxide	Aluminum forms only the Al^{3+} ion. A Roman numeral is not necessary.

2.

Name	Formula	Comments
a. Chromium(III) chloride	$CrCl_3$	Chromium(III) indicates that Cr^{3+} is present, so 3 Cl^- ions are needed for charge balance.
b. Gallium iodide	GaI_3	Gallium always forms 3+ ions, so 3 I^- ions are required for charge balance.

See Exercises 2.73 and 2.74

The common Type I and Type II ions are summarized in Fig. 2.21. Also shown in Fig. 2.21 are the common monatomic ions.

Figure 2.21 | The common cations and anions.

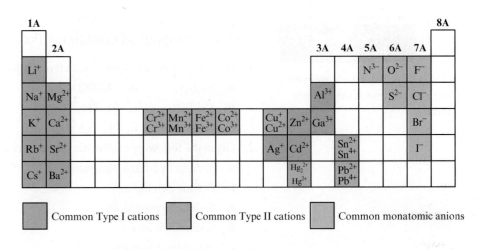

Ionic Compounds with Polyatomic Ions

We have not yet considered ionic compounds that contain polyatomic ions. For example, the compound ammonium nitrate, NH_4NO_3, contains the polyatomic ions NH_4^+ and NO_3^-. Polyatomic ions are assigned special names that *must be memorized* to name the compounds containing them. The most important polyatomic ions and their names are listed in Table 2.5.

Note in Table 2.5 that several series of anions contain an atom of a given element and different numbers of oxygen atoms. These anions are called **oxyanions**. When there are two members in such a series, the name of the one with the smaller number of oxygen atoms ends in *-ite* and the name of the one with the larger number ends in *-ate*—for example, sulfite (SO_3^{2-}) and sulfate (SO_4^{2-}). When more than two oxyanions make up a series, *hypo-* (less than) and *per-* (more than) are used as prefixes to name the members of the series with the fewest and the most oxygen atoms, respectively. The best example involves the oxyanions containing chlorine, as shown in Table 2.5.

Polyatomic ion formulas must be memorized.

Table 2.5 | Common Polyatomic Ions

Ion	Name	Ion	Name
Hg_2^{2+}	Mercury(I)	NCS^- or SCN^-	Thiocyanate
NH_4^+	Ammonium	CO_3^{2-}	Carbonate
NO_2^-	Nitrite	HCO_3^-	Hydrogen carbonate
NO_3^-	Nitrate		(bicarbonate is a widely
SO_3^{2-}	Sulfite		used common name)
SO_4^{2-}	Sulfate	ClO^- or OCl^-	Hypochlorite
HSO_4^-	Hydrogen sulfate	ClO_2^-	Chlorite
	(bisulfate is a widely	ClO_3^-	Chlorate
	used common name)	ClO_4^-	Perchlorate
OH^-	Hydroxide	$C_2H_3O_2^-$	Acetate
CN^-	Cyanide	MnO_4^-	Permanganate
PO_4^{3-}	Phosphate	$Cr_2O_7^{2-}$	Dichromate
HPO_4^{2-}	Hydrogen phosphate	CrO_4^{2-}	Chromate
$H_2PO_4^-$	Dihydrogen phosphate	O_2^{2-}	Peroxide
		$C_2O_4^{2-}$	Oxalate
		$S_2O_3^{2-}$	Thiosulfate

Naming Compounds Containing Polyatomic Ions

1. Give the systematic name for each of the following compounds:
   a. Na_2SO_4 d. $Mn(OH)_2$
   b. KH_2PO_4 e. Na_2SO_3
   c. $Fe(NO_3)_3$ f. Na_2CO_3
2. Given the following systematic names, write the formula for each compound:
   a. Sodium hydrogen carbonate
   b. Cesium perchlorate
   c. Sodium hypochlorite
   d. Sodium selenate
   e. Potassium bromate

Solution

1.

Formula	Name	Comments
a. Na_2SO_4	Sodium sulfate	
b. KH_2PO_4	Potassium dihydrogen phosphate	
c. $Fe(NO_3)_3$	Iron(III) nitrate	Transition metal—name must contain a Roman numeral. The Fe^{3+} ion balances three NO_3^- ions.
d. $Mn(OH)_2$	Manganese(II) hydroxide	Transition metal—name must contain a Roman numeral. The Mn^{2+} ion balances three OH^- ions.
e. Na_2SO_3	Sodium sulfite	
f. Na_2CO_3	Sodium carbonate	

2.

Name	Formula	Comments
a. Sodium hydrogen carbonate	$NaHCO_3$	Often called sodium bicarbonate.
b. Cesium perchlorate	$CsClO_4$	
c. Sodium hypochlorite	$NaOCl$	
d. Sodium selenate	Na_2SeO_4	Atoms in the same group, like sulfur and selenium, often form similar ions that are named similarly. Thus SeO_4^{2-} is selenate, like SO_4^{2-} (sulfate).
e. Potassium bromate	$KBrO_3$	As above, BrO_3^- is bromate, like ClO_3^- (chlorate).

See Exercises 2.75 and 2.76

Binary Covalent Compounds (Type III)

In *binary covalent compounds*, the element names follow the same rules as for binary ionic compounds.

Binary covalent compounds are formed between *two nonmetals*. Although these compounds do not contain ions, they are named very similarly to binary ionic compounds.

| Table 2.6 | Prefixes Used to Indicate Number in Chemical Names |

Prefix	Number Indicated
mono-	1
di-	2
tri-	3
tetra-	4
penta-	5
hexa-	6
hepta-	7
octa-	8
nona-	9
deca-	10

Naming Binary Covalent Compounds

1. The first element in the formula is named first, using the full element name.
2. The second element is named as if it were an anion.
3. Prefixes are used to denote the numbers of atoms present. These prefixes are given in Table 2.6.
4. The prefix *mono-* is never used for naming the first element. For example, CO is called carbon monoxide, *not* monocarbon monoxide.

To see how these rules apply, we will now consider the names of the several covalent compounds formed by nitrogen and oxygen:

Compound	Systematic Name	Common Name
N_2O	Dinitrogen monoxide	Nitrous oxide
NO	Nitrogen monoxide	Nitric oxide
NO_2	Nitrogen dioxide	
N_2O_3	Dinitrogen trioxide	
N_2O_4	Dinitrogen tetroxide	
N_2O_5	Dinitrogen pentoxide	

Notice from the preceding examples that to avoid awkward pronunciations, we often drop the final *o* or *a* of the prefix when the element begins with a vowel. For example, N_2O_4 is called dinitrogen tetroxide, *not* dinitrogen tetr*a*oxide, and CO is called carbon monoxide, *not* carbon mon*o*oxide.

Some compounds are always referred to by their common names. Three examples are water, ammonia, and hydrogen peroxide. The systematic names for H_2O, NH_3, and H_2O_2 are never used.

Naming Type III Binary Compounds

1. Name each of the following compounds:
   a. PCl_5 **b.** PCl_3 **c.** SO_2
2. From the following systematic names, write the formula for each compound:
   a. Sulfur hexafluoride
   b. Sulfur trioxide
   c. Carbon dioxide

Solution

1.

Formula	Name
a. PCl_5	Phosphorus pentachloride
b. PCl_3	Phosphorus trichloride
c. SO_2	Sulfur dioxide

2.

Name	Formula
a. Sulfur hexafluoride	SF_6
b. Sulfur trioxide	SO_3
c. Carbon dioxide	CO_2

See Exercises 2.77 and 2.78

Figure 2.22 | A flowchart for naming binary compounds.

Figure 2.23 | Overall strategy for naming chemical compounds.

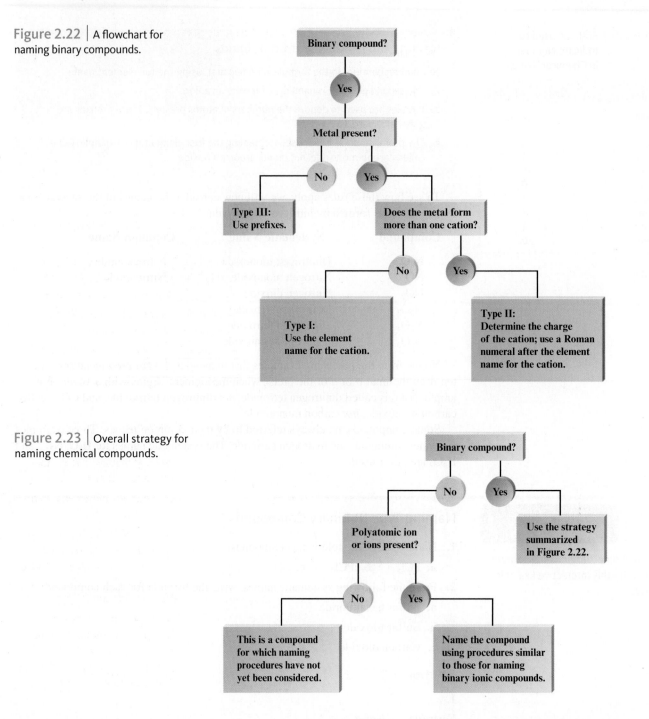

The rules for naming binary compounds are summarized in Fig. 2.22. Prefixes to indicate the number of atoms are used only in Type III binary compounds (those containing two nonmetals). An overall strategy for naming compounds is given in Fig. 2.23.

Interactive Example 2.9

Sign in at http://login.cengagebrain.com to try this Interactive Example in **OWL**.

Naming Various Types of Compounds

1. Give the systematic name for each of the following compounds:

   a. P_4O_{10} **b.** Nb_2O_5 **c.** Li_2O_2 **d.** $Ti(NO_3)_4$

2. Given the following systematic names, write the formula for each compound:

    a. Vanadium(V) fluoride

    b. Dioxygen difluoride

    c. Rubidium peroxide

    d. Gallium oxide

Solution

1.

Compound	Name	Comments
a. P_4O_{10}	Tetraphosphorus decaoxide	Binary covalent compound (Type III), so prefixes are used. The *a* in *deca-* is sometimes dropped.
b. Nb_2O_5	Niobium(V) oxide	Type II binary compound containing Nb^{5+} and O^{2-} ions. Niobium is a transition metal and requires a Roman numeral.
c. Li_2O_2	Lithium peroxide	Type I binary compound containing the Li^+ and O_2^{2-} (peroxide) ions.
d. $Ti(NO_3)_4$	Titanium(IV) nitrate	Not a binary compound. Contains the Ti^{4+} and NO_3^- ions. Titanium is a transition metal and requires a Roman numeral.

2.

Name	Chemical Formula	Comments
a. Vanadium(V) fluoride	VF_5	The compound contains V^{5+} ions and requires five F^- ions for charge balance.
b. Dioxygen difluoride	O_2F_2	The prefix *di-* indicates two of each atom.
c. Rubidium peroxide	Rb_2O_2	Because rubidium is in Group 1A, it forms only 1+ ions. Thus two Rb^+ ions are needed to balance the 2− charge on the peroxide ion (O_2^{2-}).
d. Gallium oxide	Ga_2O_3	Because gallium is in Group 3A, like aluminum, it forms only 3+ ions. Two Ga^{3+} ions are required to balance the charge on three O^{2-} ions.

See Exercises 2.79, 2.83, and 2.84

Acids

Acids can be recognized by the hydrogen that appears first in the formula.

When dissolved in water, certain molecules produce a solution containing free H^+ ions (protons). These substances, **acids**, will be discussed in detail in Chapters 4, 14, and 15. Here we will simply present the rules for naming acids.

An acid is a molecule in which one or more H^+ ions are attached to an anion. The rules for naming acids depend on whether the anion contains oxygen. If the name of the *anion ends in -ide,* the acid is named with the prefix *hydro-* and the suffix *-ic.* For example, when gaseous HCl is dissolved in water, it forms hydrochloric acid.

Similarly, HCN and H_2S dissolved in water are called hydrocyanic and hydrosulfuric acids, respectively.

When the *anion contains oxygen,* the acidic name is formed from the root name of the anion with a suffix of *-ic* or *-ous,* depending on the name of the anion.

1. If the anion name ends in *-ate,* the suffix *-ic* is added to the root name. For example, H_2SO_4 contains the sulfate anion (SO_4^{2-}) and is called sulfuric acid; H_3PO_4 contains the phosphate anion (PO_4^{3-}) and is called phosphoric acid; and $HC_2H_3O_2$ contains the acetate ion ($C_2H_3O_2^-$) and is called acetic acid.
2. If the anion has an *-ite* ending, the *-ite* is replaced by *-ous.* For example, H_2SO_3, which contains sulfite (SO_3^{2-}), is named sulfurous acid; and HNO_2, which contains nitrite (NO_2^-), is named nitrous acid.

The application of these rules can be seen in the names of the acids of the oxyanions of chlorine:

Acid	Anion	Name
$HClO_4$	Perchlor*ate*	Perchlor*ic* acid
$HClO_3$	Chlor*ate*	Chlor*ic* acid
$HClO_2$	Chlor*ite*	Chlor*ous* acid
$HClO$	Hypochlor*ite*	Hypochlor*ous* acid

The names of the most important acids are given in Tables 2.7 and 2.8. An overall strategy for naming acids is shown in Fig. 2.24.

Table 2.7 | Names of Acids* That Do Not Contain Oxygen

Acid	Name
HF	Hydrofluoric acid
HCl	Hydrochloric acid
HBr	Hydrobromic acid
HI	Hydroiodic acid
HCN	Hydrocyanic acid
H_2S	Hydrosulfuric acid

*Note that these acids are aqueous solutions containing these substances.

Table 2.8 | Names of Some Oxygen-Containing Acids

Acid	Name
HNO_3	Nitric acid
HNO_2	Nitrous acid
H_2SO_4	Sulfuric acid
H_2SO_3	Sulfurous acid
H_3PO_4	Phosphoric acid
$HC_2H_3O_2$	Acetic acid

Critical Thinking

In this chapter, you have learned a systematic way to name chemical compounds. What if all compounds had only common names? What problems would this cause?

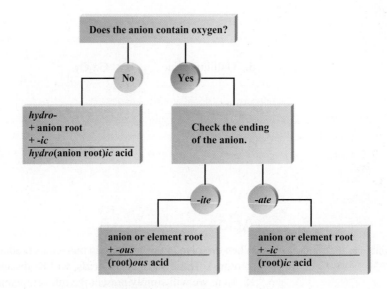

Figure 2.24 | A flowchart for naming acids. An acid is best considered as one or more H^+ ions attached to an anion.

For review

Key terms

Section 2.2
law of conservation of mass
law of definite proportion
law of multiple proportions

Section 2.3
atomic masses
atomic weights
Avogadro's hypothesis

Section 2.4
cathode-ray tubes
electrons
radioactivity
nuclear atom
nucleus

Section 2.5
proton
neutron
isotopes
atomic number
mass number

Section 2.6
chemical bond
covalent bond
molecule
chemical formula
structural formula
space-filling model
ball-and-stick model
ion
cation
anion
ionic bond
ionic solid
polyatomic ion

Section 2.7
periodic table
metal
nonmetal
group (family)
alkali metals
alkaline earth metals
halogens
noble gases
period

Fundamental laws

› Conservation of mass
› Definite proportion
› Multiple proportions

Dalton's atomic theory

› All elements are composed of atoms.
› All atoms of a given element are identical.
› Chemical compounds are formed when atoms combine.
› Atoms are not changed in chemical reactions, but the way they are bound together changes.

Early atomic experiments and models

› Thomson model
› Millikan experiment
› Rutherford experiment
› Nuclear model

Atomic structure

› Small, dense nucleus contains protons and neutrons.
  › Protons—positive charge
  › Neutrons—no charge
› Electrons reside outside the nucleus in the relatively large remaining atomic volume.
  › Electrons—negative charge, small mass (1/1840 of proton)
› Isotopes have the same atomic number but different mass numbers.

Atoms combine to form molecules by sharing electrons to form covalent bonds.

› Molecules are described by chemical formulas.
› Chemical formulas show number and type of atoms.
  › Structural formula
  › Ball-and-stick model
  › Space-filling model

Formation of ions

› Cation—formed by loss of an electron, positive charge
› Anion—formed by gain of an electron, negative charge
› Ionic bonds—formed by interaction of cations and anions

The periodic table organizes elements in order of increasing atomic number.

› Elements with similar properties are in columns, or groups.
› Metals are in the majority and tend to form cations.
› Nonmetals tend to form anions.

Key terms

Compounds are named using a system of rules depending on the type of compound.

> Binary compounds
>> Type I—contain a metal that always forms the same cation
>> Type II—contain a metal that can form more than one cation
>> Type III—contain two nonmetals
> Compounds containing a polyatomic ion

Review questions *Answers to the Review Questions can be found on the Student website (accessible from* **www.cengagebrain.com**).

1. Use Dalton's atomic theory to account for each of the following.
   a. the law of conservation of mass
   b. the law of definite proportion
   c. the law of multiple proportions

2. What evidence led to the conclusion that cathode rays had a negative charge?

3. What discoveries were made by J. J. Thomson, Henri Becquerel, and Lord Rutherford? How did Dalton's model of the atom have to be modified to account for these discoveries?

4. Consider Ernest Rutherford's α-particle bombardment experiment illustrated in Fig. 2.12. How did the results of this experiment lead Rutherford away from the plum pudding model of the atom to propose the nuclear model of the atom?

5. Do the proton and the neutron have exactly the same mass? How do the masses of the proton and neutron compare to the mass of the electron? Which particles make the greatest contribution to the mass of an atom?

Which particles make the greatest contribution to the chemical properties of an atom?

6. What is the distinction between atomic number and mass number? Between mass number and atomic mass?

7. Distinguish between the terms *family* and *period* in connection with the periodic table. For which of these terms is the term *group* also used?

8. The compounds $AlCl_3$, $CrCl_3$, and ICl_3 have similar formulas, yet each follows a different set of rules to name it. Name these compounds, and then compare and contrast the nomenclature rules used in each case.

9. When metals react with nonmetals, an ionic compound generally results. What is the predicted general formula for the compound formed between an alkali metal and sulfur? Between an alkaline earth metal and nitrogen? Between aluminum and a halogen?

10. How would you name $HBrO_4$, KIO_3, $NaBrO_2$, and HIO? Refer to Table 2.5 and the acid nomenclature discussion in the text.

Active Learning Questions

These questions are designed to be used by groups of students in class.

1. Which of the following is true about an individual atom? Explain.
   a. An individual atom should be considered to be a solid.
   b. An individual atom should be considered to be a liquid.
   c. An individual atom should be considered to be a gas.
   d. The state of the atom depends on which element it is.
   e. An individual atom cannot be considered to be a solid, liquid, or gas.
   Justify your choice, and for choices you did not pick, explain what is wrong with them.

2. How would you go about finding the number of "chalk molecules" it takes to write your name on the board? Provide an explanation of all you would need to do and a sample calculation.

3. These questions concern the work of J. J. Thomson.
   a. From Thomson's work, which particles do you think he would feel are most important for the formation of compounds (chemical changes) and why?
   b. Of the remaining two subatomic particles, which do you place second in importance for forming compounds and why?
   c. Propose three models that explain Thomson's findings and evaluate them. To be complete you should include Thomson's findings.

4. Heat is applied to an ice cube in a closed container until only steam is present. Draw a representation of this process, assuming you can see it at an extremely high level of magnification. What happens to the size of the molecules? What happens to the total mass of the sample?

5. You have a chemical in a sealed glass container filled with air. The setup is sitting on a balance as shown below. The chemical is ignited by means of a magnifying glass focusing sunlight on the reactant. After the chemical has completely burned, which of the following is true? Explain your answer.

250.0 g

   a. The balance will read less than 250.0 g.
   b. The balance will read 250.0 g.
   c. The balance will read greater than 250.0 g.
   d. Cannot be determined without knowing the identity of the chemical.

6. The formula of water is H_2O. Which of the following is indicated by this formula? Explain your answer.
   a. The mass of hydrogen is twice that of oxygen in each molecule.
   b. There are two hydrogen atoms and one oxygen atom per water molecule.
   c. The mass of oxygen is twice that of hydrogen in each molecule.
   d. There are two oxygen atoms and one hydrogen atom per water molecule.

7. You may have noticed that when water boils, you can see bubbles that rise to the surface of the water. Which of the following is inside these bubbles? Explain.
   a. air
   b. hydrogen and oxygen gas
   c. oxygen gas
   d. water vapor
   e. carbon dioxide gas

8. One of the best indications of a useful theory is that it raises more questions for further experimentation than it originally answered. Does this apply to Dalton's atomic theory? Give examples.

9. Dalton assumed that all atoms of the same element were identical in all their properties. Explain why this assumption is not valid.

10. Evaluate each of the following as an acceptable name for water:
    a. dihydrogen oxide   c. hydrogen hydroxide
    b. hydroxide hydride   d. oxygen dihydride

11. Why do we call $Ba(NO_3)_2$ barium nitrate, but we call $Fe(NO_3)_2$ iron(II) nitrate?

12. Why is calcium dichloride not the correct systematic name for $CaCl_2$?

13. The common name for NH_3 is ammonia. What would be the systematic name for NH_3? Support your answer.

14. Which (if any) of the following can be determined by knowing the number of protons in a neutral element? Explain your answer.
    a. the number of neutrons in the neutral element
    b. the number of electrons in the neutral element
    c. the name of the element

15. Which of the following explain how an ion is formed? Explain your answer.
    a. adding or subtracting protons to/from an atom
    b. adding or subtracting neutrons to/from an atom
    c. adding or subtracting electrons to/from an atom

A blue question or exercise number indicates that the answer to that question or exercise appears at the back of this book and a solution appears in the *Solutions Guide*, as found on PowerLecture.

Questions

16. What refinements had to be made in Dalton's atomic theory to account for Gay-Lussac's results on the combining volumes of gases?

17. When hydrogen is burned in oxygen to form water, the composition of water formed does not depend on the amount of oxygen reacted. Interpret this in terms of the law of definite proportion.

18. The two most reactive families of elements are the halogens and the alkali metals. How do they differ in their reactivities?

19. Explain the law of conservation of mass, the law of definite proportion, and the law of multiple proportions.

20. Section 2.3 describes the postulates of Dalton's atomic theory. With some modifications, these postulates hold up very well regarding how we view elements, compounds, and chemical reactions today. Answer the following questions concerning Dalton's atomic theory and the modifications made today.
    a. The atom can be broken down into smaller parts. What are the smaller parts?
    b. How are atoms of hydrogen identical to each other, and how can they be different from each other?
    c. How are atoms of hydrogen different from atoms of helium? How can H atoms be similar to He atoms?
    d. How is water different from hydrogen peroxide (H_2O_2) even though both compounds are composed of only hydrogen and oxygen?
    e. What happens in a chemical reaction, and why is mass conserved in a chemical reaction?

21. The contributions of J. J. Thomson and Ernest Rutherford led the way to today's understanding of the structure of the atom. What were their contributions?

22. What is the modern view of the structure of the atom?

23. The number of protons in an atom determines the identity of the atom. What does the number and arrangement of the electrons in an atom determine? What does the number of neutrons in an atom determine?

24. If the volume of a proton were similar to the volume of an electron, how will the densities of these two particles compare to each other?

25. For lighter, stable isotopes, the ratio of the mass number to the atomic number is close to a certain value. What is the value? What happens to the value of the mass number to atomic number ratio as stable isotopes become heavier?

26. List some characteristic properties that distinguish the metallic elements from the nonmetallic elements.

27. Consider the elements of Group 4A (the "carbon family"): C, Si, Ge, Sn, and Pb. What is the trend in metallic character as one goes down this group? What is the trend in metallic character going from left to right across a period in the periodic table?

28. Distinguish between the following terms.

    a. molecule versus ion

    b. covalent bonding versus ionic bonding

    c. molecule versus compound

    d. anion versus cation

29. Label the type of bonding for each of the following.

    a.   b.

    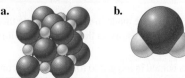

30. The vitamin niacin (nicotinic acid, $C_6H_5NO_2$) can be isolated from a variety of natural sources such as liver, yeast, milk, and whole grain. It also can be synthesized from commercially available materials. From a nutritional point of view, which source of nicotinic acid is best for use in a multivitamin tablet? Why?

31. Which of the following statements is(are) *true?* For the false statements, correct them.

    a. Most of the known elements are metals.

    b. Element 118 should be a nonmetal.

    c. Hydrogen has mostly metallic properties.

    d. A family of elements is also known as a period of elements.

    e. When an alkaline earth metal, A, reacts with a halogen, X, the formula of the covalent compound formed should be A_2X.

32. Each of the following compounds has three possible names listed for it. For each compound, what is the correct name and why aren't the other names used?

    a. N_2O: nitrogen oxide, nitrogen(I) oxide, dinitrogen monoxide

    b. Cu_2O: copper oxide, copper(I) oxide, dicopper monoxide

    c. Li_2O: lithium oxide, lithium(I) oxide, dilithium monoxide

Exercises

In this section similar exercises are paired.

Development of the Atomic Theory

33. When mixtures of gaseous H_2 and gaseous Cl_2 react, a product forms that has the same properties regardless of the relative amounts of H_2 and Cl_2 used.

    a. How is this result interpreted in terms of the law of definite proportion?

    b. When a volume of H_2 reacts with an equal volume of Cl_2 at the same temperature and pressure, what volume of product having the formula HCl is formed?

34. Observations of the reaction between nitrogen gas and hydrogen gas show us that 1 volume of nitrogen reacts with 3 volumes of hydrogen to make 2 volumes of gaseous product, as shown below:

    Determine the formula of the product and justify your answer.

35. A sample of chloroform is found to contain 12.0 g of carbon, 106.4 g of chlorine, and 1.01 g of hydrogen. If a second sample of chloroform is found to contain 30.0 g of carbon, what is the total mass of chloroform in the second sample?

36. A sample of H_2SO_4 contains 2.02 g of hydrogen, 32.07 g of sulfur, and 64.00 g of oxygen. How many grams of sulfur and grams of oxygen are present in a second sample of H_2SO_4 containing 7.27 g of hydrogen?

37. Hydrazine, ammonia, and hydrogen azide all contain only nitrogen and hydrogen. The mass of hydrogen that combines with 1.00 g of nitrogen for each compound is 1.44×10^{-1} g, 2.16×10^{-1} g, and 2.40×10^{-2} g, respectively. Show how these data illustrate the law of multiple proportions.

38. Consider 100.0-g samples of two different compounds consisting only of carbon and oxygen. One compound contains 27.2 g of carbon and the other has 42.9 g of carbon. How can these data support the law of multiple proportions if 42.9 is not a multiple of 27.2? Show that these data support the law of multiple proportions.

39. The three most stable oxides of carbon are carbon monoxide (CO), carbon dioxide (CO_2), and carbon suboxide (C_3O_2). The molecules can be represented as

    Explain how these molecules illustrate the law of multiple proportions.

40. Two elements, R and Q, combine to form two binary compounds. In the first compound, 14.0 g of R combines with 3.00 g of Q. In the second compound, 7.00 g of R combines with 4.50 g of Q. Show that these data are in accord with the law of multiple proportions. If the formula of the second compound is RQ, what is the formula of the first compound?

41. In Section 1.1 of the text, the concept of a chemical reaction was introduced with the example of the decomposition of water, represented as follows:

Use ideas from Dalton's atomic theory to explain how the above representation illustrates the law of conservation of mass.

42. In a combustion reaction, 46.0 g of ethanol reacts with 96.0 g of oxygen to produce water and carbon dioxide. If 54.0 g of water is produced, what mass of carbon dioxide is produced?

43. Early tables of atomic weights (masses) were generated by measuring the mass of a substance that reacts with 1.00 g of oxygen. Given the following data and taking the atomic mass of hydrogen as 1.00, generate a table of relative atomic masses for oxygen, sodium, and magnesium.

Element	Mass That Combines with 1.00 g Oxygen	Assumed Formula
Hydrogen	0.126 g	HO
Sodium	2.875 g	NaO
Magnesium	1.500 g	MgO

How do your values compare with those in the periodic table? How do you account for any differences?

44. Indium oxide contains 4.784 g of indium for every 1.000 g of oxygen. In 1869, when Mendeleev first presented his version of the periodic table, he proposed the formula In_2O_3 for indium oxide. Before that time it was thought that the formula was InO. What values for the atomic mass of indium are obtained using these two formulas? Assume that oxygen has an atomic mass of 16.00.

The Nature of the Atom

45. From the information in this chapter on the mass of the proton, the mass of the electron, and the sizes of the nucleus and the atom, calculate the densities of a hydrogen nucleus and a hydrogen atom.

46. If you wanted to make an accurate scale model of the hydrogen atom and decided that the nucleus would have a diameter of 1 mm, what would be the diameter of the entire model?

47. In an experiment it was found that the total charge on an oil drop was 5.93×10^{-18} C. How many negative charges does the drop contain?

48. A chemist in a galaxy far, far away performed the Millikan oil drop experiment and got the following results for the charges on various drops. Use these data to calculate the charge of the electron in zirkombs.

2.56×10^{-12} zirkombs 7.68×10^{-12} zirkombs
3.84×10^{-12} zirkombs 6.40×10^{-13} zirkombs

49. What are the symbols of the following metals: sodium, radium, iron, gold, manganese, lead?

50. What are the symbols of the following nonmetals: fluorine, chlorine, bromine, sulfur, oxygen, phosphorus?

51. Give the names of the metals that correspond to the following symbols: Sn, Pt, Hg, Mg, K, Ag.

52. Give the names of the nonmetals that correspond to the following symbols: As, I, Xe, He, C, Si.

53. a. Classify the following elements as metals or nonmetals:

Mg   Si   Rn
Ti   Ge   Eu
Au   B   Am
Bi   At   Br

b. The distinction between metals and nonmetals is really not a clear one. Some elements, called *metalloids*, are intermediate in their properties. Which of these elements would you reclassify as metalloids? What other elements in the periodic table would you expect to be metalloids?

54. a. List the noble gas elements. Which of the noble gases has only radioactive isotopes? (This situation is indicated on most periodic tables by parentheses around the mass of the element. See inside front cover.)

b. Which lanthanide element has only radioactive isotopes?

55. For each of the following sets of elements, label each as either noble gases, halogens, alkali metals, alkaline earth metals, or transition metals.

a. Ti, Fe, Ag
b. Mg, Sr, Ba
c. Li, K, Rb
d. Ne, Kr, Xe
e. F, Br, I

56. Identify the elements that correspond to the following atomic numbers. Label each as either a noble gas, a halogen, an alkali metal, an alkaline earth metal, a transition metal, a lanthanide metal, or an actinide metal.

a. 17   **e.** 2
b. 4   **f.** 92
c. 63   **g.** 55
d. 72

57. Write the atomic symbol (^A_ZX) for each of the following isotopes.

a. $Z = 8$, number of neutrons = 9
b. the isotope of chlorine in which $A = 37$
c. $Z = 27, A = 60$
d. number of protons = 26, number of neutrons = 31
e. the isotope of I with a mass number of 131
f. $Z = 3$, number of neutrons = 4

58. Write the atomic symbol (^A_ZX) for each of the isotopes described below.

a. number of protons = 27, number of neutrons = 31
b. the isotope of boron with mass number 10
c. $Z = 12, A = 23$
d. atomic number 53, number of neutrons = 79
e. $Z = 20$, number of neutrons = 27
f. number of protons = 29, mass number 65

59. Write the symbol of each atom using the $_{Z}^{A}X$ format.

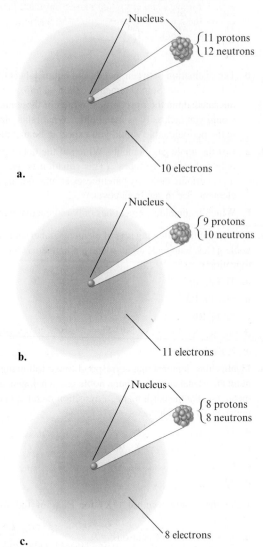

a.

Nucleus — { 11 protons
12 neutrons

10 electrons

b.

Nucleus — { 9 protons
10 neutrons

11 electrons

c.

Nucleus — { 8 protons
8 neutrons

8 electrons

60. For carbon-14 and carbon-12, how many protons and neutrons are in each nucleus? Assuming neutral atoms, how many electrons are present in an atom of carbon-14 and in an atom of carbon-12?

61. How many protons and neutrons are in the nucleus of each of the following atoms? In a neutral atom of each element, how many electrons are present?

a. ^{79}Br   d. ^{133}Cs

b. ^{81}Br   e. ^{3}H

c. ^{239}Pu   f. ^{56}Fe

62. What number of protons and neutrons are contained in the nucleus of each of the following atoms? Assuming each atom is uncharged, what number of electrons are present?

a. $_{92}^{235}$U   d. $_{82}^{208}$Pb

b. $_{13}^{27}$Al   e. $_{37}^{86}$Rb

c. $_{26}^{57}$Fe   f. $_{20}^{41}$Ca

63. For each of the following ions, indicate the number of protons and electrons the ion contains.

a. Ba^{2+} e. Co^{3+}

b. Zn^{2+} f. Te^{2-}

c. N^{3-} g. Br^{-}

d. Rb^{+}

64. How many protons, neutrons, and electrons are in each of the following atoms or ions?

a. $_{12}^{24}Mg$ d. $_{27}^{59}Co^{3+}$ g. $_{34}^{79}Se^{2-}$

b. $_{12}^{24}Mg^{2+}$ e. $_{27}^{59}Co$ h. $_{28}^{63}Ni$

c. $_{27}^{59}Co^{2+}$ f. $_{34}^{79}Se$ i. $_{28}^{59}Ni^{2+}$

65. What is the symbol for an ion with 63 protons, 60 electrons, and 88 neutrons? If an ion contains 50 protons, 68 neutrons, and 48 electrons, what is its symbol?

66. What is the symbol of an ion with 16 protons, 18 neutrons, and 18 electrons? What is the symbol for an ion that has 16 protons, 16 neutrons, and 18 electrons?

67. Complete the following table:

Symbol	Number of Protons in Nucleus	Number of Neutrons in Nucleus	Number of Electrons	Net Charge
$_{92}^{238}U$				
	20	20		2+
	23	28	20	
$_{39}^{89}Y$				
	35	44	36	
	15	16		3−

68. Complete the following table:

Symbol	Number of Protons in Nucleus	Number of Neutrons in Nucleus	Number of Electrons	Net Charge
$_{26}^{53}Fe^{2+}$				
	26	33		3+
	85	125	86	
	13	14	10	
		76	54	2−

69. Would you expect each of the following atoms to gain or lose electrons when forming ions? What ion is the most likely in each case?

a. Ra   c. P   e. Br

b. In   d. Te   f. Rb

70. For each of the following atomic numbers, use the periodic table to write the formula (including the charge) for the simple *ion* that the element is most likely to form in ionic compounds.

a. 13   c. 56   e. 87

b. 34   d. 7   f. 35

Nomenclature

71. Name the compounds in parts a–d and write the formulas for the compounds in parts e–h.

a. NaBr   e. strontium fluoride
b. Rb_2O f. aluminum selenide
c. CaS   g. potassium nitride
d. AlI_3 h. magnesium phosphide

72. Name the compounds in parts a–d and write the formulas for the compounds in parts e–h.

a. Hg_2O e. tin(II) nitride
b. $FeBr_3$ f. cobalt(III) iodide
c. CoS   g. mercury(II) oxide
d. $TiCl_4$ h. chromium(VI) sulfide

73. Name each of the following compounds:

a. CsF   c. Ag_2S e. TiO_2
b. Li_3N d. MnO_2 f. Sr_3P_2

74. Write the formula for each of the following compounds:

a. zinc chloride   d. aluminum sulfide
b. tin(IV) fluoride   e. mercury(I) selenide
c. calcium nitride   f. silver iodide

75. Name each of the following compounds:

a. $BaSO_3$ c. $KMnO_4$
b. $NaNO_2$ d. $K_2Cr_2O_7$

76. Write the formula for each of the following compounds:

a. chromium(III) hydroxide   c. lead(IV) carbonate
b. magnesium cyanide   d. ammonium acetate

77. Name each of the following compounds:

a.   c. SO_2
                               d. P_2S_5
   ● O
   ● N

b.
   ● I
   ● Cl

78. Write the formula for each of the following compounds:

a. diboron trioxide   c. dinitrogen monoxide
b. arsenic pentafluoride   d. sulfur hexachloride

79. Name each of the following compounds:

a. CuI   f. S_4N_4
b. CuI_2 g. $SeCl_4$
c. CoI_2 h. NaOCl
d. Na_2CO_3 i. $BaCrO_4$
e. $NaHCO_3$ j. NH_4NO_3

80. Name each of the following compounds. Assume the acids are dissolved in water.

a. $HC_2H_3O_2$ g. H_2SO_4
b. NH_4NO_2 h. Sr_3N_2
c. Co_2S_3 i. $Al_2(SO_3)_3$
d. ICl   j. SnO_2
e. $Pb_3(PO_4)_2$ k. Na_2CrO_4
f. $KClO_3$ l. HClO

81. Elements in the same family often form oxyanions of the same general formula. The anions are named in a similar fashion. What are the names of the oxyanions of selenium and tellurium: SeO_4^{2-}, SeO_3^{2-}, TeO_4^{2-}, TeO_3^{2-}?

82. Knowing the names of similar chlorine oxyanions and acids, deduce the names of the following: IO^-, IO_2^-, IO_3^-, IO_4^-, HIO, HIO_2, HIO_3, HIO_4.

83. Write the formula for each of the following compounds:

a. sulfur difluoride
b. sulfur hexafluoride
c. sodium dihydrogen phosphate
d. lithium nitride
e. chromium(III) carbonate
f. tin(II) fluoride
g. ammonium acetate
h. ammonium hydrogen sulfate
i. cobalt(III) nitrate
j. mercury(I) chloride
k. potassium chlorate
l. sodium hydride

84. Write the formula for each of the following compounds:

a. chromium(VI) oxide
b. disulfur dichloride
c. nickel(II) fluoride
d. potassium hydrogen phosphate
e. aluminum nitride
f. ammonia
g. manganese(IV) sulfide
h. sodium dichromate
i. ammonium sulfite
j. carbon tetraiodide

85. Write the formula for each of the following compounds:

a. sodium oxide   h. copper(I) chloride
b. sodium peroxide   i. gallium arsenide
c. potassium cyanide   j. cadmium selenide
d. copper(II) nitrate   k. zinc sulfide
e. selenium tetrabromide   l. nitrous acid
f. iodous acid   m. diphosphorus pentoxide
g. lead(IV) sulfide

86. Write the formula for each of the following compounds:

a. ammonium hydrogen phosphate
b. mercury(I) sulfide
c. silicon dioxide
d. sodium sulfite
e. aluminum hydrogen sulfate
f. nitrogen trichloride
g. hydrobromic acid
h. bromous acid
i. perbromic acid
j. potassium hydrogen sulfide
k. calcium iodide
l. cesium perchlorate

87. Name the acids illustrated below.

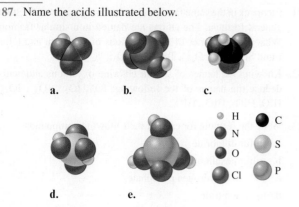

a.   **b.**   **c.**

H	C
N	S
O	P
Cl	

d.   **e.**

88. Each of the following compounds is incorrectly named. What is wrong with each name, and what is the correct name for each compound?

   a. $FeCl_3$, iron chloride

   b. NO_2, nitrogen(IV) oxide

   c. CaO, calcium(II) monoxide

   d. Al_2S_3, dialuminum trisulfide

   e. $Mg(C_2H_3O_2)_2$, manganese diacetate

   f. $FePO_4$, iron(II) phosphide

   g. P_2S_5, phosphorus sulfide

   h. Na_2O_2, sodium oxide

   i. HNO_3, nitrate acid

   j. H_2S, sulfuric acid

Additional Exercises

89. Chlorine has two natural isotopes: $^{37}_{17}Cl$ and $^{35}_{17}Cl$. Hydrogen reacts with chlorine to form the compound HCl. Would a given amount of hydrogen react with different masses of the two chlorine isotopes? Does this conflict with the law of definite proportion? Why or why not?

90. What are the symbols for the following nonmetal elements that are most often present in compounds studied in organic chemistry: carbon, hydrogen, oxygen, nitrogen, phosphorus, sulfur? Predict a stable isotope for each of these elements.

91. Four Fe^{2+} ions are key components of hemoglobin, the protein that transports oxygen in the blood. Assuming that these ions are $^{53}Fe^{2+}$, how many protons and neutrons are present in each nucleus, and how many electrons are present in each ion?

92. Which of the following statements is(are) *true?* For the false statements, correct them.

   a. All particles in the nucleus of an atom are charged.

   b. The atom is best described as a uniform sphere of matter in which electrons are embedded.

   c. The mass of the nucleus is only a very small fraction of the mass of the entire atom.

   d. The volume of the nucleus is only a very small fraction of the total volume of the atom.

   e. The number of neutrons in a neutral atom must equal the number of electrons.

93. The isotope of an unknown element, X, has a mass number of 79. The most stable ion of the isotope has 36 electrons and forms a binary compound with sodium having a formula of Na_2X. Which of the following statements is(are) *true?* For the false statements, correct them.

   a. The binary compound formed between X and fluorine will be a covalent compound.

   b. The isotope of X contains 38 protons.

   c. The isotope of X contains 41 neutrons.

   d. The identity of X is strontium, Sr.

94. For each of the following ions, indicate the total number of protons and electrons in the ion. For the positive ions in the list, predict the formula of the simplest compound formed between each positive ion and the oxide ion. Name the compounds. For the negative ions in the list, predict the formula of the simplest compound formed between each negative ion and the aluminum ion. Name the compounds.

   a. Fe^{2+} **e.** S^{2-}

   b. Fe^{3+} **f.** P^{3-}

   c. Ba^{2+} **g.** Br^{-}

   d. Cs^{+} **h.** N^{3-}

95. The formulas and common names for several substances are given below. Give the systematic names for these substances.

   a. sugar of lead $Pb(C_2H_3O_2)_2$

   b. blue vitrol $CuSO_4$

   c. quicklime CaO

   d. Epsom salts $MgSO_4$

   e. milk of magnesia $Mg(OH)_2$

   f. gypsum $CaSO_4$

   g. laughing gas N_2O

96. Identify each of the following elements:

   a. a member of the same family as oxygen whose most stable ion contains 54 electrons

   b. a member of the alkali metal family whose most stable ion contains 36 electrons

   c. a noble gas with 18 protons in the nucleus

   d. a halogen with 85 protons and 85 electrons

97. An element's most stable ion forms an ionic compound with bromine, having the formula XBr_2. If the ion of element X has a mass number of 230 and has 86 electrons, what is the identity of the element, and how many neutrons does it have?

98. A certain element has only two naturally occurring isotopes: one with 18 neutrons and the other with 20 neutrons. The element forms $1-$ charged ions when in ionic compounds. Predict the identity of the element. What number of electrons does the $1-$ charged ion have?

99. The designations 1A through 8A used for certain families of the periodic table are helpful for predicting the charges on ions in binary ionic compounds. In these compounds, the metals generally take on a positive charge equal to the family number, while the nonmetals take on a negative charge equal to the family number minus eight. Thus the compound between sodium and chlorine contains Na^{+} ions and Cl^{-} ions and has the formula NaCl. Predict the formula and the name of the binary compound formed from the following pairs of elements.

   a. Ca and N   **e.** Ba and I

   b. K and O   **f.** Al and Se

   c. Rb and F   **g.** Cs and P

   d. Mg and S   **h.** In and Br

100. By analogy with phosphorus compounds, name the following: Na_3AsO_4, H_3AsO_4, $Mg_3(SbO_4)_2$.

101. Identify each of the following elements. Give the number of protons and neutrons in each nucleus.

a. $^{31}_{15}X$ c. $^{39}_{19}X$

b. $^{127}_{53}X$ d. $^{173}_{70}X$

102. In a reaction, 34.0 g of chromium(III) oxide reacts with 12.1 g of aluminum to produce chromium and aluminum oxide. If 23.3 g of chromium is produced, what mass of aluminum oxide is produced?

ChemWork Problems

These multiconcept problems (and additional ones) are found interactively online with the same type of assistance a student would get from an instructor.

103. Complete the following table.

Atom/Ion	Protons	Neutrons	Electrons
$^{120}_{50}Sn$			
$^{25}_{12}Mg^{2+}$			
$^{56}_{26}Fe^{2+}$			
$^{79}_{34}Se$			
$^{35}_{17}Cl$			
$^{63}_{29}Cu$			

104. Which of the following is(are) correct?

a. $^{40}Ca^{2+}$ contains 20 protons and 18 electrons.

b. Rutherford created the cathode-ray tube and was the founder of the charge-to-mass ratio of an electron.

c. An electron is heavier than a proton.

d. The nucleus contains protons, neutrons, and electrons.

105. What are the formulas of the compounds that correspond to the names given in the following table?

Compound Name	Formula
Carbon tetrabromide	
Cobalt(II) phosphate	
Magnesium chloride	
Nickel(II) acetate	
Calcium nitrate	

106. What are the names of the compounds that correspond to the formulas given in the following table?

Formula	Compound Name
$Co(NO_2)_2$	
AsF_5	
$LiCN$	
K_2SO_3	
Li_3N	
$PbCrO_4$	

107. Complete the following table to predict whether the given atom will gain or lose electrons in forming the ion most likely to form when in ionic compounds.

Atom	Gain (G) or Lose (L) Electrons	Ion Formed
K		
Cs		
Br		
S		
Se		

108. Which of the following statements is(are) correct?

a. The symbols for the elements magnesium, aluminum, and xenon are Mn, Al, and Xe, respectively.

b. The elements P, As, and Bi are in the same family on the periodic table.

c. All of the following elements are expected to gain electrons to form ions in ionic compounds: Ga, Se, and Br.

d. The elements Co, Ni, and Hg are all transition elements.

e. The correct name for TiO_2 is titanium dioxide.

Challenge Problems

109. The elements in one of the groups in the periodic table are often called the coinage metals. Identify the elements in this group based on your own experience.

110. Reaction of 2.0 L of hydrogen gas with 1.0 L of oxygen gas yields 2.0 L of water vapor. All gases are at the same temperature and pressure. Show how these data support the idea that oxygen gas is a diatomic molecule. Must we consider hydrogen to be a diatomic molecule to explain these results?

111. A combustion reaction involves the reaction of a substance with oxygen gas. The complete combustion of any hydrocarbon (binary compound of carbon and hydrogen) produces carbon dioxide and water as the only products. Octane is a hydrocarbon that is found in gasoline. Complete combustion of octane produces 8 L of carbon dioxide for every 9 L of water vapor (both measured at the same temperature and pressure). What is the ratio of carbon atoms to hydrogen atoms in a molecule of octane?

112. A chemistry instructor makes the following claim: "Consider that if the nucleus were the size of a grape, the electrons would be about 1 *mile* away on average." Is this claim reasonably accurate? Provide mathematical support.

113. The early alchemists used to do an experiment in which water was boiled for several days in a sealed glass container. Eventually, some solid residue would appear in the bottom of the flask, which was interpreted to mean that some of the water in the flask had been converted into "earth." When Lavoisier repeated this experiment, he found that the water weighed the same before and after heating, and the mass of the flask plus the solid residue equaled the original mass of the flask. Were the alchemists correct? Explain what really happened. (This experiment is described in the article by A. F. Scott in *Scientific American,* January 1984.)

114. Consider the chemical reaction as depicted below. Label as much as you can using the terms *atom*, *molecule*, *element*, *compound*, *ionic*, *gas*, and *solid*.

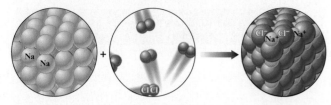

115. Each of the following statements is true, but Dalton might have had trouble explaining some of them with his atomic theory. Give explanations for the following statements.

   a. The space-filling models for ethyl alcohol and dimethyl ether are shown below.

   ● C
   ● O
   ● H

   These two compounds have the same composition by mass (52% carbon, 13% hydrogen, and 35% oxygen), yet the two have different melting points, boiling points, and solubilities in water.

   b. Burning wood leaves an ash that is only a small fraction of the mass of the original wood.

   c. Atoms can be broken down into smaller particles.

   d. One sample of lithium hydride is 87.4% lithium by mass, while another sample of lithium hydride is 74.9% lithium by mass. However, the two samples have the same chemical properties.

116. You have two distinct gaseous compounds made from element X and element Y. The mass percents are as follows:

   Compound I: 30.43% X, 69.57% Y
   Compound II: 63.64% X, 36.36% Y

   In their natural standard states, element X and element Y exist as gases. (Monatomic? Diatomic? Triatomic? That is for you to determine.) When you react "gas X" with "gas Y" to make the products, you get the following data (all at the same pressure and temperature):

   1 volume "gas X" + 2 volumes "gas Y" ⟶
   2 volumes compound I
   2 volumes "gas X" + 1 volume "gas Y" ⟶
   2 volumes compound II

   Assume the simplest possible formulas for reactants and products in the chemical equations above. Then, determine the relative atomic masses of element X and element Y.

117. A single molecule has a mass of 7.31×10^{-23} g. Provide an example of a real molecule that can have this mass. Assume the elements that make up the molecule are made of light isotopes where the number of protons equals the number of neutrons in the nucleus of each element.

118. You take three compounds, each consisting of two elements (X, Y, and/or Z), and decompose them to their respective elements. To determine the relative masses of X, Y, and Z, you collect and weigh the elements, obtaining the following data:

Elements in Compound	Masses of Elements
1. X and Y	X = 0.4 g, Y = 4.2 g
2. Y and Z	Y = 1.4 g, Z = 1.0 g
3. X and Y	X = 2.0 g, Y = 7.0 g

   a. What are the assumptions needed to solve this problem?

   b. What are the relative masses of X, Y, and Z?

   c. What are the chemical formulas of the three compounds?

   d. If you decompose 21 g of compound XY, how much of each element is present?

Integrative Problems

These problems require the integration of multiple concepts to find the solutions.

119. What is the systematic name of Ta_2O_5? If the charge on the metal remained constant and then sulfur was substituted for oxygen, how would the formula change? What is the difference in the total number of protons between Ta_2O_5 and its sulfur analog?

120. A binary ionic compound is known to contain a cation with 51 protons and 48 electrons. The anion contains one-third the number of protons as the cation. The number of electrons in the anion is equal to the number of protons plus 1. What is the formula of this compound? What is the name of this compound?

121. Using the information in Table 2.1, answer the following questions. In an ion with an unknown charge, the total mass of all the electrons was determined to be 2.55×10^{-26} g, while the total mass of its protons was 5.34×10^{-23} g. What is the identity and charge of this ion? What is the symbol and mass number of a neutral atom whose total mass of its electrons is 3.92×10^{-26} g, while its neutrons have a mass of 9.35×10^{-23} g?

Marathon Problem

This problem is designed to incorporate several concepts and techniques into one situation.

122. You have gone back in time and are working with Dalton on a table of relative masses. Following are his data.

   0.602 g gas A reacts with 0.295 g gas B
   0.172 g gas B reacts with 0.401 g gas C
   0.320 g gas A reacts with 0.374 g gas C

   a. Assuming simplest formulas (AB, BC, and AC), construct a table of relative masses for Dalton.

   b. Knowing some history of chemistry, you tell Dalton that if he determines the volumes of the gases reacted at constant temperature and pressure, he need not assume simplest formulas. You collect the following data:

   6 volumes gas A + 1 volume gas B → 4 volumes product
   1 volume gas B + 4 volumes gas C → 4 volumes product
   3 volumes gas A + 2 volumes gas C → 6 volumes product

   Write the simplest balanced equations, and find the actual relative masses of the elements. Explain your reasoning.

Chapter 3

Stoichiometry

Fireworks provide a spectacular example of chemical reactions. (Daff/Dreamstime.com)

C hemical reactions have a profound effect on our lives. There are many examples: Food is converted to energy in the human body; nitrogen and hydrogen are combined to form ammonia, which is used as a fertilizer; fuels and plastics are produced from petroleum; the starch in plants is synthesized from carbon dioxide and water using energy from sunlight; human insulin is produced in laboratories by bacteria; cancer is induced in humans by substances from our environment; and so on, in a seemingly endless list. The central activity of chemistry is to understand chemical changes such as these, and the study of reactions occupies a central place in this book. We will examine why reactions occur, how fast they occur, and the specific pathways they follow.

In this chapter we will consider the quantities of materials consumed and produced in chemical reactions. This area of study is called **chemical stoichiometry** (pronounced stoy·kē·om′·uh·trē). To understand chemical stoichiometry, you must first understand the concept of relative atomic masses.

3.1 | Counting by Weighing

Suppose you work in a candy store that sells gourmet jelly beans by the bean. People come in and ask for 50 beans, 100 beans, 1000 beans, and so on, and you have to count them out—a tedious process at best. As a good problem solver, you try to come up with a better system. It occurs to you that it might be far more efficient to buy a scale and count the jelly beans by weighing them. How can you count jelly beans by weighing them? What information about the individual beans do you need to know?

Assume that all of the jelly beans are identical and that each has a mass of 5 g. If a customer asks for 1000 jelly beans, what mass of jelly beans would be required? Each bean has a mass of 5 g, so you would need 1000 beans × 5 g/bean, or 5000 g (5 kg). It takes just a few seconds to weigh out 5 kg of jelly beans. It would take much longer to count out 1000 of them.

In reality, jelly beans are not identical. For example, let's assume that you weigh 10 beans individually and get the following results:

Jelly beans can be counted by weighing.

Bean	Mass	Bean	Mass
1	5.1 g	6	5.0 g
2	5.2 g	7	5.0 g
3	5.0 g	8	5.1 g
4	4.8 g	9	4.9 g
5	4.9 g	10	5.0 g

Can we count these nonidentical beans by weighing? Yes. The key piece of information we need is the *average mass* of the jelly beans. Let's compute the average mass for our 10-bean sample.

$$\text{Average mass} = \frac{\text{total mass of beans}}{\text{number of beans}}$$

$$= \frac{5.1\,g + 5.2\,g + 5.0\,g + 4.8\,g + 4.9\,g + 5.0\,g + 5.0\,g + 5.1\,g + 4.9\,g + 5.0\,g}{10}$$

$$= \frac{50.0}{10} = 5.0\,g$$

The average mass of a jelly bean is 5.0 g. Thus, to count out 1000 beans, we need to weigh out 5000 g of beans. This sample of beans, in which the beans have an average mass of 5.0 g, can be treated exactly like a sample where all of the beans are identical. Objects do not need to have identical masses to be counted by weighing. We simply need to know the average mass of the objects. For purposes of counting, the objects *behave as though they were all identical*, as though they each actually had the average mass.

We count atoms in exactly the same way. Because atoms are so small, we deal with samples of matter that contain huge numbers of atoms. Even if we could see the atoms, it would not be possible to count them directly. Thus we determine the number of atoms in a given sample by finding its mass. However, just as with jelly beans, to relate the mass to a number of atoms, we must know the average mass of the atoms.

3.2 | Atomic Masses

As we saw in Chapter 2, the first quantitative information about atomic masses came from the work of Dalton, Gay-Lussac, Lavoisier, Avogadro, and Berzelius. By observing the proportions in which elements combine to form various compounds, nineteenth-century chemists calculated relative atomic masses. The modern system of atomic masses, instituted in 1961, is based on ^{12}C ("carbon twelve") as the standard. In this system, ^{12}C *is assigned a mass of exactly 12 atomic mass units* (u), and the masses of all other atoms are given relative to this standard.

The most accurate method currently available for comparing the masses of atoms involves the use of the **mass spectrometer**. In this instrument, diagramed in Fig. 3.1, atoms or molecules are passed into a beam of high-speed electrons, which knock electrons off the atoms or molecules being analyzed and change them into positive ions. An applied electric field then accelerates these ions into a magnetic field. Because an accelerating ion produces its own magnetic field, an interaction with the applied magnetic field occurs, which tends to change the path of the ion. The amount of path deflection for each ion depends on its mass—the most massive ions are deflected the smallest amount—which causes the ions to separate, as shown in Fig. 3.1. A comparison of the positions where the ions hit the detector plate gives very accurate values of their relative masses. For example, when ^{12}C and ^{13}C are analyzed in a mass spectrometer, the ratio of their masses is found to be

$$\frac{\text{Mass } ^{13}C}{\text{Mass } ^{12}C} = 1.0836129$$

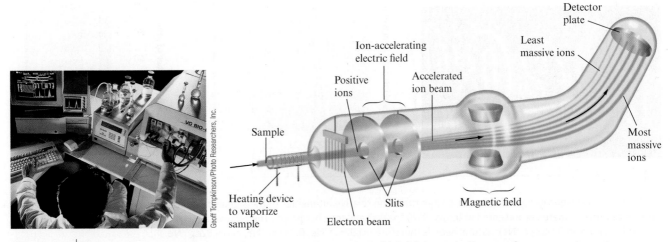

Figure 3.1 | (left) A scientist injecting a sample into a mass spectrometer. (right) Schematic diagram of a mass spectrometer.

It is much easier to weigh out 600 hex nuts than to count them one by one.

Since the atomic mass unit is defined such that the mass of ^{12}C is *exactly* 12 atomic mass units, then on this same scale,

$$\text{Mass of } ^{13}C = (1.0836129)(12 \text{ u}) = 13.003355 \text{ u}$$

Exact number
by definition

The masses of other atoms can be determined in a similar fashion.

The mass for each element is given in the table inside the front cover of this text. This value, even though it is actually a mass, is sometimes called the *atomic weight* for each element.

Look at the value of the atomic mass of carbon given in this table. You might expect to see 12, since we said the system of atomic masses is based on ^{12}C. However, the number given for carbon is not 12 but 12.01. Why? The reason for this apparent discrepancy is that the carbon found on earth (natural carbon) is a mixture of the isotopes ^{12}C, ^{13}C, and ^{14}C. All three isotopes have six protons, but they have six, seven, and eight neutrons, respectively. Because natural carbon is a mixture of isotopes, the atomic mass we use for carbon is an *average value* reflecting the average of the isotopes composing it.

The average atomic mass for carbon is computed as follows: It is known that natural carbon is composed of 98.89% ^{12}C atoms and 1.11% ^{13}C atoms. The amount of ^{14}C is negligibly small at this level of precision. Using the masses of ^{12}C (exactly 12 u) and ^{13}C (13.003355 u), we can calculate the average atomic mass for natural carbon as follows:

98.89% of 12 u + 1.11% of 13.0034 u =
$$(0.9889)(12 \text{ u}) + (0.0111)(13.0034 \text{ u}) = 12.01 \text{ u}$$

In this text we will call the average mass for an element the **average atomic mass** or, simply, the *atomic mass* for that element.

Even though natural carbon does not contain a single atom with mass 12.01, for stoichiometric purposes, we can consider carbon to be composed of only one type of atom with a mass of 12.01. This enables us to count atoms of natural carbon by weighing a sample of carbon.

Recall from Section 3.1 that counting by weighing works if you know the *average* mass of the units being counted. Counting by weighing works just the same for atoms as for jelly beans. For natural carbon with an average mass of 12.01 atomic mass units, to obtain 1000 atoms would require weighing out 12,010 atomic mass units of natural carbon (a mixture of ^{12}C and ^{13}C).

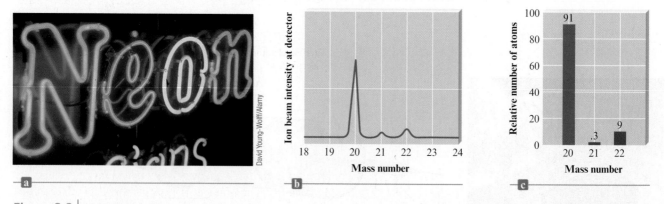

Figure 3.2 | (a) Neon gas glowing in a discharge tube. The relative intensities of the signals recorded when natural neon is injected into a mass spectrometer, represented in terms of (b) "peaks" and (c) a bar graph. The relative areas of the peaks are 0.9092 (^{20}Ne), 0.00257 (^{21}Ne), and 0.0882 (^{22}Ne); natural neon is therefore 90.92% ^{20}Ne, 0.257% ^{21}Ne, and 8.82% ^{22}Ne.

As in the case of carbon, the mass for each element listed in the table inside the front cover of the text is an average value based on the isotopic composition of the naturally occurring element. For instance, the mass listed for hydrogen (1.008) is the average mass for natural hydrogen, which is a mixture of ^1H and ^2H (deuterium). *No* atom of hydrogen actually has the mass 1.008.

In addition to being useful for determining accurate mass values for individual atoms, the mass spectrometer is used to determine the isotopic composition of a natural element. For example, when a sample of natural neon is injected into a mass spectrometer, the mass spectrum shown in Fig. 3.2 is obtained. The areas of the "peaks" or the heights of the bars indicate the relative abundances of $^{20}_{10}Ne$, $^{21}_{10}Ne$, and $^{22}_{10}Ne$ atoms.

Example 3.1

The Average Mass of an Element

When a sample of natural copper is vaporized and injected into a mass spectrometer, the results shown in Fig. 3.3 are obtained. Use these data to compute the average mass of natural copper. (The mass values for ^{63}Cu and ^{65}Cu are 62.93 u and 64.93 u, respectively.)

Solution

Where are we going?

To calculate the average mass of natural copper

What do we know?

> ^{63}Cu mass = 62.93 u
> ^{65}Cu mass = 64.93 u

How do we get there?

As shown by the graph, of every 100 atoms of natural copper, 69.09 are ^{63}Cu and 30.91 are ^{65}Cu. Thus the mass of 100 atoms of natural copper is

$$(69.09 \text{ atoms})\left(62.93 \frac{u}{atom}\right) + (30.91 \text{ atoms})\left(64.93 \frac{u}{atom}\right) = 6355 \text{ u}$$

The *average mass* of a copper atom is

$$\frac{6355 \text{ u}}{100 \text{ atoms}} = 63.55 \text{ u/atom}$$

This mass value is used in doing calculations involving the reactions of copper and is the value given in the table inside the front cover of this book.

Reality Check | When you finish a calculation, you should always check whether your answer makes sense. In this case our answer of 63.55 u is between the masses of the atoms that make up natural copper. This makes sense. The answer could not be smaller than 62.93 u or larger than 64.93 u.

See Exercises 3.37 and 3.38

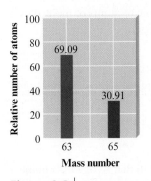

Copper nugget.

Figure 3.3 | Mass spectrum of natural copper.

3.3 | The Mole

Because samples of matter typically contain so many atoms, a unit of measure called the *mole* has been established for use in counting atoms. For our purposes, it is most convenient to define the **mole** (abbreviated mol) as *the number equal to the number of carbon atoms in exactly 12 g of pure ^{12}C*. Techniques such as mass spectrometry, which count atoms very precisely, have been used to determine this number as 6.02214×10^{23}

Figure 3.4 | One-mole samples of several elements.

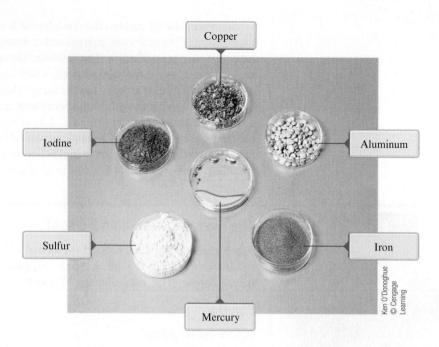

Copper

Iodine

Aluminum

Sulfur

Iron

Mercury

Ken O'Donoghue
© Cengage Learning

The SI definition of the mole is the amount of a substance that contains as many entities as there are in exactly 12 g of carbon-12.

Avogadro's number is 6.022×10^{23}. One mole of anything is 6.022×10^{23} units of that substance.

(6.022×10^{23} will be sufficient for our purposes). This number is called **Avogadro's number** to honor his contributions to chemistry. One mole of something consists of 6.022×10^{23} units of that substance. Just as a dozen eggs is 12 eggs, a mole of eggs is 6.022×10^{23} eggs.

The magnitude of the number 6.022×10^{23} is very difficult to imagine. To give you some idea, 1 mole of seconds represents a span of time 4 million times as long as the earth has already existed, and 1 mole of marbles is enough to cover the entire earth to a depth of 50 miles! However, since atoms are so tiny, a mole of atoms or molecules is a perfectly manageable quantity to use in a reaction (Fig. 3.4).

Critical Thinking

What if you were offered $1 million to count from 1 to 6×10^{23} at a rate of one number each second?
Determine your hourly wage. Would you do it? Could you do it?

How do we use the mole in chemical calculations? Recall that Avogadro's number is defined as the number of atoms in exactly 12 g of ^{12}C. This means that 12 g of ^{12}C contains 6.022×10^{23} atoms. It also means that a 12.01-g sample of natural carbon contains 6.022×10^{23} atoms (a mixture of ^{12}C, ^{13}C, and ^{14}C atoms, with an average atomic mass of 12.01). Since the ratio of the masses of the samples (12 g/12.01 g) is the same as the ratio of the masses of the individual components (12 u/12.01 u), the two samples contain the *same number* of atoms (6.022×10^{23}).

To be sure this point is clear, think of oranges with an average mass of 0.5 lb each and grapefruit with an average mass of 1.0 lb each. Any two sacks for which the sack of grapefruit weighs twice as much as the sack of oranges will contain the same number of pieces of fruit. The same idea extends to atoms. Compare natural carbon (average mass of 12.01) and natural helium (average mass of 4.003). A sample of 12.01 g of natural carbon contains the same number of atoms as 4.003 g of natural helium. Both samples contain 1 mole of atoms (6.022×10^{23}). Table 3.1 gives more examples that illustrate this basic idea.

Table 3.1 | Comparison of 1-Mole Samples of Various Elements

Element	Number of Atoms Present	Mass of Sample (g)
Aluminum	6.022×10^{23}	26.98
Copper	6.022×10^{23}	63.55
Iron	6.022×10^{23}	55.85
Sulfur	6.022×10^{23}	32.07
Iodine	6.022×10^{23}	126.9
Mercury	6.022×10^{23}	200.6

The mass of 1 mole of an element is equal to its atomic mass in grams.

Thus the mole is defined such that a sample of a natural element with a mass equal to the element's atomic mass expressed in grams contains 1 mole of atoms. This definition also fixes the relationship between the atomic mass unit and the gram. Since 6.022×10^{23} atoms of carbon (each with a mass of 12 u) have a mass of 12 g, then

$$(6.022 \times 10^{23} \text{ atoms})\left(\frac{12 \text{ u}}{\text{atom}}\right) = 12 \text{ g}$$

and

$$6.022 \times 10^{23} \text{ u} = 1 \text{ g}$$
↑
Exact number

This relationship can be used to derive the unit factor needed to convert between atomic mass units and grams.

Critical Thinking

What if you discovered Avogadro's number was not 6.02×10^{23} but 3.01×10^{23}? Would this affect the relative masses given on the periodic table? If so, how? If not, why not?

Interactive Example 3.2

Sign in at http://login.cengagebrain.com to try this Interactive Example in **OWL**.

Determining the Mass of a Sample of Atoms

Americium is an element that does not occur naturally. It can be made in very small amounts in a device known as a *particle accelerator.* Compute the mass in grams of a sample of americium containing six atoms.

Solution

Where are we going?

To calculate the mass of six americium atoms

What do we know?

› Mass of 1 atom of Am = 243 u (from the periodic table inside the front cover)

How do we get there?

The mass of six atoms is

$$6 \text{ atoms} \times 243 \frac{\text{u}}{\text{atom}} = 1.46 \times 10^3 \text{ u}$$

Using the relationship

$$6.022 \times 10^{23} \text{ u} = 1 \text{ g}$$

we write the conversion factor for converting atomic mass units to grams:

$$\frac{1 \text{ g}}{6.022 \times 10^{23} \text{ u}}$$

The mass of six americium atoms in grams is

$1.242 \times 10^{-20} \rightarrow 2.42 \times 10^{-21}$ g

$$1.46 \times 10^3 \text{ u} \times \frac{1 \text{ g}}{6.022 \times 10^{23} \text{ u}} = 2.42 \times 10^{-21} \text{ g}$$

Reality Check | Since this sample contains only six atoms, the mass should be very small as the amount 2.42×10^{-21} g indicates.

See Exercise 3.45

To do chemical calculations, you must understand what the mole means and how to determine the number of moles in a given mass of a substance. These procedures are illustrated in Examples 3.3 and 3.4.

Determining Moles of Atoms

Aluminum (Al) is a metal with a high strength-to-mass ratio and a high resistance to corrosion; thus it is often used for structural purposes. Compute both the number of moles of atoms and the number of atoms in a 10.0-g sample of aluminum.

(left) Pure aluminum. (right) Aluminum alloys are used for many products used in our kitchens.

Solution

Where are we going?

To calculate the moles and number of atoms in a sample of Al

What do we know?

> Sample contains 10.0 g of Al
> Mass of 1 mole (6.022×10^{23} atoms) of Al = 26.93 g

How do we get there?

We can calculate the number of moles of Al in a 10.0-g sample as follows:

$$10.0 \text{ g Al} \times \frac{1 \text{ mol Al}}{26.98 \text{ g Al}} = 0.371 \text{ mol Al atoms}$$

The number of atoms in 10.0 g (0.371 mole) of aluminum is

$$0.371 \text{ mol Al} \times \frac{6.022 \times 10^{23} \text{ atoms}}{1 \text{ mol Al}} = 2.23 \times 10^{23} \text{ atoms}$$

Reality Check | One mole of Al has a mass of 26.98 g and contains 6.022×10^{23} atoms. Our sample is 10.0 g, which is roughly 1/3 of 26.98. Thus the calculated amount should be on the order of 1/3 of 6×10^{23}, which it is.

See Exercise 3.46

Interactive Example 3.4

Calculating Numbers of Atoms

A silicon chip used in an integrated circuit of a microcomputer has a mass of 5.68 mg. How many silicon (Si) atoms are present in the chip?

Solution

Where are we going?

To calculate the atoms of Si in the chip

What do we know?

> The chip has 5.68 mg of Si
> Mass of 1 mole (6.022×10^{23} atoms) of Si = 28.09 g

How do we get there?

The strategy for doing this problem is to convert from milligrams of silicon to grams of silicon, then to moles of silicon, and finally to atoms of silicon:

Always check to see if your answer is sensible.

Paying careful attention to units and making sure the answer is reasonable can help you detect an inverted conversion factor or a number that was incorrectly entered in your calculator.

$$5.68 \text{ mg Si} \times \frac{1 \text{ g Si}}{1000 \text{ mg Si}} = 5.68 \times 10^{-3} \text{ g Si}$$

$$5.68 \times 10^{-3} \text{ g Si} \times \frac{1 \text{ mol Si}}{28.09 \text{ g Si}} = 2.02 \times 10^{-4} \text{ mol Si}$$

$$2.02 \times 10^{-4} \text{ mol Si} \times \frac{6.022 \times 10^{23} \text{ atoms}}{1 \text{ mol Si}} = 1.22 \times 10^{20} \text{ atoms}$$

Reality Check | Note that 5.68 mg of silicon is clearly much less than 1 mole of silicon (which has a mass of 28.09 g), so the final answer of 1.22×10^{20} atoms (compared with 6.022×10^{23} atoms) is in the right direction.

See Exercise 3.47

Interactive Example 3.5

Fragments of cobalt metal.

Russ Lappa/SPL/Photo Researchers, Inc.

Calculating the Number of Moles and Mass

Cobalt (Co) is a metal that is added to steel to improve its resistance to corrosion. Calculate both the number of moles in a sample of cobalt containing 5.00×10^{20} atoms and the mass of the sample.

Solution

Where are we going?

To calculate the number of moles and the mass of a sample of Co

What do we know?

> Sample contains 5.00×10^{20} atoms of Co

How do we get there?

Note that the sample of 5.00×10^{20} atoms of cobalt is less than 1 mole (6.022×10^{23} atoms) of cobalt. What fraction of a mole it represents can be determined as follows:

$$5.00 \times 10^{20} \text{ atoms Co} \times \frac{1 \text{ mol Co}}{6.022 \times 10^{23} \text{ atoms Co}} = 8.30 \times 10^{-4} \text{ mol Co}$$

Since the mass of 1 mole of cobalt atoms is 58.93 g, the mass of 5.00×10^{20} atoms can be determined as follows:

$$8.30 \times 10^{-4} \text{ mol Co} \times \frac{58.93 \text{ g Co}}{1 \text{ mol Co}} = 4.89 \times 10^{-2} \text{ g Co}$$

Reality Check | In this case the sample contains 5×10^{20} atoms, which is approximately 1/1000 of a mole. Thus the sample should have a mass of about $(1/1000)(58.93) \cong 0.06$. Our answer of ~0.05 makes sense.

See Exercise 3.48

3.4 | Molar Mass

A chemical compound is, ultimately, a collection of atoms. For example, methane (the major component of natural gas) consists of molecules that each contain one carbon and four hydrogen atoms (CH_4). How can we calculate the mass of 1 mole of methane; that is, what is the mass of 6.022×10^{23} CH_4 molecules? Since each CH_4 molecule contains one carbon atom and four hydrogen atoms, 1 mole of CH_4 molecules contains 1 mole of carbon atoms and 4 moles of hydrogen atoms. The mass of 1 mole of methane can be found by summing the masses of carbon and hydrogen present:

$$\text{Mass of 1 mol C} = 12.01 \text{ g}$$
$$\text{Mass of 4 mol H} = \underline{4 \times 1.008 \text{ g}}$$
$$\text{Mass of 1 mol CH}_4 = 16.04 \text{ g}$$

In this case, the term 12.01 limits the number of significant figures.

Because 16.04 g represents the mass of 1 mole of methane molecules, it makes sense to call it the *molar mass* for methane. Thus the **molar mass** of a substance is *the mass in grams of 1 mole of the compound*. Traditionally, the term *molecular weight* has been used for this quantity. However, we will use molar mass exclusively in this text. The molar mass of a known substance is obtained by summing the masses of the component atoms as we did for methane.

A substance's molar mass is the mass in grams of 1 mole of the substance.

Methane is a molecular compound—its components are molecules. Many substances are ionic—they contain simple ions or polyatomic ions. Examples are NaCl (contains Na^+ and Cl^-) and $CaCO_3$ (contains Ca^{2+} and CO_3^{2-}). Because ionic compounds do not contain molecules, we need a special name for the fundamental unit of these materials. Instead of molecule, we use the term *formula unit*. Thus $CaCO_3$ is the formula unit for calcium carbonate, and NaCl is the formula unit for sodium chloride.

Interactive Example 3.6

Sign in at http://login.cengagebrain.com to try this Interactive Example in OWL.

Calculating Molar Mass I

Juglone, a dye known for centuries, is produced from the husks of black walnuts. It is also a natural herbicide (weed killer) that kills off competitive plants around the black walnut tree but does not affect grass and other noncompetitive plants. The formula for juglone is $C_{10}H_6O_3$.

a. Calculate the molar mass of juglone.

b. A sample of 1.56×10^{-2} g of pure juglone was extracted from black walnut husks. How many moles of juglone does this sample represent?

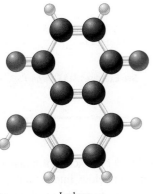

Juglone

Solution

a. The molar mass is obtained by summing the masses of the component atoms. In 1 mole of juglone, there are 10 moles of carbon atoms, 6 moles of hydrogen atoms, and 3 moles of oxygen atoms:

$$
\begin{aligned}
10\ \text{C:} \quad & 10 \times 12.01\ \text{g} = 120.1\ \text{g} \\
6\ \text{H:} \quad & 6 \times 1.008\ \text{g} = 6.048\ \text{g} \\
3\ \text{O:} \quad & 3 \times 16.00\ \text{g} = \underline{48.00\ \text{g}} \\
& \text{Mass of 1 mol } C_{10}H_6O_3 = 174.1\ \text{g}
\end{aligned}
$$

The mass of 1 mole of juglone is 174.1 g, which is the molar mass.

b. The mass of 1 mole of this compound is 174.1 g; thus 1.56×10^{-2} g is much less than a mole. The exact fraction of a mole can be determined as follows:

$$
1.56 \times 10^{-2}\ \text{g juglone} \times \frac{1\ \text{mol juglone}}{174.1\ \text{g juglone}} = 8.96 \times 10^{-5}\ \text{mol juglone}
$$

See Exercises 3.51 through 3.54

A calcite (CaCO$_3$) crystal.

Calculating Molar Mass II

Calcium carbonate ($CaCO_3$), also called *calcite,* is the principal mineral found in limestone, marble, chalk, pearls, and the shells of marine animals such as clams.

a. Calculate the molar mass of calcium carbonate.

b. A certain sample of calcium carbonate contains 4.86 moles. What is the mass in grams of this sample? What is the mass of the CO_3^{2-} ions present?

Solution

a. Calcium carbonate is an ionic compound composed of Ca^{2+} and CO_3^{2-} ions. In 1 mole of calcium carbonate, there are 1 mole of Ca^{2+} ions and 1 mole of CO_3^{2-} ions. The molar mass is calculated by summing the masses of the components:

$$
\begin{aligned}
1\ Ca^{2+}: \quad & 1 \times 40.08\ \text{g} = 40.08\ \text{g} \\
1\ CO_3^{2-}: & \\
1\ \text{C:} \quad & 1 \times 12.01\ \text{g} = 12.01\ \text{g} \\
3\ \text{O:} \quad & 3 \times 16.00\ \text{g} = \underline{48.00\ \text{g}} \\
& \text{Mass of 1 mol } CaCO_3 = 100.09\ \text{g}
\end{aligned}
$$

Thus the mass of 1 mole of $CaCO_3$ (1 mole of Ca^{2+} plus 1 mole of CO_3^{2-}) is 100.09 g. This is the molar mass.

b. The mass of 1 mole of $CaCO_3$ is 100.09 g. The sample contains nearly 5 moles, or close to 500 g. The exact amount is determined as follows:

$$
4.86\ \text{mol CaCO}_3 \times \frac{100.09\ \text{g CaCO}_3}{1\ \text{mol CaCO}_3} = 486\ \text{g CaCO}_3
$$

To find the mass of carbonate ions (CO_3^{2-}) present in this sample, we must realize that 4.86 moles of $CaCO_3$ contains 4.86 moles of Ca^{2+} ions and 4.86 moles of CO_3^{2-} ions. The mass of 1 mole of CO_3^{2-} ions is

$$
\begin{aligned}
1\ \text{C:} \quad & 1 \times 12.01 = 12.01\ \text{g} \\
3\ \text{O:} \quad & 3 \times 16.00 = \underline{48.00\ \text{g}} \\
& \text{Mass of 1 mol } CO_3^{2-} = 60.01\ \text{g}
\end{aligned}
$$

Thus the mass of 4.86 moles of CO_3^{2-} ions is

$$4.86 \ \text{mol } CO_3^{2-} \times \frac{60.01 \ \text{g } CO_3^{2-}}{1 \ \text{mol } CO_3^{2-}} = 292 \ \text{g } CO_3^{2-}$$

See Exercises 3.55 through 3.58

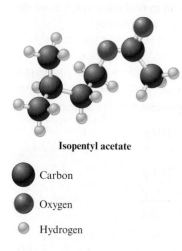

Isopentyl acetate is released when a bee stings.

Kenneth Lorenzen

Isopentyl acetate

● Carbon

● Oxygen

○ Hydrogen

To show the correct number of significant figures in each calculation, we round after each step. In your calculations, always carry extra significant figures through to the end, then round.

Molar Mass and Numbers of Molecules

Isopentyl acetate ($C_7H_{14}O_2$) is the compound responsible for the scent of bananas. A molecular model of isopentyl acetate is shown in the margin below. Interestingly, bees release about 1 μg (1×10^{-6} g) of this compound when they sting. The resulting scent attracts other bees to join the attack. How many molecules of isopentyl acetate are released in a typical bee sting? How many atoms of carbon are present?

Solution

Where are we going?

To calculate the number of molecules of isopentyl acetate and the number of carbon atoms in a bee sting

What do we know?

> Mass of isopentyl acetate in a typical bee sting is 1 microgram = 1×10^{-6} g

How do we get there?

Since we are given a mass of isopentyl acetate and want to find the number of molecules, we must first compute the molar mass of $C_7H_{14}O_2$:

$$7 \ \text{mol C} \times 12.01 \frac{\text{g}}{\text{mol}} = \ 84.07 \ \text{g C}$$

$$14 \ \text{mol H} \times 1.008 \frac{\text{g}}{\text{mol}} = \ 14.11 \ \text{g H}$$

$$2 \ \text{mol O} \times 16.00 \frac{\text{g}}{\text{mol}} = \ \underline{32.00 \ \text{g O}}$$

$$130.18 \ \text{g}$$

This means that 1 mole of isopentyl acetate (6.022×10^{23} molecules) has a mass of 130.18 g.

To find the number of molecules released in a sting, we must first determine the number of moles of isopentyl acetate in 1×10^{-6} g:

$$1 \times 10^{-6} \ \text{g } C_7H_{14}O_2 \times \frac{1 \ \text{mol } C_7H_{14}O_2}{130.18 \ \text{g } C_7H_{14}O_2} = 8 \times 10^{-9} \ \text{mol } C_7H_{14}O_2$$

Since 1 mole is 6.022×10^{23} units, we can determine the number of molecules:

$$8 \times 10^{-9} \ \text{mol } C_7H_{14}O_2 \times \frac{6.022 \times 10^{23} \ \text{molecules}}{1 \ \text{mol } C_7H_{14}O_2} = 5 \times 10^{15} \ \text{molecules}$$

To determine the number of carbon atoms present, we must multiply the number of molecules by 7, since each molecule of isopentyl acetate contains seven carbon atoms:

$$5 \times 10^{15} \ \text{molecules} \times \frac{7 \ \text{carbon atoms}}{\text{molecule}} = 4 \times 10^{16} \ \text{carbon atoms}$$

Note: In keeping with our practice of always showing the correct number of significant figures, we have rounded after each step. However, if extra digits are carried throughout this problem, the final answer rounds to 3×10^{16}.

See Exercises 3.59 through 3.64

3.5 | Learning to Solve Problems

One of the great rewards of studying chemistry is to become a good problem solver. Being able to solve complex problems is a talent that will serve you well in all walks of life. It is our purpose in this text to help you learn to solve problems in a flexible, creative way based on understanding the fundamental ideas of chemistry. We call this approach **conceptual problem solving**.

The ultimate goal is to be able to solve new problems (that is, problems you have not seen before) on your own. In this text we will provide problems and offer solutions by explaining how to think about the problems. While the answers to these problems are important, it is perhaps even more important to understand the process—the thinking necessary to get the answer. Although at first we will be solving the problem for you, do not take a passive role. While studying the solution, it is crucial that you interactively think through the problem with us. Do not skip the discussion and jump to the answer. Usually, the solution will involve asking a series of questions. Make sure that you understand each step in the process. This active approach should apply to problems outside of chemistry as well. For example, imagine riding with someone in a car to an unfamiliar destination. If your goal is simply to have the other person get you to that destination, you will probably not pay much attention to how to get there (passive), and if you have to find this same place in the future on your own, you probably will not be able to do it. If, however, your goal is to learn how to get there, you would pay attention to distances, signs, and turns (active). This is how you should read the solutions in the text (and the text in general).

While actively studying our solutions to problems is helpful, at some point you will need to know how to think through these problems on your own. If we help you too much as you solve a problem, you won't really learn effectively. If we always "drive," you won't interact as meaningfully with the material. Eventually you need to learn to drive yourself. We will provide more help at the beginning of the text and less as we proceed to later chapters.

There are two fundamentally different ways you might use to approach a problem. One way emphasizes memorization. We might call this the "pigeonholing method." In this approach, the first step is to label the problem—to decide in which pigeonhole it fits. The pigeonholing method requires that we provide you with a set of steps that you memorize and store in the appropriate slot for each different problem you encounter. The difficulty with this method is that it requires a new pigeonhole each time a problem is changed by even a small amount.

Consider the driving analogy again. Suppose you have memorized how to drive from your house to the grocery store. Do you know how to drive back from the grocery store to your house? Not necessarily. If you have only memorized the directions and do not understand fundamental principles such as "I traveled north to get to the store, so my house is south of the store," you may find yourself stranded. In a more complicated example, suppose you know how to get from your house to the store (and back) and from your house to the library (and back). Can you get from the library to the store without having to go back home? Probably not if you have only memorized directions and you do not have a "big picture" of where your house, the store, and the library are relative to one another.

The second approach is conceptual problem solving, in which we help you get the "big picture"—a real understanding of the situation. This approach to problem solving looks within the problem for a solution. In this method we assume that the problem is a new one, and we let the problem guide us as we solve it. In this approach we ask a series of questions as we proceed and use our knowledge of fundamental principles to answer these questions. Learning this approach requires some patience, but the reward for learning to solve problems this way is that we become an effective solver of any new problem that confronts us in daily life or in our work in any field. In summary, instead of looking outside the problem for a memorized solution, we will look inside the problem and let the problem help us as we proceed to a solution.

Pigeonholes can be used for sorting and classifying objects like mail.

John Humble/The Image Bank/Getty Images

As we have seen in problems we have already considered, there are several organizing principles to help you become a creative problem solver. Although we have been using these ideas in earlier problems, let's review and expand on them. Because as we progress in our study of chemistry the problems become more complicated, we will need to rely on this approach even more.

1. We need to read the problem and decide on the final goal. Then we sort through the facts given, focusing on the key words and often drawing a diagram of the problem. In this part of the analysis we need to state the problem as simply and as visually as possible. We could summarize this entire process as "**Where are we going?**"

2. In order to reach our final goal, we need to decide where to start. For example, in a stoichiometry problem we always start with the chemical reaction. Then we ask a series of questions as we proceed, such as, "What are the reactants and products?" "What is the balanced equation?" "What are the amounts of the reactants?" and so on. Our understanding of the fundamental principles of chemistry will enable us to answer each of these simple questions and eventually will lead us to the final solution. We might summarize this process as "**How do we get there?**"

3. Once we get the solution of the problem, then we ask ourselves, "Does it make sense?" That is, does our answer seem reasonable? We call this the **Reality Check**. It always pays to check your answer.

Using a conceptual approach to problem solving will enable you to develop real confidence as a problem solver. You will no longer panic when you see a problem that is different in some ways from those you have solved in the past. Although you might be frustrated at times as you learn this method, we guarantee that it will pay dividends later and should make your experience with chemistry a positive one that will prepare you for any career you choose.

To summarize, one of our major goals in this text is to help you become a creative problem solver. We will do this by, at first, giving you lots of guidance in how to solve problems. We will "drive," but we hope you will be paying attention instead of just "riding along." As we move forward, we will gradually shift more of the responsibility to you. As you gain confidence in letting the problem guide you, you will be amazed at how effective you can be at solving some really complex problems—just like the ones you will confront in "real life."

3.6 | Percent Composition of Compounds

There are two common ways of describing the composition of a compound: in terms of the numbers of its constituent atoms and in terms of the percentages (by mass) of its elements. We can obtain the mass percents of the elements from the formula of the compound by comparing the mass of each element present in 1 mole of the compound to the total mass of 1 mole of the compound.

For example, for ethanol, which has the formula C_2H_5OH, the mass of each element present and the molar mass are obtained as follows:

$$\text{Mass of C} = 2 \text{ mol} \times 12.01 \frac{\text{g}}{\text{mol}} = 24.02 \text{ g}$$

$$\text{Mass of H} = 6 \text{ mol} \times 1.008 \frac{\text{g}}{\text{mol}} = 6.048 \text{ g}$$

$$\text{Mass of O} = 1 \text{ mol} \times 16.00 \frac{\text{g}}{\text{mol}} = \underline{16.00 \text{ g}}$$

$$\text{Mass of 1 mol } C_2H_5OH = 46.07 \text{ g}$$

The **mass percent** (often called the *weight percent*) of carbon in ethanol can be computed by comparing the mass of carbon in 1 mole of ethanol to the total mass of 1 mole of ethanol and multiplying the result by 100:

$$\text{Mass percent of C} = \frac{\text{mass of C in 1 mol } C_2H_5OH}{\text{mass of 1 mol } C_2H_5OH} \times 100\%$$

$$= \frac{24.02 \text{ g}}{46.07 \text{ g}} \times 100\% = 52.14\%$$

The mass percents of hydrogen and oxygen in ethanol are obtained in a similar manner:

$$\text{Mass percent of H} = \frac{\text{mass of H in 1 mol } C_2H_5OH}{\text{mass of 1 mol } C_2H_5OH} \times 100\%$$

$$= \frac{6.048 \text{ g}}{46.07 \text{ g}} \times 100\% = 13.13\%$$

$$\text{Mass percent of O} = \frac{\text{mass of O in 1 mol } C_2H_5OH}{\text{mass of 1 mol } C_2H_5OH} \times 100\%$$

$$= \frac{16.00 \text{ g}}{46.07 \text{ g}} \times 100\% = 34.73\%$$

Reality Check | Notice that the percentages add up to 100.00%; this provides a check that the calculations are correct.

Interactive Example 3.9

Sign in at http://login.cengagebrain.com to try this Interactive Example in **OWL**.

Calculating Mass Percent

Carvone is a substance that occurs in two forms having different arrangements of the atoms but the same molecular formula ($C_{10}H_{14}O$) and mass. One type of carvone gives caraway seeds their characteristic smell, and the other type is responsible for the smell of spearmint oil. Compute the mass percent of each element in carvone.

Solution

Where are we going?

To find the mass percent of each element in carvone

What do we know?

> Molecular formula is $C_{10}H_{14}O$

What information do we need to find the mass percent?

> Mass of each element (we'll use 1 mole of carvone)
> Molar mass of carvone

How do we get there?

What is the mass of each element in 1 mole of $C_{10}H_{14}O$?

$$\text{Mass of C in 1 mol} = 10 \text{ mol} \times 12.01 \frac{\text{g}}{\text{mol}} = 120.1 \text{ g}$$

$$\text{Mass of H in 1 mol} = 14 \text{ mol} \times 1.008 \frac{\text{g}}{\text{mol}} = 14.11 \text{ g}$$

$$\text{Mass of O in 1 mol} = 1 \text{ mol} \times 16.00 \frac{\text{g}}{\text{mol}} = 16.00 \text{ g}$$

What is the molar mass of $C_{10}H_{14}O$?

$$120.1 \text{ g} + 14.11 \text{ g} + 16.00 \text{ g} = 150.2 \text{ g}$$
$$C_{10} \quad + \quad H_{14} \quad + \quad O \quad = C_{10}H_{14}O$$

Carvone

What is the mass percent of each element?
We find the fraction of the total mass contributed by each element and convert it to a percentage:

> Mass percent of C $= \dfrac{120.1 \text{ g C}}{150.2 \text{ g C}_{10}\text{H}_{14}\text{O}} \times 100\% = 79.96\%$

> Mass percent of H $= \dfrac{14.11 \text{ g H}}{150.2 \text{ g C}_{10}\text{H}_{14}\text{O}} \times 100\% = 9.394\%$

> Mass percent of O $= \dfrac{16.00 \text{ g O}}{150.2 \text{ g C}_{10}\text{H}_{14}\text{O}} \times 100\% = 10.65\%$

Reality Check | Sum the individual mass percent values—they should total to 100% within round-off errors. In this case, the percentages add up to 100.00%.

See Exercises 3.73 and 3.74

3.7 | Determining the Formula of a Compound

When a new compound is prepared, one of the first items of interest is the formula of the compound. This is most often determined by taking a weighed sample of the compound and either decomposing it into its component elements or reacting it with oxygen to produce substances such as CO_2, H_2O, and N_2, which are then collected and weighed. A device for doing this type of analysis is shown in Fig. 3.5. The results of such analyses provide the mass of each type of element in the compound, which can be used to determine the mass percent of each element.

We will see how information of this type can be used to compute the formula of a compound. Suppose a substance has been prepared that is composed of carbon, hydrogen, and nitrogen. When 0.1156 g of this compound is reacted with oxygen, 0.1638 g of carbon dioxide (CO_2) and 0.1676 g of water (H_2O) are collected. Assuming that all the carbon in the compound is converted to CO_2, we can determine the mass of carbon originally present in the 0.1156-g sample. To do this, we must use the fraction (by mass) of carbon in CO_2. The molar mass of CO_2 is

$$C: \quad 1 \text{ mol} \times 12.01\frac{\text{g}}{\text{mol}} = 12.01 \text{ g}$$

$$O: \quad 2 \text{ mol} \times 16.00\frac{\text{g}}{\text{mol}} = 32.00 \text{ g}$$

$$\text{Molar mass of } CO_2 = 44.01 \text{ g/mol}$$

CO_2

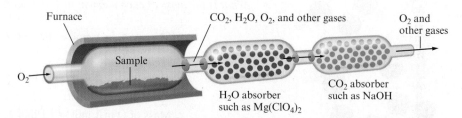

Figure 3.5 | A schematic diagram of the combustion device used to analyze substances for carbon and hydrogen. The sample is burned in the presence of excess oxygen, which converts all its carbon to carbon dioxide and all its hydrogen to water. These products are collected by absorption using appropriate materials, and their amounts are determined by measuring the increase in masses of the absorbents.

The fraction of carbon present by mass is

$$\frac{\text{Mass of C}}{\text{Total mass of CO}_2} = \frac{12.01 \text{ g C}}{44.01 \text{ g CO}_2}$$

This factor can now be used to determine the mass of carbon in 0.1638 g of CO_2:

$$0.1638 \text{ g CO}_2 \times \frac{12.01 \text{ g C}}{44.01 \text{ g CO}_2} = 0.04470 \text{ g C}$$

Remember that this carbon originally came from the 0.1156-g sample of unknown compound. Thus the mass percent of carbon in this compound is

$$\frac{0.04470 \text{ g C}}{0.1156 \text{ g compound}} \times 100\% = 38.67\% \text{ C}$$

The same procedure can be used to find the mass percent of hydrogen in the unknown compound. We assume that all the hydrogen present in the original 0.1156 g of compound was converted to H_2O. The molar mass of H_2O is 18.02 g, and the fraction of hydrogen by mass in H_2O is

$$\frac{\text{Mass of H}}{\text{Mass of H}_2\text{O}} = \frac{2.016 \text{ g H}}{18.02 \text{ g H}_2\text{O}}$$

H_2O

Therefore, the mass of hydrogen in 0.1676 g of H_2O is

$$0.1676 \text{ g H}_2\text{O} \times \frac{2.016 \text{ g H}}{18.02 \text{ g H}_2\text{O}} = 0.01875 \text{ g H}$$

The mass percent of hydrogen in the compound is

$$\frac{0.01875 \text{ g H}}{0.1156 \text{ g compound}} \times 100\% = 16.22\% \text{ H}$$

The unknown compound contains only carbon, hydrogen, and nitrogen. So far we have determined that it is 38.67% carbon and 16.22% hydrogen. The remainder must be nitrogen:

$$100.00\% - (\underset{\underset{\% \text{ C}}{\uparrow}}{38.67\%} + \underset{\underset{\% \text{ H}}{\uparrow}}{16.22\%}) = 45.11\% \text{ N}$$

We have determined that the compound contains 38.67% carbon, 16.22% hydrogen, and 45.11% nitrogen. Next we use these data to obtain the formula.

Since the formula of a compound indicates the *numbers* of atoms in the compound, we must convert the masses of the elements to numbers of atoms. The easiest way to do this is to work with 100.00 g of the compound. In the present case, 38.67% carbon by mass means 38.67 g of carbon per 100.00 g of compound; 16.22% hydrogen means 16.22 g of hydrogen per 100.00 g of compound; and so on. To determine the formula, we must calculate the number of carbon atoms in 38.67 g of carbon, the number of hydrogen atoms in 16.22 g of hydrogen, and the number of nitrogen atoms in 45.11 g of nitrogen. We can do this as follows:

$$38.67 \text{ g C} \times \frac{1 \text{ mol C}}{12.01 \text{ g C}} = 3.220 \text{ mol C}$$

$$16.22 \text{ g H} \times \frac{1 \text{ mol H}}{1.008 \text{ g H}} = 16.09 \text{ mol H}$$

$$45.11 \text{ g N} \times \frac{1 \text{ mol N}}{14.01 \text{ g N}} = 3.220 \text{ mol N}$$

Thus 100.00 g of this compound contains 3.220 moles of carbon atoms, 16.09 moles of hydrogen atoms, and 3.220 moles of nitrogen atoms.

We can find the smallest *whole-number ratio* of atoms in this compound by dividing each of the mole values above by the smallest of the three:

$$C: \quad \frac{3.220}{3.220} = 1.000 = 1$$

$$H: \quad \frac{16.09}{3.220} = 4.997 = 5$$

$$N: \quad \frac{3.220}{3.220} = 1.000 = 1$$

Thus the formula might well be CH_5N. However, it also could be $C_2H_{10}N_2$ or $C_3H_{15}N_3$, and so on—that is, some multiple of the smallest whole-number ratio. Each of these alternatives also has the correct relative numbers of atoms. That is, any molecule that can be represented as $(CH_5N)_n$, where n is an integer, has the **empirical formula** CH_5N. To be able to specify the exact formula of the molecule involved, the **molecular formula**, we must know the molar mass.

Suppose we know that this compound with empirical formula CH_5N has a molar mass of 31.06 g/mol. How do we determine which of the possible choices represents the molecular formula? Since the molecular formula is always a whole-number multiple of the empirical formula, we must first find the empirical formula mass for CH_5N:

$$1\ C: \quad 1 \times 12.01\ g = 12.01\ g$$
$$5\ H: \quad 5 \times 1.008\ g = 5.040\ g$$
$$1\ N: \quad 1 \times 14.01\ g = 14.01\ g$$

$$\text{Formula mass of } CH_5N = 31.06\ \text{g/mol}$$

This is the same as the known molar mass of the compound. Thus in this case the empirical formula and the molecular formula are the same; this substance consists of molecules with the formula CH_5N. It is quite common for the empirical and molecular formulas to be different; some examples where this is the case are shown in Fig. 3.6.

Molecular formula = (empirical formula)$_n$, where n is an integer.

Numbers very close to whole numbers, such as 9.92 and 1.08, should be rounded to whole numbers. Numbers such as 2.25, 4.33, and 2.72 should not be rounded to whole numbers.

Problem-Solving Strategy

Empirical Formula Determination

> Since mass percentage gives the number of grams of a particular element per 100 g of compound, base the calculation on 100 g of compound. Each percent will then represent the mass in grams of that element.

> Determine the number of moles of each element present in 100 g of compound using the atomic masses of the elements present.

> Divide each value of the number of moles by the smallest of the values. If each resulting number is a whole number (after appropriate rounding), these numbers represent the subscripts of the elements in the empirical formula.

> If the numbers obtained in the previous step are not whole numbers, multiply each number by an integer so that the results are all whole numbers.

Critical Thinking

One part of the problem-solving strategy for empirical formula determination is to base the calculation on 100 g of compound. What if you chose a mass other than 100 g? Would this work? What if you chose to base the calculation on 100 moles of compound? Would this work?

Figure 3.6 | Examples of substances whose empirical and molecular formulas differ. Notice that molecular formula = (empirical formula)$_n$, where n is an integer.

$C_6H_6 = (CH)_6$ $S_8 = (S)_8$ $C_6H_{12}O_6 = (CH_2O)_6$

Problem-Solving Strategy

Determining Molecular Formula from Empirical Formula
- Obtain the empirical formula.
- Compute the mass corresponding to the empirical formula.
- Calculate the ratio:

$$\frac{\text{Molar mass}}{\text{Empirical formula mass}}$$

- The integer from the previous step represents the number of empirical formula units in one molecule. When the empirical formula subscripts are multiplied by this integer, the molecular formula results. This procedure is summarized by the equation:

$$\text{Molecular formula} = \text{empirical formula} \times \frac{\text{molar mass}}{\text{empirical formula mass}}$$

Interactive Example 3.10

Sign in at http://login.cengagebrain.com to try this Interactive Example in **OWL**.

Determining Empirical and Molecular Formulas I

Determine the empirical and molecular formulas for a compound that gives the following percentages on analysis (in mass percents):

71.65% Cl   24.27% C   4.07% H

The molar mass is known to be 98.96 g/mol.

Solution

Where are we going?

To find the empirical and molecular formulas for the given compound

What do we know?
- Percent of each element
- Molar mass of the compound is 98.96 g/mol

What information do we need to find the empirical formula?
- Mass of each element in 100.00 g of compound
- Moles of each element

How do we get there?

What is the mass of each element in 100.00 g of compound?

| Cl | 71.65 g | C | 24.27 g | H | 4.07 g |

What are the moles of each element in 100.00 g of compound?

$$71.65 \text{ g Cl} \times \frac{1 \text{ mol Cl}}{35.45 \text{ g Cl}} = 2.021 \text{ mol Cl}$$

$$24.27 \text{ g C} \times \frac{1 \text{ mol C}}{12.01 \text{ g C}} = 2.021 \text{ mol C}$$

$$4.07 \text{ g H} \times \frac{1 \text{ mol H}}{1.008 \text{ g H}} = 4.04 \text{ mol H}$$

What is the empirical formula for the compound?
Dividing each mole value by 2.021 (the smallest number of moles present), we find the empirical formula $ClCH_2$.

What is the molecular formula for the compound?
Compare the empirical formula mass to the molar mass.

Empirical formula mass = 49.48 g/mol (Confirm this!)

Molar mass is given = 98.96 g/mol

$$\frac{\text{Molar mass}}{\text{Empirical formula mass}} = \frac{98.96 \text{ g/mol}}{49.48 \text{ g/mol}} = 2$$

$$\text{Molecular formula} = (ClCH_2)_2 = Cl_2C_2H_4$$

> This substance is composed of molecules with the formula $Cl_2C_2H_4$.

Note: The method we use here allows us to determine the molecular formula of a compound but not its structural formula. The compound $Cl_2C_2H_4$ is called *dichloroethane*. There are two forms of this compound, shown in Fig. 3.7. The form at the bottom was formerly used as an additive in leaded gasoline.

See Exercises 3.87 and 3.88

Figure 3.7 | The two forms of dichloroethane.

Determining Empirical and Molecular Formulas II

A white powder is analyzed and found to contain 43.64% phosphorus and 56.36% oxygen by mass. The compound has a molar mass of 283.88 g/mol. What are the compound's empirical and molecular formulas?

Solution

Where are we going?
To find the empirical and molecular formulas for the given compound

What do we know?
> Percent of each element
> Molar mass of the compound is 283.88 g/mol

What information do we need to find the empirical formula?
> Mass of each element in 100.00 g of compound
> Moles of each element

How do we get there?

What is the mass of each element in 100.00 g of compound?

$$P \qquad 43.64 \text{ g} \qquad O \qquad 56.36 \text{ g}$$

What are the moles of each element in 100.00 g of compound?

$$43.64 \text{ g P} \times \frac{1 \text{ mol P}}{30.97 \text{ g P}} = 1.409 \text{ mol P}$$

$$56.36 \text{ g O} \times \frac{1 \text{ mol O}}{16.00 \text{ g O}} = 3.523 \text{ mol O}$$

What is the empirical formula for the compound?
Dividing each mole value by the smaller one gives

$$\frac{1.409}{1.409} = 1 \text{ P} \quad \text{and} \quad \frac{3.523}{1.409} = 2.5 \text{ O}$$

This yields the formula $PO_{2.5}$. Since compounds must contain whole numbers of atoms, the empirical formula should contain only whole numbers. To obtain the simplest set of whole numbers, we multiply both numbers by 2 to give the empirical formula P_2O_5.

What is the molecular formula for the compound?
Compare the empirical formula mass to the molar mass.

Empirical formula mass = 141.94 g/mol (Confirm this!)

Molar mass is given = 283.88 g/mol

$$\frac{\text{Molar mass}}{\text{Empirical formula mass}} = \frac{283.88}{141.94} = 2$$

❯ The molecular formula is $(P_2O_5)_2$, or P_4O_{10}.

Note: The structural formula for this interesting compound is given in Fig. 3.8.

See Exercise 3.89

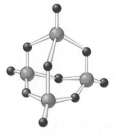

Figure 3.8 | The structure of P_4O_{10}. Note that some of the oxygen atoms act as "bridges" between the phosphorus atoms. This compound has a great affinity for water and is often used as a desiccant, or drying agent.

In Examples 3.10 and 3.11 we found the molecular formula by comparing the empirical formula mass with the molar mass. There is an alternate way to obtain the molecular formula. For example, in Example 3.10 we know the molar mass of the compound is 98.96 g/mol. This means that 1 mole of the compound weighs 98.96 g. Since we also know the mass percentages of each element, we can compute the mass of each element present in 1 mole of compound:

$$\text{Chlorine:} \quad \frac{71.65 \text{ g Cl}}{100.0 \text{ g compound}} \times \frac{98.96 \text{ g}}{\text{mol}} = \frac{70.90 \text{ g Cl}}{\text{mol compound}}$$

$$\text{Carbon:} \quad \frac{24.27 \text{ g C}}{100.0 \text{ g compound}} \times \frac{98.96 \text{ g}}{\text{mol}} = \frac{24.02 \text{ g C}}{\text{mol compound}}$$

$$\text{Hydrogen:} \quad \frac{4.07 \text{ g H}}{100.0 \text{ g compound}} \times \frac{98.96 \text{ g}}{\text{mol}} = \frac{4.03 \text{ g H}}{\text{mol compound}}$$

Now we can compute moles of atoms present per mole of compound:

$$\text{Chlorine:} \quad \frac{70.90 \text{ g Cl}}{\text{mol compound}} \times \frac{1 \text{ mol Cl}}{35.45 \text{ g Cl}} = \frac{2.000 \text{ mol Cl}}{\text{mol compound}}$$

$$\text{Carbon:} \quad \frac{24.02 \text{ g C}}{\text{mol compound}} \times \frac{1 \text{ mol C}}{12.01 \text{ g C}} = \frac{2.000 \text{ mol C}}{\text{mol compound}}$$

$$\text{Hydrogen:} \quad \frac{4.03 \text{ g H}}{\text{mol compound}} \times \frac{1 \text{ mol H}}{1.008 \text{ g H}} = \frac{4.00 \text{ mol H}}{\text{mol compound}}$$

Thus 1 mole of the compound contains 2 moles of Cl atoms, 2 moles of C atoms, and 4 moles of H atoms, and the molecular formula is $Cl_2C_2H_4$, as obtained in Example 3.10.

> **Problem-Solving Strategy**
>
> **Determining Molecular Formula from Mass Percent and Molar Mass**
>
> ❯ Using the mass percentages and the molar mass, determine the mass of each element present in 1 mole of compound.
>
> ❯ Determine the number of moles of each element present in 1 mole of compound.
>
> ❯ The integers from the previous step represent the subscripts in the molecular formula.

Interactive Example 3.12

Sign in at http://login.cengagebrain.com to try this Interactive Example in **OWL**.

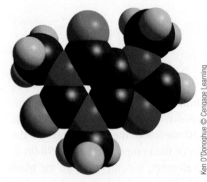

Computer-generated molecule of caffeine.

Ken O'Donoghue © Cengage Learning

Determining a Molecular Formula

Caffeine, a stimulant found in coffee, tea, and chocolate, contains 49.48% carbon, 5.15% hydrogen, 28.87% nitrogen, and 16.49% oxygen by mass and has a molar mass of 194.2 g/mol. Determine the molecular formula of caffeine.

Solution

Where are we going?

To find the molecular formula for caffeine

What do we know?

❯ Percent of each element

  ❯ 49.48% C   ❯ 28.87% N
  ❯ 5.15% H   ❯ 16.49% O

❯ Molar mass of caffeine is 194.2 g/mol

What information do we need to find the molecular formula?

❯ Mass of each element (in 1 mole of caffeine)

❯ Mole of each element (in 1 mole of caffeine)

How do we get there?

What is the mass of each element in 1 mole (194.2 g) of caffeine?

$$\frac{49.48 \text{ g C}}{100.0 \text{ g caffeine}} \times \frac{194.2 \text{ g}}{\text{mol}} = \frac{96.09 \text{ g C}}{\text{mol caffeine}}$$

$$\frac{5.15 \text{ g H}}{100.0 \text{ g caffeine}} \times \frac{194.2 \text{ g}}{\text{mol}} = \frac{10.0 \text{ g H}}{\text{mol caffeine}}$$

$$\frac{28.87 \text{ g N}}{100.0 \text{ g caffeine}} \times \frac{194.2 \text{ g}}{\text{mol}} = \frac{56.07 \text{ g N}}{\text{mol caffeine}}$$

$$\frac{16.49 \text{ g O}}{100.0 \text{ g caffeine}} \times \frac{194.2 \text{ g}}{\text{mol}} = \frac{32.02 \text{ g O}}{\text{mol caffeine}}$$

What are the moles of each element in 1 mole of caffeine?

Carbon: $\dfrac{96.09 \text{ g C}}{\text{mol caffeine}} \times \dfrac{1 \text{ mol C}}{12.01 \text{ g C}} = \dfrac{8.001 \text{ mol C}}{\text{mol caffeine}}$

Hydrogen: $\dfrac{10.0 \text{ g H}}{\text{mol caffeine}} \times \dfrac{1 \text{ mol H}}{1.008 \text{ g H}} = \dfrac{9.92 \text{ mol H}}{\text{mol caffeine}}$

Nitrogen: $\dfrac{56.07 \text{ g N}}{\text{mol caffeine}} \times \dfrac{1 \text{ mol N}}{14.01 \text{ g N}} = \dfrac{4.002 \text{ mol N}}{\text{mol caffeine}}$

Oxygen: $\dfrac{32.02 \text{ g O}}{\text{mol caffeine}} \times \dfrac{1 \text{ mol O}}{16.00 \text{ g O}} = \dfrac{2.001 \text{ mol O}}{\text{mol caffeine}}$

Rounding the numbers to integers gives the molecular formula for caffeine: $C_8H_{10}N_4O_2$.

See Exercise 3.90

3.8 | Chemical Equations

Chemical Reactions

A chemical change involves a reorganization of the atoms in one or more substances. For example, when the methane (CH_4) in natural gas combines with oxygen (O_2) in the air and burns, carbon dioxide (CO_2) and water (H_2O) are formed. This process is represented by a **chemical equation** with the **reactants** (here methane and oxygen) on the left side of an arrow and the **products** (carbon dioxide and water) on the right side:

$$CH_4 + O_2 \longrightarrow CO_2 + H_2O$$

Reactants   Products

Notice that the atoms have been reorganized. *Bonds have been broken, and new ones have been formed.* It is important to recognize that *in a chemical reaction, atoms are neither created nor destroyed.* All atoms present in the reactants must be accounted for among the products. In other words, there must be the same number of each type of atom on the products side and on the reactants side of the arrow. Making sure that this rule is obeyed is called **balancing a chemical equation** for a reaction.

The equation (shown above) for the reaction between CH_4 and O_2 is not balanced. We can see this from the following representation of the reaction:

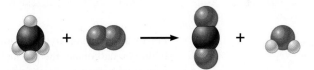

Notice that the number of oxygen atoms (in O_2) on the left of the arrow is two, while on the right there are three O atoms (in CO_2 and H_2O). Also, there are four hydrogen atoms (in CH_4) on the left and only two (in H_2O) on the right. Remember that a chemical reaction is simply a rearrangement of the atoms (a change in the way they are organized). Atoms are neither created nor destroyed in a chemical reaction. Thus the reactants and products must occur in numbers that give the same number of each type of atom among both the reactants and products. Simple trial and error will allow us to figure this out for the reaction of methane with oxygen. The needed numbers of molecules are

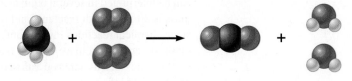

Notice that now we have the same number of each type of atom represented among the reactants and the products.

We can represent the preceding situation in a shorthand manner by the following chemical equation:

$$CH_4 + 2O_2 \longrightarrow CO_2 + 2H_2O$$

We can check that the equation is balanced by comparing the number of each type of atom on both sides:

$$CH_4 + 2O_2 \longrightarrow CO_2 + 2H_2O$$

1 C 4 H   4 O   1 C   2 O   4 H   2 O

To summarize, we have

Reactants	Products
1 C	1 C
4 H	4 H
4 O	4 O

The Meaning of a Chemical Equation

The chemical equation for a reaction gives two important types of information: the nature of the reactants and products and the relative numbers of each.

The reactants and products in a specific reaction must be identified by experiment. Besides specifying the compounds involved in the reaction, the equation often gives the *physical states* of the reactants and products:

State	Symbol
Solid	(s)
Liquid	(l)
Gas	(g)
Dissolved in water (in aqueous solution)	(aq)

For example, when hydrochloric acid in aqueous solution is added to solid sodium hydrogen carbonate, the products carbon dioxide gas, liquid water, and sodium chloride (which dissolves in the water) are formed:

$$HCl(aq) + NaHCO_3(s) \longrightarrow CO_2(g) + H_2O(l) + NaCl(aq)$$

The relative numbers of reactants and products in a reaction are indicated by the *coefficients* in the balanced equation. (The coefficients can be determined because we know that the same number of each type of atom must occur on both sides of the equation.) For example, the balanced equation

$$CH_4(g) + 2O_2(g) \longrightarrow CO_2(g) + 2H_2O(g)$$

can be interpreted in several equivalent ways, as shown in Table 3.2. Note that the total mass is 80 g for both reactants and products. We expect the mass to remain constant, since chemical reactions involve only a rearrangement of atoms. Atoms, and therefore mass, are conserved in a chemical reaction.

From this discussion you can see that a balanced chemical equation gives you a great deal of information.

Table 3.2 │ Information Conveyed by the Balanced Equation for the Combustion of Methane

Reactants		Products
$CH_4(g) + 2O_2(g)$	$\longrightarrow$	$CO_2(g) + 2H_2O(g)$
1 molecule + 2 molecules	$\longrightarrow$	1 molecule + 2 molecules
1 mole + 2 moles	$\longrightarrow$	1 mole + 2 moles
6.022×10^{23} molecules + 2 (6.022×10^{23} molecules)	$\longrightarrow$	6.022×10^{23} molecules + 2 (6.022×10^{23} molecules)
16 g + 2 (32 g)		44 g + 2 (18 g)
80 g reactants	$\longrightarrow$	80 g products

3.9 │ Balancing Chemical Equations

An unbalanced chemical equation is of limited use. Whenever you see an equation, you should ask yourself whether it is balanced. The principle that lies at the heart of the balancing process is that atoms are conserved in a chemical reaction. The same number of each type of atom must be found among the reactants and products. It is also important to recognize that the identities of the reactants and products of a reaction are determined by experimental observation. For example, when liquid ethanol is burned in the presence of sufficient oxygen gas, the products are always carbon dioxide and water. When the equation for this reaction is balanced, the *identities* of the reactants and products must not be changed. The formulas of the compounds must never be changed in balancing a chemical equation. That is, the subscripts in a formula cannot be changed, nor can atoms be added or subtracted from a formula.

Critical Thinking

What if a friend was balancing chemical equations by changing the values of the subscripts instead of using the coefficients? How would you explain to your friend that this was the wrong thing to do?

In balancing equations, start with the most complicated molecule.

Most chemical equations can be balanced by inspection, that is, by trial and error. It is always best to start with the most complicated molecules (those containing the greatest number of atoms). For example, consider the reaction of ethanol with oxygen, given by the unbalanced equation

$$C_2H_5OH(l) + O_2(g) \longrightarrow CO_2(g) + H_2O(g)$$

which can be represented by the following molecular models:

Notice that the carbon and hydrogen atoms are not balanced. There are two carbon atoms on the left and one on the right, and there are six hydrogens on the left and two on the right. We need to find the correct numbers of reactants and products so that we have the same number of all types of atoms among the reactants and products. We will balance the equation "by inspection" (a systematic trial-and-error procedure).

The most complicated molecule here is C_2H_5OH. We will begin by balancing the products that contain the atoms in C_2H_5OH. Since C_2H_5OH contains two carbon atoms, we place the coefficient 2 before the CO_2 to balance the carbon atoms:

$$\underset{\text{2 C atoms}}{C_2H_5OH(l)} + O_2(g) \longrightarrow \underset{\text{2 C atoms}}{2CO_2(g)} + H_2O(g)$$

Since C_2H_5OH contains six hydrogen atoms, the hydrogen atoms can be balanced by placing a 3 before the H_2O:

$$\underset{(5+1)\,H}{C_2H_5OH(l)} + O_2(g) \longrightarrow 2CO_2(g) + \underset{(3\times2)\,H}{3H_2O(g)}$$

Last, we balance the oxygen atoms. Note that the right side of the preceding equation contains seven oxygen atoms, whereas the left side has only three. We can correct this by putting a 3 before the O_2 to produce the balanced equation:

$$\underbrace{\underset{1\,O}{C_2H_5OH(l)} + \underset{6\,O}{3O_2(g)}}_{7\,O} \longrightarrow \underbrace{\underset{(2\times2)\,O}{2CO_2(g)} + \underset{3\,O}{3H_2O(g)}}_{7\,O}$$

Now we check:

$$C_2H_5OH(l) + 3O_2(g) \longrightarrow 2CO_2(g) + 3H_2O(g)$$

2 C atoms	2 C atoms
6 H atoms	6 H atoms
7 O atoms	7 O atoms

The equation is balanced.

The balanced equation can be represented as follows:

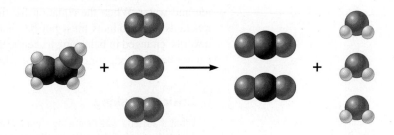

You can see that all the elements balance.

Problem-Solving Strategy

Writing and Balancing the Equation for a Chemical Reaction

1. Determine what reaction is occurring. What are the reactants, the products, and the physical states involved?

2. Write the *unbalanced* equation that summarizes the reaction described in Step 1.

3. Balance the equation by inspection, starting with the most complicated molecule(s). Determine what coefficients are necessary so that the same number of each type of atom appears on both reactant and product sides. Do not change the identities (formulas) of any of the reactants or products.

Critical Thinking

One part of the problem-solving strategy for balancing chemical equations is "starting with the most complicated molecule." What if you started with a different molecule? Could you still eventually balance the chemical equation? How would this approach be different from the suggested technique?

Balancing a Chemical Equation I

Chromium compounds exhibit a variety of bright colors. When solid ammonium dichromate, $(NH_4)_2Cr_2O_7$, a vivid orange compound, is ignited, a spectacular reaction occurs, as shown in the two photographs. Although the reaction is actually somewhat more complex, let's assume here that the products are solid chromium(III) oxide, nitrogen gas (consisting of N_2 molecules), and water vapor. Balance the equation for this reaction.

Solution

1. From the description given, the reactant is solid ammonium dichromate, $(NH_4)_2Cr_2O_7(s)$, and the products are nitrogen gas, $N_2(g)$, water vapor, $H_2O(g)$, and solid chromium(III) oxide, $Cr_2O_3(s)$. The formula for chromium(III) oxide can be determined by recognizing that the Roman numeral III means that Cr^{3+} ions are present. For a neutral compound, the formula must then be Cr_2O_3, since each oxide ion is O^{2-}.

2. The unbalanced equation is

$$(NH_4)_2Cr_2O_7(s) \rightarrow Cr_2O_3(s) + N_2(g) + H_2O(g)$$

3. Note that nitrogen and chromium are balanced (two nitrogen atoms and two chromium atoms on each side), but hydrogen and oxygen are not. A coefficient of 4 for H_2O balances the hydrogen atoms:

$$\underset{(4 \times 2)\,H}{(NH_4)_2Cr_2O_7(s)} \rightarrow Cr_2O_3(s) + N_2(g) + \underset{(4 \times 2)\,H}{4H_2O(g)}$$

Note that in balancing the hydrogen we also have balanced the oxygen, since there are seven oxygen atoms in the reactants and in the products.

Reality Check

$$2\,N,\ 8\,H,\ 2\,Cr,\ 7\,O \rightarrow 2\,N,\ 8\,H,\ 2\,Cr,\ 7\,O$$

$$\underset{\text{atoms}}{\text{Reactant}} \qquad\qquad \underset{\text{atoms}}{\text{Product}}$$

The equation is balanced.

Decomposition of ammonium dichromate.

Photos: Ken O'Donoghue © Cengage Learning

See Exercises 3.95 through 3.98

Balancing a Chemical Equation II

At 1000°C, ammonia gas, $NH_3(g)$, reacts with oxygen gas to form gaseous nitric oxide, $NO(g)$, and water vapor. This reaction is the first step in the commercial production of nitric acid by the Ostwald process. Balance the equation for this reaction.

Solution

1, 2. The unbalanced equation for the reaction is

$$NH_3(g) + O_2(g) \rightarrow NO(g) + H_2O(g)$$

3. Because all the molecules in this equation are of about equal complexity, where we start in balancing it is rather arbitrary. Let's begin by balancing the hydrogen. A coefficient of 2 for NH_3 and a coefficient of 3 for H_2O give six atoms of hydrogen on both sides:

$$2NH_3(g) + O_2(g) \rightarrow NO(g) + 3H_2O(g)$$

The nitrogen can be balanced with a coefficient of 2 for NO:

$$2NH_3(g) + O_2(g) \rightarrow 2NO(g) + 3H_2O(g)$$

Finally, note that there are two atoms of oxygen on the left and five on the right. The oxygen can be balanced with a coefficient of $\frac{5}{2}$ for O_2:

$$2NH_3(g) + \frac{5}{2}O_2(g) \rightarrow 2NO(g) + 3H_2O(g)$$

However, the usual custom is to have whole-number coefficients. We simply multiply the entire equation by 2.

$$4NH_3(g) + 5O_2(g) \rightarrow 4NO(g) + 6H_2O(g)$$

Reality Check | There are 4 N, 12 H, and 10 O on both sides, so the equation is balanced.

We can represent this balanced equation visually as

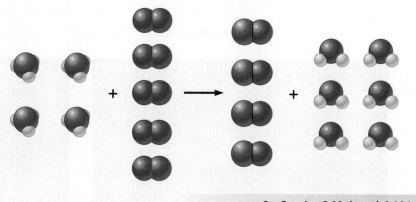

See Exercises 3.99 through 3.104

3.10 | Stoichiometric Calculations: Amounts of Reactants and Products

As we have seen in previous sections of this chapter, the coefficients in chemical equations represent *numbers* of molecules, not masses of molecules. However, when a reaction is to be run in a laboratory or chemical plant, the amounts of substances needed

Chemical connections
High Mountains—Low Octane

The next time you visit a gas station, take a moment to note the octane rating that accompanies the grade of gasoline that you are purchasing. The gasoline is priced according to its octane rating—a measure of the fuel's antiknock properties. In a conventional internal combustion engine, gasoline vapors and air are drawn into the combustion cylinder on the downward stroke of the piston. This air–fuel mixture is compressed on the upward piston stroke (compression stroke), and a spark from the sparkplug ignites the mix. The rhythmic combustion of the air–fuel mix occurring sequentially in several cylinders furnishes the power to propel the vehicle down the road. Excessive heat and pressure (or poor-quality fuel) within the cylinder may cause the premature combustion

of the mixture—commonly known as engine "knock" or "ping." Over time, this engine knock can damage the engine, resulting in inefficient performance and costly repairs.

A consumer typically is faced with three choices of gasoline, with octane ratings of 87 (regular), 89 (midgrade), and 93 (premium). But if you happen to travel or live in the higher elevations of the Rocky Mountain states, you might be surprised to find different octane ratings at the gasoline pumps. The reason for this provides a lesson in stoichiometry. At higher elevations the air is less dense—the volume of oxygen per unit volume of air is smaller. Most engines are designed to achieve a 14:1 oxygen-to-fuel ratio in the cylinder prior to combustion. If less oxygen is

available, then less fuel is required to achieve this optimal ratio. In turn, the lower volumes of oxygen and fuel result in a lower pressure in the cylinder. Because high pressure tends to promote knocking, the lower pressure within engine cylinders at higher elevations promotes a more controlled combustion of the air–fuel mixture, and therefore octane requirements are lower. While consumers in the Rocky Mountain states can purchase three grades of gasoline, the octane ratings of these fuel blends are different from those in the rest of the United States. In Denver, Colorado, regular gasoline is 85 octane, midgrade is 87 octane, and premium is 91 octane—2 points lower than gasoline sold in most of the rest of the country.

cannot be determined by counting molecules directly. Counting is always done by weighing. In this section we will see how chemical equations can be used to determine the *masses* of reacting chemicals.

To develop the principles for dealing with the stoichiometry of reactions, we will consider the reaction of propane with oxygen to produce carbon dioxide and water. We will consider the question: *"What mass of oxygen will react with 96.1 g of propane?"* In doing stoichiometry, the first thing we must do is *write the balanced chemical equation* for the reaction. In this case the balanced equation is

$$C_3H_8(g) + 5O_2(g) \longrightarrow 3CO_2(g) + 4H_2O(g)$$

which can be visualized as

Before doing any calculations involving a chemical reaction, be sure the equation for the reaction is balanced.

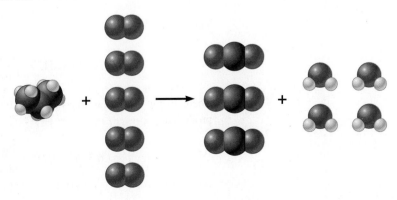

This equation means that 1 mole of C_3H_8 reacts with 5 moles of O_2 to produce 3 moles of CO_2 and 4 moles of H_2O. To use this equation to find the masses of reactants and products, we must be able to convert between masses and moles of substances. Thus we must first ask: *"How many moles of propane are present in 96.1 g of propane?"* The molar mass of propane to three significant figures is 44.1 (that is, $3 \times 12.01 + 8 \times 1.008$). The moles of propane can be calculated as follows:

$$96.1 \text{ g } C_3H_8 \times \frac{1 \text{ mol } C_3H_8}{44.1 \text{ g } C_3H_8} = 2.18 \text{ mol } C_3H_8$$

Next we must take into account the fact that each mole of propane reacts with 5 moles of oxygen. The best way to do this is to use the balanced equation to construct a **mole ratio**. In this case we want to convert from moles of propane to moles of oxygen. From the balanced equation, we see that 5 moles of O_2 are required for each mole of C_3H_8, so the appropriate ratio is

$$\frac{5 \text{ mol } O_2}{1 \text{ mol } C_3H_8}$$

Multiplying the number of moles of C_3H_8 by this factor gives the number of moles of O_2 required:

$$2.18 \text{ mol } C_3H_8 \times \frac{5 \text{ mol } O_2}{1 \text{ mol } C_3H_8} = 10.9 \text{ mol } O_2$$

Notice that the mole ratio is set up so that the moles of C_3H_8 cancel out, and the units that result are moles of O_2.

Since the original question asked for the mass of oxygen needed to react with 96.1 g of propane, the 10.9 moles of O_2 must be converted to *grams*. Since the molar mass of O_2 is 32.0 g/mol,

$$10.9 \text{ mol } O_2 \times \frac{32.0 \text{ g } O_2}{1 \text{ mol } O_2} = 349 \text{ g } O_2$$

Therefore, 349 g of oxygen are required to burn 96.1 g of propane.

This example can be extended by asking: *"What mass of carbon dioxide is produced when 96.1 g of propane are combusted with oxygen?"* In this case we must convert between moles of propane and moles of carbon dioxide. This can be accomplished by looking at the balanced equation, which shows that 3 moles of CO_2 are produced for each mole of C_3H_8 reacted. The mole ratio needed to convert from moles of propane to moles of carbon dioxide is

$$\frac{3 \text{ mol } CO_2}{1 \text{ mol } C_3H_8}$$

The conversion is

$$2.18 \text{ mol } C_3H_8 \times \frac{3 \text{ mol } CO_2}{1 \text{ mol } C_3H_8} = 6.54 \text{ mol } CO_2$$

Then, using the molar mass of CO_2 (44.0 g/mol), we calculate the mass of CO_2 produced:

$$6.54 \text{ mol } CO_2 \times \frac{44.0 \text{ g } CO_2}{1 \text{ mol } CO_2} = 288 \text{ g } CO_2$$

We will now summarize the sequence of steps needed to carry out stoichiometric calculations.

96.1 g C$_3$H$_8$ $\dfrac{\text{1 mol C}_3\text{H}_8}{\text{44.1 g C}_3\text{H}_8}$ 2.18 mol C$_3$H$_8$ $\dfrac{\text{3 mol CO}_2}{\text{1 mol C}_3\text{H}_8}$ 6.54 mol CO$_2$

$\dfrac{\text{44.0 g CO}_2}{\text{1 mol CO}_2}$ 288 g CO$_2$

Critical Thinking

Your lab partner has made the observation that you always take the mass of chemicals in lab, but then you use mole ratios to balance the equation. "Why not use the masses in the equation?" your partner asks. What if your lab partner decided to balance equations by using masses as coefficients? Is this even possible? Why or why not?

Problem-Solving Strategy

Calculating Masses of Reactants and Products in Chemical Reactions

1. Balance the equation for the reaction.
2. Convert the known mass of the reactant or product to moles of that substance.
3. Use the balanced equation to set up the appropriate mole ratios.
4. Use the appropriate mole ratios to calculate the number of moles of the desired reactant or product.
5. Convert from moles back to grams if required by the problem.

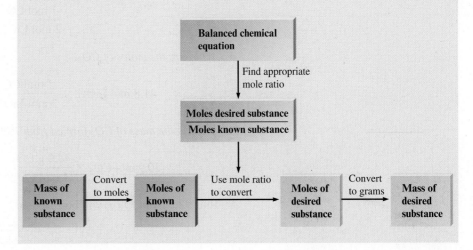

Chemical Stoichiometry I

Solid lithium hydroxide is used in space vehicles to remove exhaled carbon dioxide from the living environment by forming solid lithium carbonate and liquid water. What mass of gaseous carbon dioxide can be absorbed by 1.00 kg of lithium hydroxide?

Solution

Where are we going?

To find the mass of CO_2 absorbed by 1.00 kg LiOH

What do we know?

> Chemical reaction

$$LiOH(s) + CO_2(g) \longrightarrow Li_2CO_3(s) + H_2O(l)$$

> 1.00 kg LiOH

What information do we need to find the mass of CO_2?

> Balanced equation for the reaction

How do we get there?

1. *What is the balanced equation?*

$$2LiOH(s) + CO_2(g) \rightarrow Li_2CO_3(s) + H_2O(l)$$

2. *What are the moles of LiOH?*
   To find the moles of LiOH, we need to know the molar mass.

   What is the molar mass for LiOH?

$$6.941 + 16.00 + 1.008 = 23.95 \text{ g/mol}$$

   Now we use the molar mass to find the moles of LiOH:

$$1.00 \text{ kg LiOH} \times \frac{1000 \text{ g LiOH}}{1 \text{ kg LiOH}} \times \frac{1 \text{ mol LiOH}}{23.95 \text{ g LiOH}} = 41.8 \text{ mol LiOH}$$

3. *What is the mole ratio between CO_2 and LiOH in the balanced equation?*

$$\frac{1 \text{ mol CO}_2}{2 \text{ mol LiOH}}$$

4. *What are the moles of CO_2?*

$$41.8 \text{ mol LiOH} \times \frac{1 \text{ mol CO}_2}{2 \text{ mol LiOH}} = 20.9 \text{ mol CO}_2$$

5. *What is the mass of CO_2 formed from 1.00 kg LiOH?*

$$20.9 \text{ mol CO}_2 \times \frac{44.0 \text{ g CO}_2}{1 \text{ mol CO}_2} = 9.20 \times 10^2 \text{ g CO}_2$$

> Thus 920. g of $CO_2(g)$ will be absorbed by 1.00 kg of LiOH(s).

See Exercises 3.105 and 3.106

Interactive
Example 3.16

Sign in at http://login.cengagebrain
.com to try this Interactive Example
in **OWL**.

Chemical Stoichiometry II

Baking soda ($NaHCO_3$) is often used as an antacid. It neutralizes excess hydrochloric acid secreted by the stomach:

$$NaHCO_3(s) + HCl(aq) \longrightarrow NaCl(aq) + H_2O(l) + CO_2(aq)$$

Milk of magnesia, which is an aqueous suspension of magnesium hydroxide, is also used as an antacid:

$$Mg(OH)_2(s) + 2HCl(aq) \longrightarrow 2H_2O(l) + MgCl_2(aq)$$

Which is the more effective antacid per gram, $NaHCO_3$ or $Mg(OH)_2$?

Solution

Where are we going?

To compare the acid neutralizing power of $NaHCO_3$ and $Mg(OH)_2$ per gram

What do we know?

> Balanced equations for the reactions
> 1.00 g $NaHCO_3$
> 1.00 g $Mg(OH)_2$

How do we get there?

For $NaHCO_3$

1. *What is the balanced equation?*

$$NaHCO_3(s) + HCl(aq) \longrightarrow NaCl(aq) + H_2O(l) + CO_2(aq)$$

2. *What are the moles of $NaHCO_3$ in 1.00 g?*
   To find the moles of $NaHCO_3$, we need to know the molar mass (84.01 g/mol).

$$1.00 \text{ g } NaHCO_3 \times \frac{1 \text{ mol } NaHCO_3}{84.01 \text{ g } NaHCO_3} = 1.19 \times 10^{-2} \text{ mol } NaHCO_3$$

3. *What is the mole ratio between HCl and $NaHCO_3$ in the balanced equation?*

$$\frac{1 \text{ mol HCl}}{1 \text{ mol } NaHCO_3}$$

4. *What are the moles of HCl?*

$$1.19 \times 10^{-2} \text{ mol } NaHCO_3 \times \frac{1 \text{ mol HCl}}{1 \text{ mol } NaHCO_3} = 1.19 \times 10^{-2} \text{ mol HCl}$$

Thus 1.00 g of $NaHCO_3$ will neutralize 1.19×10^{-2} mole of HCl.

For $Mg(OH)_2$

1. *What is the balanced equation?*

$$Mg(OH)_2(s) + 2HCl(aq) \longrightarrow 2H_2O(l) + MgCl_2(aq)$$

2. *What are the moles of $Mg(OH)_2$ in 1.00 g?*
   To find the moles of $Mg(OH)_2$, we need to know the molar mass (58.32 g/mol).

$$1.00 \text{ g } Mg(OH)_2 \times \frac{1 \text{ mol } Mg(OH)_2}{58.32 \text{ g } Mg(OH)_2} = 1.71 \times 10^{-2} \text{ mol } Mg(OH)_2$$

Milk of magnesia contains a suspension of $Mg(OH)_2(s)$.

Charles D. Winters

3. *What is the mole ratio between HCl and Mg(OH)₂ in the balanced equation?*

$$\frac{2 \text{ mol HCl}}{1 \text{ mol Mg(OH)}_2}$$

4. *What are the moles of HCl?*

$$1.71 \times 10^{-2} \text{ mol } \cancel{\text{Mg(OH)}_2} \times \frac{2 \text{ mol HCl}}{1 \text{ mol } \cancel{\text{Mg(OH)}_2}} = 3.42 \times 10^{-2} \text{ mol HCl}$$

Thus 1.00 g of Mg(OH)₂ will neutralize 3.42×10^{-2} mole of HCl.

› Since 1.00 g NaHCO₃ neutralizes 1.19×10^{-2} mole of HCl and 1.00 g Mg(OH)₂ neutralizes 3.42×10^{-2} mole of HCl, Mg(OH)₂ is the more effective antacid.

See Exercises 3.107 and 3.108

3.11 | The Concept of Limiting Reactant

Suppose you have a part-time job in a sandwich shop. One very popular sandwich is always made as follows:

2 slices bread + 3 slices meat + 1 slice cheese ⟶ sandwich

Assume that you come to work one day and find the following quantities of ingredients:

8 slices bread

9 slices meat

5 slices cheese

How many sandwiches can you make? What will be left over?

To solve this problem, let's see how many sandwiches we can make with each component:

Bread: $8 \text{ slices bread} \times \dfrac{1 \text{ sandwich}}{2 \text{ slices bread}} = 4 \text{ sandwiches}$

Meat: $9 \text{ slices meat} \times \dfrac{1 \text{ sandwich}}{3 \text{ slices meat}} = 3 \text{ sandwiches}$

Cheese: $5 \text{ slices cheese} \times \dfrac{1 \text{ sandwich}}{1 \text{ slice cheese}} = 5 \text{ sandwiches}$

How many sandwiches can you make? The answer is three. When you run out of meat, you must stop making sandwiches. The meat is the limiting ingredient (Fig. 3.9).

What do you have left over? Making three sandwiches requires six pieces of bread. You started with eight slices, so you have two slices of bread left. You also used three pieces of cheese for the three sandwiches, so you have two pieces of cheese left.

In this example, the ingredient present in the largest number (the meat) was actually the component that limited the number of sandwiches you could make. This situation arose because each sandwich required three slices of meat—more than the quantity required of any other ingredient.

Figure 3.9 | Making sandwiches.

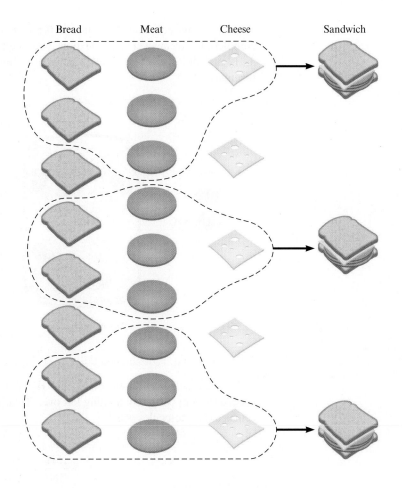

Bread   Meat   Cheese   Sandwich

When molecules react with each other to form products, considerations very similar to those involved in making sandwiches arise. We can illustrate these ideas with the reaction of $N_2(g)$ and $H_2(g)$ to form $NH_3(g)$:

$$N_2(g) + 3H_2(g) \longrightarrow 2NH_3(g)$$

Consider the following container of $N_2(g)$ and $H_2(g)$:

What will this container look like if the reaction between N_2 and H_2 proceeds to completion? To answer this question, you need to remember that each N_2 requires 3 H_2 molecules to form 2 NH_3. To make things clear, we will circle groups of reactants:

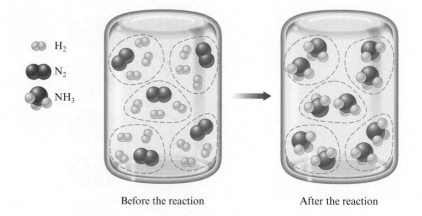

Before the reaction   After the reaction

In this case, the mixture of N_2 and H_2 contained just the number of molecules needed to form NH_3 with nothing left over. That is, the ratio of the number of H_2 molecules to N_2 molecules was

$$\frac{15H_2}{5N_2} = \frac{3H_2}{1N_2}$$

This ratio exactly matches the numbers in the balanced equation

$$3H_2(g) + N_2(g) \longrightarrow 2NH_3(g)$$

This type of mixture is called a **stoichiometric mixture**—one that contains the relative amounts of reactants that match the numbers in the balanced equation. In this case all reactants will be consumed to form products.

Now consider another container of $N_2(g)$ and $H_2(g)$:

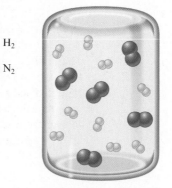

What will the container look like if the reaction between $N_2(g)$ and $H_2(g)$ proceeds to completion? Remember that each N_2 requires 3 H_2. Circling groups of reactants, we have

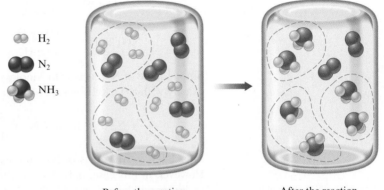

Before the reaction   After the reaction

In this case, the hydrogen (H_2) is limiting. That is, the H_2 molecules are used up before all the N_2 molecules are consumed. In this situation the amount of hydrogen limits the amount of product (ammonia) that can form—hydrogen is the limiting reactant. Some N_2 molecules are left over in this case because the reaction runs out of H_2 molecules first. To determine how much product can be formed from a given mixture of reactants, we have to look for the reactant that is limiting—the one that runs out first and thus limits the amount of product that can form. In some cases, the mixture of reactants might be stoichiometric—that is, all reactants run out at the same time. In general, however, you cannot assume that a given mixture of reactants is a stoichiometric mixture, so you must determine whether one of the reactants is limiting. The reactant that runs out first and thus limits the amounts of products that can form is called the **limiting reactant**.

To this point we have considered examples where the numbers of reactant molecules could be counted. In "real life" you can't count the molecules directly—you can't see them, and even if you could, there would be far too many to count. Instead, you must count by weighing. We must therefore explore how to find the limiting reactant, given the masses of the reactants.

A. Determination of Limiting Reactant Using Reactant Quantities

There are two ways to determine the limiting reactant in a chemical reaction. One involves comparing the moles of reactants to see which runs out first. We will consider this approach here.

In the laboratory or chemical plant, we work with much larger quantities than the few molecules of the preceding example. Therefore, we must learn to deal with limiting reactants using moles. The ideas are exactly the same, except that we are using moles of molecules instead of individual molecules. For example, suppose 25.0 kg of nitrogen and 5.00 kg of hydrogen are mixed and reacted to form ammonia. How do we calculate the mass of ammonia produced when this reaction is run to completion (until one of the reactants is completely consumed)?

As in the preceding example, we must use the balanced equation

$$N_2(g) + 3H_2(g) \longrightarrow 2NH_3(g)$$

to determine whether nitrogen or hydrogen is the limiting reactant and then to determine the amount of ammonia that is formed. We first calculate the moles of reactants present:

$$25.0 \text{ kg N}_2 \times \frac{1000 \text{ g N}_2}{1 \text{ kg N}_2} \times \frac{1 \text{ mol N}_2}{28.0 \text{ g N}_2} = 8.93 \times 10^2 \text{ mol N}_2$$

$$5.00 \text{ kg H}_2 \times \frac{1000 \text{ g H}_2}{1 \text{ kg H}_2} \times \frac{1 \text{ mol H}_2}{2.016 \text{ g H}_2} = 2.48 \times 10^3 \text{ mol H}_2$$

Since 1 mole of N_2 reacts with 3 moles of H_2, the number of moles of H_2 that will react exactly with 8.93×10^2 moles of N_2 is

$$8.93 \times 10^2 \text{ mol N}_2 \times \frac{3 \text{ mol H}_2}{1 \text{ mol N}_2} = 2.68 \times 10^3 \text{ mol H}_2$$

Thus 8.93×10^2 moles of N_2 requires 2.68×10^3 moles of H_2 to react completely. However, in this case, only 2.48×10^3 moles of H_2 are present. This means that the hydrogen will be consumed before the nitrogen. Thus hydrogen is the *limiting reactant* in this particular situation, and we must use the amount of hydrogen to compute the quantity of ammonia formed:

$$2.48 \times 10^3 \text{ mol H}_2 \times \frac{2 \text{ mol NH}_3}{3 \text{ mol H}_2} = 1.65 \times 10^3 \text{ mol NH}_3$$

Converting moles to kilograms gives

$$1.65 \times 10^3 \text{ mol NH}_3 \times \frac{17.0 \text{ g NH}_3}{1 \text{ mol NH}_3} = 2.80 \times 10^4 \text{ g NH}_3 = 28.0 \text{ kg NH}_3$$

Note that to determine the limiting reactant, we could have started instead with the given amount of hydrogen and calculated the moles of nitrogen required:

$$2.48 \times 10^3 \text{ mol H}_2 \times \frac{1 \text{ mol N}_2}{3 \text{ mol H}_2} = 8.27 \times 10^2 \text{ mol N}_2$$

Thus 2.48×10^3 moles of H_2 requires 8.27×10^2 moles of N_2. Since 8.93×10^2 moles of N_2 are actually present, the nitrogen is in excess. The hydrogen will run out first, and thus again we find that hydrogen limits the amount of ammonia formed.

A related but simpler way to determine which reactant is limiting is to compare the mole ratio of the substances required by the balanced equation with the mole ratio of reactants actually present. For example, in this case the mole ratio of H_2 to N_2 required by the balanced equation is

$$\frac{3 \text{ mol H}_2}{1 \text{ mol N}_2}$$

That is,

$$\frac{\text{mol H}_2}{\text{mol N}_2} \text{(required)} = \frac{3}{1} = 3$$

In this experiment we have 2.48×10^3 moles of H_2 and 8.93×10^2 moles of N_2. Thus the ratio

$$\frac{\text{mol H}_2}{\text{mol N}_2} \text{(actual)} = \frac{2.48 \times 10^3}{8.93 \times 10^2} = 2.78$$

Since 2.78 is less than 3, the actual mole ratio of H_2 to N_2 is too small, and H_2 must be limiting. If the actual H_2 to N_2 mole ratio had been greater than 3, then the H_2 would have been in excess and the N_2 would be limiting.

Always determine which reactant is limiting.

B. Determination of Limiting Reactant Using Quantities of Products Formed

A second method for determining which reactant in a chemical reaction is limiting is to consider the amounts of products that can be formed by completely consuming each reactant. The reactant that produces the smallest amount of product must run out first and thus be limiting. To see how this works, consider again the reaction of 25.0 kg (8.93×10^2 moles) of nitrogen with 5.00 kg (2.48×10^3 moles) of hydrogen.

We will now use these amounts of reactants to determine how much NH_3 would form. Since 1 mole of N_2 forms 2 moles of NH_3, the amount of NH_3 that would be formed if all of the N_2 was used up is calculated as follows:

$$8.93 \times 10^2 \text{ mol N}_2 \times \frac{2 \text{ mol NH}_3}{1 \text{ mol N}_2} = 1.79 \times 10^3 \text{ mol NH}_3$$

Next we will calculate how much NH_3 would be formed if the H_2 was completely used up:

$$2.48 \times 10^3 \text{ mol H}_2 \times \frac{2 \text{ mol NH}_3}{3 \text{ mol H}_2} = 1.65 \times 10^3 \text{ mol NH}_3$$

Because a smaller amount of NH_3 is produced from the H_2 than from the N_2, the amount of H_2 must be limiting.

Thus because the H_2 is the limiting reactant, the amount of NH_3 that can form is 1.65×10^3 moles. Converting moles to kilograms gives:

$$1.65 \times 10^3 \text{ mol NH}_3 \times \frac{17 \text{ g NH}_3}{1 \text{ mol NH}_3} = 2.80 \times 10^4 \text{ g NH}_3 = 28.0 \text{ kg NH}_3$$

Interactive Example 3.17

Sign in at http://login.cengagebrain.com to try this Interactive Example in **OWL**.

Stoichiometry: Limiting Reactant

Nitrogen gas can be prepared by passing gaseous ammonia over solid copper(II) oxide at high temperatures. The other products of the reaction are solid copper and water vapor. If a sample containing 18.1 g of NH_3 is reacted with 90.4 g of CuO, which is the limiting reactant? How many grams of N_2 will be formed?

Solution

Where are we going?

To find the limiting reactant
To find the mass of N_2 produced

What do we know?

› The chemical reaction

$$NH_3(g) + CuO(s) \longrightarrow N_2(g) + Cu(s) + H_2O(g)$$

› 18.1 g NH_3
› 90.4 g CuO

What information do we need?

› Balanced equation for the reaction
› Moles of NH_3
› Moles of CuO

How do we get there?

To find the limiting reactant

What is the balanced equation?

$$2NH_3(g) + 3CuO(s) \longrightarrow N_2(g) + 3Cu(s) + 3H_2O(g)$$

What are the moles of NH_3 and CuO?

To find the moles, we need to know the molar masses.

$$NH_3 \quad 17.03 \text{ g/mol}$$
$$CuO \quad 79.55 \text{ g/mol}$$

$$18.1 \text{ g NH}_3 \times \frac{1 \text{ mol NH}_3}{17.03 \text{ g NH}_3} = 1.06 \text{ mol NH}_3$$

$$90.4 \text{ g CuO} \times \frac{1 \text{ mol CuO}}{79.55 \text{ g CuO}} = 1.14 \text{ mol CuO}$$

A. First we will determine the limiting reactant by comparing the moles of reactants to see which one is consumed first.

What is the mole ratio between NH_3 and CuO in the balanced equation?

$$\frac{3 \text{ mol CuO}}{2 \text{ mol NH}_3}$$

How many moles of CuO are required to react with 1.06 moles of NH_3?

$$1.06 \text{ mol NH}_3 \times \frac{3 \text{ mol CuO}}{2 \text{ mol NH}_3} = 1.59 \text{ mol CuO}$$

> Thus 1.59 moles of CuO are required to react with 1.06 moles of NH_3. Since only 1.14 moles of CuO are actually present, the amount of CuO is limiting; CuO will run out before NH_3 does. We can verify this conclusion by comparing the mole ratio of CuO and NH_3 required by the balanced equation:

$$\frac{\text{mol CuO}}{\text{mol NH}_3} \text{(required)} = \frac{3}{2} = 1.5$$

with the mole ratio actually present:

$$\frac{\text{mol CuO}}{\text{mol NH}_3} \text{(actual)} = \frac{1.14}{1.06} = 1.08$$

> Since the actual ratio is too small (less than 1.5), CuO is the limiting reactant.

B. Alternatively we can determine the limiting reactant by computing the moles of N_2 that would be formed by complete consumption of NH_3 and CuO:

$$1.06 \text{ mol NH}_3 \times \frac{1 \text{ mol N}_2}{2 \text{ mol NH}_3} = 0.530 \text{ mol N}_2$$

$$1.14 \text{ mol CuO} \times \frac{1 \text{ mol N}_2}{3 \text{ mol CuO}} = 0.380 \text{ mol N}_2$$

As before, we see that the CuO is limiting since it produces the smaller amount of N_2.

To find the mass of N_2 produced

What are the moles of N_2 formed?

Because CuO is the limiting reactant, we must use the amount of CuO to calculate the amount of N_2 formed.

What is the mole ratio between N_2 and CuO in the balanced equation?

$$\frac{1 \text{ mol } N_2}{3 \text{ mol CuO}}$$

What are the moles of N_2?

$$1.14 \text{ mol CuO} \times \frac{1 \text{ mol } N_2}{3 \text{ mol CuO}} = 0.380 \text{ mol } N_2$$

What mass of N_2 is produced?
Using the molar mass of N_2 (28.02 g/mol), we can calculate the mass of N_2 produced:

$$■ \ 0.380 \text{ mol } N_2 \times \frac{28.02 \text{ g } N_2}{1 \text{ mol } N_2} = 10.6 \text{ g } N_2$$

See Exercises 3.117 through 3.122

The amount of a product formed when the limiting reactant is completely consumed is called the **theoretical yield** of that product. In Example 3.17, 10.6 g of nitrogen represent the theoretical yield. This is the *maximum amount* of nitrogen that can be produced from the quantities of reactants used. Actually, the amount of product predicted by the theoretical yield is seldom obtained because of side reactions (other reactions that involve one or more of the reactants or products) and other complications. The *actual yield* of product is often given as a percentage of the theoretical yield. This is called the **percent yield**:

Percent yield is important as an indicator of the efficiency of a particular laboratory or industrial reaction.

$$\frac{\text{Actual yield}}{\text{Theoretical yield}} \times 100\% = \text{percent yield}$$

For example, if the reaction considered in Example 3.17 actually gave 6.63 g of nitrogen instead of the predicted 10.6 g, the percent yield of nitrogen would be

$$\frac{6.63 \text{ g } N_2}{10.6 \text{ g } N_2} \times 100\% = 62.5\%$$

Calculating Percent Yield

Methanol (CH_3OH), also called *methyl alcohol,* is the simplest alcohol. It is used as a fuel in race cars and is a potential replacement for gasoline. Methanol can be manufactured by combining gaseous carbon monoxide and hydrogen. Suppose 68.5 kg $CO(g)$ is reacted with 8.60 kg $H_2(g)$. Calculate the theoretical yield of methanol. If 3.57×10^4 g CH_3OH is actually produced, what is the percent yield of methanol?

Solution

Where are we going?

To calculate the theoretical yield of methanol
To calculate the percent yield of methanol

What do we know?

❯ The chemical reaction

$$H_2(g) + CO(g) \longrightarrow CH_3OH(l)$$

❯ 68.5 kg $CO(g)$

❯ 8.60 kg $H_2(g)$

❯ 3.57×10^4 g CH_3OH is produced

Methanol

What information do we need?

› Balanced equation for the reaction
› Moles of H_2
› Moles of CO
› Which reactant is limiting
› Amount of CH_3OH produced

How do we get there?

To find the limiting reactant

What is the balanced equation?

$$2H_2(g) + CO(g) \longrightarrow CH_3OH(l)$$

What are the moles of H_2 and CO?

To find the moles, we need to know the molar masses.

$$H_2 \quad 2.016 \text{ g/mol}$$
$$CO \quad 28.02 \text{ g/mol}$$

$$68.5 \text{ kg CO} \times \frac{1000 \text{ g CO}}{1 \text{ kg CO}} \times \frac{1 \text{ mol CO}}{28.02 \text{ g CO}} = 2.44 \times 10^3 \text{ mol CO}$$

$$8.60 \text{ kg H}_2 \times \frac{1000 \text{ g H}_2}{1 \text{ kg H}_2} \times \frac{1 \text{ mol H}_2}{2.016 \text{ g H}_2} = 4.27 \times 10^3 \text{ mol H}_2$$

A. Determination of Limiting Reactant Using Reactant Quantities

What is the mole ratio between H_2 and CO in the balanced equation?

$$\frac{2 \text{ mol H}_2}{1 \text{ mol CO}}$$

How does the actual mole ratio compare to the stoichiometric ratio?

To determine which reactant is limiting, we compare the mole ratio of H_2 and CO required by the balanced equation

$$\frac{\text{mol H}_2}{\text{mol CO}} \text{ (required)} = \frac{2}{1} = 2$$

with the actual mole ratio

$$\frac{\text{mol H}_2}{\text{mol CO}} \text{ (actual)} = \frac{4.27 \times 10^3}{2.44 \times 10^3} = 1.75$$

› Since the actual mole ratio of H_2 to CO is smaller than the required ratio, H_2 *is limiting.*

B. Determination of Limiting Reactant Using Quantities of Products Formed

We can also determine the limiting reactant by calculating the amounts of CH_3OH formed by complete consumption of $CO(g)$ and $H_2(g)$:

$$2.44 \times 10^3 \text{ mol CO} \times \frac{1 \text{ mol CH}_3OH}{1 \text{ mol CO}} = 2.44 \times 10^3 \text{ mol CH}_3OH$$

$$4.27 \times 10^3 \text{ mol H}_2 \times \frac{1 \text{ mol CH}_3OH}{2 \text{ mol H}_2} = 2.14 \times 10^3 \text{ mol CH}_3OH$$

Since complete consumption of the H_2 produces the smaller amount of CH_3OH, the H_2 is the limiting reactant as we determined above.

To calculate the theoretical yield of methanol

What are the moles of CH_3OH formed?
We must use the amount of H_2 and the mole ratio between H_2 and CH_3OH to determine the maximum amount of methanol that can be produced:

$$4.27 \times 10^3 \ \text{mol } H_2 \times \frac{1 \ \text{mol } CH_3OH}{2 \ \text{mol } H_2} = 2.14 \times 10^3 \ \text{mol } CH_3OH$$

What is the theoretical yield of CH_3OH in grams?

$$2.14 \times 10^3 \ \text{mol } CH_3OH \times \frac{32.04 \ \text{g } CH_3OH}{1 \ \text{mol } CH_3OH} = 6.86 \times 10^4 \ \text{g } CH_3OH$$

> Thus, from the amount of reactants given, the maximum amount of CH_3OH that can be formed is 6.86×10^4 g. This is the *theoretical yield.*

What is the percent yield of CH_3OH?

> $$\frac{\text{Actual yield (grams)}}{\text{Theoretical yield (grams)}} \times 100 = \frac{3.57 \times 10^4 \ \text{g } CH_3OH}{6.86 \times 10^4 \ \text{g } CH_3OH} \times 100\% = 52.0\%$$

See Exercises 3.123 and 3.124

Problem-Solving Strategy

Solving a Stoichiometry Problem Involving Masses of Reactants and Products

1. Write and balance the equation for the reaction.
2. Convert the known masses of substances to moles.
3. Determine which reactant is limiting.
4. Using the amount of the limiting reactant and the appropriate mole ratios, compute the number of moles of the desired product.
5. Convert from moles to grams, using the molar mass.

This process is summarized in the diagram below:

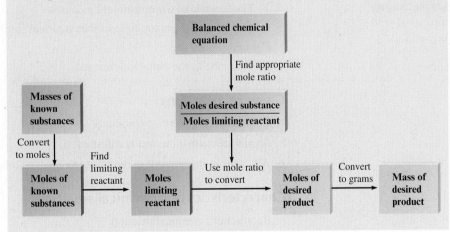

For review

Key terms

chemical stoichiometry

Section 3.2
mass spectrometer
average atomic mass

Section 3.3
mole
Avogadro's number

Section 3.4
molar mass

Section 3.5
conceptual problem solving

Section 3.6
mass percent

Section 3.7
empirical formula
molecular formula

Section 3.8
chemical equation
reactants
products
balancing a chemical equation

Section 3.10
mole ratio

Section 3.11
stoichiometric mixture
limiting reactant
theoretical yield
percent yield

Stoichiometry

> Deals with the amounts of substances consumed and/or produced in a chemical reaction.
> We count atoms by measuring the mass of the sample.
> To relate mass and the number of atoms, the average atomic mass is required.

Mole

> The amount of carbon atoms in exactly 12 g of pure ^{12}C
> 6.022×10^{23} units of a substance
> The mass of 1 mole of an element = the atomic mass in grams

Molar mass

> Mass (g) of 1 mole of a compound or element
> Obtained for a compound by finding the sum of the average masses of its constituent atoms

Percent composition

> The mass percent of each element in a compound
> $\text{Mass percent} = \dfrac{\text{mass of element in 1 mole of substance}}{\text{mass of 1 mole of substance}} \times 100\%$

Empirical formula

> The simplest whole-number ratio of the various types of atoms in a compound
> Can be obtained from the mass percent of elements in a compound

Molecular formula

> For molecular substances:
>> The formula of the constituent molecules
>> Always an integer multiple of the empirical formula
> For ionic substances:
>> The same as the empirical formula

Chemical reactions

> Reactants are turned into products.
> Atoms are neither created nor destroyed.
> All of the atoms present in the reactants must also be present in the products.

Characteristics of a chemical equation

> Represents a chemical reaction
> Reactants on the left side of the arrow, products on the right side
> When balanced, gives the relative numbers of reactant and product molecules or ions

Stoichiometry calculations

> Amounts of reactants consumed and products formed can be determined from the balanced chemical equation.

> The limiting reactant is the one consumed first, thus limiting the amount of product that can form.

Yield

> The theoretical yield is the maximum amount that can be produced from a given amount of the limiting reactant.

> The actual yield, the amount of product actually obtained, is always less than the theoretical yield.

> Percent yield $= \dfrac{\text{actual yield (g)}}{\text{theoretical yield (g)}} \times 100\%$

Review questions *Answers to the Review Questions can be found on the Student website (accessible from* **www.cengagebrain.com**).

1. Explain the concept of "counting by weighing" using marbles as your example.

2. Atomic masses are relative masses. What does this mean?

3. The atomic mass of boron (B) is given in the periodic table as 10.81, yet no single atom of boron has a mass of 10.81 u. Explain.

4. What three conversion factors and in what order would you use them to convert the mass of a compound into atoms of a particular element in that compound—for example, from 1.00 g aspirin ($C_9H_8O_4$) to number of hydrogen atoms in the 1.00-g sample?

5. Fig. 3.5 illustrates a schematic diagram of a combustion device used to analyze organic compounds. Given that a certain amount of a compound containing carbon, hydrogen, and oxygen is combusted in this device, explain how the data relating to the mass of CO_2 produced and the mass of H_2O produced can be manipulated to determine the empirical formula.

6. What is the difference between the empirical and molecular formulas of a compound? Can they ever be the same? Explain.

7. Consider the hypothetical reaction between A_2 and AB pictured below.

What is the balanced equation? If 2.50 moles of A_2 are reacted with excess AB, what amount (moles) of product will form? If the mass of AB is 30.0 u and the mass of A_2 are 40.0 u, what is the mass of the product? If 15.0 g of AB is reacted, what mass of A_2 is required to react with all of the AB, and what mass of product is formed?

8. What is a limiting reactant problem? Explain the method you are going to use to solve limiting reactant problems.

9. Consider the following mixture of $SO_2(g)$ and $O_2(g)$.

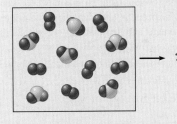

O_2
SO_2

If $SO_2(g)$ and $O_2(g)$ react to form $SO_3(g)$, draw a representation of the product mixture assuming the reaction goes to completion. What is the limiting reactant in the reaction? If 96.0 g of SO_2 react with 32.0 g O_2, what mass of product will form?

10. Why is the actual yield of a reaction often less than the theoretical yield?

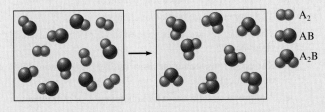

A_2
AB
A_2B

Active Learning Questions

These questions are designed to be used by groups of students in class.

1. The following are actual student responses to the question: Why is it necessary to balance chemical equations?

   a. The chemicals will not react until you have added the correct mole ratios.

   b. The correct products will not be formed unless the right amount of reactants have been added.

   c. A certain number of products cannot be formed without a certain number of reactants.

   d. The balanced equation tells you how much reactant you need and allows you to predict how much product you'll make.

   e. A mole-to-mole ratio must be established for the reaction to occur as written.

   Justify the best choice, and for choices you did not pick, explain what is wrong with them.

2. What information do we get from a chemical formula? From a chemical equation?

3. You are making cookies and are missing a key ingredient—eggs. You have most of the other ingredients needed to make the cookies, except you have only 1.33 cups of butter and no eggs. You note that the recipe calls for two cups of butter and three eggs (plus the other ingredients) to make six dozen cookies. You call a friend and have him bring you some eggs.

   a. What number of eggs do you need?

   b. If you use all the butter (and get enough eggs), what number of cookies will you make?

   Unfortunately, your friend hangs up before you tell him how many eggs you need. When he arrives, he has a surprise for you—to save time, he has broken them all in a bowl for you. You ask him how many he brought, and he replies, "I can't remember." You weigh the eggs and find that they weigh 62.1 g. Assuming that an average egg weighs 34.21 g,

   a. What quantity of butter is needed to react with all the eggs?

   b. What number of cookies can you make?

   c. Which will you have left over, eggs or butter?

   d. What quantity is left over?

4. Nitrogen gas (N_2) and hydrogen gas (H_2) react to form ammonia gas (NH_3).

   Consider the mixture of N_2 (⬤⬤) and H_2 (◯◯) in a closed container as illustrated below:

   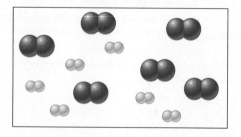

   Assuming the reaction goes to completion, draw a representation of the product mixture. Explain how you arrived at this representation.

5. For the preceding question, which of the following equations best represents the reaction?

   a. $6N_2 + 6H_2 \longrightarrow 4NH_3 + 4N_2$

   b. $N_2 + H_2 \longrightarrow NH_3$

   c. $N + 3H \longrightarrow NH_3$

   d. $N_2 + 3H_2 \longrightarrow 2NH_3$

   e. $2N_2 + 6H_2 \longrightarrow 4NH_3$

   Justify your choice, and for choices you did not pick, explain what is wrong with them.

6. You know that chemical A reacts with chemical B. You react 10.0 g A with 10.0 g B. What information do you need to determine the amount of product that will be produced? Explain.

7. A new grill has a mass of 30.0 kg. You put 3.0 kg of charcoal in the grill. You burn all the charcoal and the grill has a mass of 30.0 kg. What is the mass of the gases given off? Explain.

8. Consider an iron bar on a balance as shown.

   As the iron bar rusts, which of the following is true? Explain your answer.

   a. The balance will read less than 75.0 g.

   b. The balance will read 75.0 g.

   c. The balance will read greater than 75.0 g.

   d. The balance will read greater than 75.0 g, but if the bar is removed, the rust is scraped off, and the bar replaced, the balance will read 75.0 g.

9. You may have noticed that water sometimes drips from the exhaust of a car as it is running. Is this evidence that there is at least a small amount of water originally present in the gasoline? Explain.

Questions 10 and 11 deal with the following situation: You react chemical A with chemical B to make one product. It takes 100 g of A to react completely with 20 g of B.

10. What is the mass of the product?

   a. less than 10 g

   b. between 20 and 100 g

   c. between 100 and 120 g

   d. exactly 120 g

   e. more than 120 g

11. What is true about the chemical properties of the product?

   a. The properties are more like chemical A.

   b. The properties are more like chemical B.

   c. The properties are an average of those of chemical A and chemical B.

   d. The properties are not necessarily like either chemical A or chemical B.

   e. The properties are more like chemical A or more like chemical B, but more information is needed.

   Justify your choice, and for choices you did not pick, explain what is wrong with them.

12. Is there a difference between a homogeneous mixture of hydrogen and oxygen in a 2:1 mole ratio and a sample of water vapor? Explain.

13. Chlorine exists mainly as two isotopes, ^{37}Cl and ^{35}Cl. Which is more abundant? How do you know?

14. The average mass of a carbon atom is 12.011. Assuming you could pick up one carbon atom, estimate the chance that you would randomly get one with a mass of 12.011. Support your answer.

15. Can the subscripts in a chemical formula be fractions? Explain. Can the coefficients in a balanced chemical equation be fractions? Explain. Changing the subscripts of chemicals can balance the equations mathematically. Why is this unacceptable?

16. Consider the equation $2A + B \longrightarrow A_2B$. If you mix 1.0 mole of A with 1.0 mole of B, what amount (moles) of A_2B can be produced?

17. According to the law of conservation of mass, mass cannot be gained or destroyed in a chemical reaction. Why can't you simply add the masses of two reactants to determine the total mass of product?

18. Which of the following pairs of compounds have the same *empirical* formula?

   a. acetylene, C_2H_2, and benzene, C_6H_6

   b. ethane, C_2H_6, and butane, C_4H_{10}

   c. nitrogen dioxide, NO_2, and dinitrogen tetroxide, N_2O_4

   d. diphenyl ether, $C_{12}H_{10}O$, and phenol, C_6H_5OH

19. Atoms of three different elements are represented by O, □, and Δ. Which compound is left over when three molecules of OΔ and three molecules of □□Δ react to form O□Δ and OΔΔ?

20. In chemistry, what is meant by the term "mole"? What is the importance of the mole concept?

21. Which (if any) of the following is (are) *true* regarding the limiting reactant in a chemical reaction?

   a. The limiting reactant has the lowest coefficient in a balanced equation.

   b. The limiting reactant is the reactant for which you have the fewest number of moles.

   c. The limiting reactant has the lowest ratio of moles available/coefficient in the balanced equation.

   d. The limiting reactant has the lowest ratio of coefficient in the balanced equation/moles available.

   Justify your choice. For those you did not choose, explain why they are incorrect.

22. Consider the equation $3A + B \rightarrow C + D$. You react 4 moles of A with 2 moles of B. Which of the following is true?

   a. The limiting reactant is the one with the higher molar mass.

   b. A is the limiting reactant because you need 6 moles of A and have 4 moles.

   c. B is the limiting reactant because you have fewer moles of B than A.

   d. B is the limiting reactant because three A molecules react with each B molecule.

   e. Neither reactant is limiting.

   Justify your choice. For those you did not choose, explain why they are incorrect.

Questions

23. Reference Section 3.2 to find the atomic masses of ^{12}C and ^{13}C, the relative abundance of ^{12}C and ^{13}C in natural carbon, and the average mass (in u) of a carbon atom. If you had a sample of natural carbon containing exactly 10,000 atoms, determine the number of ^{12}C and ^{13}C atoms present. What would be the average mass (in u) and the total mass (in u) of the carbon atoms in this 10,000-atom sample? If you had a sample of natural carbon containing 6.0221×10^{23} atoms, determine the number of ^{12}C and ^{13}C atoms present. What would be the average mass (in u) and the total mass (in u) of this 6.0221×10^{23} atom sample? Given that 1 g $= 6.0221 \times 10^{23}$ u, what is the total mass of 1 mole of natural carbon in units of grams?

24. Avogadro's number, molar mass, and the chemical formula of a compound are three useful conversion factors. What unit conversions can be accomplished using these conversion factors?

25. If you had a mole of U.S. dollar bills and equally distributed the money to all of the people of the world, how rich would every person be? Assume a world population of 7 billion.

26. Describe 1 mole of CO_2 in as many ways as you can.

27. Which of the following compounds have the same empirical formulas?

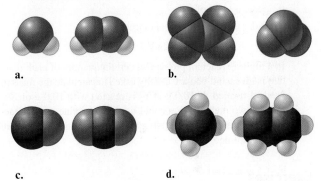

   a.   **b.**

   c.   **d.**

28. What is the difference between the molar mass and the empirical formula mass of a compound? When are these masses the same, and when are they different? When different, how is the molar mass related to the empirical formula mass?

29. How is the mass percent of elements in a compound different for a 1.0-g sample versus a 100.-g sample versus a 1-mole sample of the compound?

30. A balanced chemical equation contains a large amount of information. What information is given in a balanced equation?

31. The reaction of an element X with element Y is represented in the following diagram. Which of the equations best describes this reaction?

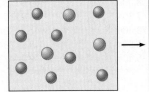 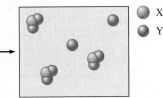

   ○ X
   ● Y

a. $3X + 8Y \rightarrow X_3Y_8$

b. $3X + 6Y \rightarrow X_3Y_6$

c. $X + 2Y \rightarrow XY_2$

d. $3X + 8Y \rightarrow 3XY_2 + 2Y$

32. Hydrogen gas and oxygen gas react to form water, and this reaction can be depicted as follows:

Explain why this equation is not balanced, and draw a picture of the balanced equation.

33. What is the theoretical yield for a reaction, and how does this quantity depend on the limiting reactant?

34. What does it mean to say a reactant is present "in excess" in a process? Can the limiting reactant be present in excess? Does the presence of an excess of a reactant affect the mass of products expected for a reaction?

35. Consider the following generic reaction:

$$A_2B_2 + 2C \longrightarrow 2CB + 2A$$

What steps and information are necessary to perform the following determinations assuming that 1.00×10^4 molecules of A_2B_2 are reacted with excess C?

a. mass of CB produced

b. atoms of A produced

c. moles of C reacted

d. percent yield of CB

36. Consider the following generic reaction:

$$Y_2 + 2XY \longrightarrow 2XY_2$$

In a limiting reactant problem, a certain quantity of each reactant is given and you are usually asked to calculate the mass of product formed. If 10.0 g of Y_2 is reacted with 10.0 g of XY, outline two methods you could use to determine which reactant is limiting (runs out first) and thus determines the mass of product formed.

Exercises

In this section similar exercises are paired.

Atomic Masses and the Mass Spectrometer

37. An element consists of 1.40% of an isotope with mass 203.973 u, 24.10% of an isotope with mass 205.9745 u, 22.10% of an isotope with mass 206.9759 u, and 52.40% of an isotope with mass 207.9766 u. Calculate the average atomic mass, and identify the element.

38. An element "X" has five major isotopes, which are listed below along with their abundances. What is the element?

Isotope	Percent Natural Abundance	Mass (u)
^{46}X	8.00%	45.95232
^{47}X	7.30%	46.951764
^{48}X	73.80%	47.947947
^{49}X	5.50%	48.947841
^{50}X	5.40%	49.944792

39. The element rhenium (Re) has two naturally occurring isotopes, ^{185}Re and ^{187}Re, with an average atomic mass of 186.207 u. Rhenium is 62.60% ^{187}Re, and the atomic mass of ^{187}Re is 186.956 u. Calculate the mass of ^{185}Re.

40. Assume silicon has three major isotopes in nature as shown in the table below. Fill in the missing information.

Isotope	Mass (u)	Abundance
^{28}Si	27.98	_____
^{29}Si	_____	4.70%
^{30}Si	29.97	3.09%

41. The element europium exists in nature as two isotopes: ^{151}Eu has a mass of 150.9196 u and ^{153}Eu has a mass of 152.9209 u. The average atomic mass of europium is 151.96 u. Calculate the relative abundance of the two europium isotopes.

42. The element silver (Ag) has two naturally occurring isotopes: ^{109}Ag and ^{107}Ag with a mass of 106.905 u. Silver consists of 51.82% ^{107}Ag and has an average atomic mass of 107.868 u. Calculate the mass of ^{109}Ag.

43. The mass spectrum of bromine (Br_2) consists of three peaks with the following characteristics:

Mass (u)	Relative Size
157.84	0.2534
159.84	0.5000
161.84	0.2466

How do you interpret these data?

44. The stable isotopes of iron are ^{54}Fe, ^{56}Fe, ^{57}Fe, and ^{58}Fe. The mass spectrum of iron looks like the following:

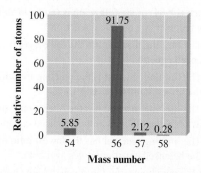

Use the data on the mass spectrum to estimate the average atomic mass of iron, and compare it to the value given in the table inside the front cover of this book.

Moles and Molar Masses

45. Calculate the mass of 500. atoms of iron (Fe).

46. What number of Fe atoms and what amount (moles) of Fe atoms are in 500.0 g of iron?

47. Diamond is a natural form of pure carbon. What number of atoms of carbon are in a 1.00-carat diamond (1.00 carat = 0.200 g)?

48. A diamond contains 5.0×10^{21} atoms of carbon. What amount (moles) of carbon and what mass (grams) of carbon are in this diamond?

49. Aluminum metal is produced by passing an electric current through a solution of aluminum oxide (Al_2O_3) dissolved in molten cryolite (Na_3AlF_6). Calculate the molar masses of Al_2O_3 and Na_3AlF_6.

50. The Freons are a class of compounds containing carbon, chlorine, and fluorine. While they have many valuable uses, they have been shown to be responsible for depletion of the ozone in the upper atmosphere. In 1991, two replacement compounds for Freons went into production: HFC-134a (CH_2FCF_3) and HCFC-124 ($CHClFCF_3$). Calculate the molar masses of these two compounds.

51. Calculate the molar mass of the following substances.

a.
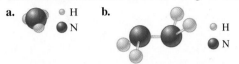 ● H
● N

b.
● H
● N

c. $(NH_4)_2Cr_2O_7$

52. Calculate the molar mass of the following substances.

a.
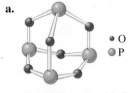 ● O
● P

b. $Ca_3(PO_4)_2$ c. Na_2HPO_4

53. What amount (moles) of compound is present in 1.00 g of each of the compounds in Exercise 51?

54. What amount (moles) of compound is present in 1.00 g of each of the compounds in Exercise 52?

55. What mass of compound is present in 5.00 moles of each of the compounds in Exercise 51?

56. What mass of compound is present in 5.00 moles of each of the compounds in Exercise 52?

57. What mass of nitrogen is present in 5.00 moles of each of the compounds in Exercise 51?

58. What mass of phosphorus is present in 5.00 moles of each of the compounds in Exercise 52?

59. What number of molecules (or formula units) are present in 1.00 g of each of the compounds in Exercise 51?

60. What number of molecules (or formula units) are present in 1.00 g of each of the compounds in Exercise 52?

61. What number of atoms of nitrogen are present in 1.00 g of each of the compounds in Exercise 51?

62. What number of atoms of phosphorus are present in 1.00 g of each of the compounds in Exercise 52?

63. Freon-12 (CCl_2F_2) is used as a refrigerant in air conditioners and as a propellant in aerosol cans. Calculate the number of molecules of Freon-12 in 5.56 mg of Freon-12. What is the mass of chlorine in 5.56 mg of Freon-12?

64. Bauxite, the principal ore used in the production of aluminum, has a molecular formula of $Al_2O_3 \cdot 2H_2O$. The $\cdot H_2O$ in the formula are called waters of hydration. Each formula unit of the compound contains two water molecules.

a. What is the molar mass of bauxite?

b. What is the mass of aluminum in 0.58 mole of bauxite?

c. How many atoms of aluminum are in 0.58 mole of bauxite?

d. What is the mass of 2.1×10^{24} formula units of bauxite?

65. What amount (moles) is represented by each of these samples?

a. 150.0 g Fe_2O_3

b. 10.0 mg NO_2

c. 1.5×10^{16} molecules of BF_3

66. What amount (moles) is represented by each of these samples?

a. 20.0 mg caffeine, $C_8H_{10}N_4O_2$

b. 2.72×10^{21} molecules of ethanol, C_2H_5OH

c. 1.50 g of dry ice, CO_2

67. What number of atoms of nitrogen are present in 5.00 g of each of the following?

a. glycine, $C_2H_5O_2N$

b. magnesium nitride

c. calcium nitrate

d. dinitrogen tetroxide

68. Complete the following table.

Mass of Sample	Moles of Sample	Molecules in Sample	Total Atoms in Sample
4.24 g C_6H_6	⸺	⸺	⸺
⸺	0.224 mol H_2O	⸺	⸺
⸺	⸺	2.71×10^{22} molecules CO_2	⸺
⸺	⸺	⸺	3.35×10^{22} total atoms in CH_3OH sample

69. Ascorbic acid, or vitamin C ($C_6H_8O_6$), is an essential vitamin. It cannot be stored by the body and must be present in the diet. What is the molar mass of ascorbic acid? Vitamin C tablets are taken as a dietary supplement. If a typical tablet contains 500.0 mg vitamin C, what amount (moles) and what number of molecules of vitamin C does it contain?

70. The molecular formula of acetylsalicylic acid (aspirin), one of the most commonly used pain relievers, is $C_9H_8O_4$.

a. Calculate the molar mass of aspirin.

b. A typical aspirin tablet contains 500. mg $C_9H_8O_4$. What amount (moles) of $C_9H_8O_4$ molecules and what number of molecules of acetylsalicylic acid are in a 500.-mg tablet?

71. Chloral hydrate ($C_2H_3Cl_3O_2$) is a drug formerly used as a sedative and hypnotic. It is the compound used to make "Mickey Finns" in detective stories.

a. Calculate the molar mass of chloral hydrate.

b. What amount (moles) of $C_2H_3Cl_3O_2$ molecules are in 500.0 g chloral hydrate?

c. What is the mass in grams of 2.0×10^{-2} mole of chloral hydrate?

d. What number of chlorine atoms are in 5.0 g chloral hydrate?

e. What mass of chloral hydrate would contain 1.0 g Cl?

f. What is the mass of exactly 500 molecules of chloral hydrate?

72. Dimethylnitrosamine, $(CH_3)_2N_2O$, is a carcinogenic (cancer-causing) substance that may be formed in foods, beverages, or gastric juices from the reaction of nitrite ion (used as a food preservative) with other substances.

a. What is the molar mass of dimethylnitrosamine?

b. How many moles of $(CH_3)_2N_2O$ molecules are present in 250 mg dimethylnitrosamine?

c. What is the mass of 0.050 mole of dimethylnitrosamine?

d. How many atoms of hydrogen are in 1.0 mole of dimethylnitrosamine?

e. What is the mass of 1.0×10^6 molecules of dimethylnitrosamine?

f. What is the mass in grams of one molecule of dimethylnitrosamine?

Percent Composition

73. Calculate the percent composition by mass of the following compounds that are important starting materials for synthetic polymers:

a. $C_3H_4O_2$ (acrylic acid, from which acrylic plastics are made)

b. $C_4H_6O_2$ (methyl acrylate, from which Plexiglas is made)

c. C_3H_3N (acrylonitrile, from which Orlon is made)

74. In 1987 the first substance to act as a superconductor at a temperature above that of liquid nitrogen (77 K) was discovered. The approximate formula of this substance is $YBa_2Cu_3O_7$. Calculate the percent composition by mass of this material.

75. The percent by mass of nitrogen for a compound is found to be 46.7%. Which of the following could be this species?

76. Arrange the following substances in order of increasing mass percent of carbon.

a. caffeine, $C_8H_{10}N_4O_2$

b. sucrose, $C_{12}H_{22}O_{11}$

c. ethanol, C_2H_5OH

77. Fungal laccase, a blue protein found in wood-rotting fungi, is 0.390% Cu by mass. If a fungal laccase molecule contains four copper atoms, what is the molar mass of fungal laccase?

78. Hemoglobin is the protein that transports oxygen in mammals. Hemoglobin is 0.347% Fe by mass, and each hemoglobin molecule contains four iron atoms. Calculate the molar mass of hemoglobin.

Empirical and Molecular Formulas

79. Express the composition of each of the following compounds as the mass percents of its elements.

a. formaldehyde, CH_2O

b. glucose, $C_6H_{12}O_6$

c. acetic acid, $HC_2H_3O_2$

80. Considering your answer to Exercise 79, which type of formula, empirical or molecular, can be obtained from elemental analysis that gives percent composition?

81. Give the empirical formula for each of the compounds represented below.

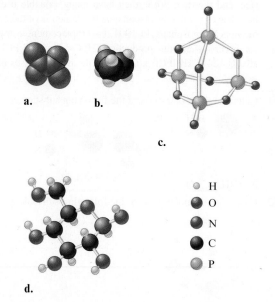

a.   b.

c.

d.

	H
	O
	N
	C
	P

82. Determine the molecular formulas to which the following empirical formulas and molar masses pertain.

a. SNH (188.35 g/mol)   c. CoC_4O_4 (341.94 g/mol)

b. $NPCl_2$ (347.64 g/mol)   d. SN (184.32 g/mol)

83. A compound that contains only carbon, hydrogen, and oxygen is 48.64% C and 8.16% H by mass. What is the empirical formula of this substance?

84. The most common form of nylon (nylon-6) is 63.68% carbon, 12.38% nitrogen, 9.80% hydrogen, and 14.14% oxygen. Calculate the empirical formula for nylon-6.

85. There are two binary compounds of mercury and oxygen. Heating either of them results in the decomposition of the compound, with oxygen gas escaping into the atmosphere while leaving a residue of pure mercury. Heating 0.6498 g of one of the compounds leaves a residue of 0.6018 g. Heating 0.4172 g of the other compound results in a mass loss of 0.016 g. Determine the empirical formula of each compound.

86. A sample of urea contains 1.121 g N, 0.161 g H, 0.480 g C, and 0.640 g O. What is the empirical formula of urea?

87. A compound containing only sulfur and nitrogen is 69.6% S by mass; the molar mass is 184 g/mol. What are the empirical and molecular formulas of the compound?

88. Determine the molecular formula of a compound that contains 26.7% P, 12.1% N, and 61.2% Cl, and has a molar mass of 580 g/mol.

89. A compound contains 47.08% carbon, 6.59% hydrogen, and 46.33% chlorine by mass; the molar mass of the compound is 153 g/mol. What are the empirical and molecular formulas of the compound?

90. Maleic acid is an organic compound composed of 41.39% C, 3.47% H, and the rest oxygen. If 0.129 mole of maleic acid has a mass of 15.0 g, what are the empirical and molecular formulas of maleic acid?

91. One of the components that make up common table sugar is fructose, a compound that contains only carbon, hydrogen, and oxygen. Complete combustion of 1.50 g of fructose produced 2.20 g of carbon dioxide and 0.900 g of water. What is the empirical formula of fructose?

92. A compound contains only C, H, and N. Combustion of 35.0 mg of the compound produces 33.5 mg CO_2 and 41.1 mg H_2O. What is the empirical formula of the compound?

93. Cumene is a compound containing only carbon and hydrogen that is used in the production of acetone and phenol in the chemical industry. Combustion of 47.6 mg cumene produces some CO_2 and 42.8 mg water. The molar mass of cumene is between 115 and 125 g/mol. Determine the empirical and molecular formulas.

94. A compound contains only carbon, hydrogen, and oxygen. Combustion of 10.68 mg of the compound yields 16.01 mg CO_2 and 4.37 mg H_2O. The molar mass of the compound is 176.1 g/mol. What are the empirical and molecular formulas of the compound?

Balancing Chemical Equations

95. Give the balanced equation for each of the following chemical reactions:
   a. Glucose ($C_6H_{12}O_6$) reacts with oxygen gas to produce gaseous carbon dioxide and water vapor.
   b. Solid iron(III) sulfide reacts with gaseous hydrogen chloride to form solid iron(III) chloride and hydrogen sulfide gas.
   c. Carbon disulfide liquid reacts with ammonia gas to produce hydrogen sulfide gas and solid ammonium thiocyanate (NH_4SCN).

96. Give the balanced equation for each of the following.
   a. The combustion of ethanol (C_2H_5OH) forms carbon dioxide and water vapor. A combustion reaction refers to a reaction of a substance with oxygen gas.
   b. Aqueous solutions of lead(II) nitrate and sodium phosphate are mixed, resulting in the precipitate formation of lead(II) phosphate with aqueous sodium nitrate as the other product.
   c. Solid zinc reacts with aqueous HCl to form aqueous zinc chloride and hydrogen gas.
   d. Aqueous strontium hydroxide reacts with aqueous hydrobromic acid to produce water and aqueous strontium bromide.

97. A common demonstration in chemistry courses involves adding a tiny speck of manganese(IV) oxide to a concentrated hydrogen peroxide (H_2O_2) solution. Hydrogen peroxide decomposes quite spectacularly under these conditions to produce oxygen gas and steam (water vapor). Manganese(IV) oxide is a catalyst for the decomposition of hydrogen peroxide and is not consumed in the reaction. Write the balanced equation for the decomposition reaction of hydrogen peroxide.

98. Iron oxide ores, commonly a mixture of FeO and Fe_2O_3, are given the general formula Fe_3O_4. They yield elemental iron when heated to a very high temperature with either carbon monoxide or elemental hydrogen. Balance the following equations for these processes:

$$Fe_3O_4(s) + H_2(g) \longrightarrow Fe(s) + H_2O(g)$$
$$Fe_3O_4(s) + CO(g) \longrightarrow Fe(s) + CO_2(g)$$

99. Balance the following equations:
   a. $Ca(OH)_2(aq) + H_3PO_4(aq) \rightarrow H_2O(l) + Ca_3(PO_4)_2(s)$
   b. $Al(OH)_3(s) + HCl(aq) \rightarrow AlCl_3(aq) + H_2O(l)$
   c. $AgNO_3(aq) + H_2SO_4(aq) \rightarrow Ag_2SO_4(s) + HNO_3(aq)$

100. Balance each of the following chemical equations.
   a. $KO_2(s) + H_2O(l) \rightarrow KOH(aq) + O_2(g) + H_2O_2(aq)$
   b. $Fe_2O_3(s) + HNO_3(aq) \rightarrow Fe(NO_3)_3(aq) + H_2O(l)$
   c. $NH_3(g) + O_2(g) \rightarrow NO(g) + H_2O(g)$
   d. $PCl_5(l) + H_2O(l) \rightarrow H_3PO_4(aq) + HCl(g)$
   e. $CaO(s) + C(s) \rightarrow CaC_2(s) + CO_2(g)$
   f. $MoS_2(s) + O_2(g) \rightarrow MoO_3(s) + SO_2(g)$
   g. $FeCO_3(s) + H_2CO_3(aq) \rightarrow Fe(HCO_3)_2(aq)$

101. Balance the following equations representing combustion reactions:

   a.
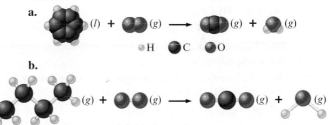
   $\bullet$ H $\bullet$ C $\bullet$ O

   b.

   c. $C_{12}H_{22}O_{11}(s) + O_2(g) \rightarrow CO_2(g) + H_2O(g)$
   d. $Fe(s) + O_2(g) \rightarrow Fe_2O_3(s)$
   e. $FeO(s) + O_2(g) \rightarrow Fe_2O_3(s)$

102. Balance the following equations:
   a. $Cr(s) + S_8(s) \rightarrow Cr_2S_3(s)$
   b. $NaHCO_3(s) \xrightarrow{\text{Heat}} Na_2CO_3(s) + CO_2(g) + H_2O(g)$
   c. $KClO_3(s) \xrightarrow{\text{Heat}} KCl(s) + O_2(g)$
   d. $Eu(s) + HF(g) \rightarrow EuF_3(s) + H_2(g)$

103. Silicon is produced for the chemical and electronics industries by the following reactions. Give the balanced equation for each reaction.
   a. $SiO_2(s) + C(s) \xrightarrow[\text{arc furnace}]{\text{Electric}} Si(s) + CO(g)$
   b. Liquid silicon tetrachloride is reacted with very pure solid magnesium, producing solid silicon and solid magnesium chloride.
   c. $Na_2SiF_6(s) + Na(s) \rightarrow Si(s) + NaF(s)$

104. Glass is a mixture of several compounds, but a major constituent of most glass is calcium silicate, $CaSiO_3$. Glass can be etched by treatment with hydrofluoric acid; HF attacks the calcium silicate of the glass, producing gaseous and water-soluble products (which can be removed by washing the glass). For example, the volumetric glassware in chemistry laboratories is

often graduated by using this process. Balance the following equation for the reaction of hydrofluoric acid with calcium silicate.

$$CaSiO_3(s) + HF(aq) \longrightarrow CaF_2(aq) + SiF_4(g) + H_2O(l)$$

Reaction Stoichiometry

105. Over the years, the thermite reaction has been used for welding railroad rails, in incendiary bombs, and to ignite solid-fuel rocket motors. The reaction is

$$Fe_2O_3(s) + 2Al(s) \longrightarrow 2Fe(l) + Al_2O_3(s)$$

What masses of iron(III) oxide and aluminum must be used to produce 15.0 g iron? What is the maximum mass of aluminum oxide that could be produced?

106. The reaction between potassium chlorate and red phosphorus takes place when you strike a match on a matchbox. If you were to react 52.9 g of potassium chlorate ($KClO_3$) with excess red phosphorus, what mass of tetraphosphorus decaoxide (P_4O_{10}) could be produced?

$$KClO_3(s) + P_4(s) \longrightarrow P_4O_{10}(s) + KCl(s) \quad \text{(unbalanced)}$$

107. The reusable booster rockets of the U.S. space shuttle employ a mixture of aluminum and ammonium perchlorate for fuel. A possible equation for this reaction is

$$3Al(s) + 3NH_4ClO_4(s) \longrightarrow$$
$$Al_2O_3(s) + AlCl_3(s) + 3NO(g) + 6H_2O(g)$$

What mass of NH_4ClO_4 should be used in the fuel mixture for every kilogram of Al?

108. One of relatively few reactions that takes place directly between two solids at room temperature is

$$Ba(OH)_2 \cdot 8H_2O(s) + NH_4SCN(s) \longrightarrow$$
$$Ba(SCN)_2(s) + H_2O(l) + NH_3(g)$$

In this equation, the $\cdot 8H_2O$ in $Ba(OH)_2 \cdot 8H_2O$ indicates the presence of eight water molecules. This compound is called barium hydroxide octahydrate.

  a. Balance the equation.

  b. What mass of ammonium thiocyanate (NH_4SCN) must be used if it is to react completely with 6.5 g barium hydroxide octahydrate?

109. Elixirs such as Alka-Seltzer use the reaction of sodium bicarbonate with citric acid in aqueous solution to produce a fizz:

$$3NaHCO_3(aq) + C_6H_8O_7(aq) \longrightarrow$$
$$3CO_2(g) + 3H_2O(l) + Na_3C_6H_5O_7(aq)$$

  a. What mass of $C_6H_8O_7$ should be used for every 1.0×10^2 mg $NaHCO_3$?

  b. What mass of $CO_2(g)$ could be produced from such a mixture?

110. Aspirin ($C_9H_8O_4$) is synthesized by reacting salicylic acid ($C_7H_6O_3$) with acetic anhydride ($C_4H_6O_3$). The balanced equation is

$$C_7H_6O_3 + C_4H_6O_3 \longrightarrow C_9H_8O_4 + HC_2H_3O_2$$

  a. What mass of acetic anhydride is needed to completely consume 1.00×10^2 g salicylic acid?

  b. What is the maximum mass of aspirin (the theoretical yield) that could be produced in this reaction?

111. Bacterial digestion is an economical method of sewage treatment. The reaction

$$5CO_2(g) + 55NH_4^+(aq) + 76O_2(g) \xrightarrow{\text{bacteria}}$$
$$\underset{\text{bacterial tissue}}{C_5H_7O_2N(s)} + 54NO_2^-(aq) + 52H_2O(l) + 109H^+(aq)$$

is an intermediate step in the conversion of the nitrogen in organic compounds into nitrate ions. What mass of bacterial tissue is produced in a treatment plant for every 1.0×10^4 kg of wastewater containing 3.0% NH_4^+ ions by mass? Assume that 95% of the ammonium ions are consumed by the bacteria.

112. Phosphorus can be prepared from calcium phosphate by the following reaction:

$$2Ca_3(PO_4)_2(s) + 6SiO_2(s) + 10C(s) \longrightarrow$$
$$6CaSiO_3(s) + P_4(s) + 10CO(g)$$

Phosphorite is a mineral that contains $Ca_3(PO_4)_2$ plus other non-phosphorus-containing compounds. What is the maximum amount of P_4 that can be produced from 1.0 kg of phosphorite if the phorphorite sample is 75% $Ca_3(PO_4)_2$ by mass? Assume an excess of the other reactants.

113. Coke is an impure form of carbon that is often used in the industrial production of metals from their oxides. If a sample of coke is 95% carbon by mass, determine the mass of coke needed to react completely with 1.0 ton of copper(II) oxide.

$$2CuO(s) + C(s) \longrightarrow 2Cu(s) + CO_2(g)$$

114. The space shuttle environmental control system handles excess CO_2 (which the astronauts breathe out; it is 4.0% by mass of exhaled air) by reacting it with lithium hydroxide, LiOH, pellets to form lithium carbonate, Li_2CO_3, and water. If there are seven astronauts on board the shuttle, and each exhales 20. L of air per minute, how long could clean air be generated if there were 25,000 g of LiOH pellets available for each shuttle mission? Assume the density of air is 0.0010 g/mL.

Limiting Reactants and Percent Yield

115. Consider the reaction between $NO(g)$ and $O_2(g)$ represented below.

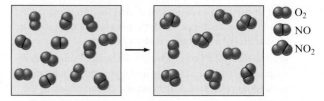

What is the balanced equation for this reaction, and what is the limiting reactant?

116. Consider the following reaction:

$$4NH_3(g) + 5O_2(g) \longrightarrow 4NO(g) + 6H_2O(g)$$

If a container were to have 10 molecules of O_2 and 10 molecules of NH_3 initially, how many total molecules (reactants plus products) would be present in the container after this reaction goes to completion?

117. Ammonia is produced from the reaction of nitrogen and hydrogen according to the following balanced equation:

$$N_2(g) + 3H_2(g) \longrightarrow 2NH_3(g)$$

  a. What is the maximum mass of ammonia that can be produced from a mixture of 1.00×10^3 g N_2 and 5.00×10^2 g H_2?

b. What mass of which starting material would remain unreacted?

118. Consider the following unbalanced equation:

$$Ca_3(PO_4)_2(s) + H_2SO_4(aq) \longrightarrow CaSO_4(s) + H_3PO_4(aq)$$

What masses of calcium sulfate and phosphoric acid can be produced from the reaction of 1.0 kg calcium phosphate with 1.0 kg concentrated sulfuric acid (98% H_2SO_4 by mass)?

119. Hydrogen peroxide is used as a cleansing agent in the treatment of cuts and abrasions for several reasons. It is an oxidizing agent that can directly kill many microorganisms; it decomposes on contact with blood, releasing elemental oxygen gas (which inhibits the growth of anaerobic microorganisms); and it foams on contact with blood, which provides a cleansing action. In the laboratory, small quantities of hydrogen peroxide can be prepared by the action of an acid on an alkaline earth metal peroxide, such as barium peroxide:

$$BaO_2(s) + 2HCl(aq) \longrightarrow H_2O_2(aq) + BaCl_2(aq)$$

What mass of hydrogen peroxide should result when 1.50 g barium peroxide is treated with 25.0 mL hydrochloric acid solution containing 0.0272 g HCl per mL? What mass of which reagent is left unreacted?

120. Silver sulfadiazine burn-treating cream creates a barrier against bacterial invasion and releases antimicrobial agents directly into the wound. If 25.0 g Ag_2O is reacted with 50.0 g $C_{10}H_{10}N_4SO_2$, what mass of silver sulfadiazine, $AgC_{10}H_9N_4SO_2$, can be produced, assuming 100% yield?

$$Ag_2O(s) + 2C_{10}H_{10}N_4SO_2(s) \longrightarrow 2AgC_{10}H_9N_4SO_2(s) + H_2O(l)$$

121. Hydrogen cyanide is produced industrially from the reaction of gaseous ammonia, oxygen, and methane:

$$2NH_3(g) + 3O_2(g) + 2CH_4(g) \longrightarrow 2HCN(g) + 6H_2O(g)$$

If 5.00×10^3 kg each of NH_3, O_2, and CH_4 are reacted, what mass of HCN and of H_2O will be produced, assuming 100% yield?

122. Acrylonitrile (C_3H_3N) is the starting material for many synthetic carpets and fabrics. It is produced by the following reaction.

$$2C_3H_6(g) + 2NH_3(g) + 3O_2(g) \longrightarrow 2C_3H_3N(g) + 6H_2O(g)$$

If 15.0 g C_3H_6, 10.0 g O_2, and 5.00 g NH_3 are reacted, what mass of acrylonitrile can be produced, assuming 100% yield?

123. The reaction of ethane gas (C_2H_6) with chlorine gas produces C_2H_5Cl as its main product (along with HCl). In addition, the reaction invariably produces a variety of other minor products, including $C_2H_4Cl_2$, $C_2H_3Cl_3$, and others. Naturally, the production of these minor products reduces the yield of the main product. Calculate the percent yield of C_2H_5Cl if the reaction of 300. g of ethane with 650. g of chlorine produced 490. g of C_2H_5Cl.

124. DDT, an insecticide harmful to fish, birds, and humans, is produced by the following reaction:

$$2C_6H_5Cl + C_2HOCl_3 \longrightarrow C_{14}H_9Cl_5 + H_2O$$
chlorobenzene   chloral   DDT

In a government lab, 1142 g of chlorobenzene is reacted with 485 g of chloral.

a. What mass of DDT is formed, assuming 100% yield?

b. Which reactant is limiting? Which is in excess?

c. What mass of the excess reactant is left over?

d. If the actual yield of DDT is 200.0 g, what is the percent yield?

125. Bornite (Cu_3FeS_3) is a copper ore used in the production of copper. When heated, the following reaction occurs:

$$2Cu_3FeS_3(s) + 7O_2(g) \longrightarrow 6Cu(s) + 2FeO(s) + 6SO_2(g)$$

If 2.50 metric tons of bornite is reacted with excess O_2 and the process has an 86.3% yield of copper, what mass of copper is produced?

126. Consider the following unbalanced reaction:

$$P_4(s) + F_2(g) \longrightarrow PF_3(g)$$

What mass of F_2 is needed to produce 120. g of PF_3 if the reaction has a 78.1% yield?

Additional Exercises

127. In using a mass spectrometer, a chemist sees a peak at a mass of 30.0106. Of the choices $^{12}C_2{}^1H_6$, $^{12}C^1H_2{}^{16}O$, and $^{14}N^{16}O$, which is responsible for this peak? Pertinent masses are ^1H, 1.007825; ^{16}O, 15.994915; and ^{14}N, 14.003074.

128. Boron consists of two isotopes, ^{10}B and ^{11}B. Chlorine also has two isotopes, ^{35}Cl and ^{37}Cl. Consider the mass spectrum of BCl_3. How many peaks would be present, and what approximate mass would each peak correspond to in the BCl_3 mass spectrum?

129. A given sample of a xenon fluoride compound contains molecules of the type XeF_n, where n is some whole number. Given that 9.03×10^{20} molecules of XeF_n weigh 0.368 g, determine the value for n in the formula.

130. Aspartame is an artificial sweetener that is 160 times sweeter than sucrose (table sugar) when dissolved in water. It is marketed as NutraSweet. The molecular formula of aspartame is $C_{14}H_{18}N_2O_5$.

a. Calculate the molar mass of aspartame.

b. What amount (moles) of molecules are present in 10.0 g aspartame?

c. Calculate the mass in grams of 1.56 mole of aspartame.

d. What number of molecules are in 5.0 mg aspartame?

e. What number of atoms of nitrogen are in 1.2 g aspartame?

f. What is the mass in grams of 1.0×10^9 molecules of aspartame?

g. What is the mass in grams of one molecule of aspartame?

131. Anabolic steroids are performance enhancement drugs whose use has been banned from most major sporting activities. One anabolic steroid is fluoxymesterone ($C_{20}H_{29}FO_3$). Calculate the percent composition by mass of fluoxymesterone.

132. Many cereals are made with high moisture content so that the cereal can be formed into various shapes before it is dried. A cereal product containing 58% H_2O by mass is produced at the rate of 1000. kg/h. What mass of water must be evaporated per hour if the final product contains only 20.% water?

133. The compound adrenaline contains 56.79% C, 6.56% H, 28.37% O, and 8.28% N by mass. What is the empirical formula for adrenaline?

134. Adipic acid is an organic compound composed of 49.31% C, 43.79% O, and the rest hydrogen. If the molar mass of adipic acid is 146.1 g/mol, what are the empirical and molecular formulas for adipic acid?

135. Vitamin B_{12}, cyanocobalamin, is essential for human nutrition. It is concentrated in animal tissue but not in higher plants. Although nutritional requirements for the vitamin are quite low, people who abstain completely from animal products may develop a deficiency anemia. Cyanocobalamin is the form used in vitamin supplements. It contains 4.34% cobalt by mass. Calculate the molar mass of cyanocobalamin, assuming that there is one atom of cobalt in every molecule of cyanocobalamin.

136. Some bismuth tablets, a medication used to treat upset stomachs, contain 262 mg of bismuth subsalicylate, $C_7H_5BiO_4$, per tablet. Assuming two tablets are digested, calculate the mass of bismuth consumed.

137. The empirical formula of styrene is CH; the molar mass of styrene is 104.14 g/mol. What number of H atoms are present in a 2.00-g sample of styrene?

138. Terephthalic acid is an important chemical used in the manufacture of polyesters and plasticizers. It contains only C, H, and O. Combustion of 19.81 mg terephthalic acid produces 41.98 mg CO_2 and 6.45 mg H_2O. If 0.250 mole of terephthalic acid has a mass of 41.5 g, determine the molecular formula for terephthalic acid.

139. A sample of a hydrocarbon (a compound consisting of only carbon and hydrogen) contains 2.59×10^{23} atoms of hydrogen and is 17.3% hydrogen by mass. If the molar mass of the hydrocarbon is between 55 and 65 g/mol, what amount (moles) of compound is present, and what is the mass of the sample?

140. A binary compound between an unknown element E and hydrogen contains 91.27% E and 8.73% H by mass. If the formula of the compound is E_3H_8, calculate the atomic mass of E.

141. A 0.755-g sample of hydrated copper(II) sulfate

$$CuSO_4 \cdot xH_2O$$

was heated carefully until it had changed completely to anhydrous copper(II) sulfate ($CuSO_4$) with a mass of 0.483 g. Determine the value of x. [This number is called the *number of waters of hydration* of copper(II) sulfate. It specifies the number of water molecules per formula unit of $CuSO_4$ in the hydrated crystal.]

142. ABS plastic is a tough, hard plastic used in applications requiring shock resistance. The polymer consists of three monomer units: acrylonitrile (C_3H_3N), butadiene (C_4H_6), and styrene (C_8H_8).

   a. A sample of ABS plastic contains 8.80% N by mass. It took 0.605 g of Br_2 to react completely with a 1.20-g sample of ABS plastic. Bromine reacts 1:1 (by moles) with the butadiene molecules in the polymer and nothing else. What is the percent by mass of acrylonitrile and butadiene in this polymer?

   b. What are the relative numbers of each of the monomer units in this polymer?

143. A sample of LSD (D-lysergic acid diethylamide, $C_{24}H_{30}N_3O$) is added to some table salt (sodium chloride) to form a mixture. Given that a 1.00-g sample of the mixture undergoes combustion to produce 1.20 g of CO_2, what is the mass percent of LSD in the mixture?

144. Methane (CH_4) is the main component of marsh gas. Heating methane in the presence of sulfur produces carbon disulfide and hydrogen sulfide as the only products.

   a. Write the balanced chemical equation for the reaction of methane and sulfur.

   b. Calculate the theoretical yield of carbon disulfide when 120. g of methane is reacted with an equal mass of sulfur.

145. A potential fuel for rockets is a combination of B_5H_9 and O_2. The two react according to the following balanced equation:

$$2B_5H_9(l) + 12O_2(g) \longrightarrow 5B_2O_3(s) + 9H_2O(g)$$

If one tank in a rocket holds 126 g B_5H_9 and another tank holds 192 g O_2, what mass of water can be produced when the entire contents of each tank react together?

146. A 0.4230-g sample of impure sodium nitrate was heated, converting all the sodium nitrate to 0.2864 g of sodium nitrite and oxygen gas. Determine the percent of sodium nitrate in the original sample.

147. An iron ore sample contains Fe_2O_3 plus other impurities. A 752-g sample of impure iron ore is heated with excess carbon, producing 453 g of pure iron by the following reaction:

$$Fe_2O_3(s) + 3C(s) \longrightarrow 2Fe(s) + 3CO(g)$$

What is the mass percent of Fe_2O_3 in the impure iron ore sample? Assume that Fe_2O_3 is the only source of iron and that the reaction is 100% efficient.

148. Commercial brass, an alloy of Zn and Cu, reacts with hydrochloric acid as follows:

$$Zn(s) + 2HCl(aq) \longrightarrow ZnCl_2(aq) + H_2(g)$$

(Cu does not react with HCl.) When 0.5065 g of a certain brass alloy is reacted with excess HCl, 0.0985 g $ZnCl_2$ is eventually isolated.

   a. What is the composition of the brass by mass?

   b. How could this result be checked without changing the above procedure?

149. Vitamin A has a molar mass of 286.4 g/mol and a general molecular formula of C_xH_yE, where E is an unknown element. If vitamin A is 83.86% C and 10.56% H by mass, what is the molecular formula of vitamin A?

150. You have seven closed containers, each with equal masses of chlorine gas (Cl_2). You add 10.0 g of sodium to the first sample, 20.0 g of sodium to the second sample, and so on (adding 70.0 g of sodium to the seventh sample). Sodium and chlorine react to form sodium chloride according to the equation

$$2Na(s) + Cl_2(g) \longrightarrow 2NaCl(s)$$

After each reaction is complete, you collect and measure the amount of sodium chloride formed. A graph of your results is shown below.

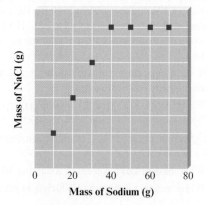

Answer the following questions:

a. Explain the shape of the graph.

b. Calculate the mass of NaCl formed when 20.0 g of sodium is used.

c. Calculate the mass of Cl_2 in each container.

d. Calculate the mass of NaCl formed when 50.0 g of sodium is used.

e. Identify the leftover reactant, and determine its mass for parts b and d above.

151. A substance X_2Z has the composition (by mass) of 40.0% X and 60.0% Z. What is the composition (by mass) of the compound XZ_2?

ChemWork Problems

These multiconcept problems (and additional ones) are found interactively online with the same type of assistance a student would get from an instructor.

152. Consider samples of phosphine (PH_3), water (H_2O), hydrogen sulfide (H_2S), and hydrogen fluoride (HF), each with a mass of 119 g. Rank the compounds from the least to the greatest number of hydrogen atoms contained in the samples.

153. Calculate the number of moles for each compound in the following table.

Compound	Mass	Moles
Magnesium phosphate	326.4 g	_____
Calcium nitrate	303.0 g	_____
Potassium chromate	141.6 g	_____
Dinitrogen pentoxide	406.3 g	_____

154. Arrange the following substances in order of increasing mass percent of nitrogen.

a. NO
b. N_2O
c. NH_3
d. SNH

155. Para-cresol, a substance used as a disinfectant and in the manufacture of several herbicides, is a molecule that contains the elements carbon, hydrogen, and oxygen. Complete combustion of a 0.345-g sample of p-cresol produced 0.983 g carbon dioxide and 0.230 g water. Determine the empirical formula for p-cresol.

156. A compound with molar mass 180.1 g/mol has the following composition by mass:

C	40.0%
H	6.70%
O	53.3%

Determine the empirical and molecular formulas of the compound.

157. Which of the following statements about chemical equations is(are) true?

a. When balancing a chemical equation, you can never change the coefficient in front of any chemical formula.

b. The coefficients in a balanced chemical equation refer to the number of grams of reactants and products.

c. In a chemical equation, the reactants are on the right and the products are on the left.

d. When balancing a chemical equation, you can never change the subscripts of any chemical formula.

e. In chemical reactions, matter is neither created nor destroyed so a chemical equation must have the same number of atoms on both sides of the equation.

158. Consider the following unbalanced chemical equation for the combustion of pentane (C_5H_{12}):

$$C_5H_{12}(l) + O_2(g) \longrightarrow CO_2(g) + H_2O(l)$$

If 20.4 g of pentane are burned in excess oxygen, what mass of water can be produced, assuming 100% yield?

159. Sulfur dioxide gas reacts with sodium hydroxide to form sodium sulfite and water. The *unbalanced* chemical equation for this reaction is given below:

$$SO_2(g) + NaOH(s) \longrightarrow Na_2SO_3(s) + H_2O(l)$$

Assuming you react 38.3 g sulfur dioxide with 32.8 g sodium hydroxide and assuming that the reaction goes to completion, calculate the mass of each product formed.

Challenge Problems

160. Gallium arsenide, GaAs, has gained widespread use in semiconductor devices that convert light and electrical signals in fiber-optic communications systems. Gallium consists of 60.% ^{69}Ga and 40.% ^{71}Ga. Arsenic has only one naturally occurring isotope, ^{75}As. Gallium arsenide is a polymeric material, but its mass spectrum shows fragments with the formulas GaAs and Ga_2As_2. What would the distribution of peaks look like for these two fragments?

161. Consider the following data for three binary compounds of hydrogen and nitrogen:

	% H (by Mass)	% N (by Mass)
I	17.75	82.25
II	12.58	87.42
III	2.34	97.66

When 1.00 L of each gaseous compound is decomposed to its elements, the following volumes of $H_2(g)$ and $N_2(g)$ are obtained:

	H_2 (L)	N_2 (L)
I	1.50	0.50
II	2.00	1.00
III	0.50	1.50

Use these data to determine the molecular formulas of compounds I, II, and III and to determine the relative values for the atomic masses of hydrogen and nitrogen.

162. Natural rubidium has the average mass of 85.4678 u and is composed of isotopes ^{85}Rb (mass = 84.9117 u) and ^{87}Rb. The ratio of atoms ^{85}Rb/^{87}Rb in natural rubidium is 2.591. Calculate the mass of ^{87}Rb.

163. A compound contains only carbon, hydrogen, nitrogen, and oxygen. Combustion of 0.157 g of the compound produced 0.213 g CO_2 and 0.0310 g H_2O. In another experiment, it is

found that 0.103 g of the compound produces 0.0230 g NH_3. What is the empirical formula of the compound? *Hint:* Combustion involves reacting with excess O_2. Assume that all the carbon ends up in CO_2 and all the hydrogen ends up in H_2O. Also assume that all the nitrogen ends up in the NH_3 in the second experiment.

164. Nitric acid is produced commercially by the Ostwald process, represented by the following equations:

$$4NH_3(g) + 5O_2(g) \longrightarrow 4NO(g) + 6H_2O(g)$$
$$2NO(g) + O_2(g) \longrightarrow 2NO_2(g)$$
$$3NO_2(g) + H_2O(l) \longrightarrow 2HNO_3(aq) + NO(g)$$

What mass of NH_3 must be used to produce 1.0×10^6 kg HNO_3 by the Ostwald process? Assume 100% yield in each reaction, and assume that the NO produced in the third step is not recycled.

165. When the supply of oxygen is limited, iron metal reacts with oxygen to produce a mixture of FeO and Fe_2O_3. In a certain experiment, 20.00 g iron metal was reacted with 11.20 g oxygen gas. After the experiment, the iron was totally consumed, and 3.24 g oxygen gas remained. Calculate the amounts of FeO and Fe_2O_3 formed in this experiment.

166. A 9.780-g gaseous mixture contains ethane (C_2H_6) and propane (C_3H_8). Complete combustion to form carbon dioxide and water requires 1.120 mole of oxygen gas. Calculate the mass percent of ethane in the original mixture.

167. Zinc and magnesium metal each reacts with hydrochloric acid to make chloride salts of the respective metals, and hydrogen gas. A 10.00-g mixture of zinc and magnesium produces 0.5171 g of hydrogen gas upon being mixed with an excess of hydrochloric acid. Determine the percent magnesium by mass in the original mixture.

168. A gas contains a mixture of $NH_3(g)$ and $N_2H_4(g)$, both of which react with $O_2(g)$ to form $NO_2(g)$ and $H_2O(g)$. The gaseous mixture (with an initial mass of 61.00 g) is reacted with 10.00 moles O_2, and after the reaction is complete, 4.062 moles of O_2 remains. Calculate the mass percent of $N_2H_4(g)$ in the original gaseous mixture.

169. Consider a gaseous binary compound with a molar mass of 62.09 g/mol. When 1.39 g of this compound is completely burned in excess oxygen, 1.21 g of water is formed. Determine the formula of the compound. Assume water is the only product that contains hydrogen.

170. A 2.25-g sample of scandium metal is reacted with excess hydrochloric acid to produce 0.1502 g hydrogen gas. What is the formula of the scandium chloride produced in the reaction?

171. In the production of printed circuit boards for the electronics industry, a 0.60-mm layer of copper is laminated onto an insulating plastic board. Next, a circuit pattern made of a chemically resistant polymer is printed on the board. The unwanted copper is removed by chemical etching, and the protective polymer is finally removed by solvents. One etching reaction is

$$Cu(NH_3)_4Cl_2(aq) + 4NH_3(aq) + Cu(s) \longrightarrow 2Cu(NH_3)_4Cl(aq)$$

A plant needs to manufacture 10,000 printed circuit boards, each 8.0×16.0 cm in area. An average of 80.% of the copper is removed from each board (density of copper = 8.96 g/cm³).

What masses of $Cu(NH_3)_4Cl_2$ and NH_3 are needed to do this? Assume 100% yield.

172. The aspirin substitute, acetaminophen ($C_8H_9O_2N$), is produced by the following three-step synthesis:

I. $C_6H_5O_3N(s) + 3H_2(g) + HCl(aq) \longrightarrow$
$$C_6H_8ONCl(s) + 2H_2O(l)$$

II. $C_6H_8ONCl(s) + NaOH(aq) \longrightarrow$
$$C_6H_7ON(s) + H_2O(l) + NaCl(aq)$$

III. $C_6H_7ON(s) + C_4H_6O_3(l) \longrightarrow$
$$C_8H_9O_2N(s) + HC_2H_3O_2(l)$$

The first two reactions have percent yields of 87% and 98% by mass, respectively. The overall reaction yields 3 moles of acetaminophen product for every 4 moles of $C_6H_5O_3N$ reacted.

a. What is the percent yield by mass for the overall process?

b. What is the percent yield by mass of Step III?

173. An element X forms both a dichloride (XCl_2) and a tetrachloride (XCl_4). Treatment of 10.00 g XCl_2 with excess chlorine forms 12.55 g XCl_4. Calculate the atomic mass of X, and identify X.

174. When $M_2S_3(s)$ is heated in air, it is converted to $MO_2(s)$. A 4.000-g sample of $M_2S_3(s)$ shows a decrease in mass of 0.277 g when it is heated in air. What is the average atomic mass of M?

175. When aluminum metal is heated with an element from Group 6A of the periodic table, an ionic compound forms. When the experiment is performed with an unknown Group 6A element, the product is 18.56% Al by mass. What is the formula of the compound?

176. Consider a mixture of potassium chloride and potassium nitrate that is 43.2% potassium by mass. What is the percent KCl by mass of the original mixture?

177. Ammonia reacts with O_2 to form either $NO(g)$ or $NO_2(g)$ according to these unbalanced equations:

$$NH_3(g) + O_2(g) \longrightarrow NO(g) + H_2O(g)$$
$$NH_3(g) + O_2(g) \longrightarrow NO_2(g) + H_2O(g)$$

In a certain experiment 2.00 moles of $NH_3(g)$ and 10.00 moles of $O_2(g)$ are contained in a closed flask. After the reaction is complete, 6.75 moles of $O_2(g)$ remains. Calculate the number of moles of $NO(g)$ in the product mixture: (*Hint:* You cannot do this problem by adding the balanced equations because you cannot assume that the two reactions will occur with equal probability.)

178. You take 1.00 g of an aspirin tablet (a compound consisting solely of carbon, hydrogen, and oxygen), burn it in air, and collect 2.20 g CO_2 and 0.400 g H_2O. You know that the molar mass of aspirin is between 170 and 190 g/mol. Reacting 1 mole of salicylic acid with 1 mole of acetic anhydride ($C_4H_6O_3$) gives you 1 mole of aspirin and 1 mole of acetic acid ($C_2H_4O_2$). Use this information to determine the molecular formula of salicylic acid.

Integrative Problems

These problems require the integration of multiple concepts to find the solutions.

179. With the advent of techniques such as scanning tunneling microscopy, it is now possible to "write" with individual atoms by manipulating and arranging atoms on an atomic surface.

a. If an image is prepared by manipulating iron atoms and their total mass is 1.05×10^{-20} g, what number of iron atoms were used?

b. If the image is prepared on a platinum surface that is exactly 20 platinum atoms high and 14 platinum atoms wide, what is the mass (grams) of the atomic surface?

c. If the atomic surface were changed to ruthenium atoms and the same surface mass as determined in part b is used, what number of ruthenium atoms is needed to construct the surface?

180. Tetrodotoxin is a toxic chemical found in fugu pufferfish, a popular but rare delicacy in Japan. This compound has an LD_{50} (the amount of substance that is lethal to 50.% of a population sample) of 10. μg per kg of body mass. Tetrodotoxin is 41.38% carbon by mass, 13.16% nitrogen by mass, and 5.37% hydrogen by mass, with the remaining amount consisting of oxygen. What is the empirical formula of tetrodotoxin? If three molecules of tetrodotoxin have a mass of 1.59×10^{-21} g, what is the molecular formula of tetrodotoxin? What number of molecules of tetrodotoxin would be the LD_{50} dosage for a person weighing 165 lb?

181. An ionic compound MX_3 is prepared according to the following unbalanced chemical equation.

$$M + X_2 \longrightarrow MX_3$$

A 0.105-g sample of X_2 contains 8.92×10^{20} molecules. The compound MX_3 consists of 54.47% X by mass. What are the identities of M and X, and what is the correct name for MX_3? Starting with 1.00 g each of M and X_2, what mass of MX_3 can be prepared?

182. The compound As_2I_4 is synthesized by reaction of arsenic metal with arsenic triiodide. If a solid cubic block of arsenic ($d = 5.72$ g/cm^3) that is 3.00 cm on edge is allowed to react with 1.01×10^{24} molecules of arsenic triiodide, what mass of As_2I_4 can be prepared? If the percent yield of As_2I_4 was 75.6%, what mass of As_2I_4 was actually isolated?

Marathon Problems

These problems are designed to incorporate several concepts and techniques into one situation.

183. A 2.077-g sample of an element, which has an atomic mass between 40 and 55, reacts with oxygen to form 3.708 g of an oxide. Determine the formula of the oxide (and identify the element).

184. Consider the following balanced chemical equation:

$$A + 5B \longrightarrow 3C + 4D$$

a. Equal masses of A and B are reacted. Complete each of the following with either "A is the limiting reactant because _____"; "B is the limiting reactant because _____"; or "we cannot determine the limiting reactant because _____."

   i. If the molar mass of A is greater than the molar mass of B, then

   ii. If the molar mass of B is greater than the molar mass of A, then

b. The products of the reaction are carbon dioxide (C) and water (D). Compound A has a similar molar mass to carbon dioxide. Compound B is a diatomic molecule. Identify compound B, and support your answer.

c. Compound A is a hydrocarbon that is 81.71% carbon by mass. Determine its empirical and molecular formulas.

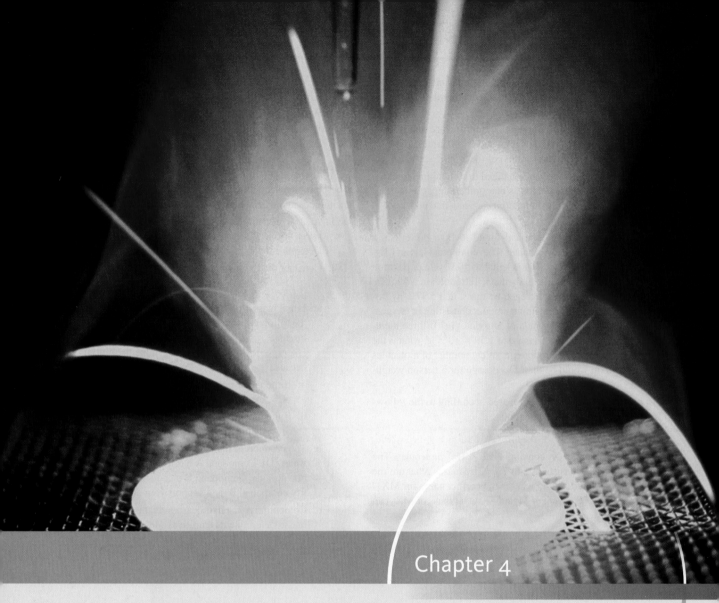

Chapter 4

Types of Chemical Reactions and Solution Stoichiometry

Sodium reacts violently when water is dripped on it. (Charles D. Winters)

Much of the chemistry that affects each of us occurs among substances dissolved in water. For example, virtually all the chemistry that makes life possible occurs in an aqueous environment. Also, various medical tests involve aqueous reactions, depending heavily on analyses of blood and other body fluids. In addition to the common tests for sugar, cholesterol, and iron, analyses for specific chemical markers allow detection of many diseases before obvious symptoms occur.

Aqueous chemistry is also important in our environment. In recent years, contamination of the groundwater by substances such as chloroform and nitrates has been widely publicized. Water is essential for life, and the maintenance of an ample supply of clean water is crucial to all civilization.

To understand the chemistry that occurs in such diverse places as the human body, the atmosphere, the groundwater, the oceans, the local water treatment plant, your hair as you shampoo it, and so on, we must understand how substances dissolved in water react with each other.

However, before we can understand solution reactions, we need to discuss the nature of solutions in which water is the dissolving medium, or *solvent*. These solutions are called **aqueous solutions**. In this chapter we will study the nature of materials after they are dissolved in water and various types of reactions that occur among these substances. You will see that the procedures developed in Chapter 3 to deal with chemical reactions work very well for reactions that take place in aqueous solutions. To understand the types of reactions that occur in aqueous solutions, we must first explore the types of species present. This requires an understanding of the nature of water.

4.1 | Water, the Common Solvent

Water is one of the most important substances on the earth. It is essential for sustaining the reactions that keep us alive, but it also affects our lives in many indirect ways. Water helps moderate the earth's temperature; it cools automobile engines, nuclear power plants, and many industrial processes; it provides a means of transportation on the earth's surface and a medium for the growth of a myriad of creatures we use as food; and much more.

One of the most valuable properties of water is its ability to dissolve many different substances. For example, salt "disappears" when you sprinkle it into the water used to cook vegetables, as does sugar when you add it to your iced tea. In each case the "disappearing" substance is obviously still present—you can taste it. What happens when a solid dissolves? To understand this process, we need to consider the nature of water. Liquid water consists of a collection of H_2O molecules. An individual H_2O molecule is "bent" or V-shaped, with an H—O—H angle of approximately 105 degrees:

The O—H bonds in the water molecule are covalent bonds formed by electron sharing between the oxygen and hydrogen atoms. However, the electrons of the bond are not shared equally between these atoms. For reasons we will discuss in later chapters, oxygen has a greater attraction for electrons than does hydrogen. If the electrons were shared equally between the two atoms, both would be electrically neutral because, on average, the number of electrons around each would equal the number of protons in that nucleus. However, because the oxygen atom has a greater attraction for electrons, the shared electrons tend to spend more time close to the oxygen than to either of the

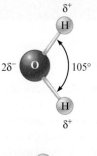

Figure 4.1 (top) The water molecule is polar. (bottom) A space-filling model of the water molecule.

hydrogens. Thus the oxygen atom gains a slight excess of negative charge, and the hydrogen atoms become slightly positive. This is shown in Fig. 4.1, where δ (delta) indicates a *partial* charge (*less than one unit of charge*). Because of this unequal charge distribution, water is said to be a **polar molecule**. It is this polarity that gives water its great ability to dissolve compounds.

A schematic of an ionic solid dissolving in water is shown in Fig. 4.2. Note that the "positive ends" of the water molecules are attracted to the negatively charged anions and that the "negative ends" are attracted to the positively charged cations. This process is called **hydration**. The hydration of its ions tends to cause a salt to "fall apart" in the water, or to dissolve. The strong forces present among the positive and negative ions of the solid are replaced by strong water–ion interactions.

It is very important to recognize that when ionic substances (salts) dissolve in water, they break up into the *individual* cations and anions. For instance, when ammonium nitrate (NH_4NO_3) dissolves in water, the resulting solution contains NH_4^+ and NO_3^- ions moving around independently. This process can be represented as

$$NH_4NO_3(s) \xrightarrow{\text{H}_2\text{O}(l)} NH_4^+(aq) + NO_3^-(aq)$$

where (*aq*) designates that the ions are hydrated by unspecified numbers of water molecules.

The **solubility** of ionic substances in water varies greatly. For example, sodium chloride is quite soluble in water, whereas silver chloride (contains Ag^+ and Cl^- ions) is only very slightly soluble. The differences in the solubilities of ionic compounds in water typically depend on the relative attractions of the ions for each other (these forces hold the solid together) and the attractions of the ions for water molecules (which cause the solid to disperse [dissolve] in water). Solubility is a complex topic that we will explore in much more detail in Chapter 11. However, the most important thing to remember at this point is that when an ionic solid does dissolve in water, the ions become hydrated and are dispersed (move around independently).

Water also dissolves many nonionic substances. Ethanol (C_2H_5OH), for example, is very soluble in water. Wine, beer, and mixed drinks are aqueous solutions of ethanol and other substances. Why is ethanol so soluble in water? The answer lies in the

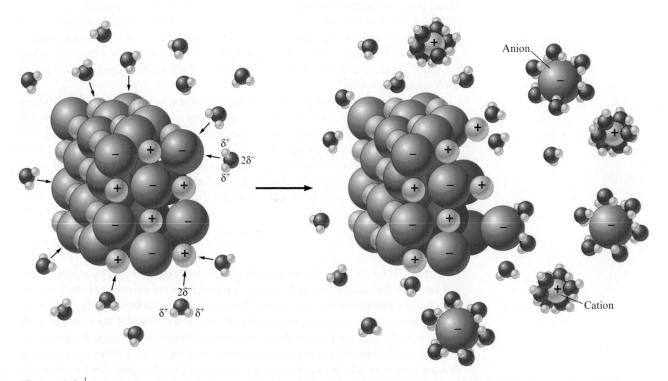

Figure 4.2 Polar water molecules interact with the positive and negative ions of a salt, assisting in the dissolving process.

Figure 4.3 | (a) The ethanol molecule contains a polar O—H bond similar to those in the water molecule. (b) The polar water molecule interacts strongly with the polar O—H bond in ethanol. This is a case of "like dissolving like."

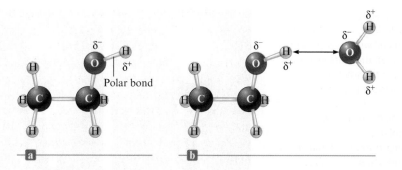

structure of the alcohol molecules, which is shown in Fig. 4.3(a). The molecule contains a polar O—H bond like those in water, which makes it very compatible with water. The interaction of water with ethanol is represented in Fig. 4.3(b).

Many substances do not dissolve in water. Pure water will not, for example, dissolve animal fat, because fat molecules are nonpolar and do not interact effectively with polar water molecules. In general, polar and ionic substances are expected to be more soluble in water than nonpolar substances. "Like dissolves like" is a useful rule for predicting solubility. We will explore the basis for this generalization when we discuss the details of solution formation in Chapter 11.

Critical Thinking

What if no ionic solids were soluble in water? How would this affect the way reactions occur in aqueous solutions?

4.2 | The Nature of Aqueous Solutions: Strong and Weak Electrolytes

As we discussed in Chapter 1, a solution is a homogeneous mixture. It is the same throughout (the first sip of a cup of coffee is the same as the last), but its composition can be varied by changing the amount of dissolved substances (you can make weak or strong coffee). In this section we will consider what happens when a substance, the **solute**, is dissolved in liquid water, the **solvent**.

One useful property for characterizing a solution is its **electrical conductivity**, its ability to conduct an electric current. This characteristic can be checked conveniently by using an apparatus like the ones shown in Fig. 4.4. If the solution in the container conducts electricity, the bulb lights. Pure water is not an electrical conductor. However, some aqueous solutions conduct current very efficiently, and the bulb shines very brightly; these solutions contain **strong electrolytes**. Other solutions conduct only a small current, and the bulb glows dimly; these solutions contain **weak electrolytes**. Some solutions permit no current to flow, and the bulb remains unlit; these solutions contain **nonelectrolytes**.

The basis for the conductivity properties of solutions was first correctly identified by Svante Arrhenius (1859–1927), then a Swedish graduate student in physics, who carried out research on the nature of solutions at the University of Uppsala in the early 1880s. Arrhenius came to believe that the conductivity of solutions arose from the presence of ions, an idea that was at first scorned by the majority of the scientific establishment. However, in the late 1890s when atoms were found to contain charged particles, the ionic theory suddenly made sense and became widely accepted.

As Arrhenius postulated, the extent to which a solution can conduct an electric current depends directly on the number of ions present. Some materials, such as sodium chloride, readily produce ions in aqueous solution and thus are strong

An *electrolyte* is a substance that when dissolved in water produces a solution that can conduct electricity.

Figure 4.4 | Electrical conductivity of aqueous solutions. The circuit will be completed and will allow current to flow only when there are charge carriers (ions) in the solution. *Note:* Water molecules are present but not shown in these pictures. (a) A hydrochloric acid solution, which is a strong electrolyte, contains ions that readily conduct the current and give a brightly lit bulb. (b) An acetic acid solution, which is a weak electrolyte, contains only a few ions and does not conduct as much current as a strong electrolyte. The bulb is only dimly lit. (c) A sucrose solution, which is a nonelectrolyte, contains no ions and does not conduct a current. The bulb remains unlit.

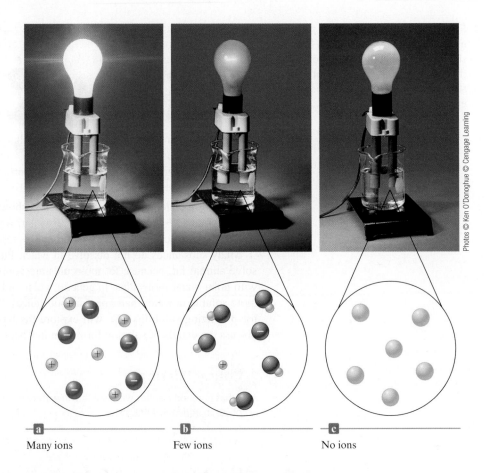

Many ions   Few ions   No ions

electrolytes. Other substances, such as acetic acid, produce relatively few ions when dissolved in water and are weak electrolytes. A third class of materials, such as sugar, form virtually no ions when dissolved in water and are nonelectrolytes.

Strong Electrolytes

Strong electrolytes are substances that are completely ionized when they are dissolved in water, as represented in Fig. 4.4(a). We will consider several classes of strong electrolytes: (1) soluble salts, (2) strong acids, and (3) strong bases.

As shown in Fig. 4.2, a salt consists of an array of cations and anions that separate and become hydrated when the salt dissolves. For example, when NaCl dissolves in water, it produces hydrated Na^+ and Cl^- ions in the solution (Fig. 4.5). Virtually no

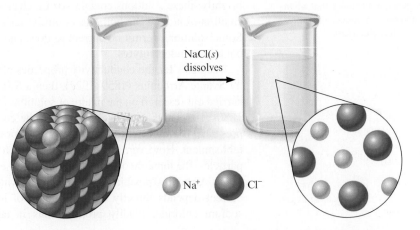

Na^+ Cl^-

Figure 4.5 | When solid NaCl dissolves, the Na^+ and Cl^- ions are randomly dispersed in the water.

 H^+

Cl^-

Figure 4.6 | HCl(*aq*) is completely ionized.

The Arrhenius definition of an acid is a substance that produces H^+ ions in solution.

Strong electrolytes dissociate (ionize) completely in aqueous solution.

Perchloric acid, $HClO_4(aq)$, is another strong acid.

NaCl units are present. Thus NaCl is a strong electrolyte. It is important to recognize that these aqueous solutions contain millions of water molecules that we will not include in our molecular-level drawings.

One of Arrhenius's most important discoveries concerned the nature of acids. Acidity was first associated with the sour taste of citrus fruits. In fact, the word *acid* comes directly from the Latin word *acidus,* meaning "sour." The mineral acids sulfuric acid (H_2SO_4) and nitric acid (HNO_3), so named because they were originally obtained by the treatment of minerals, were discovered around 1300.

Although acids were known for hundreds of years before the time of Arrhenius, no one had recognized their essential nature. In his studies of solutions, Arrhenius found that when the substances HCl, HNO_3, and H_2SO_4 were dissolved in water, they behaved as strong electrolytes. He postulated that this was the result of ionization reactions in water, for example:

$$HCl \xrightarrow{H_2O} H^+(aq) + Cl^-(aq)$$

$$HNO_3 \xrightarrow{H_2O} H^+(aq) + NO_3^-(aq)$$

$$H_2SO_4 \xrightarrow{H_2O} H^+(aq) + HSO_4^-(aq)$$

Thus Arrhenius proposed that an ***acid is a substance that produces H^+ ions (protons) when it is dissolved in water.***

We now understand that the polar nature of water plays a very important role in causing acids to produce H^+ in solution. In fact, it is most appropriate to represent the "ionization" of an acid as follows:

$$HA(aq) + H_2O(l) \longrightarrow H_3O^+(aq) + A^-(aq)$$

which emphasizes the important role of water in this process. We will have much more to say about this process in Chapter 14.

Studies of conductivity show that when HCl, HNO_3, and H_2SO_4 are placed in water, *virtually every molecule ionizes*. These substances are strong electrolytes and are thus called **strong acids**. All three are very important chemicals, and much more will be said about them as we proceed. However, at this point the following facts are important:

Sulfuric acid, nitric acid, and hydrochloric acid are aqueous solutions and should be written in chemical equations as $H_2SO_4(aq)$, $HNO_3(aq)$, and HCl(*aq*), respectively, although they often appear without the (*aq*) symbol.

A strong acid is one that completely dissociates into its ions. Thus, if 100 molecules of HCl are dissolved in water, 100 H^+ ions and 100 Cl^- ions are produced. Virtually no HCl molecules exist in aqueous solutions (Fig. 4.6).

Sulfuric acid is a special case. The formula H_2SO_4 indicates that this acid can produce two H^+ ions per molecule when dissolved in water. However, only the first H^+ ion is completely dissociated. The second H^+ ion can be pulled off under certain conditions, which we will discuss later. Thus an aqueous solution of H_2SO_4 contains mostly H^+ ions and HSO_4^- ions.

Another important class of strong electrolytes consists of the **strong bases**, soluble ionic compounds containing the hydroxide ion (OH^-). When these compounds are dissolved in water, the cations and OH^- ions separate and move independently. Solutions containing bases have a bitter taste and a slippery feel. The most common basic solutions are those produced when solid sodium hydroxide (NaOH) or potassium hydroxide (KOH) is dissolved in water to produce ions, as follows (Fig. 4.7):

$$NaOH(s) \xrightarrow{H_2O} Na^+(aq) + OH^-(aq)$$

$$KOH(s) \xrightarrow{H_2O} K^+(aq) + OH^-(aq)$$

OH^-

Na^+

Figure 4.7 | An aqueous solution of sodium hydroxide.

Chemical connections

Arrhenius: A Man with Solutions

Science is a human endeavor, subject to human frailties and governed by personalities, politics, and prejudices. One of the best illustrations of the often bumpy path of the advancement of scientific knowledge is the story of Swedish chemist Svante Arrhenius.

When Arrhenius began studies toward his doctorate at the University of Uppsala around 1880, he chose to investigate the passage of electricity through solutions, a mystery that had baffled scientists for a century. The first experiments had been done in the 1770s by Cavendish, who compared the conductivity of salt solution with that of rain water using his own physiologic reaction to the electric shocks he received! Arrhenius had an array of instruments to measure electric current, but the process of carefully weighing, measuring, and recording data from a multitude of experiments was a tedious one.

After his long series of experiments was performed, Arrhenius quit his laboratory bench and returned to his country home to try to formulate a model that could account for his data. He wrote, "I got the idea in the night of the 17th of May in the year 1883, and I could not sleep that night until I had worked through the whole problem." His idea was that ions were responsible for conducting electricity through a solution.

Back at Uppsala, Arrhenius took his doctoral dissertation containing the new theory to his advisor, Professor Cleve, an eminent chemist and the discoverer of the elements holmium and thulium. Cleve's uninterested response was what Arrhenius had expected. It was in keeping with Cleve's resistance to new ideas—he had not even accepted Mendeleev's periodic table, introduced 10 years earlier.

It is a long-standing custom that before a doctoral degree is granted, the dissertation must be defended before a panel of professors. Although this procedure is still followed at most universities today, the problems are usually worked out in private with the evaluating professors before the actual defense. However, when Arrhenius did it, the dissertation defense was an open debate, which could be rancorous and humiliating. Knowing that it would be unwise to antagonize his professors, Arrhenius downplayed his convictions about his new theory as he defended his dissertation. His diplomacy paid off: He was awarded his degree, albeit reluctantly, because the professors still did not believe his model and considered him to be a marginal scientist, at best.

Such a setback could have ended his scientific career, but Arrhenius was a crusader; he was determined to see his theory triumph. He promptly embarked on a political campaign, enlisting the aid of several prominent scientists, to get his theory accepted.

Svante August Arrhenius.

Ultimately, the ionic theory triumphed. Arrhenius's fame spread, and honors were heaped on him, culminating in the Nobel Prize in chemistry in 1903. Not one to rest on his laurels, Arrhenius turned to new fields, including astronomy; he formulated a new theory that the solar system may have come into being through the collision of stars. His exceptional versatility led him to study the use of serums to fight disease, energy resources and conservation, and the origin of life.

Additional insight on Arrhenius and his scientific career can be obtained from his address on receiving the Willard Gibbs Award. See *Journal of the American Chemical Society* 36 (1912): 353.

Weak Electrolytes

Weak electrolytes dissociate (ionize) only to a small extent in aqueous solution.

Weak electrolytes are substances that exhibit a small degree of ionization in water. That is, they produce relatively few ions when dissolved in water, as shown in Fig. 4.4(b). The most common weak electrolytes are weak acids and weak bases.

The main acidic component of vinegar is acetic acid ($HC_2H_3O_2$). The formula is written to indicate that acetic acid has two chemically distinct types of hydrogen

Figure 4.8 | Acetic acid ($HC_2H_3O_2$) exists in water mostly as undissociated molecules. Only a small percentage of the molecules are ionized.

atoms. Formulas for acids are often written with the acidic hydrogen atom or atoms (any that will produce H^+ ions in solution) listed first. If any nonacidic hydrogens are present, they are written later in the formula. Thus the formula $HC_2H_3O_2$ indicates one acidic and three nonacidic hydrogen atoms. The dissociation reaction for acetic acid in water can be written as follows:

$$HC_2H_3O_2(aq) + H_2O(l) \rightleftharpoons H_3O^+(aq) + C_2H_3O_2^-(aq)$$

Acetic acid is very different from the strong acids because only about 1% of its molecules dissociate in aqueous solutions at typical concentrations. For example, in a solution containing 0.1 mole of $HC_2H_3O_2$ per liter, for every 100 molecules of $HC_2H_3O_2$ originally dissolved in water, approximately 99 molecules of $HC_2H_3O_2$ remain intact (Fig. 4.8). That is, only one molecule out of every 100 dissociates (to produce one H^+ ion and one $C_2H_3O_2^-$ ion). The double arrow indicates the reaction can occur in either direction.

Because acetic acid is a weak electrolyte, it is called a **weak acid**. Any acid, such as acetic acid, that *dissociates (ionizes) only to a slight extent in aqueous solutions is called a weak acid.* We will explore the subject of weak acids in detail in Chapter 14.

The most common weak base is ammonia (NH_3). When ammonia is dissolved in water, it reacts as follows:

$$NH_3(aq) + H_2O(l) \rightleftharpoons NH_4^+(aq) + OH^-(aq)$$

The solution is *basic* because OH^- ions are produced. Ammonia is called a **weak base** because *the resulting solution is a weak electrolyte*; that is, very few ions are formed. In fact, in a solution containing 0.1 mole of NH_3 per liter, for every 100 molecules of NH_3 originally dissolved, only one NH_4^+ ion and one OH^- ion are produced; 99 molecules of NH_3 remain unreacted (Fig. 4.9). The double arrow indicates the reaction can occur in either direction.

Nonelectrolytes

Nonelectrolytes are substances that dissolve in water but do not produce any ions, as shown in Fig. 4.4(c). An example of a nonelectrolyte is ethanol (see Fig. 4.3 for the structural formula). When ethanol dissolves, entire C_2H_5OH molecules are dispersed in the water. Since the molecules do not break up into ions, the resulting solution does not conduct an electric current. Another common nonelectrolyte is table sugar (sucrose, $C_{12}H_{22}O_{11}$), which is very soluble in water but which produces no ions when it dissolves. The sucrose molecules remain intact.

Figure 4.9 | The reaction of NH_3 in water.

4.3 | The Composition of Solutions

Chemical reactions often take place when two solutions are mixed. To perform stoichiometric calculations in such cases, we must know two things: (1) the *nature of the reaction*, which depends on the exact forms the chemicals take when dissolved, and (2) the *amounts of chemicals* present in the solutions, usually expressed as concentrations.

The concentration of a solution can be described in many different ways, as we will see in Chapter 11. At this point we will consider only the most commonly used expression of concentration, **molarity** (*M*), which is defined as *moles of solute per volume of solution in liters*:

$$M = \text{molarity} = \frac{\text{moles of solute}}{\text{liters of solution}}$$

A solution that is 1.0 molar (written as 1.0 *M*) contains 1.0 mole of solute per liter of solution.

Calculation of Molarity I

Calculate the molarity of a solution prepared by dissolving 11.5 g of solid NaOH in enough water to make 1.50 L of solution.

Solution

Where are we going?

To find the molarity of NaOH solution

What do we know?

> 11.5 g NaOH

> 1.50 L solution

What information do we need to find molarity?

> Moles solute

> Molarity $= \dfrac{\text{mol solute}}{\text{L solution}}$

How do we get there?

What are the moles of NaOH (40.00 g/mol)?

$$11.5 \text{ g NaOH} \times \frac{1 \text{ mol NaOH}}{40.00 \text{ g NaOH}} = 0.288 \text{ mol NaOH}$$

What is the molarity of the solution?

$$\blacksquare \text{ Molarity} = \frac{\text{mol solute}}{\text{L solution}} = \frac{0.288 \text{ mol NaOH}}{1.50 \text{ L solution}} = 0.192 \ M \text{ NaOH}$$

Reality Check | The units are correct for molarity.

See Exercises 4.27 and 4.28

Calculation of Molarity II

Calculate the molarity of a solution prepared by dissolving 1.56 g of gaseous HCl in enough water to make 26.8 mL of solution.

Solution

Where are we going?

To find the molarity of HCl solution

What do we know?

> 1.56 g HCl

> 26.8 mL solution

What information do we need to find molarity?

> Moles solute

> Molarity $= \dfrac{\text{mol solute}}{\text{L solution}}$

How do we get there?

What are the moles of HCl (36.46 g/mol)?

$$1.56 \text{ g HCl} \times \frac{1 \text{ mol HCl}}{36.46 \text{ g HCl}} = 4.28 \times 10^{-2} \text{ mol HCl}$$

What is the volume of solution (in liters)?

$$26.8 \text{ mL} \times \frac{1 \text{ L}}{1000 \text{ mL}} = 2.68 \times 10^{-2} \text{ L}$$

What is the molarity of the solution?

$$\blacksquare \text{ Molarity} = \frac{4.28 \times 10^{-2} \text{ mol HCl}}{2.68 \times 10^{-2} \text{ L solution}} = 1.60 \text{ } M \text{ HCl}$$

Reality Check | The units are correct for molarity.

See Exercises 4.27 and 4.28

Concentration of Ions I

Give the concentration of each type of ion in the following solutions:

a. 0.50 M Co(NO$_3$)$_2$

b. 1 M Fe(ClO$_4$)$_3$

Solution

Where are we going?

To find the molarity of each ion in the solution

What do we know?

> 0.50 M Co(NO$_3$)$_2$

> 1 M Fe(ClO$_4$)$_3$

What information do we need to find the molarity of each ion?

> Moles of each ion

How do we get there?

For Co(NO$_3$)$_2$

What is the balanced equation for dissolving the ions?

$$Co(NO_3)_2(s) \xrightarrow{\text{H}_2\text{O}} Co^{2+}(aq) + 2NO_3^-(aq)$$

What is the molarity for each ion?

$\blacksquare$ Co^{2+} 1 $\times$ 0.50 M = 0.50 M Co^{2+}

$\blacksquare$ NO$_3^-$ 2 $\times$ 0.50 M = 1.0 M NO$_3^-$

For Fe(ClO$_4$)$_3$

What is the balanced equation for dissolving the ions?

$$Fe(ClO_4)_3(s) \xrightarrow{\text{H}_2\text{O}} Fe^{3+}(aq) + 3ClO_4^-(aq)$$

What is the molarity for each ion?

$\blacksquare$ Fe^{3+} 1 $\times$ 1 M = 1 M Fe^{3+}

$\blacksquare$ ClO$_4^-$ 3 $\times$ 1 M = 3 M ClO$_4^-$

See Exercises 4.29 and 4.30

An aqueous solution of Co(NO$_3$)$_2$.

Often chemists need to determine the number of moles of solute present in a given volume of a solution of known molarity. The procedure for doing this is easily derived from the definition of molarity. If we multiply the molarity of a solution by the volume (in liters) of a particular sample of the solution, we get the moles of solute present in that sample:

$$M = \frac{\text{moles of solute}}{\text{liters of solution}}$$

$$\text{Liters of solution} \times \text{molarity} = \text{liters of solution} \times \frac{\text{moles of solute}}{\text{liters of solution}} = \text{moles of solute}$$

This procedure is demonstrated in Examples 4.4 and 4.5.

Interactive Example 4.4

Sign in at http://login.cengagebrain .com to try this Interactive Example in **OWL**.

Concentration of Ions II

Calculate the number of moles of Cl^- ions in 1.75 L of $1.0 \times 10^{-3}\ M$ $ZnCl_2$.

Solution

Where are we going?

To find the moles of Cl^- ion in the solution

What do we know?

> $1.0 \times 10^{-3}\ M\ ZnCl_2$

> 1.75 L

What information do we need to find moles of Cl^-?

> Balanced equation for dissolving $ZnCl_2$

How do we get there?

What is the balanced equation for dissolving the ions?

$$ZnCl_2(s) \xrightarrow{\text{H}_2\text{O}} Zn^{2+}(aq) + 2Cl^-(aq)$$

What is the molarity of Cl^- ion in the solution?

$$2 \times (1.0 \times 10^{-3}\ M) = 2.0 \times 10^{-3}\ M\ Cl^-$$

How many moles of Cl^-?

- 1.75 L solution $\times$ $2.0 \times 10^{-3}\ M\ Cl^-$ = 1.75 L solution $\times \dfrac{2.0 \times 10^{-3}\ \text{mol } Cl^-}{\text{L solution}}$

$$= 3.5 \times 10^{-3}\ \text{mol } Cl^-$$

See Exercise 4.31

Interactive Example 4.5

Sign in at http://login.cengagebrain .com to try this Interactive Example in **OWL**.

Concentration and Volume

Typical blood serum is about 0.14 M NaCl. What volume of blood contains 1.0 mg of NaCl?

Solution

Where are we going?

To find the volume of blood containing 1.0 mg of NaCl

What do we know?

> 0.14 M NaCl

> 1.0 mg NaCl

What information do we need to find volume of blood containing 1.0 mg of NaCl?

› Moles of NaCl (in 1.0 mg)

How do we get there?

What are the moles of NaCl (58.44 g/mol)?

$$1.0 \text{ mg NaCl} \times \frac{1 \text{ g NaCl}}{1000 \text{ mg NaCl}} \times \frac{1 \text{ mol NaCl}}{58.44 \text{ g NaCl}} = 1.7 \times 10^{-5} \text{ mol NaCl}$$

What volume of 0.14 M NaCl contains 1.0 mg (1.7 × 10⁻⁵ mole) of NaCl?
There is some volume, call it *V*, that when multiplied by the molarity of this solution will yield 1.7×10^{-5} mole of NaCl. That is,

$$V \times \frac{0.14 \text{ mol NaCl}}{\text{L solution}} = 1.7 \times 10^{-5} \text{ mol NaCl}$$

We want to solve for the volume:

$$V = \frac{1.7 \times 10^{-5} \text{ mol NaCl}}{\dfrac{0.14 \text{ mol NaCl}}{\text{L solution}}} = 1.2 \times 10^{-4} \text{ L solution}$$

▪ Thus 0.12 mL of blood contains 1.7×10^{-5} mole of NaCl or 1.0 mg of NaCl.

See Exercises 4.33 and 4.34

A **standard solution** is a *solution whose concentration is accurately known.* Standard solutions, often used in chemical analysis, can be prepared as shown in Fig. 4.10 and in Example 4.6.

Interactive Example 4.6

Solutions of Known Concentration

To analyze the alcohol content of a certain wine, a chemist needs 1.00 L of an aqueous 0.200-*M* $K_2Cr_2O_7$ (potassium dichromate) solution. How much solid $K_2Cr_2O_7$ must be weighed out to make this solution?

Solution

Where are we going?

To find the mass of $K_2Cr_2O_7$ required for the solution

Figure 4.10 | Steps involved in the preparation of a standard aqueous solution. (a) Put a weighed amount of a substance (the solute) into the volumetric flask, and add a small quantity of water. (b) Dissolve the solid in the water by gently swirling the flask (*with the stopper in place*). (c) Add more water (with gentle swirling) until the level of the solution just reaches the mark etched on the neck of the flask. Then mix the solution thoroughly by inverting the flask several times.

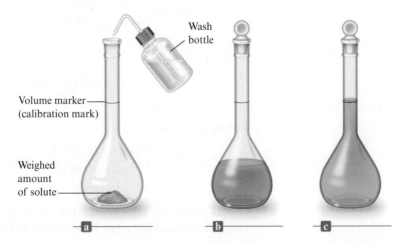

What do we know?

> 1.00 L of 0.200 M $K_2Cr_2O_7$ is required

What information do we need to find the mass of $K_2Cr_2O_7$?

> Moles of $K_2Cr_2O_7$ in the required solution

How do we get there?

What are the moles of $K_2Cr_2O_7$ required?

$$M \times V = \text{mol}$$

$$1.00 \text{ L solution} \times \frac{0.200 \text{ mol } K_2Cr_2O_7}{\text{L solution}} = 0.200 \text{ mol } K_2Cr_2O_7$$

What mass of $K_2Cr_2O_7$ is required for the solution?

$$0.200 \text{ mol } K_2Cr_2O_7 \times \frac{294.20 \text{ g } K_2Cr_2O_7}{\text{mol } K_2Cr_2O_7} = 58.8 \text{ g } K_2Cr_2O_7$$

■ To make 1.00 L of 0.200 M $K_2Cr_2O_7$, the chemist must weigh out 58.8 g $K_2Cr_2O_7$, transfer it to a 1.00-L volumetric flask, and add distilled water to the mark on the flask.

See Exercises 4.35a,c and 4.36c,e

Dilution

To save time and space in the laboratory, routinely used solutions are often purchased or prepared in concentrated form (called *stock solutions*). Water is then added to achieve the molarity desired for a particular solution. This process is called **dilution**. For example, the common acids are purchased as concentrated solutions and diluted as needed. A typical dilution calculation involves determining how much water must be added to an amount of stock solution to achieve a solution of the desired concentration. The key to doing these calculations is to remember that

Dilution with water does not alter the numbers of moles of solute present.

Moles of solute after dilution = moles of solute before dilution

because only water (no solute) is added to accomplish the dilution.

For example, suppose we need to prepare 500. mL of 1.00 M acetic acid ($HC_2H_3O_2$) from a 17.4-M stock solution of acetic acid. What volume of the stock solution is required? The first step is to determine the number of moles of acetic acid in the final solution by multiplying the volume by the molarity (remembering that the volume must be changed to liters):

$$500. \text{ mL solution} \times \frac{1 \text{ L solution}}{1000 \text{ mL solution}} \times \frac{1.00 \text{ mol } HC_2H_3O_2}{\text{L solution}} = 0.500 \text{ mol } HC_2H_3O_2$$

Thus we need to use a volume of 17.4 M acetic acid that contains 0.500 mole of $HC_2H_3O_2$. That is,

$$V \times \frac{17.4 \text{ mol } HC_2H_3O_2}{\text{L solution}} = 0.500 \text{ mol } HC_2H_3O_2$$

Solving for V gives

$$V = \frac{0.500 \text{ mol } HC_2H_3O_2}{\dfrac{17.4 \text{ mol } HC_2H_3O_2}{\text{L solution}}} = 0.0287 \text{ L or } 28.7 \text{ mL solution}$$

Thus to make 500 mL of a 1.00-M acetic acid solution, we can take 28.7 mL of 17.4 M acetic acid and dilute it to a total volume of 500 mL with distilled water.

A dilution procedure typically involves two types of glassware: a pipet and a volumetric flask. A *pipet* is a device for accurately measuring and transferring a given volume of solution. There are two common types of pipets: *volumetric* (or *transfer*) *pipets* and *measuring pipets* (Fig. 4.11). Volumetric pipets come in specific sizes, such as 5 mL, 10 mL, 25 mL, and so on. Measuring pipets are used to measure volumes for which a volumetric pipet is not available. For example, we would use a measuring pipet as shown in Fig. 4.12 on page 153 to deliver 28.7 mL of 17.4 *M* acetic acid into a 500-mL volumetric flask and then add water to the mark to perform the dilution described above.

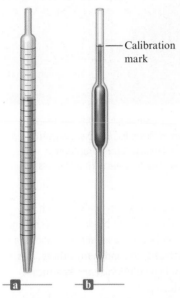

Calibration mark

a   **b**

Figure 4.11 | (a) A measuring pipet is graduated and can be used to measure various volumes of liquid accurately. (b) A volumetric (transfer) pipet is designed to measure one volume accurately. When filled to the mark, it delivers the volume indicated on the pipet.

When diluting an acid, "Do what you oughta, always add acid to water."

Sign in at http://login.cengagebrain .com to try this Interactive Example in **OWL**.

Interactive Example 4.7

Concentration and Volume

What volume of 16 *M* sulfuric acid must be used to prepare 1.5 L of a 0.10-*M* H_2SO_4 solution?

Solution

Where are we going?

To find the volume of H_2SO_4 required to prepare the solution

What do we know?

> 1.5 L of 0.10 *M* H_2SO_4 is required
> We have 16 *M* H_2SO_4

What information do we need to find the volume of H_2SO_4?

> Moles of H_2SO_4 in the required solution

How do we get there?

What are the moles of H_2SO_4 required?

$$M \times V = \text{mol}$$

$$1.5 \text{ L solution} \times \frac{0.10 \text{ mol } H_2SO_4}{\text{L solution}} = 0.15 \text{ mol } H_2SO_4$$

What volume of 16 M H_2SO_4 contains 0.15 mole of H_2SO_4?

$$V \times \frac{16 \text{ mol } H_2SO_4}{\text{L solution}} = 0.15 \text{ mol } H_2SO_4$$

Solving for *V* gives

$$V = \frac{0.15 \text{ mol } H_2SO_4}{\dfrac{16 \text{ mol } H_2SO_4}{1 \text{ L solution}}} = 9.4 \times 10^{-3} \text{ L or } 9.4 \text{ mL solution}$$

■ To make 1.5 L of 0.10 *M* H_2SO_4 using 16 *M* H_2SO_4, we must take 9.4 mL of the concentrated acid and dilute it with water to 1.5 L. The correct way to do this is to add the 9.4 mL of acid to about 1 L of distilled water and then dilute to 1.5 L by adding more water.

See Exercises 4.35b,d, and 4.36a,b,d

As noted earlier, the central idea in performing the calculations associated with dilutions is to recognize that the moles of solute are not changed by the dilution. Another way to express this condition is by the following equation:

$$M_1V_1 = M_2V_2$$

Chemical connections

Tiny Laboratories

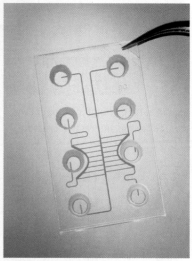

One of the major impacts of modern technology is to make things smaller. The best example is the computer. Calculations that 30 years ago required a machine the size of a large room now can be carried out on a hand-held calculator. This tendency toward miniaturization is also having a major impact on the science of chemical analysis. Using the techniques of computer chip makers, researchers are now constructing minuscule laboratories on the surface of a tiny chip made of silicon, glass, or plastic (see photo). Instead of electrons, 10^{-6} to 10^{-9} L of liquids moves between reaction chambers on the chip through tiny capillaries. The chips typically contain no moving parts. Instead of conventional pumps, the chip-based laboratories use voltage differences to move liquids that contain ions from one reaction chamber to another.

Microchip laboratories have many advantages. They require only tiny amounts of sample. This is especially advantageous for expensive, difficult-to-prepare materials or in cases such as criminal investigations, where only small amounts of evidence may exist. The chip laboratories also minimize contamination because they represent a "closed system" once the material has been introduced to the chip. In addition, the chips can be made to be disposable to prevent cross-contamination of different samples.

The chip laboratories present some difficulties not found in macroscopic laboratories. The main problem concerns the large surface area of the capillaries and reaction chambers relative to the sample volume. Molecules or biological cells in the sample solution encounter so much "wall" that they may undergo unwanted reactions with the wall materials. Glass seems to present the least of these problems, and the walls of silicon chip laboratories can be protected by formation of relatively inert silicon dioxide. Because plastic is inexpensive, it seems a good choice for disposable chips, but plastic also is the most reactive with the samples and the least durable of the available materials.

PerkinElmer, Inc. is working toward creating a miniature chemistry laboratory about the size of a toaster that can be used with "plug-in" chip-based laboratories. Various chips would be furnished with the unit that would be appropriate for different types of analyses. The entire unit would be connected to a computer to collect and analyze the data. There is even the possibility that these "laboratories" could be used in the home to perform analyses such as blood sugar and blood cholesterol and to check for the presence of bacteria such as *E. coli* and many others. This would revolutionize the healthcare industry.

Plastic chips such as this one made by PerkinElmer, Inc. are being used to perform laboratory procedures traditionally done with test tubes.

Adapted from "The Incredible Shrinking Laboratory," by Corinna Wu, as appeared in *Science News*, Vol. 154, August 15, 1998, p. 104.

where M_1 and V_1 represent the molarity and volume of the original solution (before dilution) and M_2 and V_2 represent the molarity and volume of the diluted solution. This equation makes sense because

$$M_1 \times V_1 = \text{mol solute before dilution}$$
$$= \text{mol solute after dilution} = M_2 \times V_2$$

Repeat Example 4.7 using the equation $M_1V_1 = M_2V_2$. Note that in doing so,

$$M_1 = 16\,M \qquad M_2 = 0.10\,M \qquad V_2 = 1.5\,\text{L}$$

Figure 4.12 | (a) A measuring pipet is used to transfer 28.7 mL of 17.4 M acetic acid solution to a volumetric flask. (b) Water is added to the flask to the calibration mark. (c) The resulting solution is 1.00 M acetic acid.

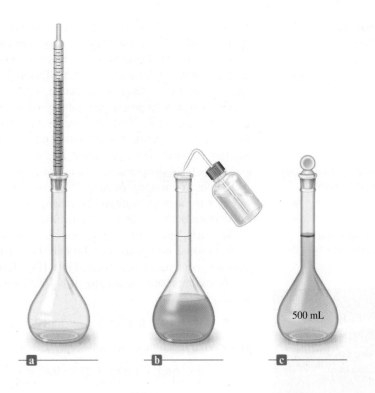

and V_1 is the unknown quantity sought. The equation $M_1V_1 = M_2V_2$ always holds for a dilution. This equation will be easy for you to remember if you understand where it comes from.

4.4 | Types of Chemical Reactions

Although we have considered many reactions so far in this text, we have examined only a tiny fraction of the millions of possible chemical reactions. To make sense of all these reactions, we need some system for grouping reactions into classes. Although there are many different ways to do this, we will use the system most commonly used by practicing chemists:

> **Types of Solution Reactions**
>
> ❯ Precipitation reactions
> ❯ Acid–base reactions
> ❯ Oxidation–reduction reactions

Virtually all reactions can be put into one of these classes. We will define and illustrate each type in the following sections.

4.5 | Precipitation Reactions

When two solutions are mixed, an insoluble substance sometimes forms; that is, a solid forms and separates from the solution. Such a reaction is called a **precipitation reaction**, and the solid that forms is called a **precipitate**. For example, a precipitation reaction occurs when an aqueous solution of potassium chromate, $K_2CrO_4(aq)$, which is yellow, is added to a colorless aqueous solution containing barium nitrate,

Figure 4.13 │ When yellow aqueous potassium chromate is added to a colorless barium nitrate solution, yellow barium chromate precipitates.

$Ba(NO_3)_2(aq)$. As shown in Fig. 4.13, when these solutions are mixed, a yellow solid forms. What is the equation that describes this chemical change? To write the equation, we must know the identities of the reactants and products. The reactants have already been described: $K_2CrO_4(aq)$ and $Ba(NO_3)_2(aq)$. Is there some way we can predict the identities of the products? In particular, what is the yellow solid?

The best way to predict the identity of this solid is to think carefully about what products are possible. To do this, we need to know what species are present in the solution after the two reactant solutions are mixed. First, let's think about the nature of each reactant solution. The designation $Ba(NO_3)_2(aq)$ means that barium nitrate (a white solid) has been dissolved in water. Notice that barium nitrate contains the Ba^{2+} and NO_3^- ions. *Remember: In virtually every case, when a solid containing ions dissolves in water, the ions separate* and move around independently. That is, $Ba(NO_3)_2(aq)$ does not contain $Ba(NO_3)_2$ units; it contains separated Ba^{2+} and NO_3^- ions [Fig. 4.14(a)].

Similarly, since solid potassium chromate contains the K^+ and CrO_4^{2-} ions, an aqueous solution of potassium chromate (which is prepared by dissolving solid K_2CrO_4 in water) contains these separated ions [Fig. 4.14(b)].

We can represent the mixing of $K_2CrO_4(aq)$ and $Ba(NO_3)_2(aq)$ in two ways. First, we can write

$$K_2CrO_4(aq) + Ba(NO_3)_2(aq) \longrightarrow \text{products}$$

However, a much more accurate representation is

$$\underbrace{2K^+(aq) + CrO_4^{2-}(aq)}_{\substack{\text{The ions in} \\ K_2CrO_4(aq)}} + \underbrace{Ba^{2+}(aq) + 2NO_3^-(aq)}_{\substack{\text{The ions in} \\ Ba(NO_3)_2(aq)}} \longrightarrow \text{products}$$

Thus the mixed solution contains the ions:

$$K^+ \qquad CrO_4^{2-} \qquad Ba^{2+} \qquad NO_3^-$$

as illustrated in Fig. 4.15(a).

How can some or all of these ions combine to form a yellow solid? This is not an easy question to answer. In fact, predicting the products of a chemical reaction is one of the hardest things a beginning chemistry student is asked to do. Even an experienced chemist, when confronted with a new reaction, is often not sure what will happen. The chemist tries to think of the various possibilities, considers the likelihood of

Figure 4.14 │ Reactant solutions: (a) $Ba(NO_3)_2(aq)$ and (b) $K_2CrO_4(aq)$.

K⁺ ○

Ba²⁺ ●

NO₃⁻

CrO₄²⁻

Figure 4.15 | The reaction of $K_2CrO_4(aq)$ and $Ba(NO_3)_2(aq)$. (a) The molecular-level "picture" of the mixed solution before any reaction has occurred. (b) The molecular-level "picture" of the solution after the reaction has occurred to form $BaCrO_4(s)$. *Note:* $BaCrO_4(s)$ is not molecular. It actually contains Ba^{2+} and CrO_4^{2-} ions packed together in a lattice. (c) A photo of the solution after the reaction has occurred, showing the solid $BaCrO_4$ on the bottom.

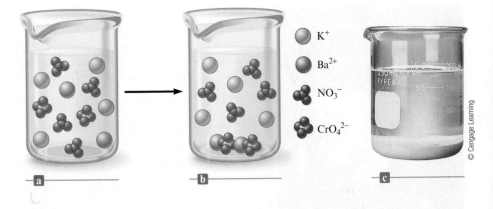

K⁺

Ba²⁺

NO₃⁻

CrO₄²⁻

each possibility, and then makes a prediction (an educated guess). Only after identifying each product *experimentally* is the chemist sure what reaction has taken place. However, an educated guess is very useful because it provides a place to start. It tells us what kinds of products we are most likely to find. We already know some things that will help us predict the products of the above reaction.

1. When ions form a solid compound, the compound must have a zero net charge. Thus the products of this reaction must contain *both anions and cations*. For example, K^+ and Ba^{2+} could not combine to form the solid, nor could CrO_4^{2-} and NO_3^-.

2. Most ionic materials contain only two types of ions: one type of cation and one type of anion (for example, NaCl, KOH, Na_2SO_4, K_2CrO_4, $Co(NO_3)_2$, NH_4Cl, Na_2CO_3).

The possible combinations of a given cation and a given anion from the list of ions K^+, CrO_4^{2-}, Ba^{2+}, and NO_3^- are

$$K_2CrO_4 \qquad KNO_3 \qquad BaCrO_4 \qquad Ba(NO_3)_2$$

Which of these possibilities is most likely to represent the yellow solid? We know it's not K_2CrO_4 or $Ba(NO_3)_2$. They are the reactants. They were present (dissolved) in the separate solutions that were mixed. The only real possibilities for the solid that formed are

$$KNO_3 \quad \text{and} \quad BaCrO_4$$

To decide which of these most likely represents the yellow solid, we need more facts. An experienced chemist knows that the K^+ ion and the NO_3^- ion are both colorless. Thus, if the solid is KNO_3, it should be white, not yellow. On the other hand, the CrO_4^{2-} ion is yellow (note in Fig. 4.14 that $K_2CrO_4(aq)$ is yellow). Thus the yellow solid is almost certainly $BaCrO_4$. Further tests show that this is the case.

So far we have determined that one product of the reaction between $K_2CrO_4(aq)$ and $Ba(NO_3)_2(aq)$ is $BaCrO_4(s)$, but what happened to the K^+ and NO_3^- ions? The answer is that these ions are left dissolved in the solution; KNO_3 does not form a solid when the K^+ and NO_3^- ions are present in this much water. In other words, if we took solid KNO_3 and put it in the same quantity of water as is present in the mixed solution, it would dissolve. Thus, when we mix $K_2CrO_4(aq)$ and $Ba(NO_3)_2(aq)$, $BaCrO_4(s)$ forms, but KNO_3 is left behind in solution (we write it as $KNO_3(aq)$). Thus the overall equation for this precipitation reaction using the formulas of the reactants and products is

$$K_2CrO_4(aq) + Ba(NO_3)_2(aq) \longrightarrow BaCrO_4(s) + 2KNO_3(aq)$$

As long as water is present, the KNO_3 remains dissolved as separated ions. (See Fig. 4.15 to help visualize what is happening in this reaction. Note the solid $BaCrO_4$ on the bottom of the container, while the K^+ and NO_3^- ions remain dispersed in the solution.) If we removed the solid $BaCrO_4$ and then evaporated the water, white solid KNO_3

Figure 4.16 | Precipitation of silver chloride by mixing solutions of silver nitrate and potassium chloride. The K^+ and NO_3^- ions remain in solution.

would be obtained; the K^+ and NO_3^- ions would assemble themselves into solid KNO_3 when the water is removed.

Now let's consider another example. When an aqueous solution of silver nitrate is added to an aqueous solution of potassium chloride, a white precipitate forms (Fig. 4.16). We can represent what we know so far as

$$AgNO_3(aq) + KCl(aq) \longrightarrow \text{unknown white solid}$$

Remembering that when ionic substances dissolve in water, the ions separate, we can write

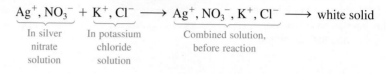

Since we know the white solid must contain both positive and negative ions, the possible compounds that can be assembled from this collection of ions are

$$AgNO_3 \qquad KCl \qquad AgCl \qquad KNO_3$$

Since $AgNO_3$ and KCl are the substances dissolved in the two reactant solutions, we know that they do not represent the white solid product. Therefore, the only real possibilities are

$$AgCl \quad \text{and} \quad KNO_3$$

From the first example considered, we know that KNO_3 is quite soluble in water. Thus solid KNO_3 will not form when the reactant solids are mixed. The product must be $AgCl(s)$ (which can be proved by experiment to be true). The overall equation for the reaction now can be written

$$AgNO_3(aq) + KCl(aq) \longrightarrow AgCl(s) + KNO_3(aq)$$

Figure 4.17 shows the result of mixing aqueous solutions of $AgNO_3$ and KCl, including a microscopic visualization of the reaction.

Notice that in these two examples we had to apply both concepts (solids must have a zero net charge) and facts (KNO_3 is very soluble in water, CrO_4^{2-} is yellow, and so on). Doing chemistry requires both understanding ideas and remembering key information. Predicting the identity of the solid product in a precipitation reaction requires knowledge of the solubilities of common ionic substances. As an aid in predicting the products of precipitation reactions, some simple solubility rules are given in Table 4.1. You should memorize these rules.

Table 4.1 | Simple Rules for the Solubility of Salts in Water

1. Most nitrate (NO_3^-) salts are soluble.
2. Most salts containing the alkali metal ions (Li^+, Na^+, K^+, Cs^+, Rb^+) and the ammonium ion (NH_4^+) are soluble.
3. Most chloride, bromide, and iodide salts are soluble. Notable exceptions are salts containing the ions Ag^+, Pb^{2+}, and Hg_2^{2+}.
4. Most sulfate salts are soluble. Notable exceptions are $BaSO_4$, $PbSO_4$, Hg_2SO_4, and $CaSO_4$.
5. Most hydroxides are only slightly soluble. The important soluble hydroxides are $NaOH$ and KOH. The compounds $Ba(OH)_2$, $Sr(OH)_2$, and $Ca(OH)_2$ are marginally soluble.
6. Most sulfide (S^{2-}), carbonate (CO_3^{2-}), chromate (CrO_4^{2-}), and phosphate (PO_4^{3-}) salts are only slightly soluble, except for those containing the cations in Rule 2.

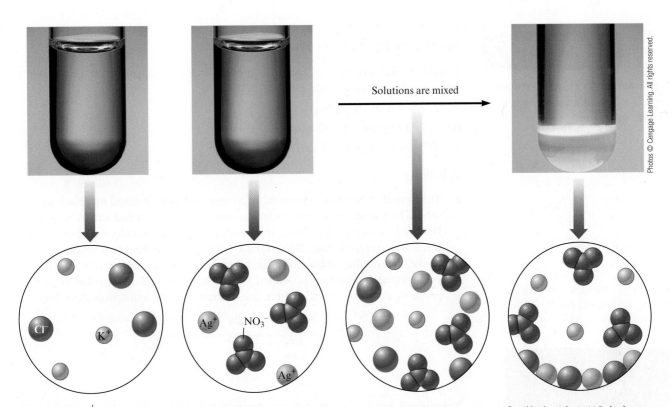

Solutions are mixed

Cl⁻ K⁺

Ag⁺ NO₃⁻ Ag⁺

Figure 4.17 │ Photos and accompanying molecular-level representations illustrating the reaction of KCl(*aq*) with AgNO₃(*aq*) to form AgCl(*s*). Note that it is not possible to have a photo of the mixed solution before the reaction occurs, because it is an imaginary step that we use to help visualize the reaction. Actually, the reaction occurs immediately when the two solutions are mixed.

The phrase *slightly soluble* used in the solubility rules in Table 4.1 means that the tiny amount of solid that dissolves is not noticeable. The solid appears to be insoluble to the naked eye. Thus the terms *insoluble* and *slightly soluble* are often used interchangeably.

Note that the information in Table 4.1 allows us to predict that AgCl is the white solid formed when solutions of AgNO₃ and KCl are mixed. Rules 1 and 2 indicate that KNO₃ is soluble, and Rule 3 states that AgCl is insoluble.

When solutions containing ionic substances are mixed, it will be helpful in determining the products if you think in terms of *ion interchange*. For example, in the preceding discussion we considered the results of mixing AgNO₃(*aq*) and KCl(*aq*). In determining the products, we took the cation from one reactant and combined it with the anion of the other reactant:

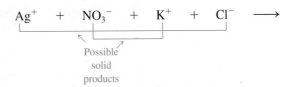

$$Ag^+ \;+\; NO_3^- \;+\; K^+ \;+\; Cl^- \longrightarrow$$

Possible
solid
products

The solubility rules in Table 4.1 allow us to predict whether either product forms as a solid.

The key to dealing with the chemistry of an aqueous solution is first to *focus on the actual components of the solution before any reaction occurs* and then to figure out how these components will react with each other. Example 4.8 illustrates this process for three different reactions.

To begin, focus on the ions in solution before any reaction occurs.

Interactive Example 4.8

Sign in at http://login.cengagebrain
.com to try this Interactive Example
in **OWL**.

Predicting Reaction Products

Using the solubility rules in Table 4.1, predict what will happen when the following pairs of solutions are mixed.

a. $KNO_3(aq)$ and $BaCl_2(aq)$

b. $Na_2SO_4(aq)$ and $Pb(NO_3)_2(aq)$

c. $KOH(aq)$ and $Fe(NO_3)_3(aq)$

Solution

a. The formula $KNO_3(aq)$ represents an aqueous solution obtained by dissolving solid KNO_3 in water to form a solution containing the hydrated ions $K^+(aq)$ and $NO_3^-(aq)$. Likewise, $BaCl_2(aq)$ represents a solution formed by dissolving solid $BaCl_2$ in water to produce $Ba^{2+}(aq)$ and $Cl^-(aq)$. When these two solutions are mixed, the resulting solution contains the ions K^+, NO_3^-, Ba^{2+}, and Cl^-. All ions are hydrated, but the (aq) is omitted for simplicity. To look for possible solid products, combine the cation from one reactant with the anion from the other:

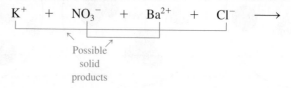

$$K^+ \ + \ NO_3^- \ + \ Ba^{2+} \ + \ Cl^- \longrightarrow$$

Possible solid products

Note from Table 4.1 that the rules predict that both KCl and $Ba(NO_3)_2$ are soluble in water. Thus no precipitate forms when $KNO_3(aq)$ and $BaCl_2(aq)$ are mixed. All the ions remain dissolved in solution. No chemical reaction occurs.

b. Using the same procedures as in part a, we find that the ions present in the combined solution before any reaction occurs are Na^+, SO_4^{2-}, Pb^{2+}, and NO_3^-. The possible salts that could form precipitates are

$$Na^+ \ + \ SO_4^{2-} \ + \ Pb^{2+} \ + \ NO_3^- \longrightarrow$$

The compound $NaNO_3$ is soluble, but $PbSO_4$ is insoluble (see Rule 4 in Table 4.1). When these solutions are mixed, $PbSO_4$ will precipitate from the solution. The balanced equation is

$$Na_2SO_4(aq) + Pb(NO_3)_2(aq) \longrightarrow PbSO_4(s) + 2NaNO_3(aq)$$

c. The combined solution (before any reaction occurs) contains the ions K^+, OH^-, Fe^{3+}, and NO_3^-. The salts that might precipitate are KNO_3 and $Fe(OH)_3$. The solubility rules in Table 4.1 indicate that both K^+ and NO_3^- salts are soluble. However, $Fe(OH)_3$ is only slightly soluble (Rule 5) and hence will precipitate. The balanced equation is

$$3KOH(aq) + Fe(NO_3)_3(aq) \longrightarrow Fe(OH)_3(s) + 3KNO_3(aq)$$

See Exercises 4.45 and 4.46

Solid $Fe(OH)_3$ forms when aqueous KOH and $Fe(NO_3)_3$ are mixed.

4.6 | Describing Reactions in Solution

In this section we will consider the types of equations used to represent reactions in solution. For example, when we mix aqueous potassium chromate with aqueous barium nitrate, a reaction occurs to form a precipitate ($BaCrO_4$) and dissolved potassium nitrate. So far we have written the overall or **formula equation** for this reaction:

$$K_2CrO_4(aq) + Ba(NO_3)_2(aq) \longrightarrow BaCrO_4(s) + 2KNO_3(aq)$$

Although the formula equation shows the reactants and products of the reaction, it does not give a correct picture of what actually occurs in solution. As we have seen, aqueous solutions of potassium chromate, barium nitrate, and potassium nitrate contain individual ions, not collections of ions, as implied by the formula equation. Thus the **complete ionic equation**

$$2K^+(aq) + CrO_4^{2-}(aq) + Ba^{2+}(aq) + 2NO_3^-(aq) \longrightarrow$$
$$BaCrO_4(s) + 2K^+(aq) + 2NO_3^-(aq)$$

better represents the actual forms of the reactants and products in solution. *In a complete ionic equation, all substances that are strong electrolytes are represented as ions.*

> A strong electrolyte is a substance that completely breaks apart into ions when dissolved in water.

The complete ionic equation reveals that only some of the ions participate in the reaction. The K^+ and NO_3^- ions are present in solution both before and after the reaction. The ions that do not participate directly in the reaction are called **spectator ions**. The ions that participate in this reaction are the Ba^{2+} and CrO_4^{2-} ions, which combine to form solid $BaCrO_4$:

$$Ba^{2+}(aq) + CrO_4^{2-}(aq) \longrightarrow BaCrO_4(s)$$

> Net ionic equations include only those components that undergo changes in the reaction.

This equation, called the **net ionic equation**, includes only those solution components directly involved in the reaction. Chemists usually write the net ionic equation for a reaction in solution because it gives the actual forms of the reactants and products and includes only the species that undergo a change.

Three Types of Equations Are Used to Describe Reactions in Solution

> The **formula equation** gives the overall reaction stoichiometry but not necessarily the actual forms of the reactants and products in solution.

> The **complete ionic equation** represents as ions all reactants and products that are strong electrolytes.

> The **net ionic equation** includes only those solution components undergoing a change. Spectator ions are not included.

Interactive Example 4.9

Sign in at http://login.cengagebrain .com to try this Interactive Example in **OWL**.

Writing Equations for Reactions

For each of the following reactions, write the formula equation, the complete ionic equation, and the net ionic equation.

a. Aqueous potassium chloride is added to aqueous silver nitrate to form a silver chloride precipitate plus aqueous potassium nitrate.

b. Aqueous potassium hydroxide is mixed with aqueous iron(III) nitrate to form a precipitate of iron(III) hydroxide and aqueous potassium nitrate.

Solution

a. Formula Equation

$$KCl(aq) + AgNO_3(aq) \longrightarrow AgCl(s) + KNO_3(aq)$$

Complete Ionic Equation
(*Remember:* Any ionic compound dissolved in water will be present as the separated ions.)

$$K^+(aq) + Cl^-(aq) + Ag^+(aq) + NO_3^-(aq) \longrightarrow AgCl(s) + K^+(aq) + NO_3^-(aq)$$

Spectator ion   Spectator ion   Solid, not written as separate ions   Spectator ion   Spectator ion

Canceling the spectator ions

$$K^+(aq) + Cl^-(aq) + Ag^+(aq) + \cancel{NO_3^-}(aq) \longrightarrow AgCl(s) + \cancel{K^+}(aq) + \cancel{NO_3^-}(aq)$$

gives the following net ionic equation.

Net Ionic Equation

$$Cl^-(aq) + Ag^+(aq) \longrightarrow AgCl(s)$$

b. **Formula Equation**

$$3KOH(aq) + Fe(NO_3)_3(aq) \longrightarrow Fe(OH)_3(s) + 3KNO_3(aq)$$

Complete Ionic Equation

$$3K^+(aq) + 3OH^-(aq) + Fe^{3+}(aq) + 3NO_3^-(aq) \longrightarrow$$
$$Fe(OH)_3(s) + 3K^+(aq) + 3NO_3^-(aq)$$

Net Ionic Equation

$$3OH^-(aq) + Fe^{3+}(aq) \longrightarrow Fe(OH)_3(s)$$

See Exercises 4.47 through 4.52

4.7 | Stoichiometry of Precipitation Reactions

In Chapter 3 we covered the principles of chemical stoichiometry: the procedures for calculating quantities of reactants and products involved in a chemical reaction. Recall that in performing these calculations we first convert all quantities to moles and then use the coefficients of the balanced equation to assemble the appropriate mole ratios. In cases where reactants are mixed, we must determine which reactant is limiting, since the reactant that is consumed first will limit the amounts of products formed. *These same principles apply to reactions that take place in solutions.* However, two points about solution reactions need special emphasis. The first is that it is sometimes difficult to tell immediately what reaction will occur when two solutions are mixed. Usually we must do some thinking about the various possibilities and then decide what probably will happen. The first step in this process *always* should be to write down the species that are actually present in the solution, as we did in Section 4.5. The second special point about solution reactions is that to obtain the moles of reactants we must use the volume of the solution and its molarity. This procedure was covered in Section 4.3.

We will introduce stoichiometric calculations for reactions in solution in Example 4.10.

Critical Thinking

What if all ionic solids were soluble in water? How would this affect stoichiometry calculations for reactions in aqueous solution?

Sign in at http://login.cengagebrain.com to try this Interactive Example in **OWL**.

Interactive Example 4.10

Determining the Mass of Product Formed I

Calculate the mass of solid NaCl that must be added to 1.50 L of a 0.100-*M* $AgNO_3$ solution to precipitate all the Ag^+ ions in the form of AgCl.

Solution

Where are we going?

To find the mass of solid NaCl required to precipitate the Ag^+

What do we know?

> 1.50 L of 0.100 *M* $AgNO_3$

What information do we need to find the mass of NaCl?

> Moles of Ag^+ in the solution

How do we get there?

What are the ions present in the combined solution?

$$Ag^+ \qquad NO_3^- \qquad Na^+ \qquad Cl^-$$

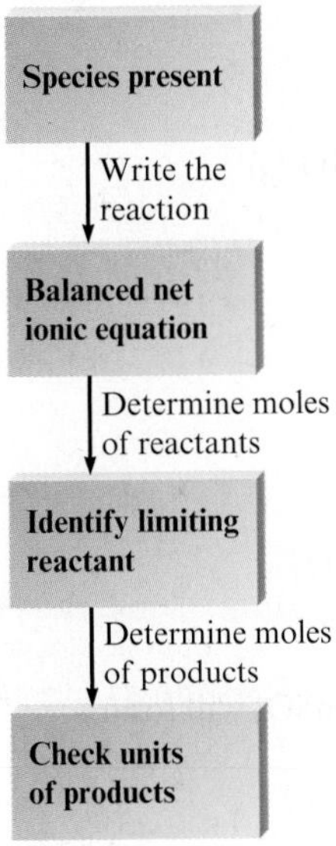

What is the balanced net ionic equation for the reaction?
Note from Table 4.1 that $NaNO_3$ is soluble and that AgCl is insoluble. Therefore, solid AgCl forms according to the following net ionic equation:

$$Ag^+(aq) + Cl^-(aq) \longrightarrow AgCl(s)$$

What are the moles of Ag^+ ions present in the solution?

$$1.50 \, \cancel{L} \times \frac{0.100 \, \text{mol } Ag^+}{\cancel{L}} = 0.150 \, \text{mol } Ag^+$$

How many moles of Cl^- are required to react with all the Ag^+?
Because Ag^+ and Cl^- react in a 1:1 ratio, 0.150 mole of Cl^- and thus 0.150 mole of NaCl are required.

What mass of NaCl is required?

$$\blacksquare \quad 0.150 \, \cancel{\text{mol NaCl}} \times \frac{58.44 \, \text{g NaCl}}{\cancel{\text{mol NaCl}}} = 8.77 \, \text{g NaCl}$$

See Exercise 4.55

[Sidebar flowchart:]

Species present

↓ Write the reaction

Balanced net ionic equation

↓ Determine moles of reactants

Identify limiting reactant

↓ Determine moles of products

Check units of products

Notice from Example 4.10 that the procedures for doing stoichiometric calculations for solution reactions are very similar to those for other types of reactions. It is useful to think in terms of the following steps for reactions in solution.

Problem-Solving Strategy

Solving Stoichiometry Problems for Reactions in Solution

1. Identify the species present in the combined solution, and determine what reaction occurs.
2. Write the balanced net ionic equation for the reaction.
3. Calculate the moles of reactants.
4. Determine which reactant is limiting.
5. Calculate the moles of product or products, as required.
6. Convert to grams or other units, as required.

Determining the Mass of Product Formed II

When aqueous solutions of Na_2SO_4 and $Pb(NO_3)_2$ are mixed, $PbSO_4$ precipitates. Calculate the mass of $PbSO_4$ formed when 1.25 L of 0.0500 M $Pb(NO_3)_2$ and 2.00 L of 0.0250 M Na_2SO_4 are mixed.

Solution

Where are we going?

To find the mass of solid $PbSO_4$ formed

What do we know?

> 1.25 L of 0.0500 M $Pb(NO_3)_2$
> 2.00 L of 0.0250 M Na_2SO_4
> Chemical reaction

$$Pb^{2+}(aq) + SO_4{}^{2-}(aq) \longrightarrow PbSO_4(s)$$

What information do we need?

> The limiting reactant

How do we get there?

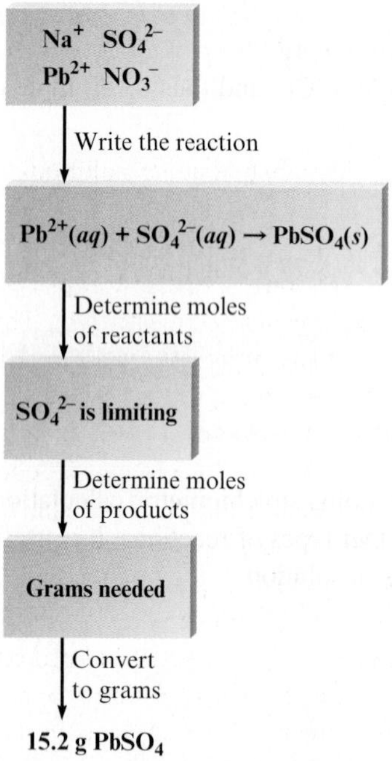

1. *What are the ions present in the combined solution?*

$$Na^+ \qquad SO_4{}^{2-} \qquad Pb^{2+} \qquad NO_3^-$$

What is the reaction?
Since $NaNO_3$ is soluble and $PbSO_4$ is insoluble, solid $PbSO_4$ will form.

2. *What is the balanced net ionic equation for the reaction?*

$$Pb^{2+}(aq) + SO_4{}^{2-}(aq) \longrightarrow PbSO_4(s)$$

3. *What are the moles of reactants present in the solution?*

$$1.25 \ L \times \frac{0.0500 \ \text{mol} \ Pb^{2+}}{L} = 0.0625 \ \text{mol} \ Pb^{2+}$$

$$2.00 \ L \times \frac{0.0250 \ \text{mol} \ SO_4{}^{2-}}{L} = 0.0500 \ \text{mol} \ SO_4{}^{2-}$$

4. *Which reactant is limiting?*
Because Pb^{2+} and $SO_4{}^{2-}$ react in a 1:1 ratio, the amount of $SO_4{}^{2-}$ will be limiting (0.0500 mol $SO_4{}^{2-}$ is less than 0.0625 mole of Pb^{2+}).

5. *What number of moles of $PbSO_4$ will be formed?*
Since $SO_4{}^{2-}$ is limiting, only 0.0500 mole of solid $PbSO_4$ will be formed.

6. *What mass of $PbSO_4$ will be formed?*

$$\blacksquare \ 0.0500 \ \text{mol} \ PbSO_4 \times \frac{303.3 \ g \ PbSO_4}{1 \ \text{mol} \ PbSO_4} = 15.2 \ g \ PbSO_4$$

See Exercises 4.57 and 4.58

Na⁺ SO₄²⁻
Pb²⁺ NO₃⁻

↓ Write the reaction

$Pb^{2+}(aq) + SO_4{}^{2-}(aq) \rightarrow PbSO_4(s)$

↓ Determine moles of reactants

$SO_4{}^{2-}$ is limiting

↓ Determine moles of products

Grams needed

↓ Convert to grams

15.2 g PbSO₄

4.8 | Acid–Base Reactions

Earlier in this chapter we considered Arrhenius's concept of acids and bases: An acid is a substance that produces H^+ ions when dissolved in water, and a base is a substance that produces OH^- ions. Although these ideas are fundamentally correct, it is convenient to have a more general definition of a base, which includes substances that do not contain OH^- ions. Such a definition was provided by Johannes N. Brønsted (1879–1947) and Thomas M. Lowry (1874–1936), who defined acids and bases as follows:

An **acid** is a proton donor.

A **base** is a proton acceptor.

The Brønsted–Lowry concept of acids and bases will be discussed in detail in Chapter 14.

How do we know when to expect an acid–base reaction? One of the most difficult tasks for someone inexperienced in chemistry is to predict what reaction might occur when two solutions are mixed. With precipitation reactions, we found that the best way to deal with this problem is to focus on the species actually present in the mixed solution. This idea also applies to acid–base reactions. For example, when an aqueous solution of hydrogen chloride (HCl) is mixed with an aqueous solution of sodium hydroxide (NaOH), the combined solution contains the ions H^+, Cl^-, Na^+, and OH^-. The separated ions are present because HCl is a strong acid and NaOH is a strong base. How can we predict what reaction occurs, if any? First, will NaCl precipitate? From Table 4.1 we can see that NaCl is soluble in water and thus will not precipitate. Therefore, the Na^+ and Cl^- ions are spectator ions. On the other hand, because water is a nonelectrolyte, large quantities of H^+ and OH^- ions cannot coexist in solution. They react to form H_2O molecules:

$$H^+(aq) + OH^-(aq) \longrightarrow H_2O(l)$$

This is the net ionic equation for the reaction that occurs when aqueous solutions of HCl and NaOH are mixed.

Next, consider mixing an aqueous solution of acetic acid ($HC_2H_3O_2$) with an aqueous solution of potassium hydroxide (KOH). In our earlier discussion of conductivity we said that an aqueous solution of acetic acid is a weak electrolyte. This tells us that acetic acid does not dissociate into ions to any great extent. In fact, in 0.1 M $HC_2H_3O_2$ approximately 99% of the $HC_2H_3O_2$ molecules remain undissociated. However, when solid KOH is dissolved in water, it dissociates completely to produce K^+ and OH^- ions. Therefore, in the solution formed by mixing aqueous solutions of $HC_2H_3O_2$ and KOH, *before any reaction occurs*, the principal species are $HC_2H_3O_2$, K^+, and OH^-. What reaction will occur? A possible precipitation reaction could occur between K^+ and OH^-. However, we know that KOH is soluble, so precipitation does not occur. Another possibility is a reaction involving the hydroxide ion (a proton acceptor) and some proton donor. Is there a source of protons in the solution? The answer is yes—the $HC_2H_3O_2$ molecules. The OH^- ion has such a strong affinity for protons that it can strip them from the $HC_2H_3O_2$ molecules. The net ionic equation for this reaction is

$$OH^-(aq) + HC_2H_3O_2(aq) \longrightarrow H_2O(l) + C_2H_3O_2^-(aq)$$

This reaction illustrates a very important general principle: The hydroxide ion is such a strong base that for purposes of stoichiometric calculations it can be assumed to react completely with any weak acid that we will encounter. Of course, OH^- ions also react completely with the H^+ ions in solutions of strong acids.

We will now deal with the stoichiometry of acid–base reactions in aqueous solutions. The procedure is fundamentally the same as that used previously for precipitation reactions.

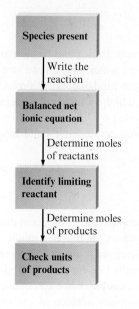

Problem-Solving Strategy

Performing Calculations for Acid–Base Reactions

1. List the species present in the combined solution *before any reaction occurs,* and decide what reaction will occur.

2. Write the balanced net ionic equation for this reaction.

3. Calculate the moles of reactants. For reactions in solution, use the volumes of the original solutions and their molarities.

4. Determine the limiting reactant where appropriate.

5. Calculate the moles of the required reactant or product.

6. Convert to grams or volume (of solution), as required.

An acid–base reaction is often called a **neutralization reaction**. When just enough base is added to react exactly with the acid in a solution, we say the acid has been *neutralized.*

Interactive Example 4.12

Sign in at http://login.cengagebrain .com to try this Interactive Example in **OWL**.

Neutralization Reactions I

What volume of a 0.100-*M* HCl solution is needed to neutralize 25.0 mL of 0.350 *M* NaOH?

Solution

Where are we going?

To find the volume of 0.100-*M* HCl required for neutralization

What do we know?

> 25 mL of 0.350 *M* NaOH

> 0.100 *M* HCl

> The chemical reaction

$$H^+(aq) + OH^-(aq) \longrightarrow H_2O(l)$$

How do we get there?

Use the Problem-Solving Strategy for Performing Calculations for Acid–Base Reactions.

1. *What are the ions present in the combined solution?*

$$H^+ \quad Cl^- \quad Na^+ \quad OH^-$$

What is the reaction?
The two possibilities are

$$Na^+(aq) + Cl^-(aq) \longrightarrow NaCl(s)$$
$$H^+(aq) + OH^-(aq) \longrightarrow H_2O(l)$$

Since we know that NaCl is soluble, the first reaction does not take place (Na^+ and Cl^- are spectator ions). However, as we have seen before, the reaction of the H^+ and OH^- ions to form H_2O does occur.

2. *What is the balanced net ionic equation for the reaction?*

$$H^+(aq) + OH^-(aq) \longrightarrow H_2O(l)$$

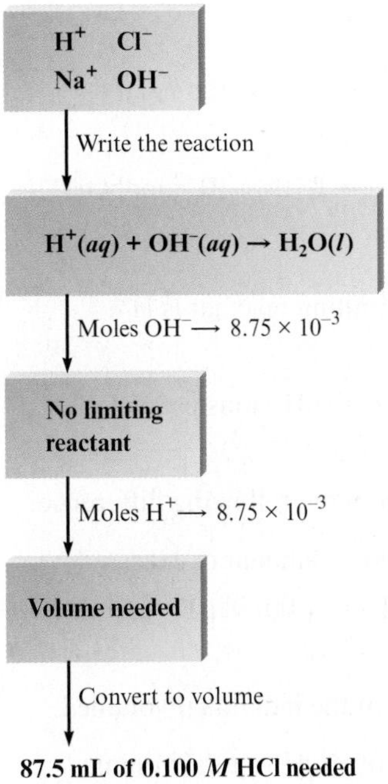

3. *What are the moles of reactant present in the solution?*

$$25.0 \text{ mL NaOH} \times \frac{1 \text{ L}}{1000 \text{ mL}} \times \frac{0.350 \text{ mol OH}^-}{\text{L NaOH}} = 8.75 \times 10^{-3} \text{ mol OH}^-$$

4. *Which reactant is limiting?*
   This problem requires the addition of just enough H^+ to react exactly with the OH^- ions present. We do not need to be concerned with limiting reactant here.

5. *What moles of H^+ are needed?*
   Since H^+ and OH^- ions react in a 1:1 ratio, 8.75×10^{-3} mole of H^+ is required to neutralize the OH^- ions present.

6. *What volume of HCl is required?*

$$V \times \frac{0.100 \text{ mol H}^+}{\text{L}} = 8.75 \times 10^{-3} \text{ mol H}^+$$

Solving for V gives

$$\blacksquare \quad V = \frac{8.75 \times 10^{-3} \text{ mol H}^+}{\dfrac{0.100 \text{ mol H}^+}{\text{L}}} = 8.75 \times 10^{-2} \text{ L}$$

See Exercises 4.69 and 4.70

Interactive Example 4.13

Sign in at http://login.cengagebrain.com to try this Interactive Example in OWL.

Neutralization Reactions II

In a certain experiment, 28.0 mL of 0.250 M HNO_3 and 53.0 mL of 0.320 M KOH are mixed. What is the concentration of H^+ or OH^- ions in excess after the reaction goes to completion?

Solution

Where are we going?
To find the concentration of H^+ or OH^- in excess after the reaction is complete

What do we know?
> 28.0 mL of 0.250 M HNO_3
> 53.0 mL of 0.320 M KOH
> The chemical reaction

$$H^+(aq) + OH^-(aq) \longrightarrow H_2O(l)$$

How do we get there?
Use the Problem-Solving Strategy for Performing Calculations for Acid–Base Reactions.

1. *What are the ions present in the combined solution?*

$$H^+ \quad NO_3^- \quad K^+ \quad OH^-$$

2. *What is the balanced net ionic equation for the reaction?*

$$H^+(aq) + OH^-(aq) \longrightarrow H_2O(l)$$

$$\boxed{\begin{array}{cc} H^+ & NO_3^- \\ K^+ & OH^- \end{array}}$$

Write the reaction

↓

$$\boxed{H^+(aq) + OH^-(aq) \rightarrow H_2O(l)}$$

Find moles H^+, OH^-

↓

$$\boxed{\begin{array}{c} \text{Limiting} \\ \text{reactant is } H^+ \end{array}}$$

Find moles OH^- that react

↓

$$\boxed{\begin{array}{c} \text{Concentration} \\ \text{of } OH^- \text{ needed} \end{array}}$$

Find excess OH^- concentration

↓

0.123 *M* OH^-

3. *What are the moles of reactant present in the solution?*

$$28.0 \text{ mL HNO}_3 \times \frac{1 \text{ L}}{1000 \text{ mL}} \times \frac{0.250 \text{ mol H}^+}{\text{L HNO}_3} = 7.00 \times 10^{-3} \text{ mol H}^+$$

$$53.0 \text{ mL KOH} \times \frac{1 \text{ L}}{1000 \text{ mL}} \times \frac{0.320 \text{ mol OH}^-}{\text{L KOH}} = 1.70 \times 10^{-2} \text{ mol OH}^-$$

4. *Which reactant is limiting?*
Since H^+ and OH^- ions react in a 1:1 ratio, the limiting reactant is H^+.

5. *What amount of OH^- will react?*
7.00×10^{-3} mole of OH^- is required to neutralize the H^+ ions present.

What amount of OH^- ions are in excess?
The amount of OH^- ions in excess is obtained from the following difference:

$$\text{Original amount} - \text{amount consumed} = \text{amount in excess}$$
$$1.70 \times 10^{-2} \text{ mol OH}^- - 7.00 \times 10^{-3} \text{ mol OH}^- = 1.00 \times 10^{-2} \text{ mol OH}^-$$

What is the volume of the combined solution?
The volume of the combined solution is the sum of the individual volumes:

$$\text{Original volume of HNO}_3 + \text{original volume of KOH} = \text{total volume}$$
$$28.0 \text{ mL} + 53.0 \text{ mL} = 81.0 \text{ mL} = 8.10 \times 10^{-2} \text{ L}$$

6. *What is the molarity of the OH^- ions in excess?*

$$\blacksquare \quad \frac{\text{mol OH}^-}{\text{L solution}} = \frac{1.00 \times 10^{-2} \text{ mol OH}^-}{8.10 \times 10^{-2} \text{ L}} = 0.123 \, M \, OH^-$$

Reality Check | This calculated molarity is less than the initial molarity, as it should be.

See Exercises 4.71 and 4.72

Acid–Base Titrations

Volumetric analysis is a technique for determining the amount of a certain substance by doing a titration. A **titration** involves delivery (from a buret) of a measured volume of a solution of known concentration (the *titrant*) into a solution containing the substance being analyzed (the *analyte*). The titrant contains a substance that reacts in a known manner with the analyte. The point in the titration where enough titrant has been added to react exactly with the analyte is called the **equivalence point** or the **stoichiometric point**. This point is often marked by an **indicator**, a substance added at the beginning of the titration that changes color at (or very near) the equivalence point. The point where the indicator *actually* changes color is called the **endpoint** of the titration. The goal is to choose an indicator such that the endpoint (where the indicator changes color) occurs exactly at the equivalence point (where just enough titrant has been added to react with all the analyte).

Ideally, the endpoint and stoichiometric point should coincide.

Requirements for a Successful Titration

❯ The exact reaction between titrant and analyte must be known (and rapid).

❯ The stoichiometric (equivalence) point must be marked accurately.

❯ The volume of titrant required to reach the stoichiometric point must be known accurately.

Figure 4.18 | The titration of an acid with a base. (a) The titrant (the base) is in the buret, and the flask contains the acid solution along with a small amount of indicator. (b) As base is added drop by drop to the acid solution in the flask during the titration, the indicator changes color, but the color disappears on mixing. (c) The stoichiometric (equivalence) point is marked by a permanent indicator color change. The volume of base added is the difference between the final and initial buret readings.

When the analyte is a base or an acid, the required titrant is a strong acid or strong base, respectively. This procedure is called an *acid–base titration.* An indicator very commonly used for acid–base titrations is phenolphthalein, which is colorless in an acidic solution and pink in a basic solution. Thus, when an acid is titrated with a base, the phenolphthalein remains colorless until after the acid is consumed and the first drop of excess base is added. In this case, the endpoint (the solution changes from colorless to pink) occurs approximately one drop of base beyond the stoichiometric point. This type of titration is illustrated in Fig. 4.18.

We will deal with the acid–base titrations only briefly here but will return to the topic of titrations and indicators in more detail in Chapter 15. The titration of an acid with a standard solution containing hydroxide ions is described in Example 4.15. In Example 4.14 we show how to determine accurately the concentration of a sodium hydroxide solution. This procedure is called *standardizing the solution.*

Interactive Example 4.14

Sign in at http://login.cengagebrain .com to try this Interactive Example in **OWL**.

Neutralization Titration

A student carries out an experiment to standardize (determine the exact concentration of) a sodium hydroxide solution. To do this, the student weighs out a 1.3009-g sample of potassium hydrogen phthalate ($KHC_8H_4O_4$, often abbreviated KHP). KHP (molar mass 204.22 g/mol) has one acidic hydrogen. The student dissolves the KHP in distilled water, adds phenolphthalein as an indicator, and titrates the resulting solution with the sodium hydroxide solution to the phenolphthalein endpoint. The difference between the final and initial buret readings indicates that 41.20 mL of the sodium hydroxide solution is required to react exactly with the 1.3009 g KHP. Calculate the concentration of the sodium hydroxide solution.

Solution

Where are we going?

To find the concentration of NaOH solution

What do we know?

> 1.3009 g $KHC_8H_4O_4$ (KHP), molar mass (204.22 g/mol)
> 41.20 mL NaOH solution to neutralize KHP
> The chemical reaction

$$HC_8H_4O_4{}^-(aq) + OH^-(aq) \longrightarrow H_2O(l) + C_8H_4O_4{}^{2-}(aq)$$

How do we get there?

Use the Problem-Solving Strategy for Performing Calculations for Acid–Base Reactions.

1. *What are the ions present in the combined solution?*

$$K^+ \qquad HC_8H_4O_4{}^- \qquad Na^+ \qquad OH^-$$

2. *What is the balanced net ionic equation for the reaction?*

$$HC_8H_4O_4{}^-(aq) + OH^-(aq) \longrightarrow H_2O(l) + C_8H_4O_4{}^{2-}(aq)$$

$HC_8H_4O_4{}^-$
Hydrogen phthalate ion

3. *What are the moles of KHP?*

$$1.3009 \text{ g } \cancel{KHC_8H_4O_4} \times \frac{1 \text{ mol } KHC_8H_4O_4}{204.22 \text{ g } \cancel{KHC_8H_4O_4}} = 6.3701 \times 10^{-3} \text{ mol } KHC_8H_4O_4$$

4. *Which reactant is limiting?*

This problem requires the addition of just enough OH^- ions to react exactly with the KHP present. We do not need to be concerned with limiting reactant here.

5. *What moles of OH^- are required?*
6.3701×10^{-3} mole of OH^- is required to neutralize the KHP present.

6. *What is the molarity of the NaOH solution?*

$$\blacksquare \text{ Molarity of NaOH} = \frac{\text{mol NaOH}}{\text{L solution}} = \frac{6.3701 \times 10^{-3} \text{ mol NaOH}}{4.120 \times 10^{-2} \text{ L}}$$
$$= 0.1546 \text{ } M$$

This standard sodium hydroxide solution can now be used in other experiments (see Example 4.15).

See Exercises 4.73 and 4.78

Critical Thinking

In Example 4.14 you determined the concentration of an aqueous solution of NaOH using phenolphthalein as an indicator. What if you used an indicator for which the endpoint of the titration occurs after the equivalence point? How would this affect your calculated concentration of NaOH?

Neutralization Analysis

An environmental chemist analyzed the effluent (the released waste material) from an industrial process known to produce the compounds carbon tetrachloride (CCl_4) and benzoic acid ($HC_7H_5O_2$), a weak acid that has one acidic hydrogen atom per molecule. A sample of this effluent weighing 0.3518 g was shaken with water, and the resulting aqueous solution required 10.59 mL of 0.1546 M NaOH for neutralization. Calculate the mass percent of $HC_7H_5O_2$ in the original sample.

Solution

Where are we going?

To find the mass percent of $HC_7H_5O_2$ in the original sample

What do we know?

> 0.3518 g effluent (original sample)

> 10.59 mL 0.1546 M NaOH for neutralization of $HC_7H_5O_2$

> The chemical reaction

$$HC_7H_5O_2(aq) + OH^-(aq) \longrightarrow H_2O(l) + C_7H_5O_2^-(aq)$$

How do we get there?

Use the Problem-Solving Strategy for Performing Calculations for Acid–Base Reactions.

1. *What are the species present in the combined solution?*

$$HC_7H_5O_2 \qquad Na^+ \qquad OH^-$$

2. *What is the balanced net ionic equation for the reaction?*

$$HC_7H_5O_2(aq) + OH^-(aq) \longrightarrow H_2O(l) + C_7H_5O_2^-(aq)$$

3. *What are the moles of OH^- required?*

$$10.59 \text{ mL NaOH} \times \frac{1 \text{ L}}{1000 \text{ mL}} \times \frac{0.1546 \text{ mol } OH^-}{\text{L NaOH}} = 1.637 \times 10^{-3} \text{ mol } OH^-$$

4. *Which reactant is limiting?*

This problem requires the addition of just enough OH^- ions to react exactly with the $HC_7H_5O_2$ present. We do not need to be concerned with limiting reactant here.

5. *What mass of $HC_7H_5O_2$ is present?*

$$1.637 \times 10^{-3} \text{ mol } HC_7H_5O_2 \times \frac{122.12 \text{ g } HC_7H_5O_2}{1 \text{ mol } HC_7H_5O_2} = 0.1999 \text{ g } HC_7H_5O_2$$

6. *What is the mass percent of the $HC_7H_5O_2$ in the effluent?*

$$\frac{0.1999 \text{ g}}{0.3518 \text{ g}} \times 100\% = 56.82\%$$

Reality Check | The calculated percent of $HC_7H_5O_2$ is less than 100%, as it should be.

See Exercise 4.77

The first step in the analysis of a complex solution is to write down the components and focus on the chemistry of each one. When a strong electrolyte is present, write it as separated ions.

In doing problems involving titrations, you must first decide what reaction is occurring. Sometimes this seems difficult because the titration solution contains several components. The key to success in doing solution reactions is to first write down all the components in the solution and focus on the chemistry of each one. We have been emphasizing this approach in dealing with the reactions between ions in solution. Make it a habit to write down the components of solutions before trying to decide what reaction(s) might take place as you attempt the end-of-chapter problems involving titrations.

4.9 | Oxidation–Reduction Reactions

We have seen that many important substances are ionic. Sodium chloride, for example, can be formed by the reaction of elemental sodium and chlorine:

$$2Na(s) + Cl_2(g) \longrightarrow 2NaCl(s)$$

In this reaction, solid sodium, which contains neutral sodium atoms, reacts with chlorine gas, which contains diatomic Cl_2 molecules, to form the ionic solid NaCl, which contains Na^+ and Cl^- ions. This process is represented in Fig. 4.19. *Reactions like this one, in which one or more electrons are transferred, are called* **oxidation–reduction reactions** *or* **redox reactions**.

Many important chemical reactions involve oxidation and reduction. Photosynthesis, which stores energy from the sun in plants by converting carbon dioxide and water to sugar, is a very important oxidation–reduction reaction. In fact, most reactions used for energy production are redox reactions. In humans, the oxidation of sugars, fats, and proteins provides the energy necessary for life. Combustion reactions, which provide

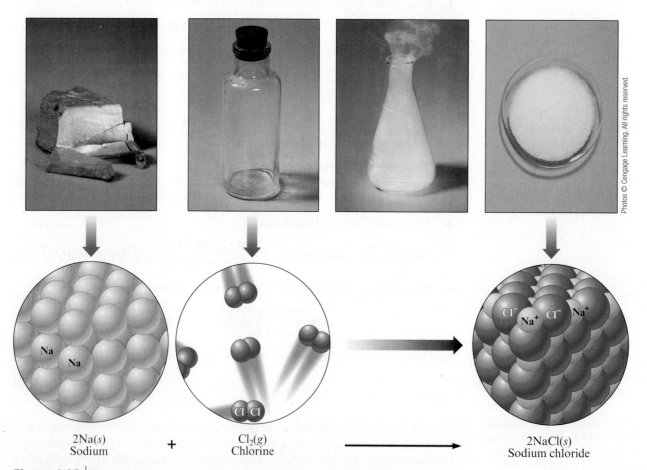

Figure 4.19 | The reaction of solid sodium and gaseous chlorine to form solid sodium chloride.

most of the energy to power our civilization, also involve oxidation and reduction. An example is the reaction of methane with oxygen:

$$CH_4(g) + 2O_2(g) \longrightarrow CO_2(g) + 2H_2O(g) + \text{energy}$$

Even though none of the reactants or products in this reaction is ionic, the reaction is still assumed to involve a transfer of electrons from carbon to oxygen. To explain this, we must introduce the concept of oxidation states.

Oxidation States

The concept of **oxidation states** (also called *oxidation numbers*) provides a way to keep track of electrons in oxidation–reduction reactions, particularly redox reactions involving covalent substances. Recall that electrons are shared by atoms in covalent bonds. The oxidation states of atoms in covalent compounds are obtained by arbitrarily assigning the electrons (which are actually shared) to particular atoms. We do this as follows: For a covalent bond between two identical atoms, the electrons are split equally between the two. In cases where two different atoms are involved (and the electrons are thus shared unequally), the shared electrons are assigned completely to the atom that has the stronger attraction for electrons. For example, recall from the discussion of the water molecule in Section 4.1 that oxygen has a greater attraction for electrons than does hydrogen. Therefore, in assigning the oxidation state of oxygen and hydrogen in H_2O, we assume that the oxygen atom actually possesses all the electrons. Recall that a hydrogen atom has one electron. Thus, in water, oxygen has formally "taken" the electrons from two hydrogen atoms. This gives the oxygen an *excess* of two electrons (its oxidation state is -2) and leaves each hydrogen with no electrons (the oxidation state of each hydrogen is thus $+1$).

We define the *oxidation states* (or *oxidation numbers*) of the atoms in a covalent compound as the imaginary charges the atoms would have if the shared electrons were divided equally between identical atoms bonded to each other or, for different atoms, were all assigned to the atom in each bond that has the greater attraction for electrons. Of course, for ionic compounds containing monatomic ions, the oxidation states of the ions are equal to the ion charges.

These considerations lead to a series of rules for assigning oxidation states that are summarized in Table 4.2. Application of these simple rules allows the assignment of oxidation states in most compounds. To apply these rules, recognize that *the sum of the*

Table 4.2 | Rules for Assigning Oxidation States

The Oxidation State of . . .	Summary	Examples
• An atom in an element is zero	Element: 0	$Na(s)$, $O_2(g)$, $O_3(g)$, $Hg(l)$
• A monatomic ion is the same as its charge	Monatomic ion: charge of ion	Na^+, Cl^-
• Fluorine is -1 in its compounds	Fluorine: -1	HF, PF_3
• Oxygen is usually -2 in its compounds Exception: peroxides (containing O_2^{2-}), in which oxygen is -1	Oxygen: -2	H_2O, CO_2
• Hydrogen is $+1$ in its covalent compounds	Hydrogen: $+1$	H_2O, HCl, NH_3

oxidation states must be zero for an electrically neutral compound. For an ion, the sum of the oxidation states must equal the charge of the ion. The principles are illustrated by Example 4.16.

It is worthwhile to note at this point that the convention is to write actual charges on ions as $n+$ or $n-$, the number being written *before* the plus or minus sign. On the other hand, oxidation states (not actual charges) are written $+n$ or $-n$, the number being written *after* the plus or minus sign.

Critical Thinking

What if the oxidation state for oxygen was defined as -1 instead of -2? What effect, if any, would it have on the oxidation state of hydrogen?

Interactive Example 4.16

Sign in at http://login.cengagebrain .com to try this Interactive Example in **OWL**.

Assigning Oxidation States

Assign oxidation states to all atoms in the following.

a. CO_2

b. SF_6

c. NO_3^-

Solution

a.   Since we have a specific rule for the oxidation state of oxygen, we will assign its value first. The oxidation state of oxygen is -2. The oxidation state of the carbon atom can be determined by recognizing that since CO_2 has no charge, the sum of the oxidation states for oxygen and carbon must be zero. Since each oxygen is -2 and there are two oxygen atoms, the carbon atom must be assigned an oxidation state of $+4$:

$$CO_2$$
$$+4 \qquad -2 \text{ for each oxygen}$$

Reality Check | We can check the assigned oxidation states by noting that when the number of atoms is taken into account, the sum is zero as required:

$$1(+4) + 2(-2) = 0$$

No. of C atoms   No. of O atoms

b.   Since we have no rule for sulfur, we first assign the oxidation state of each fluorine as -1. The sulfur must then be assigned an oxidation state of $+6$ to balance the total of -6 from the fluorine atoms:

$$SF_6$$
$$+6 \qquad -1 \text{ for each fluorine}$$

Reality Check | $+6 + 6(-1) = 0$

c.   Oxygen has an oxidation state of -2. Because the sum of the oxidation states of the three oxygens is -6 and the net charge on the NO_3^- ion is $1-$, the nitrogen must have an oxidation state of $+5$:

$$NO_3^-$$
$$+5 \qquad -2 \text{ for each oxygen}$$

Reality Check | $+5 + 3(-2) = -1$

Note that in this case the sum must be -1 (the overall charge on the ion).

See Exercises 4.79 through 4.82

Magnetite is a magnetic ore containing Fe_3O_4. Note that the compass needle points toward the ore.

We need to make one more point about oxidation states, and this can be illustrated by the compound Fe_3O_4, which is the main component in magnetite, an iron ore that accounts for the reddish color of many types of rocks and soils. To determine the oxidation states in Fe_3O_4, we first assign each oxygen atom its usual oxidation state of -2. The three iron atoms must yield a total of $+8$ to balance the total of -8 from the four oxygens. This means that each iron atom has an oxidation state of $+\frac{8}{3}$. A noninteger value for the oxidation state may seem strange because charge is expressed in whole numbers. However, although they are rare, noninteger oxidation states do occur because of the rather arbitrary way that electrons are divided up by the rules in Table 4.2. For Fe_3O_4, for example, the rules assume that all the iron atoms are equal, when in fact this compound can best be viewed as containing four O^{2-} ions, two Fe^{3+} ions, and one Fe^{2+} ion per formula unit. (Note that the "average" charge on iron works out to be $\frac{8}{3}+$, which is equal to the oxidation state we determined above.) Noninteger oxidation states should not intimidate you. They are used in the same way as integer oxidation states—for keeping track of electrons.

The Characteristics of Oxidation–Reduction Reactions

Oxidation–reduction reactions are characterized by a transfer of electrons. In some cases, the transfer occurs in a literal sense to form ions, such as in the reaction

$$2Na(s) + Cl_2(g) \longrightarrow 2NaCl(s)$$

However, sometimes the transfer is less obvious. For example, consider the combustion of methane (the oxidation state for each atom is given):

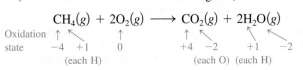

Note that the oxidation state for oxygen in O_2 is 0 because it is in elemental form. In this reaction there are no ionic compounds, but we can still describe the process in terms of a transfer of electrons. Note that carbon undergoes a change in oxidation state from -4 in CH_4 to $+4$ in CO_2. Such a change can be accounted for by a loss of eight electrons (the symbol e^- stands for an electron):

$$CH_4 \longrightarrow CO_2 + 8e^-$$
$$\quad -4 \qquad +4$$

On the other hand, each oxygen changes from an oxidation state of 0 in O_2 to -2 in H_2O and CO_2, signifying a gain of two electrons per atom. Since four oxygen atoms are involved, this is a gain of eight electrons:

$$2O_2 + 8e^- \longrightarrow CO_2 + 2H_2O$$
$$\quad 0 \qquad\qquad 4(-2) = -8$$

No change occurs in the oxidation state of hydrogen, and it is not formally involved in the electron-transfer process.

A helpful mnemonic device is OIL RIG (Oxidation Involves Loss; Reduction Involves Gain). Another common mnemonic is LEO says GER. (Loss of Electrons, Oxidation; Gain of Electrons, Reduction).

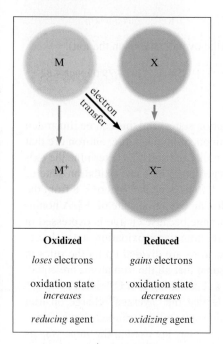

Oxidized	Reduced
loses electrons	*gains* electrons
oxidation state *increases*	oxidation state *decreases*
reducing agent	*oxidizing* agent

Figure 4.20 | A summary of an oxidation–reduction process, in which M is oxidized and X is reduced.

With this background, we can now define some important terms. **Oxidation** is an *increase* in oxidation state (a loss of electrons). **Reduction** is a *decrease* in oxidation state (a gain of electrons). Thus in the reaction

$$2Na(s) + Cl_2(g) \longrightarrow 2NaCl(s)$$
$$0 0 +1 -1$$

sodium is oxidized and chlorine is reduced. In addition, Cl_2 is called the **oxidizing agent (electron acceptor)**, and Na is called the **reducing agent (electron donor)**. These terms are summarized in Fig. 4.20.

Concerning the reaction

$$CH_4(g) + 2O_2(g) \longrightarrow CO_2(g) + 2H_2O(g)$$
$$-4 +1 0 +4 -2 +1 -2$$

we can say the following:

Methane is oxidized because there has been an increase in carbon's oxidation state (the carbon atom has formally lost electrons).

Oxygen is reduced because there has been a decrease in its oxidation state (oxygen has formally gained electrons).

CH_4 is the reducing agent.

O_2 is the oxidizing agent.

Note that when the oxidizing or reducing agent is named, the *whole compound* is specified, not just the element that undergoes the change in oxidation state.

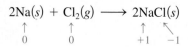

Oxidation–Reduction Reactions

Metallurgy, the process of producing a metal from its ore, always involves oxidation–reduction reactions. In the metallurgy of galena (PbS), the principal lead-containing ore, the first step is the conversion of lead sulfide to its oxide (a process called *roasting*):

$$2PbS(s) + 3O_2(g) \longrightarrow 2PbO(s) + 2SO_2(g)$$

The oxide is then treated with carbon monoxide to produce the free metal:

$$PbO(s) + CO(g) \longrightarrow Pb(s) + CO_2(g)$$

For each reaction, identify the atoms that are oxidized and reduced, and specify the oxidizing and reducing agents.

Solution

For the first reaction, we can assign the following oxidation states:

$$2PbS(s) + 3O_2(g) \longrightarrow 2PbO(s) + 2SO_2(g)$$
$$+2 -2 0 +2 -2 +4 -2 \text{ (each O)}$$

The oxidation state for the sulfur atom increases from -2 to $+4$. Thus sulfur is oxidized. The oxidation state for each oxygen atom decreases from 0 to -2. Oxygen is reduced. The oxidizing agent (that accepts the electrons) is O_2, and the reducing agent (that donates electrons) is PbS.

For the second reaction we have

$$PbO(s) + CO(g) \longrightarrow Pb(s) + CO_2(g)$$
$$+2 -2 +2 -2 0 +4 -2 \text{ (each O)}$$

Oxidation is an increase in oxidation state. *Reduction* is a decrease in oxidation state.

An oxidizing agent is reduced and a reducing agent is oxidized in a redox reaction.

Lead is reduced (its oxidation state decreases from $+2$ to 0), and carbon is oxidized (its oxidation state increases from $+2$ to $+4$). PbO is the oxidizing agent, and CO is the reducing agent.

See Exercises 4.83 and 4.84

Critical Thinking

Dalton believed that atoms were indivisible. Thomson and Rutherford helped to show that this was not true. What if atoms were indivisible? How would this affect the types of reactions you have learned about in this chapter?

4.10 | Balancing Oxidation–Reduction Equations

It is important to be able to balance oxidation–reduction reactions. One method involves the use of oxidation states (discussed in this section), and the other method (normally used for more complex reactions) involves separating the reaction into two half-reactions. We'll discuss the second method for balancing oxidation–reduction reactions in Chapter 18.

Oxidation States Method of Balancing Oxidation–Reduction Reactions

Consider the reaction between solid copper and silver ions in aqueous solution:

$$Cu(s) + Ag^+(aq) \longrightarrow Ag(s) + Cu^{2+}(aq)$$

We can tell this is a redox reaction by assigning oxidation states as follows:

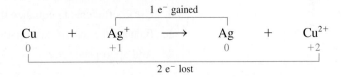

We know that in an oxidation–reduction reaction we must ultimately have equal numbers of electrons gained and lost, and we can use this principle to balance redox equations. For example, in this case, 2 Ag^+ ions must be reduced for every Cu atom oxidized:

$$Cu(s) + 2Ag^+(aq) \longrightarrow 2Ag(s) + Cu^{2+}(aq)$$

This gives us the balanced equation.

Now consider a more complex reaction:

$$H^+(aq) + Cl^-(aq) + Sn(s) + NO_3^-(aq) \longrightarrow SnCl_6^{2-}(aq) + NO_2(g) + H_2O(l)$$

To balance this equation by oxidation states, we first need to assign the oxidation states to all the atoms in the reactants and products.

$$\underset{+1}{H^+} + \underset{-1}{Cl^-} + \underset{0}{Sn} + \underset{\underset{-2}{+5}}{NO_3^-} \longrightarrow \underset{\underset{-1}{+4}}{SnCl_6^{2-}} + \underset{\underset{-2}{+4}}{NO_2} + \underset{\underset{-2}{+1}}{H_2O}$$

Note that hydrogen, chlorine, and oxygen do not change oxidation states and are not involved in electron exchange. Thus we focus our attention on Sn and N:

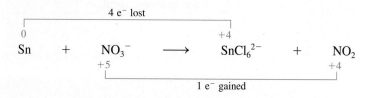

This means we need a coefficient of 4 for the N-containing species.

$$H^+ + Cl^- + Sn + 4NO_3^- \longrightarrow SnCl_6^{2-} + 4NO_2 + H_2O$$

Now we balance the rest of the equation by inspection.

Balance Cl^-:

$$H^+ + 6Cl^- + Sn + 4NO_3^- \longrightarrow SnCl_6^{2-} + 4NO_2 + H_2O$$

Balance O:

$$H^+ + 6Cl^- + Sn + 4NO_3^- \longrightarrow SnCl_6^{2-} + 4NO_2 + 4H_2O$$

Balance H:

$$8H^+ + 6Cl^- + Sn + 4NO_3^- \longrightarrow SnCl_6^{2-} + 4NO_2 + 4H_2O$$

This gives the final balanced equation. Now we will write the equation with the states included:

$$8H^+(aq) + 6Cl^-(aq) + Sn(s) + 4NO_3^-(aq) \longrightarrow SnCl_6^{2-}(aq) + 4NO_2(g) + 4H_2O(l)$$

Problem-Solving Strategy

Balancing Oxidation–Reduction Reactions by Oxidation States

1. Write the unbalanced equation.
2. Determine the oxidation states of all atoms in the reactants and products.
3. Show electrons gained and lost using "tie lines."
4. Use coefficients to equalize the electrons gained and lost.
5. Balance the rest of the equation by inspection.
6. Add appropriate states.

Example 4.18

Balancing Oxidation–Reduction Reactions

Balance the reaction between solid lead(II) oxide and ammonia gas to produce nitrogen gas, liquid water, and solid lead.

Solution

We'll use the Problem-Solving Strategy for Balancing Oxidation–Reduction Reactions by Oxidation States.

1. *What is the unbalanced equation?*

$$PbO(s) + NH_3(g) \longrightarrow N_2(g) + H_2O(l) + Pb(s)$$

2. *What are the oxidation states for each atom?*

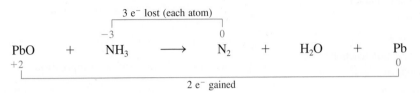

$$\overset{-2}{PbO} + \overset{-3}{NH_3} \longrightarrow \overset{0}{N_2} + \overset{-2}{H_2O} + \overset{}{Pb}$$
$$\underset{+2}{} \quad \underset{+1}{} \quad \quad \underset{+1}{} \quad \underset{0}{}$$

3. *How are electrons gained and lost?*

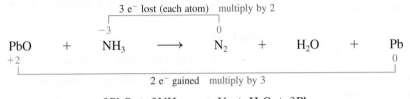

The oxidation states of all other atoms are unchanged.

4. *What coefficients are needed to equalize the electrons gained and lost?*

$$3PbO + 2NH_3 \longrightarrow N_2 + H_2O + 3Pb$$

5. *What coefficients are needed to balance the remaining elements?*
Balance O:

$$3PbO + 2NH_3 \longrightarrow N_2 + 3H_2O + 3Pb$$

All the elements are now balanced. The balanced equation with states is:

▪ $3PbO(s) + 2NH_3(g) \longrightarrow N_2(g) + 3H_2O(l) + 3Pb(s)$

See Exercises 4.87 and 4.88

For review

Key terms

aqueous solution

Section 4.1
polar molecule
hydration
solubility

Section 4.2
solute
solvent
electrical conductivity
strong electrolyte
weak electrolyte
nonelectrolyte
acid
strong acid
strong base

Chemical reactions in solution are very important in everyday life.

Water is a polar solvent that dissolves many ionic and polar substances.

Electrolytes

> Strong electrolyte: 100% dissociated to produce separate ions; strongly conducts an electric current
> Weak electrolyte: Only a small percentage of dissolved molecules produce ions; weakly conducts an electric current
> Nonelectrolyte: Dissolved substance produces no ions; does not conduct an electric current

Acids and bases

> Arrhenius model
> > Acid: produces H^+
> > Base: produces OH^-

Key terms

weak acid
weak base

Section 4.3
molarity
standard solution
dilution

Section 4.5
precipitation reaction
precipitate

Section 4.6
formula equation
complete ionic equation
spectator ions
net ionic equation

Section 4.8
acid
base
neutralization reaction
volumetric analysis
titration
stoichiometric (equivalence)
    point
indicator
endpoint

Section 4.9
oxidation–reduction (redox)
    reaction
oxidation state
oxidation
reduction
oxidizing agent (electron
    acceptor)
reducing agent (electron donor)

Acids and bases

› Brønsted–Lowry model
  › Acid: proton donor
  › Base: proton acceptor
› Strong acid: completely dissociates into separated H^+ and anions
› Weak acid: dissociates to a slight extent

Molarity

› One way to describe solution composition

$$\text{Molarity } (M) = \frac{\text{moles of solute}}{\text{volume of solution (L)}}$$

› Moles solute = volume of solution (L) × molarity
› Standard solution: molarity is accurately known

Dilution

› Solvent is added to reduce the molarity
› Moles of solute after dilution = moles of solute before dilution

$$M_1V_1 = M_2V_2$$

Types of equations that describe solution reactions

› Formula equation: All reactants and products are written as complete formulas
› Complete ionic equation: All reactants and products that are strong electrolytes are written as separated ions
› Net ionic equation: Only those compounds that undergo a change are written; spectator ions are not included

Solubility rules

› Based on experiment observation
› Help predict the outcomes of precipitation reactions

Important types of solution reactions

› Acid–base reactions: involve a transfer of H^+ ions
› Precipitation reactions: formation of a solid occurs
› Oxidation–reduction reactions: involve electron transfer

Titrations

› Measures the volume of a standard solution (titrant) needed to react with a substance in solution
› Stoichiometric (equivalence) point: the point at which the required amount of titrant has been added to exactly react with the substance being analyzed
› Endpoint: the point at which a chemical indicator changes color

Oxidation–reduction reactions

> Oxidation states are assigned using a set of rules to keep track of electron flow.

> Oxidation: increase in oxidation state (a loss of electrons)

> Reduction: decrease in oxidation state (a gain of electrons)

> Oxidizing agent: gains electrons (is reduced)

> Reducing agent: loses electrons (is oxidized)

> Equations for oxidation–reduction reactions can be balanced by the oxidation states method.

Review questions *Answers to the Review Questions can be found on the Student website (accessible from* **www.cengagebrain.com**).

1. The (*aq*) designation listed after a solute indicates the process of hydration. Using KBr(*aq*) and C_2H_5OH(*aq*) as your examples, explain the process of hydration for soluble ionic compounds and for soluble covalent compounds.

2. Characterize strong electrolytes versus weak electrolytes versus nonelectrolytes. Give examples of each. How do you experimentally determine whether a soluble substance is a strong electrolyte, weak electrolyte, or nonelectrolyte?

3. Distinguish between the terms *slightly soluble* and *weak electrolyte.*

4. Molarity is a conversion factor relating moles of solute in solution to the volume of the solution. How does one use molarity as a conversion factor to convert from moles of solute to volume of solution, and from volume of solution to moles of solute present?

5. What is a dilution? What stays constant in a dilution? Explain why the equation $M_1V_1 = M_2V_2$ works for dilution problems.

6. When the following beakers are mixed, draw a molecular-level representation of the product mixture (see Fig. 4.17).

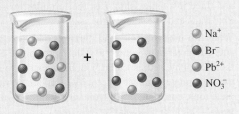

7. Differentiate between the formula equation, the complete ionic equation, and the net ionic equation. For each reaction in Question 6, write all three balanced equations.

8. What is an acid–base reaction? Strong bases are soluble ionic compounds that contain the hydroxide ion. List the strong bases. When a strong base reacts with an acid, what is always produced? Explain the terms *titration, stoichiometric point, neutralization,* and *standardization.*

9. Define the terms *oxidation, reduction, oxidizing agent,* and *reducing agent.* Given a chemical reaction, how can you tell if it is a redox reaction?

10. Consider the steps involved in balancing oxidation–reduction reactions by using oxidation states. The key to the oxidation states method is to equalize the electrons lost by the species oxidized with the electrons gained by the species reduced. First of all, how do you recognize what is oxidized and what is reduced? Second, how do you balance the electrons lost with the electrons gained? Once the electrons are balanced, what else is needed to balance the oxidation–reduction reaction?

Active Learning Questions

These questions are designed to be used by groups of students in class.

1. Assume you have a highly magnified view of a solution of HCl that allows you to "see" the HCl. Draw this magnified view. If you dropped in a piece of magnesium, the magnesium would disappear and hydrogen gas would be released. Represent this change using symbols for the elements, and write out the balanced equation.

2. You have a solution of table salt in water. What happens to the salt concentration (increases, decreases, or stays the same) as the solution boils? Draw pictures to explain your answer.

3. You have a sugar solution (solution *A*) with concentration *x*. You pour one-fourth of this solution into a beaker, and add an equivalent volume of water (solution *B*).

   a. What is the ratio of sugar in solutions *A* and *B*?

   b. Compare the volumes of solutions *A* and *B*.

   c. What is the ratio of the concentrations of sugar in solutions *A* and *B*?

4. You add an aqueous solution of lead nitrate to an aqueous solution of potassium iodide. Draw highly magnified views of each solution individually, and the mixed solution, including any product that forms. Write the balanced equation for the reaction.

5. Order the following molecules from lowest to highest oxidation state of the nitrogen atom: HNO_3, NH_4Cl, N_2O, NO_2, $NaNO_2$.

6. Why is it that when something gains electrons, it is said to be *reduced?* What is being reduced?

7. Consider separate aqueous solutions of HCl and H_2SO_4 with the same molar concentrations. You wish to neutralize an aqueous solution of NaOH. For which acid solution would you need to add more volume (in milliliters) to neutralize the base?

   a. the HCl solution

   b. the H_2SO_4 solution

   c. You need to know the acid concentrations to answer this question.

   d. You need to know the volume and concentration of the NaOH solution to answer this question.

   e. c and d

   Explain.

8. Draw molecular-level pictures to differentiate between concentrated and dilute solutions.

9. You need to make 150.0 mL of a 0.10-*M* NaCl solution. You have solid NaCl, and your lab partner has a 2.5-*M* NaCl solution. Explain how you each make the 0.10-*M* NaCl solution.

10. The exposed electrodes of a light bulb are placed in a solution of H_2SO_4 in an electrical circuit such that the light bulb is glowing. You add a dilute salt solution, and the bulb dims. Which of the following could be the salt in the solution?

   a. $Ba(NO_3)_2$ **c.** K_2SO_4

   b. $NaNO_3$ **d.** $Ca(NO_3)_2$

   Justify your choices. For those you did not choose, explain why they are incorrect.

11. You have two solutions of chemical A. To determine which has the highest concentration of A (molarity), which of the following must you know (there may be more than one answer)?

   a. the mass in grams of A in each solution

   b. the molar mass of A

   c. the volume of water added to each solution

   d. the total volume of the solution

   Explain.

12. Which of the following must be known to calculate the molarity of a salt solution (there may be more than one answer)?

   a. the mass of salt added

   b. the molar mass of the salt

   c. the volume of water added

   d. the total volume of the solution

   Explain.

A blue question or exercise number indicates that the answer to that question or exercise appears at the back of this book and a solution appears in the *Solutions Guide,* as found on PowerLecture.

Questions

13. Differentiate between what happens when the following are added to water.

   a. polar solute versus nonpolar solute

   b. KF versus $C_6H_{12}O_6$

   c. RbCl versus AgCl

   d. HNO_3 versus CO

14. A typical solution used in general chemistry laboratories is 3.0 *M* HCl. Describe, in detail, the composition of 2.0 L of a 3.0-*M* HCl solution. How would 2.0 L of a 3.0-*M* $HC_2H_3O_2$ solution differ from the same quantity of the HCl solution?

15. Which of the following statements is(are) *true?* For the false statements, correct them.

   a. A concentrated solution in water will always contain a strong or weak electrolyte.

   b. A strong electrolyte will break up into ions when dissolved in water.

   c. An acid is a strong electrolyte.

   d. All ionic compounds are strong electrolytes in water.

16. A student wants to prepare 1.00 L of a 1.00-*M* solution of NaOH (molar mass = 40.00 g/mol). If solid NaOH is available, how would the student prepare this solution? If 2.00 *M* NaOH is available, how would the student prepare the solution? To help ensure three significant figures in the NaOH molarity, to how many significant figures should the volumes and mass be determined?

17. List the formulas of three soluble bromide salts and three insoluble bromide salts. Do the same exercise for sulfate salts, hydroxide salts, and phosphate salts (list three soluble salts and three insoluble salts). List the formulas for six insoluble Pb^{2+} salts and one soluble Pb^{2+} salt.

18. When 1.0 mole of solid lead nitrate is added to 2.0 moles of aqueous potassium iodide, a yellow precipitate forms. After the precipitate settles to the bottom, does the solution above the precipitate conduct electricity? Explain. Write the complete ionic equation to help you answer this question.

19. What is an acid and what is a base? An acid–base reaction is sometimes called a proton-transfer reaction. Explain.

20. A student had 1.00 L of a 1.00-*M* acid solution. Much to the surprise of the student, it took 2.00 L of 1.00 *M* NaOH solution to react completely with the acid. Explain why it took twice as much NaOH to react with all of the acid.

   In a different experiment, a student had 10.0 mL of 0.020 *M* HCl. Again, much to the surprise of the student, it took only 5.00 mL of 0.020 *M* strong base to react completely with the HCl. Explain why it took only half as much strong base to react with all of the HCl.

21. Differentiate between the following terms.
   a. species reduced versus the reducing agent
   b. species oxidized versus the oxidizing agent
   c. oxidation state versus actual charge

22. How does one balance redox reactions by the oxidation states method?

Exercises

In this section similar exercises are paired.

Aqueous Solutions: Strong and Weak Electrolytes

23. Show how each of the following strong electrolytes "breaks up" into its component ions upon dissolving in water by drawing molecular-level pictures.
   a. NaBr   f. $FeSO_4$
   b. $MgCl_2$ g. $KMnO_4$
   c. $Al(NO_3)_3$ h. $HClO_4$
   d. $(NH_4)_2SO_4$ i. $NH_4C_2H_3O_2$ (ammonium acetate)
   e. NaOH

24. Match each name below with the following microscopic pictures of that compound in aqueous solution.

i.   ii.   iii.   iv.

   a. barium nitrate   c. potassium carbonate
   b. sodium chloride   d. magnesium sulfate

   Which picture best represents $HNO_3(aq)$? Why aren't any of the pictures a good representation of $HC_2H_3O_2(aq)$?

25. Calcium chloride is a strong electrolyte and is used to "salt" streets in the winter to melt ice and snow. Write a reaction to show how this substance breaks apart when it dissolves in water.

26. Commercial cold packs and hot packs are available for treating athletic injuries. Both types contain a pouch of water and a dry chemical. When the pack is struck, the pouch of water breaks, dissolving the chemical, and the solution becomes either hot or cold. Many hot packs use magnesium sulfate, and many cold packs use ammonium nitrate. Write reactions to show how these strong electrolytes break apart when they dissolve in water.

Solution Concentration: Molarity

27. Calculate the molarity of each of these solutions.
   a. A 5.623-g sample of $NaHCO_3$ is dissolved in enough water to make 250.0 mL of solution.
   b. A 184.6-mg sample of $K_2Cr_2O_7$ is dissolved in enough water to make 500.0 mL of solution.
   c. A 0.1025-g sample of copper metal is dissolved in 35 mL of concentrated HNO_3 to form Cu^{2+} ions and then water

is added to make a total volume of 200.0 mL. (Calculate the molarity of Cu^{2+}.)

28. A solution of ethanol (C_2H_5OH) in water is prepared by dissolving 75.0 mL of ethanol (density = 0.79 g/cm^3) in enough water to make 250.0 mL of solution. What is the molarity of the ethanol in this solution?

29. Calculate the concentration of all ions present in each of the following solutions of strong electrolytes.
   a. 0.100 mole of $Ca(NO_3)_2$ in 100.0 mL of solution
   b. 2.5 moles of Na_2SO_4 in 1.25 L of solution
   c. 5.00 g of NH_4Cl in 500.0 mL of solution
   d. 1.00 g K_3PO_4 in 250.0 mL of solution

30. Calculate the concentration of all ions present in each of the following solutions of strong electrolytes.
   a. 0.0200 mole of sodium phosphate in 10.0 mL of solution
   b. 0.300 mole of barium nitrate in 600.0 mL of solution
   c. 1.00 g of potassium chloride in 0.500 L of solution
   d. 132 g of ammonium sulfate in 1.50 L of solution

31. Which of the following solutions of strong electrolytes contains the largest number of moles of chloride ions: 100.0 mL of 0.30 M $AlCl_3$, 50.0 mL of 0.60 M $MgCl_2$, or 200.0 mL of 0.40 M NaCl?

32. Which of the following solutions of strong electrolytes contains the largest number of ions: 100.0 mL of 0.100 M NaOH, 50.0 mL of 0.200 M $BaCl_2$, or 75.0 mL of 0.150 M Na_3PO_4?

33. What mass of NaOH is contained in 250.0 mL of a 0.400 M sodium hydroxide solution?

34. If 10. g of $AgNO_3$ is available, what volume of 0.25 M $AgNO_3$ solution can be prepared?

35. Describe how you would prepare 2.00 L of each of the following solutions.
   a. 0.250 M NaOH from solid NaOH
   b. 0.250 M NaOH from 1.00 M NaOH stock solution
   c. 0.100 M K_2CrO_4 from solid K_2CrO_4
   d. 0.100 M K_2CrO_4 from 1.75 M K_2CrO_4 stock solution

36. How would you prepare 1.00 L of a 0.50-M solution of each of the following?
   a. H_2SO_4 from "concentrated" (18 M) sulfuric acid
   b. HCl from "concentrated" (12 M) reagent
   c. $NiCl_2$ from the salt $NiCl_2 \cdot 6H_2O$
   d. HNO_3 from "concentrated" (16 M) reagent
   e. Sodium carbonate from the pure solid

37. A solution is prepared by dissolving 10.8 g ammonium sulfate in enough water to make 100.0 mL of stock solution. A 10.00-mL sample of this stock solution is added to 50.00 mL of water. Calculate the concentration of ammonium ions and sulfate ions in the final solution.

38. A solution was prepared by mixing 50.00 mL of 0.100 M HNO_3 and 100.00 mL of 0.200 M HNO_3. Calculate the molarity of the final solution of nitric acid.

39. Calculate the sodium ion concentration when 70.0 mL of 3.0 M sodium carbonate is added to 30.0 mL of 1.0 M sodium bicarbonate.

40. Suppose 50.0 mL of 0.250 *M* $CoCl_2$ solution is added to 25.0 mL of 0.350 *M* $NiCl_2$ solution. Calculate the concentration, in moles per liter, of each of the ions present after mixing. Assume that the volumes are additive.

41. A standard solution is prepared for the analysis of fluoxymesterone ($C_{20}H_{29}FO_3$), an anabolic steroid. A stock solution is first prepared by dissolving 10.0 mg of fluoxymesterone in enough water to give a total volume of 500.0 mL. A 100.0-μL aliquot (portion) of this solution is diluted to a final volume of 100.0 mL. Calculate the concentration of the final solution in terms of molarity.

42. A stock solution containing Mn^{2+} ions was prepared by dissolving 1.584 g pure manganese metal in nitric acid and diluting to a final volume of 1.000 L. The following solutions were then prepared by dilution:

For solution *A*, 50.00 mL of stock solution was diluted to 1000.0 mL.
For solution *B*, 10.00 mL of solution *A* was diluted to 250.0 mL.
For solution *C*, 10.00 mL of solution *B* was diluted to 500.0 mL.

Calculate the concentrations of the stock solution and solutions *A*, *B*, and *C*.

Precipitation Reactions

43. On the basis of the general solubility rules given in Table 4.1, predict which of the following substances are likely to be soluble in water.

a. aluminum nitrate
b. magnesium chloride
c. rubidium sulfate
d. nickel(II) hydroxide
e. lead(II) sulfide
f. magnesium hydroxide
g. iron(III) phosphate

44. On the basis of the general solubility rules given in Table 4.1, predict which of the following substances are likely to be soluble in water.

a. zinc chloride
b. lead(II) nitrate
c. lead(II) sulfate
d. sodium iodide
e. cobalt(III) sulfide
f. chromium(III) hydroxide
g. magnesium carbonate
h. ammonium carbonate

45. When the following solutions are mixed together, what precipitate (if any) will form?

a. $FeSO_4(aq) + KCl(aq)$
b. $Al(NO_3)_3(aq) + Ba(OH)_2(aq)$
c. $CaCl_2(aq) + Na_2SO_4(aq)$
d. $K_2S(aq) + Ni(NO_3)_2(aq)$

46. When the following solutions are mixed together, what precipitate (if any) will form?

a. $Hg_2(NO_3)_2(aq) + CuSO_4(aq)$
b. $Ni(NO_3)_2(aq) + CaCl_2(aq)$

c. $K_2CO_3(aq) + MgI_2(aq)$
d. $Na_2CrO_4(aq) + AlBr_3(aq)$

47. For the reactions in Exercise 45, write the balanced formula equation, complete ionic equation, and net ionic equation. If no precipitate forms, write "No reaction."

48. For the reactions in Exercise 46, write the balanced formula equation, complete ionic equation, and net ionic equation. If no precipitate forms, write "No reaction."

49. Write the balanced formula and net ionic equation for the reaction that occurs when the contents of the two beakers are added together. What colors represent the spectator ions in each reaction?

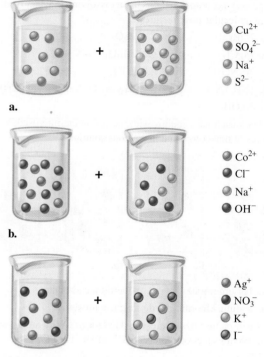

a.

b.

c.

50. Give an example how each of the following insoluble ionic compounds could be produced using a precipitation reaction. Write the balanced formula equation for each reaction.

a. $Fe(OH)_3(s)$ c. $PbSO_4(s)$
b. $Hg_2Cl_2(s)$ d. $BaCrO_4(s)$

51. Write net ionic equations for the reaction, if any, that occurs when aqueous solutions of the following are mixed.

a. ammonium sulfate and barium nitrate
b. lead(II) nitrate and sodium chloride
c. sodium phosphate and potassium nitrate
d. sodium bromide and rubidium chloride
e. copper(II) chloride and sodium hydroxide

52. Write net ionic equations for the reaction, if any, that occurs when aqueous solutions of the following are mixed.

a. chromium(III) chloride and sodium hydroxide
b. silver nitrate and ammonium carbonate
c. copper(II) sulfate and mercury(I) nitrate
d. strontium nitrate and potassium iodide

53. Separate samples of a solution of an unknown soluble ionic compound are treated with KCl, Na_2SO_4, and NaOH. A precipitate forms only when Na_2SO_4 is added. Which cations could be present in the unknown soluble ionic compound?

54. A sample may contain any or all of the following ions: Hg_2^{2+}, Ba^{2+}, and Mn^{2+}.

  a. No precipitate formed when an aqueous solution of NaCl was added to the sample solution.

  b. No precipitate formed when an aqueous solution of Na_2SO_4 was added to the sample solution.

  c. A precipitate formed when the sample solution was made basic with NaOH.

  Which ion or ions are present in the sample solution?

55. What mass of Na_2CrO_4 is required to precipitate all of the silver ions from 75.0 mL of a 0.100-M solution of $AgNO_3$?

56. What volume of 0.100 M Na_3PO_4 is required to precipitate all the lead(II) ions from 150.0 mL of 0.250 M $Pb(NO_3)_2$?

57. What mass of solid aluminum hydroxide can be produced when 50.0 mL of 0.200 M $Al(NO_3)_3$ is added to 200.0 mL of 0.100 M KOH?

58. What mass of barium sulfate can be produced when 100.0 mL of a 0.100-M solution of barium chloride is mixed with 100.0 mL of a 0.100-M solution of iron(III) sulfate?

59. What mass of solid AgBr is produced when 100.0 mL of 0.150 M $AgNO_3$ is added to 20.0 mL of 1.00 M NaBr?

60. What mass of silver chloride can be prepared by the reaction of 100.0 mL of 0.20 M silver nitrate with 100.0 mL of 0.15 M calcium chloride? Calculate the concentrations of each ion remaining in solution after precipitation is complete.

61. A 100.0-mL aliquot of 0.200 M aqueous potassium hydroxide is mixed with 100.0 mL of 0.200 M aqueous magnesium nitrate.

  a. Write a balanced chemical equation for any reaction that occurs.

  b. What precipitate forms?

  c. What mass of precipitate is produced?

  d. Calculate the concentration of each ion remaining in solution after precipitation is complete.

62. The drawings below represent aqueous solutions. Solution A is 2.00 L of a 2.00-M aqueous solution of copper(II) nitrate. Solution B is 2.00 L of a 3.00-M aqueous solution of potassium hydroxide.

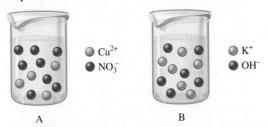

  A   B

  a. Draw a picture of the solution made by mixing solutions A and B together after the precipitation reaction takes place. Make sure this picture shows the correct relative volume compared to solutions A and B, and the correct relative number of ions, along with the correct relative amount of solid formed.

b. Determine the concentrations (in M) of all ions left in solution (from part a) and the mass of solid formed.

63. A 1.42-g sample of a pure compound, with formula M_2SO_4, was dissolved in water and treated with an excess of aqueous calcium chloride, resulting in the precipitation of all the sulfate ions as calcium sulfate. The precipitate was collected, dried, and found to weigh 1.36 g. Determine the atomic mass of M, and identify M.

64. You are given a 1.50-g mixture of sodium nitrate and sodium chloride. You dissolve this mixture into 100 mL of water and then add an excess of 0.500 M silver nitrate solution. You produce a white solid, which you then collect, dry, and measure. The white solid has a mass of 0.641 g.

  a. If you had an extremely magnified view of the solution (to the atomic-molecular level), list the species you would see (include charges, if any).

  b. Write the balanced net ionic equation for the reaction that produces the solid. Include phases and charges.

  c. Calculate the percent sodium chloride in the original unknown mixture.

Acid–Base Reactions

65. Write the balanced formula, complete ionic, and net ionic equations for each of the following acid–base reactions.

  a. $HClO_4(aq) + Mg(OH)_2(s) \rightarrow$

  b. $HCN(aq) + NaOH(aq) \rightarrow$

  c. $HCl(aq) + NaOH(aq) \rightarrow$

66. Write the balanced formula, complete ionic, and net ionic equations for each of the following acid–base reactions.

  a. $HNO_3(aq) + Al(OH)_3(s) \rightarrow$

  b. $HC_2H_3O_2(aq) + KOH(aq) \rightarrow$

  c. $Ca(OH)_2(aq) + HCl(aq) \rightarrow$

67. Write the balanced formula equation for the acid–base reactions that occur when the following are mixed.

  a. potassium hydroxide (aqueous) and nitric acid

  b. barium hydroxide (aqueous) and hydrochloric acid

  c. perchloric acid [$HClO_4(aq)$] and solid iron(III) hydroxide

  d. solid silver hydroxide and hydrobromic acid

  e. aqueous strontium hydroxide and hydroiodic acid

68. What acid and what base would react in aqueous solution so that the following salts appear as products in the formula equation? Write the balanced formula equation for each reaction.

  a. potassium perchlorate

  b. cesium nitrate

  c. calcium iodide

69. What volume of each of the following acids will react completely with 50.00 mL of 0.200 M NaOH?

  a. 0.100 M HCl

  b. 0.150 M HNO_3

  c. 0.200 M $HC_2H_3O_2$ (1 acidic hydrogen)

70. What volume of each of the following bases will react completely with 25.00 mL of 0.200 M HCl?

  a. 0.100 M NaOH   **c.** 0.250 M KOH

  b. 0.0500 M $Sr(OH)_2$

71. Hydrochloric acid (75.0 mL of 0.250 M) is added to 225.0 mL of 0.0550 M $Ba(OH)_2$ solution. What is the concentration of the excess H^+ or OH^- ions left in this solution?

72. A student mixes four reagents together, thinking that the solutions will neutralize each other. The solutions mixed together are 50.0 mL of 0.100 M hydrochloric acid, 100.0 mL of 0.200 M of nitric acid, 500.0 mL of 0.0100 M calcium hydroxide, and 200.0 mL of 0.100 M rubidium hydroxide. Did the acids and bases exactly neutralize each other? If not, calculate the concentration of excess H^+ or OH^- ions left in solution.

73. A 25.00-mL sample of hydrochloric acid solution requires 24.16 mL of 0.106 M sodium hydroxide for complete neutralization. What is the concentration of the original hydrochloric acid solution?

74. A 10.00-mL sample of vinegar, an aqueous solution of acetic acid ($HC_2H_3O_2$), is titrated with 0.5062 M NaOH, and 16.58 mL is required to reach the equivalence point.
  a. What is the molarity of the acetic acid?
  b. If the density of the vinegar is 1.006 g/cm^3, what is the mass percent of acetic acid in the vinegar?

75. What volume of 0.0200 M calcium hydroxide is required to neutralize 35.00 mL of 0.0500 M nitric acid?

76. A 30.0-mL sample of an unknown strong base is neutralized after the addition of 12.0 mL of a 0.150 M HNO_3 solution. If the unknown base concentration is 0.0300 M, give some possible identities for the unknown base.

77. A student titrates an unknown amount of potassium hydrogen phthalate ($KHC_8H_4O_4$, often abbreviated KHP) with 20.46 mL of a 0.1000-M NaOH solution. KHP (molar mass = 204.22 g/mol) has one acidic hydrogen. What mass of KHP was titrated (reacted completely) by the sodium hydroxide solution?

78. The concentration of a certain sodium hydroxide solution was determined by using the solution to titrate a sample of potassium hydrogen phthalate (abbreviated as KHP). KHP is an acid with one acidic hydrogen and a molar mass of 204.22 g/mol. In the titration, 34.67 mL of the sodium hydroxide solution was required to react with 0.1082 g KHP. Calculate the molarity of the sodium hydroxide.

Oxidation–Reduction Reactions

79. Assign oxidation states for all atoms in each of the following compounds.
  a. $KMnO_4$
  b. NiO_2
  c. $Na_4Fe(OH)_6$
  d. $(NH_4)_2HPO_4$
  e. P_4O_6
  f. Fe_3O_4
  g. $XeOF_4$
  h. SF_4
  i. CO
  j. $C_6H_{12}O_6$

80. Assign oxidation states for all atoms in each of the following compounds.
  a. UO_2^{2+}
  b. As_2O_3
  c. $NaBiO_3$
  d. As_4
  e. $HAsO_2$
  f. $Mg_2P_2O_7$
  g. $Na_2S_2O_3$
  h. Hg_2Cl_2
  i. $Ca(NO_3)_2$

81. Assign the oxidation state for nitrogen in each of the following.
  a. Li_3N
  b. NH_3
  c. N_2H_4
  d. NO
  e. N_2O
  f. NO_2
  g. NO_2^-
  h. NO_3^-
  i. N_2

82. Assign oxidation numbers to all the atoms in each of the following.
  a. $SrCr_2O_7$
  b. $CuCl_2$
  c. O_2
  d. H_2O_2
  e. $MgCO_3$
  f. Ag
  g. $PbSO_3$
  h. PbO_2
  i. $Na_2C_2O_4$
  j. CO_2
  k. $(NH_4)_2Ce(SO_4)_3$
  l. Cr_2O_3

83. Specify which of the following are oxidation–reduction reactions, and identify the oxidizing agent, the reducing agent, the substance being oxidized, and the substance being reduced.
  a. $Cu(s) + 2Ag^+(aq) \rightarrow 2Ag(s) + Cu^{2+}(aq)$
  b. $HCl(g) + NH_3(g) \rightarrow NH_4Cl(s)$
  c. $SiCl_4(l) + 2H_2O(l) \rightarrow 4HCl(aq) + SiO_2(s)$
  d. $SiCl_4(l) + 2Mg(s) \rightarrow 2MgCl_2(s) + Si(s)$
  e. $Al(OH)_4^-(aq) \rightarrow AlO_2^-(aq) + 2H_2O(l)$

84. Specify which of the following equations represent oxidation–reduction reactions, and indicate the oxidizing agent, the reducing agent, the species being oxidized, and the species being reduced.
  a. $CH_4(g) + H_2O(g) \rightarrow CO(g) + 3H_2(g)$
  b. $2AgNO_3(aq) + Cu(s) \rightarrow Cu(NO_3)_2(aq) + 2Ag(s)$
  c. $Zn(s) + 2HCl(aq) \rightarrow ZnCl_2(aq) + H_2(g)$
  d. $2H^+(aq) + 2CrO_4^{2-}(aq) \rightarrow Cr_2O_7^{2-}(aq) + H_2O(l)$

85. Consider the reaction between sodium metal and fluorine (F_2) gas to form sodium fluoride. Using oxidation states, how many electrons would each sodium atom lose, and how many electrons would each fluorine atom gain? How many sodium atoms are needed to react with one fluorine molecule? Write a balanced equation for this reaction.

86. Consider the reaction between oxygen (O_2) gas and magnesium metal to form magnesium oxide. Using oxidation states, how many electrons would each oxygen atom gain, and how many electrons would each magnesium atom lose? How many magnesium atoms are needed to react with one oxygen molecule? Write a balanced equation for this reaction.

87. Balance each of the following oxidation–reduction reactions by using the oxidation states method.
  a. $C_2H_6(g) + O_2(g) \rightarrow CO_2(g) + H_2O(g)$
  b. $Mg(s) + HCl(aq) \rightarrow Mg^{2+}(aq) + Cl^-(aq) + H_2(g)$
  c. $Co^{3+}(aq) + Ni(s) \rightarrow Co^{2+}(aq) + Ni^{2+}(aq)$
  d. $Zn(s) + H_2SO_4(aq) \rightarrow ZnSO_4(aq) + H_2(g)$

88. Balance each of the following oxidation–reduction reactions by using the oxidation states method.
  a. $Cl_2(g) + Al(s) \rightarrow Al^{3+}(aq) + Cl^-(aq)$
  b. $O_2(g) + H_2O(l) + Pb(s) \rightarrow Pb(OH)_2(s)$
  c. $H^+(aq) + MnO_4^-(aq) + Fe^{2+}(aq) \rightarrow$
  $Mn^{2+}(aq) + Fe^{3+}(aq) + H_2O(l)$

Additional Exercises

89. You wish to prepare 1 L of a 0.02-M potassium iodate solution. You require that the final concentration be within 1% of 0.02 M and that the concentration must be known accurately to the fourth decimal place. How would you prepare this solution? Specify the glassware you would use, the accuracy needed for the balance, and the ranges of acceptable masses of KIO_3 that can be used.

90. The figures below are molecular-level representations of four aqueous solutions of the same solute. Arrange the solutions from most to least concentrated.

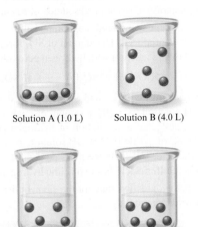

Solution A (1.0 L)   Solution B (4.0 L)

Solution C (2.0 L)   Solution D (2.0 L)

91. An average human being has about 5.0 L of blood in his or her body. If an average person were to eat 32.0 g of sugar (sucrose, $C_{12}H_{22}O_{11}$, 342.30 g/mol), and all that sugar were dissolved into the bloodstream, how would the molarity of the blood sugar change?

92. A 230.-mL sample of a 0.275-M $CaCl_2$ solution is left on a hot plate overnight; the following morning, the solution is 1.10 M. What volume of water evaporated from the 0.275 M $CaCl_2$ solution?

93. Using the general solubility rules given in Table 4.1, name three reagents that would form precipitates with each of the following ions in aqueous solution. Write the net ionic equation for each of your suggestions.

   a. chloride ion   **d.** sulfate ion

   b. calcium ion   **e.** mercury(I) ion, Hg_2^{2+}

   c. iron(III) ion   **f.** silver ion

94. Consider a 1.50-g mixture of magnesium nitrate and magnesium chloride. After dissolving this mixture in water, 0.500 M silver nitrate is added dropwise until precipitate formation is complete. The mass of the white precipitate formed is 0.641 g.

   a. Calculate the mass percent of magnesium chloride in the mixture.

   b. Determine the minimum volume of silver nitrate that must have been added to ensure complete formation of the precipitate.

95. A 1.00-g sample of an alkaline earth metal chloride is treated with excess silver nitrate. All of the chloride is recovered as 1.38 g of silver chloride. Identify the metal.

96. A mixture contains only NaCl and $Al_2(SO_4)_3$. A 1.45-g sample of the mixture is dissolved in water and an excess of NaOH is added, producing a precipitate of $Al(OH)_3$. The precipitate is filtered, dried, and weighed. The mass of the precipitate is 0.107 g. What is the mass percent of $Al_2(SO_4)_3$ in the sample?

97. The thallium (present as Tl_2SO_4) in a 9.486-g pesticide sample was precipitated as thallium(I) iodide. Calculate the mass percent of Tl_2SO_4 in the sample if 0.1824 g of TlI was recovered.

98. A mixture contains only NaCl and $Fe(NO_3)_3$. A 0.456-g sample of the mixture is dissolved in water, and an excess of NaOH is added, producing a precipitate of $Fe(OH)_3$. The precipitate is filtered, dried, and weighed. Its mass is 0.107 g. Calculate the following.

   a. the mass of iron in the sample

   b. the mass of $Fe(NO_3)_3$ in the sample

   c. the mass percent of $Fe(NO_3)_3$ in the sample

99. A student added 50.0 mL of an NaOH solution to 100.0 mL of 0.400 M HCl. The solution was then treated with an excess of aqueous chromium(III) nitrate, resulting in formation of 2.06 g of precipitate. Determine the concentration of the NaOH solution.

100. Some of the substances commonly used in stomach antacids are MgO, $Mg(OH)_2$, and $Al(OH)_3$.

   a. Write a balanced equation for the neutralization of hydrochloric acid by each of these substances.

   b. Which of these substances will neutralize the greatest amount of 0.10 M HCl per gram?

101. Acetylsalicylic acid is the active ingredient in aspirin. It took 35.17 mL of 0.5065 M sodium hydroxide to react completely with 3.210 g of acetylsalicylic acid. Acetylsalicylic acid has one acidic hydrogen. What is the molar mass of acetylsalicylic acid?

102. When hydrochloric acid reacts with magnesium metal, hydrogen gas and aqueous magnesium chloride are produced. What volume of 5.0 M HCl is required to react completely with 3.00 g of magnesium?

103. A 2.20-g sample of an unknown acid (empirical formula = $C_3H_4O_3$) is dissolved in 1.0 L of water. A titration required 25.0 mL of 0.500 M NaOH to react completely with all the acid present. Assuming the unknown acid has one acidic proton per molecule, what is the molecular formula of the unknown acid?

104. Carminic acid, a naturally occurring red pigment extracted from the cochineal insect, contains only carbon, hydrogen, and oxygen. It was commonly used as a dye in the first half of the nineteenth century. It is 53.66% C and 4.09% H by mass. A titration required 18.02 mL of 0.0406 M NaOH to neutralize 0.3602 g carminic acid. Assuming that there is only one acidic hydrogen per molecule, what is the molecular formula of carminic acid?

105. Chlorisondamine chloride ($C_{14}H_{20}Cl_6N_2$) is a drug used in the treatment of hypertension. A 1.28-g sample of a medication containing the drug was treated to destroy the organic material and to release all the chlorine as chloride ion. When the filtered solution containing chloride ion was treated with an excess of silver nitrate, 0.104 g silver chloride was recovered. Calculate the mass percent of chlorisondamine chloride in the medication, assuming the drug is the only source of chloride.

106. Saccharin ($C_7H_5NO_3S$) is sometimes dispensed in tablet form. Ten tablets with a total mass of 0.5894 g were dissolved in water. The saccharin was oxidized to convert all the sulfur to sulfate ion, which was precipitated by adding an excess of barium chloride solution. The mass of $BaSO_4$ obtained was 0.5032 g. What is the average mass of saccharin per tablet? What is the average mass percent of saccharin in the tablets?

107. Douglasite is a mineral with the formula $2KCl \cdot FeCl_2 \cdot 2H_2O$. Calculate the mass percent of douglasite in a 455.0-mg sample if it took 37.20 mL of a 0.1000-M $AgNO_3$ solution to precipitate all the Cl^- as AgCl. Assume the douglasite is the only source of chloride ion.

108. Many oxidation–reduction reactions can be balanced by inspection. Try to balance the following reactions by inspection. In each reaction, identify the substance reduced and the substance oxidized.

  a. $Al(s) + HCl(aq) \rightarrow AlCl_3(aq) + H_2(g)$
  b. $CH_4(g) + S(s) \rightarrow CS_2(l) + H_2S(g)$
  c. $C_3H_8(g) + O_2(g) \rightarrow CO_2(g) + H_2O(l)$
  d. $Cu(s) + Ag^+(aq) \rightarrow Ag(s) + Cu^{2+}(aq)$

109. The blood alcohol (C_2H_5OH) level can be determined by titrating a sample of blood plasma with an acidic potassium dichromate solution, resulting in the production of $Cr^{3+}(aq)$ and carbon dioxide. The reaction can be monitored because the dichromate ion ($Cr_2O_7^{2-}$) is orange in solution, and the Cr^{3+} ion is green. The balanced equation is

$$16H^+(aq) + 2Cr_2O_7^{2-}(aq) + C_2H_5OH(aq) \longrightarrow$$
$$4Cr^{3+}(aq) + 2CO_2(g) + 11H_2O(l)$$

This reaction is an oxidation–reduction reaction. What species is reduced, and what species is oxidized? How many electrons are transferred in the balanced equation above?

ChemWork Problems

These multiconcept problems (and additional ones) are found interactively online with the same type of assistance a student would get from an instructor.

110. Calculate the concentration of all ions present when 0.160 g of $MgCl_2$ is dissolved in 100.0 mL of solution.

111. A solution is prepared by dissolving 0.6706 g oxalic acid ($H_2C_2O_4$) in enough water to make 100.0 mL of solution. A 10.00-mL aliquot (portion) of this solution is then diluted to a final volume of 250.0 mL. What is the final molarity of the oxalic acid solution?

112. For the following chemical reactions, determine the precipitate produced when the two reactants listed below are mixed together. Indicate "none" if no precipitate will form.

	Formula of Precipitate
$Sr(NO_3)_2(aq) + K_3PO_4(aq) \longrightarrow$	_____ (s)
$K_2CO_3(aq) + AgNO_3(aq) \longrightarrow$	_____ (s)
$NaCl(aq) + KNO_3(aq) \longrightarrow$	_____ (s)
$KCl(aq) + AgNO_3(aq) \longrightarrow$	_____ (s)
$FeCl_3(aq) + Pb(NO_3)_2(aq) \longrightarrow$	_____ (s)

113. What volume of 0.100 M NaOH is required to precipitate all of the nickel(II) ions from 150.0 mL of a 0.249-M solution of $Ni(NO_3)_2$?

114. A 500.0-mL sample of 0.200 M sodium phosphate is mixed with 400.0 mL of 0.289 M barium chloride. What is the mass of the solid produced?

115. A 450.0-mL sample of a 0.257-M solution of silver nitrate is mixed with 400.0 mL of 0.200 M calcium chloride. What is the concentration of Cl^- in solution after the reaction is complete?

116. The zinc in a 1.343-g sample of a foot powder was precipitated as $ZnNH_4PO_4$. Strong heating of the precipitate yielded 0.4089 g $Zn_2P_2O_7$. Calculate the mass percent of zinc in the sample of foot powder.

117. A 50.00-mL sample of aqueous $Ca(OH)_2$ requires 34.66 mL of a 0.944-M nitric acid for neutralization. Calculate the concentration (molarity) of the original solution of calcium hydroxide.

118. When organic compounds containing sulfur are burned, sulfur dioxide is produced. The amount of SO_2 formed can be determined by the reaction with hydrogen peroxide:

$$H_2O_2(aq) + SO_2(g) \longrightarrow H_2SO_4(aq)$$

The resulting sulfuric acid is then titrated with a standard NaOH solution. A 1.302-g sample of coal is burned and the SO_2 is collected in a solution of hydrogen peroxide. It took 28.44 mL of a 0.1000-M NaOH solution to titrate the resulting sulfuric acid. Calculate the mass percent of sulfur in the coal sample. Sulfuric acid has two acidic hydrogens.

119. Assign the oxidation state for the element listed in each of the following compounds:

	Oxidation State
S in $MgSO_4$	_____
Pb in $PbSO_4$	_____
O in O_2	_____
Ag in Ag	_____
Cu in $CuCl_2$	_____

Challenge Problems

120. A 10.00-g sample consisting of a mixture of sodium chloride and potassium sulfate is dissolved in water. This aqueous mixture then reacts with excess aqueous lead(II) nitrate to form 21.75 g of solid. Determine the mass percent of sodium chloride in the original mixture.

121. The units of parts per million (ppm) and parts per billion (ppb) are commonly used by environmental chemists. In general, 1 ppm means 1 part of solute for every 10^6 parts of solution. Mathematically, by mass:

$$ppm = \frac{\mu g\ solute}{g\ solution} = \frac{mg\ solute}{kg\ solution}$$

In the case of very dilute aqueous solutions, a concentration of 1.0 ppm is equal to 1.0 μg of solute per 1.0 mL, which equals 1.0 g solution. Parts per billion is defined in a similar fashion. Calculate the molarity of each of the following aqueous solutions.

  a. 5.0 ppb Hg in H_2O
  b. 1.0 ppb $CHCl_3$ in H_2O
  c. 10.0 ppm As in H_2O
  d. 0.10 ppm DDT ($C_{14}H_9Cl_5$) in H_2O

122. In the spectroscopic analysis of many substances, a series of standard solutions of known concentration are measured to generate a calibration curve. How would you prepare standard solutions containing 10.0, 25.0, 50.0, 75.0, and 100. ppm of copper from a commercially produced 1000.0-ppm solution? Assume each solution has a final volume of 100.0 mL. (See Exercise 121 for definitions.)

123. In most of its ionic compounds, cobalt is either Co(II) or Co(III). One such compound, containing chloride ion and waters of hydration, was analyzed, and the following results were obtained. A 0.256-g sample of the compound was dissolved in water, and excess silver nitrate was added. The silver chloride was filtered, dried, and weighed, and it had a mass of 0.308 g. A second sample of 0.416 g of the compound was dissolved in water, and an excess of sodium hydroxide was added. The hydroxide salt was filtered and heated in a flame, forming cobalt(III) oxide. The mass of cobalt(III) oxide formed was 0.145 g.

  a. What is the percent composition, by mass, of the compound?

  b. Assuming the compound contains one cobalt ion per formula unit, what is the formula?

  c. Write balanced equations for the three reactions described.

124. Polychlorinated biphenyls (PCBs) have been used extensively as dielectric materials in electrical transformers. Because PCBs have been shown to be potentially harmful, analysis for their presence in the environment has become very important. PCBs are manufactured according to the following generic reaction:

$$C_{12}H_{10} + nCl_2 \rightarrow C_{12}H_{10-n}Cl_n + nHCl$$

This reaction results in a mixture of PCB products. The mixture is analyzed by decomposing the PCBs and then precipitating the resulting Cl^- as AgCl.

  a. Develop a general equation that relates the average value of n to the mass of a given mixture of PCBs and the mass of AgCl produced.

  b. A 0.1947-g sample of a commercial PCB yielded 0.4791 g of AgCl. What is the average value of n for this sample?

125. Consider the reaction of 19.0 g of zinc with excess silver nitrite to produce silver metal and zinc nitrite. The reaction is stopped before all the zinc metal has reacted and 29.0 g of solid metal is present. Calculate the mass of each metal in the 29.0-g mixture.

126. A mixture contains only sodium chloride and potassium chloride. A 0.1586-g sample of the mixture was dissolved in water. It took 22.90 mL of 0.1000 M $AgNO_3$ to completely precipitate all the chloride present. What is the composition (by mass percent) of the mixture?

127. You are given a solid that is a mixture of Na_2SO_4 and K_2SO_4. A 0.205-g sample of the mixture is dissolved in water. An excess of an aqueous solution of $BaCl_2$ is added. The $BaSO_4$ that is formed is filtered, dried, and weighed. Its mass is 0.298 g. What mass of SO_4^{2-} ion is in the sample? What is the mass percent of SO_4^{2-} ion in the sample? What are the percent compositions by mass of Na_2SO_4 and K_2SO_4 in the sample?

128. Zinc and magnesium metal each react with hydrochloric acid according to the following equations:

$$Zn(s) + 2HCl(aq) \longrightarrow ZnCl_2(aq) + H_2(g)$$
$$Mg(s) + 2HCl(aq) \longrightarrow MgCl_2(aq) + H_2(g)$$

A 10.00-g mixture of zinc and magnesium is reacted with the stoichiometric amount of hydrochloric acid. The reaction mixture is then reacted with 156 mL of 3.00 M silver nitrate to produce the maximum possible amount of silver chloride.

  a. Determine the percent magnesium by mass in the original mixture.

  b. If 78.0 mL of HCl was added, what was the concentration of the HCl?

129. You made 100.0 mL of a lead(II) nitrate solution for lab but forgot to cap it. The next lab session you noticed that there was only 80.0 mL left (the rest had evaporated). In addition, you forgot the initial concentration of the solution. You decide to take 2.00 mL of the solution and add an excess of a concentrated sodium chloride solution. You obtain a solid with a mass of 3.407 g. What was the concentration of the original lead(II) nitrate solution?

130. Consider reacting copper(II) sulfate with iron. Two possible reactions can occur, as represented by the following equations.

copper(II) sulfate(aq) + iron(s) $\longrightarrow$
  copper(s) + iron(II) sulfate(aq)

copper(II) sulfate(aq) + iron(s) $\longrightarrow$
  copper(s) + iron(III) sulfate(aq)

You place 87.7 mL of a 0.500-M solution of copper(II) sulfate in a beaker. You then add 2.00 g of iron filings to the copper(II) sulfate solution. After one of the above reactions occurs, you isolate 2.27 g of copper. Which equation above describes the reaction that occurred? Support your answer.

131. Consider an experiment in which two burets, Y and Z, are simultaneously draining into a beaker that initially contained 275.0 mL of 0.300 M HCl. Buret Y contains 0.150 M NaOH and buret Z contains 0.250 M KOH. The stoichiometric point in the titration is reached 60.65 minutes after Y and Z were started simultaneously. The total volume in the beaker at the stoichiometric point is 655 mL. Calculate the flow rates of burets Y and Z. Assume the flow rates remain constant during the experiment.

132. Complete and balance each acid–base reaction.

  a. $H_3PO_4(aq) + NaOH(aq) \rightarrow$
    Contains three acidic hydrogens

  b. $H_2SO_4(aq) + Al(OH)_3(s) \rightarrow$
    Contains two acidic hydrogens

  c. $H_2Se(aq) + Ba(OH)_2(aq) \rightarrow$
    Contains two acidic hydrogens

  d. $H_2C_2O_4(aq) + NaOH(aq) \rightarrow$
    Contains two acidic hydrogens

133. What volume of 0.0521 M $Ba(OH)_2$ is required to neutralize exactly 14.20 mL of 0.141 M H_3PO_4? Phosphoric acid contains three acidic hydrogens.

134. A 10.00-mL sample of sulfuric acid from an automobile battery requires 35.08 mL of 2.12 M sodium hydroxide solution for complete neutralization. What is the molarity of the sulfuric acid? Sulfuric acid contains two acidic hydrogens.

135. A 0.500-L sample of H_2SO_4 solution was analyzed by taking a 100.0-mL aliquot and adding 50.0 mL of 0.213 M NaOH. After the reaction occurred, an excess of OH^- ions remained in the solution. The excess base required 13.21 mL of 0.103 M

HCl for neutralization. Calculate the molarity of the original sample of H_2SO_4. Sulfuric acid has two acidic hydrogens.

136. A 6.50-g sample of a diprotic acid requires 137.5 mL of a 0.750 M NaOH solution for complete neutralization. Determine the molar mass of the acid.

137. Citric acid, which can be obtained from lemon juice, has the molecular formula $C_6H_8O_7$. A 0.250-g sample of citric acid dissolved in 25.0 mL of water requires 37.2 mL of 0.105 M NaOH for complete neutralization. What number of acidic hydrogens per molecule does citric acid have?

138. A stream flows at a rate of 5.00×10^4 liters per second (L/s) upstream of a manufacturing plant. The plant discharges 3.50×10^3 L/s of water that contains 65.0 ppm HCl into the stream. (See Exercise 121 for definitions.)

a. Calculate the stream's total flow rate downstream from this plant.

b. Calculate the concentration of HCl in ppm downstream from this plant.

c. Further downstream, another manufacturing plant diverts 1.80×10^4 L/s of water from the stream for its own use. This plant must first neutralize the acid and does so by adding lime:

$$CaO(s) + 2H^+(aq) \longrightarrow Ca^{2+}(aq) + H_2O(l)$$

What mass of CaO is consumed in an 8.00-h work day by this plant?

d. The original stream water contained 10.2 ppm Ca^{2+}. Although no calcium was in the waste water from the first plant, the waste water of the second plant contains Ca^{2+} from the neutralization process. If 90.0% of the water used by the second plant is returned to the stream, calculate the concentration of Ca^{2+} in ppm downstream of the second plant.

139. It took 25.06 ±0.05 mL of a sodium hydroxide solution to titrate a 0.4016-g sample of KHP (see Exercise 77). Calculate the concentration and uncertainty in the concentration of the sodium hydroxide solution. (See Appendix 1.5.) Neglect any uncertainty in the mass.

Integrative Problems

These problems require the integration of multiple concepts to find the solutions.

140. Tris(pentafluorophenyl)borane, commonly known by its acronym BARF, is frequently used to initiate polymerization of ethylene or propylene in the presence of a catalytic transition metal compound. It is composed solely of C, F, and B; it is 42.23% C and 55.66% F by mass.

a. What is the empirical formula of BARF?

b. A 2.251-g sample of BARF dissolved in 347.0 mL of solution produces a 0.01267-M solution. What is the molecular formula of BARF?

141. In a 1-L beaker, 203 mL of 0.307 M ammonium chromate was mixed with 137 mL of 0.269 M chromium(III) nitrite to produce ammonium nitrite and chromium(III) chromate. Write the balanced chemical equation for the reaction occurring

here. If the percent yield of the reaction was 88.0%, what mass of chromium(III) chromate was isolated?

142. The vanadium in a sample of ore is converted to VO^{2+}. The VO^{2+} ion is subsequently titrated with MnO_4^- in acidic solution to form $V(OH)_4^+$ and manganese(II) ion. The unbalanced titration reaction is

$$MnO_4^-(aq) + VO^{2+}(aq) + H_2O(l) \longrightarrow$$
$$V(OH)_4^+(aq) + Mn^{2+}(aq) + H^+(aq)$$

To titrate the solution, 26.45 mL of 0.02250 M MnO_4^- was required. If the mass percent of vanadium in the ore was 58.1%, what was the mass of the ore sample? *Hint:* Balance the titration reaction by the oxidation states method.

143. The unknown acid H_2X can be neutralized completely by OH^- according to the following (unbalanced) equation:

$$H_2X(aq) + OH^-(aq) \longrightarrow X^{2-}(aq) + H_2O(l)$$

The ion formed as a product, X^{2-}, was shown to have 36 total electrons. What is element X? Propose a name for H_2X. To completely neutralize a sample of H_2X, 35.6 mL of 0.175 M OH^- solution was required. What was the mass of the H_2X sample used?

Marathon Problems

These problems are designed to incorporate several concepts and techniques into one situation.

144. Three students were asked to find the identity of the metal in a particular sulfate salt. They dissolved a 0.1472-g sample of the salt in water and treated it with excess barium chloride, resulting in the precipitation of barium sulfate. After the precipitate had been filtered and dried, it weighed 0.2327 g.

Each student analyzed the data independently and came to different conclusions. Pat decided that the metal was titanium. Chris thought it was sodium. Randy reported that it was gallium. What formula did each student assign to the sulfate salt?

Look for information on the sulfates of gallium, sodium, and titanium in this text and reference books such as the *CRC Handbook of Chemistry and Physics*. What further tests would you suggest to determine which student is most likely correct?

145. You have two 500.0-mL aqueous solutions. Solution A is a solution of a metal nitrate that is 8.246% nitrogen by mass. The ionic compound in solution B consists of potassium, chromium, and oxygen; chromium has an oxidation state of +6 and there are 2 potassiums and 1 chromium in the formula. The masses of the solutes in each of the solutions are the same. When the solutions are added together, a blood-red precipitate forms. After the reaction has gone to completion, you dry the solid and find that it has a mass of 331.8 g.

a. Identify the ionic compounds in solution A and solution B.

b. Identify the blood-red precipitate.

c. Calculate the concentration (molarity) of all ions in the original solutions.

d. Calculate the concentration (molarity) of all ions in the final solution.

Chapter 5

Gases

The artistry of this sunset over Lamarck Col in California's Sierra Nevada mountains results from reflections on the clouds in our gaseous atmosphere. (Jerry Dodrill/Aurora Photos/Getty Images)

Matter exists in three distinct physical states: gas, liquid, and solid. Although relatively few substances exist in the gaseous state under typical conditions, gases are very important. For example, we live immersed in a gaseous solution. The earth's atmosphere is a mixture of gases that consists mainly of elemental nitrogen (N_2) and oxygen (O_2). The atmosphere both supports life and acts as a waste receptacle for the exhaust gases that accompany many industrial processes. The chemical reactions of these waste gases in the atmosphere lead to various types of pollution, including smog and acid rain. The gases in the atmosphere also shield us from harmful radiation from the sun and keep the earth warm by reflecting heat radiation back toward the earth. In fact, there is now great concern that an increase in atmospheric carbon dioxide, a product of the combustion of fossil fuels, is causing a dangerous warming of the earth.

In this chapter we will look carefully at the properties of gases. First we will see how measurements of gas properties lead to various types of laws—statements that show how the properties are related to each other. Then we will construct a model to explain why gases behave as they do. This model will show how the behavior of the individual particles of a gas leads to the observed properties of the gas itself (a collection of many, many particles).

The study of gases provides an excellent example of the scientific method in action. It illustrates how observations lead to natural laws, which in turn can be accounted for by models.

5.1 | Pressure

A gas uniformly fills any container, is easily compressed, and mixes completely with any other gas. One of the most obvious properties of a gas is that it exerts pressure on its surroundings. For example, when you blow up a balloon, the air inside pushes against the elastic sides of the balloon and keeps it firm.

As mentioned earlier, the gases most familiar to us form the earth's atmosphere. The pressure exerted by this gaseous mixture that we call air can be dramatically demonstrated by the experiment shown in Fig. 5.1. A small volume of water is placed in a

As a gas, water occupies 1200 times as much space as it does as a liquid at 25°C and atmospheric pressure.

Charles D. Winters

Figure 5.1 | The pressure exerted by the gases in the atmosphere can be demonstrated by boiling water in a large metal can (a) and then turning off the heat and sealing the can. As the can cools, the water vapor condenses, lowering the gas pressure inside the can. This causes the can to crumple (b).

190

Vacuum

$h = 760$ mm Hg for standard atmosphere

Figure 5.2 | A torricellian barometer. The tube, completely filled with mercury, is inverted in a dish of mercury. Mercury flows out of the tube until the pressure of the column of mercury (shown by the black arrow) "standing on the surface" of the mercury in the dish is equal to the pressure of the air (shown by the purple arrows) on the rest of the surface of the mercury in the dish.

Soon after Torricelli died, a German physicist named Otto von Guericke invented an air pump. In a famous demonstration for the King of Prussia in 1663, Guericke placed two hemispheres together, pumped the air out of the resulting sphere through a valve, and showed that teams of horses could not pull the hemispheres apart. Then, after secretly opening the air valve, Guericke easily separated the hemispheres by hand. The King of Prussia was so impressed that he awarded Guericke a lifetime pension!

metal can, and the water is boiled, which fills the can with steam. The can is then sealed and allowed to cool. Why does the can collapse as it cools? It is the atmospheric pressure that crumples the can. When the can is cooled after being sealed so that no air can flow in, the water vapor (steam) condenses to a very small volume of liquid water. As a gas, the water filled the can, but when it is condensed to a liquid, the liquid does not come close to filling the can. The H_2O molecules formerly present as a gas are now collected in a very small volume of liquid, and there are very few molecules of gas left to exert pressure outward and counteract the air pressure. As a result, the pressure exerted by the gas molecules in the atmosphere smashes the can.

A device to measure atmospheric pressure, the **barometer**, was invented in 1643 by an Italian scientist named Evangelista Torricelli (1608–1647), who had been a student of Galileo. Torricelli's barometer is constructed by filling a glass tube with liquid mercury and inverting it in a dish of mercury (Fig. 5.2). Notice that a large quantity of mercury stays in the tube. In fact, at sea level the height of this column of mercury averages 760 mm. Why does this mercury stay in the tube, seemingly in defiance of gravity? Figure 5.2 illustrates how the pressure exerted by the atmospheric gases on the surface of mercury in the dish keeps the mercury in the tube.

Atmospheric pressure results from the mass of the air being pulled toward the center of the earth by gravity—in other words, it results from the weight of the air. Changing weather conditions cause the atmospheric pressure to vary, so the height of the column of Hg supported by the atmosphere at sea level varies; it is not always 760 mm. The meteorologist who says a "low" is approaching means that the atmospheric pressure is going to decrease. This condition often occurs in conjunction with a storm.

Atmospheric pressure also varies with altitude. For example, when Torricelli's experiment is done in Breckenridge, Colorado (elevation 9600 feet), the atmosphere supports a column of mercury only about 520 mm high because the air is "thinner." That is, there is less air pushing down on the earth's surface at Breckenridge than at sea level.

Units of Pressure

Because instruments used for measuring pressure, such as the **manometer** (Fig. 5.3), often contain mercury, the most commonly used units for pressure are based on the height of the mercury column (in millimeters) that the gas pressure can support. The

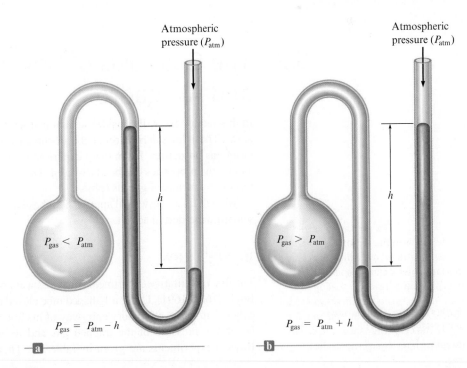

Figure 5.3 | A simple manometer, a device for measuring the pressure of a gas in a container. The pressure of the gas is given by h (the difference in mercury levels) in units of torr (equivalent to mm Hg). (a) Gas pressure = atmospheric pressure − h. (b) Gas pressure = atmospheric pressure + h.

Atmospheric pressure (P_{atm})

$P_{gas} < P_{atm}$

h

$P_{gas} = P_{atm} - h$

a

Atmospheric pressure (P_{atm})

$P_{gas} > P_{atm}$

h

$P_{gas} = P_{atm} + h$

b

Checking tire pressure.

unit **mm Hg** (millimeter of mercury) is often called the **torr** in honor of Torricelli. The terms *torr* and *mm Hg* are used interchangeably by chemists. A related unit for pressure is the **standard atmosphere** (abbreviated atm):

$$1 \text{ standard atmosphere} = 1 \text{ atm} = 760 \text{ mm Hg} = 760 \text{ torr}$$

However, since pressure is defined as force per unit area,

$$\text{Pressure} = \frac{\text{force}}{\text{area}}$$

the fundamental units of pressure involve units of force divided by units of area. In the SI system, the unit of force is the newton (N) and the unit of area is meters squared (m^2). (For a review of the SI system, see Chapter 1.) Thus the unit of pressure in the SI system is newtons per meter squared (N/m^2) and is called the **pascal** (Pa). In terms of pascals, the standard atmosphere is

$$1 \text{ standard atmosphere} = 101,325 \text{ Pa}$$

Thus 1 atmosphere is about 10^5 pascals. Since the pascal is so small, and since it is not commonly used in the United States, we will use it sparingly in this book. However, converting from torrs or atmospheres to pascals is straightforward, as shown in Example 5.1.

Interactive Example 5.1

Sign in at http://login.cengagebrain.com to try this Interactive Example in **OWL**.

1 atm = 760 mm Hg
     = 760 torr
     = 101,325 Pa
     = 29.92 in Hg
     = 14.7 lb/in²

Pressure Conversions

The pressure of a gas is measured as 49 torr. Represent this pressure in both atmospheres and pascals.

Solution

$$49 \text{ torr} \times \frac{1 \text{ atm}}{760 \text{ torr}} = 6.4 \times 10^{-2} \text{ atm}$$

$$6.4 \times 10^{-2} \text{ atm} \times \frac{101,325 \text{ Pa}}{1 \text{ atm}} = 6.5 \times 10^{3} \text{ Pa}$$

See Exercises 5.37 and 5.38

5.2 | The Gas Laws of Boyle, Charles, and Avogadro

In this section we will consider several mathematical laws that relate the properties of gases. These laws derive from experiments involving careful measurements of the relevant gas properties. From these experimental results, the mathematical relationships among the properties can be discovered. These relationships are often represented pictorially by means of graphs (plots).

We will take a historical approach to these laws to give you some perspective on the scientific method in action.

Boyle's Law

Boyle's law: $V \propto 1/P$ at constant temperature

Graphing is reviewed in Appendix 1.3.

The first quantitative experiments on gases were performed by an Irish chemist, Robert Boyle (1627–1691). Using a J-shaped tube closed at one end (Fig. 5.4), which he reportedly set up in the multistory entryway of his house, Boyle studied the relationship between the pressure of the trapped gas and its volume. Representative values from Boyle's experiments are given in Table 5.1. These data show that the product of the

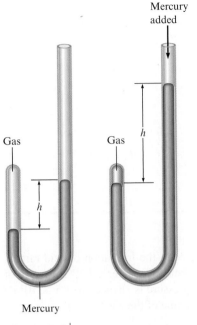

Figure 5.4 | A J-tube similar to the one used by Boyle. When mercury is added to the tube, pressure on the trapped gas is increased, resulting in a decreased volume.

Table 5.1 | Actual Data from Boyle's Experiment

Pressure (in Hg)	Volume (in³)	Pressure × Volume (in Hg × in³)
117.5	12.0	14.1×10^2
87.2	16.0	14.0×10^2
70.7	20.0	14.1×10^2
58.8	24.0	14.1×10^2
44.2	32.0	14.1×10^2
35.3	40.0	14.1×10^2
29.1	48.0	14.0×10^2

pressure and volume for the trapped air sample is constant within the accuracies of Boyle's measurements (note the third column in Table 5.1). This behavior can be represented by the equation

$$PV = k$$

which is called **Boyle's law**, where k is a constant for a given sample of air at a specific temperature.

It is convenient to represent the data in Table 5.1 by using two different plots. The first type of plot, P versus V, forms a curve called a *hyperbola* [Fig. 5.5(a)]. Looking at this plot, note that as the pressure drops by about half (from 58.8 to 29.1), the volume doubles (from 24.0 to 48.0). In other words, there is an *inverse relationship* between pressure and volume. The second type of plot can be obtained by rearranging Boyle's law to give

$$V = \frac{k}{P} = k\frac{1}{P}$$

which is the equation for a straight line of the type

$$y = mx + b$$

where m represents the slope and b is the intercept of the straight line. In this case, $y = V$, $x = 1/P$, $m = k$, and $b = 0$. Thus a plot of V versus $1/P$ using Boyle's data gives a straight line with an intercept of zero [Fig. 5.5(b)].

In the three centuries since Boyle carried out his studies, the sophistication of measuring techniques has increased tremendously. The results of highly accurate measurements show that Boyle's law holds precisely only at very low pressures. Measurements at higher pressures reveal that PV is not constant but varies as the pressure is varied. Results for several gases at pressures below 1 atm are shown in Fig. 5.6. Note the very small changes that occur in the product PV as the pressure is changed at these low pressures. Such changes become more significant at much higher pressures, where the complex nature of the dependence of PV on pressure becomes more obvious. We will discuss these deviations and the reasons for them in detail in Section 5.8. *A gas that strictly obeys Boyle's law is called an **ideal gas**.* We will describe the characteristics of an ideal gas more completely in Section 5.3.

One common use of Boyle's law is to predict the new volume of a gas when the pressure is changed (at constant temperature), or vice versa. Because deviations from Boyle's law are so slight at pressures close to 1 atm, in our calculations we will assume that gases obey Boyle's law (unless stated otherwise).

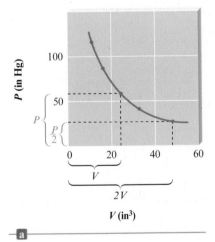

a

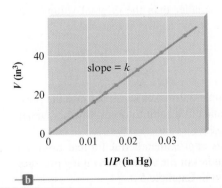

b

Figure 5.5 | Plotting Boyle's data from Table 5.1. (a) A plot of P versus V shows that the volume doubles as the pressure is halved. (b) A plot of V versus $1/P$ gives a straight line. The slope of this line equals the value of the constant k.

Figure 5.6 | A plot of PV versus P for several gases at pressures below 1 atm. An ideal gas is expected to have a constant value of PV, as shown by the dotted line. Carbon dioxide shows the largest change in PV, and this change is actually quite small: PV changes from about 22.39 L · atm at 0.25 atm to 22.26 L · atm at 1.00 atm. Thus Boyle's law is a good approximation at these relatively low pressures.

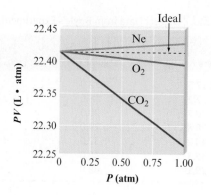

Interactive Example 5.2

Sign in at http://login.cengagebrain.com to try this Interactive Example in **OWL**.

Boyle's Law I

Sulfur dioxide (SO_2), a gas that plays a central role in the formation of acid rain, is found in the exhaust of automobiles and power plants. Consider a 1.53-L sample of gaseous SO_2 at a pressure of 5.6×10^3 Pa. If the pressure is changed to 1.5×10^4 Pa at a constant temperature, what will be the new volume of the gas?

Solution

Where are we going?

To calculate the new volume of gas

What do we know?

> $P_1 = 5.6 \times 10^3$ Pa   > $P_2 = 1.5 \times 10^4$ Pa
> $V_1 = 1.53$ L   > $V_2 = ?$

What information do we need?

> Boyle's law

$$PV = k$$

Boyle's law also can be written as

$$P_1V_1 = P_2V_2$$

5.6×10^3 Pa 1.5×10^4 Pa

$V = 1.53$ L $V = ?$

As pressure increases, the volume of SO_2 decreases.

How do we get there?

What is Boyle's law (in a form useful with our knowns)?

$$P_1V_1 = P_2V_2$$

What is V_2?

$$V_2 = \frac{P_1V_1}{P_2} = \frac{5.6 \times 10^3 \text{ Pa} \times 1.53 \text{ L}}{1.5 \times 10^4 \text{ Pa}} = 0.57 \text{ L}$$

> The new volume will be 0.57 L.

Reality Check | The new volume (0.57 L) is smaller than the original volume. As pressure increases, the volume should decrease, so our answer is reasonable.

See Exercise 5.43

The fact that the volume decreases in Example 5.2 makes sense because the pressure was increased. To help eliminate errors, make it a habit to check whether an answer to a problem makes physical (common!) sense.

We mentioned before that Boyle's law is only approximately true for real gases. To determine the significance of the deviations, studies of the effect of changing pressure on the volume of a gas are often done, as shown in Example 5.3.

Example 5.3

Boyle's Law II

In a study to see how closely gaseous ammonia obeys Boyle's law, several volume measurements were made at various pressures, using 1.0 mole of NH_3 gas at a temperature of 0°C. Using the results listed below, calculate the Boyle's law constant for NH_3 at the various pressures.

Experiment	Pressure (atm)	Volume (L)
1	0.1300	172.1
2	0.2500	89.28
3	0.3000	74.35
4	0.5000	44.49
5	0.7500	29.55
6	1.000	22.08

Solution

To determine how closely NH_3 gas follows Boyle's law under these conditions, we calculate the value of k (in L · atm) for each set of values:

Experiment	1	2	3	4	5	6
$k = PV$	22.37	22.32	22.31	22.25	22.16	22.08

Although the deviations from true Boyle's law behavior are quite small at these low pressures, note that the value of k changes regularly in one direction as the pressure is increased. Thus, to calculate the "ideal" value of k for NH_3, we can plot PV versus P (Fig. 5.7), and extrapolate (extend the line beyond the experimental points) back to zero pressure, where, for reasons we will discuss later, a gas behaves most ideally. The value of k obtained by this extrapolation is 22.41 L · atm. Notice that this is the same value obtained from similar plots for the gases CO_2, O_2, and Ne at 0°C (see Fig. 5.6).

See Exercise 5.125

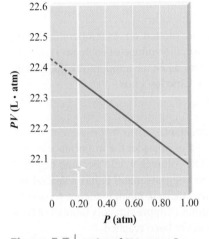

Figure 5.7 | A plot of PV versus P for 1 mole of ammonia. The dashed line shows the extrapolation of the data to zero pressure to give the "ideal" value of PV of 22.41 L · atm.

Charles's Law

In the century following Boyle's findings, scientists continued to study the properties of gases. One of these scientists was a French physicist, Jacques Charles (1746–1823), who was the first person to fill a balloon with hydrogen gas and who made the first solo balloon flight. Charles found in 1787 that the volume of a gas at constant pressure increases *linearly* with the temperature of the gas. That is, a plot of the volume of a gas (at constant pressure) versus its temperature (°C) gives a straight line. This behavior is shown for samples of several gases in Fig. 5.8. The slopes of the lines in this graph are different because the samples contain different numbers of moles of gas. A very interesting feature of these plots is that the volumes of all the gases extrapolate to zero at the same temperature, −273°C. On the Kelvin temperature scale, this point is defined as 0 K, which leads to the following relationship between the Kelvin and Celsius scales:

$$K = °C + 273$$

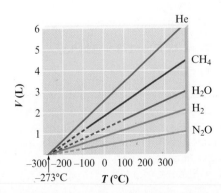

Figure 5.8 | Plots of V versus T (°C) for several gases. The solid lines represent experimental measurements of gases. The dashed lines represent extrapolation of the data into regions where these gases would become liquids or solids. Note that the samples of the various gases contain different numbers of moles.

Figure 5.9 | Plots of V versus T as in Fig. 5.8, except here the Kelvin scale is used for temperature.

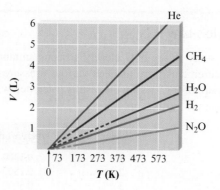

When the volumes of the gases shown in Fig. 5.8 are plotted versus temperature on the Kelvin scale, the plots in Fig. 5.9 result. In this case, the volume of each gas is *directly proportional to temperature* and extrapolates to zero when the temperature is 0 K. This behavior is represented by the equation known as **Charles's law**,

$$V = bT$$

where T is in kelvins and b is a proportionality constant.

Before we illustrate the uses of Charles's law, let us consider the importance of 0 K. At temperatures below this point, the extrapolated volumes would become negative. The fact that a gas cannot have a negative volume suggests that 0 K has a special significance. In fact, 0 K is called **absolute zero**, and there is much evidence to suggest that this temperature cannot be attained. Temperatures of approximately 0.000000001 K have been produced in laboratories, but 0 K has never been reached.

Charles's law: $V \propto T$ (expressed in K) at constant pressure.

Critical Thinking

According to Charles's law, the volume of a gas is directly related to its temperature in kelvins at constant pressure and number of moles. What if the volume of a gas was directly related to its temperature in degrees Celsius at constant pressure and number of moles? What differences would you notice?

Charles's Law

A sample of gas at 15°C and 1 atm has a volume of 2.58 L. What volume will this gas occupy at 38°C and 1 atm?

Solution

Where are we going?

To calculate the new volume of gas

What do we know?

› $T_1 = 15°C + 273 = 288$ K   › $T_2 = 38°C + 273 = 311$ K

› $V_1 = 2.58$ L   › $V_2 = ?$

What information do we need?

› Charles's law

Charles's law also can be written as

$$\frac{V_1}{T_1} = \frac{V_2}{T_2}$$

$$\frac{V}{T} = b$$

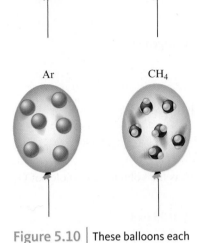

Figure 5.10 | These balloons each hold 1.0 L gas at 25°C and 1 atm. Each balloon contains 0.041 mole of gas, or 2.5×10^{22} molecules.

How do we get there?

What is Charles's law (in a form useful with our knowns)?

$$\frac{V_1}{T_1} = \frac{V_2}{T_2}$$

What is V_2?

■ $V_2 = \left(\frac{T_2}{T_1}\right) V_1 = \left(\frac{311 \text{ K}}{288 \text{ K}}\right) 2.58 \text{ L} = 2.79 \text{ L}$

Reality Check | The new volume is greater than the original volume, which makes physical sense because the gas will expand as it is heated.

See Exercise 5.44

Avogadro's Law

In Chapter 2 we noted that in 1811 the Italian chemist Avogadro postulated that equal volumes of gases at the same temperature and pressure contain the same number of "particles." This observation is called **Avogadro's law**, which is illustrated by Fig. 5.10. Stated mathematically, Avogadro's law is

$$V = an$$

where V is the volume of the gas, n is the number of moles of gas particles, and a is a proportionality constant. This equation states that *for a gas at constant temperature and pressure, the volume is directly proportional to the number of moles of gas.* This relationship is obeyed closely by gases at low pressures.

Interactive Example 5.5

Sign in at http://login.cengagebrain .com to try this Interactive Example in **OWL**.

Avogadro's law also can be written as

$$\frac{V_1}{n_1} = \frac{V_2}{n_2}$$

Avogadro's Law

Suppose we have a 12.2-L sample containing 0.50 mole of oxygen gas (O_2) at a pressure of 1 atm and a temperature of 25°C. If all this O_2 were converted to ozone (O_3) at the same temperature and pressure, what would be the volume of the ozone?

Solution

Where are we going?

To calculate the volume of the ozone produced by 0.50 mole of oxygen

What do we know?

> $n_1 = 0.50 \text{ mol } O_2$ > $n_2 = ? \text{ mol } O_3$
> $V_1 = 12.2 \text{ L } O_2$ > $V_2 = ? \text{ L } O_3$

What information do we need?

> Balanced equation
> Moles of O_3
> Avogadro's law

$$V = an$$

How do we get there?

How many moles of O_3 are produced by 0.50 mole of O_2?

What is the balanced equation?

$$3O_2(g) \longrightarrow 2O_3(g)$$

What is the mole ratio between O_3 and O_2?

$$\frac{2 \text{ mol } O_3}{3 \text{ mol } O_2}$$

Now we can calculate the moles of O_3 formed.

$$0.50 \text{ mol } O_2 \times \frac{2 \text{ mol } O_3}{3 \text{ mol } O_2} = 0.33 \text{ mol } O_3$$

What is the volume of O_3 produced?
Avogadro's law states that $V = an$, which can be rearranged to give

$$\frac{V}{n} = a$$

Since a is a constant, an alternative representation is

$$\frac{V_1}{n_1} = a = \frac{V_2}{n_2}$$

where V_1 is the volume of n_1 moles of O_2 gas and V_2 is the volume of n_2 moles of O_3 gas. In this case we have

> $n_1 = 0.50$ mol   > $n_2 = 0.33$ mol
> $V_1 = 12.2$ L   > $V_2 = ?$

- Solving for V_2 gives

$$V_2 = \left(\frac{n_2}{n_1}\right)V_1 = \left(\frac{0.33 \text{ mol}}{0.50 \text{ mol}}\right)12.2 \text{ L} = 8.1 \text{ L}$$

Reality Check | Note that the volume decreases, as it should, since fewer moles of gas molecules will be present after O_2 is converted to O_3.

See Exercises 5.45 and 5.46

5.3 | The Ideal Gas Law

We have considered three laws that describe the behavior of gases as revealed by experimental observations:

Boyle's law:	$V = \dfrac{k}{P}$	(at constant T and n)
Charles's law:	$V = bT$	(at constant P and n)
Avogadro's law:	$V = an$	(at constant T and P)

These relationships, which show how the volume of a gas depends on pressure, temperature, and number of moles of gas present, can be combined as follows:

$$V = R\left(\frac{Tn}{P}\right)$$

where R is the combined proportionality constant called the **universal gas constant**. When the pressure is expressed in atmospheres and the volume in liters, R has the value 0.08206 L · atm/K · mol. The preceding equation can be rearranged to the more familiar form of the **ideal gas law**:

$R = 0.08206\dfrac{\text{L} \cdot \text{atm}}{\text{K} \cdot \text{mol}}$

$$PV = nRT$$

The ideal gas law is an *equation of state* for a gas, where the state of the gas is its condition at a given time. A particular *state* of a gas is described by its pressure, volume, temperature, and number of moles. Knowledge of any three of these properties is enough to completely define the state of a gas, since the fourth property can then be determined from the equation for the ideal gas law.

It is important to recognize that the ideal gas law is an empirical equation—it is based on experimental measurements of the properties of gases. A gas that obeys this equation is said to behave *ideally*. The ideal gas equation is best regarded as a limiting law—it expresses behavior that real gases *approach* at low pressures and high temperatures. Therefore, an ideal gas is a hypothetical substance. However, most gases obey the ideal gas equation closely enough at pressures below 1 atm that only minimal errors result from assuming ideal behavior. Unless you are given information to the contrary, you should assume ideal gas behavior when solving problems involving gases in this text.

The ideal gas law can be used to solve a variety of problems. Example 5.6 demonstrates one type, where you are asked to find one property characterizing the state of a gas, given the other three.

The ideal gas law applies best at pressures smaller than 1 atm.

Interactive Example 5.6

Sign in at http://login.cengagebrain.com to try this Interactive Example in **OWL**.

The reaction of zinc with hydrochloric acid to produce bubbles of hydrogen gas.

Ideal Gas Law I

A sample of hydrogen gas (H_2) has a volume of 8.56 L at a temperature of 0°C and a pressure of 1.5 atm. Calculate the moles of H_2 molecules present in this gas sample.

Solution

Where are we going?
To calculate the moles of H_2

What do we know?
- $n = ?$ mol H_2
- $V = 8.56$ L
- $P = 1.5$ atm
- $T = 0°C + 273 = 273$ K

What information do we need?
- Ideal gas law
$$PV = nRT$$
- $R = 0.08206$ L · atm/K · mol

How do we get there?
How many moles of H_2 are present in the sample?
- Solve the ideal gas equation for n:

$$n = \frac{(1.5 \text{ atm})(8.56 \text{ L})}{\left(0.08206\frac{\text{L} \cdot \text{atm}}{\text{K} \cdot \text{mol}}\right)(273 \text{ K})} = 0.57 \text{ mol}$$

See Exercises 5.47 through 5.54

Gas law problems can be solved in a variety of ways. They can be classified as a Boyle's law, Charles's law, or Avogadro's law problem and solved, but now we need to remember the specific law and when it applies. The real advantage of using the ideal gas law is that it applies to virtually any problem dealing with gases and is easy to remember.

The basic assumption we make when using the ideal gas law to describe a change in state for a gas is that the equation applies equally well to both the initial and final states. In dealing with changes in state, we always *place the variables that change on one side of the equal sign and the constants on the other.* Let's see how this might work in several examples.

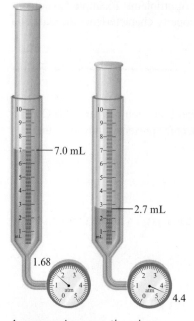

Interactive Example 5.7

Sign in at http://login.cengagebrain.com to try this Interactive Example in **OWL**.

As pressure increases, the volume decreases.

Ideal Gas Law II

Suppose we have a sample of ammonia gas with a volume of 7.0 mL at a pressure of 1.68 atm. The gas is compressed to a volume of 2.7 mL at a constant temperature. Use the ideal gas law to calculate the final pressure.

Solution

Where are we going?

To use the ideal gas equation to determine the final pressure

What do we know?

> $P_1 = 1.68$ atm   > $P_2 = ?$
> $V_1 = 7.0$ mL   > $V_2 = 2.7$ mL

What information do we need?

> Ideal gas law

$$PV = nRT$$

> $R = 0.08206$ L · atm/K · mol

How do we get there?

What are the variables that change?

$$P, V$$

What are the variables that remain constant?

$$n, R, T$$

Write the ideal gas law, collecting the change variables on one side of the equal sign and the variables that do not change on the other.

$$PV = nRT$$

Change   Remain constant

Since n and T remain the same in this case, we can write $P_1V_1 = nRT$ and $P_2V_2 = nRT$. Combining these gives

$$P_1V_1 = nRT = P_2V_2 \quad \text{or} \quad P_1V_1 = P_2V_2$$

We are given $P_1 = 1.68$ atm, $V_1 = 7.0$ mL, and $V_2 = 2.7$ mL. Solving for P_2 thus gives

$$P_2 = \left(\frac{V_1}{V_2}\right)P_1 = \left(\frac{7.0 \text{ mL}}{2.7 \text{ mL}}\right)1.68 \text{ atm} = 4.4 \text{ atm}$$

Reality Check | Does this answer make sense? The volume decreased (at constant temperature), so the pressure should increase, as the result of the calculation indicates. Note that the calculated final pressure is 4.4 atm. Most gases do not behave ideally above 1 atm. Therefore, we might find that if we *measured* the pressure of this gas sample, the observed pressure would differ slightly from 4.4 atm.

See Exercises 5.55 and 5.56

Ideal Gas Law III

A sample of methane gas that has a volume of 3.8 L at 5°C is heated to 86°C at constant pressure. Calculate its new volume.

Solution

Where are we going?

To use the ideal gas equation to determine the final volume

What do we know?

> $T_1 = 5°C + 273 = 278$ K   > $T_2 = 86°C + 273 = 359$ K
> $V_1 = 3.8$ L   > $V_2 = ?$

What information do we need?

> Ideal gas law

$$PV = nRT$$

> $R = 0.08206$ L · atm/K · mol

How do we get there?

What are the variables that change?

$$V, T$$

What are the variables that remain constant?

$$n, R, P$$

Write the ideal gas law, collecting the change variables on one side of the equal sign and the variables that do not change on the other.

$$\frac{V}{T} = \frac{nR}{P}$$

which leads to

$$\frac{V_1}{T_1} = \frac{nR}{P} \quad \text{and} \quad \frac{V_2}{T_2} = \frac{nR}{P}$$

Combining these gives

$$\frac{V_1}{T_1} = \frac{nR}{P} = \frac{V_2}{T_2} \quad \text{or} \quad \frac{V_1}{T_1} = \frac{V_2}{T_2}$$

> Solving for V_2:

$$V_2 = \frac{T_2 V_1}{T_1} = \frac{(359 \text{ K})(3.8 \text{ L})}{278 \text{ K}} = 4.9 \text{ L}$$

Reality Check | Is the answer sensible? In this case the temperature increased (at constant pressure), so the volume should increase. Thus the answer makes sense.

See Exercises 5.57 through 5.59

The problem in Example 5.8 could be described as a Charles's law problem, whereas the problem in Example 5.7 might be called a Boyle's law problem. In both cases, however, we started with the ideal gas law. The real advantage of using the ideal gas law is that it applies to virtually any problem dealing with gases and is easy to remember.

Interactive
Example 5.9

Sign in at http://login.cengagebrain
.com to try this Interactive Example
in OWL.

Ideal Gas Law IV

A sample of diborane gas (B_2H_6), a substance that bursts into flame when exposed to air, has a pressure of 345 torr at a temperature of $-15°C$ and a volume of 3.48 L. If conditions are changed so that the temperature is 36°C and the pressure is 468 torr, what will be the volume of the sample?

Solution

Where are we going?

To use the ideal gas equation to determine the final volume

What do we know?

> $T_1 = 15°C + 273 = 258\ K$ > $T_2 = 36°C + 273 = 309\ K$
> $V_1 = 3.48\ L$ > $V_2 = ?$
> $P_1 = 345\ torr$ > $P_2 = 468\ torr$

What information do we need?

> Ideal gas law

$$PV = nRT$$

> $R = 0.08206\ L \cdot atm/K \cdot mol$

How do we get there?

What are the variables that change?

$$P, V, T$$

What are the variables that remain constant?

$$n, R$$

Write the ideal gas law, collecting the change variables on one side of the equal sign and the variables that do not change on the other.

$$\frac{PV}{T} = nR$$

which leads to

$$\frac{P_1V_1}{T_1} = nR = \frac{P_2V_2}{T_2} \quad \text{or} \quad \frac{P_1V_1}{T_1} = \frac{P_2V_2}{T_2}$$

> Solving for V_2:

$$V_2 = \frac{T_2P_1V_1}{T_1P_2} = \frac{(309\ \cancel{K})(345\ \cancel{torr})(3.48\ L)}{(258\ \cancel{K})(468\ \cancel{torr})} = 3.07\ L$$

See Exercises 5.61 and 5.62

Always convert the temperature to the Kelvin scale when applying the ideal gas law.

Since the equation used in Example 5.9 involves a *ratio* of pressures, it was unnecessary to convert pressures to units of atmospheres. The units of torrs cancel. (You will obtain the same answer by inserting $P_1 = \dfrac{345}{760}$ and $P_2 = \dfrac{468}{760}$ into the equation.)

However, temperature *must always* be converted to the Kelvin scale; since this conversion involves *addition* of 273, the conversion factor does not cancel. Be careful.

One of the many other types of problems dealing with gases that can be solved using the ideal gas law is illustrated in Example 5.10.

Interactive Example 5.10

Sign in at http://login.cengagebrain.com to try this Interactive Example in **OWL**.

A Twyman–Green interferometer emits green argon laser light. Interferometers can measure extremely small distances, useful in configuring telescope mirrors.

Ideal Gas Law V

A sample containing 0.35 mole of argon gas at a temperature of 13°C and a pressure of 568 torr is heated to 56°C and a pressure of 897 torr. Calculate the change in volume that occurs.

Solution

Where are we going?

To use the ideal gas equation to determine the final volume

What do we know?

State 1	State 2
$n_1 = 0.35$ mol	$n_2 = 0.35$ mol
$P_1 = 568 \text{ torr} \times \dfrac{1 \text{ atm}}{760 \text{ torr}} = 0.747$ atm	$P_2 = 897 \text{ torr} \times \dfrac{1 \text{ atm}}{760 \text{ torr}} = 1.18$ atm
$T_1 = 13°C + 273 = 286$ K	$T_2 = 56°C + 273 = 329$ K

What information do we need?

> Ideal gas law

$$PV = nRT$$

> $R = 0.08206 \text{ L} \cdot \text{atm/K} \cdot \text{mol}$

> V_1 and V_2

How do we get there?

What is V_1?

$$V_1 = \frac{n_1 R T_1}{P_1} = \frac{(0.35 \text{ mol})(0.08206 \text{ L} \cdot \text{atm/K} \cdot \text{mol})(286 \text{ K})}{(0.747 \text{ atm})} = 11 \text{ L}$$

What is V_2?

$$V_2 = \frac{n_2 R T_2}{P_2} = \frac{(0.35 \text{ mol})(0.08206 \text{ L} \cdot \text{atm/K} \cdot \text{mol})(329 \text{ K})}{(1.18 \text{ atm})} = 8.0 \text{ L}$$

What is the change in volume ΔV?

$$\blacksquare \quad \Delta V = V_2 - V_1 = 8.0 \text{ L} - 11 \text{ L} = -3 \text{ L}$$

The *change* in volume is negative because the volume decreases.

Note: For this problem (unlike Example 5.9), the pressures must be converted from torrs to atmospheres as required by the atmospheres part of the units for R since each volume was found separately, and the conversion factor does not cancel.

See Exercise 5.63

5.4 | Gas Stoichiometry

When 273.15 K is used in this calculation, the molar volume obtained in Example 5.3 is the same value as 22.41 L.

Suppose we have 1 mole of an ideal gas at 0°C (273.2 K) and 1 atm. From the ideal gas law, the volume of the gas is given by

$$V = \frac{nRT}{P} = \frac{(1.000 \text{ mol})(0.08206 \text{ L} \cdot \text{atm/K} \cdot \text{mol})(273.2 \text{ K})}{1.000 \text{ atm}} = 22.42 \text{ L}$$

Figure 5.11 | 22.42 L of a gas would just fit into this beach ball.

STP: 0°C and 1 atm

Table 5.2 | Molar Volumes for Various Gases at 0°C and 1 atm

Gas	Molar Volume (L)
Oxygen (O_2)	22.397
Nitrogen (N_2)	22.402
Hydrogen (H_2)	22.433
Helium (He)	22.434
Argon (Ar)	22.397
Carbon dioxide (CO_2)	22.260
Ammonia (NH_3)	22.079

This volume of 22.42 L is the **molar volume** of an ideal gas (at 0°C and 1 atm). The measured molar volumes of several gases are listed in Table 5.2. Note that the molar volumes of some of the gases are very close to the ideal value, while others deviate significantly. Later in this chapter, we will discuss some of the reasons for the deviations.

The conditions 0°C and 1 atm, called **standard temperature and pressure** (abbreviated **STP**), are common reference conditions for the properties of gases. For example, the molar volume of an ideal gas is 22.42 L at STP (Fig. 5.11).

Critical Thinking

What if STP was defined as normal room temperature (22°C) and 1 atm? How would this affect the molar volume of an ideal gas? Include an explanation and a number.

Interactive Example 5.11

Sign in at http://login.cengagebrain.com to try this Interactive Example in **OWL**.

Gas Stoichiometry I

A sample of nitrogen gas has a volume of 1.75 L at STP. How many moles of N_2 are present?

Solution

We could solve this problem by using the ideal gas equation, but we can take a shortcut by using the molar volume of an ideal gas at STP. Since 1 mole of an ideal gas at STP has a volume of 22.42 L, 1.75 L N_2 at STP will contain less than 1 mole. We can find how many moles using the ratio of 1.75 L to 22.42 L:

$$1.75 \text{ L } N_2 \times \frac{1 \text{ mol } N_2}{22.42 \text{ L } N_2} = 7.81 \times 10^{-2} \text{ mol } N_2$$

See Exercises 5.65 and 5.66

Many chemical reactions involve gases. By assuming ideal behavior for these gases, we can carry out stoichiometric calculations if the pressure, volume, and temperature of the gases are known.

Interactive Example 5.12

Sign in at http://login.cengagebrain.com to try this Interactive Example in **OWL**.

Gas Stoichiometry II

Quicklime (CaO) is produced by the thermal decomposition of calcium carbonate ($CaCO_3$). Calculate the volume of CO_2 at STP produced from the decomposition of 152 g $CaCO_3$ by the reaction

$$CaCO_3(s) \longrightarrow CaO(s) + CO_2(g)$$

Solution

Where are we going?

To use stoichiometry to determine the volume of CO_2 produced

What do we know?

$$CaCO_3(s) \longrightarrow CaO(s) + CO_2(g)$$

What information do we need?

> Molar volume of a gas at STP is 22.42 L

How do we get there?

We need to use the strategy for solving stoichiometry problems that we learned in Chapter 3.

1. *What is the balanced equation?*

$$CaCO_3(s) \longrightarrow CaO(s) + CO_2(g)$$

2. *What are the moles of $CaCO_3$ (100.09 g/mol)?*

$$152 \text{ g } CaCO_3 \times \frac{1 \text{ mol } CaCO_3}{100.09 \text{ g } CaCO_3} = 1.52 \text{ mol } CaCO_3$$

3. *What is the mole ratio between CO_2 and $CaCO_3$ in the balanced equation?*

$$\frac{1 \text{ mol } CO_2}{1 \text{ mol } CaCO_3}$$

4. *What are the moles of CO_2?*

1.52 moles of CO_2, which is the same as the moles of $CaCO_3$ because the mole ratio is 1.

5. *What is the volume of CO_2 produced?*

We can compute this by using the molar volume since the sample is at STP:

$$1.52 \text{ mol } CO_2 \times \frac{22.42 \text{ L } CO_2}{1 \text{ mol } CO_2} = 34.1 \text{ L } CO_2$$

> Thus the decomposition of 152 g $CaCO_3$ produces 34.1 L CO_2 at STP.

See Exercises 5.67 through 5.70

Remember that the molar volume of an ideal gas is 22.42 L when measured at STP.

Note that in Example 5.12 the final step involved calculation of the volume of gas from the number of moles. Since the conditions were specified as STP, we were able to use the molar volume of a gas at STP. If the conditions of a problem are different from STP, the ideal gas law must be used to compute the volume.

Interactive Example 5.13

Sign in at http://login.cengagebrain.com to try this Interactive Example in **OWL**.

Gas Stoichiometry III

A sample of methane gas having a volume of 2.80 L at 25°C and 1.65 atm was mixed with a sample of oxygen gas having a volume of 35.0 L at 31°C and 1.25 atm. The mixture was then ignited to form carbon dioxide and water. Calculate the volume of CO_2 formed at a pressure of 2.50 atm and a temperature of 125°C.

Solution

Where are we going?

To determine the volume of CO_2 produced

What do we know?

	CH₄	O₂	CO₂
P	1.65 atm	1.25 atm	2.50 atm
V	2.80 L	35.0 L	?
T	25°C + 273 = 298 K	31°C + 273 = 304 K	125°C + 273 = 398 K

What information do we need?

> Balanced chemical equation for the reaction

> Ideal gas law

$$PV = nRT$$

> $R = 0.08206$ L $\cdot$ atm/K $\cdot$ mol

How do we get there?

We need to use the strategy for solving stoichiometry problems that we learned in Chapter 3.

1. *What is the balanced equation?*
   From the description of the reaction, the unbalanced equation is

$$CH_4(g) + O_2(g) \longrightarrow CO_2(g) + H_2O(g)$$

   which can be balanced to give

$$CH_4(g) + 2O_2(g) \longrightarrow CO_2(g) + 2H_2O(g)$$

2. *What is the limiting reactant?*
   We can determine this by using the ideal gas law to determine the moles for each reactant:

$$n_{CH_4} = \frac{PV}{RT} = \frac{(1.65 \text{ atm})(2.80 \text{ L})}{(0.08206 \text{ L} \cdot \text{atm/K} \cdot \text{mol})(298 \text{ K})} = 0.189 \text{ mol}$$

$$n_{O_2} = \frac{PV}{RT} = \frac{(1.25 \text{ atm})(35.0 \text{ L})}{(0.08206 \text{ L} \cdot \text{atm/K} \cdot \text{mol})(304 \text{ K})} = 1.75 \text{ mol}$$

   In the balanced equation for the combustion reaction, 1 mole of CH_4 requires 2 moles of O_2. Thus the moles of O_2 required by 0.189 mole of CH_4 can be calculated as follows:

$$0.189 \text{ mol CH}_4 \times \frac{2 \text{ mol O}_2}{1 \text{ mol CH}_4} = 0.378 \text{ mol O}_2$$

   The limiting reactant is CH_4.

3. *What are the moles of CO₂?*
   Since CH_4 is limiting, we use the moles of CH_4 to determine the moles of CO_2 produced:

$$0.189 \text{ mol CH}_4 \times \frac{1 \text{ mol CO}_2}{1 \text{ mol CH}_4} = 0.189 \text{ mol CO}_2$$

4. *What is the volume of CO₂ produced?*
   Since the conditions stated are not STP, we must use the ideal gas law to calculate the volume:

$$V = \frac{nRT}{P}$$

In this case $n = 0.189$ mol, $T = 125°C + 273 = 398$ K, $P = 2.50$ atm, and $R = 0.08206$ L · atm/K · mol. Thus

■ $V = \dfrac{(0.189 \text{ mol})(0.08206 \text{ L} \cdot \text{atm/K} \cdot \text{mol})(398 \text{ K})}{2.50 \text{ atm}} = 2.47 \text{ L}$

This represents the volume of CO_2 produced under these conditions.

See Exercises 5.71 and 5.74

Molar Mass of a Gas

One very important use of the ideal gas law is in the calculation of the molar mass (molecular weight) of a gas from its measured density. To see the relationship between gas density and molar mass, consider that the number of moles of gas, n, can be expressed as

$$n = \frac{\text{grams of gas}}{\text{molar mass}} = \frac{\text{mass}}{\text{molar mass}} = \frac{m}{\text{molar mass}}$$

Substitution into the ideal gas equation gives

$$P = \frac{nRT}{V} = \frac{(m/\text{molar mass})RT}{V} = \frac{m(RT)}{V(\text{molar mass})}$$

Density $= \dfrac{\text{mass}}{\text{volume}}$

However, m/V is the gas density, d, in units of grams per liter. Thus

$$P = \frac{dRT}{\text{molar mass}}$$

or

$$\text{Molar mass} = \frac{dRT}{P} \qquad (5.1)$$

Thus, if the density of a gas at a given temperature and pressure is known, its molar mass can be calculated.

Interactive Example 5.14

Sign in at http://login.cengagebrain .com to try this Interactive Example in **OWL**.

Gas Density/Molar Mass

The density of a gas was measured at 1.50 atm and 27°C and found to be 1.95 g/L. Calculate the molar mass of the gas.

Solution

Where are we going?

To determine the molar mass of the gas

What do we know?

> $P = 1.50$ atm
> $T = 27°C + 273 = 300.$ K
> $d = 1.95$ g/L

What information do we need?

> Molar mass $= \dfrac{dRT}{P}$

> $R = 0.08206$ L · atm/K · mol

How do we get there?

$$\blacksquare \text{ Molar mass} = \frac{dRT}{P} = \frac{\left(1.95\frac{\text{g}}{\text{L}}\right)\left(0.08206\frac{\text{L} \cdot \text{atm}}{\text{K} \cdot \text{mol}}\right)(300.\text{ K})}{1.50 \text{ atm}} = 32.0 \text{ g/mol}$$

Reality Check | These are the units expected for molar mass.

See Exercises 5.77 through 5.80

You could memorize the equation involving gas density and molar mass, but it is better simply to remember the ideal gas equation, the definition of density, and the relationship between number of moles and molar mass. You can then derive the appropriate equation when you need it. This approach ensures that you understand the concepts and means one less equation to memorize.

5.5 | Dalton's Law of Partial Pressures

Among the experiments that led John Dalton to propose the atomic theory were his studies of mixtures of gases. In 1803 Dalton summarized his observations as follows: *For a mixture of gases in a container, the total pressure exerted is the sum of the pressures that each gas would exert if it were alone.* This statement, known as **Dalton's law of partial pressures**, can be expressed as follows:

$$P_{\text{TOTAL}} = P_1 + P_2 + P_3 + \cdots$$

where the subscripts refer to the individual gases (gas 1, gas 2, and so on). The symbols P_1, P_2, P_3, and so on represent each **partial pressure**, the pressure that a particular gas would exert if it were alone in the container.

Assuming that each gas behaves ideally, the partial pressure of each gas can be calculated from the ideal gas law:

$$P_1 = \frac{n_1RT}{V}, \qquad P_2 = \frac{n_2RT}{V}, \qquad P_3 = \frac{n_3RT}{V}, \qquad \cdots$$

The total pressure of the mixture P_{TOTAL} can be represented as

$$P_{\text{TOTAL}} = P_1 + P_2 + P_3 + \cdots = \frac{n_1RT}{V} + \frac{n_2RT}{V} + \frac{n_3RT}{V} + \cdots$$

$$= (n_1 + n_2 + n_3 + \cdots)\left(\frac{RT}{V}\right)$$

$$= n_{\text{TOTAL}}\left(\frac{RT}{V}\right)$$

where n_{TOTAL} is the sum of the numbers of moles of the various gases. Thus, for a mixture of ideal gases, it is the *total number of moles of particles* that is important, not the identity or composition of the involved gas particles. This idea is illustrated in Fig. 5.12.

Figure 5.12 | The partial pressure of each gas in a mixture of gases in a container depends on the number of moles of that gas. The total pressure is the sum of the partial pressures and depends on the total moles of gas particles present, no matter what they are. Note that the volume remains constant.

This important observation indicates some fundamental characteristics of an ideal gas. The fact that the pressure exerted by an ideal gas is not affected by the identity (composition) of the gas particles reveals two things about ideal gases: (1) the volume of the individual gas particle must not be important, and (2) the forces among the particles must not be important. If these factors were important, the pressure exerted by the gas would depend on the nature of the individual particles. These observations will strongly influence the model that we will eventually construct to explain ideal gas behavior.

Interactive Example 5.15

Sign in at http://login.cengagebrain.com to try this Interactive Example in **OWL**.

Dalton's Law I

Mixtures of helium and oxygen can be used in scuba diving tanks to help prevent "the bends." For a particular dive, 46 L He at 25°C and 1.0 atm and 12 L O_2 at 25°C and 1.0 atm were pumped into a tank with a volume of 5.0 L. Calculate the partial pressure of each gas and the total pressure in the tank at 25°C.

Solution

Where are we going?
To determine the partial pressure of each gas
To determine the total pressure in the tank at 25°C

What do we know?

	He	O_2	Tank
P	1.00 atm	1.00 atm	? atm
V	46 L	12 L	5.0 L
T	25°C + 273 = 298 K	25°C + 273 = 298 K	25°C + 273 = 298 K

What information do we need?
> Ideal gas law
$$PV = nRT$$
> $R = 0.08206$ L · atm/K · mol

How do we get there?
How many moles are present for each gas?
$$n = \frac{PV}{RT}$$
$$n_{He} = \frac{(1.0 \text{ atm})(46 \text{ L})}{(0.08206 \text{ L} \cdot \text{atm/K} \cdot \text{mol})(298 \text{ K})} = 1.9 \text{ mol}$$
$$n_{O_2} = \frac{(1.0 \text{ atm})(12 \text{ L})}{(0.08206 \text{ L} \cdot \text{atm/K} \cdot \text{mol})(298 \text{ K})} = 0.49 \text{ mol}$$

What is the partial pressure for each gas in the tank?
The tank containing the mixture has a volume of 5.0 L, and the temperature is 25°C. We can use these data and the ideal gas law to calculate the partial pressure of each gas:
$$P = \frac{nRT}{V}$$
$$P_{He} = \frac{(1.9 \text{ mol})(0.08206 \text{ L} \cdot \text{atm/K} \cdot \text{mol})(298 \text{ K})}{5.0 \text{ L}} = 9.3 \text{ atm}$$
$$P_{O_2} = \frac{(0.49 \text{ mol})(0.08206 \text{ L} \cdot \text{atm/K} \cdot \text{mol})(298 \text{ K})}{5.0 \text{ L}} = 2.4 \text{ atm}$$

Chemical connections
Separating Gases

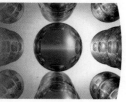

Assume you work for an oil company that owns a huge natural gas reservoir containing a mixture of methane and nitrogen gases. In fact, the gas mixture contains so much nitrogen that it is unusable as a fuel. Your job is to separate the nitrogen (N_2) from the methane (CH_4). How might you accomplish this task? You clearly need some sort of "molecular filter" that will stop the slightly larger methane molecules (size $\approx$ 430 pm) and allow the nitrogen molecules (size $\approx$ 410 pm) to pass through. To accomplish the separation of molecules so similar in size will require a very precise "filter."

The good news is that such a filter exists. Recent work by Steven Kuznicki and Valerie Bell at Engelhard Corporation in New Jersey and Michael Tsapatsis at the University of

Massachusetts has produced a "molecular sieve" in which the pore (passage) sizes can be adjusted precisely enough to separate N_2 molecules from CH_4 molecules. The material involved is a special hydrated titanosilicate (contains H_2O, Ti, Si, O, and Sr) compound patented by Engelhard known as ETS-4 (Engelhard TitanoSilicate-4). When sodium ions are substituted for the strontium ions

in ETS-4 and the new material is carefully dehydrated, a uniform and controllable pore-size reduction occurs (see figure). The researchers have shown that the material can be used to separate N_2 ($\approx$ 410 pm) from O_2 ($\approx$ 390 pm). They have also shown that it is possible to reduce the nitrogen content of natural gas from 18% to less than 5% with a 90% recovery of methane.

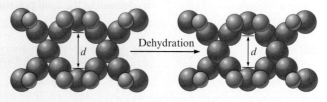

Molecular sieve framework of titanium (blue), silicon (green), and oxygen (red) atoms contracts on heating—at room temperature (left), d = 4.27 Å; at 250°C (right), d = 3.94 Å.

What is the total pressure of the mixture of gases in the tank?
The total pressure is the sum of the partial pressures:

$$\blacksquare \quad P_{\text{TOTAL}} = P_{\text{He}} + P_{O_2} = 9.3 \text{ atm} + 2.4 \text{ atm} = 11.7 \text{ atm}$$

See Exercises 5.83 and 5.84

At this point we need to define the **mole fraction**: *the ratio of the number of moles of a given component in a mixture to the total number of moles in the mixture.* The Greek lowercase letter chi (χ) is used to symbolize the mole fraction. For example, for a given component in a mixture, the mole fraction χ_1 is

$$\chi_1 = \frac{n_1}{n_{\text{TOTAL}}} = \frac{n_1}{n_1 + n_2 + n_3 + \cdots}$$

From the ideal gas equation, we know that the number of moles of a gas is directly proportional to the pressure of the gas, since

$$n = P\left(\frac{V}{RT}\right)$$

Chemical connections
The Chemistry of Air Bags

The inclusion of air bags in modern automobiles has led to a significant reduction in the number of injuries as a result of car crashes. Air bags are stored in the steering wheel and dashboard of all cars, and many autos now have additional air bags that protect the occupant's knees, head, and shoulders. In fact, some auto manufacturers now include air bags in the seat belts. Also, because deployment of an air bag can severely injure a child, all cars now have "smart" air bags that deploy with an inflation force that is proportional to the seat occupant's weight.

The term "air bag" is really a misnomer because air is not involved in the inflation process. Rather, an air bag inflates rapidly (in about 30 ms) due to the explosive production of

N$_2$ gas. Originally, sodium azide, which decomposes to produce N$_2$,

$$2NaN_3(s) \rightarrow 2Na(s) + 3N_2(g)$$

was used, but it has now been replaced by less toxic materials.

The sensing devices that trigger the air bags must react very rapidly. For example, consider a car hitting a concrete bridge abutment. When this happens, an internal accelerometer sends a message to the control module that a collision possibly is occurring. The microprocessor then analyzes the measured deceleration from several accelerometers and door pressure sensors and decides whether air bag deployment is appropriate. All this happens within 8 to 40 ms of the initial impact.

Because an air bag must provide the appropriate cushioning effect, the

bag begins to vent even as it is being filled. In fact, the maximum pressure in the bag is 5 pounds per square inch (psi), even in the middle of a collision event. Air bags represent a case where an explosive chemical reaction saves lives rather than the reverse.

Inflated air bags.

That is, for each component in the mixture,

$$n_1 = P_1\left(\frac{V}{RT}\right), \qquad n_2 = P_2\left(\frac{V}{RT}\right), \qquad \cdots$$

Therefore, we can represent the mole fraction in terms of pressures:

$$\chi_1 = \frac{n_1}{n_{\text{TOTAL}}} = \frac{\overbrace{P_1(V/RT)}^{n_1}}{\underbrace{P_1(V/RT)}_{n_1} + \underbrace{P_2(V/RT)}_{n_2} + \underbrace{P_3(V/RT)}_{n_3} + \cdots}$$

$$= \frac{(V/RT)P_1}{(V/RT)(P_1 + P_2 + P_3 + \cdots)}$$

$$= \frac{P_1}{P_1 + P_2 + P_3 + \cdots} = \frac{P_1}{P_{\text{TOTAL}}}$$

In fact, the mole fraction of each component in a mixture of ideal gases is directly related to its partial pressure:

$$\chi_2 = \frac{n_2}{n_{\text{TOTAL}}} = \frac{P_2}{P_{\text{TOTAL}}}$$

Dalton's Law II

The partial pressure of oxygen was observed to be 156 torr in air with a total atmospheric pressure of 743 torr. Calculate the mole fraction of O_2 present.

Solution

Where are we going?

To determine the mole fraction of O_2

What do we know?

> P_{O_2} = 156 torr
> P_{TOTAL} = 743 torr

How do we get there?

The mole fraction of O_2 can be calculated from the equation

$$\chi_{O_2} = \frac{P_{O_2}}{P_{TOTAL}} = \frac{156 \text{ torr}}{743 \text{ torr}} = 0.210$$

Note that the mole fraction has no units.

See Exercise 5.89

The expression for the mole fraction,

$$\chi_1 = \frac{P_1}{P_{TOTAL}}$$

can be rearranged to give

$$P_1 = \chi_1 \times P_{TOTAL}$$

That is, *the partial pressure of a particular component of a gaseous mixture is the mole fraction of that component times the total pressure.*

Dalton's Law III

The mole fraction of nitrogen in the air is 0.7808. Calculate the partial pressure of N_2 in air when the atmospheric pressure is 760. torr.

Solution

The partial pressure of N_2 can be calculated as follows:

$$P_{N_2} = \chi_{N_2} \times P_{TOTAL} = 0.7808 \times 760. \text{ torr} = 593 \text{ torr}$$

See Exercise 5.90

Collecting a Gas over Water

A mixture of gases results whenever a gas is collected by displacement of water. For example, Fig. 5.13 shows the collection of oxygen gas produced by the decomposition of solid potassium chlorate. In this situation, the gas in the bottle is a mixture of water vapor and the oxygen being collected. Water vapor is present because molecules of water escape from the surface of the liquid and collect in the space above the liquid.

Figure 5.13 | The production of oxygen by thermal decomposition of $KClO_3$. The MnO_2 is mixed with the $KClO_3$ to make the reaction faster.

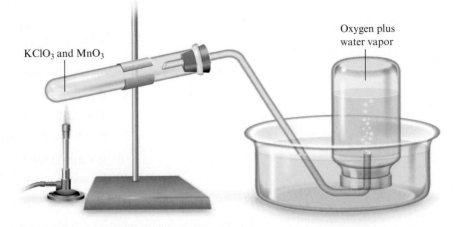

KClO₃ and MnO₃

Oxygen plus water vapor

Molecules of water also return to the liquid. When the rate of escape equals the rate of return, the number of water molecules in the vapor state remains constant, and thus the pressure of water vapor remains constant. This pressure, which depends on temperature, is called the *vapor pressure of water.*

Vapor pressure will be discussed in detail in Chapter 10. A table of water vapor pressure values is given in Section 10.8.

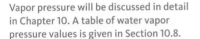
Interactive Example 5.18

Sign in at http://login.cengagebrain .com to try this Interactive Example in **OWL**.

Gas Collection over Water

A sample of solid potassium chlorate ($KClO_3$) was heated in a test tube (see Fig. 5.13) and decomposed by the following reaction:

$$2KClO_3(s) \longrightarrow 2KCl(s) + 3O_2(g)$$

The oxygen produced was collected by displacement of water at 22°C at a total pressure of 754 torr. The volume of the gas collected was 0.650 L, and the vapor pressure of water at 22°C is 21 torr. Calculate the partial pressure of O_2 in the gas collected and the mass of $KClO_3$ in the sample that was decomposed.

Solution

Where are we going?

To determine the partial pressure of O_2 in the gas collected
Calculate the mass of $KClO_3$ in the original sample

What do we know?

	Gas Collected	Water Vapor
P	754 torr	21 torr
V	0.650 L	
T	22°C + 273 = 295 K	22°C + 273 = 295 K

How do we get there?

What is the partial pressure of O_2?

$$P_{TOTAL} = P_{O_2} + P_{H_2O} = P_{O_2} + 21 \text{ torr} = 754 \text{ torr}$$

$$\blacksquare \ P_{O_2} = 754 \text{ torr} - 21 \text{ torr} = 733 \text{ torr}$$

What is the number of moles of O_2?

Now we use the ideal gas law to find the number of moles of O_2:

$$n_{O_2} = \frac{P_{O_2}V}{RT}$$

In this case, the partial pressure of the O_2 is

$$P_{O_2} = 733 \text{ torr} = \frac{733 \text{ torr}}{760 \text{ torr/atm}} = 0.964 \text{ atm}$$

To find the moles of O_2 produced, we use

$$V = 0.650 \text{ L}$$
$$T = 22°C + 273 = 295 \text{ K}$$
$$R = 0.08206 \text{ L} \cdot \text{atm/K} \cdot \text{mol}$$

$$n_{O_2} = \frac{(0.964 \text{ atm})(0.650 \text{ L})}{(0.08206 \text{ L} \cdot \text{atm/K} \cdot \text{mol})(295 \text{ K})} = 2.59 \times 10^{-2} \text{ mol}$$

How many moles of $KClO_3$ are required to produce this amount of O_2?

Use the stoichiometry problem-solving strategy:

1. *What is the balanced equation?*

$$2KClO_3(s) \longrightarrow 2KCl(s) + 3O_2(g)$$

2. *What is the mole ratio between $KClO_3$ and O_2 in the balanced equation?*

$$\frac{2 \text{ mol } KClO_3}{3 \text{ mol } O_2}$$

3. *What are the moles of $KClO_3$?*

$$2.59 \times 10^{-2} \text{ mol } O_2 \times \frac{2 \text{ mol } KClO_3}{3 \text{ mol } O_2} = 1.73 \times 10^{-2} \text{ mol } KClO_3$$

4. *What is the mass of $KClO_3$ (molar mass 122.6 g/mol) in the original sample?*

$$1.73 \times 10^{-2} \text{ mol } KClO_3 \times \frac{122.6 \text{ g } KClO_3}{1 \text{ mol } KClO_3} = 2.12 \text{ g } KClO_3$$

❯ Thus the original sample contained 2.12 g $KClO_3$.

See Exercises 5.91 through 5.93

5.6 | The Kinetic Molecular Theory of Gases

We have so far considered the behavior of gases from an experimental point of view. Based on observations from different types of experiments, we know that at pressures of less than 1 atm most gases closely approach the behavior described by the ideal gas law. Now we want to construct a model to explain this behavior.

Before we do this, let's briefly review the scientific method. Recall that a law is a way of generalizing behavior that has been observed in many experiments. Laws are very useful, since they allow us to predict the behavior of similar systems. For example, if a chemist prepares a new gaseous compound, a measurement of the gas density at known pressure and temperature can provide a reliable value for the compound's molar mass.

However, although laws summarize observed behavior, they do not tell us *why* nature behaves in the observed fashion. This is the central question for scientists. To try to answer this question, we construct theories (build models). The models in chemistry consist of speculations about what the individual atoms or molecules (microscopic particles) might be doing to cause the observed behavior of the macroscopic systems (collections of very large numbers of atoms and molecules).

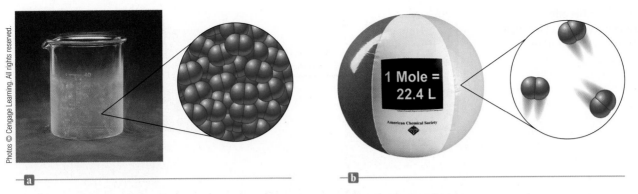

Figure 5.14 | (a) One mole of $N_2(l)$ has a volume of approximately 35 mL and a density of 0.81 g/mL. (b) One mole of $N_2(g)$ has a volume of 22.42 L (STP) and a density of 1.2×10^{-3} g/mL. Thus the ratio of the volumes of gaseous N_2 and liquid N_2 is 22.42/0.035 = 640, and the spacing of the molecules is 9 times farther apart in $N_2(g)$.

A model is considered successful if it explains the observed behavior in question and predicts correctly the results of future experiments. It is important to understand that a model can never be proved absolutely true. In fact, *any model is an approximation* by its very nature and is bound to fail at some point. Models range from the simple to the extraordinarily complex. We use simple models to predict approximate behavior and more complicated models to account very precisely for observed quantitative behavior. In this text we will stress simple models that provide an approximate picture of what might be happening and that fit the most important experimental results.

An example of this type of model is the **kinetic molecular theory (KMT)**, a simple model that attempts to explain the properties of an ideal gas. This model is based on speculations about the behavior of the individual gas particles (atoms or molecules). The postulates of the kinetic molecular theory as they relate to the particles of an ideal gas can be stated as follows:

Postulates of the Kinetic Molecular Theory

1. The particles are so small compared with the distances between them that *the volume of the individual particles can be assumed to be negligible* (zero) (Fig. 5.14).

2. *The particles are in constant motion. The collisions of the particles with the walls of the container are the cause of the pressure exerted by the gas.*

3. *The particles are assumed to exert no forces on each other;* they are assumed neither to attract nor to repel each other.

4. *The average kinetic energy of a collection of gas particles is assumed to be directly proportional to the Kelvin temperature of the gas.*

Of course, the molecules in a real gas have finite volumes and do exert forces on each other. Thus *real gases* do not conform to these assumptions. However, we will see that these postulates do indeed explain *ideal gas* behavior.

The true test of a model is how well its predictions fit the experimental observations. The postulates of the kinetic molecular model picture an ideal gas as consisting of particles having no volume and no attractions for each other, and the model assumes that the gas produces pressure on its container by collisions with the walls.

Let's consider how this model accounts for the properties of gases as summarized by the ideal gas law: $PV = nRT$.

Figure 5.15 | The effects of decreasing the volume of a sample of gas at constant temperature.

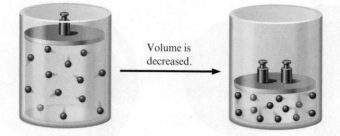

Figure 5.15 | The effects of decreasing the volume of a sample of gas at constant temperature.

Volume is decreased.

Pressure and Volume (Boyle's Law)

We have seen that for a given sample of gas at a given temperature (*n* and *T* are constant) that if the volume of a gas is decreased, the pressure increases:

$$P = \underbrace{(nRT)}_{\text{Constant}}\frac{1}{V}$$

This makes sense based on the kinetic molecular theory because a decrease in volume means that the gas particles will hit the wall more often, thus increasing pressure (Fig. 5.15).

Pressure and Temperature

From the ideal gas law, we can predict that for a given sample of an ideal gas at a constant volume, the pressure will be directly proportional to the temperature:

$$P = \underbrace{\left(\frac{nR}{V}\right)}_{\text{Constant}}T$$

The KMT accounts for this behavior because when the temperature of a gas increases, the speeds of its particles increase, the particles hitting the wall with greater force and greater frequency. Since the volume remains the same, this would result in increased gas pressure (Fig. 5.16).

Critical Thinking

You have learned the postulates of the KMT. What if we could not assume the third postulate to be true? How would this affect the measured pressure of a gas?

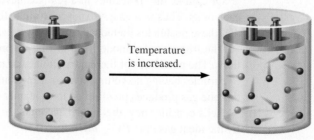

Temperature is increased.

Figure 5.16 | The effects of increasing the temperature of a sample of gas at constant volume.

Figure 5.17 | The effects of increasing the temperature of a sample of gas at constant pressure.

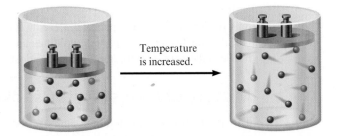

Volume and Temperature (Charles's Law)

The ideal gas law indicates that for a given sample of gas at a constant pressure, the volume of the gas is directly proportional to the temperature in kelvins:

$$V = \left(\frac{nR}{P}\right)T$$

$\underbrace{\phantom{\frac{nR}{P}}}$
↑
Constant

This can be visualized from the KMT (Fig. 5.17). When the gas is heated to a higher temperature, the speeds of its molecules increase and thus they hit the walls more often and with more force. The only way to keep the pressure constant in this situation is to increase the volume of the container. This compensates for the increased particle speeds.

Volume and Number of Moles (Avogadro's Law)

The ideal gas law predicts that the volume of a gas at a constant temperature and pressure depends directly on the number of gas particles present:

$$V = \left(\frac{RT}{P}\right)n$$

$\underbrace{\phantom{\frac{RT}{P}}}$
↑
Constant

This makes sense in terms of the KMT because an increase in the number of gas particles at the same temperature would cause the pressure to increase if the volume were held constant (Fig. 5.18). The only way to return the pressure to its original value is to increase the volume.

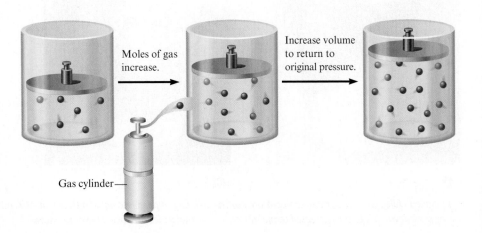

Figure 5.18 | The effects of increasing the number of moles of gas particles at constant temperature and pressure.

It is important to recognize that the volume of a gas (at constant P and T) depends only on the *number* of gas particles present. The individual volumes of the particles are not a factor because the particle volumes are so small compared with the distances between the particles (for a gas behaving ideally).

Mixture of Gases (Dalton's Law)

The observation that the total pressure exerted by a mixture of gases is the sum of the pressures of the individual gases is expected because the KMT assumes that all gas particles are independent of each other and that the volumes of the individual particles are unimportant. Thus the identities of the gas particles do not matter.

Deriving the Ideal Gas Law

We have shown qualitatively that the assumptions of the KMT successfully account for the observed behavior of an ideal gas. We can go further. By applying the principles of physics to the assumptions of the KMT, we can in effect derive the ideal gas law.

As shown in detail in Appendix 2, we can apply the definitions of velocity, momentum, force, and pressure to the collection of particles in an ideal gas and *derive* the following expression for pressure:

$$P = \frac{2}{3}\left[\frac{nN_A\left(\frac{1}{2}m\overline{u^2}\right)}{V}\right]$$

where P is the pressure of the gas, n is the number of moles of gas, N_A is Avogadro's number, m is the mass of each particle, $\overline{u^2}$ is the average of the square of the velocities of the particles, and V is the volume of the container.

The quantity $\frac{1}{2}m\overline{u^2}$ represents the average kinetic energy of a gas particle. If the average kinetic energy of an individual particle is multiplied by N_A, the number of particles in a mole, we get the average kinetic energy for a mole of gas particles:

$$(KE)_{avg} = N_A\left(\frac{1}{2}m\overline{u^2}\right)$$

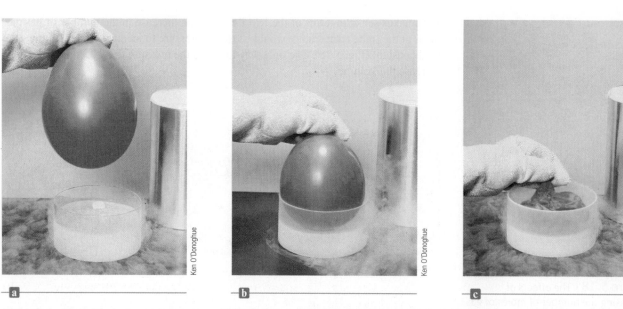

(a) A balloon filled with air at room temperature. (b) The balloon is dipped into liquid nitrogen at 77 K. (c) The balloon collapses as the molecules inside slow down due to the decreased temperature. Slower molecules produce a lower pressure.

Kinetic energy (KE) given by the equation $KE = \frac{1}{2}m\overline{u}^2$ is the energy due to the motion of a particle. We will discuss this further in Section 6.1.

Using this definition, we can rewrite the expression for pressure as

$$P = \frac{2}{3}\left[\frac{n(KE)_{avg}}{V}\right] \quad \text{or} \quad \frac{PV}{n} = \frac{2}{3}(KE)_{avg}$$

The fourth postulate of the kinetic molecular theory is that the average kinetic energy of the particles in the gas sample is directly proportional to the temperature in kelvin. Thus, since $(KE)_{avg} \propto T$, we can write

$$\frac{PV}{n} = \frac{2}{3}(KE)_{avg} \propto T \quad \text{or} \quad \frac{PV}{n} \propto T$$

Note that this expression has been *derived* from the assumptions of the kinetic molecular theory. How does it compare to the ideal gas law—the equation obtained from experiment? Compare the ideal gas law,

$$\frac{PV}{n} = RT \quad \text{From experiment}$$

with the result from the kinetic molecular theory,

$$\frac{PV}{n} \propto T \quad \text{From theory}$$

These expressions have exactly the same form if R, the universal gas constant, is considered the proportionality constant in the second case.

The agreement between the ideal gas law and the predictions of the kinetic molecular theory gives us confidence in the validity of the model. The characteristics we have assumed for ideal gas particles must agree, at least under certain conditions, with their actual behavior.

The Meaning of Temperature

We have seen from the kinetic molecular theory that the Kelvin temperature indicates the average kinetic energy of the gas particles. The exact relationship between temperature and average kinetic energy can be obtained by combining the equations:

$$\frac{PV}{n} = RT = \frac{2}{3}(KE)_{avg}$$

which yields the expression

$$(KE)_{avg} = \frac{3}{2}RT$$

This is a very important relationship. It summarizes the meaning of the Kelvin temperature of a gas: The Kelvin temperature is an index of the random motions of the particles of a gas, with higher temperature meaning greater motion. (As we will see in Chapter 10, temperature is an index of the random motions in solids and liquids as well as in gases.)

Root Mean Square Velocity

In the equation from the kinetic molecular theory, the average velocity of the gas particles is a special kind of average. The symbol $\overline{u^2}$ means the average of the *squares* of the particle velocities. The square root of $\overline{u^2}$ is called the **root mean square velocity** and is symbolized by u_{rms}:

$$u_{rms} = \sqrt{\overline{u^2}}$$

We can obtain an expression for u_{rms} from the equations

$$(KE)_{avg} = N_A(\tfrac{1}{2}m\overline{u^2}) \quad \text{and} \quad (KE)_{avg} = \frac{3}{2}RT$$

Combination of these equations gives

$$N_A(\tfrac{1}{2}m\overline{u^2}) = \frac{3}{2}RT \quad \text{or} \quad \overline{u^2} = \frac{3RT}{N_A m}$$

Taking the square root of both sides of the last equation produces

$$\sqrt{\overline{u^2}} = u_{rms} = \sqrt{\frac{3RT}{N_A m}}$$

In this expression m represents the mass in kilograms of a single gas particle. When N_A, the number of particles in a mole, is multiplied by m, the product is the mass of a *mole* of gas particles in *kilograms*. We will call this quantity M. Substituting M for $N_A m$ in the equation for u_{rms}, we obtain

$$u_{rms} = \sqrt{\frac{3RT}{M}}$$

$$R = 0.08206 \frac{\text{L} \cdot \text{atm}}{\text{K} \cdot \text{mol}}$$

$$R = 8.3145 \frac{\text{J}}{\text{K} \cdot \text{mol}}$$

Before we can use this equation, we need to consider the units for R. So far we have used 0.08206 L · atm/K · mol as the value of R. But to obtain the desired units (meters per second) for u_{rms}, R must be expressed in different units. As we will see in more detail in Chapter 6, the energy unit most often used in the SI system is the joule (J). A **joule** is defined as a kilogram meter squared per second squared (kg · m²/s²). When R is converted to include the unit of joules, it has the value 8.3145 J/K · mol. When R in these units is used in the expression $\sqrt{3RT/M}$, u_{rms} is obtained in the units of meters per second as desired.

Interactive Example 5.19

Sign in at http://login.cengagebrain.com to try this Interactive Example in OWL.

Root Mean Square Velocity

Calculate the root mean square velocity for the atoms in a sample of helium gas at 25°C.

Solution

Where are we going?

To determine the root mean square velocity for the atoms of He

What do we know?

> $T = 25°C + 273 = 298$ K

> $R = 8.3145$ J/K · mol

What information do we need?

> Root mean square velocity is $\quad u_{rms} = \sqrt{\dfrac{3RT}{M}}$

How do we get there?

What is the mass of a mole of He in kilograms?

$$M = 4.00 \frac{\text{g}}{\text{mol}} \times \frac{1 \text{ kg}}{1000 \text{ g}} = 4.00 \times 10^{-3} \text{ kg/mol}$$

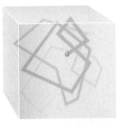

Figure 5.19 | Path of one particle in a gas. Any given particle will continuously change its course as a result of collisions with other particles, as well as with the walls of the container.

What is the root mean square velocity for the atoms of He?

$$u_{rms} = \sqrt{\frac{3\left(8.3145\frac{J}{K \cdot mol}\right)(298\ K)}{4.00 \times 10^{-3}\frac{kg}{mol}}} = \sqrt{1.86 \times 10^6 \frac{J}{kg}}$$

Since the units of J are kg · m²/s², this expression gives

$$u_{rms} = \sqrt{1.86 \times 10^6 \frac{kg \cdot m^2}{kg \cdot s^2}} = 1.36 \times 10^3\ m/s$$

Reality Check | The resulting units are appropriate for velocity.

See Exercises 5.103 and 5.104

So far we have said nothing about the range of velocities actually found in a gas sample. In a real gas, there are large numbers of collisions between particles. For example, as we will see in the next section, when an odorous gas such as ammonia is released in a room, it takes some time for the odor to permeate the air. This delay results from collisions between the NH_3 molecules and the O_2 and N_2 molecules in the air, which greatly slow the mixing process.

If the path of a particular gas particle could be monitored, it would look very erratic, something like that shown in Fig. 5.19. The average distance a particle travels between collisions in a particular gas sample is called the *mean free path*. It is typically a very small distance (1×10^{-7} m for O_2 at STP). One effect of the many collisions among gas particles is to produce a large range of velocities as the particles collide and exchange kinetic energy. Although u_{rms} for oxygen gas at STP is approximately 500 m/s, the majority of O_2 molecules do not have this velocity. The actual distribution of molecular velocities for oxygen gas at STP is shown in Fig. 5.20. This figure shows the relative number of gas molecules having each particular velocity.

We are also interested in the effect of *temperature* on the velocity distribution in a gas. Figure 5.21 shows the velocity distribution (called the Maxwell-Boltzmann distribution) for nitrogen gas at three temperatures. Note that as the temperature is increased,

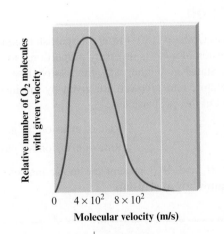

Figure 5.20 | A plot of the relative number of O_2 molecules that have a given velocity at STP.

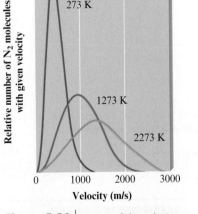

Figure 5.21 | A plot of the relative number of N_2 molecules that have a given velocity at three temperatures. Note that as the temperature increases, both the average velocity and the spread of velocities increase.

Figure 5.22 | The effusion of a gas into an evacuated chamber. The rate of effusion (the rate at which the gas is transferred across the barrier through the pin hole) is inversely proportional to the square root of the mass of the gas molecules.

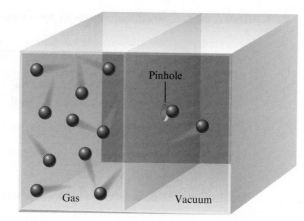

the curve peak moves toward higher values and the range of velocities becomes much larger. The peak of the curve reflects the most probable velocity (the velocity found most often as we sample the movement of the various particles in the gas). Because the kinetic energy increases with temperature, it makes sense that the peak of the curve should move to higher values as the temperature of the gas is increased.

5.7 | Effusion and Diffusion

We have seen that the postulates of the kinetic molecular theory, when combined with the appropriate physical principles, produce an equation that successfully fits the experimentally observed behavior of gases as they approach ideal behavior. Two phenomena involving gases provide further tests of this model.

Diffusion is the term used to describe the mixing of gases. When a small amount of pungent-smelling ammonia is released at the front of a classroom, it takes some time before everyone in the room can smell it, because time is required for the ammonia to mix with the air. The rate of diffusion is the rate of the mixing of gases. **Effusion** is the term used to describe the passage of a gas through a tiny orifice into an evacuated chamber, as shown in Fig. 5.22. The rate of effusion measures the speed at which the gas is transferred into the chamber.

Effusion

Thomas Graham (1805–1869), a Scottish chemist, found experimentally that the rate of effusion of a gas is inversely proportional to the square root of the mass of its particles. Stated in another way, the relative rates of effusion of two gases at the same temperature and pressure are given by the inverse ratio of the square roots of the masses of the gas particles:

In Graham's law the units for molar mass can be g/mol or kg/mol, since the units cancel in the ratio $\dfrac{\sqrt{M_2}}{\sqrt{M_1}}$

$$\frac{\text{Rate of effusion for gas 1}}{\text{Rate of effusion for gas 2}} = \frac{\sqrt{M_2}}{\sqrt{M_1}}$$

where M_1 and M_2 represent the molar masses of the gases. This equation is called **Graham's law of effusion**.

Interactive
Example 5.20

Sign in at http://login.cengagebrain
.com to try this Interactive Example
in **OWL**.

Effusion Rates

Calculate the ratio of the effusion rates of hydrogen gas (H_2) and uranium hexafluoride (UF_6), a gas used in the enrichment process to produce fuel for nuclear reactors (Fig. 5.23).

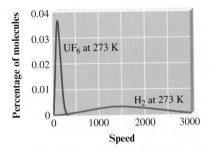

Figure 5.23 | Relative molecular speed distribution of H_2 and UF_6.

Solution

First we need to compute the molar masses: Molar mass of H_2 = 2.016 g/mol, and molar mass of UF_6 = 352.02 g/mol. Using Graham's law,

$$\frac{\text{Rate of effusion for } H_2}{\text{Rate of effusion for } UF_6} = \frac{\sqrt{M_{UF_6}}}{\sqrt{M_{H_2}}} = \sqrt{\frac{352.02}{2.016}} = 13.2$$

The effusion rate of the very light H_2 molecules is about 13 times that of the massive UF_6 molecules.

See Exercises 5.111 through 5.114

Does the kinetic molecular model for gases correctly predict the relative effusion rates of gases summarized by Graham's law? To answer this question, we must recognize that the effusion rate for a gas depends directly on the average velocity of its particles. The faster the gas particles are moving, the more likely they are to pass through the effusion orifice. This reasoning leads to the following *prediction* for two gases at the same pressure and temperature (T):

$$\frac{\text{Effusion rate for gas 1}}{\text{Effusion rate for gas 2}} = \frac{u_{rms} \text{ for gas 1}}{u_{rms} \text{ for gas 2}} = \frac{\sqrt{\dfrac{3RT}{M_1}}}{\sqrt{\dfrac{3RT}{M_2}}} = \frac{\sqrt{M_2}}{\sqrt{M_1}}$$

This equation is identical to Graham's law. Thus the kinetic molecular model does fit the experimental results for the effusion of gases.

Diffusion

Diffusion is frequently illustrated by the lecture demonstration represented in Fig. 5.24, in which two cotton plugs soaked in ammonia and hydrochloric acid are simultaneously placed at the ends of a long tube. A white ring of ammonium chloride (NH_4Cl) forms where the NH_3 and HCl molecules meet several minutes later:

$$NH_3(g) + HCl(g) \longrightarrow NH_4Cl(s)$$
White solid

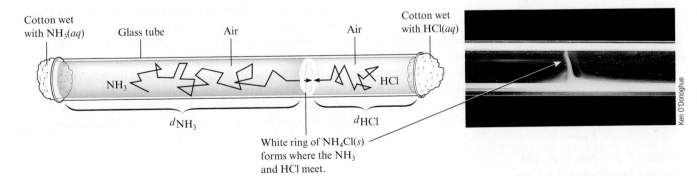

Figure 5.24 | (above left) A demonstration of the relative diffusion rates of NH_3 and HCl molecules through air. Two cotton plugs, one dipped in HCl(*aq*) and one dipped in NH_3(*aq*), are simultaneously inserted into the ends of the tube. Gaseous NH_3 and HCl vaporizing from the cotton plugs diffuse toward each other and, where they meet, react to form $NH_4Cl(s)$. (above right) When HCl(*g*) and NH_3(*g*) meet in the tube, a white ring of $NH_4Cl(s)$ forms.

As a first approximation, we might expect that the distances traveled by the two gases are related to the relative velocities of the gas molecules:

$$\frac{\text{Distance traveled by NH}_3}{\text{Distance traveled by HCl}} = \frac{u_{\text{rms}} \text{ for NH}_3}{u_{\text{rms}} \text{ for HCl}} = \sqrt{\frac{M_{\text{HCl}}}{M_{\text{NH}_3}}} = \sqrt{\frac{36.5}{17}} = 1.5$$

However, careful experiments produce an observed ratio of less than 1.5, indicating that a quantitative analysis of diffusion requires a more complex analysis.

The diffusion of the gases through the tube is surprisingly slow in light of the fact that the velocities of HCl and NH$_3$ molecules at 25°C are about 450 and 660 m/s, respectively. Why does it take several minutes for the NH$_3$ and HCl molecules to meet? The answer is that the tube contains air and thus the NH$_3$ and HCl molecules undergo many collisions with O$_2$ and N$_2$ molecules as they travel through the tube. Because so many collisions occur when gases mix, diffusion is quite complicated to describe theoretically.

5.8 | Real Gases

An ideal gas is a hypothetical concept. No gas *exactly* follows the ideal gas law, although many gases come very close at low pressures and/or high temperatures. Thus ideal gas behavior can best be thought of as the behavior *approached by real gases* under certain conditions.

We have seen that a very simple model, the kinetic molecular theory, by making some rather drastic assumptions (no interparticle interactions and zero volume for the gas particles), successfully explains ideal behavior. However, it is important that we examine real gas behavior to see how it differs from that predicted by the ideal gas law and to determine what modifications are needed in the kinetic molecular theory to explain the observed behavior. Since a model is an approximation and will inevitably fail, we must be ready to learn from such failures. In fact, we often learn more about nature from the failures of our models than from their successes.

We will examine the experimentally observed behavior of real gases by measuring the pressure, volume, temperature, and number of moles for a gas and noting how the quantity PV/nRT depends on pressure. Plots of PV/nRT versus P are shown for several gases in Fig. 5.25. For an ideal gas, PV/nRT equals 1 under all conditions, but notice that for real gases, PV/nRT approaches 1 only at very low pressures (typically below 1 atm). To illustrate the effect of temperature, PV/nRT is plotted versus P for nitrogen gas at several temperatures in Fig. 5.26. Note that the behavior of the gas appears to become more nearly ideal as the temperature is increased. The most important conclusion to be drawn from these figures is that a real gas typically exhibits behavior that is closest to ideal behavior at *low pressures* and *high temperatures*.

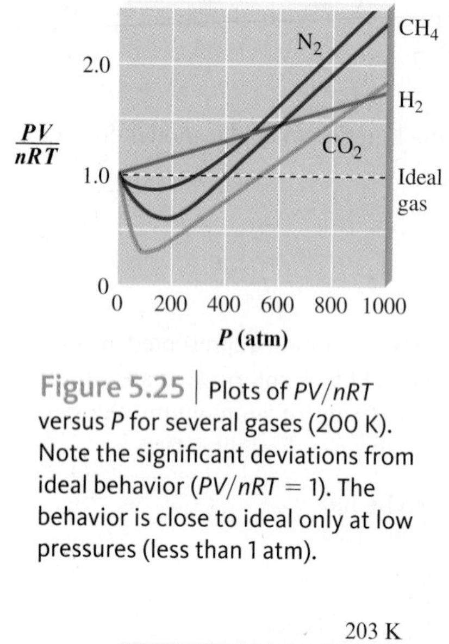

One of the most important procedures in science is correcting our models as we collect more data. We will understand more clearly how gases actually behave if we can figure out how to correct the simple model that explains the ideal gas law so that the new model fits the behavior we actually observe for gases. So the question is: How can we modify the assumptions of the kinetic molecular theory to fit the behavior of real gases? The first person to do important work in this area was Johannes van der Waals (1837–1923), a physics professor at the University of Amsterdam who in 1910 received a Nobel Prize for his work. To follow his analysis, we start with the ideal gas law,

$$P = \frac{nRT}{V}$$

Remember that this equation describes the behavior of a hypothetical gas consisting of volumeless entities that do not interact with each other. In contrast, a real gas consists of atoms or molecules that have finite volumes. Therefore, the volume available to a given particle in a real gas is less than the volume of the container because the gas

Figure 5.25 | Plots of PV/nRT versus P for several gases (200 K). Note the significant deviations from ideal behavior ($PV/nRT = 1$). The behavior is close to ideal only at low pressures (less than 1 atm).

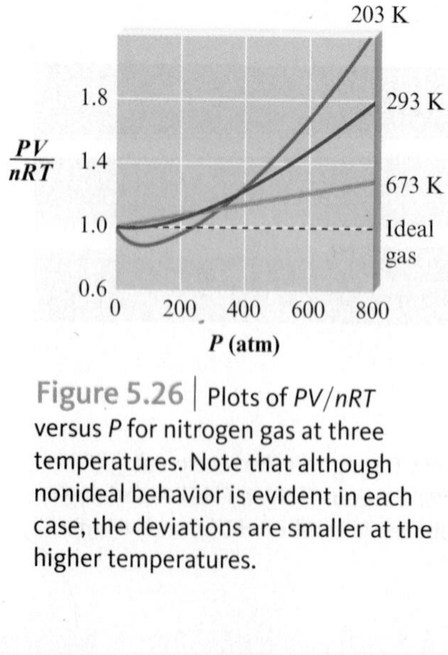

Figure 5.26 | Plots of PV/nRT versus P for nitrogen gas at three temperatures. Note that although nonideal behavior is evident in each case, the deviations are smaller at the higher temperatures.

particles themselves take up some of the space. To account for this discrepancy, van der Waals represented the actual volume as the volume of the container V minus a correction factor for the volume of the molecules nb, where n is the number of moles of gas and b is an empirical constant (one determined by fitting the equation to the experimental results). Thus the volume *actually available* to a given gas molecule is given by the difference $V - nb$.

This modification of the ideal gas equation leads to the equation

$$P' = \frac{nRT}{V - nb}$$

The volume of the gas particles has now been taken into account.

The next step is to allow for the attractions that occur among the particles in a real gas. The effect of these attractions is to make the observed pressure P_{obs} smaller than it would be if the gas particles did not interact:

$$P_{obs} = (P' - \text{correction factor}) = \left(\frac{nRT}{V - nb} - \text{correction factor} \right)$$

This effect can be understood using the following model. When gas particles come close together, attractive forces occur, which cause the particles to hit the wall very slightly less often than they would in the absence of these interactions (Fig. 5.27).

The size of the correction factor depends on the concentration of gas molecules defined in terms of moles of gas particles per liter (n/V). The higher the concentration, the more likely a pair of gas particles will be close enough to attract each other. For large numbers of particles, the number of interacting *pairs* of particles depends on the square of the number of particles and thus on the square of the concentration, or $(n/V)^2$. This can be justified as follows: In a gas sample containing N particles, there are $N - 1$ partners available for each particle (Fig. 5.28). Since the $1 \cdots 2$ pair is the same as the $2 \cdots 1$ pair, this analysis counts each pair twice. Thus, for N particles, there are $N(N - 1)/2$ pairs. If N is a very large number, $N - 1$ approximately equals N, giving $N^2/2$ possible pairs. Thus the pressure, corrected for the attractions of the particles, has the form

$$P_{obs} = P' - a\left(\frac{n}{V} \right)^2$$

where a is a proportionality constant (which includes the factor of $\frac{1}{2}$ from $N^2/2$). The value of a for a given real gas can be determined from observing the actual behavior of that gas. Inserting the corrections for both the volume of the particles and the attractions of the particles gives the equation

$$P_{obs} = \frac{nRT}{V - nb} - a\left(\frac{n}{V} \right)^2$$

Observed pressure   Volume of the container   Volume correction   Pressure correction

---

P' is corrected for the finite volume of the particles. The attractive forces have not yet been taken into account.

The attractive forces among molecules will be discussed in Chapter 10.

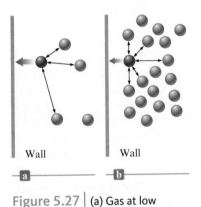

| Wall | Wall |
| a | b |

Figure 5.27 | (a) Gas at low concentration—relatively few interactions between particles. The indicated gas particle exerts a pressure on the wall close to that predicted for an ideal gas. (b) Gas at high concentration—many more interactions between particles. The indicated gas particle exerts a much lower pressure on the wall than would be expected in the absence of interactions.

We have now corrected for both the finite volume and the attractive forces of the particles.

Given particle

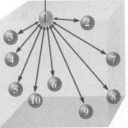

Gas sample with 10 particles

Figure 5.28 | Illustration of pairwise interactions among gas particles. In a sample with 10 particles, each particle has 9 possible partners, to give $10(9)/2 = 45$ distinct pairs. The factor of $\frac{1}{2}$ arises because when particle ① is the particle of interest, we count the ① $\cdots$ ② pair, and when particle ② is the particle of interest, we count the ② $\cdots$ ① pair. However, ① $\cdots$ ② and ② $\cdots$ ① are the same pair, which we thus have counted twice. Therefore, we must divide by 2 to get the correct number of pairs.

Figure 5.29 | The volume taken up by the gas particles themselves is less important at (a) large container volume (low pressure) than at (b) small container volume (high pressure).

P_{obs} is usually called just P.

This equation can be rearranged to give the **van der Waals equation**:

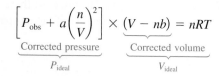

$$\underbrace{\left[P_{obs} + a\left(\frac{n}{V}\right)^2\right]}_{\substack{\text{Corrected pressure} \\ P_{ideal}}} \times \underbrace{(V - nb)}_{\substack{\text{Corrected volume} \\ V_{ideal}}} = nRT$$

The values of the weighting factors a and b are determined for a given gas by fitting experimental behavior. That is, a and b are varied until the best fit of the observed pressure is obtained under all conditions. The values of a and b for various gases are given in Table 5.3.

Experimental studies indicate that the changes van der Waals made in the basic assumptions of the kinetic molecular theory correct the major flaws in the model. First, consider the effects of volume. For a gas at low pressure (large volume), the volume of the container is very large compared with the volumes of the gas particles. That is, in this case the volume available to the gas is essentially equal to the volume of the container, and the gas behaves ideally. On the other hand, for a gas at high pressure (small container volume), the volume of the particles becomes significant so that the volume available to the gas is significantly less than the container volume. These cases are illustrated in Fig. 5.29. Note from Table 5.3 that the volume correction constant b generally increases with the size of the gas molecule, which gives further support to these arguments.

The fact that a real gas tends to behave more ideally at high temperatures also can be explained in terms of the van der Waals model. At high temperatures the particles are moving so rapidly that the effects of interparticle interactions are not very important.

The corrections to the kinetic molecular theory that van der Waals found necessary to explain real gas behavior make physical sense, which makes us confident that we understand the fundamentals of gas behavior at the particle level. This is significant because so much important chemistry takes place in the gas phase. In fact, the mixture of gases called the atmosphere is vital to our existence. In Section 5.10 we consider some of the important reactions that occur in the atmosphere.

Table 5.3 | Values of the van der Waals Constants for Some Common Gases

Gas	$a\left(\dfrac{atm \cdot L^2}{mol^2}\right)$	$b\left(\dfrac{L}{mol}\right)$
He	0.0341	0.0237
Ne	0.211	0.0171
Ar	1.35	0.0322
Kr	2.32	0.0398
Xe	4.19	0.0511
H_2	0.244	0.0266
N_2	1.39	0.0391
O_2	1.36	0.0318
Cl_2	6.49	0.0562
CO_2	3.59	0.0427
CH_4	2.25	0.0428
NH_3	4.17	0.0371
H_2O	5.46	0.0305

Critical Thinking

You have learned that no gases behave perfectly ideally, but under conditions of high temperature and low pressure (high volume), gases behave more ideally. What if all gases always behaved perfectly ideally? How would the world be different?

5.9 | Characteristics of Several Real Gases

We can understand gas behavior more completely if we examine the characteristics of several common gases. Note from Fig. 5.25 that the gases H_2, N_2, CH_4, and CO_2 show different behavior when the compressibility $\left(\dfrac{PV}{nRT}\right)$ is plotted versus P. For example, notice that the plot for $H_2(g)$ never drops below the ideal value (1.0) in contrast to all the other gases. What is special about H_2 compared to these other gases? Recall from Section 5.8 that the reason that the compressibility of a real gas falls below 1.0 is that the actual (observed) pressure is lower than the pressure expected for an ideal gas due to the intermolecular attractions that occur in real gases. This must mean that H_2 molecules have very low attractive forces for each other. This idea is borne out by looking at the van der Waals a value for H_2 in Table 5.3. Note that H_2 has the lowest value among the gases H_2, N_2, CH_4, and CO_2. Remember that the value of a reflects how much of a correction must be made to adjust the observed pressure up to the expected ideal pressure:

$$P_{ideal} = P_{observed} + a\left(\frac{n}{V}\right)^2$$

A low value for a reflects weak intermolecular forces among the gas molecules.

Also notice that although the compressibility for N_2 dips below 1.0, it does not show as much deviation as that for CH_4, which in turn does not show as much deviation as the compressibility for CO_2. Based on this behavior, we can surmise that the importance of intermolecular interactions increases in this order:

$$H_2 < N_2 < CH_4 < CO_2$$

This order is reflected by the relative a values for these gases in Table 5.3. In Section 10.1, we will see how these variations in intermolecular interactions can be explained. The main point to be made here is that real gas behavior can tell us about the relative importance of intermolecular attractions among gas molecules.

5.10 | Chemistry in the Atmosphere

The most important gases to us are those in the **atmosphere** that surrounds the earth's surface. The principal components are N_2 and O_2, but many other important gases, such as H_2O and CO_2, are also present. The average composition of the earth's atmosphere near sea level, with the water vapor removed, is shown in Table 5.4. Because of gravitational effects, the composition of the earth's atmosphere is not constant; heavier molecules tend to be near the earth's surface, and light molecules tend to migrate to higher altitudes, with some eventually escaping into space. The atmosphere is a highly complex and dynamic system, but for convenience we divide it into several layers. Fig. 5.30 shows how the temperature changes with altitude.

The chemistry occurring in the higher levels of the atmosphere is mostly determined by the effects of high-energy radiation and particles from the sun and other sources in space. In fact, the upper atmosphere serves as an important shield to prevent this high-energy radiation from reaching the earth, where it would damage the relatively fragile molecules sustaining life. In particular, the ozone in the upper atmosphere helps prevent high-energy ultraviolet radiation from penetrating to the earth. Intensive research is in progress to determine the natural factors that control the ozone concentration and how it is affected by chemicals released into the atmosphere.

The chemistry occurring in the troposphere, the layer of atmosphere closest to the earth's surface, is strongly influenced by human activities. Millions of tons of gases and particulates are released into the troposphere by our highly industrial civilization. Actually, it is amazing that the atmosphere can absorb so much material with relatively small permanent changes (so far).

Significant changes, however, are occurring. Severe **air pollution** is found around many large cities, and it is probable that long-range changes in our planet's weather are taking place. We will discuss some of the long-range effects of pollution in Chapter 6. In this section we will deal with short-term, localized effects of pollution.

Table 5.4 | Atmospheric Composition Near Sea Level (Dry Air)*

Component	Mole Fraction
N_2	0.78084
O_2	0.20948
Ar	0.00934
CO_2	0.000345
Ne	0.00001818
He	0.00000524
CH_4	0.00000168
Kr	0.00000114
H_2	0.0000005
NO	0.0000005
Xe	0.000000087

*The atmosphere contains various amounts of water vapor depending on conditions.

Figure 5.30 | The variation of temperature with altitude.

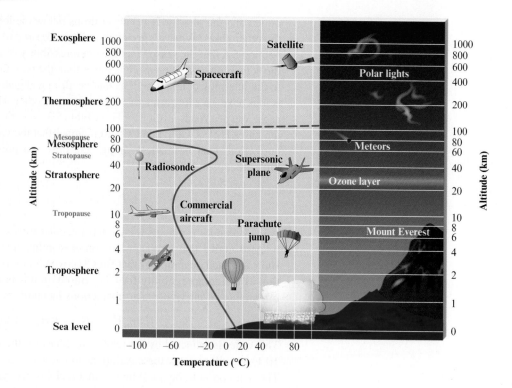

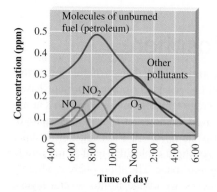

Figure 5.31 | Concentration (in molecules per million molecules of "air") for some smog components versus time of day.

(Source: P.A. Leighton, "Photochemistry of Air Pollution," in *Physical Chemistry: A Series of Monographs*, edited by Eric Hutchinson and P. Van Rysselberghe, Vol. IX. New York: Academic Press, 1961.)

The OH radical has no charge [it has one fewer electron than the hydroxide ion (OH^-)].

The two main sources of pollution are transportation and the production of electricity. The combustion of petroleum in vehicles produces CO, CO_2, NO, and NO_2, along with unburned molecules from petroleum. When this mixture is trapped close to the ground in stagnant air, reactions occur, producing chemicals that are potentially irritating and harmful to living systems.

The complex chemistry of polluted air appears to center around the nitrogen oxides (NO_x). At the high temperatures found in the gasoline and diesel engines of cars and trucks, N_2 and O_2 react to form a small quantity of NO that is emitted into the air with the exhaust gases (Fig. 5.31). This NO is immediately oxidized in air to NO_2, which, in turn, absorbs energy from sunlight and breaks up into nitric oxide and free oxygen atoms:

$$NO_2(g) \xrightarrow{\text{Radiant energy}} NO(g) + O(g)$$

Oxygen atoms are very reactive and can combine with O_2 to form *ozone:*

$$O(g) + O_2(g) \longrightarrow O_3(g)$$

Ozone is also very reactive and can react directly with other pollutants, or the ozone can absorb light and break up to form an energetically excited O_2 molecule (O_2^*) and an energetically excited oxygen atom (O^*). The latter species readily reacts with a water molecule to form two hydroxyl radicals (OH):

$$O^* + H_2O \longrightarrow 2OH$$

The hydroxyl radical is a very reactive oxidizing agent. For example, OH can react with NO_2 to form nitric acid:

$$OH + NO_2 \longrightarrow HNO_3$$

The OH radical also can react with the unburned hydrocarbons in the polluted air to produce chemicals that cause the eyes to water and burn and are harmful to the respiratory system.

The end product of this whole process is often referred to as **photochemical smog**, so called because light is required to initiate some of the reactions. The production of photochemical smog can be understood more clearly by examining as a group the reactions discussed above:

$$NO_2(g) \longrightarrow NO(g) + O(g)$$
$$O(g) + O_2(g) \longrightarrow O_3(g)$$
$$NO(g) + \tfrac{1}{2}O_2(g) \longrightarrow NO_2(g)$$

Net reaction:
$$\tfrac{3}{2}O_2(g) \longrightarrow O_3(g)$$

Note that the NO_2 molecules assist in the formation of ozone without being themselves used up. The ozone formed then leads to the formation of OH and other pollutants.

We can observe this process by analyzing polluted air at various times during a day (see Fig. 5.31). As people drive to work between 6 and 8 a.m., the amounts of NO, NO_2, and unburned molecules from petroleum increase. Later, as the decomposition of NO_2 occurs, the concentration of ozone and other pollutants builds up. Current efforts to combat the formation of photochemical smog are focused on cutting down the amounts of molecules from unburned fuel in automobile exhaust and designing engines that produce less nitric oxide.

The other major source of pollution results from burning coal to produce electricity. Much of the coal found in the Midwest contains significant quantities of sulfur, which, when burned, produces sulfur dioxide:

$$S\,(\text{in coal}) + O_2(g) \longrightarrow SO_2(g)$$

A further oxidation reaction occurs when sulfur dioxide is changed to sulfur trioxide in the air:*

$$2SO_2(g) + O_2(g) \longrightarrow 2SO_3(g)$$

The production of sulfur trioxide is significant because it can combine with droplets of water in the air to form sulfuric acid:

$$SO_3(g) + H_2O(l) \longrightarrow H_2SO_4(aq)$$

Sulfuric acid is very corrosive to both living things and building materials. Another result of this type of pollution is **acid rain**. In many parts of the northeastern United States and southeastern Canada, acid rain has caused some freshwater lakes to become too acidic to support any life (Fig. 5.32).

The problem of sulfur dioxide pollution is made more complicated by the energy crisis. As petroleum supplies dwindle and the price increases, our dependence on coal will probably grow. As supplies of low-sulfur coal are used up, high-sulfur coal will be utilized. One way to use high-sulfur coal without further harming the air quality is to remove the sulfur dioxide from the exhaust gas by means of a system called a *scrubber* before it is emitted from the power plant stack. A common method of scrubbing is to blow powdered limestone ($CaCO_3$) into the combustion chamber, where it is decomposed to lime and carbon dioxide:

$$CaCO_3(s) \longrightarrow CaO(s) + CO_2(g)$$

The lime then combines with the sulfur dioxide to form calcium sulfite:

$$CaO(s) + SO_2(g) \longrightarrow CaSO_3(s)$$

To remove the calcium sulfite and any remaining unreacted sulfur dioxide, an aqueous suspension of lime is injected into the exhaust gases to produce a *slurry* (a thick suspension) (Fig. 5.33).

Although represented here as O_2, the actual oxidant for NO is OH or an organic peroxide such as CH_3COO, formed by oxidation of organic pollutants.

Figure 5.32 | Testing for acid rain in Wilderness Lake, Colorado.

Robert Krueger/US Forest Service/Getty Images

*This reaction is very slow unless solid particles are present. See Chapter 12 for a discussion.

Figure 5.33 | A schematic diagram of the process for scrubbing sulfur dioxide from stack gases in power plants.

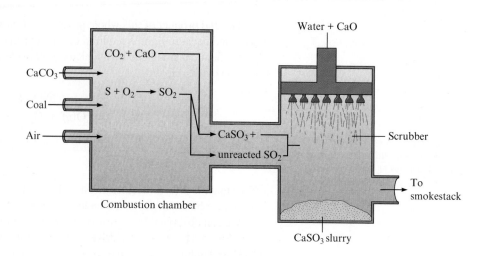

$CO_2 + CaO$

$CaCO_3$

$S + O_2 \longrightarrow SO_2$

Coal

Air

$CaSO_3 +$

unreacted SO_2

Combustion chamber

Water + CaO

Scrubber

To smokestack

$CaSO_3$ slurry

Unfortunately, there are many problems associated with scrubbing. The systems are complicated and expensive and consume a great deal of energy. The large quantities of calcium sulfite produced in the process present a disposal problem. With a typical scrubber, approximately 1 ton of calcium sulfite per year is produced per person served by the power plant. Since no use has yet been found for this calcium sulfite, it is usually buried in a landfill. As a result of these difficulties, air pollution by sulfur dioxide continues to be a major problem, one that is expensive in terms of damage to the environment and human health as well as in monetary terms.

For review

Key terms

Section 5.1
barometer
manometer
mm Hg
torr
standard atmosphere
pascal

Section 5.2
Boyle's law
ideal gas
Charles's law
absolute zero
Avogadro's law

Section 5.3
universal gas constant
ideal gas law

Section 5.4
molar volume
standard temperature and
   pressure (STP)

State of a gas

› The state of a gas can be described completely by specifying its pressure (P), volume (V), temperature (T), and the amount (moles) of gas present (n)

› Pressure

› Common units

$$1 \text{ torr} = 1 \text{ mm Hg}$$
$$1 \text{ atm} = 760 \text{ torr}$$

› SI unit: pascal

$$1 \text{ atm} = 101,325 \text{ Pa}$$

Gas laws

› Discovered by observing the properties of gases

› Boyle's law: $PV = k$

› Charles's law: $V = bT$

› Avogadro's law: $V = an$

› Ideal gas law: $PV = nRT$

› Dalton's law of partial pressures: $P_{\text{TOTAL}} = P_1 + P_2 + P_3 + \cdots$, where P_n represents the partial pressure of component n in a mixture of gases

Key terms

Kinetic molecular theory (KMT)

› Model that accounts for ideal gas behavior
› Postulates of the KMT:
  › Volume of gas particles is zero
  › No particle interactions
  › Particles are in constant motion, colliding with the container walls to produce pressure
  › The average kinetic energy of the gas particles is directly proportional to the temperature of the gas in kelvins

Gas properties

› The particles in any gas sample have a range of velocities
› The root mean square (rms) velocity for a gas represents the average of the squares of the particle velocities

$$u_{rms} = \sqrt{\frac{3RT}{M}}$$

› Diffusion: the mixing of two or more gases
› Effusion: the process in which a gas passes through a small hole into an empty chamber

Real gas behavior

› Real gases behave ideally only at high temperatures and low pressures
› Understanding how the ideal gas equation must be modified to account for real gas behavior helps us understand how gases behave on a molecular level
› Van der Waals found that to describe real gas behavior we must consider particle interactions and particle volumes

Review questions *Answers to the Review Questions can be found on the Student website (accessible from www.cengagebrain.com).*

1. Explain how a barometer and a manometer work to measure the pressure of the atmosphere or the pressure of a gas in a container.

2. What are Boyle's law, Charles's law, and Avogadro's law? What plots do you make to show a linear relationship for each law?

3. Show how Boyle's law, Charles's law, and Avogadro's law are special cases of the ideal gas law. Using the ideal gas law, determine the relationship between P and n (at constant V and T) and between P and T (at constant V and n).

4. Rationalize the following observations.
   a. Aerosol cans will explode if heated.
   b. You can drink through a soda straw.
   c. A thin-walled can will collapse when the air inside is removed by a vacuum pump.
   d. Manufacturers produce different types of tennis balls for high and low elevations.

5. Consider the following balanced equation in which gas X forms gas X_2:

$$2X(g) \longrightarrow X_2(g)$$

Equal moles of X are placed in two separate containers. One container is rigid so the volume cannot change; the other container is flexible so the volume changes to keep the internal pressure equal to the external pressure. The above reaction is run in each container. What happens to the pressure and density of the gas inside each container as reactants are converted to products?

6. Use the postulates of the kinetic molecular theory (KMT) to explain why Boyle's law, Charles's law, Avogadro's law, and Dalton's law of partial pressures hold true for ideal gases. Use the KMT to explain the P versus n (at constant V and T) relationship and the P versus T (at constant V and n) relationship.

7. Consider the following velocity distribution curves *A* and *B*.

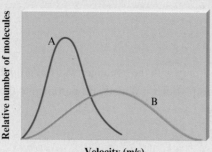

a. If the plots represent the velocity distribution of 1.0 L of He(*g*) at STP versus 1.0 L of Cl₂(*g*) at STP, which plot corresponds to each gas? Explain your reasoning.

b. If the plots represent the velocity distribution of 1.0 L of O₂(*g*) at temperatures of 273 K versus 1273 K, which plot corresponds to each temperature? Explain your reasoning. Under which temperature condition would the O₂(*g*) sample behave most ideally? Explain.

8. Briefly describe two methods one might use to find the molar mass of a newly synthesized gas for which a molecular formula was not known.

9. In the van der Waals equation, why is a term added to the observed pressure and why is a term subtracted from the container volume to correct for nonideal gas behavior?

10. Why do real gases not always behave ideally? Under what conditions does a real gas behave most ideally? Why?

Active Learning Questions

These questions are designed to be used by groups of students in class.

1. Consider the following apparatus: a test tube covered with a nonpermeable elastic membrane inside a container that is closed with a cork. A syringe goes through the cork.

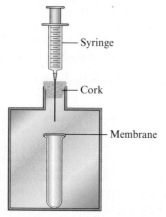

a. As you push down on the syringe, how does the membrane covering the test tube change?

b. You stop pushing the syringe but continue to hold it down. In a few seconds, what happens to the membrane?

2. Figure 5.2 shows a picture of a barometer. Which of the following statements is the best explanation of how this barometer works?

a. Air pressure outside the tube causes the mercury to move in the tube until the air pressure inside and outside the tube is equal.

b. Air pressure inside the tube causes the mercury to move in the tube until the air pressure inside and outside the tube is equal.

c. Air pressure outside the tube counterbalances the weight of the mercury in the tube.

d. Capillary action of the mercury causes the mercury to go up the tube.

e. The vacuum that is formed at the top of the tube holds up the mercury.

Justify your choice, and for the choices you did not pick, explain what is wrong with them. Pictures help!

3. The barometer below shows the level of mercury at a given atmospheric pressure. Fill all the other barometers with mercury for that same atmospheric pressure. Explain your answer.

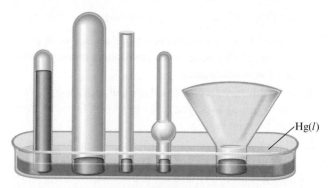

4. As you increase the temperature of a gas in a sealed, rigid container, what happens to the density of the gas? Would the results be the same if you did the same experiment in a container with a piston at constant pressure? (See Fig. 5.17.)

5. A diagram in a chemistry book shows a magnified view of a flask of air as follows:

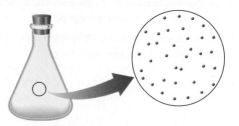

What do you suppose is between the dots (the dots represent air molecules)?

a. air

b. dust

c. pollutants

d. oxygen

e. nothing

6. If you put a drinking straw in water, place your finger over the opening, and lift the straw out of the water, some water stays in the straw. Explain.

7. A chemistry student relates the following story: I noticed my tires were a bit low and went to the gas station. As I was filling the tires, I thought about the kinetic molecular theory (KMT). I noticed the tires because the volume was low, and I realized that I was increasing both the pressure and volume of the tires. "Hmmm," I thought, "that goes against what I learned in chemistry, where I was told pressure and volume are inversely proportional." What is the fault in the logic of the chemistry student in this situation? Explain *why* we think pressure and volume to be inversely related (draw pictures and use the KMT).

8. Chemicals X and Y (both gases) react to form the gas XY, but it takes a bit of time for the reaction to occur. Both X and Y are placed in a container with a piston (free to move), and you note the volume. As the reaction occurs, what happens to the volume of the container? (See Fig. 5.18.)

9. Which statement best explains why a hot-air balloon rises when the air in the balloon is heated?

a. According to Charles's law, the temperature of a gas is directly related to its volume. Thus the volume of the balloon increases, making the density smaller. This lifts the balloon.

b. Hot air rises inside the balloon, and this lifts the balloon.

c. The temperature of a gas is directly related to its pressure. The pressure therefore increases, and this lifts the balloon.

d. Some of the gas escapes from the bottom of the balloon, thus decreasing the mass of gas in the balloon. This decreases the density of the gas in the balloon, which lifts the balloon.

e. Temperature is related to the root mean square velocity of the gas molecules. Thus the molecules are moving faster, hitting the balloon more, and thus lifting the balloon.

Justify your choice, and for the choices you did not pick, explain what is wrong with them.

10. Draw a highly magnified view of a sealed, rigid container filled with a gas. Then draw what it would look like if you cooled the gas significantly but kept the temperature above the boiling point of the substance in the container. Also draw what it would look like if you heated the gas significantly. Finally, draw what each situation would look like if you evacuated enough of the gas to decrease the pressure by a factor of 2.

11. If you release a helium balloon, it soars upward and eventually pops. Explain this behavior.

12. If you have any two gases in different containers that are the same size at the same pressure and same temperature, what is true about the moles of each gas? Why is this true?

13. Explain the following seeming contradiction: You have two gases, A and B, in two separate containers of equal volume and at equal pressure and temperature. Therefore, you must have the same number of moles of each gas. Because the two temperatures are equal, the average kinetic energies of the two samples are equal. Therefore, since the energy given such a system will be converted to translational motion (that is, move the molecules), the root mean square velocities of the two are equal, and thus the particles in each sample move, on average, with the same relative speed. Since A and B are different gases, they each must have a different molar mass. If A has a higher molar mass than B, the particles of A must be hitting the sides of the container with more force. Thus the pressure in the container of gas A must be higher than that in the container with gas B. However, one of our initial assumptions was that the pressures were equal.

14. You have a balloon covering the mouth of a flask filled with air at 1 atm. You apply heat to the bottom of the flask until the volume of the balloon is equal to that of the flask.

a. Which has more air in it, the balloon or the flask? Or do both have the same amount? Explain.

b. In which is the pressure greater, the balloon or the flask? Or is the pressure the same? Explain.

15. How does Dalton's law of partial pressures help us with our model of ideal gases? That is, what postulates of the kinetic molecular theory does it support?

16. At the same conditions of pressure and temperature, ammonia gas is less dense than air. Why is this true?

17. For each of the quantities listed below, explain which of the following properties (mass of the molecule, density of the gas sample, temperature of the gas sample, size of the molecule, and number of moles of gas) must be known to calculate the quantity.

a. average kinetic energy

b. average number of collisions per second with other gas molecules

c. average force of each impact with the wall of the container

d. root mean square velocity

e. average number of collisions with a given area of the container

f. distance between collisions

18. You have two containers each with 1 mole of xenon gas at 15°C. Container A has a volume of 3.0 L, and container B has a volume of 1.0 L. Explain how the following quantities compare between the two containers.

a. the average kinetic energy of the Xe atoms

b. the force with which the Xe atoms collide with the container walls

c. the root mean square velocity of the Xe atoms

d. the collision frequency of the Xe atoms (with other atoms)

e. the pressure of the Xe sample

19. Draw molecular-level views that show the differences among solids, liquids, and gases.

Questions

20. At room temperature, water is a liquid with a molar volume of 18 mL. At 105°C and 1 atm pressure, water is a gas and has a molar volume of over 30 L. Explain the large difference in molar volumes.

21. If a barometer were built using water ($d = 1.0$ g/cm³) instead of mercury ($d = 13.6$ g/cm³), would the column of water be higher than, lower than, or the same as the column of mercury at 1.00 atm? If the level is different, by what factor? Explain.

22. A bag of potato chips is packed and sealed in Los Angeles, California, and then shipped to Lake Tahoe, Nevada, during ski season. It is noticed that the volume of the bag of potato chips has increased upon its arrival in Lake Tahoe. What external conditions would most likely cause the volume increase?

23. Boyle's law can be represented graphically in several ways. Which of the following plots does *not* correctly represent Boyle's law (assuming constant T and n)? Explain.

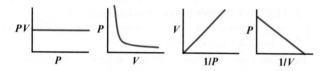

24. As weather balloons rise from the earth's surface, the pressure of the atmosphere becomes less, tending to cause the volume of the balloons to expand. However, the temperature is much lower in the upper atmosphere than at sea level. Would this temperature effect tend to make such a balloon expand or contract? Weather balloons do, in fact, expand as they rise. What does this tell you?

25. Which noble gas has the smallest density at STP? Explain.

26. Consider two different containers, each filled with 2 moles of Ne(g). One of the containers is rigid and has constant volume. The other container is flexible (like a balloon) and is capable of changing its volume to keep the external pressure and internal pressure equal to each other. If you raise the temperature in both containers, what happens to the pressure and density of the gas inside each container? Assume a constant external pressure.

27. In Example 5.11 of the text, the molar volume of N$_2$(g) at STP is given as 22.42 L/mol N$_2$. How is this number calculated? How does the molar volume of He(g) at STP compare to the molar volume of N$_2$(g) at STP (assuming ideal gas behavior)? Is the molar volume of N$_2$(g) at 1.000 atm and 25.0°C equal to, less than, or greater than 22.42 L/mol? Explain. Is the molar volume of N$_2$(g) collected over water at a total pressure of 1.000 atm and 0.0°C equal to, less than, or greater than 22.42 L/mol? Explain.

28. Consider the flasks in the following diagrams.

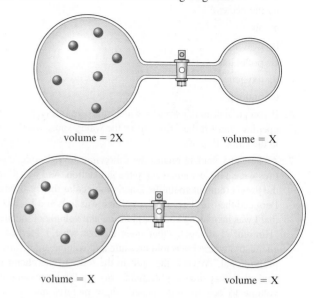

Assuming the connecting tube has negligible volume, draw what each diagram will look like after the stopcock between the two flasks is opened. Also, solve for the final pressure in each case, in terms of the original pressure. Assume temperature is constant.

29. Do all the molecules in a 1-mole sample of CH$_4$(g) have the same kinetic energy at 273 K? Do all molecules in a 1-mole sample of N$_2$(g) have the same velocity at 546 K? Explain.

30. Consider the following samples of gases at the same temperature.

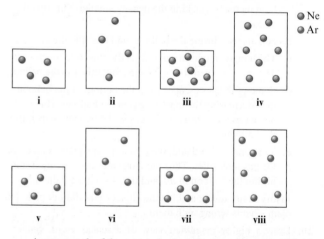

Arrange each of these samples in order from lowest to highest:

a. pressure

b. average kinetic energy

c. density

d. root mean square velocity

Note: Some samples of gases may have equal values for these attributes. Assume the larger containers have a volume twice the volume of the smaller containers, and assume the mass of an argon atom is twice the mass of a neon atom.

31. As $NH_3(g)$ is decomposed into nitrogen gas and hydrogen gas at constant pressure and temperature, the volume of the product gases collected is twice the volume of NH_3 reacted. Explain. As $NH_3(g)$ is decomposed into nitrogen gas and hydrogen gas at constant volume and temperature, the total pressure increases by some factor. Why the increase in pressure and by what factor does the total pressure increase when reactants are completely converted into products? How do the partial pressures of the product gases compare to each other and to the initial pressure of NH_3?

32. Which of the following statements is(are) *true?* For the false statements, correct them.

    a. At constant temperature, the lighter the gas molecules, the faster the average velocity of the gas molecules.

    b. At constant temperature, the heavier the gas molecules, the larger the average kinetic energy of the gas molecules.

    c. A real gas behaves most ideally when the container volume is relatively large and the gas molecules are moving relatively quickly.

    d. As temperature increases, the effect of interparticle interactions on gas behavior is increased.

    e. At constant V and T, as gas molecules are added into a container, the number of collisions per unit area increases resulting in a higher pressure.

    f. The kinetic molecular theory predicts that pressure is inversely proportional to temperature at constant volume and moles of gas.

33. From the values in Table 5.3 for the van der Waals constant a for the gases H_2, CO_2, N_2, and CH_4, predict which of these gas molecules show the strongest intermolecular attractions.

34. Without looking at a table of values, which of the following gases would you expect to have the largest value of the van der Waals constant b: H_2, N_2, CH_4, C_2H_6, or C_3H_8?

35. Figure 5.6 shows the PV versus P plot for three different gases. Which gas behaves most ideally? Explain.

36. Ideal gas particles are assumed to be volumeless and to neither attract nor repel each other. Why are these assumptions crucial to the validity of Dalton's law of partial pressures?

Exercises

In this section similar exercises are paired.

Pressure

37. Freon-12 (CF_2Cl_2) is commonly used as the refrigerant in central home air conditioners. The system is initially charged to a pressure of 4.8 atm. Express this pressure in each of the following units (1 atm = 14.7 psi).

    a. mm Hg   c. Pa
    b. torr   d. psi

38. A gauge on a compressed gas cylinder reads 2200 psi (pounds per square inch; 1 atm = 14.7 psi). Express this pressure in each of the following units.

    a. standard atmospheres
    b. megapascals (MPa)
    c. torr

39. A sealed-tube manometer (as shown below) can be used to measure pressures below atmospheric pressure. The tube above the mercury is evacuated. When there is a vacuum in the flask, the mercury levels in both arms of the U-tube are equal. If a gaseous sample is introduced into the flask, the mercury levels are different. The difference h is a measure of the pressure of the gas inside the flask. If h is equal to 6.5 cm, calculate the pressure in the flask in torr, pascals, and atmospheres.

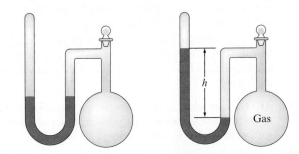

40. If the sealed-tube manometer in Exercise 39 had a height difference of 20.0 inches between the mercury levels, what is the pressure in the flask in torr and atmospheres?

41. A diagram for an open-tube manometer is shown below.

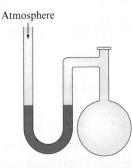

Atmosphere

If the flask is open to the atmosphere, the mercury levels are equal. For each of the following situations where a gas is contained in the flask, calculate the pressure in the flask in torr, atmospheres, and pascals.

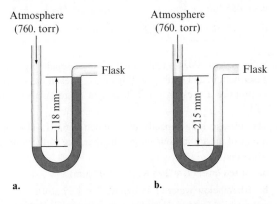

a.   b.

c. Calculate the pressures in the flask in parts a and b (in torr) if the atmospheric pressure is 635 torr.

42. a. If the open-tube manometer in Exercise 41 contains a nonvolatile silicone oil (density = 1.30 g/cm³) instead of mercury (density = 13.6 g/cm³), what are the pressures in the flask as shown in parts a and b in torr, atmospheres, and pascals?

b. What advantage would there be in using a less dense fluid than mercury in a manometer used to measure relatively small differences in pressure?

Gas Laws

43. A particular balloon is designed by its manufacturer to be inflated to a volume of no more than 2.5 L. If the balloon is filled with 2.0 L helium at sea level, is released, and rises to an altitude at which the atmospheric pressure is only 500. mm Hg, will the balloon burst? (Assume temperature is constant.)

44. A balloon is filled to a volume of 7.00×10^2 mL at a temperature of 20.0°C. The balloon is then cooled at constant pressure to a temperature of 1.00×10^2 K. What is the final volume of the balloon?

45. An 11.2-L sample of gas is determined to contain 0.50 mole of N_2. At the same temperature and pressure, how many moles of gas would there be in a 20.-L sample?

46. Consider the following chemical equation.

$$2NO_2(g) \longrightarrow N_2O_4(g)$$

If 25.0 mL NO_2 gas is completely converted to N_2O_4 gas under the same conditions, what volume will the N_2O_4 occupy?

47. Complete the following table for an ideal gas.

	P (atm)	V (L)	n (mol)	T
a.	5.00		2.00	155°C
b.	0.300	2.00		155 K
c.	4.47	25.0	2.01	
d.		2.25	10.5	75°C

48. Complete the following table for an ideal gas.

	P	V	n	T
a.	7.74×10^3 Pa	12.2 mL		25°C
b.		43.0 mL	0.421 mol	223 K
c.	455 torr		4.4×10^{-2} mol	331°C
d.	745 mm Hg	11.2 L	0.401 mol	

49. Suppose two 200.0-L tanks are to be filled separately with the gases helium and hydrogen. What mass of each gas is needed to produce a pressure of 2.70 atm in its respective tank at 24°C?

50. The average lung capacity of a human is 6.0 L. How many moles of air are in your lungs when you are in the following situations?

   a. At sea level ($T = 298$ K, $P = 1.00$ atm).

   b. 10. m below water ($T = 298$ K, $P = 1.97$ atm).

   c. At the top of Mount Everest ($T = 200.$ K, $P = 0.296$ atm).

51. The steel reaction vessel of a bomb calorimeter, which has a volume of 75.0 mL, is charged with oxygen gas to a pressure of 14.5 atm at 22°C. Calculate the moles of oxygen in the reaction vessel.

52. A 5.0-L flask contains 0.60 g O_2 at a temperature of 22°C. What is the pressure (in atm) inside the flask?

53. A 2.50-L container is filled with 175 g argon.

   a. If the pressure is 10.0 atm, what is the temperature?

   b. If the temperature is 225 K, what is the pressure?

54. A person accidentally swallows a drop of liquid oxygen, $O_2(l)$, which has a density of 1.149 g/mL. Assuming the drop has a volume of 0.050 mL, what volume of gas will be produced in the person's stomach at body temperature (37°C) and a pressure of 1.0 atm?

55. A gas sample containing 1.50 moles at 25°C exerts a pressure of 400. torr. Some gas is added to the same container and the temperature is increased to 50.°C. If the pressure increases to 800. torr, how many moles of gas were added to the container? Assume a constant-volume container.

56. A bicycle tire is filled with air to a pressure of 75 psi at a temperature of 19°C. Riding the bike on asphalt on a hot day increases the temperature of the tire to 58°C. The volume of the tire increases by 4.0%. What is the new pressure in the bicycle tire?

57. Consider two separate gas containers at the following conditions:

Container A	Container B
Contents: $SO_2(g)$	Contents: unknown gas
Pressure = P_A	Pressure = P_B
Moles of gas = 1.0 mol	Moles of gas = 2.0 mol
Volume = 1.0 L	Volume = 2.0 L
Temperature = 7°C	Temperature = 287°C

How is the pressure in container B related to the pressure in container A?

58. What will be the effect on the volume of an ideal gas if the pressure is doubled and the absolute temperature is halved?

59. A container is filled with an ideal gas to a pressure of 11.0 atm at 0°C.

   a. What will be the pressure in the container if it is heated to 45°C?

   b. At what temperature would the pressure be 6.50 atm?

   c. At what temperature would the pressure be 25.0 atm?

60. An ideal gas at 7°C is in a spherical flexible container having a radius of 1.00 cm. The gas is heated at constant pressure to 88°C. Determine the radius of the spherical container after the gas is heated. [Volume of a sphere $= (4/3)\pi r^3$.]

61. An ideal gas is contained in a cylinder with a volume of 5.0×10^2 mL at a temperature of 30.°C and a pressure of 710. torr. The gas is then compressed to a volume of 25 mL, and the temperature is raised to 820.°C. What is the new pressure of the gas?

62. A compressed gas cylinder contains 1.00×10^3 g argon gas. The pressure inside the cylinder is 2050. psi (pounds per square inch) at a temperature of 18°C. How much gas remains in the cylinder if the pressure is decreased to 650. psi at a temperature of 26°C?

63. A sealed balloon is filled with 1.00 L helium at 23°C and 1.00 atm. The balloon rises to a point in the atmosphere where the pressure is 220. torr and the temperature is −31°C. What is the change in volume of the balloon as it ascends from 1.00 atm to a pressure of 220. torr?

64. A hot-air balloon is filled with air to a volume of 4.00×10^3 m^3 at 745 torr and 21°C. The air in the balloon is then heated to 62°C, causing the balloon to expand to a volume of 4.20×10^3 m^3. What is the ratio of the number of moles of air in the heated balloon to the original number of moles of air in the balloon? (*Hint:* Openings in the balloon allow air to flow in and out. Thus the pressure in the balloon is always the same as that of the atmosphere.)

Gas Density, Molar Mass, and Reaction Stoichiometry

65. Consider the following reaction:

$$4Al(s) + 3O_2(g) \longrightarrow 2Al_2O_3(s)$$

It takes 2.00 L of pure oxygen gas at STP to react completely with a certain sample of aluminum. What is the mass of aluminum reacted?

66. A student adds 4.00 g of dry ice (solid CO_2) to an empty balloon. What will be the volume of the balloon at STP after all the dry ice sublimes (converts to gaseous CO_2)?

67. Air bags are activated when a severe impact causes a steel ball to compress a spring and electrically ignite a detonator cap. This causes sodium azide (NaN_3) to decompose explosively according to the following reaction:

$$2NaN_3(s) \longrightarrow 2Na(s) + 3N_2(g)$$

What mass of $NaN_3(s)$ must be reacted to inflate an air bag to 70.0 L at STP?

68. Concentrated hydrogen peroxide solutions are explosively decomposed by traces of transition metal ions (such as Mn or Fe):

$$2H_2O_2(aq) \longrightarrow 2H_2O(l) + O_2(g)$$

What volume of pure $O_2(g)$, collected at 27°C and 746 torr, would be generated by decomposition of 125 g of a 50.0% by mass hydrogen peroxide solution? Ignore any water vapor that may be present.

69. In 1897 the Swedish explorer Andreé tried to reach the North Pole in a balloon. The balloon was filled with hydrogen gas. The hydrogen gas was prepared from iron splints and diluted sulfuric acid. The reaction is

$$Fe(s) + H_2SO_4(aq) \longrightarrow FeSO_4(aq) + H_2(g)$$

The volume of the balloon was 4800 m^3 and the loss of hydrogen gas during filling was estimated at 20.%. What mass of iron splints and 98% (by mass) H_2SO_4 were needed to ensure the complete filling of the balloon? Assume a temperature of 0°C, a pressure of 1.0 atm during filling, and 100% yield.

70. Sulfur trioxide, SO_3, is produced in enormous quantities each year for use in the synthesis of sulfuric acid.

$$S(s) + O_2(g) \longrightarrow SO_2(g)$$
$$2SO_2(g) + O_2(g) \longrightarrow 2SO_3(g)$$

What volume of $O_2(g)$ at 350.°C and a pressure of 5.25 atm is needed to completely convert 5.00 g sulfur to sulfur trioxide?

71. A 15.0-L rigid container was charged with 0.500 atm of krypton gas and 1.50 atm of chlorine gas at 350.°C. The krypton and chlorine react to form krypton tetrachloride. What mass of krypton tetrachloride can be produced assuming 100% yield?

72. An important process for the production of acrylonitrile (C_3H_3N) is given by the following equation:

$$2C_3H_6(g) + 2NH_3(g) + 3O_2(g) \longrightarrow 2C_3H_3N(g) + 6H_2O(g)$$

A 150.-L reactor is charged to the following partial pressures at 25°C:

$$P_{C_3H_6} = 0.500 \text{ MPa}$$
$$P_{NH_3} = 0.800 \text{ MPa}$$
$$P_{O_2} = 1.500 \text{ MPa}$$

What mass of acrylonitrile can be produced from this mixture (MPa = 10^6 Pa)?

73. Consider the reaction between 50.0 mL liquid methanol, CH_3OH (density = 0.850 g/mL), and 22.8 L O_2 at 27°C and a pressure of 2.00 atm. The products of the reaction are $CO_2(g)$ and $H_2O(g)$. Calculate the number of moles of H_2O formed if the reaction goes to completion.

74. Urea (H_2NCONH_2) is used extensively as a nitrogen source in fertilizers. It is produced commercially from the reaction of ammonia and carbon dioxide:

$$2NH_3(g) + CO_2(g) \xrightarrow[\text{Pressure}]{\text{Heat}} H_2NCONH_2(s) + H_2O(g)$$

Ammonia gas at 223°C and 90. atm flows into a reactor at a rate of 500. L/min. Carbon dioxide at 223°C and 45 atm flows into the reactor at a rate of 600. L/min. What mass of urea is produced per minute by this reaction assuming 100% yield?

75. Hydrogen cyanide is prepared commercially by the reaction of methane, $CH_4(g)$, ammonia, $NH_3(g)$, and oxygen, $O_2(g)$, at high temperature. The other product is gaseous water.

  a. Write a chemical equation for the reaction.

  b. What volume of $HCN(g)$ can be obtained from the reaction of 20.0 L $CH_4(g)$, 20.0 L $NH_3(g)$, and 20.0 L $O_2(g)$? The volumes of all gases are measured at the same temperature and pressure.

76. Ethene is converted to ethane by the reaction

$$C_2H_4(g) + H_2(g) \xrightarrow{\text{Catalyst}} C_2H_6(g)$$

C_2H_4 flows into a catalytic reactor at 25.0 atm and 300.°C with a flow rate of 1000. L/min. Hydrogen at 25.0 atm and 300.°C flows into the reactor at a flow rate of 1500. L/min. If 15.0 kg C_2H_6 is collected per minute, what is the percent yield of the reaction?

77. An unknown diatomic gas has a density of 3.164 g/L at STP. What is the identity of the gas?

78. A compound has the empirical formula CHCl. A 256-mL flask, at 373 K and 750. torr, contains 0.800 g of the gaseous compound. Give the molecular formula.

79. Uranium hexafluoride is a solid at room temperature, but it boils at 56°C. Determine the density of uranium hexafluoride at 60.°C and 745 torr.

80. Given that a sample of air is made up of nitrogen, oxygen, and argon in the mole fractions 0.78 N_2, 0.21 O_2, and 0.010 Ar, what is the density of air at standard temperature and pressure?

Partial Pressure

81. Determine the partial pressure of each gas as shown in the figure below. *Note:* The relative numbers of each type of gas are depicted in the figure.

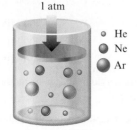

82. Consider the flasks in the following diagrams.

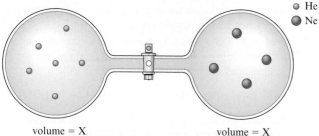

volume = X   volume = X

  a. Which is greater, the initial pressure of helium or the initial pressure of neon? How much greater?

  b. Assuming the connecting tube has negligible volume, draw what each diagram will look like after the stopcock between the two flasks is opened.

  c. Solve for the final pressure in terms of the original pressures of helium and neon. Assume temperature is constant.

  d. Solve for the final partial pressures of helium and neon in terms of their original pressures. Assume the temperature is constant.

83. A piece of solid carbon dioxide, with a mass of 7.8 g, is placed in a 4.0-L otherwise empty container at 27°C. What is the pressure in the container after all the carbon dioxide vaporizes? If 7.8 g solid carbon dioxide were placed in the same container but it already contained air at 740 torr, what would be the partial pressure of carbon dioxide and the total pressure in the container after the carbon dioxide vaporizes?

84. A mixture of 1.00 g H_2 and 1.00 g He is placed in a 1.00-L container at 27°C. Calculate the partial pressure of each gas and the total pressure.

85. Consider the flasks in the following diagram. What are the final partial pressures of H_2 and N_2 after the stopcock between the two flasks is opened? (Assume the final volume is 3.00 L.) What is the total pressure (in torr)?

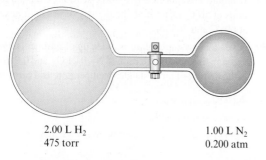

2.00 L H_2 1.00 L N_2
475 torr   0.200 atm

86. Consider the flask apparatus in Exercise 85, which now contains 2.00 L H_2 at a pressure of 360. torr and 1.00 L N_2 at an unknown pressure. If the total pressure in the flasks is 320. torr after the stopcock is opened, determine the initial pressure of N_2 in the 1.00-L flask.

87. Consider the three flasks in the diagram below. Assuming the connecting tubes have negligible volume, what is the partial pressure of each gas and the total pressure after all the stopcocks are opened?

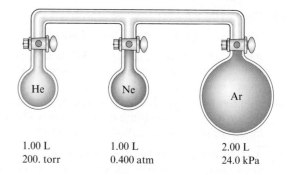

1.00 L   1.00 L   2.00 L
200. torr   0.400 atm   24.0 kPa

88. At 0°C a 1.0-L flask contains 5.0×10^{-2} mole of N_2, 1.5×10^2 mg O_2, and 5.0×10^{21} molecules of NH_3. What is the partial pressure of each gas, and what is the total pressure in the flask?

89. The partial pressure of $CH_4(g)$ is 0.175 atm and that of $O_2(g)$ is 0.250 atm in a mixture of the two gases.

  a. What is the mole fraction of each gas in the mixture?

  b. If the mixture occupies a volume of 10.5 L at 65°C, calculate the total number of moles of gas in the mixture.

  c. Calculate the number of grams of each gas in the mixture.

90. A tank contains a mixture of 52.5 g oxygen gas and 65.1 g carbon dioxide gas at 27°C. The total pressure in the tank is 9.21 atm. Calculate the partial pressures of each gas in the container.

91. Small quantities of hydrogen gas can be prepared in the laboratory by the addition of aqueous hydrochloric acid to metallic zinc.

$$Zn(s) + 2HCl(aq) \longrightarrow ZnCl_2(aq) + H_2(g)$$

Typically, the hydrogen gas is bubbled through water for collection and becomes saturated with water vapor. Suppose 240. mL of hydrogen gas is collected at 30.°C and has a total pressure of 1.032 atm by this process. What is the partial pressure of hydrogen gas in the sample? How many grams of zinc must have reacted to produce this quantity of hydrogen? (The vapor pressure of water is 32 torr at 30°C.)

92. Helium is collected over water at 25°C and 1.00 atm total pressure. What total volume of gas must be collected to obtain 0.586 g helium? (At 25°C the vapor pressure of water is 23.8 torr.)

93. At elevated temperatures, sodium chlorate decomposes to produce sodium chloride and oxygen gas. A 0.8765-g sample of impure sodium chlorate was heated until the production of oxygen gas ceased. The oxygen gas collected over water occupied 57.2 mL at a temperature of 22°C and a pressure of 734 torr. Calculate the mass percent of $NaClO_3$ in the original sample. (At 22°C the vapor pressure of water is 19.8 torr.)

94. Xenon and fluorine will react to form binary compounds when a mixture of these two gases is heated to 400°C in a nickel reaction vessel. A 100.0-mL nickel container is filled with xenon and fluorine, giving partial pressures of 1.24 atm and 10.10 atm, respectively, at a temperature of 25°C. The reaction vessel is heated to 400°C to cause a reaction to occur and then cooled to a temperature at which F_2 is a gas and the xenon fluoride compound produced is a nonvolatile solid. The remaining F_2 gas is transferred to another 100.0-mL nickel container, where the pressure of F_2 at 25°C is 7.62 atm. Assuming all of the xenon has reacted, what is the formula of the product?

95. Methanol (CH_3OH) can be produced by the following reaction:

$$CO(g) + 2H_2(g) \longrightarrow CH_3OH(g)$$

Hydrogen at STP flows into a reactor at a rate of 16.0 L/min. Carbon monoxide at STP flows into the reactor at a rate of 25.0 L/min. If 5.30 g methanol is produced per minute, what is the percent yield of the reaction?

96. In the "Méthode Champenoise," grape juice is fermented in a wine bottle to produce sparkling wine. The reaction is

$$C_6H_{12}O_6(aq) \longrightarrow 2C_2H_5OH(aq) + 2CO_2(g)$$

Fermentation of 750. mL grape juice (density = 1.0 g/cm³) is allowed to take place in a bottle with a total volume of 825 mL until 12% by volume is ethanol (C_2H_5OH). Assuming that the CO_2 is insoluble in H_2O (actually, a wrong assumption), what would be the pressure of CO_2 inside the wine bottle at 25°C? (The density of ethanol is 0.79 g/cm³.)

97. Hydrogen azide, HN_3, decomposes on heating by the following *unbalanced* equation:

$$HN_3(g) \longrightarrow N_2(g) + H_2(g)$$

If 3.0 atm of pure $HN_3(g)$ is decomposed initially, what is the final total pressure in the reaction container? What are the partial pressures of nitrogen and hydrogen gas? Assume the volume and temperature of the reaction container are constant.

98. Equal moles of sulfur dioxide gas and oxygen gas are mixed in a flexible reaction vessel and then sparked to initiate the formation of gaseous sulfur trioxide. Assuming that the reaction goes to completion, what is the ratio of the final volume of the gas mixture to the initial volume of the gas mixture if both volumes are measured at the same temperature and pressure?

99. Some very effective rocket fuels are composed of lightweight liquids. The fuel composed of dimethylhydrazine [$(CH_3)_2N_2H_2$] mixed with dinitrogen tetroxide was used to power the Lunar Lander in its missions to the moon. The two components react according to the following equation:

$$(CH_3)_2N_2H_2(l) + 2N_2O_4(l) \longrightarrow 3N_2(g) + 4H_2O(g) + 2CO_2(g)$$

If 150 g dimethylhydrazine reacts with excess dinitrogen tetroxide and the product gases are collected at 127°C in an evacuated 250-L tank, what is the partial pressure of nitrogen gas produced and what is the total pressure in the tank assuming the reaction has 100% yield?

100. The oxides of Group 2A metals (symbolized by M here) react with carbon dioxide according to the following reaction:

$$MO(s) + CO_2(g) \longrightarrow MCO_3(s)$$

A 2.85-g sample containing only MgO and CuO is placed in a 3.00-L container. The container is filled with CO_2 to a pressure of 740. torr at 20.°C. After the reaction has gone to completion, the pressure inside the flask is 390. torr at 20.°C. What is the mass percent of MgO in the mixture? Assume that only the MgO reacts with CO_2.

Kinetic Molecular Theory and Real Gases

101. Calculate the average kinetic energies of $CH_4(g)$ and $N_2(g)$ molecules at 273 K and 546 K.

102. A 100.-L flask contains a mixture of methane (CH_4) and argon gases at 25°C. The mass of argon present is 228 g and the mole fraction of methane in the mixture is 0.650. Calculate the total kinetic energy of the gaseous mixture.

103. Calculate the root mean square velocities of $CH_4(g)$ and $N_2(g)$ molecules at 273 K and 546 K.

104. Consider separate 1.0-L samples of He(g) and $UF_6(g)$, both at 1.00 atm and containing the same number of moles. What ratio of temperatures for the two samples would produce the same root mean square velocity?

105. You have a gas in a container fitted with a piston and you change one of the conditions of the gas such that a change takes place, as shown below:

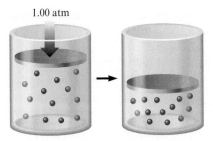

1.00 atm

State two distinct changes you can make to accomplish this, and explain why each would work.

106. You have a gas in a container fitted with a piston and you change one of the conditions of the gas such that a change takes place, as shown below:

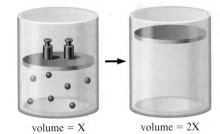

volume = X   volume = 2X

State three distinct changes you can make to accomplish this, and explain why each would work.

107. Consider a 1.0-L container of neon gas at STP. Will the average kinetic energy, average velocity, and frequency of collisions of gas molecules with the walls of the container increase, decrease, or remain the same under each of the following conditions?

   a. The temperature is increased to 100°C.
   b. The temperature is decreased to −50°C.
   c. The volume is decreased to 0.5 L.
   d. The number of moles of neon is doubled.

108. Consider two gases, A and B, each in a 1.0-L container with both gases at the same temperature and pressure. The mass of gas A in the container is 0.34 g and the mass of gas B in the container is 0.48 g.

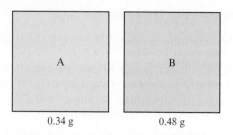

0.34 g   0.48 g

   a. Which gas sample has the most molecules present? Explain.

   b. Which gas sample has the largest average kinetic energy? Explain.

   c. Which gas sample has the fastest average velocity? Explain.

   d. How can the pressure in the two containers be equal to each other since the larger gas B molecules collide with the container walls more forcefully?

109. Consider three identical flasks filled with different gases.

Flask A: CO at 760 torr and 0°C
Flask B: N_2 at 250 torr and 0°C
Flask C: H_2 at 100 torr and 0°C

   a. In which flask will the molecules have the greatest average kinetic energy?

   b. In which flask will the molecules have the greatest average velocity?

110. Consider separate 1.0-L gaseous samples of H_2, Xe, Cl_2, and O_2 all at STP.

   a. Rank the gases in order of increasing average kinetic energy.

   b. Rank the gases in order of increasing average velocity.

   c. How can separate 1.0-L samples of O_2 and H_2 each have the same average velocity?

111. Freon-12 is used as a refrigerant in central home air conditioners. The rate of effusion of Freon-12 to Freon-11 (molar mass = 137.4 g/mol) is 1.07:1. The formula of Freon-12 is one of the following: CF_4, CF_3Cl, CF_2Cl_2, $CFCl_3$, or CCl_4. Which formula is correct for Freon-12?

112. The rate of effusion of a particular gas was measured and found to be 24.0 mL/min. Under the same conditions, the rate of effusion of pure methane (CH_4) gas is 47.8 mL/min. What is the molar mass of the unknown gas?

113. One way of separating oxygen isotopes is by gaseous diffusion of carbon monoxide. The gaseous diffusion process behaves like an effusion process. Calculate the relative rates of effusion of $^{12}C^{16}O$, $^{12}C^{17}O$, and $^{12}C^{18}O$. Name some advantages and disadvantages of separating oxygen isotopes by gaseous diffusion of carbon dioxide instead of carbon monoxide.

114. It took 4.5 minutes for 1.0 L helium to effuse through a porous barrier. How long will it take for 1.0 L Cl_2 gas to effuse under identical conditions?

115. Calculate the pressure exerted by 0.5000 mole of N_2 in a 1.0000-L container at 25.0°C

   a. using the ideal gas law.

   b. using the van der Waals equation.

   c. Compare the results.

116. Calculate the pressure exerted by 0.5000 mole of N_2 in a 10.000-L container at 25.0°C

   a. using the ideal gas law.

   b. using the van der Waals equation.

   c. Compare the results.

   d. Compare the results with those in Exercise 115.

Atmosphere Chemistry

117. Use the data in Table 5.4 to calculate the partial pressure of He in dry air assuming that the total pressure is 1.0 atm. Assuming a temperature of 25°C, calculate the number of He atoms per cubic centimeter.

118. A 1.0-L sample of air is collected at 25°C at sea level (1.00 atm). Estimate the volume this sample of air would have at an altitude of 15 km (see Fig. 5.30). At 15 km, the pressure is about 0.1 atm.

119. Write an equation to show how sulfuric acid is produced in the atmosphere.

120. Write an equation to show how sulfuric acids in acid rain reacts with marble and limestone. (Both marble and limestone are primarily calcium carbonate.)

121. Atmospheric scientists often use mixing ratios to express the concentrations of trace compounds in air. Mixing ratios are often expressed as ppmv (parts per million volume):

$$\text{ppmv of } X = \frac{\text{vol of } X \text{ at STP}}{\text{total vol of air at STP}} \times 10^6$$

On a certain November day, the concentration of carbon monoxide in the air in downtown Denver, Colorado, reached 3.0×10^2 ppmv. The atmospheric pressure at that time was 628 torr and the temperature was 0°C.

   a. What was the partial pressure of CO?

   b. What was the concentration of CO in molecules per cubic meter?

   c. What was the concentration of CO in molecules per cubic centimeter?

122. Trace organic compounds in the atmosphere are first concentrated and then measured by gas chromatography. In the concentration step, several liters of air are pumped through a tube containing a porous substance that traps organic compounds. The tube is then connected to a gas chromatograph and heated to release the trapped compounds. The organic compounds are separated in the column and the amounts are measured. In an analysis for benzene and toluene in air, a 3.00-L sample of air at 748 torr and 23°C was passed through the trap. The gas chromatography analysis showed that this air sample contained 89.6 ng benzene (C_6H_6) and 153 ng toluene (C_7H_8). Calculate the mixing ratio (see Exercise 121) and number of molecules per cubic centimeter for both benzene and toluene.

Additional Exercises

123. Draw a qualitative graph to show how the first property varies with the second in each of the following (assume 1 mole of an ideal gas and T in kelvin).

  a. PV versus V with constant T

  b. P versus T with constant V

  c. T versus V with constant P

  d. P versus V with constant T

  e. P versus $1/V$ with constant T

  f. PV/T versus P

124. At STP, 1.0 L Br_2 reacts completely with 3.0 L F_2, producing 2.0 L of a product. What is the formula of the product? (All substances are gases.)

125. A form of Boyle's law is $PV = k$ (at constant T and n). Table 5.1 contains actual data from pressure–volume experiments conducted by Robert Boyle. The value of k in most experiments is 14.1×10^2 in Hg $\cdot$ in^3. Express k in units of atm $\cdot$ L. In Example 5.3, k was determined for NH_3 at various pressures and volumes. Give some reasons why the k values differ so dramatically between Example 5.3 and Table 5.1.

126. A 2.747-g sample of manganese metal is reacted with excess HCl gas to produce 3.22 L $H_2(g)$ at 373 K and 0.951 atm and a manganese chloride compound ($MnCl_x$). What is the formula of the manganese chloride compound produced in the reaction?

127. A 1.00-L gas sample at 100.°C and 600. torr contains 50.0% helium and 50.0% xenon by mass. What are the partial pressures of the individual gases?

128. Cyclopropane, a gas that when mixed with oxygen is used as a general anesthetic, is composed of 85.7% C and 14.3% H by mass. If the density of cyclopropane is 1.88 g/L at STP, what is the molecular formula of cyclopropane?

129. The nitrogen content of organic compounds can be determined by the Dumas method. The compound in question is first reacted by passage over hot $CuO(s)$:

$$\text{Compound} \xrightarrow[\text{CuO}(s)]{\text{Hot}} N_2(g) + CO_2(g) + H_2O(g)$$

The product gas is then passed through a concentrated solution of KOH to remove the CO_2. After passage through the KOH solution, the gas contains N_2 and is saturated with water vapor. In a given experiment a 0.253-g sample of a compound produced 31.8 mL N_2 saturated with water vapor at 25°C and 726 torr. What is the mass percent of nitrogen in the compound? (The vapor pressure of water at 25°C is 23.8 torr.)

130. An organic compound containing only C, H, and N yields the following data.

  i. Complete combustion of 35.0 mg of the compound produced 33.5 mg CO_2 and 41.1 mg H_2O.

  ii. A 65.2-mg sample of the compound was analyzed for nitrogen by the Dumas method (see Exercise 129), giving 35.6 mL of dry N_2 at 740. torr and 25°C.

  iii. The effusion rate of the compound as a gas was measured and found to be 24.6 mL/min. The effusion rate of argon gas, under identical conditions, is 26.4 mL/min.

  What is the molecular formula of the compound?

131. A 15.0-L tank is filled with H_2 to a pressure of 2.00×10^2 atm. How many balloons (each 2.00 L) can be inflated to a pressure of 1.00 atm from the tank? Assume that there is no temperature change and that the tank cannot be emptied below 1.00 atm pressure.

132. A spherical glass container of unknown volume contains helium gas at 25°C and 1.960 atm. When a portion of the helium is withdrawn and adjusted to 1.00 atm at 25°C, it is found to have a volume of 1.75 cm^3. The gas remaining in the first container shows a pressure of 1.710 atm. Calculate the volume of the spherical container.

133. A 2.00-L sample of $O_2(g)$ was collected over water at a total pressure of 785 torr and 25°C. When the $O_2(g)$ was dried (water vapor removed), the gas had a volume of 1.94 L at 25°C and 785 torr. Calculate the vapor pressure of water at 25°C.

134. A 20.0-L stainless steel container at 25°C was charged with 2.00 atm of hydrogen gas and 3.00 atm of oxygen gas. A spark ignited the mixture, producing water. What is the pressure in the tank at 25°C? If the exact same experiment were performed, but the temperature was 125°C instead of 25°C, what would be the pressure in the tank?

135. Metallic molybdenum can be produced from the mineral molybdenite, MoS_2. The mineral is first oxidized in air to molybdenum trioxide and sulfur dioxide. Molybdenum trioxide is then reduced to metallic molybdenum using hydrogen gas. The balanced equations are

$$MoS_2(s) + \tfrac{7}{2}O_2(g) \longrightarrow MoO_3(s) + 2SO_2(g)$$
$$MoO_3(s) + 3H_2(g) \longrightarrow Mo(s) + 3H_2O(l)$$

Calculate the volumes of air and hydrogen gas at 17°C and 1.00 atm that are necessary to produce 1.00×10^3 kg pure molybdenum from MoS_2. Assume air contains 21% oxygen by volume, and assume 100% yield for each reaction.

136. Nitric acid is produced commercially by the Ostwald process. In the first step ammonia is oxidized to nitric oxide:

$$4NH_3(g) + 5O_2(g) \longrightarrow 4NO(g) + 6H_2O(g)$$

Assume this reaction is carried out in the apparatus diagrammed below.

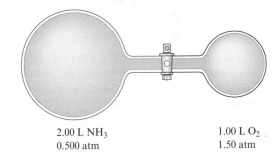

2.00 L NH_3 1.00 L O_2
0.500 atm   1.50 atm

The stopcock between the two reaction containers is opened, and the reaction proceeds using proper catalysts. Calculate the partial pressure of NO after the reaction is complete. Assume 100% yield for the reaction, assume the final container volume is 3.00 L, and assume the temperature is constant.

137. A compound contains only C, H, and N. It is 58.51% C and 7.37% H by mass. Helium effuses through a porous frit 3.20 times as fast as the compound does. Determine the empirical and molecular formulas of this compound.

138. One of the chemical controversies of the nineteenth century concerned the element beryllium (Be). Berzelius originally claimed that beryllium was a trivalent element (forming Be^{3+} ions) and that it gave an oxide with the formula Be_2O_3. This resulted in a calculated atomic mass of 13.5 for beryllium. In formulating his periodic table, Mendeleev proposed that beryllium was divalent (forming Be^{2+} ions) and that it gave an oxide with the formula BeO. This assumption gives an atomic mass of 9.0. In 1894, A. Combes (*Comptes Rendus* 1894, p. 1221) reacted beryllium with the anion $C_5H_7O_2^-$ and measured the density of the gaseous product. Combes's data for two different experiments are as follows:

	I	II
Mass	0.2022 g	0.2224 g
Volume	22.6 cm^3	26.0 cm^3
Temperature	13°C	17°C
Pressure	765.2 mm Hg	764.6 mm

If beryllium is a divalent metal, the molecular formula of the product will be $Be(C_5H_7O_2)_2$; if it is trivalent, the formula will be $Be(C_5H_7O_2)_3$. Show how Combes's data help to confirm that beryllium is a divalent metal.

139. An organic compound contains C, H, N, and O. Combustion of 0.1023 g of the compound in excess oxygen yielded 0.2766 g CO_2 and 0.0991 g H_2O. A sample of 0.4831 g of the compound was analyzed for nitrogen by the Dumas method (see Exercise 129). At STP, 27.6 mL of dry N_2 was obtained. In a third experiment, the density of the compound as a gas was found to be 4.02 g/L at 127°C and 256 torr. What are the empirical and molecular formulas of the compound?

140. Consider the following diagram:

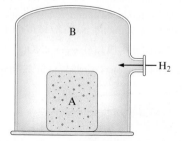

Container A (with porous walls) is filled with air at STP. It is then inserted into a large enclosed container (B), which is then flushed with $H_2(g)$. What will happen to the pressure inside container A? Explain your answer.

ChemWork Problems

These multiconcept problems (and additional ones) are found interactively online with the same type of assistance a student would get from an instructor.

141. A glass vessel contains 28 g of nitrogen gas. Assuming ideal behavior, which of the processes listed below would double the pressure exerted on the walls of the vessel?
   a. Adding 28 g of oxygen gas
   b. Raising the temperature of the container from −73°C to 127°C
   c. Adding enough mercury to fill one-half the container
   d. Adding 32 g of oxygen gas

e. Raising the temperature of the container from 30.°C to 60.°C

142. A steel cylinder contains 150.0 moles of argon gas at a temperature of 25°C and a pressure of 8.93 MPa. After some argon has been used, the pressure is 2.00 MPa at a temperature of 19°C. What mass of argon remains in the cylinder?

143. A certain flexible weather balloon contains helium gas at a volume of 855 L. Initially, the balloon is at sea level where the temperature is 25°C and the barometric pressure is 730 torr. The balloon then rises to an altitude of 6000 ft, where the pressure is 605 torr and the temperature is 15°C. What is the change in volume of the balloon as it ascends from sea level to 6000 ft?

144. A large flask with a volume of 936 mL is evacuated and found to have a mass of 134.66 g. It is then filled to a pressure of 0.967 atm at 31°C with a gas of unknown molar mass and then reweighed to give a new mass of 135.87 g. What is the molar mass of this gas?

145. A 20.0-L nickel container was charged with 0.859 atm of xenon gas and 1.37 atm of fluorine gas at 400°C. The xenon and fluorine react to form xenon tetrafluoride. What mass of xenon tetrafluoride can be produced assuming 100% yield?

146. Consider the *unbalanced* chemical equation below:

$$CaSiO_3(s) + HF(g) \longrightarrow CaF_2(aq) + SiF_4(g) + H_2O(l)$$

Suppose a 32.9-g sample of $CaSiO_3$ is reacted with 31.8 L of HF at 27.0°C and 1.00 atm. Assuming the reaction goes to completion, calculate the mass of the SiF_4 and H_2O produced in the reaction.

147. Consider separate 1.0-L gaseous samples of He, N_2, and O_2, all at STP and all acting ideally. Rank the gases in order of increasing average kinetic energy and in order of increasing average velocity.

148. Which of the following statements is(are) *true*?
   a. If the number of moles of a gas is doubled, the volume will double, assuming the pressure and temperature of the gas remain constant.
   b. If the temperature of a gas increases from 25°C to 50°C, the volume of the gas would double, assuming that the pressure and the number of moles of gas remain constant.
   c. The device that measures atmospheric pressure is called a barometer.
   d. If the volume of a gas decreases by one half, then the pressure would double, assuming that the number of moles and the temperature of the gas remain constant.

Challenge Problems

149. A chemist weighed out 5.14 g of a mixture containing unknown amounts of $BaO(s)$ and $CaO(s)$ and placed the sample in a 1.50-L flask containing $CO_2(g)$ at 30.0°C and 750. torr. After the reaction to form $BaCO_3(s)$ and $CaCO_3(s)$ was completed, the pressure of $CO_2(g)$ remaining was 230. torr. Calculate the mass percentages of $CaO(s)$ and $BaO(s)$ in the mixture.

150. A mixture of chromium and zinc weighing 0.362 g was reacted with an excess of hydrochloric acid. After all the metals in the mixture reacted, 225 mL dry of hydrogen gas was collected at 27°C and 750. torr. Determine the mass percent of Zn

in the metal sample. [Zinc reacts with hydrochloric acid to produce zinc chloride and hydrogen gas; chromium reacts with hydrochloric acid to produce chromium(III) chloride and hydrogen gas.]

151. Consider a sample of a hydrocarbon (a compound consisting of only carbon and hydrogen) at 0.959 atm and 298 K. Upon combusting the entire sample in oxygen, you collect a mixture of gaseous carbon dioxide and water vapor at 1.51 atm and 375 K. This mixture has a density of 1.391 g/L and occupies a volume four times as large as that of the pure hydrocarbon. Determine the molecular formula of the hydrocarbon.

152. You have an equimolar mixture of the gases SO_2 and O_2, along with some He, in a container fitted with a piston. The density of this mixture at STP is 1.924 g/L. Assume ideal behavior and constant temperature and pressure.

   a. What is the mole fraction of He in the original mixture?

   b. The SO_2 and O_2 react to completion to form SO_3. What is the density of the gas mixture after the reaction is complete?

153. Methane (CH_4) gas flows into a combustion chamber at a rate of 200. L/min at 1.50 atm and ambient temperature. Air is added to the chamber at 1.00 atm and the same temperature, and the gases are ignited.

   a. To ensure complete combustion of CH_4 to $CO_2(g)$ and $H_2O(g)$, three times as much oxygen as is necessary is reacted. Assuming air is 21 mole percent O_2 and 79 mole percent N_2, calculate the flow rate of air necessary to deliver the required amount of oxygen.

   b. Under the conditions in part a, combustion of methane was not complete as a mixture of $CO_2(g)$ and $CO(g)$ was produced. It was determined that 95.0% of the carbon in the exhaust gas was present in CO_2. The remainder was present as carbon in CO. Calculate the composition of the exhaust gas in terms of mole fraction of CO, CO_2, O_2, N_2, and H_2O. Assume CH_4 is completely reacted and N_2 is unreacted.

154. A steel cylinder contains 5.00 mole of graphite (pure carbon) and 5.00 moles of O_2. The mixture is ignited and all the graphite reacts. Combustion produces a mixture of CO gas and CO_2 gas. After the cylinder has cooled to its original temperature, it is found that the pressure of the cylinder has increased by 17.0%. Calculate the mole fractions of CO, CO_2, and O_2 in the final gaseous mixture.

155. The total mass that can be lifted by a balloon is given by the difference between the mass of air displaced by the balloon and the mass of the gas inside the balloon. Consider a hot-air balloon that approximates a sphere 5.00 m in diameter and contains air heated to 65°C. The surrounding air temperature is 21°C. The pressure in the balloon is equal to the atmospheric pressure, which is 745 torr.

   a. What total mass can the balloon lift? Assume that the average molar mass of air is 29.0 g/mol. (*Hint:* Heated air is less dense than cool air.)

   b. If the balloon is filled with enough helium at 21°C and 745 torr to achieve the same volume as in part a, what total mass can the balloon lift?

   c. What mass could the hot-air balloon in part a lift if it were on the ground in Denver, Colorado, where a typical atmospheric pressure is 630. torr?

156. You have a sealed, flexible balloon filled with argon gas. The atmospheric pressure is 1.00 atm and the temperature is 25°C. Assume that air has a mole fraction of nitrogen of 0.790, the rest being oxygen.

   a. Explain why the balloon would float when heated. Make sure to discuss which factors change and which remain constant, and why this matters.

   b. Above what temperature would you heat the balloon so that it would float?

157. You have a helium balloon at 1.00 atm and 25°C. You want to make a hot-air balloon with the same volume and same lift as the helium balloon. Assume air is 79.0% nitrogen and 21.0% oxygen by volume. The "lift" of a balloon is given by the difference between the mass of air displaced by the balloon and the mass of gas inside the balloon.

   a. Will the temperature in the hot-air balloon have to be higher or lower than 25°C? Explain.

   b. Calculate the temperature of the air required for the hot-air balloon to provide the same lift as the helium balloon at 1.00 atm and 25°C. Assume atmospheric conditions are 1.00 atm and 25°C.

158. We state that the ideal gas law tends to hold best at low pressures and high temperatures. Show how the van der Waals equation simplifies to the ideal gas law under these conditions.

159. You are given an unknown gaseous binary compound (that is, a compound consisting of two different elements). When 10.0 g of the compound is burned in excess oxygen, 16.3 g of water is produced. The compound has a density 1.38 times that of oxygen gas at the same conditions of temperature and pressure. Give a possible identity for the unknown compound.

160. Nitrogen gas (N_2) reacts with hydrogen gas (H_2) to form ammonia gas (NH_3). You have nitrogen and hydrogen gases in a 15.0-L container fitted with a movable piston (the piston allows the container volume to change so as to keep the pressure constant inside the container). Initially the partial pressure of each reactant gas is 1.00 atm. Assume the temperature is constant and that the reaction goes to completion.

   a. Calculate the partial pressure of ammonia in the container after the reaction has reached completion.

   b. Calculate the volume of the container after the reaction has reached completion.

Integrative Problems

These problems require the integration of multiple concepts to find the solutions.

161. In the presence of nitric acid, UO^{2+} undergoes a redox process. It is converted to UO_2^{2+} and nitric oxide (NO) gas is produced according to the following unbalanced equation:

$$H^+(aq) + NO_3^-(aq) + UO^{2+}(aq) \longrightarrow$$
$$NO(g) + UO_2^{2+}(aq) + H_2O(l)$$

If 2.55×10^2 mL NO(g) is isolated at 29°C and 1.5 atm, what amount (moles) of UO^{2+} was used in the reaction? (*Hint:* Balance the reaction by the oxidation states method.)

162. Silane, SiH_4, is the silicon analogue of methane, CH_4. It is prepared industrially according to the following equations:

$$Si(s) + 3HCl(g) \longrightarrow HSiCl_3(l) + H_2(g)$$

$$4HSiCl_3(l) \longrightarrow SiH_4(g) + 3SiCl_4(l)$$

a. If 156 mL $HSiCl_3$ ($d = 1.34$ g/mL) is isolated when 15.0 L HCl at 10.0 atm and 35°C is used, what is the percent yield of $HSiCl_3$?

b. When 156 mL $HSiCl_3$ is heated, what volume of SiH_4 at 10.0 atm and 35°C will be obtained if the percent yield of the reaction is 93.1%?

163. Solid thorium(IV) fluoride has a boiling point of 1680°C. What is the density of a sample of gaseous thorium(IV) fluoride at its boiling point under a pressure of 2.5 atm in a 1.7-L container? Which gas will effuse faster at 1680°C, thorium(IV) fluoride or uranium(III) fluoride? How much faster?

164. Natural gas is a mixture of hydrocarbons, primarily methane (CH_4) and ethane (C_2H_6). A typical mixture might have $\chi_{methane} = 0.915$ and $\chi_{ethane} = 0.085$. What are the partial pressures of the two gases in a 15.00-L container of natural gas at 20.°C and

1.44 atm? Assuming complete combustion of both gases in the natural gas sample, what is the total mass of water formed?

Marathon Problem

This problem is designed to incorporate several concepts and techniques into one situation.

165. Consider an equimolar mixture (equal number of moles) of two diatomic gases (A_2 and B_2) in a container fitted with a piston. The gases react to form one product (which is also a gas) with the formula A_xB_y. The density of the sample after the reaction is complete (and the temperature returns to its original state) is 1.50 times greater than the density of the reactant mixture.

a. Specify the formula of the product, and explain if more than one answer is possible based on the given data.

b. Can you determine the molecular formula of the product with the information given or only the empirical formula?

Thermochemistry

A burning match is an example of exothermic reaction. This double exposure shows the match lit and the match blown out. (© Caren Brinkema/Science Faction/Corbis)

Energy is the essence of our very existence as individuals and as a society. The food that we eat furnishes the energy to live, work, and play, just as the coal and oil consumed by manufacturing and transportation systems power our modern industrialized civilization.

In the past, huge quantities of carbon-based fossil fuels have been available for the taking. This abundance of fuels has led to a world society with a voracious appetite for energy, consuming millions of barrels of petroleum every day. We are now dangerously dependent on the dwindling supplies of oil, and this dependence is an important source of tension among nations in today's world. In an incredibly short time, we have moved from a period of ample and cheap supplies of petroleum to one of high prices and uncertain supplies. If our present standard of living is to be maintained, we must find alternatives to petroleum. To do this, we need to know the relationship between chemistry and energy, which we explore in this chapter.

There are additional problems with fossil fuels. The waste products from burning fossil fuels significantly affect our environment. For example, when a carbon-based fuel is burned, the carbon reacts with oxygen to form carbon dioxide, which is released into the atmosphere. Although much of this carbon dioxide is consumed in various natural processes such as photosynthesis and the formation of carbonate materials, the amount of carbon dioxide in the atmosphere is steadily increasing. This increase is significant because atmospheric carbon dioxide absorbs heat radiated from the earth's surface and radiates it back toward the earth. Since this is an important mechanism for controlling the earth's temperature, many scientists fear that an increase in the concentration of carbon dioxide will warm the earth, causing significant changes in climate. In addition, impurities in the fossil fuels react with components of the air to produce air pollution. We discussed some aspects of this problem in Chapter 5.

Just as energy is important to our society on a macroscopic scale, it is critically important to each living organism on a microscopic scale. The living cell is a miniature chemical factory powered by energy from chemical reactions. The process of cellular respiration extracts the energy stored in sugars and other nutrients to drive the various tasks of the cell. Although the extraction process is more complex and more subtle, the energy obtained from "fuel" molecules by the cell is the same as would be obtained from burning the fuel to power an internal combustion engine.

Whether it is an engine or a cell that is converting energy from one form to another, the processes are all governed by the same principles, which we will begin to explore in this chapter. Additional aspects of energy transformation will be covered in Chapter 17.

6.1 | The Nature of Energy

Although the concept of energy is quite familiar, energy itself is rather difficult to define precisely. We will define **energy** as the *capacity to do work or to produce heat.* In this chapter we will concentrate specifically on the heat transfer that accompanies chemical processes.

One of the most important characteristics of energy is that it is conserved. The **law of conservation of energy** states that *energy can be converted from one form to another but can be neither created nor destroyed.* That is, the energy of the universe is constant. Energy can be classified as either potential or kinetic energy. **Potential energy** is energy due to position or composition. For example, water behind a dam has potential energy that can be converted to work when the water flows down through turbines, thereby creating electricity. Attractive and repulsive forces also lead to

The total energy content of the universe is constant.

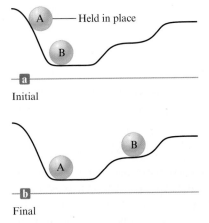

Initial

Final

Figure 6.1 | (a) In the initial positions, ball A has a higher potential energy than ball B. (b) After A has rolled down the hill, the potential energy lost by A has been converted to random motions of the components of the hill (frictional heating) and to the increase in the potential energy of B.

Heat involves a transfer of energy.

This infrared photo of a house shows where energy leaks occur. The more red the color, the more energy (heat) is leaving the house.

potential energy. The energy released when gasoline is burned results from differences in attractive forces between the nuclei and electrons in the reactants and products. The **kinetic energy** of an object is energy due to the motion of the object and depends on the mass of the object m and its velocity v: $KE = \frac{1}{2}mv^2$.

Energy can be converted from one form to another. For example, consider the two balls in Fig. 6.1(a). Ball A, because of its higher position initially, has more potential energy than ball B. When A is released, it moves down the hill and strikes B. Eventually, the arrangement shown in Fig. 6.1(b) is achieved. What has happened in going from the initial to the final arrangement? The potential energy of A has decreased, but since energy is conserved, all the energy lost by A must be accounted for. How is this energy distributed?

Initially, the potential energy of A is changed to kinetic energy as the ball rolls down the hill. Part of this kinetic energy is then transferred to B, causing it to be raised to a higher final position. Thus the potential energy of B has been increased. However, since the final position of B is lower than the original position of A, some of the energy is still unaccounted for. Both balls in their final positions are at rest, so the missing energy cannot be due to their motions. What has happened to the remaining energy?

The answer lies in the interaction between the hill's surface and the ball. As ball A rolls down the hill, some of its kinetic energy is transferred to the surface of the hill as heat. This transfer of energy is called *frictional heating*. The temperature of the hill increases very slightly as the ball rolls down.

Before we proceed further, it is important to recognize that heat and temperature are decidedly different. As we saw in Chapter 5, *temperature* is a property that reflects the random motions of the particles in a particular substance. **Heat**, on the other hand, involves the *transfer* of energy between two objects due to a temperature difference. Heat is not a substance contained by an object, although we often talk of heat as if this were true.

Note that in going from the initial to the final arrangements in Fig. 6.1, ball B gains potential energy because work was done by ball A on B. **Work** is defined as force acting over a distance. Work is required to raise B from its original position to its final one. Part of the original energy stored as potential energy in A has been transferred through work to B, thereby increasing B's potential energy. Thus there are two ways to transfer energy: through work and through heat.

In rolling to the bottom of the hill shown in Fig. 6.1, ball A will always lose the same amount of potential energy. However, the way that this energy transfer is divided between work and heat depends on the specific conditions—the **pathway**. For example, the surface of the hill might be so rough that the energy of A is expended completely through frictional heating; A is moving so slowly when it hits B that it cannot move B to the next level. In this case, no work is done. Regardless of the condition of the hill's surface, the *total energy* transferred will be constant. However, the amounts of heat and work will differ. Energy change is independent of the pathway; however, work and heat are both dependent on the pathway.

This brings us to a very important concept: the **state function** or **state property**. A state function refers to a property of the system that depends only on its *present state*. A state function (property) does not depend in any way on the system's past (or future). In other words, the value of a state function does not depend on how the system arrived at the present state; it depends only on the characteristics of the present state. This leads to a very important characteristic of a state function: A change in this function (property) in going from one state to another state is independent of the particular pathway taken between the two states.

A nonscientific analogy that illustrates the difference between a state function and a nonstate function is elevation on the earth's surface and distance between two points. In traveling from Chicago (elevation 674 ft) to Denver (elevation 5280 ft), the change in elevation is always $5280 - 674 = 4606$ ft regardless of the route taken between the two cities. The distance traveled, however, depends on how you make the trip. Thus elevation is a function that does not depend on the route (pathway), but distance is pathway dependent. Elevation is a state function and distance is not.

Courtesy, Sierra Pacific Innovations

Energy is a state function; work and heat are not.

Of the functions considered in our present example, energy is a state function, but work and heat are not state functions.

Chemical Energy

The ideas we have just illustrated using mechanical examples also apply to chemical systems. The combustion of methane, for example, is used to heat many homes in the United States:

$$CH_4(g) + 2O_2(g) \longrightarrow CO_2(g) + 2H_2O(g) + \text{energy (heat)}$$

To discuss this reaction, we divide the universe into two parts: the system and the surroundings. The **system** is the part of the universe on which we wish to focus attention; the **surroundings** include everything else in the universe. In this case we define the system as the reactants and products of the reaction. The surroundings consist of the reaction container (a furnace, for example), the room, and anything else other than the reactants and products.

When a reaction results in the evolution of heat, it is said to be **exothermic** (*exo-* is a prefix meaning "out of"); that is, energy flows *out of the system*. For example, in the combustion of methane, energy flows out of the system as heat. Reactions that absorb energy from the surroundings are said to be **endothermic**. When the heat flow is *into a system*, the process is endothermic. For example, the formation of nitric oxide from nitrogen and oxygen is endothermic:

$$N_2(g) + O_2(g) + \text{energy (heat)} \longrightarrow 2NO(g)$$

Where does the energy, released as heat, come from in an exothermic reaction? The answer lies in the difference in potential energies between the products and the reactants. Which has lower potential energy, the reactants or the products? We know that total energy is conserved and that energy flows from the system into the surroundings in an exothermic reaction. This means that *the energy gained by the surroundings must be equal to the energy lost by the system.* In the combustion of methane, the energy content of the system decreases, which means that 1 mole of CO_2 and 2 moles of H_2O molecules (the products) possess less potential energy than do 1 mole of CH_4 and 2 moles of O_2 molecules (the reactants). The heat flow into the surroundings results from a lowering of the potential energy of the reaction system. This always holds true. *In any exothermic reaction, some of the potential energy stored in the chemical bonds is being converted to thermal energy (random kinetic energy) via heat.*

The energy diagram for the combustion of methane is shown in Fig. 6.2, where Δ(PE) represents the *change* in potential energy stored in the bonds of the products as compared with the bonds of the reactants. In other words, this quantity represents the difference between the energy required to break the bonds in the reactants and the

Figure 6.2 | The combustion of methane releases the quantity of energy Δ(PE) to the surroundings via heat flow. This is an exothermic process.

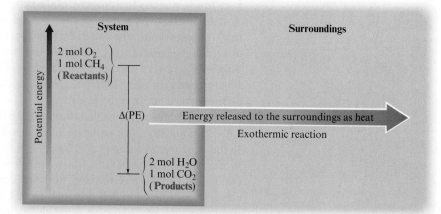

Figure 6.3 | The energy diagram for the reaction of nitrogen and oxygen to form nitric oxide. This is an endothermic process: Heat [equal in magnitude to Δ(PE)] flows into the system from the surroundings.

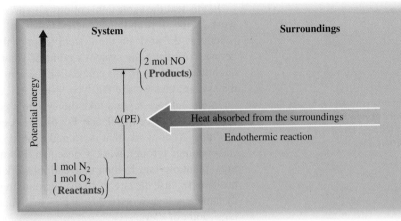

energy released when the bonds in the products are formed. In an exothermic process, the bonds in the products are stronger (on average) than those of the reactants. That is, more energy is released by forming the new bonds in the products than is consumed to break the bonds in the reactants. The net result is that the quantity of energy Δ(PE) is transferred to the surroundings through heat.

For an endothermic reaction, the situation is reversed, as shown in Fig. 6.3. Energy that flows into the system as heat is used to increase the potential energy of the system. In this case the products have higher potential energy (weaker bonds on average) than the reactants.

The study of energy and its interconversions is called **thermodynamics**. The law of conservation of energy is often called the **first law of thermodynamics** and is stated as follows: *The energy of the universe is constant.*

The **internal energy** E of a system can be defined most precisely as the sum of the kinetic and potential energies of all the "particles" in the system. The internal energy of a system can be changed by a flow of work, heat, or both. That is,

$$\Delta E = q + w$$

where ΔE represents the change in the system's internal energy, q represents heat, and w represents work.

Thermodynamic quantities always consist of two parts: a *number*, giving the magnitude of the change, and a *sign*, indicating the direction of the flow. *The sign reflects the system's point of view.* For example, if a quantity of energy flows *into* the system via heat (an endothermic process), q is equal to $+x$, where the *positive* sign indicates that the *system's energy is increasing.* On the other hand, when energy flows *out of* the system via heat (an exothermic process), q is equal to $-x$, where the *negative* sign indicates that the *system's energy is decreasing.*

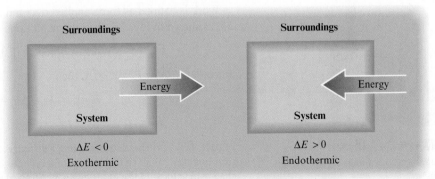

In this text the same conventions also apply to the flow of work. If the system does work on the surroundings (energy flows out of the system), w is negative. If the surroundings do work on the system (energy flows into the system), w is positive. We define work from the system's point of view to be consistent for all thermodynamic quantities. That is, in this convention the signs of both q and w reflect what happens to the system; thus we use $\Delta E = q + w$.

In this text we *always* take the system's point of view. This convention is not followed in every area of science. For example, engineers are in the business of designing machines to do work, that is, to make the system (the machine) transfer energy to its surroundings through work. Consequently, engineers define work from the surroundings' point of view. In their convention, work that flows out of the system is treated as positive because the energy of the surroundings has increased. The first law of thermodynamics is then written $\Delta E = q - w'$, where w' signifies work from the surroundings' point of view.

> The convention in this text is to take the system's point of view; $q = -x$ denotes an exothermic process, and $q = +x$ denotes an endothermic one.

Interactive Example 6.1

Sign in at http://login.cengagebrain .com to try this Interactive Example in **OWL**.

The joule (J) is the fundamental SI unit for energy:

$$J = \frac{kg \cdot m^2}{s^2}$$

One kilojoule (kJ) = 10^3 J.

Internal Energy

Calculate ΔE for a system undergoing an endothermic process in which 15.6 kJ of heat flows and where 1.4 kJ of work is done on the system.

Solution

We use the equation

$$\Delta E = q + w$$

where $q = +15.6$ kJ, since the process is endothermic, and $w = +1.4$ kJ, since work is done on the system. Thus

$$\Delta E = 15.6 \text{ kJ} + 1.4 \text{ kJ} = 17.0 \text{ kJ}$$

The system has gained 17.0 kJ of energy.

See Exercises 6.29 through 6.32

A common type of work associated with chemical processes is work done by a gas (through *expansion*) or work done to a gas (through *compression*). For example, in an automobile engine, the heat from the combustion of the gasoline expands the gases in the cylinder to push back the piston, and this motion is then translated into the motion of the car.

Suppose we have a gas confined to a cylindrical container with a movable piston as shown in Fig. 6.4, where F is the force acting on a piston of area A. Since pressure is defined as force per unit area, the pressure of the gas is

$$P = \frac{F}{A}$$

Work is defined as force applied over a distance, so if the piston moves a distance Δh, as shown in Fig. 6.4, then the work done is

$$\text{Work} = \text{force} \times \text{distance} = F \times \Delta h$$

Since $P = F/A$ or $F = P \times A$, then

$$\text{Work} = F \times \Delta h = P \times A \times \Delta h$$

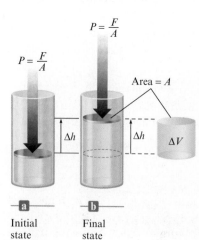

a Initial state **b** Final state

Figure 6.4 | (a) The piston, moving a distance Δh against a pressure P, does work on the surroundings. (b) Since the volume of a cylinder is the area of the base times its height, the change in volume of the gas is given by $\Delta h \times A = \Delta V$.

Since the volume of a cylinder equals the area of the piston times the height of the cylinder (see Fig. 6.4), the change in volume ΔV resulting from the piston moving a distance Δh is

$$\Delta V = \text{final volume} - \text{initial volume} = A \times \Delta h$$

Substituting $\Delta V = A \times \Delta h$ into the expression for work gives

$$\text{Work} = P \times A \times \Delta h = P\Delta V$$

This gives us the *magnitude* (size) of the work required to expand a gas ΔV against a pressure P.

What about the sign of the work? The gas (the system) is expanding, moving the piston against the pressure. Thus the system is doing work on the surroundings, so from the system's point of view the sign of the work should be negative.

For an *expanding* gas, ΔV is a positive quantity because the volume is increasing. Thus ΔV and w must have opposite signs, which leads to the equation

$$w = -P\Delta V$$

Note that for a gas expanding against an external pressure P, w is a negative quantity as required, since work flows out of the system. When a gas is *compressed*, ΔV is a negative quantity (the volume decreases), which makes w a positive quantity (work flows into the system).

w and *PΔV* have opposite signs because when the gas expands (ΔV is positive), work flows into the surroundings (*w* is negative).

Critical Thinking

You are calculating ΔE in a chemistry problem. What if you confuse the system and the surroundings? How would this affect the magnitude of the answer you calculate? The sign?

Interactive Example 6.2

Sign in at http://login.cengagebrain.com to try this Interactive Example in **OWL**.

For an ideal gas, work can occur only when its volume changes. Thus, if a gas is heated at constant volume, the pressure increases but no work occurs.

PV Work

Calculate the work associated with the expansion of a gas from 46 L to 64 L at a constant external pressure of 15 atm.

Solution

For a gas at constant pressure,

$$w = -P\Delta V$$

In this case $P = 15$ atm and $\Delta V = 64 - 46 = 18$ L. Hence

$$w = -15 \text{ atm} \times 18 \text{ L} = -270 \text{ L} \cdot \text{atm}$$

Note that since the gas expands, it does work on its surroundings.

Reality Check | Energy flows out of the gas, so w is a negative quantity.

See Exercises 6.35 through 6.37

Sign in at http://login.cengagebrain.com to try this Interactive Example in **OWL**.

In dealing with "*PV* work," keep in mind that the P in $P\Delta V$ always refers to the external pressure—the pressure that causes a compression or that resists an expansion.

Interactive Example 6.3

Internal Energy, Heat, and Work

A balloon is being inflated to its full extent by heating the air inside it. In the final stages of this process, the volume of the balloon changes from 4.00×10^6 L to 4.50×10^6 L by the addition of 1.3×10^8 J of energy as heat. Assuming that the

A propane burner is used to heat the air in a hot-air balloon.

balloon expands against a constant pressure of 1.0 atm, calculate ΔE for the process. (To convert between L · atm and J, use 1 L · atm = 101.3 J.)

Solution

Where are we going?

To calculate ΔE

What do we know?

> $V_1 = 4.00 \times 10^6$ L
> $q = +1.3 \times 10^8$ J
> $P = 1.0$ atm
> 1 L · atm $= 101.3$ J
> $V_2 = 4.50 \times 10^6$ L

What do we need?

> $\Delta E = q + w$

How do we get there?

What is the work done on the gas?

$$w = -P\Delta V$$

What is ΔV?

$$\Delta V = V_2 - V_1 = 4.50 \times 10^6 \text{ L} - 4.00 \times 10^6 \text{ L} = 5.0 \times 10^5 \text{ L}$$

What is the work?

$$w = -P\Delta V = -1.0 \text{ atm} \times 5.0 \times 10^5 \text{ L} = -5.0 \times 10^5 \text{ L} \cdot \text{atm}$$

The negative sign makes sense because the gas is expanding and doing work on the surroundings. To calculate ΔE, we must sum q and w. However, since q is given in units of J and w is given in units of L · atm, we must change the work to units of joules:

$$w = -5.0 \times 10^5 \text{ L} \cdot \text{atm} \times \frac{101.3 \text{ J}}{\text{L} \cdot \text{atm}} = -5.1 \times 10^7 \text{ J}$$

Then

> $\Delta E = q + w = (+1.3 \times 10^8 \text{ J}) + (-5.1 \times 10^7 \text{ J}) = 8 \times 10^7 \text{ J}$

Reality Check | Since more energy is added through heating than the gas expends doing work, there is a net increase in the internal energy of the gas in the balloon. Hence ΔE is positive.

See Exercises 6.38 through 6.40

6.2 | Enthalpy and Calorimetry

Enthalpy

So far we have discussed the internal energy of a system. A less familiar property of a system is its **enthalpy** H, which is defined as

$$H = E + PV$$

where E is the internal energy of the system, P is the pressure of the system, and V is the volume of the system.

Since internal energy, pressure, and volume are all state functions, *enthalpy is also a state function.* But what exactly is enthalpy? To help answer this question, consider a process carried out at constant pressure and where the only work allowed is pressure–volume work ($w = -P\Delta V$). Under these conditions, the expression

$$\Delta E = q_P + w$$

becomes

$$\Delta E = q_P - P\Delta V$$

or

$$q_P = \Delta E + P\Delta V$$

where q_P is the heat at constant pressure.

We will now relate q_P to a change in enthalpy. The definition of enthalpy is $H = E + PV$. Therefore, we can say

$$\text{Change in } H = (\text{change in } E) + (\text{change in } PV)$$

or

$$\Delta H = \Delta E + \Delta(PV)$$

Since P is constant, the change in PV is due only to a change in volume. Thus

$$\Delta(PV) = P\Delta V$$

and

$$\Delta H = \Delta E + P\Delta V$$

This expression is identical to the one we obtained for q_P:

$$q_P = \Delta E + P\Delta V$$

Thus, for a process carried out at constant pressure and where the only work allowed is that from a volume change, we have

$$\Delta H = q_P$$

At constant pressure (where only PV work is allowed), the change in enthalpy ΔH of the system is equal to the energy flow as heat. This means that for a reaction studied at constant pressure, the flow of heat is a measure of the change in enthalpy for the system. For this reason, the terms *heat of reaction* and *change in enthalpy* are used interchangeably for reactions studied at constant pressure.

For a chemical reaction, the enthalpy change is given by the equation

$$\Delta H = H_{\text{products}} - H_{\text{reactants}}$$

In a case in which the products of a reaction have a greater enthalpy than the reactants, ΔH will be positive. Thus heat will be absorbed by the system, and the reaction is endothermic. On the other hand, if the enthalpy of the products is less than that of the reactants, ΔH will be negative. In this case the overall decrease in enthalpy is achieved by the generation of heat, and the reaction is exothermic.

Marginal notes:

Enthalpy is a state function. A change in enthalpy does not depend on the pathway between two states.

Recall from the previous section that w and $P\Delta V$ have opposite signs:

$$w = -P\Delta V$$

$\Delta H = q$ only at constant pressure.

The change in enthalpy of a system has no easily interpreted meaning except at constant pressure, where $\Delta H = $ heat.

At constant pressure, exothermic means ΔH is negative; endothermic means ΔH is positive.

Sign in at http://login.cengagebrain .com to try this Interactive Example in **OWL**.

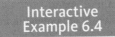

Interactive Example 6.4

Enthalpy

When 1 mole of methane (CH_4) is burned at constant pressure, 890 kJ of energy is released as heat. Calculate ΔH for a process in which a 5.8-g sample of methane is burned at constant pressure.

Solution

Where are we going?

To calculate ΔH

What do we know?

> $q_P = \Delta H = -890$ kJ/mol CH_4
> Mass $= 5.8$ g CH_4
> Molar mass $CH_4 = 16.04$ g

How do we get there?

What are the moles of CH_4 burned?

$$5.8 \text{ g CH}_4 \times \frac{1 \text{ mol CH}_4}{16.04 \text{ g CH}_4} = 0.36 \text{ mol CH}_4$$

What is ΔH?

$$\Delta H = 0.36 \text{ mol CH}_4 \times \frac{-890 \text{ kJ}}{\text{mol CH}_4} = -320 \text{ kJ}$$

Thus, when a 5.8-g sample of CH_4 is burned at constant pressure,

> $\Delta H =$ heat flow $= -320$ kJ

Reality Check | In this case, a 5.8-g sample of CH_4 is burned. Since this amount is smaller than 1 mole, less than 890 kJ will be released as heat.

See Exercises 6.45 through 6.48

Calorimetry

The device used experimentally to determine the heat associated with a chemical reaction is called a **calorimeter**. **Calorimetry**, the science of measuring heat, is based on observing the temperature change when a body absorbs or discharges energy as heat. Substances respond differently to being heated. One substance might require a great deal of heat energy to raise its temperature by one degree, whereas another will exhibit the same temperature change after absorbing relatively little heat. The **heat capacity** C of a substance, which is a measure of this property, is defined as

$$C = \frac{\text{heat absorbed}}{\text{increase in temperature}}$$

When an element or a compound is heated, the energy required will depend on the amount of the substance present (for example, it takes twice as much energy to raise the temperature of 2 g of water by one degree than it takes to raise the temperature of 1 g of water by one degree). Thus, in defining the heat capacity of a substance, the amount of substance must be specified. If the heat capacity is given *per gram* of substance, it is called the **specific heat capacity**, and its units are J/°C · g or J/K · g. If the heat capacity is given *per mole* of the substance, it is called the **molar heat capacity**, and it has the units J/°C · mol or J/K · mol. The specific heat capacities of some common substances are given in Table 6.1. Note from this table that the heat capacities of metals are very different from that of water. It takes much less energy to change the temperature of a gram of a metal by 1°C than for a gram of water.

Although the calorimeters used for highly accurate work are precision instruments, a very simple calorimeter can be used to examine the fundamentals of calorimetry. All we need are two nested Styrofoam cups with a cover through which a stirrer and thermometer can be inserted (Fig. 6.5). This device is called a "coffee-cup calorimeter."

Table 6.1 | The Specific Heat Capacities of Some Common Substances

Substance	Specific Heat Capacity (J/°C · g)
$H_2O(l)$	4.18
$H_2O(s)$	2.03
$Al(s)$	0.89
$Fe(s)$	0.45
$Hg(l)$	0.14
$C(s)$	0.71

Specific heat capacity: the energy required to raise the temperature of one gram of a substance by one degree Celsius.

Molar heat capacity: the energy required to raise the temperature of one mole of a substance by one degree Celsius.

The outer cup is used to provide extra insulation. The inner cup holds the solution in which the reaction occurs.

The measurement of heat using a simple calorimeter such as that shown in Fig. 6.5 is an example of **constant-pressure calorimetry**, since the pressure (atmospheric pressure) remains constant during the process. Constant-pressure calorimetry is used in determining the changes in enthalpy (heats of reactions) for reactions occurring in solution. Recall that under these conditions, the change in enthalpy equals the heat; that is, $\Delta H = q_P$.

For example, suppose we mix 50.0 mL of 1.0 M HCl at 25.0°C with 50.0 mL of 1.0 M NaOH also at 25°C in a calorimeter. After the reactants are mixed by stirring, the temperature is observed to increase to 31.9°C. As we saw in Section 4.8, the net ionic equation for this reaction is

$$H^+(aq) + OH^-(aq) \longrightarrow H_2O(l)$$

When these reactants (each originally at the same temperature) are mixed, the temperature of the mixed solution is observed to increase. Therefore, the chemical reaction must be releasing energy as heat. This released energy increases the random motions of the solution components, which in turn increases the temperature. The quantity of energy released can be determined from the temperature increase, the mass of solution, and the specific heat capacity of the solution. For an approximate result, we will assume that the calorimeter does not absorb or leak any heat and that the solution can be treated as if it were pure water with a density of 1.0 g/mL.

We also need to know the heat required to raise the temperature of a given amount of water by 1°C. Table 6.1 lists the specific heat capacity of water as 4.18 J/°C · g. This means that 4.18 J of energy is required to raise the temperature of 1 g of water by 1°C.

From these assumptions and definitions, we can calculate the heat (change in enthalpy) for the neutralization reaction:

Energy (as heat) released by the reaction
= energy (as heat) absorbed by the solution
= specific heat capacity × mass of solution × increase in temperature
= $s \times m \times \Delta T$

In this case the increase in temperature $(\Delta T) = 31.9°C - 25.0°C = 6.9°C$, and the mass of solution $(m) = 100.0$ mL × 1.0 g/mL = 1.0×10^2 g. Thus

$$\text{Energy released (as heat)} = s \times m \times \Delta T$$
$$= \left(4.18 \frac{J}{°C \cdot g}\right)(1.0 \times 10^2 \, g)(6.9°C)$$
$$= 2.9 \times 10^3 \, J$$

How much energy would have been released if twice these amounts of solutions had been mixed? The answer is that twice as much heat would have been produced. The heat of a reaction is an *extensive property;* it depends directly on the amount of substance, in this case on the amounts of reactants. In contrast, an *intensive property* is not related to the amount of a substance. For example, temperature is an intensive property.

Enthalpies of reaction are often expressed in terms of moles of reacting substances. The number of moles of H^+ ions consumed in the preceding experiment is

$$50.0 \, mL \times \frac{1 \, L}{1000 \, mL} \times \frac{1.0 \, mol \, H^+}{L} = 5.0 \times 10^{-2} \, mol \, H^+$$

Thus 2.9×10^3 J heat was released when 5.0×10^{-2} mole of H^+ ions reacted, or

$$\frac{2.9 \times 10^3 \, J}{5.0 \times 10^{-2} \, mol \, H^+} = 5.8 \times 10^4 \, J/mol$$

If two reactants at the same temperature are mixed and the resulting solution gets warmer, this means the reaction taking place is exothermic. An endothermic reaction cools the solution.

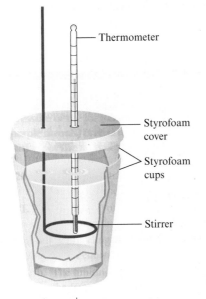

Thermometer

Styrofoam cover

Styrofoam cups

Stirrer

Figure 6.5 | A coffee-cup calorimeter made of two Styrofoam cups.

Chemical connections
Nature Has Hot Plants

The voodoo lily is a beautiful, seductive—and foul-smelling—plant. The exotic-looking lily features an elaborate reproductive mechanism—a purple spike that can reach nearly 3 feet in length and is cloaked by a hoodlike leaf. But approach to the plant reveals bad news—it smells terrible!

Despite its antisocial odor, this putrid plant has fascinated biologists for many years because of its ability to generate heat. At the peak of its metabolic activity, the plant's blossom can be as much as 15°C above its ambient temperature. To generate this much heat, the metabolic rate of the plant must be close to that of a flying hummingbird!

What's the purpose of this intense heat production? For a plant faced with limited food supplies in the very competitive tropical climate where it grows, heat production seems like a great waste of energy. The answer to this mystery is that the voodoo lily is pollinated mainly by carrion-loving insects. Thus the lily prepares a malodorous mixture of chemicals characteristic of rotting meat, which it then "cooks" off into the surrounding air to attract flesh-feeding beetles and flies. Then, once the insects enter the pollination chamber, the high temperatures there (as high as 110°F) cause the insects to remain very active to better carry out their pollination duties.

Notice that in this example we mentally keep track of the direction of the energy flow and assign the correct sign at the end of the calculation.

of heat released per 1.0 mole of H^+ ions neutralized. Thus the *magnitude* of the enthalpy change per mole for the reaction

$$H^+(aq) + OH^-(aq) \longrightarrow H_2O(l)$$

is 58 kJ/mol. Since heat is *evolved*, $\Delta H = -58$ kJ/mol.

To summarize, we have found that when an object changes temperature, the heat can be calculated from the equation

$$q = s \times m \times \Delta T$$

where s is the specific heat capacity, m is the mass, and ΔT is the change in temperature. Note that when ΔT is positive, q also has a positive sign. The object has absorbed heat, so its temperature increases.

Interactive Example 6.5

Sign in at http://login.cengagebrain.com to try this Interactive Example in OWL.

Constant-Pressure Calorimetry

When 1.00 L of 1.00 M $Ba(NO_3)_2$ solution at 25.0°C is mixed with 1.00 L of 1.00 M Na_2SO_4 solution at 25.0°C in a calorimeter, the white solid $BaSO_4$ forms, and the temperature of the mixture increases to 28.1°C. Assuming that the calorimeter absorbs only a negligible quantity of heat, the specific heat capacity of the solution is 4.18 J/°C · g, and the density of the final solution is 1.0 g/mL, calculate the enthalpy change per mole of $BaSO_4$ formed.

Solution

Where are we going?

To calculate ΔH per mole of $BaSO_4$ formed

The voodoo lily is only one of many such thermogenic (heat-producing) plants. Another interesting example is the eastern skunk cabbage, which produces enough heat to bloom inside of a snow bank by creating its own ice caves. These plants are of special interest to biologists because they provide opportunities to study metabolic reactions that are quite subtle in "normal" plants. For example, recent studies have shown that salicylic acid, the active form of aspirin, is probably very important in producing the metabolic bursts in thermogenic plants.

Besides studying the dramatic heat effects in thermogenic plants, biologists are also interested in calorimetric studies of regular plants.

For example, very precise calorimeters have been designed that can be used to study the heat produced, and thus the metabolic activities, of clumps of cells no larger than a bread crumb. Several scientists have suggested that a single calorimetric measurement taking just a few minutes on a tiny plant might be useful in predicting the growth rate of the mature plant throughout its lifetime. If true, this would provide a very efficient method for selecting the plants most likely to thrive as adults.

Because the study of the heat production by plants is an excellent way to learn about plant metabolism, this continues to be a "hot" area of research.

The voodoo lily attracts pollinating insects with its foul odor.

What do we know?

> 1.00 L of 1.00 M $Ba(NO_3)_2$
> 1.00 L of 1.00 M Na_2SO_4
> $T_{initial} = 25.0°C$
> $T_{final} = 28.1°C$
> Heat capacity of solution = 4.18 J/°C · g
> Density of final solution = 1.0 g/mL

What do we need?

> Net ionic equation for the reaction

The ions present before any reaction occurs are Ba^{2+}, NO_3^-, Na^+, and SO_4^{2-}. The Na^+ and NO_3^- ions are spectator ions, since $NaNO_3$ is very soluble in water and will not precipitate under these conditions. The net ionic equation for the reaction is therefore

$$Ba^{2+}(aq) + SO_4^{2-}(aq) \longrightarrow BaSO_4(s)$$

How do we get there?

What is ΔH?

Since the temperature increases, formation of solid $BaSO_4$ must be exothermic; ΔH is negative.

Heat evolved by the reaction

= heat absorbed by the solution

= specific heat capacity × mass of solution × increase in temperature

What is the mass of the final solution?

$$\text{Mass of solution} = 2.00 \, \cancel{L} \times \frac{1000 \, \cancel{mL}}{1 \, \cancel{L}} \times \frac{1.0 \, g}{\cancel{mL}} = 2.0 \times 10^3 \, g$$

What is the temperature increase?

$$\Delta T = T_{\text{final}} - T_{\text{initial}} = 28.1°C - 25.0°C = 3.1°C$$

How much heat is evolved by the reaction?

$$\text{Heat evolved} = (4.18 \, J/°\cancel{C} \cdot \cancel{g})(2.0 \times 10^3 \, \cancel{g})(3.1°\cancel{C}) = 2.6 \times 10^4 \, J$$

Thus

$$q = q_P = \Delta H = -2.6 \times 10^4 \, J$$

What is ΔH per mole of $BaSO_4$ formed?

Since 1.0 L of 1.0 M $Ba(NO_3)_2$ contains 1 mole of Ba^{2+} ions and 1.0 L of 1.0 M Na_2SO_4 contains 1.0 mole of SO_4^{2-} ions, 1.0 mole of solid $BaSO_4$ is formed in this experiment. Thus the enthalpy change per mole of $BaSO_4$ formed is

❭ $$\Delta H = -2.6 \times 10^4 \, J/mol = -26 \, kJ/mol$$

See Exercises 6.61 through 6.64

Constant-volume calorimetry experiments can also be performed. For example, when a photographic flashbulb flashes, the bulb becomes very hot, because the reaction of the zirconium or magnesium wire with the oxygen inside the bulb is exothermic. The reaction occurs inside the flashbulb, which is rigid and does not change volume. Under these conditions, no work is done (because the volume must change for pressure–volume work to be performed). To study the energy changes in reactions under conditions of constant volume, a "bomb calorimeter" (Fig. 6.6) is used. Weighed reactants are placed inside a rigid steel container (the "bomb") and ignited. The energy change is determined by measuring the increase in the temperature of the water and

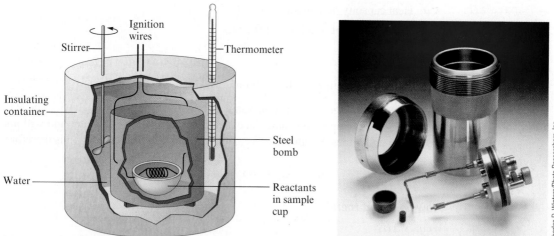

Figure 6.6 | A bomb calorimeter. The reaction is carried out inside a rigid steel "bomb" (photo of actual disassembled "bomb" shown on right), and the heat evolved is absorbed by the surrounding water and other calorimeter parts. The quantity of energy produced by the reaction can be calculated from the temperature increase.

other calorimeter parts. For a constant-volume process, the change in volume ΔV is equal to zero, so work (which is $-P\Delta V$) is also equal to zero. Therefore,

$$\Delta E = q + w = q = q_V \qquad \text{(constant volume)}$$

Suppose we wish to measure the energy of combustion of octane (C_8H_{18}), a component of gasoline. A 0.5269-g sample of octane is placed in a bomb calorimeter known to have a heat capacity of 11.3 kJ/°C. This means that 11.3 kJ of energy is required to raise the temperature of the water and other parts of the calorimeter by 1°C. The octane is ignited in the presence of excess oxygen, and the temperature increase of the calorimeter is 2.25°C. The amount of energy released is calculated as follows:

Energy released by the reaction

$= \text{temperature increase} \times \text{energy required to change the temperature by } 1°C$

$= \Delta T \times \text{heat capacity of calorimeter}$

$= 2.25°\cancel{C} \times 11.3 \text{ kJ}/°\cancel{C} = 25.4 \text{ kJ}$

This means that 25.4 kJ of energy was released by the combustion of 0.5269 g octane. The number of moles of octane is

$$0.5269 \text{ g octane} \times \frac{1 \text{ mol octane}}{114.2 \text{ g octane}} = 4.614 \times 10^{-3} \text{ mol octane}$$

Since 25.4 kJ of energy was released for 4.614×10^{-3} mole of octane, the energy released per mole is

$$\frac{25.4 \text{ kJ}}{4.614 \times 10^{-3} \text{ mol}} = 5.50 \times 10^3 \text{ kJ/mol}$$

Since the reaction is exothermic, ΔE is negative:

$$\Delta E_{\text{combustion}} = -5.50 \times 10^3 \text{ kJ/mol}$$

Note that since no work is done in this case, ΔE is equal to the heat.

$$\Delta E = q + w = q \qquad \text{since } w = 0$$

Thus $q = -5.50 \times 10^3$ kJ/mol.

| Example 6.6 | Constant-Volume Calorimetry |

Hydrogen's potential as a fuel is discussed in Section 6.6.

It has been suggested that hydrogen gas obtained by the decomposition of water might be a substitute for natural gas (principally methane). To compare the energies of combustion of these fuels, the following experiment was carried out using a bomb calorimeter with a heat capacity of 11.3 kJ/°C. When a 1.50-g sample of methane gas was burned with excess oxygen in the calorimeter, the temperature increased by 7.3°C. When a 1.15-g sample of hydrogen gas was burned with excess oxygen, the temperature increase was 14.3°C. Compare the energies of combustion (per gram) for hydrogen and methane.

Solution

Where are we going?

To calculate ΔH of combustion per gram for H_2 and CH_4

What do we know?

> 1.50 g $CH_4 \Rightarrow \Delta T = 7.3°C$

> 1.15 g $H_2 \Rightarrow \Delta T = 14.3°C$

> Heat capacity of calorimeter = 11.3 kJ/°C

What do we need?

> $\Delta E = \Delta T \times$ heat capacity of calorimeter

How do we get there?

What is the energy released for each combustion?

For CH_4, we calculate the energy of combustion for methane using the heat capacity of the calorimeter (11.3 kJ/°C) and the observed temperature increase of 7.3°C:

$$\text{Energy } \textit{released} \text{ in the combustion of 1.5 g } CH_4 = (11.3 \text{ kJ/°C})(7.3°C)$$
$$= 83 \text{ kJ}$$

$$\text{Energy } \textit{released} \text{ in the combustion of 1 g } CH_4 = \frac{83 \text{ kJ}}{1.5 \text{ g}} = 55 \text{ kJ/g}$$

For H_2,

$$\text{Energy } \textit{released} \text{ in the combustion of 1.15 g } H_2 = (11.3 \text{ kJ/°C})(14.3°C)$$
$$= 162 \text{ kJ}$$

$$\text{Energy } \textit{released} \text{ in the combustion of 1 g } H_2 = \frac{162 \text{ kJ}}{1.15 \text{ g}} = 141 \text{ kJ/g}$$

The direction of energy flow is indicated by words in this example. Using signs to designate the direction of energy flow:

$$\Delta E_{combustion} = -55 \text{ kJ/g}$$

for methane and

$$\Delta E_{combustion} = -141 \text{ kJ/g}$$

for hydrogen.

How do the energies of combustion compare?

> The energy released in the combustion of 1 g hydrogen is approximately 2.5 times that for 1 g methane, indicating that hydrogen gas is a potentially useful fuel.

See Exercises 6.67 and 6.68

6.3 | Hess's Law

Since enthalpy is a state function, the change in enthalpy in going from some initial state to some final state is independent of the pathway. This means that *in going from a particular set of reactants to a particular set of products, the change in enthalpy is the same whether the reaction takes place in one step or in a series of steps.* This principle is known as **Hess's law** and can be illustrated by examining the oxidation of nitrogen to produce nitrogen dioxide. The overall reaction can be written in one step, where the enthalpy change is represented by ΔH_1.

ΔH is not dependent on the reaction pathway.

$$N_2(g) + 2O_2(g) \longrightarrow 2NO_2(g) \qquad \Delta H_1 = 68 \text{ kJ}$$

This reaction also can be carried out in two distinct steps, with enthalpy changes designated by ΔH_2 and ΔH_3:

$$N_2(g) + O_2(g) \longrightarrow 2NO(g) \qquad\qquad \Delta H_2 = 180 \text{ kJ}$$
$$2NO(g) + O_2(g) \longrightarrow 2NO_2(g) \qquad\qquad \Delta H_3 = -112 \text{ kJ}$$

Net reaction: $N_2(g) + 2O_2(g) \longrightarrow 2NO_2(g) \qquad \Delta H_2 + \Delta H_3 = 68 \text{ kJ}$

Note that the sum of the two steps gives the net, or overall, reaction and that

$$\Delta H_1 = \Delta H_2 + \Delta H_3 = 68 \text{ kJ}$$

The principle of Hess's law is shown schematically in Fig. 6.7.

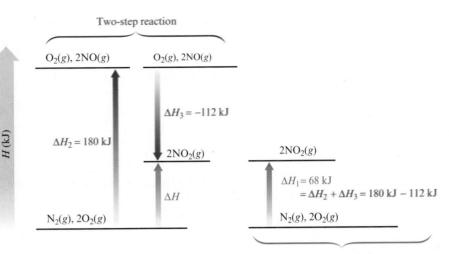

Figure 6.7 | The principle of Hess's law. The same change in enthalpy occurs when nitrogen and oxygen react to form nitrogen dioxide, regardless of whether the reaction occurs in one (red) or two (blue) steps.

Characteristics of Enthalpy Changes

To use Hess's law to compute enthalpy changes for reactions, it is important to understand two characteristics of ΔH for a reaction:

Reversing the direction of a reaction changes the sign of ΔH.

Characteristics of ΔH for a Reaction

❯ If a reaction is reversed, the sign of ΔH is also reversed.

❯ The magnitude of ΔH is directly proportional to the quantities of reactants and products in a reaction. If the coefficients in a balanced reaction are multiplied by an integer, the value of ΔH is multiplied by the same integer.

Crystals of xenon tetrafluoride, the first reported binary compound containing a noble gas element.

Both these rules follow in a straightforward way from the properties of enthalpy changes. The first rule can be explained by recalling that the *sign* of ΔH indicates the *direction* of the heat flow at constant pressure. If the direction of the reaction is reversed, the direction of the heat flow also will be reversed. To see this, consider the preparation of xenon tetrafluoride, which was the first binary compound made from a noble gas:

$$Xe(g) + 2F_2(g) \longrightarrow XeF_4(s) \qquad \Delta H = -251 \text{ kJ}$$

This reaction is exothermic, and 251 kJ of energy flows into the surroundings as heat. On the other hand, if the colorless XeF_4 crystals are decomposed into the elements, according to the equation

$$XeF_4(s) \longrightarrow Xe(g) + 2F_2(g)$$

the opposite energy flow occurs because 251 kJ of energy must be added to the system to produce this endothermic reaction. Thus, for this reaction, $\Delta H = +251$ kJ.

The second rule comes from the fact that ΔH is an extensive property, depending on the amount of substances reacting. For example, since 251 kJ of energy is evolved for the reaction

$$Xe(g) + 2F_2(g) \longrightarrow XeF_4(s)$$

then for a preparation involving twice the quantities of reactants and products, or

$$2Xe(g) + 4F_2(g) \longrightarrow 2XeF_4(s)$$

twice as much heat would be evolved:

$$\Delta H = 2(-251 \text{ kJ}) = -502 \text{ kJ}$$

Critical Thinking

What if Hess's law were not true? What are some possible repercussions this would have?

Hess's Law I

Two forms of carbon are *graphite,* the soft, black, slippery material used in "lead" pencils and as a lubricant for locks, and *diamond,* the brilliant, hard gemstone. Using the enthalpies of combustion for graphite (-394 kJ/mol) and diamond (-396 kJ/mol), calculate ΔH for the conversion of graphite to diamond:

$$C_{\text{graphite}}(s) \longrightarrow C_{\text{diamond}}(s)$$

Solution

Where are we going?

To calculate ΔH for the conversion of graphite to diamond

What do we know?
The combustion reactions are

> $C_{\text{graphite}}(s) + O_2(g) \longrightarrow CO_2(g) \qquad \Delta H = -394 \text{ kJ}$
> $C_{\text{diamond}}(s) + O_2(g) \longrightarrow CO_2(g) \qquad \Delta H = -396 \text{ kJ}$

How do we get there?

How do we combine the combustion equations to obtain the equation for the conversion of graphite to diamond?
Note that if we reverse the second reaction (which means we must change the sign of ΔH) and sum the two reactions, we obtain the desired reaction:

$$
\begin{array}{ll}
C_{\text{graphite}}(s) + O_2(g) \longrightarrow CO_2(g) & \Delta H = -394 \text{ kJ} \\
\underline{CO_2(g) \longrightarrow C_{\text{diamond}}(s) + O_2(g)} & \underline{\Delta H = -(-396 \text{ kJ})} \\
C_{\text{graphite}}(s) \longrightarrow C_{\text{diamond}}(s) & \Delta H = 2 \text{ kJ}
\end{array}
$$

> The ΔH for the conversion of graphite to diamond is

$$\Delta H = 2 \text{ kJ/mol graphite}$$

We obtain this value by summing the ΔH values for the equations as shown above. This process is endothermic since the sign of ΔH is positive.

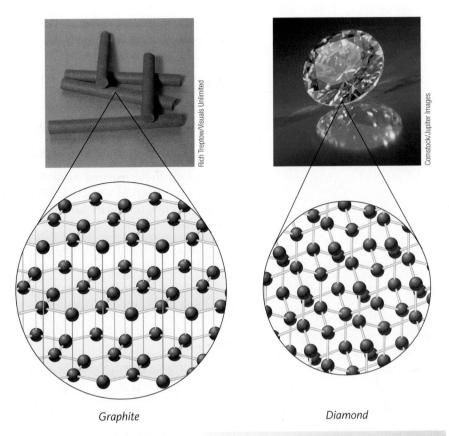

Graphite

Diamond

See Exercises 6.69 and 6.70

Interactive Example 6.8

Sign in at http://login.cengagebrain .com to try this Interactive Example in **OWL**.

Hess's Law II

Diborane (B_2H_6) is a highly reactive boron hydride that was once considered as a possible rocket fuel for the U.S. space program. Calculate ΔH for the synthesis of diborane from its elements, according to the equation

$$2B(s) + 3H_2(g) \longrightarrow B_2H_6(g)$$

using the following data:

Reaction	ΔH
(a) $2B(s) + \frac{3}{2}O_2(g) \longrightarrow B_2O_3(s)$	-1273 kJ
(b) $B_2H_6(g) + 3O_2(g) \longrightarrow B_2O_3(s) + 3H_2O(g)$	-2035 kJ
(c) $H_2(g) + \frac{1}{2}O_2(g) \longrightarrow H_2O(l)$	-286 kJ
(d) $H_2O(l) \longrightarrow H_2O(g)$	44 kJ

Solution

To obtain ΔH for the required reaction, we must somehow combine equations (a), (b), (c), and (d) to produce that reaction and add the corresponding ΔH values. This can best be done by focusing on the reactants and products of the required reaction. The reactants are $B(s)$ and $H_2(g)$, and the product is $B_2H_6(g)$. How can we obtain the correct equation? Reaction (a) has $B(s)$ as a reactant, as needed in the required equation. Thus reaction (a) will be used as it is. Reaction (b) has $B_2H_6(g)$ as a reactant, but this substance is needed as a product. Thus reaction (b) must be reversed, and the sign of ΔH must be changed accordingly. Up to this point we have

(a)

$$2B(s) + \tfrac{3}{2}O_2(g) \longrightarrow B_2O_3(s) \qquad\qquad \Delta H = -1273 \text{ kJ}$$

$-$(b)

$$B_2O_3(s) + 3H_2O(g) \longrightarrow B_2H_6(g) + 3O_2(g) \qquad\qquad \Delta H = -(-2035 \text{ kJ})$$

Sum: $B_2O_3(s) + 2B(s) + \tfrac{3}{2}O_2(g) + 3H_2O(g) \longrightarrow B_2O_3(s) + B_2H_6(g) + 3O_2(g) \qquad \Delta H = 762 \text{ kJ}$

Deleting the species that occur on both sides gives

$$2B(s) + 3H_2O(g) \longrightarrow B_2H_6(g) + \tfrac{3}{2}O_2(g) \qquad \Delta H = 762 \text{ kJ}$$

We are closer to the required reaction, but we still need to remove $H_2O(g)$ and $O_2(g)$ and introduce $H_2(g)$ as a reactant. We can do this using reactions (c) and (d). If we multiply reaction (c) and its ΔH value by 3 and add the result to the preceding equation, we have

$$2B(s) + 3H_2O(g) \longrightarrow B_2H_6(g) + \tfrac{3}{2}O_2(g) \qquad\qquad \Delta H = 762 \text{ kJ}$$

$3 \times$ (c)

$$3\left[H_2(g) + \tfrac{1}{2}O_2(g) \longrightarrow H_2O(l)\right] \qquad\qquad \Delta H = 3(-286 \text{ kJ})$$

Sum: $2B(s) + 3H_2(g) + \tfrac{3}{2}O_2(g) + 3H_2O(g) \longrightarrow B_2H_6(g) + \tfrac{3}{2}O_2(g) + 3H_2O(l) \qquad \Delta H = -96 \text{ kJ}$

We can cancel the $\tfrac{3}{2}O_2(g)$ on both sides, but we cannot cancel the H_2O because it is gaseous on one side and liquid on the other. This can be solved by adding reaction (d), multiplied by 3:

$$2B(s) + 3H_2(g) + 3H_2O(g) \longrightarrow B_2H_6(g) + 3H_2O(l) \qquad\qquad \Delta H = -96 \text{ kJ}$$

$3 \times$ (d)

$$3\left[H_2O(l) \longrightarrow H_2O(g)\right] \qquad\qquad \Delta H = 3(44 \text{ kJ})$$

$2B(s) + 3H_2(g) + 3H_2O(g) + 3H_2O(l) \longrightarrow B_2H_6(g) + 3H_2O(l) + 3H_2O(g) \qquad \Delta H = +36 \text{ kJ}$

This gives the reaction required by the problem:

$$2B(s) + 3H_2(g) \longrightarrow B_2H_6(g) \qquad \Delta H = +36 \text{ kJ}$$

Thus ΔH for the synthesis of 1 mole of diborane from the elements is $+36$ kJ.

See Exercises 6.71 through 6.76

Problem-Solving Strategy

Hess's Law

Calculations involving Hess's law typically require that several reactions be manipulated and combined to finally give the reaction of interest. In doing this procedure you should

> Work *backward* from the required reaction, using the reactants and products to decide how to manipulate the other given reactions at your disposal

> Reverse any reactions as needed to give the required reactants and products

> Multiply reactions to give the correct numbers of reactants and products

This process involves some trial and error, but it can be very systematic if you always allow the final reaction to guide you.

6.4 | Standard Enthalpies of Formation

For a reaction studied under conditions of constant pressure, we can obtain the enthalpy change using a calorimeter. However, this process can be very difficult. In fact, in some cases it is impossible, since certain reactions do not lend themselves to such study. An example is the conversion of solid carbon from its graphite form to its diamond form:

$$C_{graphite}(s) \longrightarrow C_{diamond}(s)$$

The value of ΔH for this process cannot be obtained by direct measurement in a calorimeter because the process is much too slow under normal conditions. However, as we saw in Example 6.7, ΔH for this process can be calculated from heats of combustion. This is only one example of how useful it is to be able to *calculate* ΔH values for chemical reactions. We will next show how to do this using standard enthalpies of formation.

The **standard enthalpy of formation** (ΔH_f°) of a compound is defined as the *change in enthalpy that accompanies the formation of 1 mole of a compound from its elements with all substances in their standard states.*

A *degree symbol* on a thermodynamic function, for example, ΔH°, indicates that the corresponding process has been carried out under standard conditions. The **standard state** for a substance is a precisely defined reference state. Because thermodynamic functions often depend on the concentrations (or pressures) of the substances involved, we must use a common reference state to properly compare the thermodynamic properties of two substances. This is especially important because, for most thermodynamic properties, we can measure only *changes* in the property. For example, we have no method for determining absolute values of enthalpy. We can measure enthalpy changes (ΔH values) only by performing heat-flow experiments.

The International Union of Pure and Applied Chemists (IUPAC) has adopted 1 bar (100,000 Pa) as the standard pressure instead of 1 atm (101,305 Pa). Both standards are now in wide use.

Standard state is *not* the same as the standard temperature and pressure (STP) for a gas (discussed in Section 5.4).

Conventional Definitions of Standard States

For a Compound

> The standard state of a gaseous substance is a pressure of exactly 1 atmosphere.

> For a pure substance in a condensed state (liquid or solid), the standard state is the pure liquid or solid.

> For a substance present in a solution, the standard state is a concentration of exactly 1 M.

For an Element

> The standard state of an element is the form in which the element exists under conditions of 1 atmosphere and 25°C. (The standard state for oxygen is $O_2(g)$ at a pressure of 1 atmosphere; the standard state for sodium is $Na(s)$; the standard state for mercury is $Hg(l)$; and so on.)

Several important characteristics of the definition of the enthalpy of formation will become clearer if we again consider the formation of nitrogen dioxide from the elements in their standard states:

$$\tfrac{1}{2}N_2(g) + O_2(g) \longrightarrow NO_2(g) \qquad \Delta H_f^\circ = 34 \text{ kJ/mol}$$

Note that the reaction is written so that both elements are in their standard states, and 1 mole of product is formed. Enthalpies of formation are *always* given per mole of product with the product in its standard state.

The formation reaction for methanol is written as

$$C(s) + 2H_2(g) + \tfrac{1}{2}O_2(g) \longrightarrow CH_3OH(l) \qquad \Delta H_f^\circ = -239 \text{ kJ/mol}$$

The standard state of carbon is graphite, the standard states for oxygen and hydrogen are the diatomic gases, and the standard state for methanol is the liquid.

The ΔH_f° values for some common substances are shown in Table 6.2. More values are found in Appendix 4. The importance of the tabulated ΔH_f° values is that enthalpies

Brown nitrogen dioxide gas.

Table 6.2 | Standard Enthalpies of Formation for Several Compounds at 25°C

Compound	ΔH_f° (kJ/mol)
$NH_3(g)$	−46
$NO_2(g)$	34
$H_2O(l)$	−286
$Al_2O_3(s)$	−1676
$Fe_2O_3(s)$	−826
$CO_2(g)$	−394
$CH_3OH(l)$	−239
$C_8H_{18}(l)$	−269

for many reactions can be calculated using these numbers. To see how this is done, we will calculate the standard enthalpy change for the combustion of methane:

$$CH_4(g) + 2O_2(g) \longrightarrow CO_2(g) + 2H_2O(l)$$

Enthalpy is a state function, so we can invoke Hess's law and choose *any* convenient pathway from reactants to products and then sum the enthalpy changes along the chosen pathway. A convenient pathway, shown in Fig. 6.8, involves taking the reactants apart to the respective elements in their standard states in reactions (a) and (b) and then forming the products from these elements in reactions (c) and (d). This general pathway will work for any reaction, since atoms are conserved in a chemical reaction.

Note from Fig. 6.8 that reaction (a), where methane is taken apart into its elements,

$$CH_4(g) \longrightarrow C(s) + 2H_2(g)$$

is just the reverse of the formation reaction for methane:

$$C(s) + 2H_2(g) \longrightarrow CH_4(g) \qquad \Delta H_f^\circ = -75 \text{ kJ/mol}$$

Since reversing a reaction means changing the sign of ΔH but keeping the magnitude the same, ΔH for reaction (a) is $-\Delta H_f^\circ$, or 75 kJ. Thus $\Delta H^\circ_{(a)} = 75$ kJ.

Next we consider reaction (b). Here oxygen is already an element in its standard state, so no change is needed. Thus $\Delta H^\circ_{(b)} = 0$.

The next steps, reactions (c) and (d), use the elements formed in reactions (a) and (b) to form the products. Note that reaction (c) is simply the formation reaction for carbon dioxide:

$$C(s) + O_2(g) \longrightarrow CO_2(g) \qquad \Delta H_f^\circ = -394 \text{ kJ/mol}$$

and

$$\Delta H^\circ_{(c)} = \Delta H_f^\circ \text{ for } CO_2(g) = -394 \text{ kJ}$$

Reaction (d) is the formation reaction for water:

$$H_2(g) + \tfrac{1}{2}O_2(g) \longrightarrow H_2O(l) \qquad \Delta H_f^\circ = -286 \text{ kJ/mol}$$

However, since 2 moles of water are required in the balanced equation, we must form 2 moles of water from the elements:

$$2H_2(g) + O_2(g) \longrightarrow 2H_2O(l)$$

Thus

$$\Delta H^\circ_{(d)} = 2 \times \Delta H_f^\circ \text{ for } H_2O(l) = 2(-286 \text{ kJ}) = -572 \text{ kJ}$$

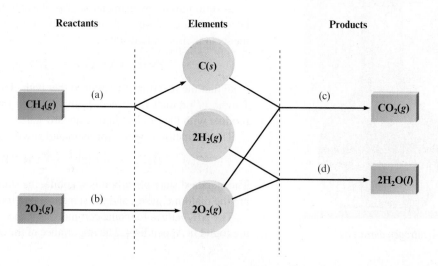

Figure 6.8 | In this pathway for the combustion of methane, the reactants are first taken apart in reactions (a) and (b) to form the constituent elements in their standard states, which are then used to assemble the products in reactions (c) and (d).

Figure 6.9 | A schematic diagram of the energy changes for the reaction $CH_4(g) + 2O_2(g) \rightarrow CO_2(g) + 2H_2O(l)$.

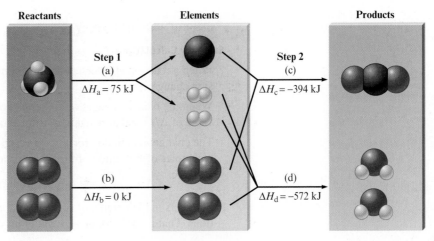

We have now completed the pathway from the reactants to the products. The change in enthalpy for the reaction is the sum of the ΔH values (including their signs) for the steps:

$$\Delta H^\circ_{\text{reaction}}$$
$$= \Delta H^\circ_{(a)} + \Delta H^\circ_{(b)} + \Delta H^\circ_{(c)} + \Delta H^\circ_{(d)}$$
$$= \left[-\Delta H^\circ_f \text{ for } CH_4(g)\right] + 0 + \left[\Delta H^\circ_f \text{ for } CO_2(g)\right] + \left[2 \times \Delta H^\circ_f \text{ for } H_2O(l)\right]$$
$$= -(-75 \text{ kJ}) + 0 + (-394 \text{ kJ}) + (-572 \text{ kJ})$$
$$= -891 \text{ kJ}$$

This process is diagramed in Fig. 6.9. Notice that the reactants are taken apart and converted to elements [not necessary for $O_2(g)$] that are then used to form products. You can see that this is a very exothermic reaction because very little energy is required to convert the reactants to the respective elements but a great deal of energy is released when these elements form the products. This is why this reaction is so useful for producing heat to warm homes and offices.

Let's examine carefully the pathway we used in this example. First, the reactants were broken down into the elements in their standard states. This process involved reversing the formation reactions and thus switching the signs of the enthalpies of formation. The products were then constructed from these elements. This involved formation reactions and thus enthalpies of formation. We can summarize this entire process as follows: The enthalpy change for a given reaction can be calculated by subtracting the enthalpies of formation of the reactants from the enthalpies of formation of the products. Remember to multiply the enthalpies of formation by integers as required by the balanced equation. This statement can be represented symbolically as follows:

$$\Delta H^\circ_{\text{reaction}} = \Sigma n_p \Delta H^\circ_f(\text{products}) - \Sigma n_r \Delta H^\circ_f(\text{reactants}) \quad (6.1)$$

where the symbol Σ (sigma) means "to take the sum of the terms," and n_p and n_r represent the moles of each product or reactant, respectively.

Elements are not included in the calculation because elements require no change in form. We have in effect *defined* the enthalpy of formation of an element in its standard state as zero, since we have chosen this as our reference point for calculating enthalpy changes in reactions.

Subtraction means to reverse the sign and add.

Elements in their standard states are not included in enthalpy calculations using ΔH°_f values.

Problem-Solving Strategy

Enthalpy Calculations

› When a reaction is reversed, the magnitude of ΔH remains the same, but its sign changes.

› When the balanced equation for a reaction is multiplied by an integer, the value of ΔH for that reaction must be multiplied by the same integer.

› The change in enthalpy for a given reaction can be calculated from the enthalpies of formation of the reactants and products:

$$\Delta H^{\circ}{}_{\text{reaction}} = \Sigma n_p \Delta H_f^{\circ}(\text{products}) - \Sigma n_r \Delta H_f^{\circ}(\text{reactants})$$

› Elements in their standard states are not included in the $\Delta H_{\text{reaction}}$ calculations. That is, ΔH_f° for an element in its standard state is zero.

Interactive Example 6.9

Sign in at http://login.cengagebrain .com to try this Interactive Example in **OWL**.

Enthalpies from Standard Enthalpies of Formation I

Using the standard enthalpies of formation listed in Table 6.2, calculate the standard enthalpy change for the overall reaction that occurs when ammonia is burned in air to form nitrogen dioxide and water. This is the first step in the manufacture of nitric acid.

$$4NH_3(g) + 7O_2(g) \longrightarrow 4NO_2(g) + 6H_2O(l)$$

Solution

We will use the pathway in which the reactants are broken down into elements in their standard states, which are then used to form the products (Fig. 6.10).

› *Decomposition of $NH_3(g)$ into elements* [reaction (a) in Fig. 6.10]. The first step is to decompose 4 moles of NH_3 into N_2 and H_2:

$$4NH_3(g) \longrightarrow 2N_2(g) + 6H_2(g)$$

The preceding reaction is 4 times the *reverse* of the formation reaction for NH_3:

$$\tfrac{1}{2}N_2(g) + \tfrac{3}{2}H_2(g) \longrightarrow NH_3(g) \qquad \Delta H_f^{\circ} = -46 \text{ kJ/mol}$$

Thus

$$\Delta H^{\circ}{}_{(a)} = 4 \text{ mol}\big[-(-46 \text{ kJ/mol})\big] = 184 \text{ kJ}$$

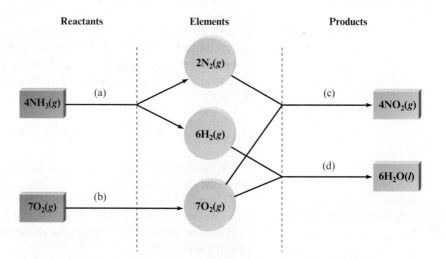

Figure 6.10 | A pathway for the combustion of ammonia.

> *Elemental oxygen* [reaction (b) in Fig. 6.10]. Since $O_2(g)$ is an element in its standard state, $\Delta H°_{(b)} = 0$.
> We now have the elements $N_2(g)$, $H_2(g)$, and $O_2(g)$, which can be combined to form the products of the overall reaction.

> *Synthesis of $NO_2(g)$ from elements* [reaction (c) in Fig. 6.10]. The overall reaction equation has 4 moles of NO_2. Thus the required reaction is 4 times the formation reaction for NO_2:

$$4 \times \left[\tfrac{1}{2}N_2(g) + O_2(g) \longrightarrow NO_2(g)\right]$$

and

$$\Delta H°_{(c)} = 4 \times \Delta H°_f \text{ for } NO_2(g)$$

From Table 6.2, $\Delta H°_f$ for $NO_2(g) = 34$ kJ/mol and

$$\Delta H°_{(c)} = 4 \text{ mol} \times 34 \text{ kJ/mol} = 136 \text{ kJ}$$

> *Synthesis of $H_2O(l)$ from elements* [reaction (d) in Fig. 6.10]. Since the overall equation for the reaction has 6 moles of $H_2O(l)$, the required reaction is 6 times the formation reaction for $H_2O(l)$:

$$6 \times \left[H_2(g) + \tfrac{1}{2}O_2(g) \longrightarrow H_2O(l)\right]$$

and

$$\Delta H°_{(d)} = 6 \times \Delta H°_f \text{ for } H_2O(l)$$

From Table 6.2, $\Delta H°_f$ for $H_2O(l) = -286$ kJ/mol and

$$\Delta H°_{(d)} = 6 \text{ mol} \left(-286 \text{ kJ/mol}\right) = -1716 \text{ kJ}$$

To summarize, we have done the following:

$$
\begin{array}{l}
4NH_3(g) \xrightarrow{\Delta H°_{(a)}} \\
7O_2(g) \xrightarrow{\Delta H°_{(b)}}
\end{array}
\left\{
\begin{array}{l}
2N_2(g) + 6H_2(g) \xrightarrow{\Delta H°_{(c)}} 4NO_2(g) \\
7O_2(g)
\end{array}
\right\} \xrightarrow{\Delta H°_{(d)}} 6H_2O(l)
$$

Elements in their
standard states

We add the $\Delta H°$ values for the steps to get $\Delta H°$ for the overall reaction:

$$
\begin{aligned}
\Delta H°_{\text{reaction}} &= \Delta H°_{(a)} + \Delta H°_{(b)} + \Delta H°_{(c)} + \Delta H°_{(d)} \\
&= \left[4 \times -\Delta H°_f \text{ for } NH_3(g)\right] + 0 + \left[4 \times \Delta H°_f \text{ for } NO_2(g)\right] \\
&\quad + \left[6 \times \Delta H°_f \text{ for } H_2O(l)\right] \\
&= \left[4 \times \Delta H°_f \text{ for } NO_2(g)\right] + \left[6 \times \Delta H°_f \text{ for } H_2O(l)\right] \\
&\quad - \left[4 \times \Delta H°_f \text{ for } NH_3(g)\right] \\
&= \Sigma n_p \Delta H°_f(\text{products}) - \Sigma n_r \Delta H°_f(\text{reactants})
\end{aligned}
$$

Remember that elemental reactants and products do not need to be included, since $\Delta H°_f$ for an element in its standard state is zero. Note that we have again obtained Equation (6.1). The final solution is

$$
\begin{aligned}
\Delta H°_{\text{reaction}} &= \left[4 \times (34 \text{ kJ})\right] + \left[6 \times (-286 \text{ kJ})\right] - \left[4 \times (-46 \text{ kJ})\right] \\
&= -1396 \text{ kJ}
\end{aligned}
$$

See Exercises 6.79 and 6.80

Now that we have shown the basis for Equation (6.1), we will make direct use of it to calculate ΔH for reactions in succeeding exercises.

The thermite reaction is one of the most energetic chemical reactions known.

Enthalpies from Standard Enthalpies of Formation II

Using enthalpies of formation, calculate the standard change in enthalpy for the thermite reaction:

$$2Al(s) + Fe_2O_3(s) \longrightarrow Al_2O_3(s) + 2Fe(s)$$

This reaction occurs when a mixture of powdered aluminum and iron(III) oxide is ignited with a magnesium fuse.

Solution

Where are we going?

To calculate ΔH for the reaction

What do we know?

> ΔH_f° for $Fe_2O_3(s) = -826$ kJ/mol
> ΔH_f° for $Al_2O_3(s) = -1676$ kJ/mol
> ΔH_f° for $Al(s) = \Delta H_f^\circ$ for $Fe(s) = 0$

What do we need?

> We use Equation (6.1):

$$\Delta H^\circ = \Sigma n_p \Delta H_f^\circ(\text{products}) - \Sigma n_r \Delta H_f^\circ(\text{reactants})$$

How do we get there?

> $\Delta H^\circ_{\text{reaction}} = \Delta H_f^\circ$ for $Al_2O_3(s) - \Delta H_f^\circ$ for $Fe_2O_3(s)$
> $= -1676$ kJ $- (-826$ kJ$) = -850.$ kJ

This reaction is so highly exothermic that the iron produced is initially molten. This process is often used as a lecture demonstration and also has been used in welding massive steel objects such as ships' propellers.

See Exercises 6.83 and 6.84

Critical Thinking

For $\Delta H_{\text{reaction}}$ calculations, we define ΔH_f° for an element in its standard state as zero. What if we define ΔH_f° for an element in its standard state as 10 kJ/mol? How would this affect your determination of $\Delta H_{\text{reaction}}$? Provide support for your answer with a sample calculation.

Example 6.11

Enthalpies from Standard Enthalpies of Formation III

Until recently, methanol (CH_3OH) was used as a fuel in high-performance engines in race cars. Using the data in Table 6.2, compare the standard enthalpy of combustion per gram of methanol with that per gram of gasoline. Gasoline is actually a mixture of compounds, but assume for this problem that gasoline is pure liquid octane (C_8H_{18}).

Solution

Where are we going?

To compare ΔH of combustion for methanol and octane

What do we know?

> Standard enthalpies of formation from Table 6.2

How do we get there?

For methanol:

What is the combustion reaction?

$$2CH_3OH(l) + 3O_2(g) \longrightarrow 2CO_2(g) + 4H_2O(l)$$

What is the $\Delta H°_{reaction}$?

Using the standard enthalpies of formation from Table 6.2 and Equation (6.1), we have

$$\Delta H°_{reaction} = 2 \times \Delta H°_f \text{ for } CO_2(g) + 4 \times \Delta H°_f \text{ for } H_2O(l)$$
$$- 2 \times \Delta H°_f \text{ for } CH_3OH(l)$$
$$= 2 \times (-394 \text{ kJ}) + 4 \times (-286 \text{ kJ}) - 2 \times (-239 \text{ kJ})$$
$$= -1.45 \times 10^3 \text{ kJ}$$

What is the enthalpy of combustion per gram?

Thus 1.45×10^3 kJ of heat is evolved when 2 moles of methanol burn. The molar mass of methanol is 32.04 g/mol. This means that 1.45×10^3 kJ of energy is produced when 64.08 g methanol burns. The enthalpy of combustion per gram of methanol is

$$\frac{-1.45 \times 10^3 \text{ kJ}}{64.08 \text{ g}} = -22.6 \text{ kJ/g}$$

For octane:

What is the combustion reaction?

$$2C_8H_{18}(l) + 25O_2(g) \longrightarrow 16CO_2(g) + 18H_2O(l)$$

What is the $\Delta H°_{reaction}$?

Using the standard enthalpies of formation from Table 6.2 and Equation (6.1), we have

$$\Delta H°_{reaction} = 16 \times \Delta H°_f \text{ for } CO_2(g) + 18 \times \Delta H°_f \text{ for } H_2O(l)$$
$$- 2 \times \Delta H°_f \text{ for } C_8H_{18}(l)$$
$$= 16 \times (-394 \text{ kJ}) + 18 \times (-286 \text{ kJ}) - 2 \times (-269 \text{ kJ})$$
$$= -1.09 \times 10^4 \text{ kJ}$$

What is the enthalpy of combustion per gram?

This is the amount of heat evolved when 2 moles of octane burn. Since the molar mass of octane is 114.22 g/mol, the enthalpy of combustion per gram of octane is

$$\frac{-1.09 \times 10^4 \text{ kJ}}{2(114.22 \text{ g})} = -47.7 \text{ kJ/g}$$

> The enthalpy of combustion per gram of octane is approximately twice that per gram of methanol. On this basis, gasoline appears to be superior to methanol for use in a racing car, where weight considerations are usually very important. Why, then, was methanol used in racing cars? The answer is that methanol burns much more smoothly than gasoline in high-performance engines, and this advantage more than compensates for its weight disadvantage.

In the cars used in the Indianapolis 500, ethanol is now used instead of methanol.

See Exercise 6.91

6.5 | Present Sources of Energy

Woody plants, coal, petroleum, and natural gas hold a vast amount of energy that originally came from the sun. By the process of photosynthesis, plants store energy that can be claimed by burning the plants themselves or the decay products that have been converted over millions of years to **fossil fuels**. Although the United States currently

Figure 6.11 | Energy sources used in the United States.

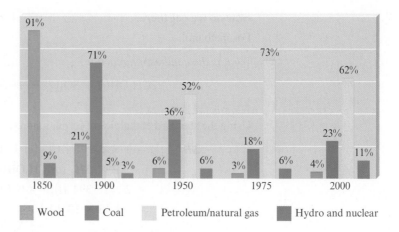

- Wood
- Coal
- Petroleum/natural gas
- Hydro and nuclear

Oil rig in Gulf of Mexico.

Keith Wood/Stone/Getty Images

Table 6.3 | Names and Formulas for Some Common Hydrocarbons

Formula	Name
CH_4	Methane
C_2H_6	Ethane
C_3H_8	Propane
C_4H_{10}	Butane
C_5H_{12}	Pentane
C_6H_{14}	Hexane
C_7H_{16}	Heptane
C_8H_{18}	Octane

depends heavily on petroleum for energy, this dependency is a relatively recent phenomenon, as shown in Fig. 6.11. In this section we will discuss some sources of energy and their effects on the environment.

Petroleum and Natural Gas

Although how they were produced is not completely understood, petroleum and natural gas were most likely formed from the remains of marine organisms that lived approximately 500 million years ago. **Petroleum** is a thick, dark liquid composed mostly of compounds called *hydrocarbons* that contain carbon and hydrogen. (Carbon is unique among elements in the extent to which it can bond to itself to form chains of various lengths.) Table 6.3 gives the formulas and names for several common hydrocarbons. **Natural gas**, usually associated with petroleum deposits, consists mostly of methane, but it also contains significant amounts of ethane, propane, and butane. Over the last several years it has become clear that there are tremendous reserves of natural gas deep in shale deposits. Estimates indicate that there may be as much as 200 trillion cubic meters of recoverable natural gas in these deposits around the globe. In the eastern United States, the 1-mile-deep Marsellus Shale deposit may contain as much as two trillion cubic meters of recoverable gas. The problem with these gas deposits is that the shale is very impermeable, and the gas does not flow out into wells on its own. A technique called hydraulic fracturing, or "fracking," is now being used to access these gas deposits. Fracking involves injecting a slurry of water, sand, and chemical additives under pressure through a well bore drilled into the shale. This produces fractures in the rock that allow the gas to flow out into wells. With the increased use of fracking has come environmental concerns, including potential contamination of ground water, risks to air quality, and possible mishandling of wastes associated with the process. However, even in view of these potential hazards, it is expected that natural gas obtained from fracking may supply as much as half of the U.S. natural gas supply by the year 2020.

The composition of petroleum varies somewhat, but it consists mostly of hydrocarbons having chains that contain from 5 to more than 25 carbons. To be used efficiently, the petroleum must be separated into fractions by boiling. The lighter molecules (having the lowest boiling points) can be boiled off, leaving the heavier ones behind. The commercial uses of various petroleum fractions are shown in Table 6.4.

The petroleum era began when the demand for lamp oil during the Industrial Revolution outstripped the traditional sources: animal fats and whale oil. In response to this increased demand, Edwin Drake drilled the first oil well in 1859 at Titusville, Pennsylvania. The petroleum from this well was refined to produce *kerosene* (fraction C_{10}–C_{18}), which served as an excellent lamp oil. *Gasoline* (fraction C_5–C_{10}) had limited use and was often discarded. However, this situation soon changed. The development of the

Table 6.4 | Uses of the Various Petroleum Fractions

Petroleum Fraction in Terms of Numbers of Carbon Atoms	Major Uses
C_5–C_{10}	Gasoline
C_{10}–C_{18}	Kerosene
	Jet fuel
C_{15}–C_{25}	Diesel fuel
	Heating oil
	Lubricating oil
>C_{25}	Asphalt

electric light decreased the need for kerosene, and the advent of the "horseless carriage" with its gasoline-powered engine signaled the birth of the gasoline age.

As gasoline became more important, new ways were sought to increase the yield of gasoline obtained from each barrel of petroleum. William Burton invented a process at Standard Oil of Indiana called *pyrolytic (high-temperature) cracking.* In this process, the heavier molecules of the kerosene fraction are heated to about 700°C, causing them to break (crack) into the smaller molecules of hydrocarbons in the gasoline fraction. As cars became larger, more efficient internal combustion engines were designed. Because of the uneven burning of the gasoline then available, these engines "knocked," producing unwanted noise and even engine damage. Intensive research to find additives that would promote smoother burning produced tetraethyl lead, $(C_2H_5)_4Pb$, a very effective "antiknock" agent.

The addition of tetraethyl lead to gasoline became a common practice, and by 1960, gasoline contained as much as 3 grams of lead per gallon. As we have discovered so often in recent years, technological advances can produce environmental problems. To prevent air pollution from automobile exhaust, catalytic converters have been added to car exhaust systems. The effectiveness of these converters, however, is destroyed by lead. The use of leaded gasoline also greatly increased the amount of lead in the environment, where it can be ingested by animals and humans. For these reasons, the use of lead in gasoline has been phased out, requiring extensive (and expensive) modifications of engines and of the gasoline refining process.

Coal

Coal was formed from the remains of plants that were buried and subjected to high pressure and heat over long periods of time. Plant materials have a high content of cellulose, a complex molecule whose empirical formula is CH_2O but whose molar mass is around 500,000 g/mol. After the plants and trees that flourished on the earth at various times and places died and were buried, chemical changes gradually lowered the oxygen and hydrogen content of the cellulose molecules. Coal "matures" through four stages: lignite, subbituminous, bituminous, and anthracite. Each stage has a higher carbon-to-oxygen and carbon-to-hydrogen ratio; that is, the relative carbon content gradually increases. Typical elemental compositions of the various coals are given in Table 6.5. The energy available from the combustion of a given mass of coal increases as the carbon content increases. Therefore, anthracite is the most valuable coal and lignite the least valuable.

Coal has variable composition depending on both its age and location.

Coal is an important and plentiful fuel in the United States, currently furnishing approximately 23% of our energy. As the supply of petroleum dwindles, the share of the energy supply from coal is expected to increase. However, coal is expensive and dangerous to mine underground, and the strip mining of fertile farmland in the Midwest or of scenic land in the West causes obvious problems. In addition, the burning of coal, especially high-sulfur coal, yields air pollutants such as sulfur dioxide, which, in turn, can lead to acid rain, as we learned in Chapter 5. However, even if coal were pure carbon, the carbon dioxide produced when it was burned would still have significant effects on the earth's climate.

Table 6.5 | Elemental Composition of Various Types of Coal

Type of Coal	Mass Percent of Each Element				
	C	H	O	N	S
Lignite	71	4	23	1	1
Subbituminous	77	5	16	1	1
Bituminous	80	6	8	1	5
Anthracite	92	3	3	1	1

The electromagnetic spectrum, including visible and infrared radiation, is discussed in Chapter 7.

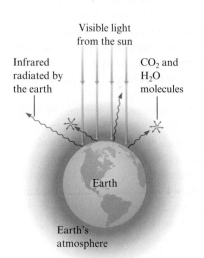

Figure 6.12 | The earth's atmosphere is transparent to visible light from the sun. This visible light strikes the earth, and part of it is changed to infrared radiation. The infrared radiation from the earth's surface is strongly absorbed by CO_2, H_2O, and other molecules present in smaller amounts (for example, CH_4 and N_2O) in the atmosphere. In effect, the atmosphere traps some of the energy, acting like the glass in a greenhouse and keeping the earth warmer than it would otherwise be.

The average temperature of the earth's surface is 298 K. It would be 255 K without the "greenhouse gases."

Effects of Carbon Dioxide on Climate

The earth receives a tremendous quantity of radiant energy from the sun, about 30% of which is reflected back into space by the earth's atmosphere. The remaining energy passes through the atmosphere to the earth's surface. Some of this energy is absorbed by plants for photosynthesis and some by the oceans to evaporate water, but most of it is absorbed by soil, rocks, and water, increasing the temperature of the earth's surface. This energy is in turn radiated from the heated surface mainly as *infrared radiation,* often called *heat radiation.*

The atmosphere, like window glass, is transparent to visible light but does not allow all the infrared radiation to pass back into space. Molecules in the atmosphere, principally H_2O and CO_2, strongly absorb infrared radiation and radiate it back toward the earth (Fig. 6.12), so a net amount of thermal energy is retained by the earth's atmosphere, causing the earth to be much warmer than it would be without its atmosphere. In a way, the atmosphere acts like the glass of a greenhouse, which is transparent to visible light but absorbs infrared radiation, thus raising the temperature inside the building. This **greenhouse effect** is seen even more spectacularly on Venus, where the dense atmosphere is thought to be responsible for the high surface temperature of that planet.

Thus the temperature of the earth's surface is controlled to a significant extent by the water content of the atmosphere as well as greenhouse gases such as carbon dioxide. The effect of atmospheric moisture (humidity) is apparent in the Midwest. In summer, when the humidity is high, the heat of the sun is retained well into the night, giving very high nighttime temperatures. On the other hand, in winter, the coldest temperatures always occur on clear nights, when the low humidity allows efficient radiation of energy back into space.

The atmosphere's water content is controlled by the water cycle (evaporation and precipitation), and the average remains constant over the years. However, as fossil fuels have been used more extensively, the carbon dioxide concentration has increased by about 16% from 1880 to 1980. Comparisons of satellite data have now produced evidence that the increase in carbon dioxide concentration has contributed to the warming of the earth's atmosphere. The data compare the same areas in both 1979 and 1997. The analysis shows that more infrared radiation was blocked by CO_2, methane, and other greenhouse gases. This *could* increase the earth's average temperature by as much as 3°C, causing dramatic changes in climate and greatly affecting the growth of food crops.

How well can we predict long-term effects? Because weather has been studied for a period of time that is minuscule compared with the age of the earth, the factors that control the earth's climate in the long range, such as changes in the sun, are not clearly understood. For example, we do not understand what causes the earth's periodic ice ages. So it is difficult to estimate the impact of the increasing carbon dioxide levels.

In fact, the variation in the earth's average temperature over the past century is somewhat confusing. In the northern latitudes during the past century, the average temperature rose by 0.8°C over a period of 60 years, then cooled by 0.5°C during the next 25 years, and finally warmed by 0.2°C in the succeeding 15 years. Such fluctuations do not match the steady increase in carbon dioxide. However, in southern latitudes and near the equator during the past century, the average temperature showed a steady rise totaling 0.4°C. This figure is in reasonable agreement with the predicted effect of the increasing carbon dioxide concentration over that period. Another significant fact is that the past 10 years constitute the warmest decade on record, although the warming has not been as great as the models predicted for the increase in carbon dioxide concentration.

Although the relationship between the carbon dioxide concentration in the atmosphere and the earth's temperature is not known at present, one thing is clear: The

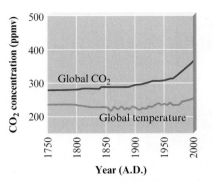

Figure 6.13 | The atmospheric CO_2 concentration and the average global temperature over the last 250 years. Note the significant increase in CO_2 concentration in the last 50 years.

(Source: National Assessment Synthesis Team, *Climate Change Impacts on the United States: The Potential Consequences of Climate, Variability and Change, Overview,* Report for the U.S. Global Change Research Program, Cambridge University Press, Cambridge, UK, p. 13, 2000.)

increase in the atmospheric concentration of carbon dioxide is quite dramatic (Fig. 6.13). We must consider the implications of this increase as we consider our future energy needs.

Methane is another greenhouse gas that is 21 times more potent than carbon dioxide. This fact is particularly significant for countries with lots of animals, because methane is produced by methanogenic archae that live in the animals' rumen. For example, sheep and cattle produce about 14% of Australia's total greenhouse emissions. To reduce this level, Australia has initiated a program to vaccinate sheep and cattle to lower the number of archae present in their digestive systems. It is hoped that this effort will reduce by 20% the amount of methane emitted by these animals.

6.6 | New Energy Sources

As we search for the energy sources of the future, we need to consider economic, climatic, and supply factors. There are several potential energy sources: the sun (solar), nuclear processes (fission and fusion), biomass (plants), and synthetic fuels. Direct use of the sun's radiant energy to heat our homes and run our factories and transportation systems seems a sensible long-term goal. But what do we do now? Conservation of fossil fuels is one obvious step, but substitutes for fossil fuels also must be found. We will discuss some alternative sources of energy here. Nuclear power will be considered in Chapter 19.

Coal Conversion

One alternative energy source involves using a traditional fuel—coal—in new ways. Since transportation costs for solid coal are high, more energy-efficient fuels are being developed from coal. One possibility is to produce a gaseous fuel. Substances like coal that contain large molecules have high boiling points and tend to be solids or thick liquids. To convert coal from a solid to a gas therefore requires reducing the size of the molecules; the coal structure must be broken down in a process called *coal gasification.* This is done by treating the coal with oxygen and steam at high temperatures to break many of the carbon–carbon bonds. These bonds are replaced by carbon–hydrogen and carbon–oxygen bonds as the coal fragments react with the water and oxygen. The process is represented in Fig. 6.14. The desired product is a mixture of carbon monoxide and hydrogen called *synthesis gas,* or **syngas**, and methane (CH_4) gas. Since all the components of this product can react with oxygen to release heat in a combustion reaction, this gas is a useful fuel.

One of the most important considerations in designing an industrial process is efficient use of energy. In coal gasification, some of the reactions are exothermic:

$$C(s) + 2H_2(g) \longrightarrow CH_4(g) \qquad \Delta H° = -75 \text{ kJ}$$
$$C(s) + \tfrac{1}{2}O_2(g) \longrightarrow CO(g) \qquad \Delta H° = -111 \text{ kJ}$$
$$C(s) + O_2(g) \longrightarrow CO_2(g) \qquad \Delta H° = -394 \text{ kJ}$$

Other gasification reactions are endothermic, for example:

$$C(s) + H_2O(g) \longrightarrow H_2(g) + CO(g) \qquad \Delta H° = 131 \text{ kJ}$$

Figure 6.14 | Coal gasification. Reaction of coal with a mixture of steam and air breaks down the large hydrocarbon molecules in the coal to smaller gaseous molecules, which can be used as fuels.

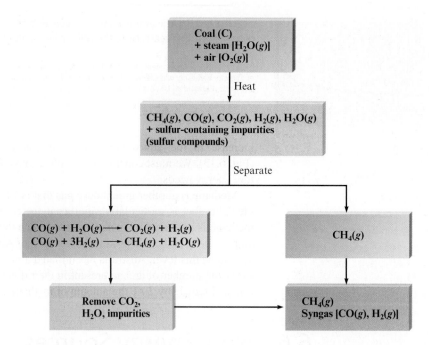

An industrial process must be energy efficient.

The main engines in the space shuttle Endeavor *used hydrogen and oxygen as fuel in the shuttle's many trips into space.*

If such conditions as the rate of feed of coal, air, and steam are carefully controlled, the correct temperature can be maintained in the process without using any external energy source. That is, an energy balance is maintained.

Presently only a few plants in the United States use syngas produced on site to produce electricity. These plants are being used to evaluate the economic feasibility of producing electrical power by coal gasification.

Although syngas can be used directly as a fuel, it is also important as a raw material to produce other fuels. For example, syngas can be converted directly to methanol:

$$CO(g) + 2H_2(g) \longrightarrow CH_3OH(l)$$

Methanol is used in the production of synthetic fibers and plastics and also can be used as a fuel. In addition, it can be converted directly to gasoline. Approximately half of South Africa's gasoline supply comes from methanol produced from syngas.

In addition to coal gasification, the formation of *coal slurries* is another new use of coal. A slurry is a suspension of fine particles in a liquid, and coal must be pulverized and mixed with water to form a slurry. The slurry can be handled, stored, and burned in ways similar to those used for *residual oil,* a heavy fuel oil from petroleum accounting for almost 15% of U.S. petroleum imports. One hope is that coal slurries might replace solid coal and residual oil as fuels for electricity-generating power plants. However, the water needed for slurries might place an unacceptable burden on water resources, especially in the western states.

Hydrogen as a Fuel

If you have ever seen a lecture demonstration where hydrogen–oxygen mixtures were ignited, you have witnessed a demonstration of hydrogen's potential as a fuel. The combustion reaction is

$$H_2(g) + \tfrac{1}{2}O_2(g) \longrightarrow H_2O(l) \qquad \Delta H° = -286 \text{ kJ}$$

As we saw in Example 6.6, the heat of combustion of $H_2(g)$ per gram is approximately 2.5 times that of natural gas. In addition, hydrogen has a real advantage over fossil fuels in that the only product of hydrogen combustion is water; fossil fuels also produce

Chemical connections
Farming the Wind

In the Midwest the wind blows across fields of corn, soybeans, wheat, and wind turbines—wind turbines? It turns out that the wind that seems to blow almost continuously across the plains is now becoming the latest cash crop. One of these new-breed wind farmers is Daniel Juhl, who recently erected 17 wind turbines on six acres of land near Woodstock, Minnesota. These turbines can generate as much as 10 megawatts (MW) of electricity, which Juhl sells to the local electrical utility.

There is plenty of untapped wind power in the United States. Wind mappers rate regions on a scale of 1 to 6 (with 6 being the best) to indicate the quality of the wind resource. Wind farms are now being developed in areas rated from 4 to 6. The farmers who own the land welcome the increased income derived from the wind blowing across their land. Economists estimate that each acre devoted to wind turbines can pay royalties to the farmers of as much as $8000 per year, or many times the revenue from growing corn on that same land. Juhl claims that farmers who construct the turbines themselves can realize as much as $20,000 per year per turbine. Globally, wind generation of electricity has nearly quadrupled in the last five years and is expected to increase by about 60%

per year in the United States. The economic feasibility of wind-generated electricity has greatly improved in the past 30 years as the wind turbines have become more efficient. Today's turbines can produce electricity that costs about the same as that from other sources. The most impressive thing about wind power is the magnitude of the supply. According to the American Wind Energy Association in Washington, D.C., the wind-power potential in the United States is comparable or larger than the energy resources under the sands of Saudi Arabia.

The biggest hurdle that must be overcome before wind power can become a significant electricity producer in the United States is construction of the transmission infrastructure—the power lines needed to move the electricity from the rural areas to the cities where most of the power is used. For example, the hundreds of turbines planned in southwest Minnesota in a development called Buffalo Ridge could supply enough electricity to

This State Line Wind Project along the Oregon–Washington border uses approximately 399 wind turbines to create enough electricity to power some 70,000 households.

power 1 million homes if transmission problems can be solved.

Another possible scenario for wind farms is to use the electrical power generated to decompose water to produce hydrogen gas that could be carried to cities by pipelines and used as a fuel. One real benefit of hydrogen is that it produces water as its only combustion product. Thus, it is essentially pollution-free.

Within a few years, wind power could be a major source of electricity. There could be a fresh wind blowing across the energy landscape of the United States in the near future.

carbon dioxide. However, even though it appears that hydrogen is a very logical choice as a major fuel for the future, there are three main problems: the cost of production, storage, and transport.

First let's look at the production problem. Although hydrogen is very abundant on the earth, virtually none of it exists as the free gas. Currently, the main source of hydrogen gas is from the treatment of natural gas with steam:

$$CH_4(g) + H_2O(g) \longrightarrow 3H_2(g) + CO(g)$$

We can calculate ΔH for this reaction using Equation (6.1):

$$\Delta H° = \Sigma n_p \Delta H_f°(\text{products}) - \Sigma n_r \Delta H_f°(\text{reactants})$$
$$= \Delta H_f° \text{ for } CO(g) - \Delta H_f° \text{ for } CH_4(g) - \Delta H_f° \text{ for } H_2O(g)$$
$$= -111 \text{ kJ} - (-75 \text{ kJ}) - (-242 \text{ kJ}) = 206 \text{ kJ}$$

Note that this reaction is highly endothermic; treating methane with steam is not an efficient way to obtain hydrogen for fuel. It would be much more economical to burn the methane directly.

A virtually inexhaustible supply of hydrogen exists in the waters of the world's oceans. However, the reaction

$$H_2O(l) \longrightarrow H_2(g) + \tfrac{1}{2}O_2(g)$$

requires 286 kJ of energy per mole of liquid water, and under current circumstances, large-scale production of hydrogen from water is not economically feasible. However, several methods for such production are currently being studied: electrolysis of water, thermal decomposition of water, thermochemical decomposition of water, and biological decomposition of water.

Electrolysis of water involves passing an electric current through it. The present cost of electricity makes the hydrogen produced by electrolysis too expensive to be competitive as a fuel. However, if in the future we develop more efficient sources of electricity, this situation could change.

Recent research at the University of Minnesota by Lanny Schmidt and his coworkers suggests that corn could be a feasible source of hydrogen. In this process the starch from the corn is fermented to produce alcohol, which is then decomposed in a special reactor at 140°C with a rhodium and cerium oxide catalyst to give hydrogen. These scientists indicate that enough hydrogen gas can be obtained from a few ounces of ethanol to generate electricity to run six 60-watt bulbs for an hour.

Thermal decomposition is another method for producing hydrogen from water. This involves heating the water to several thousand degrees, where it spontaneously decomposes into hydrogen and oxygen. However, attaining temperatures in this range would be very expensive even if a practical heat source and a suitable reaction container were available.

In the thermochemical decomposition of water, chemical reactions, as well as heat, are used to "split" water into its components. One such system involves the following reactions (the temperature required for each is given in parentheses):

$$2HI \longrightarrow I_2 + H_2 \qquad (425°C)$$
$$2H_2O + SO_2 + I_2 \longrightarrow H_2SO_4 + 2HI \qquad (90°C)$$
$$\underline{H_2SO_4 \longrightarrow SO_2 + H_2O + \tfrac{1}{2}O_2 \qquad (825°C)}$$
$$\text{Net reaction: } H_2O \longrightarrow H_2 + \tfrac{1}{2}O_2$$

Note that the HI is not consumed in the net reaction. Note also that the maximum temperature required is 825°C, a temperature that is feasible if a nuclear reactor is used as a heat source. A current research goal is to find a system for which the required temperatures are low enough that sunlight can be used as the energy source.

But what about the organisms that decompose water without the aid of electricity or high temperatures? In the process of photosynthesis, green plants absorb carbon dioxide and water and use them along with energy from the sun to produce the substances needed for growth. Scientists have studied photosynthesis for years, hoping to get answers to humanity's food and energy shortages. At present, much of this research involves attempts to modify the photosynthetic process so that plants will release hydrogen gas from water instead of using the hydrogen to produce complex compounds. Small-scale experiments have shown that under certain conditions plants do produce hydrogen gas, but the yields are far from being commercially useful. At this point the economical production of hydrogen gas remains unrealized.

Electrolysis will be discussed in Chapter 18.

The storage and transportation of hydrogen also present problems. First, on metal surfaces the H_2 molecule decomposes to atoms. Since the atoms are so small, they can migrate into the metal, causing structural changes that make it brittle. This might lead to a pipeline failure if hydrogen were pumped under high pressure.

An additional problem is the relatively small amount of energy that is available *per unit volume* of hydrogen gas. Although the energy available per gram of hydrogen is significantly greater than that per gram of methane, the energy available per given volume of hydrogen is about one-third that available from the same volume of methane. This is demonstrated in Example 6.12.

Although the use of hydrogen as a fuel solves some of the problems associated with fossil fuels, it does present some potential environmental problems of its own. Studies by John M. Eiler and his colleagues at California Institute of Technology indicate that, if hydrogen becomes a major source of energy, accidental leakage of the gas into the atmosphere could pose a threat. The Cal Tech scientists calculate that leakage could raise the concentration of H_2 in the atmosphere from its natural level of 0.5 part per million to more than 2 parts per million. As some of the H_2 eventually finds its way into the upper atmosphere, it would react with O_2 to form water, which would increase the number of ice crystals. This could lead to the destruction of some of the protective ozone because many of the chemical reactions that destroy ozone occur on the surfaces of ice crystals. However, as is the usual case with environmental issues, the situation is complicated. The scenario suggested by Eiler's team may not happen because the leaked H_2 could be consumed by soil microbes that use hydrogen as a nutrient. In fact, Eiler's studies show that 90% of the H_2 emitted into the atmosphere today from sources such as motor vehicles and forest fires is eventually absorbed by soil organisms.

The evaluation of hydrogen as a fuel illustrates how complex and interconnected the economic and environmental issues are.

Example 6.12

Enthalpies of Combustion

Compare the energy available from the combustion of a given volume of methane and the same volume of hydrogen at the same temperature and pressure.

Solution

In Example 6.6 we calculated the heat released for the combustion of methane and hydrogen: 55 kJ/g CH_4 and 141 kJ/g H_2. We also know from our study of gases that 1 mole of $H_2(g)$ has the same volume as 1 mole of $CH_4(g)$ at the same temperature and pressure (assuming ideal behavior). Thus, for molar volumes of both gases under the same conditions of temperature and pressure,

$$\frac{\text{Enthalpy of combustion of 1 molar volume of } H_2(g)}{\text{Enthalpy of combustion of 1 molar volume of } CH_4(g)}$$

$$= \frac{\text{enthalpy of combustion per mole of } H_2}{\text{enthalpy of combustion per mole of } CH_4}$$

$$= \frac{(-141 \text{ kJ/g})(2.02 \text{ g } H_2/\text{mol } H_2)}{(-55 \text{ kJ/g})(16.04 \text{ g } CH_4/\text{mol } CH_4)}$$

$$= \frac{-285}{-882} \approx \frac{1}{3}$$

Thus about three times the volume of hydrogen gas is needed to furnish the same energy as a given volume of methane.

See Exercise 6.92

Could hydrogen be considered as a potential fuel for automobiles? This is an intriguing question. The internal combustion engines in automobiles can be easily adapted to burn hydrogen. In fact, BMW is now experimenting with a fleet of cars powered by hydrogen-burning internal combustion engines. However, the primary difficulty is the storage of enough hydrogen to give an automobile a reasonable range. This is illustrated by Example 6.13.

Interactive Example 6.13

Sign in at http://login.cengagebrain
.com to try this Interactive Example
in **OWL**.

Comparing Enthalpies of Combustion

Assuming that the combustion of hydrogen gas provides three times as much energy per gram as gasoline, calculate the volume of liquid H_2 (density = 0.0710 g/mL) required to furnish the energy contained in 80.0 L (about 20 gal) of gasoline (density = 0.740 g/mL). Calculate also the volume that this hydrogen would occupy as a gas at 1.00 atm and 25°C.

Solution

Where are we going?

To calculate the volume of $H_2(l)$ required and its volume as a gas at the given conditions

What do we know?

> Density for $H_2(l)$ = 0.0710 g/mL
> 80.0 L gasoline
> Density for gasoline = 0.740 g/mL
> $H_2(g) \Rightarrow P$ = 1.00 atm, T = 25°C = 298 K

How do we get there?

What is the mass of gasoline?

$$80.0 \text{ L} \times \frac{1000 \text{ mL}}{1 \text{ L}} \times \frac{0.740 \text{ g}}{\text{mL}} = 59{,}200 \text{ g}$$

How much $H_2(l)$ is needed?
Since H_2 furnishes three times as much energy per gram as gasoline, only a third as much liquid hydrogen is needed to furnish the same energy:

$$\text{Mass of } H_2(l) \text{ needed} = \frac{59{,}200 \text{ g}}{3} = 19{,}700 \text{ g}$$

Since density = mass/volume, then volume = mass/density, and the volume of $H_2(l)$ needed is

$$V = \frac{19{,}700 \text{ g}}{0.0710 \text{ g/mL}}$$
$$= 2.77 \times 10^5 \text{ mL} = 277 \text{ L}$$

Thus 277 L of liquid H_2 is needed to furnish the same energy of combustion as 80.0 L of gasoline.

What is the volume of the $H_2(g)$?
To calculate the volume that this hydrogen would occupy as a gas at 1.00 atm and 25°C, we use the ideal gas law:

$$PV = nRT$$

In this case

$$P = 1.00 \text{ atm}, T = 273 + 25°C = 298 \text{ K, and } R = 0.08206 \text{ L} \cdot \text{atm/K} \cdot \text{mol}$$

What are the moles of $H_2(g)$?

$$n = 19,700 \text{ g H}_2 \times \frac{1 \text{ mol H}_2}{2.016 \text{ g H}_2} = 9.77 \times 10^3 \text{ mol H}_2$$

Thus

$$V = \frac{nRT}{P} = \frac{(9.77 \times 10^3 \text{ mol})(0.08206 \text{ L} \cdot \text{atm/K} \cdot \text{mol})(298 \text{ K})}{1.00 \text{ atm}}$$

$$= 2.39 \times 10^5 \text{ L} = 239,000 \text{ L}$$

› At 1 atm and 25°C, the hydrogen gas needed to replace 20 gal of gasoline occupies a volume of 239,000 L.

See Exercises 6.93 and 6.94

You can see from Example 6.13 that an automobile would need a huge tank to hold enough hydrogen gas (at 1 atm) to have a typical mileage range. Clearly, hydrogen must be stored as a liquid or in some other way. Is this feasible? Because of its very low boiling point (20 K), storage of liquid hydrogen requires a superinsulated container that can withstand high pressures. Storage in this manner would be both expensive and hazardous because of the potential for explosion. Thus storage of hydrogen in the individual automobile as a liquid does not seem practical.

Metal hydrides are discussed in Chapter 20.

A much better alternative seems to be the use of metals that absorb hydrogen to form solid metal hydrides:

$$H_2(g) + M(s) \longrightarrow MH_2(s)$$

To use this method of storage, hydrogen gas would be pumped into a tank containing the solid metal in powdered form, where it would be absorbed to form the hydride, whose volume would be little more than that of the metal alone. This hydrogen would then be available for combustion in the engine by release of $H_2(g)$ from the hydride as needed:

$$MH_2(s) \longrightarrow M(s) + H_2(g)$$

Several types of solids that absorb hydrogen to form hydrides are being studied for use in hydrogen-powered vehicles. The most likely use of hydrogen in automobiles will be to power fuel cells. Ford, Honda, and Toyota are all experimenting with cars powered by hydrogen fuel cells.

Other Energy Alternatives

Many other energy sources are being considered for future use. The western states, especially Colorado, contain huge deposits of *oil shale,* which consists of a complex carbon-based material called *kerogen* contained in porous rock formations. These deposits have the potential of being a larger energy source than the vast petroleum deposits of the Middle East. The main problem with oil shale is that the trapped fuel is not fluid and cannot be pumped. To recover the fuel, the rock must be heated to a temperature of 250°C or higher to decompose the kerogen to smaller molecules that produce gaseous and liquid products. This process is expensive and yields large quantities of waste rock, which have a negative environmental impact.

Ethanol (C_2H_5OH) is another fuel used to supplement gasoline. The most common method of producing ethanol is fermentation, a process in which sugar is changed to alcohol by the action of yeast. The sugar can come from virtually any source, including fruits and grains, although fuel-grade ethanol would probably come mostly from corn. Car engines can burn pure alcohol or *gasohol,* an alcohol–gasoline mixture (10% ethanol in gasoline), with little modification. Gasohol is now widely available in the

Castor seed oil is one of the raw materials used in the production of biofuels.

Chemical connections
Veggie Gasoline?

Gasoline usage is as high as ever, and world petroleum supplies will eventually dwindle. One possible alternative to petroleum as a source of fuels and lubricants is vegetable oil—the same vegetable oil we now use to cook french fries. Researchers believe that the oils from soybeans, corn, canola, and sunflowers all have the potential to be used in cars as well as on salads.

The use of vegetable oil for fuel is not a new idea. Rudolf Diesel reportedly used peanut oil to run one of his engines at the Paris Exposition in 1900. In addition, ethyl alcohol has been used widely as a fuel in South America and as a fuel additive in the United States.

Biodiesel, a fuel made by esterifying the fatty acids found in vegetable oil, has some real advantages over regular diesel fuel. Biodiesel produces fewer pollutants such as particulates, carbon monoxide, and complex organic molecules, and since vegetable oils have no sulfur, there is no noxious sulfur dioxide in the exhaust gases. Also, biodiesel can run in existing engines with little modification. In addition, biodiesel is much more biodegradable than petroleum-based fuels, so spills cause less environmental damage.

Of course, biodiesel also has some serious drawbacks. The main one is that it costs about three times as much as regular diesel fuel. Biodiesel also produces more nitrogen oxides in the exhaust than conventional diesel fuel and is less stable in

This promotion bus both advertises biodiesel and demonstrates its usefulness.

storage. Biodiesel also can leave more gummy deposits in engines and must be "winterized" by removing components that tend to solidify at low temperatures.

The best solution may be to use biodiesel as an additive to regular diesel fuel. One such fuel is known as B20 because it is 20% biodiesel and 80% conventional diesel fuel. B20 is especially attractive because of the higher lubricating ability of vegetable oils, thus reducing diesel engine wear.

Vegetable oils are also being looked at as replacements for motor oils and hydraulic fluids. Tests of a sunflower seed–based engine lubricant manufactured by Renewable Lubricants of Hartville, Ohio, have shown satisfactory lubricating ability while lowering particle emissions. In addition, Lou

Honary and his colleagues at the University of Northern Iowa have developed BioSOY, a vegetable oil–based hydraulic fluid for use in heavy machinery.

Veggie oil fuels and lubricants seem to have a growing market as petroleum supplies wane and environmental laws become more stringent. In Germany's Black Forest region, for example, environmental protection laws require that farm equipment use only vegetable oil fuels and lubricants. In the near future there may be veggie oil in your garage as well as in your kitchen.

Adapted from "Fill 'Er Up . . . with Veggie Oil," by Corinna Wu, as appeared in *Science News*, Vol. 154, December 5, 1998, p. 364.

United States. A fuel called E85, which is 85% ethanol and 15% gasoline, is also widely available for cars with "flex-fuel" engines. The use of pure alcohol as a motor fuel is not feasible in most of the United States because it does not vaporize easily when temperatures are low. However, pure ethanol could be a very practical fuel in warm climates. For example, in Brazil, large quantities of ethanol fuel are being produced for cars. It is hoped that in the future ethanol can be derived from sources of cellulose, such as switch grass, so that corn, which is a valuable food source, would not have to be used. However, the process to efficiently convert cellulose to ethanol has proved very difficult to develop.

Methanol (CH_3OH), an alcohol similar to ethanol, which was used successfully for many years in race cars, is now being evaluated as a motor fuel in California. A major gasoline retailer has agreed to install pumps at 25 locations to dispense a fuel that is 85% methanol and 15% gasoline for use in specially prepared automobiles. The California Energy Commission feels that methanol has great potential for providing a secure, long-term energy supply that would alleviate air quality problems. Arizona and Colorado are also considering methanol as a source of portable energy.

Another potential source of liquid fuels is oil squeezed from seeds (*seed oil*). For example, some farmers in North Dakota, South Africa, and Australia are now using sunflower oil to replace diesel fuel. Oil seeds, found in a wide variety of plants, can be processed to produce an oil composed mainly of carbon and hydrogen, which of course reacts with oxygen to produce carbon dioxide, water, and heat. It is hoped that oil-seed plants can be developed that will thrive under soil and climatic conditions unsuitable for corn and wheat. The main advantage of seed oil as a fuel is that it is renewable. Ideally, fuel would be grown just like food crops.

For review

Key terms

Section 6.1
energy
law of conservation of energy
potential energy
kinetic energy
heat
work
pathway
state function (property)
system
surroundings
exothermic
endothermic
thermodynamics
first law of thermodynamics
internal energy

Section 6.2
enthalpy
calorimeter
calorimetry
heat capacity
specific heat capacity
molar heat capacity
constant-pressure calorimetry
constant-volume calorimetry

Section 6.3
Hess's law

Energy

⟩ The capacity to do work or produce heat
⟩ Is conserved (first law of thermodynamics)
⟩ Can be converted from one form to another
⟩ Is a state function
⟩ Potential energy: stored energy
⟩ Kinetic energy: energy due to motion
⟩ The internal energy for a system is the sum of its potential and kinetic energies
⟩ The internal energy of a system can be changed by work and heat:

$$\Delta E = q + w$$

Work

⟩ Force applied over a distance
⟩ For an expanding/contracting gas
⟩ Not a state function

$$w = -P\Delta V$$

Heat

⟩ Energy flow due to a temperature difference
⟩ Exothermic: energy as heat flows out of a system
⟩ Endothermic: energy as heat flows into a system
⟩ Not a state function
⟩ Measured for chemical reactions by calorimetry

Key terms

Section 6.4
standard enthalpy of formation
standard state

Section 6.5
fossil fuels
petroleum
natural gas
coal
greenhouse effect

Section 6.6
syngas

Enthalpy

> $H = E + PV$

> Is a state function

> Hess's law: the change in enthalpy in going from a given set of reactants to a given set of products is the same whether the process takes place in one step or a series of steps

> Standard enthalpies of formation (ΔH_f°) can be used to calculate ΔH for a chemical reaction

$$\Delta H^\circ_{\text{reaction}} = \Sigma\, n_p \Delta H_f^\circ(\text{products}) - \Sigma\, n_r \Delta H^\circ(\text{reactants})$$

Energy use

> Energy sources from fossil fuels are associated with difficult supply and environmental impact issues

> The greenhouse effect results from release into the atmosphere of gases, including carbon dioxide, that strongly absorb infrared radiation, thus warming the earth

> Alternative fuels are being sought to replace fossil fuels:

> Hydrogen

> Syngas from coal

> Biofuels from plants such as corn and certain seed-producing plants

Review questions   *Answers to the Review Questions can be found on the Student website (accessible from* **www.cengagebrain.com**).

1. Define the following terms: potential energy, kinetic energy, path-dependent function, state function, system, surroundings.

2. Consider the following potential energy diagrams for two different reactions.

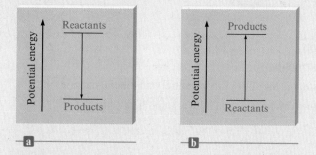

Which plot represents an exothermic reaction? In plot a, do the reactants on average have stronger or weaker bonds than the products? In plot b, reactants must gain potential energy to convert to products. How does this occur?

3. What is the first law of thermodynamics? How can a system change its internal energy, E? What are the sign conventions for thermodynamic quantities used in this text?

4. When a gas expands, what is the sign of w? Why? When a gas contracts, what is the sign of w? Why? What are the signs of q and w for the process of boiling water?

5. What is the heat gained/released at constant pressure equal to ($q_P = ?$)? What is the heat gained/released at constant volume equal to ($q_V = ?$)? Explain why ΔH is obtained directly from a coffee-cup calorimeter, whereas ΔE is obtained directly from a bomb calorimeter.

6. High-quality audio amplifiers generate large amounts of heat. To dissipate the heat and prevent damage to the electronic components, heat-radiating metal fins are used. Would it be better to make these fins out of iron or aluminum? Why? (See Table 6.1 for specific heat capacities.)

7. Explain how calorimetry works to calculate ΔH or ΔE for a reaction. Does the temperature of the calorimeter increase or decrease for an endothermic reaction? For an exothermic reaction? Explain.

8. What is Hess's law? When a reaction is reversed, what happens to the sign and magnitude of ΔH for that reversed reaction? When the coefficients in a balanced reaction are multiplied by a factor n, what happens to the sign and magnitude of ΔH for that multiplied reaction?

9. Define the standard enthalpy of formation. What are standard states for elements and for compounds? Using Hess's law, illustrate why the formula $\Delta H^\circ_{\text{reaction}} = \Sigma n_p \Delta H_f^\circ (\text{products}) - \Sigma n_r \Delta H_f^\circ (\text{reactants})$ works to calculate ΔH° for a reaction.

10. What are some of the problems associated with the world's dependence on fossil fuels? What are some alternative fuels for petroleum products?

Active Learning Questions

These questions are designed to be used by groups of students in class.

1. Objects placed together eventually reach the same temperature. When you go into a room and touch a piece of metal in that room, it feels colder than a piece of plastic. Explain.

2. What is meant by the term *lower in energy?* Which is lower in energy, a mixture of hydrogen and oxygen gases or liquid water? How do you know? Which of the two is more stable? How do you know?

3. A fire is started in a fireplace by striking a match and lighting crumpled paper under some logs. Explain all the energy transfers in this scenario using the terms *exothermic, endothermic, system, surroundings, potential energy,* and *kinetic energy* in the discussion.

4. Liquid water turns to ice. Is this process endothermic or exothermic? Explain what is occurring using the terms *system, surroundings, heat, potential energy,* and *kinetic energy* in the discussion.

5. Consider the following statements: "Heat is a form of energy, and energy is conserved. The heat lost by a system must be equal to the amount of heat gained by the surroundings. Therefore, heat is conserved." Indicate everything you think is correct in these statements. Indicate everything you think is incorrect. Correct the incorrect statements and explain.

6. Consider 5.5 L of a gas at a pressure of 3.0 atm in a cylinder with a movable piston. The external pressure is changed so that the volume changes to 10.5 L.

   a. Calculate the work done, and indicate the correct sign.

   b. Use the preceding data but consider the process to occur in two steps. At the end of the first step, the volume is 7.0 L. The second step results in a final volume of 10.5 L. Calculate the work done, and indicate the correct sign.

   c. Calculate the work done if after the first step the volume is 8.0 L and the second step leads to a volume of 10.5 L. Does the work differ from that in part b? Explain.

7. In Question 6 the work calculated for the different conditions in the various parts of the question was different even though the system had the same initial and final conditions. Based on this information, is work a state function?

   a. Explain how you know that work is not a state function.

   b. Why does the work increase with an increase in the number of steps?

   c. Which two-step process resulted in more work, when the first step had the bigger change in volume or when the second step had the bigger change in volume? Explain.

8. Explain why oceanfront areas generally have smaller temperature fluctuations than inland areas.

9. Hess's law is really just another statement of the first law of thermodynamics. Explain.

10. In the equation $w = -P\Delta V$, why is there a negative sign?

Questions

A blue question or exercise number indicates that the answer to that question or exercise appears at the back of this book and a solution appears in the *Solutions Guide,* as found on PowerLecture.

11. Consider an airplane trip from Chicago, Illinois, to Denver, Colorado. List some path-dependent functions and some state functions for the plane trip.

12. How is average bond strength related to relative potential energies of the reactants and the products?

13. Assuming gasoline is pure $C_8H_{18}(l)$, predict the signs of q and w for the process of combusting gasoline into $CO_2(g)$ and $H_2O(g)$.

14. What is the difference between ΔH and ΔE?

15. The enthalpy change for the reaction

$$CH_4(g) + 2O_2(g) \longrightarrow CO_2(g) + 2H_2O(l)$$

   is -891 kJ for the reaction *as written.*

   a. What quantity of heat is released for each mole of water formed?

   b. What quantity of heat is released for each mole of oxygen reacted?

16. For the reaction $HgO(s) \rightarrow Hg(l) + \frac{1}{2}O_2(g)$, $\Delta H = +90.7$ kJ:

   a. What quantity of heat is required to produce 1 mole of mercury by this reaction?

   b. What quantity of heat is required to produce 1 mole of oxygen gas by this reaction?

   c. What quantity of heat would be released in the following reaction as written?

$$2Hg(l) + O_2(g) \longrightarrow 2HgO(s)$$

17. The enthalpy of combustion of $CH_4(g)$ when $H_2O(l)$ is formed is -891 kJ/mol and the enthalpy of combustion of $CH_4(g)$ when $H_2O(g)$ is formed is -803 kJ/mol. Use these data and Hess's law to determine the enthalpy of vaporization for water.

18. The enthalpy change for a reaction is a state function and it is an extensive property. Explain.

19. Standard enthalpies of formation are relative values. What are ΔH_f° values relative to?

20. The combustion of methane can be represented as follows:

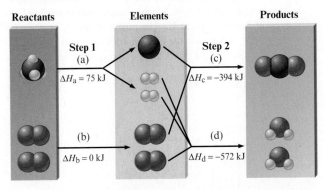

   a. Use the information given above to determine the value of ΔH for the combustion of methane to form $CO_2(g)$ and $2H_2O(l)$.

   b. What is ΔH_f° for an element in its standard state? Why is this? Use the figure above to support your answer.

c. How does ΔH for the reaction $CO_2(g) + 2H_2O(l) \rightarrow$ $CH_4(g) + O_2(g)$ compare to that of the combustion of methane? Why is this?

21. Why is it a good idea to rinse your thermos bottle with hot water before filling it with hot coffee?

22. Photosynthetic plants use the following reaction to produce glucose, cellulose, and so forth:

$$6CO_2(g) + 6H_2O(l) \xrightarrow{\text{Sunlight}} C_6H_{12}O_6(s) + 6O_2(g)$$

How might extensive destruction of forests exacerbate the greenhouse effect?

23. What is incomplete combustion of fossil fuels? Why can this be a problem?

24. Explain the advantages and disadvantages of hydrogen as an alternative fuel.

Exercises

In this section similar exercises are paired.

Potential and Kinetic Energy

25. Calculate the kinetic energy of a baseball (mass = 5.25 oz) with a velocity of 1.0×10^2 mi/h.

26. Which has the greater kinetic energy, an object with a mass of 2.0 kg and a velocity of 1.0 m/s or an object with a mass of 1.0 kg and a velocity of 2.0 m/s?

27. Consider the following diagram when answering the questions below.

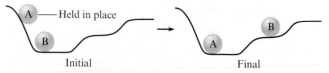

a. Compare balls A and B in terms of potential energy in both the initial and final setups.

b. Ball A has stopped moving in the figure on the right above, but energy must be conserved. What happened to the potential energy of ball A?

28. Consider the accompanying diagram. Ball A is allowed to fall and strike ball B. Assume that all of ball A's energy is transferred to ball B at point I, and that there is no loss of energy to other sources. What is the kinetic energy and the potential energy of ball B at point II? The potential energy is given by $PE = mgz$, where m is the mass in kilograms, g is the gravitational constant (9.81 m/s²), and z is the distance in meters.

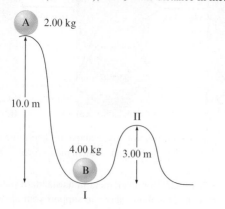

Heat and Work

29. A gas absorbs 45 kJ of heat and does 29 kJ of work. Calculate ΔE.

30. A system releases 125 kJ of heat while 104 kJ of work is done on it. Calculate ΔE.

31. Calculate ΔE for each of the following.
   a. $q = -47$ kJ, $w = +88$ kJ
   b. $q = +82$ kJ, $w = -47$ kJ
   c. $q = +47$ kJ, $w = 0$
   d. In which of these cases do the surroundings do work on the system?

32. A system undergoes a process consisting of the following two steps:

   Step 1: The system absorbs 72 J of heat while 35 J of work is done on it.

   Step 2: The system absorbs 35 J of heat while performing 72 J of work.

   Calculate ΔE for the overall process.

33. If the internal energy of a thermodynamic system is increased by 300. J while 75 J of expansion work is done, how much heat was transferred and in which direction, to or from the system?

34. Calculate the internal energy change for each of the following.
   a. One hundred (100.) joules of work is required to compress a gas. At the same time, the gas releases 23 J of heat.
   b. A piston is compressed from a volume of 8.30 L to 2.80 L against a constant pressure of 1.90 atm. In the process, there is a heat gain by the system of 350. J.
   c. A piston expands against 1.00 atm of pressure from 11.2 L to 29.1 L. In the process, 1037 J of heat is absorbed.

35. A sample of an ideal gas at 15.0 atm and 10.0 L is allowed to expand against a constant external pressure of 2.00 atm at a constant temperature. Calculate the work in units of kJ for the gas expansion. (*Hint:* Boyle's law applies.)

36. A piston performs work of 210. L · atm on the surroundings, while the cylinder in which it is placed expands from 10. L to 25 L. At the same time, 45 J of heat is transferred from the surroundings to the system. Against what pressure was the piston working?

37. Consider a mixture of air and gasoline vapor in a cylinder with a piston. The original volume is 40. cm³. If the combustion of this mixture releases 950. J of energy, to what volume will the gases expand against a constant pressure of 650. torr if all the energy of combustion is converted into work to push back the piston?

38. As a system increases in volume, it absorbs 52.5 J of energy in the form of heat from the surroundings. The piston is working against a pressure of 0.500 atm. The final volume of the system is 58.0 L. What was the initial volume of the system if the internal energy of the system decreased by 102.5 J?

39. A balloon filled with 39.1 moles of helium has a volume of 876 L at 0.0°C and 1.00 atm pressure. The temperature of the balloon is increased to 38.0°C as it expands to a volume of

998 L, the pressure remaining constant. Calculate q, w, and ΔE for the helium in the balloon. (The molar heat capacity for helium gas is 20.8 J/°C · mol.)

40. One mole of $H_2O(g)$ at 1.00 atm and 100.°C occupies a volume of 30.6 L. When 1 mole of $H_2O(g)$ is condensed to 1 mole of $H_2O(l)$ at 1.00 atm and 100.°C, 40.66 kJ of heat is released. If the density of $H_2O(l)$ at this temperature and pressure is 0.996 g/cm^3, calculate ΔE for the condensation of 1 mole of water at 1.00 atm and 100.°C.

Properties of Enthalpy

41. One of the components of polluted air is NO. It is formed in the high-temperature environment of internal combustion engines by the following reaction:

$$N_2(g) + O_2(g) \longrightarrow 2NO(g) \qquad \Delta H = 180 \text{ kJ}$$

Why are high temperatures needed to convert N_2 and O_2 to NO?

42. The reaction

$$SO_3(g) + H_2O(l) \longrightarrow H_2SO_4(aq)$$

is the last step in the commercial production of sulfuric acid. The enthalpy change for this reaction is -227 kJ. In designing a sulfuric acid plant, is it necessary to provide for heating or cooling of the reaction mixture? Explain.

43. Are the following processes exothermic or endothermic?
   a. When solid KBr is dissolved in water, the solution gets colder.
   b. Natural gas (CH_4) is burned in a furnace.
   c. When concentrated H_2SO_4 is added to water, the solution gets very hot.
   d. Water is boiled in a teakettle.

44. Are the following processes exothermic or endothermic?
   a. the combustion of gasoline in a car engine
   b. water condensing on a cold pipe
   c. $CO_2(s) \longrightarrow CO_2(g)$
   d. $F_2(g) \longrightarrow 2F(g)$

45. The overall reaction in a commercial heat pack can be represented as

$$4Fe(s) + 3O_2(g) \longrightarrow 2Fe_2O_3(s) \qquad \Delta H = -1652 \text{ kJ}$$

   a. How much heat is released when 4.00 moles of iron are reacted with excess O_2?
   b. How much heat is released when 1.00 mole of Fe_2O_3 is produced?
   c. How much heat is released when 1.00 g iron is reacted with excess O_2?
   d. How much heat is released when 10.0 g Fe and 2.00 g O_2 are reacted?

46. Consider the following reaction:

$$2H_2(g) + O_2(g) \longrightarrow 2H_2O(l) \qquad \Delta H = -572 \text{ kJ}$$

   a. How much heat is evolved for the production of 1.00 mole of $H_2O(l)$?
   b. How much heat is evolved when 4.03 g hydrogen are reacted with excess oxygen?
   c. How much heat is evolved when 186 g oxygen are reacted with excess hydrogen?

   d. The total volume of hydrogen gas needed to fill the *Hindenburg* was 2.0×10^8 L at 1.0 atm and 25°C. How much heat was evolved when the *Hindenburg* exploded, assuming all of the hydrogen reacted?

47. Consider the combustion of propane:

$$C_3H_8(g) + 5O_2(g) \longrightarrow 3CO_2(g) + 4H_2O(l) \qquad \Delta H = -2221 \text{ kJ}$$

Assume that all the heat in Example 6.3 comes from the combustion of propane. What mass of propane must be burned to furnish this amount of energy assuming the heat transfer process is 60.% efficient?

48. Consider the following reaction:

$$CH_4(g) + 2O_2(g) \longrightarrow CO_2(g) + 2H_2O(l) \qquad \Delta H = -891 \text{ kJ}$$

Calculate the enthalpy change for each of the following cases:
   a. 1.00 g methane is burned in excess oxygen.
   b. 1.00×10^3 L methane gas at 740. torr and 25°C are burned in excess oxygen.

49. For the process $H_2O(l) \longrightarrow H_2O(g)$ at 298 K and 1.0 atm, ΔH is more positive than ΔE by 2.5 kJ/mol. What does the 2.5 kJ/mol quantity represent?

50. For the following reactions at constant pressure, predict if $\Delta H > \Delta E$, $\Delta H < \Delta E$, or $\Delta H = \Delta E$.
   a. $2HF(g) \longrightarrow H_2(g) + F_2(g)$
   b. $N_2(g) + 3H_2(g) \longrightarrow 2NH_3(g)$
   c. $4NH_3(g) + 5O_2(g) \longrightarrow 4NO(g) + 6H_2O(g)$

Calorimetry and Heat Capacity

51. Consider the substances in Table 6.1. Which substance requires the largest amount of energy to raise the temperature of 25.0 g of the substance from 15.0°C to 37.0°C? Calculate the energy. Which substance in Table 6.1 has the largest temperature change when 550. g of the substance absorbs 10.7 kJ of energy? Calculate the temperature change.

52. The specific heat capacity of silver is 0.24 J/°C · g.
   a. Calculate the energy required to raise the temperature of 150.0 g Ag from 273 K to 298 K.
   b. Calculate the energy required to raise the temperature of 1.0 mole of Ag by 1.0°C (called the *molar heat capacity* of silver).
   c. It takes 1.25 kJ of energy to heat a sample of pure silver from 12.0°C to 15.2°C. Calculate the mass of the sample of silver.

53. A 5.00-g sample of one of the substances listed in Table 6.1 was heated from 25.2°C to 55.1°C, requiring 133 J to do so. Which substance was it?

54. It takes 585 J of energy to raise the temperature of 125.6 g mercury from 20.0°C to 53.5°C. Calculate the specific heat capacity and the molar heat capacity of mercury.

55. A 30.0-g sample of water at 280. K is mixed with 50.0 g water at 330. K. Calculate the final temperature of the mixture assuming no heat loss to the surroundings.

56. A biology experiment requires the preparation of a water bath at 37.0°C (body temperature). The temperature of the cold tap water is 22.0°C, and the temperature of the hot tap water is

55.0°C. If a student starts with 90.0 g cold water, what mass of hot water must be added to reach 37.0°C?

57. A 5.00-g sample of aluminum pellets (specific heat capacity = 0.89 J/°C · g) and a 10.00-g sample of iron pellets (specific heat capacity = 0.45 J/°C · g) are heated to 100.0°C. The mixture of hot iron and aluminum is then dropped into 97.3 g water at 22.0°C. Calculate the final temperature of the metal and water mixture, assuming no heat loss to the surroundings.

58. Hydrogen gives off 120. J/g of energy when burned in oxygen, and methane gives off 50. J/g under the same circumstances. If a mixture of 5.0 g hydrogen and 10. g methane is burned, and the heat released is transferred to 50.0 g water at 25.0°C, what final temperature will be reached by the water?

59. A 150.0-g sample of a metal at 75.0°C is added to 150.0 g H$_2$O at 15.0°C. The temperature of the water rises to 18.3°C. Calculate the specific heat capacity of the metal, assuming that all the heat lost by the metal is gained by the water.

60. A 110.-g sample of copper (specific heat capacity = 0.20 J/°C · g) is heated to 82.4°C and then placed in a container of water at 22.3°C. The final temperature of the water and copper is 24.9°C. What is the mass of the water in the container, assuming that all the heat lost by the copper is gained by the water?

61. In a coffee-cup calorimeter, 50.0 mL of 0.100 M AgNO$_3$ and 50.0 mL of 0.100 M HCl are mixed to yield the following reaction:

$$Ag^+(aq) + Cl^-(aq) \longrightarrow AgCl(s)$$

The two solutions were initially at 22.60°C, and the final temperature is 23.40°C. Calculate the heat that accompanies this reaction in kJ/mol of AgCl formed. Assume that the combined solution has a mass of 100.0 g and a specific heat capacity of 4.18 J/°C · g.

62. In a coffee-cup calorimeter, 100.0 mL of 1.0 M NaOH and 100.0 mL of 1.0 M HCl are mixed. Both solutions were originally at 24.6°C. After the reaction, the final temperature is 31.3°C. Assuming that all the solutions have a density of 1.0 g/cm^3 and a specific heat capacity of 4.18 J/°C · g, calculate the enthalpy change for the neutralization of HCl by NaOH. Assume that no heat is lost to the surroundings or to the calorimeter.

63. A coffee-cup calorimeter initially contains 125 g water at 24.2°C. Potassium bromide (10.5 g), also at 24.2°C, is added to the water, and after the KBr dissolves, the final temperature is 21.1°C. Calculate the enthalpy change for dissolving the salt in J/g and kJ/mol. Assume that the specific heat capacity of the solution is 4.18 J/°C · g and that no heat is transferred to the surroundings or to the calorimeter.

64. In a coffee-cup calorimeter, 1.60 g NH$_4$NO$_3$ is mixed with 75.0 g water at an initial temperature of 25.00°C. After dissolution of the salt, the final temperature of the calorimeter contents is 23.34°C. Assuming the solution has a heat capacity of 4.18 J/°C · g and assuming no heat loss to the calorimeter, calculate the enthalpy change for the dissolution of NH$_4$NO$_3$ in units of kJ/mol.

65. Consider the dissolution of CaCl$_2$:

$$CaCl_2(s) \longrightarrow Ca^{2+}(aq) + 2Cl^-(aq) \qquad \Delta H = -81.5 \text{ kJ}$$

An 11.0-g sample of CaCl$_2$ is dissolved in 125 g water, with both substances at 25.0°C. Calculate the final temperature of the solution assuming no heat loss to the surroundings and assuming the solution has a specific heat capacity of 4.18 J/°C · g.

66. Consider the reaction

$$2HCl(aq) + Ba(OH)_2(aq) \longrightarrow BaCl_2(aq) + 2H_2O(l)$$
$$\Delta H = -118 \text{ kJ}$$

Calculate the heat when 100.0 mL of 0.500 M HCl is mixed with 300.0 mL of 0.100 M Ba(OH)$_2$. Assuming that the temperature of both solutions was initially 25.0°C and that the final mixture has a mass of 400.0 g and a specific heat capacity of 4.18 J/°C · g, calculate the final temperature of the mixture.

67. The heat capacity of a bomb calorimeter was determined by burning 6.79 g methane (energy of combustion = −802 kJ/mol CH$_4$) in the bomb. The temperature changed by 10.8°C.

   a. What is the heat capacity of the bomb?

   b. A 12.6-g sample of acetylene, C$_2$H$_2$, produced a temperature increase of 16.9°C in the same calorimeter. What is the energy of combustion of acetylene (in kJ/mol)?

68. The combustion of 0.1584 g benzoic acid increases the temperature of a bomb calorimeter by 2.54°C. Calculate the heat capacity of this calorimeter. (The energy released by combustion of benzoic acid is 26.42 kJ/g.) A 0.2130-g sample of vanillin (C$_8$H$_8$O$_3$) is then burned in the same calorimeter, and the temperature increases by 3.25°C. What is the energy of combustion per gram of vanillin? Per mole of vanillin?

Hess's Law

69. The enthalpy of combustion of solid carbon to form carbon dioxide is −393.7 kJ/mol carbon, and the enthalpy of combustion of carbon monoxide to form carbon dioxide is −283.3 kJ/mol CO. Use these data to calculate ΔH for the reaction

$$2C(s) + O_2(g) \longrightarrow 2CO(g)$$

70. Combustion reactions involve reacting a substance with oxygen. When compounds containing carbon and hydrogen are combusted, carbon dioxide and water are the products. Using the enthalpies of combustion for C$_4$H$_4$ (−2341 kJ/mol), C$_4$H$_8$ (−2755 kJ/mol), and H$_2$ (−286 kJ/mol), calculate ΔH for the reaction

$$C_4H_4(g) + 2H_2(g) \longrightarrow C_4H_8(g)$$

71. Given the following data

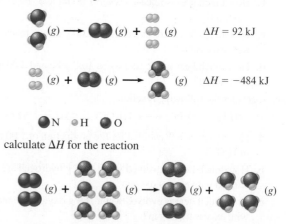

calculate ΔH for the reaction

On the basis of the enthalpy change, is this a useful reaction for the synthesis of ammonia?

72. Given the following data

$$2ClF(g) + O_2(g) \longrightarrow Cl_2O(g) + F_2O(g) \qquad \Delta H = 167.4 \text{ kJ}$$
$$2ClF_3(g) + 2O_2(g) \longrightarrow Cl_2O(g) + 3F_2O(g) \qquad \Delta H = 341.4 \text{ kJ}$$
$$2F_2(g) + O_2(g) \longrightarrow 2F_2O(g) \qquad \Delta H = -43.4 \text{ kJ}$$

calculate ΔH for the reaction

$$ClF(g) + F_2(g) \longrightarrow ClF_3(g)$$

73. Given the following data

$$2O_3(g) \longrightarrow 3O_2(g) \qquad \Delta H = -427 \text{ kJ}$$
$$O_2(g) \longrightarrow 2O(g) \qquad \Delta H = 495 \text{ kJ}$$
$$NO(g) + O_3(g) \longrightarrow NO_2(g) + O_2(g) \qquad \Delta H = -199 \text{ kJ}$$

calculate ΔH for the reaction

$$NO(g) + O(g) \longrightarrow NO_2(g)$$

74. Calculate ΔH for the reaction

$$N_2H_4(l) + O_2(g) \longrightarrow N_2(g) + 2H_2O(l)$$

given the following data:

$$2NH_3(g) + 3N_2O(g) \longrightarrow 4N_2(g) + 3H_2O(l) \qquad \Delta H = -1010. \text{ kJ}$$
$$N_2O(g) + 3H_2(g) \longrightarrow N_2H_4(l) + H_2O(l) \qquad \Delta H = -317 \text{ kJ}$$
$$2NH_3(g) + \tfrac{1}{2}O_2(g) \longrightarrow N_2H_4(l) + H_2O(l) \qquad \Delta H = -143 \text{ kJ}$$
$$H_2(g) + \tfrac{1}{2}O_2(g) \longrightarrow H_2O(l) \qquad \Delta H = -286 \text{ kJ}$$

75. Given the following data

$$Ca(s) + 2C(graphite) \longrightarrow CaC_2(s) \qquad \Delta H = -62.8 \text{ kJ}$$
$$Ca(s) + \tfrac{1}{2}O_2(g) \longrightarrow CaO(s) \qquad \Delta H = -635.5 \text{ kJ}$$
$$CaO(s) + H_2O(l) \longrightarrow Ca(OH)_2(aq) \qquad \Delta H = -653.1 \text{ kJ}$$
$$C_2H_2(g) + \tfrac{5}{2}O_2(g) \longrightarrow 2CO_2(g) + H_2O(l) \qquad \Delta H = -1300. \text{ kJ}$$
$$C(graphite) + O_2(g) \longrightarrow CO_2(g) \qquad \Delta H = -393.5 \text{ kJ}$$

calculate ΔH for the reaction

$$CaC_2(s) + 2H_2O(l) \longrightarrow Ca(OH)_2(aq) + C_2H_2(g)$$

76. Given the following data

$$P_4(s) + 6Cl_2(g) \longrightarrow 4PCl_3(g) \qquad \Delta H = -1225.6 \text{ kJ}$$
$$P_4(s) + 5O_2(g) \longrightarrow P_4O_{10}(s) \qquad \Delta H = -2967.3 \text{ kJ}$$
$$PCl_3(g) + Cl_2(g) \longrightarrow PCl_5(g) \qquad \Delta H = -84.2 \text{ kJ}$$
$$PCl_3(g) + \tfrac{1}{2}O_2(g) \longrightarrow Cl_3PO(g) \qquad \Delta H = -285.7 \text{ kJ}$$

calculate ΔH for the reaction

$$P_4O_{10}(s) + 6PCl_5(g) \longrightarrow 10Cl_3PO(g)$$

Standard Enthalpies of Formation

77. Give the definition of the standard enthalpy of formation for a substance. Write separate reactions for the formation of NaCl, H_2O, $C_6H_{12}O_6$, and $PbSO_4$ that have $\Delta H°$ values equal to $\Delta H_f°$ for each compound.

78. Write reactions for which the enthalpy change will be

a. $\Delta H_f°$ for solid aluminum oxide.

b. the standard enthalpy of combustion of liquid ethanol, $C_2H_5OH(l)$.

c. the standard enthalpy of neutralization of sodium hydroxide solution by hydrochloric acid.

d. $\Delta H_f°$ for gaseous vinyl chloride, $C_2H_3Cl(g)$.

e. the enthalpy of combustion of liquid benzene, $C_6H_6(l)$.

f. the enthalpy of solution of solid ammonium bromide.

79. Use the values of $\Delta H_f°$ in Appendix 4 to calculate $\Delta H°$ for the following reactions.

a.

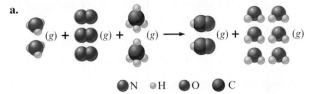

● N ○ H ● O ● C

b. $Ca_3(PO_4)_2(s) + 3H_2SO_4(l) \longrightarrow 3CaSO_4(s) + 2H_3PO_4(l)$

c. $NH_3(g) + HCl(g) \longrightarrow NH_4Cl(s)$

80. Use the values of $\Delta H_f°$ in Appendix 4 to calculate $\Delta H°$ for the following reactions. (See Exercise 79.)

a.

b. $SiCl_4(l) + 2H_2O(l) \longrightarrow SiO_2(s) + 4HCl(aq)$

c. $MgO(s) + H_2O(l) \longrightarrow Mg(OH)_2(s)$

81. The Ostwald process for the commercial production of nitric acid from ammonia and oxygen involves the following steps:

$$4NH_3(g) + 5O_2(g) \longrightarrow 4NO(g) + 6H_2O(g)$$
$$2NO(g) + O_2(g) \longrightarrow 2NO_2(g)$$
$$3NO_2(g) + H_2O(l) \longrightarrow 2HNO_3(aq) + NO(g)$$

a. Use the values of $\Delta H_f°$ in Appendix 4 to calculate the value of $\Delta H°$ for each of the preceding reactions.

b. Write the overall equation for the production of nitric acid by the Ostwald process by combining the preceding equations. (Water is also a product.) Is the overall reaction exothermic or endothermic?

82. Calculate $\Delta H°$ for each of the following reactions using the data in Appendix 4:

$$4Na(s) + O_2(g) \longrightarrow 2Na_2O(s)$$
$$2Na(s) + 2H_2O(l) \longrightarrow 2NaOH(aq) + H_2(g)$$
$$2Na(s) + CO_2(g) \longrightarrow Na_2O(s) + CO(g)$$

Explain why a water or carbon dioxide fire extinguisher might not be effective in putting out a sodium fire.

83. The reusable booster rockets of the space shuttle use a mixture of aluminum and ammonium perchlorate as fuel. A possible reaction is

$$3Al(s) + 3NH_4ClO_4(s) \longrightarrow$$
$$Al_2O_3(s) + AlCl_3(s) + 3NO(g) + 6H_2O(g)$$

Calculate $\Delta H°$ for this reaction.

84. The space shuttle *Orbiter* utilizes the oxidation of methylhydrazine by dinitrogen tetroxide for propulsion:

$$4N_2H_3CH_3(l) + 5N_2O_4(l) \longrightarrow 12H_2O(g) + 9N_2(g) + 4CO_2(g)$$

Calculate $\Delta H°$ for this reaction.

85. Consider the reaction

$$2ClF_3(g) + 2NH_3(g) \longrightarrow N_2(g) + 6HF(g) + Cl_2(g)$$
$$\Delta H° = -1196 \text{ kJ}$$

Calculate $\Delta H_f°$ for $ClF_3(g)$.

86. The standard enthalpy of combustion of ethene gas, $C_2H_4(g)$, is -1411.1 kJ/mol at 298 K. Given the following enthalpies of formation, calculate $\Delta H_f°$ for $C_2H_4(g)$.

$CO_2(g)$	-393.5 kJ/mol
$H_2O(l)$	-285.8 kJ/mol

Energy Consumption and Sources

87. Water gas is produced from the reaction of steam with coal:

$$C(s) + H_2O(g) \longrightarrow H_2(g) + CO(g)$$

Assuming that coal is pure graphite, calculate $\Delta H°$ for this reaction.

88. Syngas can be burned directly or converted to methanol. Calculate $\Delta H°$ for the reaction

$$CO(g) + 2H_2(g) \longrightarrow CH_3OH(l)$$

89. Ethanol (C_2H_5OH) has been proposed as an alternative fuel. Calculate the standard enthalpy of combustion per gram of liquid ethanol.

90. Methanol (CH_3OH) has also been proposed as an alternative fuel. Calculate the standard enthalpy of combustion per gram of liquid methanol, and compare this answer to that for ethanol in Exercise 89.

91. Some automobiles and buses have been equipped to burn propane (C_3H_8). Compare the amounts of energy that can be obtained per gram of $C_3H_8(g)$ and per gram of gasoline, assuming that gasoline is pure octane, $C_8H_{18}(l)$. (See Example 6.11.) Look up the boiling point of propane. What disadvantages are there to using propane instead of gasoline as a fuel?

92. Acetylene (C_2H_2) and butane (C_4H_{10}) are gaseous fuels with enthalpies of combustion of -49.9 kJ/g and -49.5 kJ/g, respectively. Compare the energy available from the combustion of a given volume of acetylene to the combustion energy from the same volume of butane at the same temperature and pressure.

93. Assume that 4.19×10^6 kJ of energy is needed to heat a home. If this energy is derived from the combustion of methane (CH_4), what volume of methane, measured at STP, must be burned? ($\Delta H°_{combustion}$ for $CH_4 = -891$ kJ/mol)

94. The complete combustion of acetylene, $C_2H_2(g)$, produces 1300. kJ of energy per mole of acetylene consumed. How many grams of acetylene must be burned to produce enough heat to raise the temperature of 1.00 gal water by 10.0°C if the process is 80.0% efficient? Assume the density of water is 1.00 g/cm³.

Additional Exercises

95. It has been determined that the body can generate 5500 kJ of energy during one hour of strenuous exercise. Perspiration is the body's mechanism for eliminating this heat. What mass of water would have to be evaporated through perspiration to rid

the body of the heat generated during 2 hours of exercise? (The heat of vaporization of water is 40.6 kJ/mol.)

96. One way to lose weight is to exercise! Walking briskly at 4.0 miles per hour for an hour consumes about 400 kcal of energy. How many hours would you have to walk at 4.0 miles per hour to lose one pound of body fat? One gram of body fat is equivalent to 7.7 kcal of energy. There are 454 g in 1 lb.

97. Three gas-phase reactions were run in a constant-pressure piston apparatus as shown in the following illustration. For each reaction, give the balanced reaction and predict the sign of w (the work done) for the reaction.

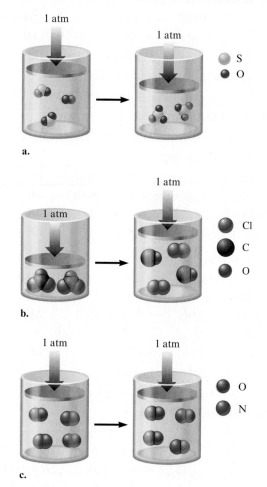

If just the balanced reactions were given, how could you predict the sign of w for a reaction?

98. Nitrogen gas reacts with hydrogen gas to form ammonia gas. Consider the reaction between nitrogen and hydrogen as depicted below:

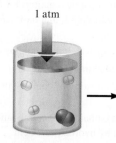

a. Draw what the container will look like after the reaction has gone to completion. Assume a constant pressure of 1 atm.

b. Is the sign of work positive or negative, or is the value of work equal to zero for the reaction? Explain your answer.

99. Combustion of table sugar produces $CO_2(g)$ and $H_2O(l)$. When 1.46 g table sugar is combusted in a constant-volume (bomb) calorimeter, 24.00 kJ of heat is liberated.

a. Assuming that table sugar is pure sucrose, $C_{12}H_{22}O_{11}(s)$, write the balanced equation for the combustion reaction.

b. Calculate ΔE in kJ/mol $C_{12}H_{22}O_{11}$ for the combustion reaction of sucrose.

c. Calculate ΔH in kJ/mol $C_{12}H_{22}O_{11}$ for the combustion reaction of sucrose at 25°C.

100. Consider the following changes:

a. $N_2(g) \longrightarrow N_2(l)$

b. $CO(g) + H_2O(g) \longrightarrow H_2(g) + CO_2(g)$

c. $Ca_3P_2(s) + 6H_2O(l) \longrightarrow 3Ca(OH)_2(s) + 2PH_3(g)$

d. $2CH_3OH(l) + 3O_2(g) \longrightarrow 2CO_2(g) + 4H_2O(l)$

e. $I_2(s) \longrightarrow I_2(g)$

At constant temperature and pressure, in which of these changes is work done by the system on the surroundings? By the surroundings on the system? In which of them is no work done?

101. Consider the following cyclic process carried out in two steps on a gas:

Step 1: 45 J of heat is added to the gas, and 10. J of expansion work is performed.

Step 2: 60. J of heat is removed from the gas as the gas is compressed back to the initial state.

Calculate the work for the gas compression in Step 2.

102. Calculate $\Delta H°$ for the reaction

$$2K(s) + 2H_2O(l) \longrightarrow 2KOH(aq) + H_2(g)$$

A 5.00-g chunk of potassium is dropped into 1.00 kg water at 24.0°C. What is the final temperature of the water after the preceding reaction occurs? Assume that all the heat is used to raise the temperature of the water. (Never run this reaction. It is very dangerous; it bursts into flame!)

103. The enthalpy of neutralization for the reaction of a strong acid with a strong base is −56 kJ/mol water produced. How much energy will be released when 200.0 mL of 0.400 M HNO_3 is mixed with 150.0 mL of 0.500 M KOH?

104. When 1.00 L of 2.00 M Na_2SO_4 solution at 30.0°C is added to 2.00 L of 0.750 M $Ba(NO_3)_2$ solution at 30.0°C in a calorimeter, a white solid ($BaSO_4$) forms. The temperature of the mixture increases to 42.0°C. Assuming that the specific heat capacity of the solution is 6.37 J/°C · g and that the density of the final solution is 2.00 g/mL, calculate the enthalpy change per mole of $BaSO_4$ formed.

105. If a student performs an endothermic reaction in a calorimeter, how does the calculated value of ΔH differ from the actual value if the heat exchanged with the calorimeter is not taken into account?

106. In a bomb calorimeter, the reaction vessel is surrounded by water that must be added for each experiment. Since the amount of water is not constant from experiment to

experiment, the mass of water must be measured in each case. The heat capacity of the calorimeter is broken down into two parts: the water and the calorimeter components. If a calorimeter contains 1.00 kg water and has a total heat capacity of 10.84 kJ/°C, what is the heat capacity of the calorimeter components?

107. The bomb calorimeter in Exercise 106 is filled with 987 g water. The initial temperature of the calorimeter contents is 23.32°C. A 1.056-g sample of benzoic acid ($\Delta E_{comb} = -26.42$ kJ/g) is combusted in the calorimeter. What is the final temperature of the calorimeter contents?

108. Consider the two space shuttle fuel reactions in Exercises 83 and 84. Which reaction produces more energy per kilogram of reactant mixture (stoichiometric amounts)?

109. Consider the following equations:

$$3A + 6B \longrightarrow 3D \qquad \Delta H = -403 \text{ kJ/mol}$$
$$E + 2F \longrightarrow A \qquad \Delta H = -105.2 \text{ kJ/mol}$$
$$C \longrightarrow E + 3D \qquad \Delta H = 64.8 \text{ kJ/mol}$$

Suppose the first equation is reversed and multiplied by $\frac{1}{6}$, the second and third equations are divided by 2, and the three adjusted equations are added. What is the net reaction and what is the overall heat of this reaction?

110. Given the following data

$$Fe_2O_3(s) + 3CO(g) \longrightarrow 2Fe(s) + 3CO_2(g) \qquad \Delta H° = -23 \text{ kJ}$$
$$3Fe_2O_3(s) + CO(g) \longrightarrow 2Fe_3O_4(s) + CO_2(g) \qquad \Delta H° = -39 \text{ kJ}$$
$$Fe_3O_4(s) + CO(g) \longrightarrow 3FeO(s) + CO_2(g) \qquad \Delta H° = 18 \text{ kJ}$$

calculate $\Delta H°$ for the reaction

$$FeO(s) + CO(g) \longrightarrow Fe(s) + CO_2(g)$$

111. At 298 K, the standard enthalpies of formation for $C_2H_2(g)$ and $C_6H_6(l)$ are 227 kJ/mol and 49 kJ/mol, respectively.

a. Calculate $\Delta H°$ for

$$C_6H_6(l) \longrightarrow 3C_2H_2(g)$$

b. Both acetylene (C_2H_2) and benzene (C_6H_6) can be used as fuels. Which compound would liberate more energy per gram when combusted in air?

112. Using the following data, calculate the standard heat of formation of $ICl(g)$ in kJ/mol:

$$Cl_2(g) \longrightarrow 2Cl(g) \qquad \Delta H° = 242.3 \text{ kJ}$$
$$I_2(g) \longrightarrow 2I(g) \qquad \Delta H° = 151.0 \text{ kJ}$$
$$ICl(g) \longrightarrow I(g) + Cl(g) \qquad \Delta H° = 211.3 \text{ kJ}$$
$$I_2(s) \longrightarrow I_2(g) \qquad \Delta H° = 62.8 \text{ kJ}$$

113. A sample of nickel is heated to 99.8°C and placed in a coffee-cup calorimeter containing 150.0 g water at 23.5°C. After the metal cools, the final temperature of metal and water mixture is 25.0°C. If the specific heat capacity of nickel is 0.444 J/°C · g, what mass of nickel was originally heated? Assume no heat loss to the surroundings.

114. Quinone is an important type of molecule that is involved in photosynthesis. The transport of electrons mediated by quinone in certain enzymes allows plants to take water, carbon dioxide, and the energy of sunlight to create glucose. A 0.1964-g sample of quinone ($C_6H_4O_2$) is burned in a bomb calorimeter with a heat capacity of 1.56 kJ/°C. The tempera-

ture of the calorimeter increases by 3.2°C. Calculate the energy of combustion of quinone per gram and per mole.

115. Calculate $\Delta H°$ for each of the following reactions, which occur in the atmosphere.

a. $C_2H_4(g) + O_3(g) \longrightarrow CH_3CHO(g) + O_2(g)$
b. $O_3(g) + NO(g) \longrightarrow NO_2(g) + O_2(g)$
c. $SO_3(g) + H_2O(l) \longrightarrow H_2SO_4(aq)$
d. $2NO(g) + O_2(g) \longrightarrow 2NO_2(g)$

ChemWork Problems

These multiconcept problems (and additional ones) are found interactively online with the same type of assistance a student would get from an instructor.

116. Consider a balloon filled with helium at the following conditions.

313 g He
1.00 atm
1910. L
Molar Heat Capacity = 20.8 J/°C · mol

The temperature of this balloon is decreased by 41.6°C as the volume decreases to 1643 L, with the pressure remaining constant. Determine q, w, and ΔE (in kJ) for the compression of the balloon.

117. In which of the following systems is(are) work done by the surroundings on the system? Assume pressure and temperature are constant.

a. $2SO_2(g) + O_2(g) \longrightarrow 2SO_3(g)$
b. $CO_2(s) \longrightarrow CO_2(g)$
c. $4NH_3(g) + 7O_2(g) \longrightarrow 4NO_2(g) + 6H_2O(g)$
d. $N_2O_4(g) \longrightarrow 2NO_2(g)$
e. $CaCO_3(s) \longrightarrow CaCO(s) + CO_2(g)$

118. Which of the following processes are exothermic?

a. $N_2(g) \longrightarrow 2N(g)$
b. $H_2O(l) \longrightarrow H_2O(s)$
c. $Cl_2(g) \longrightarrow 2Cl(g)$
d. $2H_2(g) + O_2(g) \longrightarrow 2H_2O(g)$
e. $O_2(g) \longrightarrow 2O(g)$

119. Consider the reaction

$$B_2H_6(g) + 3O_2(g) \longrightarrow B_2O_3(s) + 3H_2O(g) \qquad \Delta H = -2035 \text{ kJ}$$

Calculate the amount of heat released when 54.0 g of diborane is combusted.

120. A swimming pool, 10.0 m by 4.0 m, is filled with water to a depth of 3.0 m at a temperature of 20.2°C. How much energy is required to raise the temperature of the water to 24.6°C?

121. In a coffee-cup calorimeter, 150.0 mL of 0.50 M HCl is added to 50.0 mL of 1.00 M NaOH to make 200.0 g solution at an initial temperature of 48.2°C. If the enthalpy of neutralization for the reaction between a strong acid and a strong base is −56 kJ/mol, calculate the final temperature of the calorimeter contents. Assume the specific heat capacity of the solution is 4.184 J/g · °C and assume no heat loss to the surroundings.

122. Calculate ΔH for the reaction

$$N_2H_4(l) + O_2(g) \longrightarrow N_2(g) + 2H_2O(l)$$

given the following data:

Equation	ΔH (kJ)
$2NH_3(g) + 3N_2O(g) \longrightarrow 4N_2(g) + 3H_2O(l)$	−1010
$N_2O(g) + 3H_2(g) \longrightarrow N_2H_4(l) + H_2O(l)$	−317
$2NH_3(g) + \frac{1}{2}O_2(g) \longrightarrow N_2H_4(l) + H_2O(l)$	−143
$H_2(g) + \frac{1}{2}O_2(g) \longrightarrow H_2O(l)$	−286

123. Which of the following substances have an enthalpy of formation equal to zero?

a. $Cl_2(g)$
b. $H_2(g)$
c. $N_2(l)$
d. $Cl(g)$

Challenge Problems

124. Consider 2.00 moles of an ideal gas that are taken from state A ($P_A = 2.00$ atm, $V_A = 10.0$ L) to state B ($P_B = 1.00$ atm, $V_B = 30.0$ L) by two different pathways:

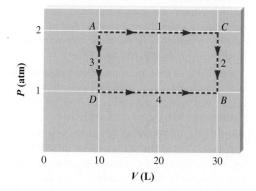

These pathways are summarized on the following graph of P versus V:

Calculate the work (in units of J) associated with the two pathways. Is work a state function? Explain.

125. Calculate w and ΔE when 1 mole of a liquid is vaporized at its boiling point (80.°C) and 1.00 atm pressure. ΔH_{vap} for the liquid is 30.7 kJ/mol at 80.°C.

126. The sun supplies energy at a rate of about 1.0 kilowatt per square meter of surface area (1 watt = 1 J/s). The plants in an agricultural field produce the equivalent of 20. kg sucrose

($C_{12}H_{22}O_{11}$) per hour per hectare (1 ha = 10,000 m²). Assuming that sucrose is produced by the reaction

$$12CO_2(g) + 11H_2O(l) \longrightarrow C_{12}H_{22}O_{11}(s) + 12O_2(g)$$
$$\Delta H = 5640 \text{ kJ}$$

calculate the percentage of sunlight used to produce the sucrose—that is, determine the efficiency of photosynthesis.

127. The best solar panels currently available are about 19% efficient in converting sunlight to electricity. A typical home will use about 40. kWh of electricity per day (1 kWh = 1 kilowatt hour; 1 kW = 1000 J/s). Assuming 8.0 hours of useful sunlight per day, calculate the minimum solar panel surface area necessary to provide all of a typical home's electricity. (See Exercise 126 for the energy rate supplied by the sun.)

128. On Easter Sunday, April 3, 1983, nitric acid spilled from a tank car near downtown Denver, Colorado. The spill was neutralized with sodium carbonate:

$$2HNO_3(aq) + Na_2CO_3(s) \longrightarrow 2NaNO_3(aq) + H_2O(l) + CO_2(g)$$

a. Calculate $\Delta H°$ for this reaction. Approximately 2.0×10^4 gal nitric acid was spilled. Assume that the acid was an aqueous solution containing 70.0% HNO_3 by mass with a density of 1.42 g/cm³. What mass of sodium carbonate was required for complete neutralization of the spill, and what quantity of heat was evolved? ($\Delta H_f°$ for $NaNO_3(aq) = -467$ kJ/mol)

b. According to *The Denver Post* for April 4, 1983, authorities feared that dangerous air pollution might occur during the neutralization. Considering the magnitude of $\Delta H°$, what was their major concern?

129. A piece of chocolate cake contains about 400 Calories. A nutritional Calorie is equal to 1000 calories (thermochemical calories), which is equal to 4.184 kJ. How many 8-in-high steps must a 180-lb man climb to expend the 400 Cal from the piece of cake? See Exercise 28 for the formula for potential energy.

130. The standard enthalpy of formation of $H_2O(l)$ at 298 K is -285.8 kJ/mol. Calculate the change in internal energy for the following process at 298 K and 1 atm:

$$H_2O(l) \longrightarrow H_2(g) + \tfrac{1}{2}O_2(g) \qquad \Delta E° = ?$$

(*Hint:* Using the ideal gas equation, derive an expression for work in terms of n, R, and T.)

131. You have a 1.00-mole sample of water at $-30.°C$ and you heat it until you have gaseous water at $140.°C$. Calculate q for the entire process. Use the following data.

Specific heat capacity of ice = 2.03 J/°C · g
Specific heat capacity of water = 4.18 J/°C · g
Specific heat capacity of steam = 2.02 J/°C · g

$H_2O(s) \longrightarrow H_2O(l) \qquad \Delta H_{fusion} = 6.02$ kJ/mol (at 0°C)
$H_2O(l) \longrightarrow H_2O(g) \qquad \Delta H_{vaporization} = 40.7$ kJ/mol (at 100.°C)

132. A 500.0-g sample of an element at 195°C is dropped into an ice–water mixture; 109.5 g ice melts and an ice–water mixture remains. Calculate the specific heat of the element. See Exercise 131 for pertinent information.

Integrative Problems

These problems require the integration of multiple concepts to find the solutions.

133. The preparation of $NO_2(g)$ from $N_2(g)$ and $O_2(g)$ is an endothermic reaction:

$$N_2(g) + O_2(g) \longrightarrow NO_2(g) \text{ (unbalanced)}$$

The enthalpy change of reaction for the balanced equation (with lowest whole-number coefficients) is $\Delta H = 67.7$ kJ. If 2.50×10^2 mL $N_2(g)$ at $100.°C$ and 3.50 atm and 4.50×10^2 mL $O_2(g)$ at $100.°C$ and 3.50 atm are mixed, what amount of heat is necessary to synthesize the maximum yield of $NO_2(g)$?

134. Nitromethane, CH_3NO_2, can be used as a fuel. When the liquid is burned, the (unbalanced) reaction is mainly

$$CH_3NO_2(l) + O_2(g) \longrightarrow CO_2(g) + N_2(g) + H_2O(g)$$

a. The standard enthalpy change of reaction ($\Delta H°_{rxn}$) for the balanced reaction (with lowest whole-number coefficients) is -1288.5 kJ. Calculate $\Delta H_f°$ for nitromethane.

b. A 15.0-L flask containing a sample of nitromethane is filled with O_2 and the flask is heated to $100.°C$. At this temperature, and after the reaction is complete, the total pressure of all the gases inside the flask is 950. torr. If the mole fraction of nitrogen ($\chi_{nitrogen}$) is 0.134 after the reaction is complete, what mass of nitrogen was produced?

135. A cubic piece of uranium metal (specific heat capacity = 0.117 J/°C · g) at 200.0°C is dropped into 1.00 L deuterium oxide ("heavy water," specific heat capacity = 4.211 J/°C · g) at 25.5°C. The final temperature of the uranium and deuterium oxide mixture is 28.5°C. Given the densities of uranium (19.05 g/cm³) and deuterium oxide (1.11 g/mL), what is the edge length of the cube of uranium?

Marathon Problems

These problems are designed to incorporate several concepts and techniques into one situation.

136. Consider a sample containing 5.00 moles of a monatomic ideal gas that is taken from state A to state B by the following two pathways:

Pathway one:
$$P_A = 3.00 \text{ atm} \xrightarrow{\;1\;} P_C = 3.00 \text{ atm}$$
$$V_A = 15.0 \text{ L} \qquad\quad V_C = 55.0 \text{ L}$$
$$\xrightarrow{\;2\;} P_B = 6.00 \text{ atm}$$
$$V_B = 20.0 \text{ L}$$

Pathway two:
$$P_A = 3.00 \text{ atm} \xrightarrow{\;3\;} P_D = 6.00 \text{ atm}$$
$$V_A = 15.0 \text{ L} \qquad\quad V_D = 15.0 \text{ L}$$
$$\xrightarrow{\;4\;} P_B = 6.00 \text{ atm}$$
$$V_B = 20.0 \text{ L}$$

For each step, assume that the external pressure is constant and equals the final pressure of the gas for that step. Calculate q, w, ΔE, and ΔH for each step in kJ, and calculate overall values for each pathway. Explain how the overall values for the two pathways illustrate that ΔE and ΔH are state functions, whereas q

and *w* are path functions. *Hint:* In a more rigorous study of thermochemistry, it can be shown that for an ideal gas:

$$\Delta E = nC_v\Delta T \text{ and}$$

$$\Delta H = nC_p\Delta T$$

where C_v is the molar heat capacity at constant volume and C_p is the molar heat capacity at constant pressure. In addition, for a monotomic ideal gas, $C_v = \frac{3}{2}R$ and $C_p = \frac{5}{2}R$.

137. A gaseous hydrocarbon reacts completely with oxygen gas to form carbon dioxide and water vapor. Given the following data, determine ΔH_f° for the hydrocarbon:

$$\Delta H^\circ_{\text{reaction}} = -2044.5 \text{ kJ/mol hydrocarbon}$$
$$\Delta H_f^\circ(CO_2) = -393.5 \text{ kJ/mol}$$
$$\Delta H_f^\circ(H_2O) = -242 \text{ kJ/mol}$$

Density of CO_2 and H_2O product mixture at 1.00 atm, 200.°C = 0.751 g/L.

The density of the hydrocarbon is less than the density of Kr at the same conditions.

Chapter 7

Atomic Structure and Periodicity

The line spectra of a number of gases. Each spectrum is unique and allows the identification of the elements. (Ted Kinsman/Photo Researchers, Inc.)

In the past 200 years, a great deal of experimental evidence has accumulated to support the atomic model. This theory has proved to be both extremely useful and physically reasonable. When atoms were first suggested by the Greek philosophers Democritus and Leucippus about 400 B.C., the concept was based mostly on intuition. In fact, for the following 20 centuries, no convincing experimental evidence was available to support the existence of atoms. The first real scientific data were gathered by Lavoisier and others from quantitative measurements of chemical reactions. The results of these stoichiometric experiments led John Dalton to propose the first systematic atomic theory. Dalton's theory, although crude, has stood the test of time extremely well.

Once we came to "believe in" atoms, it was logical to ask: What is the nature of an atom? Does an atom have parts, and if so, what are they? In Chapter 2 we considered some of the experiments most important for shedding light on the nature of the atom. Now we will see how the atomic theory has evolved to its present state.

One of the most striking things about the chemistry of the elements is the periodic repetition of properties. There are several groups of elements that show great similarities in chemical behavior. As we saw in Chapter 2, these similarities led to the development of the periodic table of the elements. In this chapter we will see that the modern theory of atomic structure accounts for periodicity in terms of the electron arrangements in atoms.

However, before we examine atomic structure, we must consider the revolution that took place in physics in the first 30 years of the twentieth century. During that time, experiments were carried out, the results of which could not be explained by the theories of classical physics developed by Isaac Newton and many others who followed him. A radical new theory called *quantum mechanics* was developed to account for the behavior of light and atoms. This "new physics" provides many surprises for humans who are used to the macroscopic world, but it seems to account flawlessly (within the bounds of necessary approximations) for the behavior of matter.

As the first step in our exploration of this revolution in science, we will consider the properties of light, more properly called *electromagnetic radiation*.

7.1 | Electromagnetic Radiation

One of the ways that energy travels through space is by **electromagnetic radiation**. The light from the sun, the energy used to cook food in a microwave oven, the X rays used by dentists, and the radiant heat from a fireplace are all examples of electromagnetic radiation. Although these forms of radiant energy seem quite different, they all exhibit the same type of wavelike behavior and travel at the speed of light in a vacuum.

Waves have three primary characteristics: wavelength, frequency, and speed. **Wavelength** (symbolized by the lowercase Greek letter lambda, λ) is the *distance between two consecutive peaks or troughs in a wave* (Fig. 7.1). The **frequency** (symbolized by the lowercase Greek letter nu, ν) is defined as the *number of waves (cycles) per second that pass a given point in space*. Since all types of electromagnetic radiation travel at the speed of light, short-wavelength radiation must have a high frequency. You can see this in Fig. 7.1, where three waves are shown traveling between two points at constant speed. Note that the wave with the shortest wavelength (λ_3) has the highest frequency and the wave with the longest wavelength (λ_1) has the lowest

Wavelength, λ, and frequency, ν, are inversely related.

Figure 7.1 | The nature of waves. Many of the properties of ocean waves are the same as those of light waves. Note that the radiation with the shortest wavelength has the highest frequency.

$\nu_1 = 4$ cycles/second $= 4$ hertz

$\nu_2 = 8$ cycles/second $= 8$ hertz

$\nu_3 = 16$ cycles/second $= 16$ hertz

c = speed of light
$= 2.9979 \times 10^8$ m/s

frequency. This implies an inverse relationship between wavelength and frequency, that is, $\lambda \propto 1/\nu$, or

$$\lambda\nu = c$$

where λ is the wavelength in meters, ν is the frequency in cycles per second, and c is the speed of light (2.9979×10^8 m/s). In the SI system, cycles are understood, and the unit per second becomes 1/s, or s^{-1}, which is called the *hertz* (abbreviated Hz).

Electromagnetic radiation is classified as shown in Fig. 7.2. Radiation provides an important means of energy transfer. For example, the energy from the sun reaches the earth mainly in the form of visible and ultraviolet radiation, whereas the glowing coals of a fireplace transmit heat energy by infrared radiation. In a microwave oven, the water molecules in food absorb microwave radiation, which increases their motions. This energy is then transferred to other types of molecules via collisions, causing an increase in the food's temperature. As we proceed in the study of chemistry, we will consider many of the classes of electromagnetic radiation and the ways in which they affect matter.

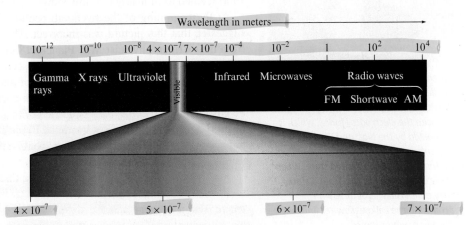

Figure 7.2 | Classification of electromagnetic radiation.

Interactive Example 7.1

Sign in at http://login.cengagebrain .com to try this Interactive Example in OWL.

When a strontium salt is dissolved in methanol (with a little water) and ignited, it gives a brilliant red flame. The red color is produced by emission of light when electrons, excited by the energy of the burning methanol, fall back to their ground states.

Frequency of Electromagnetic Radiation

The brilliant red colors seen in fireworks are due to the emission of light with wavelengths around 650 nm when strontium salts such as $Sr(NO_3)_2$ and $SrCO_3$ are heated. (This can be easily demonstrated in the lab by dissolving one of these salts in methanol that contains a little water and igniting the mixture in an evaporating dish.) Calculate the frequency of red light of wavelength 6.50×10^2 nm.

Solution

We can convert wavelength to frequency using the equation

$$\lambda\nu = c \quad \text{or} \quad \nu = \frac{c}{\lambda}$$

where $c = 2.9979 \times 10^8$ m/s. In this case $\lambda = 6.50 \times 10^2$ nm. Changing the wavelength to meters, we have

$$6.50 \times 10^2 \text{ nm} \times \frac{1 \text{ m}}{10^9 \text{ nm}} = 6.50 \times 10^{-7} \text{ m}$$

and

$$\nu = \frac{c}{\lambda} = \frac{2.9979 \times 10^8 \text{ m/s}}{6.50 \times 10^{-7} \text{ m}} = 4.61 \times 10^{14} \text{ s}^{-1} = 4.61 \times 10^{14} \text{ Hz}$$

See Exercises 7.39 and 7.40

7.2 | The Nature of Matter

It is probably fair to say that at the end of the nineteenth century, physicists were feeling rather smug. Theories could explain phenomena as diverse as the motions of the planets and the dispersion of visible light by a prism. Rumor has it that students were being discouraged from pursuing physics as a career because it was felt that all the major problems had been solved, or at least described in terms of the current physical theories.

At the end of the nineteenth century, the idea prevailed that matter and energy were distinct. Matter was thought to consist of particles, whereas energy in the form of light (electromagnetic radiation) was described as a wave. Particles were things that had mass and whose position in space could be specified. Waves were described as massless and delocalized; that is, their position in space could not be specified. It also was assumed that there was no intermingling of matter and light. Everything known before 1900 seemed to fit neatly into this view.

At the beginning of the twentieth century, however, certain experimental results suggested that this picture was incorrect. The first important advance came in 1900 from the German physicist Max Planck (1858–1947). Studying the radiation profiles emitted by solid bodies heated to incandescence, Planck found that the results could not be explained in terms of the physics of his day, which held that matter could absorb or emit any quantity of energy. Planck could account for these observations only by postulating that energy can be gained or lost only in *whole-number multiples* of the quantity $h\nu$, where h is a constant called **Planck's constant**, determined by experiment to have the value 6.626×10^{-34} J · s. That is, the change in energy for a system, ΔE, can be represented by the equation

$$\Delta E = nh\nu$$

where n is an integer (1, 2, 3, . . .), h is Planck's constant, and ν is the frequency of the electromagnetic radiation absorbed or emitted.

When alternating current at 110 volts is applied to a dill pickle, a glowing discharge occurs. The current flowing between the electrodes (forks), which is supported by the Na^+ and Cl^- ions present, apparently causes some sodium atoms to form in an excited state. When these atoms relax to the ground state, they emit visible light at 589 nm, producing the yellow glow reminiscent of sodium vapor lamps.

Planck's result was a real surprise. It had always been assumed that the energy of matter was continuous, which meant that the transfer of any quantity of energy was possible. Now it seemed clear that energy is in fact **quantized** and can occur only in discrete units of size $h\nu$. Each of these small "packets" of energy is called a *quantum*. A system can transfer energy only in whole quanta. Thus energy seems to have particulate properties.

Energy can be gained or lost only in integer multiples of $h\nu$.

Planck's constant $= 6.626 \times 10^{-34}$ J $\cdot$ s.

The Energy of a Photon

The blue color in fireworks is often achieved by heating copper(I) chloride (CuCl) to about 1200°C. Then the compound emits blue light having a wavelength of 450 nm. What is the increment of energy (the quantum) that is emitted at 4.50×10^2 nm by CuCl?

Solution

The quantum of energy can be calculated from the equation

$$\Delta E = h\nu$$

The frequency ν for this case can be calculated as follows:

$$\nu = \frac{c}{\lambda} = \frac{2.9979 \times 10^8 \text{ m/s}}{4.50 \times 10^{-7} \text{ m}} = 6.66 \times 10^{14} \text{ s}^{-1}$$

So

$$\Delta E = h\nu = (6.626 \times 10^{-34} \text{ J} \cdot \text{s})(6.66 \times 10^{14} \text{ s}^{-1}) = 4.41 \times 10^{-19} \text{ J}$$

A sample of CuCl emitting light at 450 nm can lose energy only in increments of 4.41×10^{-19} J, the size of the quantum in this case.

See Exercises 7.41 and 7.42

The next important development in the knowledge of atomic structure came when Albert Einstein (Fig. 7.3) proposed that electromagnetic radiation is itself quantized. Einstein suggested that electromagnetic radiation can be viewed as a stream of "particles" called **photons**. The energy of each photon is given by the expression

$$E_{\text{photon}} = h\nu = \frac{hc}{\lambda}$$

Figure 7.3 | Albert Einstein (1879–1955) was born in Germany. Nothing in his early development suggested genius; even at the age of 9 he did not speak clearly, and his parents feared that he might be handicapped. When asked what profession Einstein should follow, his school principal replied, "It doesn't matter; he'll never make a success of anything." When he was 10, Einstein entered the Luitpold Gymnasium (high school), which was typical of German schools of that time in being harshly disciplinarian. There he developed a deep suspicion of authority and a skepticism that encouraged him to question and doubt—valuable qualities in a scientist. In 1905, while a patent clerk in Switzerland, Einstein published a paper explaining the photoelectric effect via the quantum theory. For this revolutionary thinking, he received a Nobel Prize in 1921. Highly regarded by this time, he worked in Germany until 1933, when Hitler's persecution of the Jews forced him to come to the United States. He worked at the Institute for Advanced Studies in Princeton, New Jersey, until his death in 1955.

Einstein was undoubtedly the greatest physicist of our age. Even if someone else had derived the theory of relativity, his other work would have ensured his ranking as the second greatest physicist of his time. Our concepts of space and time were radically changed by ideas he first proposed when he was 26 years old. From then until the end of his life, he attempted unsuccessfully to find a single unifying theory that would explain all physical events.

Photo Researchers, Inc./Getty Images

Chemical connections
Fireworks

The art of using mixtures of chemicals to produce explosives is an ancient one. Black powder—a mixture of potassium nitrate, charcoal, and sulfur—was being used in China well before 1000 A.D. and has been used subsequently through the centuries in military explosives, in construction blasting, and for fireworks. The du Pont Company, now a major chemical manufacturer, started out as a manufacturer of black powder. In fact, the founder, Eleuthère du Pont, learned the manufacturing technique from none other than Lavoisier.

Before the nineteenth century, fireworks were confined mainly to rockets and loud bangs. Orange and yellow colors came from the presence of charcoal and iron filings. However, with the great advances in chemistry in the nineteenth century, new compounds found their way into fireworks. Salts of copper, strontium, and barium added brilliant colors. Magnesium and aluminum metals gave a dazzling white light. Fireworks, in fact, have changed very little since then.

How do fireworks produce their brilliant colors and loud bangs? Actually, only a handful of different chemicals are responsible for most of the spectacular effects. To produce the noise and flashes, an oxidizer (an oxidizing agent) and a fuel (a reducing agent) are used. A common mixture involves potassium perchlorate

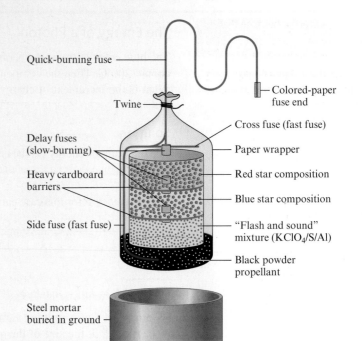

A typical aerial shell used in fireworks displays. Time-delayed fuses cause a shell to explode in stages. In this case a red starburst occurs first, followed by a blue starburst, and finally a flash and loud report.
(Reprinted with permission from *Chemical & Engineering News*, June 29, 1981, p. 24. Copyright © 1981, American Chemical Society.)

($KClO_4$) as the oxidizer and aluminum and sulfur as the fuel. The perchlorate oxidizes the fuel in a very exothermic reaction, which produces a brilliant flash, due to the aluminum, and a loud report from the rapidly expanding gases produced. For a color effect, an element with a colored emission spectrum is included. Recall that the electrons in atoms can be raised to higher-energy orbitals when the atoms absorb energy. The excited atoms can then release this excess energy by emitting light of specific wavelengths, often in the visible region. In fireworks, the energy to excite the electrons comes from the reaction between the oxidizer and fuel.

Yellow colors in fireworks are due to the 589-nm emission of sodium ions. Red colors come from strontium salts emitting at 606 nm and from

where h is Planck's constant, ν is the frequency of the radiation, and λ is the wavelength of the radiation.

The Photoelectric Effect

Einstein arrived at this conclusion through his analysis of the **photoelectric effect** (for which he later was awarded the Nobel Prize). The photoelectric effect refers to the

636 to 688 nm. This red color is familiar from highway safety flares. Barium salts give a green color in fireworks, due to a series of emission lines between 505 and 535 nm. A really good blue color, however, is hard to obtain. Copper salts give a blue color, emitting in the 420- to 460-nm region. But difficulties occur because the oxidizing agent, potassium chlorate ($KClO_3$), reacts with copper salts to form copper chlorate, a highly explosive compound that is dangerous to store. (The use of $KClO_3$ in fireworks has been largely abandoned because of its explosive hazards.) Paris green, a copper salt containing arsenic, was once used extensively but is now considered to be too toxic.

In recent years the colors produced by fireworks have become more intense because of the formation of metal chlorides during the burning process. These gaseous metal chloride molecules produce colors much more brilliant than do the metal atoms by themselves. For example, strontium chloride produces a much brighter red than do strontium atoms. Thus, chlorine-donating compounds are now included in many fireworks shells.

A typical aerial shell is shown in the diagram. The shell is launched from a mortar (a steel cylinder) using black powder as the propellant. Time-delayed fuses are used to fire the shell in stages. A list of chemicals commonly used in fireworks is given in the table.

Although you might think that the chemistry of fireworks is simple, the achievement of the vivid white flashes and the brilliant colors requires complex combinations of chemicals. For example, because the white flashes produce high flame temperatures, the colors tend to wash out. Thus oxidizers such as $KClO_4$ are commonly used with fuels that produce relatively low flame temperatures. An added difficulty, however, is that perchlorates are very sensitive to accidental ignition and are therefore quite hazardous. Another problem arises from the use of sodium salts. Because sodium produces an extremely bright yellow emission, sodium salts cannot be used when other colors are desired. Carbon-based fuels also give a yellow flame that masks other colors, and this limits the use of organic compounds as fuels. You can see that the manufacture of

Fireworks in Washington, D.C.

fireworks that produce the desired effects and are also safe to handle requires careful selection of chemicals. And, of course, there is still the dream of a deep blue flame.

Chemicals Commonly Used in the Manufacture of Fireworks

Oxidizers	Fuels	Special Effects
Potassium nitrate	Aluminum	Red flame: strontium nitrate, strontium carbonate
Potassium chlorate	Magnesium	Green flame: barium nitrate, barium chlorate
Potassium perchlorate	Titanium	Blue flame: copper carbonate, copper sulfate, copper oxide
Ammonium perchlorate	Charcoal	Yellow flame: sodium oxalate, cryolite (Na_3AlF_6)
Barium nitrate	Sulfur	White flame: magnesium, aluminum
Barium chlorate	Antimony sulfide	Gold sparks: iron filings, charcoal
Strontium nitrate	Dextrin	White sparks: aluminum, magnesium, aluminum–magnesium alloy, titanium
	Red gum	Whistle effect: potassium benzoate or sodium salicylate
	Polyvinyl chloride	White smoke: mixture of potassium nitrate and sulfur
		Colored smoke: mixture of potassium chlorate, sulfur, and organic dye

phenomenon in which electrons are emitted from the surface of a metal when light strikes it. The following observations characterize the photoelectric effect.

1. Studies in which the frequency of the light is varied show that no electrons are emitted by a given metal below a specific threshold frequency, ν_0.
2. For light with frequency lower than the threshold frequency, no electrons are emitted regardless of the intensity of the light.

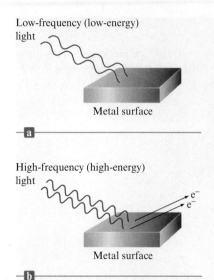

Low-frequency (low-energy) light

Metal surface

a

High-frequency (high-energy) light

Metal surface

b

Figure 7.4 | The photoelectric effect. (a) Light with frequency less than the threshold frequency produces no electrons. (b) Light with frequency higher than the threshold frequency causes electrons to be emitted from the metal.

3. For light with frequency greater than the threshold frequency, the number of electrons emitted increases with the intensity of the light.
4. For light with frequency greater than the threshold frequency, the kinetic energy of the emitted electrons increases linearly with the frequency of the light.

These observations can be explained by assuming that electromagnetic radiation is quantized (consists of photons), and that the threshold frequency represents the minimum energy required to remove the electron from the metal's surface.

$$\text{Minimum energy required to remove an electron} = E_0 = h\nu_0$$

Because a photon with energy less than E_0 ($\nu < \nu_0$) cannot remove an electron, light with a frequency less than the threshold frequency produces no electrons (Fig. 7.4). On the other hand, for light where $\nu > \nu_0$, the energy in excess of that required to remove the electron is given to the electron as kinetic energy (KE):

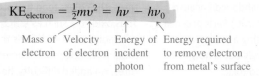

$$KE_{electron} = \tfrac{1}{2}mv^2 = h\nu - h\nu_0$$

Mass of Velocity   Energy of   Energy required
electron of electron incident   to remove electron
photon   from metal's surface

Because in this picture the intensity of light is a measure of the number of photons present in a given part of the beam, a greater intensity means that more photons are available to release electrons (as long as $\nu > \nu_0$ for the radiation).

In a related development, Einstein derived the famous equation

$$E = mc^2$$

in his *special theory of relativity* published in 1905. The main significance of this equation is that *energy has mass.* This is more apparent if we rearrange the equation in the following form:

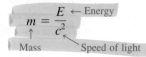

$$m = \frac{E}{c^2}$$

Energy

Mass   Speed of light

Note that the *apparent mass* of a photon depends on its wavelength. A photon does not have mass in a classical sense.

Using this form of the equation, we can calculate the mass associated with a given quantity of energy. For example, we can calculate the *apparent* mass of a photon. For electromagnetic radiation of wavelength λ, the energy of each photon is given by the expression

$$E_{photon} = \frac{hc}{\lambda}$$

Then the apparent mass of a photon of light with wavelength λ is given by

$$m = \frac{E}{c^2} = \frac{hc/\lambda}{c^2} = \frac{h}{\lambda c}$$

The photon behaves as if it has mass under certain circumstances. In 1922 American physicist Arthur Compton (1892–1962) performed experiments involving collisions of X rays and electrons that showed that photons do exhibit the apparent mass calculated from the preceding equation. However, it is clear that photons do not have mass in the classical sense. A photon has mass only in a relativistic sense—it has no rest mass.

We can summarize the important conclusions from the work of Planck and Einstein as follows:

Energy is quantized. It can occur only in discrete units called quanta.

Electromagnetic radiation, which was previously thought to exhibit only wave properties, seems to show certain characteristics of particulate matter as well. This phenomenon is sometimes referred to as the **dual nature of light** (Fig. 7.5).

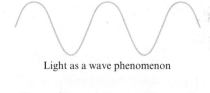

Light as a wave phenomenon

Light as a stream of photons

Figure 7.5 | Electromagnetic radiation exhibits wave properties and particulate properties. The energy of each photon of the radiation is related to the wavelength and frequency by the equation $E_{photon} = h\nu = hc/\lambda$.

Thus light, which previously was thought to be purely wavelike, was found to have certain characteristics of particulate matter. But is the opposite also true? That is, does matter that is normally assumed to be particulate exhibit wave properties? This question was raised in 1923 by a young French physicist named Louis de Broglie (1892–1987). To see how de Broglie supplied the answer to this question, recall that the relationship between mass and wavelength for electromagnetic radiation is $m = h/\lambda c$. For a particle with velocity v, the corresponding expression is

Do not confuse ν (frequency) with v (velocity).

$$m = \frac{h}{\lambda v}$$

Rearranging to solve for λ, we have

$$\lambda = \frac{h}{mv}$$

This equation, called *de Broglie's equation,* allows us to calculate the wavelength for a particle, as shown in Example 7.3.

$ball: \lambda = \frac{E}{V}$

Calculations of Wavelength

Compare the wavelength for an electron (mass = 9.11×10^{-31} kg) traveling at a speed of 1.0×10^7 m/s with that for a ball (mass = 0.10 kg) traveling at 35 m/s.

Solution

We use the equation $\lambda = h/mv$, where

$$h = 6.626 \times 10^{-34} \text{ J} \cdot \text{s} \quad \text{or} \quad 6.626 \times 10^{-34} \text{ kg} \cdot \text{m}^2/\text{s}$$

since

$$1 \text{ J} = 1 \text{ kg} \cdot \text{m}^2/\text{s}^2$$

For the electron,

$$\lambda_e = \frac{6.626 \times 10^{-34} \dfrac{\text{kg} \cdot \text{m} \cdot \text{m}}{\text{s}}}{(9.11 \times 10^{-31} \text{ kg})(1.0 \times 10^7 \text{ m/s})} = 7.27 \times 10^{-11} \text{ m}$$

For the ball,

$$\lambda_b = \frac{6.626 \times 10^{-34} \dfrac{\text{kg} \cdot \text{m} \cdot \text{m}}{\text{s}}}{(0.10 \text{ kg})(35 \text{ m/s})} = 1.9 \times 10^{-34} \text{ m}$$

See Exercises 7.53 through 7.56

Notice from Example 7.3 that the wavelength associated with the ball is incredibly short. On the other hand, the wavelength of the electron, although still quite small, happens to be on the same order as the spacing between the atoms in a typical crystal. This is important because, as we will see presently, it provides a means for testing de Broglie's equation.

Diffraction results when light is scattered from a regular array of points or lines. You may have noticed the diffraction of light from the ridges and grooves of a compact disc. The colors result because the various wavelengths of visible light are not all scattered in the same way. The colors are "separated," giving the same effect as light passing through a prism. Just as a regular arrangement of ridges and grooves produces diffraction, so does a regular array of atoms or ions in a crystal, as shown in the

Figure 7.6 | A diffraction pattern of a beryl crystal. (a) A light area results from constructive interference of the waves. (b) A dark area arises from destructive interference of the waves.

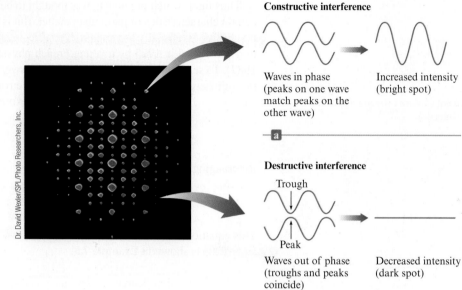

Constructive interference

Waves in phase (peaks on one wave match peaks on the other wave)

Increased intensity (bright spot)

a

Destructive interference

Trough

Peak

Waves out of phase (troughs and peaks coincide)

Decreased intensity (dark spot)

b

photograph on the next page. For example, when X rays are directed onto a crystal of a particular nickel/titanium alloy, the scattered radiation produces a **diffraction pattern** of bright spots and dark areas on a photographic plate (Fig. 7.6). This occurs because the scattered light can interfere constructively (the peaks and troughs of the beams are in phase) to produce a bright area [Fig. 7.6(a)] or destructively (the peaks and troughs are out of phase) to produce a dark spot [Fig. 7.6(b)].

A diffraction pattern can be explained only in terms of waves. Thus, this phenomenon provides a test for the postulate that particles such as electrons have wavelengths. As we saw in Example 7.3, an electron with a velocity of 10^7 m/s (easily achieved by acceleration of the electron in an electric field) has a wavelength of about 10^{-10} m, which is roughly the distance between the ions in a crystal such as sodium chloride. This is important because diffraction occurs most efficiently when the spacing between the scattering points is about the same as the wavelength of the wave being diffracted. Thus, if electrons really do have an associated wavelength, a crystal should diffract electrons. An experiment to test this idea was carried out in 1927 by C. J. Davisson and L. H. Germer at Bell Laboratories. When they directed a beam of electrons at a nickel crystal, they observed a diffraction pattern similar to that seen from the diffraction of X rays. This result verified de Broglie's relationship, at least for electrons. Larger chunks of matter, such as balls, have such small wavelengths (see Example 7.3) that they are impossible to verify experimentally. However, we believe that all matter obeys de Broglie's equation.

Now we have come full circle. Electromagnetic radiation, which at the turn of the twentieth century was thought to be a pure waveform, was found to possess particulate properties. Conversely, electrons, which were thought to be particles, were found to have a wavelength associated with them. The significance of these results is that matter and energy are not distinct. Energy is really a form of matter, and all matter shows the same types of properties. That is, *all matter exhibits both particulate and wave properties.* Large pieces of matter, such as baseballs, exhibit predominantly particulate properties. The associated wavelength is so small that it is not observed. Very small "bits of matter," such as photons, while showing some particulate properties, exhibit predominantly wave properties. Pieces of matter with intermediate mass, such as electrons, show clearly both the particulate and wave properties of matter.

7.3 | The Atomic Spectrum of Hydrogen

A beautiful rainbow.

As we saw in Chapter 2, key information about the atom came from several experiments carried out in the early twentieth century, in particular Thomson's discovery of the electron and Rutherford's discovery of the nucleus. Another important experiment was the study of the emission of light by excited hydrogen atoms. When a sample of hydrogen gas receives a high-energy spark, the H_2 molecules absorb energy, and some of the H—H bonds are broken. The resulting hydrogen atoms are *excited;* that is, they contain excess energy, which they release by emitting light of various wavelengths to produce what is called the *emission spectrum* of the hydrogen atom.

To understand the significance of the hydrogen emission spectrum, we must first describe the **continuous spectrum** that results when white light is passed through a prism [Fig. 7.7(a)]. This spectrum, like the rainbow produced when sunlight is dispersed by raindrops, contains *all* the wavelengths of visible light. In contrast, when the hydrogen emission spectrum in the visible region is passed through a prism [Fig. 7.7(b)], we see only a few lines, each of which corresponds to a discrete wavelength. The hydrogen emission spectrum is called a **line spectrum**.

What is the significance of the line spectrum of hydrogen? It indicates that *only certain energies are allowed for the electron in the hydrogen atom*. In other words, the energy of the electron in the hydrogen atom is *quantized*. This observation ties in perfectly with the postulates of Max Planck discussed in Section 7.2. Changes in energy between discrete energy levels in hydrogen will produce only certain wavelengths of

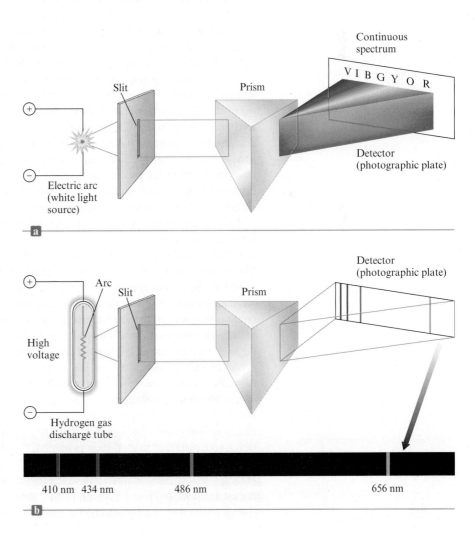

Figure 7.7 | (a) A continuous spectrum containing all wavelengths of visible light (indicated by the initial letters of the colors of the rainbow). (b) The hydrogen line spectrum contains only a few discrete wavelengths.

Spectrum adapted by permission from C. W. Keenan, D. C. Kleinfelter, and J. H. Wood, *General College Chemistry*, Sixth Edition (New York: Harper & Row, 1980).

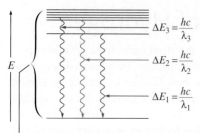

Various energy levels
in the hydrogen atom

Figure 7.8 | A change between two
discrete energy levels emits a photon
of light.

emitted light (Fig. 7.8). For example, a given change in energy from a high to a lower level would give a wavelength of light that can be calculated from Planck's equation:

$$\Delta E = h\nu = \frac{hc}{\lambda}$$

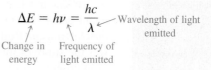

The discrete line spectrum of hydrogen shows that only certain energies are possible; that is, the electron energy levels are quantized. In contrast, if any energy level were allowed, the emission spectrum would be continuous.

Critical Thinking

We now have evidence that electron energy levels in the atoms are quantized. Some of this evidence is discussed in this chapter. What if energy levels in atoms were not quantized? What are some differences we would notice?

7.4 | The Bohr Model

In 1913, a Danish physicist named Niels Bohr (1885–1962), aware of the experimental results we have just discussed, developed a **quantum model** for the hydrogen atom. Bohr proposed that the *electron in a hydrogen atom moves around the nucleus only in certain allowed circular orbits.* He calculated the radii for these allowed orbits by using the theories of classical physics and by making some new assumptions.

From classical physics Bohr knew that a particle in motion tends to move in a straight line and can be made to travel in a circle only by application of a force toward the center of the circle. Thus Bohr reasoned that the tendency of the revolving electron to fly off the atom must be just balanced by its attraction for the positively charged nucleus. But classical physics also decreed that a charged particle under acceleration should radiate energy. Since an electron revolving around the nucleus constantly changes its direction, it is constantly accelerating. Therefore, the electron should emit light and lose energy—and thus be drawn into the nucleus. This, of course, does not correlate with the existence of stable atoms.

Clearly, an atomic model based solely on the theories of classical physics was untenable. Bohr also knew that the correct model had to account for the experimental spectrum of hydrogen, which showed that only certain electron energies were allowed. The experimental data were absolutely clear on this point. Bohr found that his model would fit the experimental results if he assumed that the angular momentum of the electron (angular momentum equals the product of mass, velocity, and orbital radius) could occur only in certain increments. It was not clear why this should be true, but with this assumption, Bohr's model gave hydrogen atom energy levels consistent with the hydrogen emission spectrum. The model is represented pictorially in Fig. 7.9.

Although we will not show the derivation here, the most important equation to come from Bohr's model is the expression for the *energy levels available to the electron in the hydrogen atom:*

The J in Equation (7.1) stands for joules.

$$E = -2.178 \times 10^{-18} \text{ J}\left(\frac{Z^2}{n^2}\right) \tag{7.1}$$

in which n is an integer (the larger the value of n, the larger the orbit radius) and Z is the nuclear charge. Using Equation (7.1), Bohr was able to calculate hydrogen atom energy levels that exactly matched the values obtained by experiment.

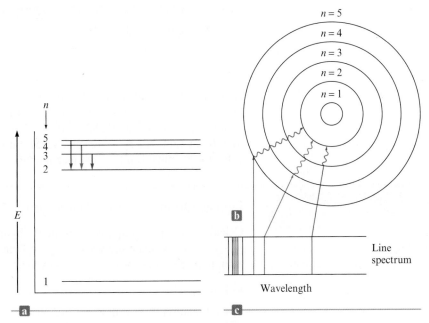

Figure 7.9 | Electronic transitions in the Bohr model for the hydrogen atom. (a) An energy-level diagram for electronic transitions. (b) An orbit-transition diagram, which accounts for the experimental spectrum. (Note that the orbits shown are schematic. They are not drawn to scale.) (c) The resulting line spectrum on a photographic plate is shown. Note that the lines in the visible region of the spectrum correspond to transitions from higher levels to the $n = 2$ level.

Niels Hendrik David Bohr (1885–1962) as a boy lived in the shadow of his younger brother Harald, who played on the 1908 Danish Olympic Soccer Team and later became a distinguished mathematician. In school, Bohr received his poorest marks in composition and struggled with writing during his entire life. In fact, he wrote so poorly that he was forced to dictate his Ph.D. thesis to his mother. Nevertheless, Bohr was a brilliant physicist. After receiving his Ph.D. in Denmark, he constructed a quantum model for the hydrogen atom by the time he was 27. Even though his model later proved to be incorrect, Bohr remained a central figure in the drive to understand the atom. He was awarded the Nobel Prize in physics in 1922.

The negative sign in Equation (7.1) simply means that the energy of the electron bound to the nucleus is lower than it would be if the electron were at an infinite distance ($n = \infty$) from the nucleus, where there is no interaction and the energy is zero:

$$E = -2.178 \times 10^{-18} \text{ J} \left(\frac{Z^2}{\infty} \right) = 0$$

The energy of the electron in any orbit is negative relative to this reference state.

Equation (7.1) can be used to calculate the change in energy of an electron when the electron changes orbits. For example, suppose an electron in level $n = 6$ of an excited hydrogen atom falls back to level $n = 1$ as the hydrogen atom returns to its lowest possible energy state, its **ground state**. We use Equation (7.1) with $Z = 1$, since the hydrogen nucleus contains a single proton. The energies corresponding to the two states are as follows:

For $n = 6$: $E_6 = -2.178 \times 10^{-18} \text{ J} \left(\frac{1^2}{6^2} \right) = -6.050 \times 10^{-20} \text{ J}$

For $n = 1$: $E_1 = -2.178 \times 10^{-18} \text{ J} \left(\frac{1^2}{1^2} \right) = -2.178 \times 10^{-18} \text{ J}$

Note that for $n = 1$ the electron has a more negative energy than it does for $n = 6$, which means that the electron is more tightly bound in the smallest allowed orbit.

The change in energy, ΔE, when the electron falls from $n = 6$ to $n = 1$ is

$$\Delta E = \text{energy of final state} - \text{energy of initial state}$$
$$= E_1 - E_6 = (-2.178 \times 10^{-18} \text{ J}) - (-6.050 \times 10^{-20} \text{ J})$$
$$= -2.117 \times 10^{-18} \text{ J}$$

The negative sign for the *change* in energy indicates that the atom has *lost* energy and is now in a more stable state. The energy is carried away from the atom by the production (emission) of a photon.

The wavelength of the emitted photon can be calculated from the equation

$$\Delta E = h \left(\frac{c}{\lambda} \right) \quad \text{or} \quad \lambda = \frac{hc}{\Delta E}$$

where ΔE represents the change in energy of the atom, which equals the energy of the emitted photon. We have

$$\lambda = \frac{hc}{\Delta E} = \frac{(6.626 \times 10^{-34} \text{ J} \cdot \text{s})(2.9979 \times 10^8 \text{ m/s})}{2.117 \times 10^{-18} \text{ J}} = 9.383 \times 10^{-8} \text{ m}$$

Note that for this calculation the absolute value of ΔE is used (we have not included the negative sign). In this case we indicate the direction of energy flow by saying that a photon of wavelength 9.383×10^{-8} m has been *emitted* from the hydrogen atom. Simply plugging the negative value of ΔE into the equation would produce a negative value for λ, which is physically meaningless.

Interactive Example 7.4

Sign in at http://login.cengagebrain.com to try this Interactive Example in OWL.

Energy Quantization in Hydrogen

Calculate the energy required to excite the hydrogen electron from level $n = 1$ to level $n = 2$. Also calculate the wavelength of light that must be absorbed by a hydrogen atom in its ground state to reach this excited state.*

Solution

Using Equation (7.1) with $Z = 1$, we have

$$E_1 = -2.178 \times 10^{-18} \text{ J}\left(\frac{1^2}{1^2}\right) = -2.178 \times 10^{-18} \text{ J}$$

$$E_2 = -2.178 \times 10^{-18} \text{ J}\left(\frac{1^2}{2^2}\right) = -5.445 \times 10^{-19} \text{ J}$$

$$\Delta E = E_2 - E_1 = (-5.445 \times 10^{-19} \text{ J}) - (-2.178 \times 10^{-18} \text{ J}) = 1.633 \times 10^{-18} \text{ J}$$

The positive value for ΔE indicates that the system has gained energy. The wavelength of light that must be *absorbed* to produce this change is

$$\lambda = \frac{hc}{\Delta E} = \frac{(6.626 \times 10^{-34} \text{ J} \cdot \text{s})(2.9979 \times 10^8 \text{ m/s})}{1.633 \times 10^{-18} \text{ J}}$$

$$= 1.216 \times 10^{-7} \text{ m}$$

Note from Fig. 7.2 that the light required to produce the transition from the $n = 1$ to $n = 2$ level in hydrogen lies in the ultraviolet region.

See Exercises 7.57 and 7.58

At this time we must emphasize two important points about the Bohr model:

1. The model correctly fits the quantized energy levels of the hydrogen atom and postulates only certain allowed circular orbits for the electron.
2. As the electron becomes more tightly bound, its energy becomes more negative relative to the zero-energy reference state (corresponding to the electron being at infinite distance from the nucleus). As the electron is brought closer to the nucleus, energy is released from the system.

Using Equation (7.1), we can derive a general equation for the electron moving from one level (n_{initial}) to another level (n_{final}):

$$\Delta E = \text{energy of level } n_{\text{final}} - \text{energy of level } n_{\text{initial}}$$
$$= E_{\text{final}} - E_{\text{initial}}$$
$$= (-2.178 \times 10^{-18} \text{ J})\left(\frac{1^2}{n_{\text{final}}^2}\right) - (-2.178 \times 10^{-18} \text{ J})\left(\frac{1^2}{n_{\text{initial}}^2}\right)$$
$$= -2.178 \times 10^{-18} \text{ J}\left(\frac{1}{n_{\text{final}}^2} - \frac{1}{n_{\text{initial}}^2}\right) \quad (7.2)$$

*After this example we will no longer show cancellation marks. However, the same process for canceling units applies throughout this text.

Chemical connections
0.035 Femtometer Is a Big Deal

The long-accepted value for the radius of the proton is 0.8768 fm (fm = 10^{-15} m). However, new experiments suggest that this value might be about 4% too large. This doesn't sound like much, but if proven correct it will raise havoc in the scientific community.

The possible problem with the currently accepted value of the proton radius results from work done by Randolf Pohl and his coworkers at the Max Planck Institute of Quantum Optics in Garching, Germany. These scientists created an exotic form of hydrogen in which they replaced the hydrogen electron by a muon. A muon is a particle that has the same charge as an electron but is 200 times more massive than the electron. Because of its greater mass, the muon has orbitals with smaller average radii than those of an electron. The German team attempted to excite the muon to higher energy levels using laser pulses. They set their laser to detect muon transitions assuming the proton radius was in the range from 0.87 to 0.91 femtometers (fm), the value expected from current theories. After years of failing to see the expected energy transitions, the German scientists were within weeks of shutting down the experiment when they decided to assume a smaller value for the proton radius. They saw the expected transition when they tuned their laser at a value corresponding to a proton radius of 0.84184 fm. Although this value doesn't seem too different from 0.8768 fm, it has tremendous implications. A change in the radius of the proton will affect the value of the charge density of the proton, a value that affects the values of all of the fundamental physical constants since all of these constants are interrelated. Thus, more experiments are needed to determine the correct value of the proton radius. If the new, smaller value proves to be correct, it has important implications for the fundamental theory (quantum electrodynamics) of matter. Perhaps this will lead to a better understanding of nature.

Equation (7.2) can be used to calculate the energy change between *any* two energy levels in a hydrogen atom, as shown in Example 7.5.

Example 7.5	Electron Energies

Calculate the energy required to remove the electron from a hydrogen atom in its ground state.

Solution

Removing the electron from a hydrogen atom in its ground state corresponds to taking the electron from $n_{\text{initial}} = 1$ to $n_{\text{final}} = \infty$. Thus

$$\Delta E = -2.178 \times 10^{-18} \text{ J}\left(\frac{1}{n_{\text{final}}^{2}} - \frac{1}{n_{\text{initial}}^{2}}\right)$$

$$= -2.178 \times 10^{-18} \text{ J}\left(\frac{1}{\infty} - \frac{1}{1^{2}}\right)$$

$$= -2.178 \times 10^{-18} \text{ J}(0 - 1) = 2.178 \times 10^{-18} \text{ J}$$

The energy required to remove the electron from a hydrogen atom in its ground state is 2.178×10^{-18} J.

See Exercises 7.65 and 7.66

Although Bohr's model fits the energy levels for hydrogen, it is a fundamentally incorrect model for the hydrogen atom.

At first Bohr's model appeared to be very promising. The energy levels calculated by Bohr closely agreed with the values obtained from the hydrogen emission spectrum. However, when Bohr's model was applied to atoms other than hydrogen, it did not work at all. Although some attempts were made to adapt the model using elliptical orbits, it was concluded that Bohr's model is fundamentally incorrect. The model is, however, very important historically, because it showed that the observed quantization of energy in atoms could be explained by making rather simple assumptions. Bohr's model paved the way for later theories. It is important to realize, however, that the current theory of atomic structure is in no way derived from the Bohr model. Electrons do *not* move around the nucleus in circular orbits, as we shall see later in this chapter.

7.5 | The Quantum Mechanical Model of the Atom

By the mid-1920s it had become apparent that the Bohr model could not be made to work. A totally new approach was needed. Three physicists were at the forefront of this effort: Werner Heisenberg (1901–1976), Louis de Broglie (1892–1987), and Erwin Schrödinger (1887–1961). The approach they developed became known as *wave mechanics* or, more commonly, *quantum mechanics.* As we have already seen, de Broglie originated the idea that the electron, previously considered to be a particle, also shows wave properties. Pursuing this line of reasoning, Schrödinger, an Austrian physicist, decided to attack the problem of atomic structure by giving emphasis to the wave properties of the electron. To Schrödinger and de Broglie, the electron bound to the nucleus seemed similar to a **standing wave**, and they began research on a wave mechanical description of the atom.

The most familiar example of standing waves occurs in association with musical instruments such as guitars or violins, where a string attached at both ends vibrates to produce a musical tone. The waves are described as "standing" because they are stationary; the waves do not travel along the length of the string. The motions of the string can be explained as a combination of simple waves of the type shown in Fig. 7.10. The dots in this figure indicate the nodes, or points of zero lateral (sideways) displacement, for a given wave. Note that there are limitations on the allowed wavelengths of the standing wave. Each end of the string is fixed, so there is always a node at each end. This means that there must be a whole number of *half* wavelengths in any of the allowed motions of the string (see Fig. 7.10). Standing waves can be illustrated using the wave generator shown in the photo.

A similar situation results when the electron in the hydrogen atom is imagined to be a standing wave. As shown in Fig. 7.11, only certain circular orbits have a circumference into which a whole number of wavelengths of the standing electron wave will "fit." All other orbits would produce destructive interference of the standing electron wave and are not allowed. This seemed like a possible explanation for the observed quantization of the hydrogen atom, so Schrödinger worked out a model for the hydrogen atom in which the electron was assumed to behave as a standing wave.

It is important to recognize that Schrödinger could not be sure that this idea would work. The test had to be whether or not the model would correctly fit the experimental

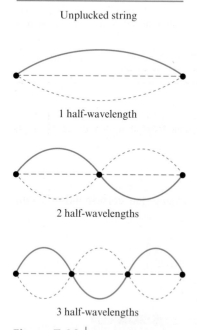

Unplucked string

1 half-wavelength

2 half-wavelengths

3 half-wavelengths

Figure 7.10 | The standing waves caused by the vibration of a guitar string fastened at both ends. Each dot represents a node (a point of zero displacement).

Wave-generating apparatus.

David Blackburn

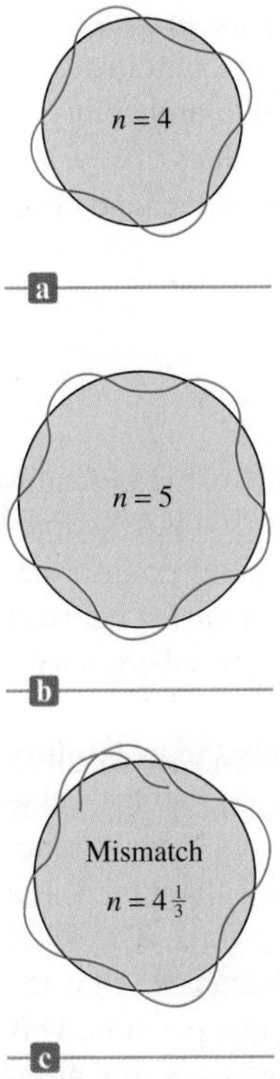

data on hydrogen and other atoms. The physical principles for describing standing waves were well known in 1925 when Schrödinger decided to treat the electron in this way. His mathematical treatment is too complicated to be detailed here. However, the form of Schrödinger's equation is

$$\hat{H}\psi = E\psi$$

where ψ, called the **wave function**, is a function of the coordinates (x, y, and z) of the electron's position in three-dimensional space and $\hat{H}$ represents a set of mathematical instructions called an *operator*. In this case, the operator contains mathematical terms that produce the total energy of the atom when they are applied to the wave function. E represents the total energy of the atom (the sum of the potential energy due to the attraction between the proton and electron and the kinetic energy of the moving electron). When this equation is analyzed, many solutions are found. Each solution consists of a wave function, ψ, that is characterized by a particular value of E. A specific wave function is often called an **orbital**.

To illustrate the most important ideas of the **quantum (wave) mechanical model** of the atom, we will first concentrate on the wave function corresponding to the lowest energy for the hydrogen atom. This wave function is called the $1s$ orbital. The first point of interest is to explore the meaning of the word *orbital*. As we will see, this is not a trivial matter. One thing is clear: An orbital is *not* a Bohr orbit. The electron in the hydrogen $1s$ orbital is not moving around the nucleus in a circular orbit. How, then, is the electron moving? The answer is quite surprising: *We do not know.* The wave function gives us no information about the detailed pathway of the electron. This is somewhat disturbing. When we solve problems involving the motions of particles in the macroscopic world, we are able to predict their pathways. For example, when two billiard balls with known velocities collide, we can predict their motions after the collision. However, we cannot predict the electron's motion from the $1s$ orbital function. Does this mean that the theory is wrong? Not necessarily: We have already learned that an electron does not behave much like a billiard ball, so we must examine the situation closely before we discard the theory.

To help us understand the nature of an orbital, we need to consider a principle discovered by Werner Heisenberg, one of the primary developers of quantum mechanics. Heisenberg's mathematical analysis led him to a surprising conclusion: *There is a fundamental limitation to just how precisely we can know both the position and momentum of a particle at a given time.* This is a statement of the **Heisenberg uncertainty principle**. Stated mathematically, the uncertainty principle is

$$\Delta x \cdot \Delta(mv) \geq \frac{h}{4\pi}$$

where Δx is the uncertainty in a particle's position, $\Delta(mv)$ is the uncertainty in a particle's momentum, and h is Planck's constant. Thus the minimum uncertainty in the product $\Delta x \cdot \Delta(mv)$ is $h/4\pi$. What this equation really says is that the more accurately we know a particle's position, the less accurately we can know its momentum, and vice versa. This limitation is so small for large particles such as baseballs or billiard balls that it is unnoticed. However, for a small particle such as the electron, the limitation becomes quite important. Applied to the electron, the uncertainty principle implies that we cannot know the exact motion of the electron as it moves around the nucleus. It is therefore not appropriate to assume that the electron is moving around the nucleus in a well-defined orbit, as in the Bohr model.

The Physical Meaning of a Wave Function

Given the limitations indicated by the uncertainty principle, what then is the physical meaning of a wave function for an electron? That is, what is an atomic orbital? Although the wave function itself has no easily visualized meaning, the square of the function does have a definite physical significance. *The square of the function*

Figure 7.11 | The hydrogen electron visualized as a standing wave around the nucleus. The circumference of a particular circular orbit would have to correspond to a whole number of wavelengths, as shown in (a) and (b), or else destructive interference occurs, as shown in (c). This is consistent with the fact that only certain electron energies are allowed; the atom is quantized. (Although this idea encouraged scientists to use a wave theory, it does not mean that the electron really travels in circular orbits.)

Probability is the likelihood, or odds, that something will occur.

indicates the probability of finding an electron near a particular point in space. For example, suppose we have two positions in space, one defined by the coordinates x_1, y_1, and z_1 and the other by the coordinates x_2, y_2, and z_2. The relative probability of finding the electron at positions 1 and 2 is given by substituting the values of x, y, and z for the two positions into the wave function, squaring the function value, and computing the following ratio:

$$\frac{[\psi(x_1, y_1, z_1)]^2}{[\psi(x_2, y_2, z_2)]^2} = \frac{N_1}{N_2}$$

The quotient N_1/N_2 is the ratio of the probabilities of finding the electron at positions 1 and 2. For example, if the value of the ratio N_1/N_2 is 100, the electron is 100 times more likely to be found at position 1 than at position 2. The model gives no information concerning when the electron will be at either position or how it moves between the positions. This vagueness is consistent with the concept of the Heisenberg uncertainty principle.

The square of the wave function is most conveniently represented as a **probability distribution**, in which the intensity of color is used to indicate the probability value near a given point in space. The probability distribution for the hydrogen $1s$ wave function (orbital) is shown in Fig. 7.12(a). The best way to think about this diagram is as a three-dimensional time exposure with the electron as a tiny moving light. The more times the electron visits a particular point, the darker the negative becomes. Thus the darkness of a point indicates the probability of finding an electron at that position. This diagram is also known as an *electron density map; electron density* and *electron probability* mean the same thing. When a chemist uses the term *atomic orbital,* he or she is probably picturing an electron density map of this type.

Another way of representing the electron probability distribution for the $1s$ wave function is to calculate the probability at points along a line drawn outward in any direction from the nucleus. The result is shown in Fig. 7.12(b). Note that the probability of finding the electron at a particular position is greatest close to the nucleus and drops off rapidly as the distance from the nucleus increases. We are also interested in knowing the *total* probability of finding the electron in the hydrogen atom at a particular *distance* from the nucleus. Imagine that the space around the hydrogen nucleus is made up of a series of thin, spherical shells (rather like layers in an onion), as shown in Fig. 7.13(a). When the total probability of finding the electron in each spherical shell is plotted versus the distance from the nucleus, the plot in Fig. 7.13(b) is obtained. This graph is called the **radial probability distribution**.

The maximum in the curve occurs because of two opposing effects. The probability of finding an electron at a particular position is greatest near the nucleus, but the volume of the spherical shell increases with distance from the nucleus. Therefore, as we move away from the nucleus, the probability of finding the electron at a given position

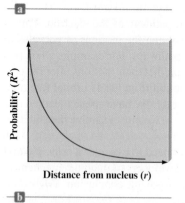

Figure 7.12 | (a) The probability distribution for the hydrogen $1s$ orbital in three-dimensional space. (b) The probability of finding the electron at points along a line drawn from the nucleus outward in any direction for the hydrogen $1s$ orbital.

Figure 7.13 | (a) Cross section of the hydrogen $1s$ orbital probability distribution divided into successive thin spherical shells. (b) The radial probability distribution. A plot of the total probability of finding the electron in each thin spherical shell as a function of distance from the nucleus.

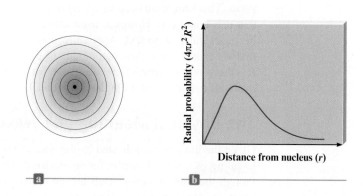

$1\,\text{Å} = 10^{-10}$ m; the angstrom is most often used as the unit for atomic radius because of its convenient size. Another convenient unit is the picometer:

$$1\,\text{pm} = 10^{-12}\,\text{m}$$

decreases, but we are summing more positions. Thus the total probability increases to a certain radius and then decreases as the electron probability at each position becomes very small. For the hydrogen $1s$ orbital, the maximum radial probability (the distance at which the electron is most likely to be found) occurs at a distance of 5.29×10^{-2} nm or 0.529 Å from the nucleus. Interestingly, this is exactly the radius of the innermost orbit in the Bohr model. Note that in Bohr's model the electron is assumed to have a circular path and so is *always* found at this distance. In the quantum mechanical model, the specific electron motions are unknown, and this is the *most probable* distance at which the electron is found.

One more characteristic of the hydrogen $1s$ orbital that we must consider is its size. As we can see from Fig. 7.12, the size of this orbital cannot be defined precisely, since the probability never becomes zero (although it drops to an extremely small value at large values of r). So, in fact, the hydrogen $1s$ orbital has no distinct size. However, it is useful to have a definition of relative orbital size. *The definition most often used by chemists to describe the size of the hydrogen $1s$ orbital is the radius of the sphere that encloses 90% of the total electron probability.* That is, 90% of the time the electron is inside this sphere.

So far we have described only the lowest-energy wave function in the hydrogen atom, the $1s$ orbital. Hydrogen has many other orbitals, which we will describe in the next section. However, before we proceed, we should summarize what we have said about the meaning of an atomic orbital. An orbital is difficult to define precisely at an introductory level. Technically, an orbital is a wave function. However, it is usually most helpful to picture an orbital as a three-dimensional electron density map. That is, an electron "in" a particular atomic orbital is assumed to exhibit the electron probability indicated by the orbital map.

7.6 | Quantum Numbers

When we solve the Schrödinger equation for the hydrogen atom, we find many wave functions (orbitals) that satisfy it. Each of these orbitals is characterized by a series of numbers called **quantum numbers**, which describe various properties of the orbital:

The **principal quantum number** (n) has integral values: 1, 2, 3, ... The principal quantum number is related to the size and energy of the orbital. As n increases, the orbital becomes larger and the electron spends more time farther from the nucleus. An increase in n also means higher energy, because the electron is less tightly bound to the nucleus, and the energy is less negative.

The **angular momentum quantum number** (ℓ) has integral values from 0 to $n - 1$ for each value of n. This quantum number is related to the shape of atomic orbitals. The value of ℓ for a particular orbital is commonly assigned a letter: $\ell = 0$ is called s; $\ell = 1$ is called p; $\ell = 2$ is called d; $\ell = 3$ is called f. This system arises from early spectral studies and is summarized in Table 7.1.

The **magnetic quantum number** (m_ℓ) has integral values between ℓ and $-\ell$, including zero. The value of m_ℓ is related to the orientation of the orbital in space relative to the other orbitals in the atom.

Table 7.1 | The Angular Momentum Quantum Numbers and Corresponding Letters Used to Designate Atomic Orbitals

Value of ℓ	0	1	2	3	4
Letter Used	s	p	d	f	g

Number of Orbitals per Subshell

$s = 1$
$p = 3$
$d = 5$
$f = 7$
$g = 9$

Table 7.2 | Quantum Numbers for the First Four Levels of Orbitals in the Hydrogen Atom

n	ℓ	Sublevel Designation	m_ℓ	Number of Orbitals
1	0	1*s*	0	1
2	0	2*s*	0	1
	1	2*p*	−1, 0, +1	3
3	0	3*s*	0	1
	1	3*p*	−1, 0, 1	3
	2	3*d*	−2, −1, 0, 1, 2	5
4	0	4*s*	0	1
	1	4*p*	−1, 0, 1	3
	2	4*d*	−2, −1, 0, 1, 2	5
	3	4*f*	−3, −2, −1, 0, 1, 2, 3	7

$n = 1, 2, 3, \ldots$
$\ell = 0, 1, \ldots (n - 1)$
$m_\ell = -\ell, \ldots 0, \ldots +\ell$

The first four levels of orbitals in the hydrogen atom are listed with their quantum numbers in Table 7.2. Note that each set of orbitals with a given value of ℓ (sometimes called a **subshell**) is designated by giving the value of n and the letter for ℓ. Thus an orbital where $n = 2$ and $\ell = 1$ is symbolized as 2*p*. There are three 2*p* orbitals, which have different orientations in space. We will describe these orbitals in the next section.

Interactive Example 7.6

Sign in at http://login.cengagebrain.com to try this Interactive Example in OWL.

Electron Subshells

For principal quantum level $n = 5$, determine the number of allowed subshells (different values of ℓ), and give the designation of each.

Solution

For $n = 5$, the allowed values of ℓ run from 0 to 4 ($n - 1 = 5 - 1$). Thus, the subshells and their designations are

$\ell = 0$	$\ell = 1$	$\ell = 2$	$\ell = 3$	$\ell = 4$
5*s*	5*p*	5*d*	5*f*	5*g*

See Exercises 7.71 and 7.73

7.7 | Orbital Shapes and Energies

We have seen that the meaning of an orbital is represented most clearly by a probability distribution. Each orbital in the hydrogen atom has a unique probability distribution. We also saw that another means of representing an orbital is by the surface that surrounds 90% of the total electron probability. These three types of representations for the hydrogen 1*s*, 2*s*, and 3*s* orbitals are shown in Fig. 7.14. Note the characteristic spherical shape of each of the *s* orbitals. Note also that the 2*s* and 3*s* orbitals contain areas of high probability separated by areas of zero probability. These latter areas are called **nodal surfaces**, or simply **nodes**. The number of nodes increases as n increases. For *s* orbitals, the number of nodes is given by $n - 1$. For our purposes, however, we will think of *s* orbitals only in terms of their overall spherical shape, which becomes larger as the value of n increases.

Figure 7.14 | Three representations of the hydrogen 1s, 2s, and 3s orbitals. (a) The square of the wave function. (b) "Slices" of the three-dimensional electron density. (c) The surfaces that contain 90% of the total electron probability (the "sizes" of the orbitals).

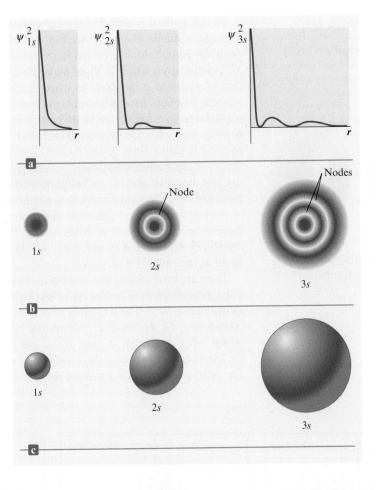

The two types of representations for the 2p orbitals (there are no 1p orbitals) are shown in Fig. 7.15. Note that the p orbitals are not spherical like s orbitals but have two *lobes* separated by a node at the nucleus. The p orbitals are labeled according to the axis of the *xyz* coordinate system along which the lobes lie. For example, the 2p orbital with lobes centered along the x axis is called the $2p_x$ orbital.

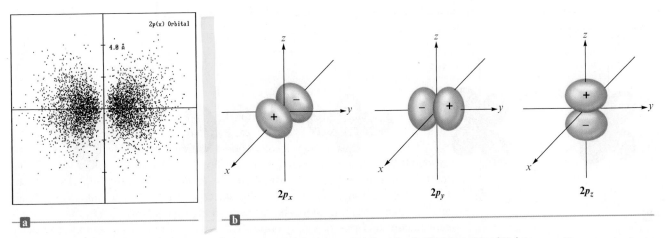

Figure 7.15 | Representation of the 2p orbitals. (a) The electron probability distribution for a 2p orbital. (Generated from a program by Robert Allendoerfer on Project SERAPHIM disk PC 2402; reprinted with permission.) (b) The boundary surface representations of all three 2p orbitals. Note that the signs inside the surface indicate the phases (signs) of the orbital in that region of space.

n value
↓
2*p*_x ← orientation in space
↑
ℓ value

Figure 7.16 | A cross section of the electron probability distribution for a 3*p* orbital.

At this point it is useful to remember that mathematical functions have signs. For example, a simple sine wave (see Fig. 7.1) oscillates from positive to negative and repeats this pattern. Atomic orbital functions also have signs. The functions for *s* orbitals are positive everywhere in three-dimensional space. That is, when the *s* orbital function is evaluated at any point in space, it results in a positive number. In contrast, the *p* orbital functions have different signs in different regions of space. For example, the *p*_z orbital has a positive sign in all the regions of space in which *z* is positive and has a negative sign when *z* is negative. This behavior is indicated in Fig. 7.15(b) by the positive and negative signs inside their boundary surfaces. It is important to understand that these are mathematical signs, not charges. Just as a sine wave has alternating positive and negative phases, so too *p* orbitals have positive and negative phases. The phases of the *p*_x, *p*_y, and *p*_z orbitals are indicated in Fig. 7.15(b).

As you might expect from our discussion of the *s* orbitals, the 3*p* orbitals have a more complex probability distribution than that of the 2*p* orbitals (Fig. 7.16), but they can still be represented by the same boundary surface shapes. The surfaces just grow larger as the value of *n* increases.

There are no *d* orbitals that correspond to principal quantum levels *n* = 1 and *n* = 2. The *d* orbitals (ℓ = 2) first occur in level *n* = 3. The five 3*d* orbitals have the shapes shown in Fig. 7.17. The *d* orbitals have two different fundamental shapes. Four of the orbitals (d_{xz}, d_{yz}, d_{xy}, and $d_{x^2-y^2}$) have four lobes centered in the plane indicated in the orbital label. Note that d_{xy} and $d_{x^2-y^2}$ are both centered in the *xy* plane; however, the lobes of $d_{x^2-y^2}$ lie *along* the *x* and *y* axes, while the lobes of d_{xy} lie *between* the axes. The fifth orbital, d_{z^2}, has a unique shape with two lobes along the *z* axis and a belt

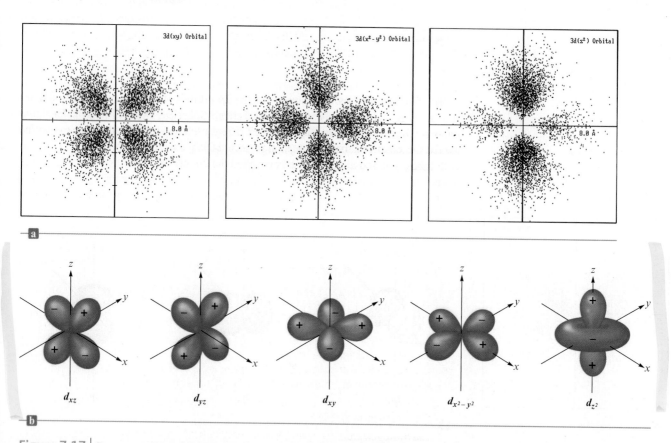

Figure 7.17 | Representation of the 3*d* orbitals. (a) Electron density plots of selected 3*d* orbitals. (Generated from a program by Robert Allendoerfer on Project SERAPHIM disk PC 2402; reprinted with permission.) (b) The boundary surfaces of all five 3*d* orbitals, with the signs (phases) indicated.

Figure 7.18 | Representation of the 4f orbitals in terms of their boundary surfaces.

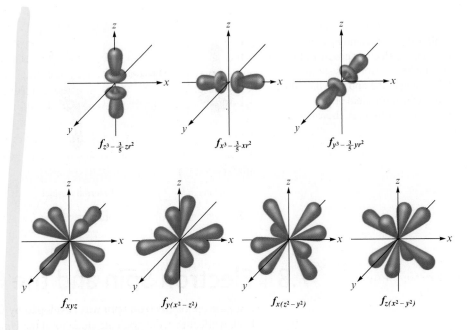

$f_{z^3 - \frac{3}{5}zr^2}$ $f_{x^3 - \frac{3}{5}xr^2}$ $f_{y^3 - \frac{3}{5}yr^2}$

f_{xyz} $f_{y(x^2-z^2)}$ $f_{x(z^2-y^2)}$ $f_{z(x^2-y^2)}$

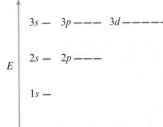

Figure 7.19 | Orbital energy levels for the hydrogen atom.

centered in the *xy* plane. The *d* orbitals for levels $n > 3$ look like the 3*d* orbitals but have larger lobes.

The *f* orbitals first occur in level $n = 4$, and as might be expected, they have shapes even more complex than those of the *d* orbitals. Figure 7.18 shows representations of the 4*f* orbitals ($\ell = 3$) along with their designations. These orbitals are not involved in the bonding in any of the compounds we will consider in this text. Their shapes and labels are simply included for completeness.

So far we have talked about the shapes of the hydrogen atomic orbitals but not about their energies. For the hydrogen atom, the energy of a particular orbital is determined by its value of *n*. Thus *all* orbitals with the same value of *n* have the *same energy*—they are said to be **degenerate**. This is shown in Fig. 7.19, where the energies for the orbitals in the first three quantum levels for hydrogen are shown.

Hydrogen's single electron can occupy any of its atomic orbitals. However, in the lowest energy state, the *ground state,* the electron resides in the 1*s* orbital. If energy is put into the atom, the electron can be transferred to a higher-energy orbital, producing an *excited state.*

Let's Review | *A Summary of the Hydrogen Atom*

❯ In the quantum (wave) mechanical model, the electron is viewed as a standing wave. This representation leads to a series of wave functions (orbitals) that describe the possible energies and spatial distributions available to the electron.

❯ In agreement with the Heisenberg uncertainty principle, the model cannot specify the detailed electron motions. Instead, the square of the wave function represents the probability distribution of the electron in that orbital. This allows us to picture orbitals in terms of probability distributions, or electron density maps.

❯ The size of an orbital is arbitrarily defined as the surface that contains 90% of the total electron probability.

❯ The hydrogen atom has many types of orbitals. In the ground state, the single electron resides in the 1s orbital. The electron can be excited to higher-energy orbitals if energy is put into the atom.

Figure 7.20 | A picture of the spinning electron. Spinning in one direction, the electron produces the magnetic field oriented as shown in (a). Spinning in the opposite direction, it gives a magnetic field of the opposite orientation, as shown in (b).

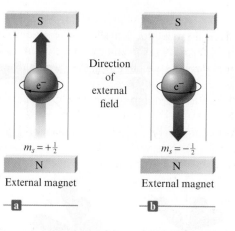

7.8 | Electron Spin and the Pauli Principle

The concept of **electron spin** was developed by Samuel Goudsmit and George Uhlenbeck while they were graduate students at the University of Leyden in the Netherlands. They found that a fourth quantum number (in addition to n, ℓ, and m_ℓ) was necessary to account for the details of the emission spectra of atoms. The spectral data indicate that the electron has a magnetic moment with two possible orientations when the atom is placed in an external magnetic field. Since they knew from classical physics that a spinning charge produces a magnetic moment, it seemed reasonable to assume that the electron could have two spin states, thus producing the two oppositely directed magnetic moments (Fig. 7.20). The new quantum number adopted to describe this phenomenon, called the **electron spin quantum number** (m_s), can have only one of two values, $+\frac{1}{2}$ and $-\frac{1}{2}$. We can interpret this to mean that the electron can spin in one of two opposite directions, although other interpretations also have been suggested.

For our purposes, the main significance of electron spin is connected with the postulate of Austrian physicist Wolfgang Pauli (1900–1958): In a given atom no two electrons can have the same set of four quantum numbers (n, ℓ, m_ℓ, and m_s). This is called the **Pauli exclusion principle**. Since electrons in the same orbital have the same values of n, ℓ, and m_ℓ, this postulate says that they must have different values of m_s. Then, since only two values of m_s are allowed, *an orbital can hold only two electrons, and they must have opposite spins*. This principle will have important consequences as we use the atomic model to account for the electron arrangements of the atoms in the periodic table.

7.9 | Polyelectronic Atoms

The quantum mechanical model gives a description of the hydrogen atom that agrees very well with experimental data. However, the model would not be very useful if it did not account for the properties of all the other atoms as well.

To see how the model applies to **polyelectronic atoms**, that is, atoms with more than one electron, let's consider helium, which has two protons in its nucleus and two electrons:

Three energy contributions must be considered in the description of the helium atom: (1) the kinetic energy of the electrons as they move around the nucleus, (2) the potential energy of attraction between the nucleus and the electrons, and (3) the potential energy of repulsion between the two electrons.

Although the helium atom can be readily described in terms of the quantum mechanical model, the Schrödinger equation that results cannot be solved exactly. The difficulty arises in dealing with the repulsions between the electrons. Since the electron pathways are unknown, the electron repulsions cannot be calculated exactly. This is called the *electron correlation problem.*

The electron correlation problem occurs with all polyelectronic atoms. To treat these systems using the quantum mechanical model, we must make approximations. Most commonly, the approximation used is to treat each electron as if it were moving in a *field of charge that is the net result of the nuclear attraction and the average repulsions of all the other electrons.*

For example, consider the sodium atom, which has 11 electrons:

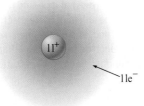

Now let's single out the outermost electron and consider the forces this electron feels. The electron clearly is attracted to the highly charged nucleus. However, the electron also feels the repulsions caused by the other 10 electrons. The net effect is that the electron is not bound nearly as tightly to the nucleus as it would be if the other electrons were not present. We say that the electron is *screened* or *shielded* from the nuclear charge by the repulsions of the other electrons.

This picture of polyelectronic atoms leads to *hydrogen-like orbitals* for these atoms. They have the same general shapes as the orbitals for hydrogen, but their sizes and energies are different. The differences occur because of the interplay between nuclear attraction and the electron repulsions.

One especially important difference between polyelectronic atoms and the hydrogen atom is that for hydrogen all the orbitals in a given principal quantum level have the same energy (they are said to be *degenerate*). This is not the case for polyelectronic atoms, where we find that for a given principal quantum level the orbitals vary in energy as follows:

$$E_{ns} < E_{np} < E_{nd} < E_{nf}$$

In other words, when electrons are placed in a particular quantum level, they "prefer" the orbitals in the order *s*, *p*, *d*, and then *f*. Why does this happen? Although the concept of orbital energies is a complicated matter, we can qualitatively understand why the 2*s* orbital has a lower energy than the 2*p* orbital in a polyelectronic atom by looking at the probability profiles of these orbitals (Fig. 7.21). Notice that the 2*p* orbital has its maximum probability closer to the nucleus than for the 2*s*. This might lead us to predict that the 2*p* would be preferable (lower energy) to the 2*s* orbital. However, notice the small hump of electron density that occurs in the 2*s* profile very near the nucleus. This means that although an electron in the 2*s* orbital spends most of its time a little farther from the nucleus than does an electron in the 2*p* orbital, it spends a small but very significant amount of time very near the nucleus. We say that the 2*s* electron *penetrates* to the nucleus more than once in the 2*p* orbital. This *penetration effect* causes an electron in a 2*s* orbital to be attracted to the nucleus more strongly than an electron in a 2*p* orbital. That is, the 2*s* orbital is lower in energy than the 2*p* orbitals in a polyelectronic atom.

The same thing happens in the other principal quantum levels as well. Figure 7.22 shows the radial probability profiles for the 3*s*, 3*p*, and 3*d* orbitals. Note again the hump in the 3*s* profile very near the nucleus. The innermost hump for the 3*p* is farther

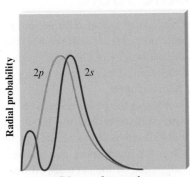

Figure 7.21 | A comparison of the radial probability distributions of the 2*s* and 2*p* orbitals.

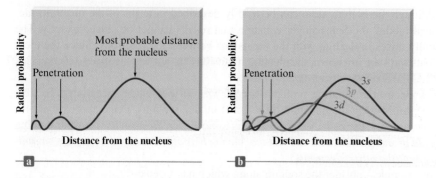

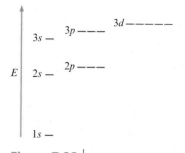

Figure 7.22 | (a) The radial probability distribution for an electron in a 3s orbital. Although a 3s electron is mostly found far from the nucleus, there is a small but significant probability (shown by the arrows) of its being found close to the nucleus. The 3s electron penetrates the shield of inner electrons. (b) The radial probability distribution for the 3s, 3p, and 3d orbitals. The arrows indicate that the s orbital (red arrow) allows greater electron penetration than the p orbital (yellow arrow) does; the d orbital allows minimal electron penetration.

Figure 7.23 | The orders of the energies of the orbitals in the first three levels of polyelectronic atoms.

out, which causes the energy of the 3s orbital to be lower than that of the 3p. Notice that the 3d orbital has its maximum probability closer to the nucleus than either the 3s or 3p does, but its absence of probability near the nucleus causes it to be highest in energy of the three orbitals. The relative energies of the orbitals for $n = 3$ are

$$E_{3s} < E_{3p} < E_{3d}$$

In general, the more effectively an orbital allows its electron to penetrate the shielding electrons to be close to the nuclear charge, the lower is the energy of that orbital.

A summary diagram of the orders of the orbital energies for polyelectronic atoms is represented in Fig. 7.23. We will use these orbitals in Section 7.11 to show how the electrons are arranged in polyelectronic atoms.

Critical Thinking

What if Bohr's model was correct? How would this affect the radial probability profiles as shown in Figs. 7.21 and 7.22?

7.10 | The History of the Periodic Table

The modern periodic table contains a tremendous amount of useful information. In this section we will discuss the origin of this valuable tool; later we will see how the quantum mechanical model for the atom explains the periodicity of chemical properties. Certainly the greatest triumph of the quantum mechanical model is its ability to account for the arrangement of the elements in the periodic table.

The periodic table was originally constructed to represent the patterns observed in the chemical properties of the elements. As chemistry progressed during the eighteenth and nineteenth centuries, it became evident that the earth is composed of a great many elements with very different properties. Things are much more complicated than the simple model of earth, air, fire, and water suggested by the ancients. At first, the array of elements and properties was bewildering. Gradually, however, patterns were noticed.

The first chemist to recognize patterns was Johann Dobereiner (1780–1849), who found several groups of three elements that have similar properties, for example,

Figure 7.24 | Dmitri Ivanovich Mendeleev (1834–1907), born in Siberia as the youngest of 17 children, taught chemistry at the University of St. Petersburg. In 1860 Mendeleev heard the Italian chemist Cannizzaro lecture on a reliable method for determining the correct atomic masses of the elements. This important development paved the way for Mendeleev's own brilliant contribution to chemistry—the periodic table. In 1861 Mendeleev returned to St. Petersburg, where he wrote a book on organic chemistry. Later Mendeleev also wrote a book on inorganic chemistry, and he was struck by the fact that the systematic approach characterizing organic chemistry was lacking in inorganic chemistry. In attempting to systematize inorganic chemistry, he eventually arranged the elements in the form of the periodic table.

Mendeleev was a versatile genius who was interested in many fields of science. He worked on many problems associated with Russia's natural resources, such as coal, salt, and various metals. Being particularly interested in the petroleum industry, he visited the United States in 1876 to study the Pennsylvania oil fields. His interests also included meteorology and hot-air balloons. In 1887 he made an ascent in a balloon to study a total eclipse of the sun.

chlorine, bromine, and iodine. However, as Dobereiner attempted to expand this model of *triads* (as he called them) to the rest of the known elements, it became clear that it was severely limited.

The next notable attempt was made by the English chemist John Newlands, who in 1864 suggested that elements should be arranged in *octaves,* based on the idea that certain properties seemed to repeat for every eighth element in a way similar to the musical scale, which repeats for every eighth tone. Even though this model managed to group several elements with similar properties, it was not generally successful.

The present form of the periodic table was conceived independently by two chemists: the German Julius Lothar Meyer (1830–1895) and Dmitri Ivanovich Mendeleev (1834–1907), a Russian (Fig. 7.24). Usually Mendeleev is given most of the credit, because it was he who emphasized how useful the table could be in predicting the existence and properties of still unknown elements. For example, in 1872 when Mendeleev first published his table (Fig. 7.25), the elements gallium, scandium, and germanium were unknown. Mendeleev correctly predicted the existence and properties of

TABELLE II

REIHEN	GRUPPE I. — R^2O	GRUPPE II. — RO	GRUPPE III. — R^2O^3	GRUPPE IV. RH^4 RO^2	GRUPPE V. RH^3 R^2O^5	GRUPPE VI. RH^2 RO^3	GRUPPE VII. RH R^2O^7	GRUPPE VIII. — RO^4
1	H=1							
2	Li=7	Be=9,4	B=11	C=12	N=14	O=16	F=19	
3	Na=23	Mg=24	Al=27,3	Si=28	P=31	S=32	Cl=35,5	
4	K=39	Ca=40	=44	Ti=48	V=51	Cr=52	Mn=55	Fe=56, Co=59, Ni=59, Cu=63.
5	(Cu=63)	Zn=65	=68	=72	As=75	Se=78	Br=80	
6	Rb=85	Sr=87	?Yt=88	Zr=90	Nb=94	Mo=96	=100	Ru=104, Rh=104, Pd=106, Ag=108.
7	(Ag=108)	Cd=112	In=113	Sn=118	Sb=122	Te=125	J=127	
8	Cs=133	Ba=137	?Di=138	?Ce=140	—	—	—	
9	(—)		—	—	—	—	—	Os=195, Ir=197, Pt=198, Au=199.
10	—	—	?Er=178	?La=180	Ta=182	W=184	—	
11	(Au=199)	Hg=200	Tl=204	Pb=207	Bi=208	—	—	— — — —
12	—	—	—	Th=231	—	U=240	—	

Figure 7.25 | Mendeleev's early periodic table, published in 1872. Note the spaces left for missing elements with atomic masses 44, 68, 72, and 100.

(From *Annalen der Chemie und Pharmacie,* VIII, Supplementary Volume for 1872, page 511.)

Table 7.3 | Comparison of the Properties of Germanium as Predicted by Mendeleev and as Actually Observed

Properties of Germanium	Predicted in 1871	Observed in 1886
Atomic weight	72	72.3
Density	5.5 g/cm³	5.47 g/cm³
Specific heat	0.31 J/(°C · g)	0.32 J/(°C · g)
Melting point	Very high	960°C
Oxide formula	RO₂	GeO₂
Oxide density	4.7 g/cm³	4.70 g/cm³
Chloride formula	RCl₄	GeCl₄
Boiling point of chloride	100°C	86°C

Table 7.4 | Predicted Properties of Elements 113 and 114

Property	Element 113	Element 114
Chemically like	Thallium	Lead
Atomic mass	297	298
Density	16 g/mL	14 g/mL
Melting point	430°C	70°C
Boiling point	1100°C	150°C

these elements from gaps in his periodic table. The data for germanium (which Mendeleev called *ekasilicon*) are shown in Table 7.3. Note the excellent agreement between the actual values and Mendeleev's predictions, which were based on the properties of other members in the group of elements similar to germanium.

Using his table, Mendeleev also was able to correct several values for atomic masses. For example, the original atomic mass of 76 for indium was based on the assumption that indium oxide had the formula InO. This atomic mass placed indium, which has metallic properties, among the nonmetals. Mendeleev assumed the atomic mass was probably incorrect and proposed that the formula of indium oxide was really In₂O₃. Based on this correct formula, indium has an atomic mass of approximately 113, placing the element among the metals. Mendeleev also corrected the atomic masses of beryllium and uranium.

Because of its obvious usefulness, Mendeleev's periodic table was almost universally adopted, and it remains one of the most valuable tools at the chemist's disposal. For example, it is still used to predict the properties of elements recently discovered, as shown in Table 7.4.

A current version of the periodic table is shown inside the front cover of this book. The only fundamental difference between this table and that of Mendeleev is that it lists the elements in order by atomic number rather than by atomic mass. The reason for this will become clear later in this chapter as we explore the electron arrangements of the atom. Another recent format of the table is discussed in the following section.

7.11 | The Aufbau Principle and the Periodic Table

We can use the quantum mechanical model of the atom to show how the electron arrangements in the hydrogen-like atomic orbitals of the various atoms account for the organization of the periodic table. Our main assumption here is that all atoms have the same type of orbitals as have been described for the hydrogen atom. As protons are added one by one to the nucleus to build up the elements, electrons are similarly added to these hydrogen-like orbitals. This is called the **aufbau principle**.

Hydrogen has one electron, which occupies the $1s$ orbital in its ground state. The configuration for hydrogen is written as $1s^1$, which can be represented by the following *orbital diagram*:

Aufbau is German for "building up."

H ($Z = 1$)
He ($Z = 2$)
Li ($Z = 3$)
Be ($Z = 4$)
B ($Z = 5$)
etc.
(Z = atomic number)

$$\text{H: } 1s^1 \quad \begin{array}{ccc} 1s & 2s & 2p \\ \boxed{\uparrow} & \boxed{} & \boxed{} \end{array}$$

Chemical connections
The Chemistry of Copernicium

Although element 112 is not exactly a newborn (it was discovered in 1996), it has just been given a name. The International Union of Pure and Applied Chemistry announced on February 19, 2010, that the official name of 112 is Copernicium (pronounced koh-pur-NEE-see-um) and has the symbol Cn. The name honors Nicolaus Copernicus, who is best known for first recognizing that the planetary system is sun-centered rather than Earth-centered.

Now scientists from the Paul Sherrer Institute in Switzerland and the Joint Institute for Nuclear Research in Russia have studied the properties of Cn. By firing a beam of ^{48}Ca projectiles at a target of ^{242}Pu (doped with Nd_2O_3), these scientists have produced just two atoms of Cn. Given the relatively long lifetime of Cn atoms (a few seconds), the researchers have been able to compare the properties of Cn, Hg, and Ar atoms. When these atoms were

injected into a series of gold-covered detectors, it was found that, similar to Hg (and in contrast to Ar), Cn atoms were "mildly" volatile and readily formed bonds with gold. Thus, as predicted by the periodic table, the element Cn, which has an expected electron configuration of $[Rn]7s^26d^{10}$, fits with the Zn, Cd, and Hg group of elements, all of which have the $ns^2(n-1)d^{10}$ configuration.

The arrow represents an electron spinning in a particular direction.

The next element, *helium,* has two electrons. Since two electrons with opposite spins can occupy an orbital, according to the Pauli exclusion principle, the electrons for helium are in the $1s$ orbital with opposite spins, producing a $1s^2$ configuration:

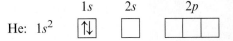

Lithium has three electrons, two of which can go into the $1s$ orbital before the orbital is filled. Since the $1s$ orbital is the only orbital for $n = 1$, the third electron will occupy the lowest-energy orbital with $n = 2$, or the $2s$ orbital, giving a $1s^22s^1$ configuration:

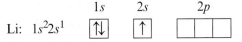

The next element, *beryllium,* has four electrons, which occupy the $1s$ and $2s$ orbitals:

Boron has five electrons, four of which occupy the $1s$ and $2s$ orbitals. The fifth electron goes into the second type of orbital with $n = 2$, the $2p$ orbitals:

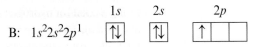

Since all the $2p$ orbitals have the same energy (are degenerate), it does not matter which $2p$ orbital the electron occupies.

For an atom with unfilled subshells, the lowest energy is achieved by electrons occupying separate orbitals with parallel spins, as far as allowed by the Pauli exclusion principle.

[Ne] is shorthand for $1s^2 2s^2 2p^6$.

Sodium metal is so reactive that it is stored in kerosene to protect it from the oxygen in the air.

Carbon is the next element and has six electrons. Two electrons occupy the $1s$ orbital, two occupy the $2s$ orbital, and two occupy $2p$ orbitals. Since there are three $2p$ orbitals with the same energy, the mutually repulsive electrons will occupy *separate* $2p$ orbitals. This behavior is summarized by **Hund's rule** (named for the German physicist F. H. Hund): The lowest energy configuration for an atom is the one having the maximum number of unpaired electrons allowed by the Pauli principle in a particular set of degenerate orbitals. By convention, the unpaired electrons are represented as having parallel spins (with spin "up").

The configuration for carbon could be written $1s^2 2s^2 2p^1 2p^1$ to indicate that the electrons occupy separate $2p$ orbitals. However, the configuration is usually given as $1s^2 2s^2 2p^2$, and it is understood that the electrons are in different $2p$ orbitals. The orbital diagram for carbon is

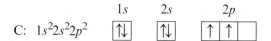

C: $1s^2 2s^2 2p^2$

Note that the unpaired electrons in the $2p$ orbitals are shown with parallel spins.

The configuration for *nitrogen,* which has seven electrons, is $1s^2 2s^2 2p^3$. The three electrons in the $2p$ orbitals occupy separate orbitals with parallel spins:

N: $1s^2 2s^2 2p^3$

The configuration for *oxygen,* which has eight electrons, is $1s^2 2s^2 2p^4$. One of the $2p$ orbitals is now occupied by a pair of electrons with opposite spins, as required by the Pauli exclusion principle:

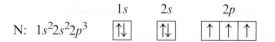

O: $1s^2 2s^2 2p^4$

The orbital diagrams and electron configurations for *fluorine* (nine electrons) and *neon* (ten electrons) are as follows:

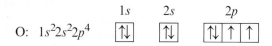

F: $1s^2 2s^2 2p^5$

Ne: $1s^2 2s^2 2p^6$

With neon, the orbitals with $n = 1$ and $n = 2$ are now completely filled.

For *sodium,* the first ten electrons occupy the $1s$, $2s$, and $2p$ orbitals, and the eleventh electron must occupy the first orbital with $n = 3$, the $3s$ orbital. The electron configuration for sodium is $1s^2 2s^2 2p^6 3s^1$. To avoid writing the inner-level electrons, this configuration is often abbreviated as $[Ne]3s^1$, where [Ne] represents the electron configuration of neon, $1s^2 2s^2 2p^6$. *prior element*

The next element, *magnesium,* has the configuration $1s^2 2s^2 2p^6 3s^2$, or $[Ne]3s^2$. Then the next six elements, *aluminum* through *argon,* have configurations obtained by filling the $3p$ orbitals one electron at a time. Figure 7.26 summarizes the electron configurations of the first 18 elements by giving the number of electrons in the type of orbital occupied last.

At this point it is useful to introduce the concept of **valence electrons**, *the electrons in the outermost principal quantum level of an atom.* The valence electrons of the nitrogen atom, for example, are the $2s$ and $2p$ electrons. For the sodium atom, the valence electron is the electron in the $3s$ orbital, and so on. Valence electrons are the most important electrons to chemists because they are involved in bonding, as we will see in the next two chapters. The inner electrons are known as **core electrons**.

Figure 7.26 | The electron configurations in the type of orbital occupied last for the first 18 elements.

H $1s^1$								He $1s^2$
Li $2s^1$	Be $2s^2$		B $2p^1$	C $2p^2$	N $2p^3$	O $2p^4$	F $2p^5$	Ne $2p^6$
Na $3s^1$	Mg $3s^2$		Al $3p^1$	Si $3p^2$	P $3p^3$	S $3p^4$	Cl $3p^5$	Ar $3p^6$

A vial containing potassium metal. The sealed vial contains an inert gas to protect the potassium from reacting with oxygen.

Calcium metal.

Chromium is often used to plate bumpers and hood ornaments, such as this statue of Mercury found on a 1929 Buick.

Note in Fig. 7.26 that a very important pattern is developing: The elements in the same group (vertical column of the periodic table) have the same valence electron configuration. Remember that Mendeleev originally placed the elements in groups based on similarities in chemical properties. Now we understand the reason behind these groupings. Elements with the same valence electron configuration show similar chemical behavior.

The element after argon is *potassium*. Since the 3p orbitals are fully occupied in argon, we might expect the next electron to go into a 3d orbital (recall that for $n = 3$ the orbitals are 3s, 3p, and 3d). However, the chemistry of potassium is clearly very similar to that of lithium and sodium, indicating that the last electron in potassium occupies the 4s orbital instead of one of the 3d orbitals, a conclusion confirmed by many types of experiments. The electron configuration of potassium is

$$\text{K:} \quad 1s^2 2s^2 2p^6 3s^2 3p^6 4s^1 \quad \text{or} \quad [\text{Ar}]4s^1$$

The next element is *calcium*:

$$\text{Ca:} \quad [\text{Ar}]4s^2$$

The next element, *scandium*, begins a series of 10 elements (scandium through zinc) called the **transition metals**, whose configurations are obtained by adding electrons to the five 3d orbitals. The configuration of scandium is

$$\text{Sc:} \quad [\text{Ar}]4s^2 3d^1$$

That of *titanium* is

$$\text{Ti:} \quad [\text{Ar}]4s^2 3d^2$$

And that of *vanadium* is

$$\text{V:} \quad [\text{Ar}]4s^2 3d^3$$

Chromium is the next element. The expected configuration is $[\text{Ar}]4s^2 3d^4$. However, the observed configuration is

$$\text{Cr:} \quad [\text{Ar}]4s^1 3d^5$$

The explanation for this configuration of chromium is beyond the scope of this book. In fact, chemists are still disagreeing over the exact cause of this anomaly. Note, however, that the observed configuration has both the 4s and 3d orbitals half-filled. This is a good way to remember the correct configuration.

The next four elements, *manganese* through *nickel*, have the expected configurations:

Mn:	$[\text{Ar}]4s^2 3d^5$	Co:	$[\text{Ar}]4s^2 3d^7$
Fe:	$[\text{Ar}]4s^2 3d^6$	Ni:	$[\text{Ar}]4s^2 3d^8$

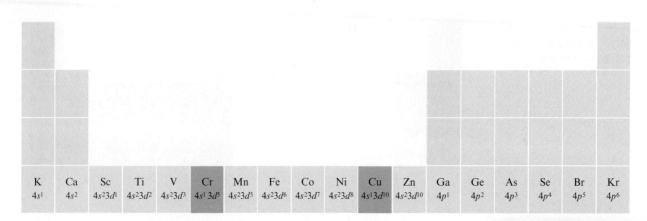

K	Ca	Sc	Ti	V	Cr	Mn	Fe	Co	Ni	Cu	Zn	Ga	Ge	As	Se	Br	Kr
$4s^1$	$4s^2$	$4s^23d^1$	$4s^23d^2$	$4s^23d^3$	$4s^1\,3d^5$	$4s^23d^5$	$4s^23d^6$	$4s^23d^7$	$4s^23d^8$	$4s^13d^{10}$	$4s^23d^{10}$	$4p^1$	$4p^2$	$4p^3$	$4p^4$	$4p^5$	$4p^6$

Figure 7.27 | Valence electron configurations for potassium through krypton. The transition metals (scandium through zinc) have the general configuration $[Ar]4s^23d^n$, except for chromium and copper.

The configuration for *copper* is expected to be $[Ar]4s^23d^9$. However, the observed configuration is

$$\text{Cu:}\quad [Ar]4s^13d^{10}$$

In this case, a half-filled $4s$ orbital and a filled set of $3d$ orbitals characterize the actual configuration.

Zinc has the expected configuration:

$$\text{Zn:}\quad [Ar]4s^23d^{10}$$

The configurations of the transition metals are shown in Fig. 7.27. After that, the next six elements, *gallium* through *krypton,* have configurations that correspond to filling the $4p$ orbitals (Fig. 7.27).

The entire periodic table is represented in Fig. 7.28 in terms of which orbitals are being filled. The valence electron configurations are given in Fig. 7.29. From these two figures, note the following additional points:

The $(n + 1)s$ orbitals fill before the nd orbitals.

1. The $(n + 1)s$ orbitals always fill before the nd orbitals. For example, the $5s$ orbitals fill in rubidium and strontium before the $4d$ orbitals fill in the second row of

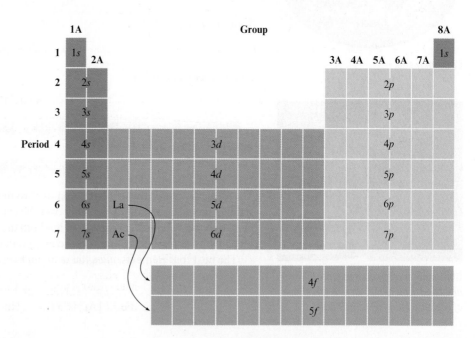

Figure 7.28 | The orbitals being filled for elements in various parts of the periodic table. Note that in going along a horizontal row (a period), the $(n + 1)s$ orbital fills before the nd orbital. The group labels indicate the number of valence electrons (*ns* plus *np* electrons) for the elements in each group.

Transition metals

	Representative Elements		d-Transition Elements											Representative Elements					Noble Gases
	1A 1 ns^1	**Group numbers**																	**8A** 18 ns^2np^6
1	1 H $1s^1$	**2A** 2 ns^2											**3A** 13 ns^2np^1	**4A** 14 ns^2np^2	**5A** 15 ns^2np^3	**6A** 16 ns^2np^4	**7A** 17 ns^2np^5	2 He $1s^2$
2	3 Li $2s^1$	4 Be $2s^2$											5 B $2s^22p^1$	6 C $2s^22p^2$	7 N $2s^22p^3$	8 O $2s^22p^4$	9 F $2s^22p^5$	10 Ne $2s^22p^6$
3	11 Na $3s^1$	12 Mg $3s^2$	3	4	5	6	7	8	9	10	11	12	13 Al $3s^23p^1$	14 Si $3s^23p^2$	15 P $3s^23p^3$	16 S $3s^23p^4$	17 Cl $3s^23p^5$	18 Ar $3s^23p^6$
4	19 K $4s^1$	20 Ca $4s^2$	21 Sc $4s^23d^1$	22 Ti $4s^23d^2$	23 V $4s^23d^3$	24 Cr $4s^13d^5$	25 Mn $4s^23d^5$	26 Fe $4s^23d^6$	27 Co $4s^23d^7$	28 Ni $4s^23d^8$	29 Cu $4s^13d^{10}$	30 Zn $4s^23d^{10}$	31 Ga $4s^24p^1$	32 Ge $4s^24p^2$	33 As $4s^24p^3$	34 Se $4s^24p^4$	35 Br $4s^24p^5$	36 Kr $4s^24p^6$
5	37 Rb $5s^1$	38 Sr $5s^2$	39 Y $5s^24d^1$	40 Zr $5s^24d^2$	41 Nb $5s^14d^4$	42 Mo $5s^14d^5$	43 Tc $5s^24d^5$	44 Ru $5s^14d^7$	45 Rh $5s^14d^8$	46 Pd $4d^{10}$	47 Ag $5s^14d^{10}$	48 Cd $5s^24d^{10}$	49 In $5s^25p^1$	50 Sn $5s^25p^2$	51 Sb $5s^25p^3$	52 Te $5s^25p^4$	53 I $5s^25p^5$	54 Xe $5s^25p^6$
6	55 Cs $6s^1$	56 Ba $6s^2$	57 La* $6s^25d^1$	72 Hf $4f^{14}6s^25d^2$	73 Ta $6s^25d^3$	74 W $6s^25d^4$	75 Re $6s^25d^5$	76 Os $6s^25d^6$	77 Ir $6s^25d^7$	78 Pt $6s^15d^9$	79 Au $6s^15d^{10}$	80 Hg $6s^25d^{10}$	81 Tl $6s^26p^1$	82 Pb $6s^26p^2$	83 Bi $6s^26p^3$	84 Po $6s^26p^4$	85 At $6s^26p^5$	86 Rn $6s^26p^6$
7	87 Fr $7s^1$	88 Ra $7s^2$	89 Ac** $7s^26d^1$	104 Rf $7s^26d^2$	105 Db $7s^26d^3$	106 Sg $7s^26d^4$	107 Bh $7s^26d^5$	108 Hs $7s^26d^6$	109 Mt $7s^26d^7$	110 Ds $7s^26d^8$	111 Rg $7s^16d^{10}$	112 Cn $7s^26d^{10}$	113 Uut $7s^27p^1$	114 Fl $7s^27p^2$	115 Uup $7s^27p^3$	116 Lv $7s^27p^4$	117 Uus $7s^27p^5$	118 Uuo $7s^27p^6$

Period number, highest occupied electron level

f-Transition Elements

*Lanthanides

58 Ce $6s^24f^15d^1$	59 Pr $6s^24f^35d^0$	60 Nd $6s^24f^45d^0$	61 Pm $6s^24f^55d^0$	62 Sm $6s^24f^65d^0$	63 Eu $6s^24f^75d^0$	64 Gd $6s^24f^75d^1$	65 Tb $6s^24f^95d^0$	66 Dy $6s^24f^{10}5d^0$	67 Ho $6s^24f^{11}5d^0$	68 Er $6s^24f^{12}5d^0$	69 Tm $6s^24f^{13}5d^0$	70 Yb $6s^24f^{14}5d^0$	71 Lu $6s^24f^{14}5d^1$

**Actinides

90 Th $7s^25f^06d^2$	91 Pa $7s^25f^26d^1$	92 U $7s^25f^36d^1$	93 Np $7s^25f^46d^1$	94 Pu $7s^25f^66d^0$	95 Am $7s^25f^76d^0$	96 Cm $7s^25f^76d^1$	97 Bk $7s^25f^96d^0$	98 Cf $7s^25f^{10}6d^0$	99 Es $7s^25f^{11}6d^0$	100 Fm $7s^25f^{12}6d^0$	101 Md $7s^25f^{13}6d^0$	102 No $7s^25f^{14}6d^0$	103 Lr $7s^25f^{14}6d^1$

Figure 7.29 | The periodic table with atomic symbols, atomic numbers, and partial electron configurations.

transition metals (yttrium through cadmium). This early filling of the s orbitals can be explained by the penetration effect. For example, the 4s orbital allows for so much more penetration to the vicinity of the nucleus that it becomes lower in energy than the 3d orbital. Thus, the 4s fills before the 3d. The same things can be said about the 5s and 4d, the 6s and 5d, and the 7s and 6d orbitals.

2. After lanthanum, which has the configuration $[Xe]6s^25d^1$, a group of 14 elements called the **lanthanide series**, or the **lanthanides**, occurs. This series of elements corresponds to the filling of the seven 4f orbitals. Note that sometimes an electron occupies a 5d orbital instead of a 4f orbital. This occurs because the energies of the 4f and 5d orbitals are very similar.

Lanthanides are elements in which the 4f orbitals are being filled.

3. After actinium, which has the configuration $[Rn]7s^26d^1$, a group of 14 elements called the **actinide series**, or the **actinides**, occurs. This series corresponds to the filling of the seven 5f orbitals. Note that sometimes one or two electrons occupy the 6d orbitals instead of the 5f orbitals, because these orbitals have very similar energies.

Actinides are elements in which the 5f orbitals are being filled.

The group label tells the total number of valence electrons for that group.

4. The group labels for Groups 1A, 2A, 3A, 4A, 5A, 6A, 7A, and 8A indicate the *total number* of valence electrons for the atoms in these groups. For example, all the elements in Group 5A have the configuration ns^2np^3. (The *d* electrons fill one period late and are usually not counted as valence electrons.) The meaning of the group labels for the transition metals is not as clear as for the Group A elements, and these will not be used in this text.

5. The groups labeled 1A, 2A, 3A, 4A, 5A, 6A, 7A, and 8A are often called the **main-group**, or **representative**, **elements**. Every member of these groups has the same valence electron configuration.

Critical Thinking

You have learned that each orbital is allowed two electrons, and this pattern is evident on the periodic table. What if each orbital was allowed three electrons? How would this change the appearance of the periodic table?

The International Union of Pure and Applied Chemistry (IUPAC), a body of scientists organized to standardize scientific conventions, has recommended a new form for the periodic table, which the American Chemical Society has adopted (see the blue numbers in Fig. 7.29). In this new version, the group number indicates the number of *s*, *p*, and *d* electrons added since the last noble gas. We will not use the new format in this book, but you should be aware that the familiar periodic table may be soon replaced by this or a similar format.

The results considered in this section are very important. We have seen that the quantum mechanical model can be used to explain the arrangement of the elements in the periodic table. This model allows us to understand that the similar chemistry exhibited by the members of a given group arises from the fact that they all have the same valence electron configuration. Only the principal quantum number of the valence orbitals changes in going down a particular group.

When an electron configuration is given in this text, the orbitals are listed in the order in which they fill.

It is important to be able to give the electron configuration for each of the main-group elements. This is most easily done by using the periodic table. If you understand how the table is organized, it is not necessary to memorize the order in which the orbitals fill. Review Figs. 7.28 and 7.29 to make sure that you understand the correspondence between the orbitals and the periods and groups.

Cr: $[Ar]4s^13d^5$
Cu: $[Ar]4s^13d^{10}$

Predicting the configurations of the transition metals (3*d*, 4*d*, and 5*d* elements), the lanthanides (4*f* elements), and the actinides (5*f* elements) is somewhat more difficult because there are many exceptions of the type encountered in the first-row transition metals (the 3*d* elements). You should memorize the configurations of chromium and copper, the two exceptions in the first-row transition metals, since these elements are often encountered.

Interactive Example 7.7

Sign in at http://login.cengagebrain.com to try this Interactive Example in **OWL**.

Electron Configurations

Give the electron configurations for sulfur (S), cadmium (Cd), hafnium (Hf), and radium (Ra) using the periodic table inside the front cover of this book.

Solution

Sulfur is element 16 and resides in Period 3, where the 3*p* orbitals are being filled (Fig. 7.30). Since sulfur is the fourth among the "3*p* elements," it must have four 3*p* electrons. Its configuration is

$$S: \quad 1s^22s^22p^63s^23p^4 \quad \text{or} \quad [Ne]3s^23p^4$$

Figure 7.30 | The positions of the elements considered in Example 7.7.

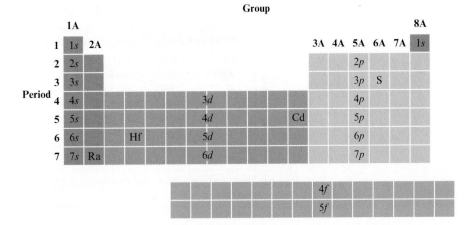

Cadmium is element 48 and is located in Period 5 at the end of the 4*d* transition metals (Fig. 7.30). It is the tenth element in the series and thus has 10 electrons in the 4*d* orbitals, in addition to the 2 electrons in the 5*s* orbital. The configuration is

$$\text{Cd:} \quad 1s^2 2s^2 2p^6 3s^2 3p^6 4s^2 3d^{10} 4p^6 5s^2 4d^{10} \quad \text{or} \quad [\text{Kr}]5s^2 4d^{10}$$

Hafnium is element 72 and is found in Period 6 (Fig. 7.30). Note that it occurs just after the lanthanide series. Thus the 4*f* orbitals are already filled. Hafnium is the second member of the 5*d* transition series and has two 5*d* electrons. The configuration is

$$\text{Hf:} \quad 1s^2 2s^2 2p^6 3s^2 3p^6 4s^2 3d^{10} 4p^6 5s^2 4d^{10} 5p^6 6s^2 4f^{14} 5d^2 \quad \text{or} \quad [\text{Xe}]6s^2 4f^{14} 5d^2$$

Radium is element 88 and is in Period 7 (and Group 2A) (Fig. 7.30). Thus radium has two electrons in the 7*s* orbital, and the configuration is

$$\text{Ra:} \quad 1s^2 2s^2 2p^6 3s^2 3p^6 4s^2 3d^{10} 4p^6 5s^2 4d^{10} 5p^6 6s^2 4f^{14} 5d^{10} 6p^6 7s^2 \quad \text{or} \quad [\text{Rn}]7s^2$$

See Exercises 7.85 through 7.88

7.12 | Periodic Trends in Atomic Properties

We have developed a fairly complete picture of polyelectronic atoms. Although the model is rather crude because the nuclear attractions and electron repulsions are simply lumped together, it is very successful in accounting for the periodic table of elements. We will next use the model to account for the observed trends in several important atomic properties: ionization energy, electron affinity, and atomic size.

Ionization Energy

Ionization energy results in the formation of a **positive** ion.

Ionization energy is the energy required to remove an electron from a gaseous atom or ion:

$$X(g) \longrightarrow X^+(g) + e^-$$

where the atom or ion is assumed to be in its ground state.

To introduce some of the characteristics of ionization energy, we will consider the energy required to remove several electrons in succession from aluminum in the gaseous state. The ionization energies are

$$\text{Al}(g) \longrightarrow \text{Al}^+(g) + e^- \qquad I_1 = 580 \text{ kJ/mol}$$
$$\text{Al}^+(g) \longrightarrow \text{Al}^{2+}(g) + e^- \qquad I_2 = 1815 \text{ kJ/mol}$$
$$\text{Al}^{2+}(g) \longrightarrow \text{Al}^{3+}(g) + e^- \qquad I_3 = 2740 \text{ kJ/mol}$$
$$\text{Al}^{3+}(g) \longrightarrow \text{Al}^{4+}(g) + e^- \qquad I_4 = 11,600 \text{ kJ/mol}$$

Table 7.5 | Successive Ionization Energies (kJ/mol) for the Elements in Period 3

Element	I_1	I_2	I_3	I_4	I_5	I_6	I_7
Na	495	4560					
Mg	735	1445	7730				
Al	580	1815	2740	11,600			
Si	780	1575	3220	4350	16,100		
P	1060	1890	2905	4950	6270	21,200	
S	1005	2260	3375	4565	6950	8490	27,000
Cl	1255	2295	3850	5160	6560	9360	11,000
Ar	1527	2665	3945	5770	7230	8780	12,000

Core electrons*

General decrease

*Note the large jump in ionization energy in going from removal of valence electrons to removal of core electrons.

—————————————— General increase ——————————————⟶

Setting the aluminum cap on the Washington Monument in 1884. At that time, aluminum was regarded as a precious metal.

Table 7.6 | First Ionization Energies for the Alkali Metals and Noble Gases

Atom	I_1 (kJ/mol)
Group 1A	
Li	520
Na	495
K	419
Rb	409
Cs	382
Group 8A	
He	2377
Ne	2088
Ar	1527
Kr	1356
Xe	1176
Rn	1042

Several important points can be illustrated from these results. In a stepwise ionization process, it is always the highest-energy electron (the one bound least tightly) that is removed first. The **first ionization energy** I_1 is the energy required to remove the highest-energy electron of an atom. The first electron removed from the aluminum atom comes from the $3p$ orbital (Al has the electron configuration $[Ne]3s^23p^1$). The second electron comes from the $3s$ orbital (since Al^+ has the configuration $[Ne]3s^2$). Note that the value of I_1 is considerably smaller than the value of I_2, the **second ionization energy**.

This makes sense for several reasons. The primary factor is simply charge. Note that the first electron is removed from a neutral atom (Al), whereas the second electron is removed from a $1+$ ion (Al^+). The increase in positive charge binds the electrons more firmly, and the ionization energy increases. The same trend shows up in the third (I_3) and fourth (I_4) ionization energies, where the electron is removed from the Al^{2+} and Al^{3+} ions, respectively.

The increase in successive ionization energies for an atom also can be interpreted using our simple model for polyelectronic atoms. The increase in ionization energy from I_1 to I_2 makes sense because the first electron is removed from a $3p$ orbital that is higher in energy than the $3s$ orbital from which the second electron is removed. The largest jump in ionization energy by far occurs in going from the third ionization energy (I_3) to the fourth (I_4). This is so because I_4 corresponds to removing a core electron (Al^{3+} has the configuration $1s^22s^22p^6$), and core electrons are bound much more tightly than valence electrons.

Table 7.5 gives the values of ionization energies for all the Period 3 elements. Note the large jump in energy in each case in going from removal of valence electrons to removal of core electrons.

The values of the first ionization energies for the elements in the first six periods of the periodic table are graphed in Fig. 7.31. Note that in general as we go *across a period from left to right, the first ionization energy increases.* This is consistent with the idea that electrons added in the same principal quantum level do not completely shield the increasing nuclear charge caused by the added protons. Thus electrons in the same principal quantum level are generally more strongly bound as we move to the right on the periodic table, and there is a general increase in ionization energy values as electrons are added to a given principal quantum level.

On the other hand, *first ionization energy decreases in going down a group.* This can be seen most clearly by focusing on the Group 1A elements (the alkali metals) and the Group 8A elements (the noble gases), as shown in Table 7.6. The main reason for the decrease in ionization energy in going down a group is that the electrons being removed are, on average, farther from the nucleus. As n increases, the size of the orbital increases, and the electron is easier to remove.

In Fig. 7.31 we see that there are some discontinuities in ionization energy in going across a period. For example, for Period 2, discontinuities occur in going from

Figure 7.31 | The values of first ionization energy for the elements in the first six periods. In general, ionization energy decreases in going down a group. For example, note the decrease in values for Group 1A and Group 8A. In general, ionization energy increases in going left to right across a period. For example, note the sharp increase going across Period 2 from lithium through neon.

First ionization energy increases across a period and decreases down a group.

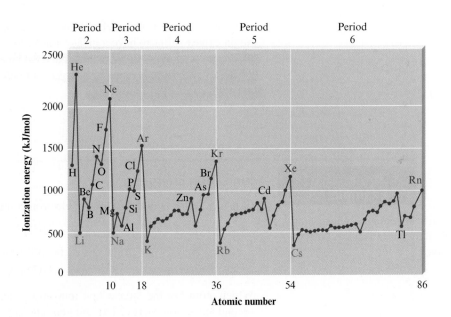

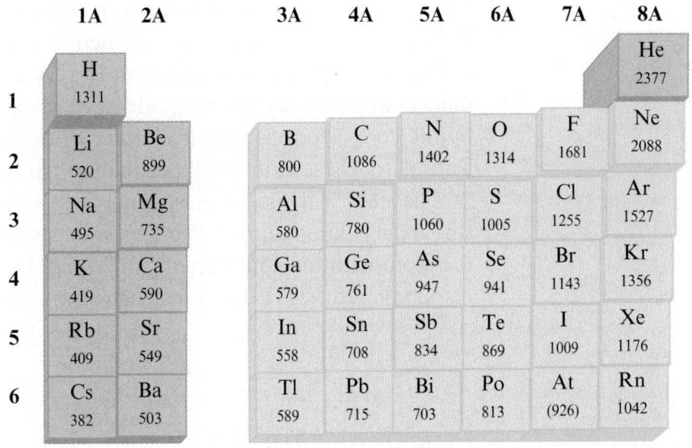

Figure 7.32 | Trends in ionization energies (kJ/mol) for the representative elements.

beryllium to boron and from nitrogen to oxygen. These exceptions to the normal trend can be explained in terms of electron repulsions. The decrease in ionization energy in going from beryllium to boron reflects the fact that the electrons in the filled $2s$ orbital provide some shielding for electrons in the $2p$ orbital from the nuclear charge. The decrease in ionization energy in going from nitrogen to oxygen reflects the extra electron repulsions in the doubly occupied oxygen $2p$ orbital.

The ionization energies for the representative elements are summarized in Fig. 7.32.

Example 7.8

Trends in Ionization Energies

The first ionization energy for phosphorus is 1060 kJ/mol, and that for sulfur is 1005 kJ/mol. Why?

Solution

Phosphorus and sulfur are neighboring elements in Period 3 of the periodic table and have the following valence electron configurations: Phosphorus is $3s^2 3p^3$, and sulfur is $3s^2 3p^4$.

Ordinarily, the first ionization energy increases as we go across a period, so we might expect sulfur to have a greater ionization energy than phosphorus. However, in this case the fourth p electron in sulfur must be placed in an already occupied orbital. The electron–electron repulsions that result cause this electron to be more easily removed than might be expected.

See Exercises 7.113 and 7.114

Interactive Example 7.9

Sign in at http://login.cengagebrain.com to try this Interactive Example in **OWL**.

Ionization Energies

Consider atoms with the following electron configurations:

$$1s^2 2s^2 2p^6$$
$$1s^2 2s^2 2p^6 3s^1$$
$$1s^2 2s^2 2p^6 3s^2$$

Which atom has the largest first ionization energy, and which one has the smallest second ionization energy? Explain your choices.

Solution

The atom with the largest value of I_1 is the one with the configuration $1s^2 2s^2 2p^6$ (this is the neon atom), because this element is found at the right end of Period 2. Since the $2p$ electrons do not shield each other very effectively, I_1 will be relatively large. The other configurations given include $3s$ electrons. These electrons are effectively shielded by the core electrons and are farther from the nucleus than the $2p$ electrons in neon. Thus I_1 for these atoms will be smaller than for neon.

The atom with the smallest value of I_2 is the one with the configuration $1s^2 2s^2 2p^6 3s^2$ (the magnesium atom). For magnesium, both I_1 and I_2 involve valence electrons. For the atom with the configuration $1s^2 2s^2 2p^6 3s^1$ (sodium), the second electron lost (corresponding to I_2) is a core electron (from a $2p$ orbital).

See Exercises 7.115 and 7.116

Electron Affinity

Electron affinity is associated with the production of a **negative** ion.

Electron affinity *is the energy change associated with the addition of an electron to a gaseous atom:*

$$X(g) + e^- \longrightarrow X^-(g)$$

The sign convention for electron affinity values follows the convention for energy changes used in Chapter 6.

Because two different conventions have been used, there is a good deal of confusion in the chemical literature about the signs for electron affinity values. Electron affinity has been defined in many textbooks as the energy *released* when an electron is added to a gaseous atom. This convention requires that a positive sign be attached to an exothermic addition of an electron to an atom, which opposes normal thermodynamic conventions. Therefore, in this book we define electron affinity as a *change* in energy, which means that if the addition of the electron is exothermic, the corresponding value for electron affinity will carry a negative sign.

Figure 7.33 shows the electron affinity values for the atoms among the first 20 elements that form stable, isolated negative ions—that is, the atoms that undergo the addition of an electron as shown above. As expected, all these elements have negative (exothermic) electron affinities. Note that the *more negative* the energy, the greater the quantity of energy released. Although electron affinities generally become more negative from left to right across a period, there are several exceptions to this rule in each period. The dependence of electron affinity on atomic number can be explained by considering the changes in electron repulsions as a function of electron configurations. For example, the fact that

Figure 7.33 | The electron affinity values for atoms among the first 20 elements that form stable, isolated X^- ions. The lines shown connect adjacent elements. The absence of a line indicates missing elements (He, Be, N, Ne, Mg, and Ar) whose atoms do not add an electron exothermically and thus do not form stable, isolated X^- ions.

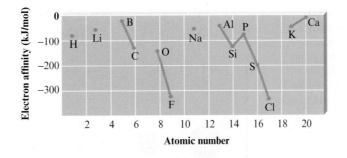

the nitrogen atom does not form a stable, isolated $N^-(g)$ ion, whereas carbon forms $C^-(g)$, reflects the difference in the electron configurations of these atoms. An electron added to nitrogen $(1s^22s^22p^3)$ to form the $N^-(g)$ ion $(1s^22s^22p^4)$ would have to occupy a $2p$ orbital that already contains one electron. The extra repulsion between the electrons in this doubly occupied orbital causes $N^-(g)$ to be unstable. When an electron is added to carbon $(1s^22s^22p^2)$ to form the $C^-(g)$ ion $(1s^22s^22p^3)$, no such extra repulsions occur.

In contrast to the nitrogen atom, the oxygen atom can add one electron to form the stable $O^-(g)$ ion. Presumably oxygen's greater nuclear charge compared with that of nitrogen is sufficient to overcome the repulsion associated with putting a second electron into an already occupied $2p$ orbital. However, it should be noted that a second electron *cannot* be added to an oxygen atom $[O^-(g) + e^- \nrightarrow O^{2-}(g)]$ to form an isolated oxide ion. This outcome seems strange in view of the many stable oxide compounds (MgO, Fe_2O_3, and so on) that are known. As we will discuss in detail in Chapter 8, the O^{2-} ion is stabilized in ionic compounds by the large attractions that occur among the positive ions and the oxide ions.

When we go down a group, electron affinity should become more positive (less energy released), since the electron is added at increasing distances from the nucleus. Although this is generally the case, the changes in electron affinity in going down most groups are relatively small, and numerous exceptions occur. This behavior is demonstrated by the electron affinities of the Group 7A elements (the halogens) shown in Table 7.7. Note that the range of values is quite small compared with the changes that typically occur across a period. Also note that although chlorine, bromine, and iodine show the expected trend, the energy released when an electron is added to fluorine is smaller than might be expected. This smaller energy release has been attributed to the small size of the $2p$ orbitals. Because the electrons must be very close together in these orbitals, there are unusually large electron–electron repulsions. In the other halogens with their larger orbitals, the repulsions are not as severe.

Table 7.7 | Electron Affinities of the Halogens

Atom	Electron Affinity (kJ/mol)
F	−327.8
Cl	−348.7
Br	−324.5
I	−295.2

Atomic Radius

Just as the size of an orbital cannot be specified exactly, neither can the size of an atom. We must make some arbitrary choices to obtain values for **atomic radii**. These values can be obtained by measuring the distances between atoms in chemical compounds. For example, in the bromine molecule, the distance between the two nuclei is known to be 228 pm. The bromine atomic radius is assumed to be half this distance, or 114 pm (Fig. 7.34). These radii are often called *covalent atomic radii* because of the way they are determined (from the distances between atoms in covalent bonds).

Figure 7.34 | The radius of an atom (r) is defined as half the distance between the nuclei in a molecule consisting of identical atoms.

For nonmetallic atoms that do not form diatomic molecules, the atomic radii are estimated from their various covalent compounds. The radii for metal atoms (called *metallic radii*) are obtained from half the distance between metal atoms in solid metal crystals.

The values of the atomic radii for the representative elements are shown in Fig. 7.35. Note that these values are significantly smaller than might be expected from the 90% electron density volumes of isolated atoms, because when atoms form bonds, their electron "clouds" interpenetrate. However, these values form a self-consistent data set that can be used to discuss the trends in atomic radii.

Figure 7.35 | Atomic radii (in picometers) for selected atoms. Note that atomic radius decreases going across a period and increases going down a group. The values for the noble gases are estimated, because data from bonded atoms are lacking.

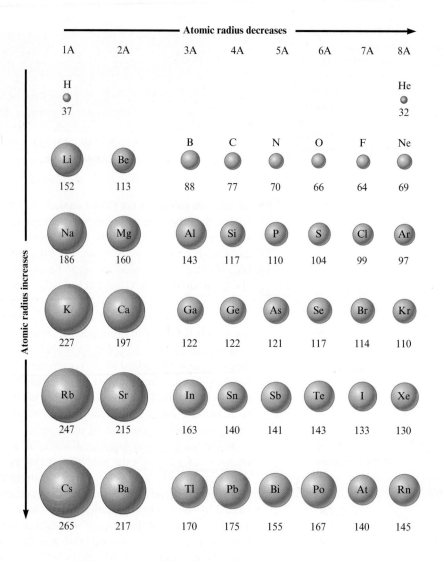

Note from Fig. 7.35 that the atomic radii decrease in going from left to right across a period. This decrease can be explained in terms of the increasing effective nuclear charge (decreasing shielding) in going from left to right. This means that the valence electrons are drawn closer to the nucleus, decreasing the size of the atom.

Atomic radius increases down a group, because of the increases in the orbital sizes in successive principal quantum levels.

Interactive Example 7.10

Sign in at http://login.cengagebrain .com to try this Interactive Example in **OWL**.

Trends in Radii

Predict the trend in radius for the following ions: Be^{2+}, Mg^{2+}, Ca^{2+}, and Sr^{2+}.

Solution

All these ions are formed by removing two electrons from an atom of a Group 2A element. In going from beryllium to strontium, we are going down the group, so the sizes increase:

$$Be^{2+} < Mg^{2+} < Ca^{2+} < Sr^{2+}$$

↑ Smallest radius   ↑ Largest radius

See Exercises 7.105, 7.106, and 7.109

7.13 | The Properties of a Group: The Alkali Metals

We have seen that the periodic table originated as a way to portray the systematic properties of the elements. Mendeleev was primarily responsible for first showing its usefulness in correlating and predicting the elemental properties. In this section we will summarize much of the information available from the table. We also will illustrate the usefulness of the table by discussing the properties of a representative group, the alkali metals.

Information Contained in the Periodic Table

1. The essence of the periodic table is that the groups of representative elements exhibit similar chemical properties that change in a regular way. The quantum mechanical model of the atom has allowed us to understand the basis for the similarity of properties in a group—that each group member has the same valence electron configuration. *It is the number and type of valence electrons that primarily determine an atom's chemistry.*

2. One of the most valuable types of information available from the periodic table is the electron configuration of any representative element. If you understand the organization of the table, you will not need to memorize electron configurations for these elements. Although the predicted electron configurations for transition metals are sometimes incorrect, this is not a serious problem. You should, however, memorize the configurations of two exceptions, chromium and copper, since these $3d$ transition elements are found in many important compounds.

3. As we mentioned in Chapter 2, certain groups in the periodic table have special names. These are summarized in Fig. 7.36. Groups are often referred to by these names, so you should learn them.

4. The most basic division of the elements in the periodic table is into metals and nonmetals. The most important chemical property of a metal atom is the tendency to give up one or more electrons to form a positive ion; metals tend to have low ionization energies. The metallic elements are found on the left side of the table, as shown in Fig. 7.36. The most chemically reactive metals are found on the lower left-hand portion of the table, where the ionization energies are smallest. The most distinctive chemical property of a nonmetal atom is the ability to gain one or more electrons to form an anion when reacting with a metal. Thus nonmetals are elements with large ionization energies and the most negative electron affinities. The nonmetals are found on the right side of the table, with the most reactive ones in the upper right-hand corner, except for the noble gas elements, which are quite unreactive. The division into metals and nonmetals shown in Fig. 7.36 is only approximate. Many elements along the division line exhibit both metallic and non-metallic properties under certain circumstances. These elements are often called **metalloids**, or sometimes **semimetals**.

The Alkali Metals

The metals of Group 1A, the alkali metals, illustrate very well the relationships among the properties of the elements in a group. Lithium, sodium, potassium, rubidium, cesium, and francium are the most chemically reactive of the metals. We will not discuss francium here because it occurs in nature in only very small quantities. Although hydrogen is found in Group 1A of the periodic table, it behaves as a nonmetal, in contrast to the other members of that group. The fundamental reason for hydrogen's nonmetallic character is its very small size (see Fig. 7.35). The electron in the small $1s$ orbital is bound tightly to the nucleus.

Metals and nonmetals were first discussed in Chapter 2.

Hydrogen will be discussed further in Chapter 20.

Figure 7.36 | Special names for groups in the periodic table.

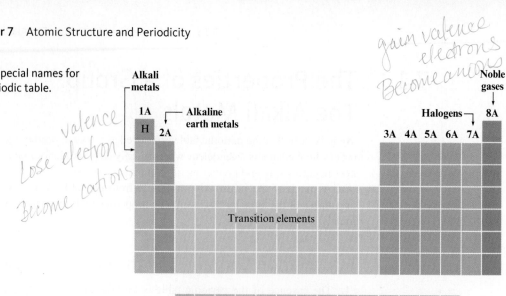

gain valence electrons Become anions

Lose valence electron Become cations

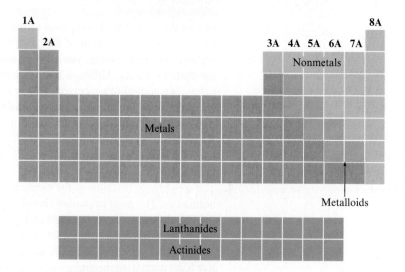

Some important properties of the first five alkali metals are shown in Table 7.8. The data in Table 7.8 show that in going down the group, the first ionization energy decreases and the atomic radius increases. This agrees with the general trends discussed in Section 7.12.

The overall increase in density in going down Group 1A is typical of all groups. This occurs because atomic mass generally increases more rapidly than atomic size. Thus there is more mass per unit volume for each succeeding element.

Other groups will be discussed in Chapters 20 and 21.

Table 7.8 | Properties of Five Alkali Metals

Element	Valence Electron Configuration	Density at 25°C (g/cm³)	mp (°C)	bp (°C)	First Ionization Energy (kJ/mol)	Atomic (covalent) Radius (pm)	Ionic (M⁺) Radius (pm)
Li	$2s^1$	0.53	180	1330	520	152	60
Na	$3s^1$	0.97	98	892	495	186	95
K	$4s^1$	0.86	64	760	419	227	133
Rb	$5s^1$	1.53	39	668	409	247	148
Cs	$6s^1$	1.87	29	690	382	265	169

Chemical connections

Potassium—Too Much of a Good Thing Can Kill You

Potassium is widely recognized as an essential element. In fact, our daily requirement for potassium is more than twice that for sodium. Because most foods contain potassium, serious deficiency of this element in humans is rare. However, potassium deficiency can be caused by kidney malfunction or by the use of certain diuretics. Potassium deficiency leads to muscle weakness, irregular heartbeat, and depression.

Potassium is found in the fluids of the body as the K^+ ion, and its presence is essential to the operation of our nervous system. The passage of impulses along the nerves requires the flow of K^+ (and Na^+) through channels in the membranes of the nerve cells. Failure of this ion flow prevents nerve transmissions and results in death. For example, the black mamba snake kills its victims by injecting a venom that blocks the potassium channels in the nerve cells.

Although a steady intake of potassium is essential to preserve life, ironically, too much potassium can be lethal. In fact, the deadly ingredient in the drug mixture used for executing

The black mamba snake's venom kills by blocking the potassium channels in the nerve cells of victims.

criminals is potassium chloride. Injection of a large amount of a potassium chloride solution produces an excess of K^+ ion in the fluids surrounding the cells and prevents the essential flow of K^+ out the cells to allow nerve impulses to occur. This causes the heart to stop beating.

Unlike other forms of execution, death by lethal injection of potassium chloride does not harm the organs of the body. Thus condemned criminals who are executed in this manner could potentially donate their organs for transplants. However, this idea is very controversial.

Oxidation–reduction reactions were discussed in Chapter 4.

The smooth decrease in melting point and boiling point in going down Group 1A is not typical; in most other groups, more complicated behavior occurs. Note that the melting point of cesium is only 29°C. Cesium can be melted readily using only the heat from your hand. This is very unusual—metals typically have rather high melting points. For example, tungsten melts at 3410°C. The only other metals with low melting points are mercury (mp −38°C) and gallium (mp 30°C).

The chemical property most characteristic of a metal is the ability to lose its valence electrons. The Group 1A elements are very reactive. They have low ionization energies and react with nonmetals to form ionic solids. A typical example involves the reaction of sodium with chlorine to form sodium chloride:

$$2Na(s) + Cl_2(g) \longrightarrow 2NaCl(s)$$

where sodium chloride contains Na^+ and Cl^- ions. This is an oxidation–reduction reaction in which chlorine oxidizes sodium. In the reactions between metals and

Potassium reacts vigorously with water.

nonmetals, it is typical for the nonmetal to behave as the oxidizing agent and the metal to behave as the reducing agent, as shown by the following reactions:

$$2Na(s) + S(s) \longrightarrow Na_2S(s)$$
Contains Na^+ and S^{2-} ions

$$6Li(s) + N_2(g) \longrightarrow 2Li_3N(s)$$
Contains Li^+ and N^{3-} ions

$$2Na(s) + O_2(g) \longrightarrow Na_2O_2(s)$$
Contains Na^+ and $O_2{}^{2-}$ ions

For reactions of the types just shown, the relative reducing powers of the alkali metals can be predicted from the first ionization energies listed in Table 7.8. Since it is much easier to remove an electron from a cesium atom than from a lithium atom, cesium should be the better reducing agent. The expected trend in reducing ability is

$$Cs > Rb > K > Na > Li$$

This order is observed experimentally for direct reactions between the solid alkali metals and nonmetals. However, this is not the order for reducing ability found when the alkali metals react in aqueous solution. For example, the reduction of water by an alkali metal is very vigorous and exothermic:

$$2M(s) + 2H_2O(l) \longrightarrow H_2(g) + 2M^+(aq) + 2OH^-(aq) + \text{energy}$$

The order of reducing abilities observed for this reaction for the first three group members is

$$Li > K > Na$$

In the gas phase, potassium loses an electron more easily than sodium, and sodium more easily than lithium. Thus it is surprising that lithium is the best reducing agent toward water.

This reversal occurs because the formation of the M^+ ions in aqueous solution is strongly influenced by the hydration of these ions by the polar water molecules. The *hydration energy* of an ion represents the change in energy that occurs when water molecules attach to the M^+ ion. The hydration energies for the Li^+, Na^+, and K^+ ions (Table 7.9) indicate that the process is exothermic in each case. However, nearly twice as much energy is released by the hydration of the Li^+ ion as for the K^+ ion. This difference is caused by size effects; the Li^+ ion is much smaller than the K^+ ion, and thus its *charge density* (charge per unit volume) is also much greater. This means that the polar water molecules are more strongly attracted to the small Li^+ ion. Because the Li^+ ion is so strongly hydrated, its formation from the lithium atom occurs more readily than the formation of the K^+ ion from the potassium atom. Although a potassium atom in the gas phase loses its valence electron more readily than a lithium atom in the gas phase, the opposite is true in aqueous solution. This anomaly is an example of the importance of the polarity of the water molecule in aqueous reactions.

There is one more surprise involving the highly exothermic reactions of the alkali metals with water. Experiments show that in water lithium is the best reducing agent, so we might expect that lithium should react the most vigorously with water. However, this is not true. Sodium and potassium react much more vigorously. Why is this so? The answer lies in the relatively high melting point of lithium. When sodium and potassium react with water, the heat evolved causes them to melt, giving a larger area of contact with water. Lithium, on the other hand, does not melt under these conditions and reacts more slowly. This illustrates the important principle (which we will discuss in detail in Chapter 12) that the energy change for a reaction and the rate at which it occurs are not necessarily related.

In this section we have seen that the trends in atomic properties summarized by the periodic table can be a great help in understanding the chemical behavior of the elements. This fact will be emphasized over and over as we proceed in our study of chemistry.

Table 7.9 | Hydration Energies for Li^+, Na^+, and K^+ Ions

Ion	Hydration Energy (kJ/mol)
Li^+	-510
Na^+	-402
K^+	-314

For review

Key terms

Electromagnetic radiation

> Characterized by its wavelength (λ), frequency (ν), and speed ($c = 2.9979 \times 10^8$ m/s)

$$\lambda\nu = c$$

> Can be viewed as a stream of "particles" called photons, each with energy $h\nu$, where h is Planck's constant (6.626×10^{-34} J · s)

Photoelectric effect

> When light strikes a metal surface, electrons are emitted
> Analysis of the kinetic energy and numbers of the emitted electrons led Einstein to suggest that electromagnetic radiation can be viewed as a stream of photons

Hydrogen spectrum

> The emission spectrum of hydrogen shows discrete wavelengths
> Indicates that hydrogen has discrete energy levels

Bohr model of the hydrogen atom

> Using the data from the hydrogen spectrum and assuming angular momentum to be quantized, Bohr devised a model in which the electron traveled in circular orbits
> Although an important pioneering effort, this model proved to be entirely incorrect

Wave (quantum) mechanical model

> An electron is described as a standing wave
> The square of the wave function (often called an orbital) gives a probability distribution for the electron position
> The exact position of the electron is never known, which is consistent with the Heisenberg uncertainty principle: It is impossible to know accurately both the position and the momentum of a particle simultaneously
> Probability maps are used to define orbital shapes
> Orbitals are characterized by the quantum numbers n, ℓ, and m_ℓ

Electron spin

> Described by the spin quantum number m_s, which can have values of $\pm\frac{1}{2}$
> Pauli exclusion principle: No two electrons in a given atom can have the same set of quantum numbers n, ℓ, m_ℓ, and m_s
> Only two electrons with opposite spins can occupy a given orbital

Periodic table

> By populating the orbitals from the wave mechanical model (the aufbau principle), the form of the periodic table can be explained
> According to the wave mechanical model, atoms in a given group have the same valence (outer) electron configuration
> The trends in properties such as ionization energies and atomic radii can be explained in terms of the concepts of nuclear attraction, electron repulsions, shielding, and penetration

Key terms

Section 7.11
aufbau principle
Hund's rule
valence electrons
core electrons
transition metals
lanthanide series
actinide series
main-group elements
    (representative elements)

Section 7.12
first ionization energy
second ionization energy
electron affinity
atomic radii

Section 7.13
metalloids (semimetals)

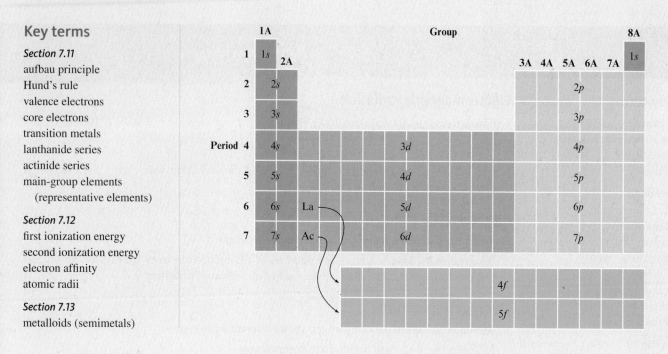

Review questions   *Answers to the Review Questions can be found on the Student website (accessible from www.cengagebrain.com).*

1. Four types of electromagnetic radiation (EMR) are ultraviolet, microwaves, gamma rays, and visible. All of these types of EMR can be characterized by wavelength, frequency, photon energy, and speed of travel. Define these terms, and rank the four types of electromagnetic radiation in order of increasing wavelength, frequency, photon energy, and speed.

2. Characterize the Bohr model of the atom. In the Bohr model, what do we mean when we say something is quantized? How does the Bohr model of the hydrogen atom explain the hydrogen emission spectrum? Why is the Bohr model fundamentally incorrect?

3. What experimental evidence supports the quantum theory of light? Explain the wave-particle duality of all matter. For what size particles must one consider both the wave and the particle properties?

4. List the most important ideas of the quantum mechanical model of the atom. Include in your discussion the terms or names *wave function, orbital, Heisenberg uncertainty principle, de Broglie, Schrödinger,* and *probability distribution.*

5. What are quantum numbers? What information do we get from the quantum numbers n, ℓ, and m_ℓ? We define a spin quantum number (m_s), but do we know that an electron literally spins?

6. How do $2p$ orbitals differ from each other? How do $2p$ and $3p$ orbitals differ from each other? What is a nodal surface in an atomic orbital? What is wrong with $1p$, $1d$, $2d$, $1f$, $2f$, and $3f$ orbitals? Explain what we mean when

we say that a $4s$ electron is more penetrating than a $3d$ electron.

7. Four blocks of elements in a periodic table refer to various atomic orbitals being filled. What are the four blocks and the corresponding orbitals? How do you get the energy ordering of the atomic orbitals from the periodic table? What is the aufbau principle? Hund's rule? The Pauli exclusion principle? There are two common exceptions to the ground-state electron configuration for elements 1–36 as predicted by the periodic table. What are they?

8. What is the difference between core electrons and valence electrons? Why do we emphasize the valence electrons in an atom when discussing atomic properties? What is the relationship between valence electrons and elements in the same group of the periodic table?

9. Using the element phosphorus as an example, write the equation for a process in which the energy change will correspond to the ionization energy and to the electron affinity.

    Explain why the first ionization energy tends to increase as one proceeds from left to right across a period. Why is the first ionization energy of aluminum lower than that of magnesium, and the first ionization energy of sulfur lower than that of phosphorus?

    Why do the successive ionization energies of an atom always increase? Note the successive ionization energies for silicon given in Table 7.5. Would you expect to see any large jumps between successive ionization energies of silicon as you removed all the electrons, one by one, beyond those shown in the table?

10. The radius trend and the ionization energy trend are exact opposites. Does this make sense? Define electron affinity. Electron affinity values are both exothermic (negative) and endothermic (positive). However, ionization energy values are always endothermic (positive). Explain.

Active Learning Questions

These questions are designed to be used by groups of students in class.

1. What does it mean for something to have wavelike properties? Particulate properties? Electromagnetic radiation can be discussed in terms of both particles and waves. Explain the experimental verification for each of these views.

2. Defend and criticize Bohr's model. Why was it reasonable that such a model was proposed, and what evidence was there that it "works"? Why do we no longer "believe" in it?

3. The first four ionization energies for the elements X and Y are shown below. The units are not kJ/mol.

	X	Y
First	170	200
Second	350	400
Third	1800	3500
Fourth	2500	5000

Identify the elements X and Y. There may be more than one correct answer, so explain completely.

4. Compare the first ionization energy of helium to its second ionization energy, remembering that both electrons come from the $1s$ orbital. Explain the difference without using actual numbers from the text.

5. Which has the larger second ionization energy, lithium or beryllium? Why?

6. Explain why a graph of ionization energy versus atomic number (across a row) is not linear. Where are the exceptions? Why are there exceptions?

7. Without referring to your text, predict the trend of second ionization energies for the elements sodium through argon. Compare your answer with Table 7.5. Explain any differences.

8. Account for the fact that the line that separates the metals from the nonmetals on the periodic table is diagonal downward to the right instead of horizontal or vertical.

9. Explain *electron* from a quantum mechanical perspective, including a discussion of atomic radii, probabilities, and orbitals.

10. Choose the best response for the following. The ionization energy for the chlorine atom is equal in magnitude to the electron affinity for

a. the Cl atom.   **d.** the F atom.

b. the Cl^- ion.   **e.** none of these.

c. the Cl^+ ion.

Explain each choice. Justify your choice, and for the choices you did not select, explain what is incorrect about them.

11. Consider the following statement: "The ionization energy for the potassium atom is negative, because when K loses an electron to become K^+, it achieves a noble gas electron configuration." Indicate everything that is correct in this statement. Indicate everything that is incorrect. Correct the incorrect information, and explain.

12. In going across a row of the periodic table, electrons are added and ionization energy generally increases. In going down a column of the periodic table, electrons are also being added but ionization energy decreases. Explain.

13. How does probability fit into the description of the atom?

14. What is meant by an *orbital*?

15. Explain the difference between the probability density distribution for an orbital and its radial probability.

16. Is the following statement true or false? The hydrogen atom has a $3s$ orbital. Explain.

17. Which is higher in energy: the $2s$ or $2p$ orbital in hydrogen? Is this also true for helium? Explain.

18. Prove mathematically that it is more energetically favorable for a fluorine atom to take an electron from a sodium atom than for a fluorine atom to take an electron from another fluorine atom.

A blue question or exercise number indicates that the answer to that question or exercise appears at the back of the book and a solution appears in the *Solutions Guide,* as found on PowerLecture.

Questions

19. What type of relationship (direct or inverse) exists between wavelength, frequency, and photon energy? What does a photon energy unit of a joule equal? $E = \lambda \nu$ $E = 2.18 \times 10^5$ *[handwritten]* 18

20. What do we mean by the *frequency* of electromagnetic radiation? Is the frequency the same as the *speed* of the electromagnetic radiation?   *[handwritten]* cycle/s = s⁻¹ = Hz   so light/EMR ≠ frequency

21. Explain the photoelectric effect.   *[handwritten]* v > h·v₀ add KE to e   v < hv₀ - oe's move,

22. Describe briefly why the study of electromagnetic radiation has been important to our understanding of the arrangement of electrons in atoms.   *[handwritten]* wave vs. particulate properties discovered

23. How does the wavelength of a fast-pitched baseball compare to the wavelength of an electron traveling at 1/10 the speed of light? What is the significance of this comparison? See Example 7.3.   *[handwritten]* 2.9979 × 10⁸ → 2.9979 × 10⁷ m/s

24. The following is an energy-level diagram for electronic transitions in the Bohr hydrogen atom.

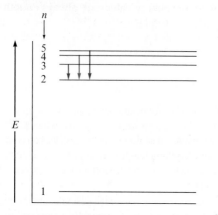

   a. Explain why the energy levels get closer together as they increase. Provide mathematical support for this.
   b. Verify that the colors given in the diagram are correct. Provide mathematical support.

25. The Bohr model only works for one electron species. Why do we discuss it in this text (what's good about it)?

26. We can represent both probability and radial probability versus distance from the nucleus for a hydrogen $1s$ orbital as depicted below.

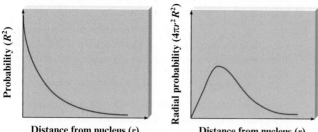

   What does each graph tell us about the electron in a hydrogen $1s$ orbital? Describe the significance of the radial probability distribution.

27. Consider the representations of the p and d atomic orbitals in Figs. 7.15 and 7.17. What do the $+$ and $-$ signs indicate?

28. The periodic table consists of four blocks of elements that correspond to s, p, d, and f orbitals being filled. After f orbitals come g and h orbitals. In theory, if a g block and an h block of elements existed, how long would the rows of g and h elements be in this theoretical periodic table?

29. Many times the claim is made that subshells half-filled with electrons are particularly stable. Can you suggest a possible physical basis for this claim?

30. Diagonal relationships in the periodic table exist as well as the vertical relationships. For example, Be and Al are similar in some of their properties, as are B and Si. Rationalize why these diagonal relationships hold for properties such as size, ionization energy, and electron affinity.

31. Elements with very large ionization energies also tend to have highly exothermic electron affinities. Explain. Which group of elements would you expect to be an exception to this statement?

32. The changes in electron affinity as one goes down a group in the periodic table are not nearly as large as the variations in ionization energies. Why?

33. Why is it much harder to explain the line spectra of polyelectronic atoms and ions than it is to explain the line spectra of hydrogen and hydrogen-like ions?

34. Scientists use emission spectra to confirm the presence of an element in materials of unknown composition. Why is this possible?

35. Does the minimization of electron–electron repulsions correlate with Hund's rule?

36. In the hydrogen atom, what is the physical significance of the state for which $n = \infty$ and $E = 0$?

37. The work function is the energy required to remove an electron from an atom on the surface of a metal. How does this definition differ from that for ionization energy?

38. Many more anhydrous lithium salts are hygroscopic (readily absorb water) than are those of the other alkali metals. Explain.

Exercises

In this section similar exercises are paired.

Light and Matter

39. The laser in an audio CD player uses light with a wavelength of 7.80×10^2 nm. Calculate the frequency of this light.

40. An FM radio station broadcasts at 99.5 MHz. Calculate the wavelength of the corresponding radio waves.

41. Microwave radiation has a wavelength on the order of 1.0 cm. Calculate the frequency and the energy of a single photon of this radiation. Calculate the energy of an Avogadro's number of photons (called an *einstein*) of this radiation.

42. A photon of ultraviolet (UV) light possesses enough energy to mutate a strand of human DNA. What is the energy of a single UV photon and a mole of UV photons having a wavelength of 25 nm?

43. Octyl methoxycinnamate and oxybenzone are common ingredients in sunscreen applications. These compounds work by absorbing ultraviolet (UV) B light (wavelength 280–320 nm), the UV light most associated with sunburn symptoms. What frequency range of light do these compounds absorb?

44. Human color vision is "produced" by the nervous system based on how three different cone receptors interact with photons of light in the eye. These three different types of cones interact with photons of different frequency light, as indicated in the following chart:

Cone Type	Range of Light Frequency Detected
S	$6.00–7.49 \times 10^{14}$ s^{-1}
M	$4.76–6.62 \times 10^{14}$ s^{-1}
L	$4.28–6.00 \times 10^{14}$ s^{-1}

What wavelength ranges (and corresponding colors) do the three types of cones detect?

45. Consider the following waves representing electromagnetic radiation:

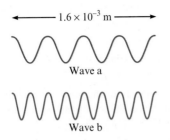

$\longleftarrow 1.6 \times 10^{-3}$ m $\longrightarrow$

Wave a

Wave b

Which wave has the longer wavelength? Calculate the wavelength. Which wave has the higher frequency and larger photon energy? Calculate these values. Which wave has the greater velocity? What type of electromagnetic radiation does each wave represent?

46. One type of electromagnetic radiation has a frequency of 107.1 MHz, another type has a wavelength of 2.12×10^{-10} m, and another type of electromagnetic radiation has photons with energy equal to 3.97×10^{-19} J/photon. Identify each type of electromagnetic radiation and place them in order of increasing photon energy and increasing frequency.

47. Carbon absorbs energy at a wavelength of 150. nm. The total amount of energy emitted by a carbon sample is 1.98×10^5 J. Calculate the number of carbon atoms present in the sample, assuming that each atom emits one photon.

48. X rays have wavelengths on the order of 1×10^{-10} m. Calculate the energy of 1.0×10^{-10} m X rays in units of kilojoules per mole of X rays. AM radio waves have wavelengths on the order of 1×10^4 m. Calculate the energy of 1.0×10^4 m radio waves in units of kilojoules per mole of radio waves. Consider that the bond energy of a carbon–carbon single bond found in organic compounds is 347 kJ/mol. Would X rays and/or radio waves be able to disrupt organic compounds by breaking carbon–carbon single bonds?

49. The work function of an element is the energy required to remove an electron from the surface of the solid element. The work function for lithium is 279.7 kJ/mol (that is, it takes 279.7 kJ of energy to remove one mole of electrons from one mole of Li atoms on the surface of Li metal). What is the maximum wavelength of light that can remove an electron from an atom on the surface of lithium metal?

50. It takes 208.4 kJ of energy to remove 1 mole of electrons from an atom on the surface of rubidium metal. How much energy does it take to remove a single electron from an atom on the surface of solid rubidium? What is the maximum wavelength of light capable of doing this?

51. It takes 7.21×10^{-19} J of energy to remove an electron from an iron atom. What is the maximum wavelength of light that can do this?

52. Ionization energy is the energy required to remove an electron from an atom in the gas phase. The ionization energy of gold is 890.1 kJ/mol. Is light with a wavelength of 225 nm capable of ionizing a gold atom (removing an electron) in the gas phase?

53. Calculate the de Broglie wavelength for each of the following.
    a. an electron with a velocity 10.% of the speed of light
    b. a tennis ball (55 g) served at 35 m/s ($\sim$80 mi/h)

54. Neutron diffraction is used in determining the structures of molecules.
    a. Calculate the de Broglie wavelength of a neutron moving at 1.00% of the speed of light.
    b. Calculate the velocity of a neutron with a wavelength of 75 pm (1 pm = 10^{-12} m).

55. A particle has a velocity that is 90.% of the speed of light. If the wavelength of the particle is 1.5×10^{-15} m, calculate the mass of the particle.

56. Calculate the velocities of electrons with de Broglie wavelengths of 1.0×10^2 nm and 1.0 nm, respectively.

Hydrogen Atom: The Bohr Model

57. Calculate the wavelength of light emitted when each of the following transitions occur in the hydrogen atom. What type of electromagnetic radiation is emitted in each transition?
    a. $n = 3 \rightarrow n = 2$
    b. $n = 4 \rightarrow n = 2$
    c. $n = 2 \rightarrow n = 1$

58. Calculate the wavelength of light emitted when each of the following transitions occur in the hydrogen atom. What type of electromagnetic radiation is emitted in each transition?
    a. $n = 4 \rightarrow n = 3$
    b. $n = 5 \rightarrow n = 4$
    c. $n = 5 \rightarrow n = 3$

59. Using vertical lines, indicate the transitions from Exercise 57 on an energy-level diagram for the hydrogen atom (see Fig. 7.9).

60. Using vertical lines, indicate the transitions from Exercise 58 on an energy-level diagram for the hydrogen atom (see Fig. 7.9).

61. Calculate the longest and shortest wavelengths of light emitted by electrons in the hydrogen atom that begin in the $n = 6$ state and then fall to states with smaller values of n.

62. Assume that a hydrogen atom's electron has been excited to the $n = 5$ level. How many different wavelengths of light can be emitted as this excited atom loses energy?

63. Does a photon of visible light ($\lambda \approx 400$ to 700 nm) have sufficient energy to excite an electron in a hydrogen atom from the $n = 1$ to the $n = 5$ energy state? from the $n = 2$ to the $n = 6$ energy state?

64. An electron is excited from the $n = 1$ ground state to the $n = 3$ state in a hydrogen atom. Which of the following statements are true? Correct the false statements to make them true.
    a. It takes more energy to ionize (completely remove) the electron from $n = 3$ than from the ground state.
    b. The electron is farther from the nucleus on average in the $n = 3$ state than in the $n = 1$ state.
    c. The wavelength of light emitted if the electron drops from $n = 3$ to $n = 2$ will be shorter than the wavelength of light emitted if the electron falls from $n = 3$ to $n = 1$.
    d. The wavelength of light emitted when the electron returns to the ground state from $n = 3$ will be the same as the wavelength of light absorbed to go from $n = 1$ to $n = 3$.
    e. For $n = 3$, the electron is in the first excited state.

65. Calculate the maximum wavelength of light capable of removing an electron for a hydrogen atom from the energy state characterized by $n = 1$, by $n = 2$.

66. Consider an electron for a hydrogen atom in an excited state. The maximum wavelength of electromagnetic radiation that can completely remove (ionize) the electron from the H atom is 1460 nm. What is the initial excited state for the electron ($n = $?)?

67. An excited hydrogen atom with an electron in the $n = 5$ state emits light having a frequency of 6.90×10^{14} s^{-1}. Determine the principal quantum level for the final state in this electronic transition.

68. An excited hydrogen atom emits light with a wavelength of 397.2 nm to reach the energy level for which $n = 2$. In which principal quantum level did the electron begin?

Quantum Mechanics, Quantum Numbers, and Orbitals

69. Using the Heisenberg uncertainty principle, calculate Δx for each of the following.
    a. an electron with $\Delta v = 0.100$ m/s
    b. a baseball (mass = 145 g) with $\Delta v = 0.100$ m/s
    c. How does the answer in part a compare with the size of a hydrogen atom?
    d. How does the answer in part b correspond to the size of a baseball?

70. The Heisenberg uncertainty principle can be expressed in the form

$$\Delta E \cdot \Delta t \geq \frac{h}{4\pi}$$

where E represents energy and t represents time. Show that the units for this form are the same as the units for the form used in this chapter:

$$\Delta x \cdot \Delta(mv) \geq \frac{h}{4\pi}$$

71. What are the possible values for the quantum numbers n, ℓ, and m_ℓ?

72. Identify each of the following orbitals and determine the n and l quantum numbers. Explain your answers.
    a.

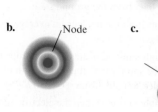

    b.

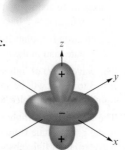

Node

    c.

73. Which of the following sets of quantum numbers are not allowed in the hydrogen atom? For the sets of quantum numbers that are incorrect, state what is wrong in each set.
    a. $n = 3$, $\ell = 2$, $m_\ell = 2$
    b. $n = 4$, $\ell = 3$, $m_\ell = 4$
    c. $n = 0$, $\ell = 0$, $m_\ell = 0$
    d. $n = 2$, $\ell = -1$, $m_\ell = 1$

74. Which of the following sets of quantum numbers are not allowed? For each incorrect set, state why it is incorrect.
    a. $n = 3$, $\ell = 3$, $m_\ell = 0$, $m_s = -\frac{1}{2}$
    b. $n = 4$, $\ell = 3$, $m_\ell = 2$, $m_s = -\frac{1}{2}$
    c. $n = 4$, $\ell = 1$, $m_\ell = 1$, $m_s = +\frac{1}{2}$
    d. $n = 2$, $\ell = 1$, $m_\ell = -1$, $m_s = -1$
    e. $n = 5$, $\ell = -4$, $m_\ell = 2$, $m_s = +\frac{1}{2}$
    f. $n = 3$, $\ell = 1$, $m_\ell = 2$, $m_s = -\frac{1}{2}$

75. What is the physical significance of the value of ψ^2 at a particular point in an atomic orbital?

76. In defining the sizes of orbitals, why must we use an arbitrary value, such as 90% of the probability of finding an electron in that region?

Polyelectronic Atoms

77. Total radial probability distributions for the helium, neon, and argon atoms are shown in the following graph. How can one interpret the shapes of these curves in terms of electron configurations, quantum numbers, and nuclear charges?

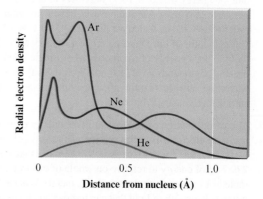

78. The relative orbital levels for the hydrogen atom can be represented as

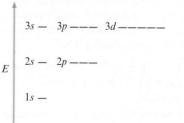

$3s$ — $3p$ ——— $3d$ ——————

$2s$ — $2p$ ———

E

$1s$ —

Draw the relative orbital energy levels for atoms with more than one electron, and explain your answer. Also explain how the following radial probability distributions support your answer.

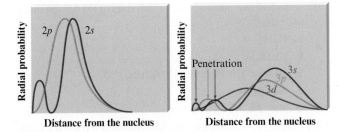

Distance from the nucleus   **Distance from the nucleus**

79. How many orbitals in an atom can have the designation $5p$, $3d_{z^2}$, $4d$, $n = 5$, $n = 4$?

80. How many electrons in an atom can have the designation $1p$, $6d_{x^2-y^2}$, $4f$, $7p_y$, $2s$, $n = 3$?

81. Give the maximum number of electrons in an atom that can have these quantum numbers:
    a. $n = 4$
    b. $n = 5, m_\ell = +1$
    c. $n = 5, m_s = +\frac{1}{2}$
    d. $n = 3, \ell = 2$
    e. $n = 2, \ell = 1$

82. Give the maximum number of electrons in an atom that can have these quantum numbers:
    a. $n = 0, \ell = 0, m_\ell = 0$
    b. $n = 2, \ell = 1, m_\ell = -1, m_s = -\frac{1}{2}$
    c. $n = 3, m_s = +\frac{1}{2}$
    d. $n = 2, \ell = 2$
    e. $n = 1, \ell = 0, m_\ell = 0$

83. Draw atomic orbital diagrams representing the ground-state electron configuration for each of the following elements.
    a. Na
    b. Co
    c. Kr
    How many unpaired electrons are present in each element?

84. For elements 1–36, there are two exceptions to the filling order as predicted from the periodic table. Draw the atomic orbital diagrams for the two exceptions and indicate how many unpaired electrons are present.

85. The elements Si, Ga, As, Ge, Al, Cd, S, and Se are all used in the manufacture of various semiconductor devices. Write the expected electron configuration for these atoms.

86. The elements Cu, O, La, Y, Ba, Tl, and Bi are all found in high-temperature ceramic superconductors. Write the expected electron configuration for these atoms.

87. Write the expected electron configurations for each of the following atoms: Sc, Fe, P, Cs, Eu, Pt, Xe, Br.

88. Write the expected electron configurations for each of the following atoms: Cl, Sb, Sr, W, Pb, Cf.

89. Write the expected ground-state electron configuration for the following:
    a. the element with one unpaired $5p$ electron that forms a covalent with compound fluorine
    b. the (as yet undiscovered) alkaline earth metal after radium

    c. the noble gas with electrons occupying $4f$ orbitals
    d. the first-row transition metal with the most unpaired electrons

90. Using only the periodic table inside the front cover of the text, write the expected ground-state electron configurations for
    a. the third element in Group 5A.
    b. element number 116.
    c. an element with three unpaired $5d$ electrons.
    d. the halogen with electrons in the $6p$ atomic orbitals.

91. Given the valence electron orbital level diagram and the description, identify the element or ion.
    a. a ground-state atom

    $3s$ $3p$

    [↑↓]   [↑↓][↑][↑]

    b. an atom in an excited state (assume two electrons occupy the $1s$ orbital)

    $2s$ $2p$

    [↑]   [↑↓][↑][↑]

    c. a ground-state ion with a charge of -1

    $4s$ $4p$

    [↑↓]   [↑↓][↑↓][↑]

92. Identify the following elements.
    a. An excited state of this element has the electron configuration $1s^2 2s^2 2p^5 3s^1$.
    b. The ground-state electron configuration is $[Ne]3s^2 3p^4$.
    c. An excited state of this element has the electron configuration $[Kr]5s^2 4d^6 5p^2 6s^1$.
    d. The ground-state electron configuration contains three unpaired $6p$ electrons.

93. In the ground state of mercury, Hg,
    a. how many electrons occupy atomic orbitals with $n = 3$?
    b. how many electrons occupy d atomic orbitals?
    c. how many electrons occupy p_z atomic orbitals?
    d. how many electrons have spin "up" ($m_s = +\frac{1}{2}$)?

94. In the ground state of element 115, Uup,
    a. how many electrons have $n = 5$ as one of their quantum numbers?
    b. how many electrons have $\ell = 3$ as one of their quantum numbers?
    c. how many electrons have $m_\ell = 1$ as one of their quantum numbers?
    d. how many electrons have $m_s = -\frac{1}{2}$ as one of their quantum numbers?

95. Give a possible set of values of the four quantum numbers for all the electrons in a boron atom and a nitrogen atom if each is in the ground state.

96. Give a possible set of values of the four quantum numbers for the $4s$ and $3d$ electrons in titanium.

97. Valence electrons are those electrons in the outermost principal quantum level (highest n level) of an atom in its ground state. Groups 1A to 8A have from 1 to 8 valence electrons. For each group of the representative elements (1A–8A), give the number of valence electrons, the general valence electron configuration, a sample element in that group, and the specific valence electron configuration for that element.

98. How many valence electrons do each of the following elements have, and what are the specific valence electrons for each element?

 a. Ca
 b. O
 c. element 117
 d. In
 e. Ar
 f. Bi

99. A certain oxygen atom has the electron configuration $1s^2 2s^2 2p_x^2 2p_y^2$. How many unpaired electrons are present? Is this an excited state of oxygen? In going from this state to the ground state, would energy be released or absorbed?

100. Which of the following electron configurations correspond to an excited state? Identify the atoms and write the ground-state electron configuration where appropriate.

 a. $1s^2 2s^2 3p^1$
 b. $1s^2 2s^2 2p^6$
 c. $1s^2 2s^2 2p^4 3s^1$
 d. $[Ar]4s^2 3d^5 4p^1$

 How many unpaired electrons are present in each of these species?

101. Which of elements 1–36 have two unpaired electrons in the ground state?

102. Which of elements 1–36 have one unpaired electron in the ground state?

103. One bit of evidence that the quantum mechanical model is "correct" lies in the magnetic properties of matter. Atoms with unpaired electrons are attracted by magnetic fields and thus are said to exhibit *paramagnetism*. The degree to which this effect is observed is directly related to the number of unpaired electrons present in the atom. Consider the ground-state electron configurations for Li, N, Ni, Te, Ba, and Hg. Which of these atoms would be expected to be paramagnetic, and how many unpaired electrons are present in each paramagnetic atom?

104. How many unpaired electrons are present in each of the following in the ground state: O, O^+, O^-, Os, Zr, S, F, Ar?

The Periodic Table and Periodic Properties

105. Arrange the following groups of atoms in order of increasing size.

 a. Te, S, Se
 b. K, Br, Ni
 c. Ba, Si, F

106. Arrange the following groups of atoms in order of increasing size.

 a. Rb, Na, Be
 b. Sr, Se, Ne
 c. Fe, P, O

107. Arrange the atoms in Exercise 105 in order of increasing first ionization energy.

108. Arrange the atoms in Exercise 106 in order of increasing first ionization energy.

109. In each of the following sets, which atom or ion has the smallest radius?

 a. H, He
 b. Cl, In, Se
 c. element 120, element 119, element 116
 d. Nb, Zn, Si
 e. Na^-, Na, Na^+

110. In each of the following sets, which atom or ion has the smallest ionization energy?

 a. Ca, Sr, Ba
 b. K, Mn, Ga
 c. N, O, F
 d. S^{2-}, S, S^{2+}
 e. Cs, Ge, Ar

111. Element 106 has been named seaborgium, Sg, in honor of Glenn Seaborg, discoverer of the first transuranium element.

 a. Write the expected electron configuration for element 106.
 b. What other element would be most like element 106 in its properties?
 c. Predict the formula for a possible oxide and a possible oxyanion of element 106.

112. Predict some of the properties of element 117 (the symbol is Uus, following conventions proposed by the International Union of Pure and Applied Chemistry, or IUPAC).

 a. What will be its electron configuration?
 b. What element will it most resemble chemically?
 c. What will be the formula of the neutral binary compounds it forms with sodium, magnesium, carbon, and oxygen?
 d. What oxyanions would you expect Uus to form?

113. The first ionization energies of As and Se are 0.947 and 0.941 MJ/mol, respectively. Rationalize these values in terms of electron configurations.

114. Rank the elements Be, B, C, N, and O in order of increasing first ionization energy. Explain your reasoning.

115. Consider the following ionization energies for aluminum:

$$Al(g) \longrightarrow Al^+(g) + e^- \quad I_1 = 580 \text{ kJ/mol}$$
$$Al^+(g) \longrightarrow Al^{2+}(g) + e^- \quad I_2 = 1815 \text{ kJ/mol}$$
$$Al^{2+}(g) \longrightarrow Al^{3+}(g) + e^- \quad I_3 = 2740 \text{ kJ/mol}$$
$$Al^{3+}(g) \longrightarrow Al^{4+}(g) + e^- \quad I_4 = 11{,}600 \text{ kJ/mol}$$

 a. Account for the trend in the values of the ionization energies.
 b. Explain the large increase between I_3 and I_4.

116. The following graph plots the first, second, and third ionization energies for Mg, Al, and Si.

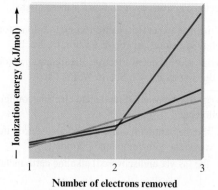

Without referencing the text, which plot corresponds to which element? In one of the plots, there is a huge jump in energy between I_2 and I_3, unlike in the other two plots. Explain this phenomenon.

117. For each of the following pairs of elements

(C and N)   (Ar and Br)

pick the atom with

   a. more favorable (exothermic) electron affinity.

   b. higher ionization energy.

   c. larger size.

118. For each of the following pairs of elements

(Mg and K)   (F and Cl)

pick the atom with

   a. more favorable (exothermic) electron affinity.

   b. higher ionization energy.

   c. larger size.

119. The electron affinities of the elements from aluminum to chlorine are -44, -120, -74, -200.4, and -384.7 kJ/mol, respectively. Rationalize the trend in these values.

120. In the second row of the periodic table, Be, N, and Ne all have endothermic (unfavorable) electron affinities, whereas the other second-row elements have exothermic (favorable) electron affinities. Rationalize why Be, N, and Ne have unfavorable electron affinities.

121. Order the atoms in each of the following sets from the least exothermic electron affinity to the most.

   a. S, Se   b. F, Cl, Br, I

122. Order the atoms in each of the following sets from the least exothermic electron affinity to the most.

   a. N, O, F   b. Al, Si, P

123. The electron affinity for sulfur is more exothermic than that for oxygen. How do you account for this?

124. Which has the more negative electron affinity, the oxygen atom or the O^- ion? Explain your answer.

125. Write equations corresponding to the following.

   a. the fourth ionization energy of Se

   b. the electron affinity of S^-

   c. the electron affinity of Fe^{3+}

   d. the ionization energy of Mg

126. Using data from the text, determine the following values (justify your answer):

   a. the electron affinity of Mg^{2+}

   b. the ionization energy of Cl^-

   c. the electron affinity of Cl^+

   d. the ionization energy of Mg^- (Electron affinity of Mg = 230 kJ/mol)

Alkali Metals

127. An ionic compound of potassium and oxygen has the empirical formula KO. Would you expect this compound to be potassium(II) oxide or potassium peroxide? Explain.

128. Give the name and formula of each of the binary compounds formed from the following elements.

   a. Li and N

   b. Na and Br

   c. K and S

129. Cesium was discovered in natural mineral waters in 1860 by R. W. Bunsen and G. R. Kirchhoff using the spectroscope they invented in 1859. The name came from the Latin *caesius* ("sky blue") because of the prominent blue line observed for this element at 455.5 nm. Calculate the frequency and energy of a photon of this light.

130. The bright yellow light emitted by a sodium vapor lamp consists of two emission lines at 589.0 and 589.6 nm. What are the frequency and the energy of a photon of light at each of these wavelengths? What are the energies in kJ/mol?

131. Does the information on alkali metals in Table 7.8 of the text confirm the general periodic trends in ionization energy and atomic radius? Explain.

132. Predict the atomic number of the next alkali metal after francium and give its ground-state electron configuration.

133. Complete and balance the equations for the following reactions.

   a. $Li(s) + N_2(g) \rightarrow$

   b. $Rb(s) + S(s) \rightarrow$

134. Complete and balance the equations for the following reactions.

   a. $Cs(s) + H_2O(l) \rightarrow$

   b. $Na(s) + Cl_2(g) \rightarrow$

Additional Exercises

135. "Lithium" is often prescribed as a mood-stabilizing drug. Do you think the "lithium" prescribed is in the elemental form? What is the more likely form of lithium to be prescribed as a drug?

136. A carbon–oxygen double bond in a certain organic molecule absorbs radiation that has a frequency of 6.0×10^{13} s^{-1}.

   a. What is the wavelength of this radiation?

   b. To what region of the spectrum does this radiation belong?

   c. What is the energy of this radiation per photon? per mole of photons?

   d. A carbon–oxygen bond in a different molecule absorbs radiation with frequency equal to 5.4×10^{13} s^{-1}. Is this radiation more or less energetic?

137. Photogray lenses incorporate small amounts of silver chloride in the glass of the lens. When light hits the AgCl particles, the following reaction occurs:

$$AgCl \xrightarrow{hv} Ag + Cl$$

The silver metal that is formed causes the lenses to darken. The enthalpy change for this reaction is 3.10×10^2 kJ/mol. Assuming all this energy must be supplied by light, what is the maximum wavelength of light that can cause this reaction?

138. A certain microwave oven delivers 750. watts (J/s) of power to a coffee cup containing 50.0 g water at 25.0°C. If the wavelength of microwaves in the oven is 9.75 cm, how long does it take, and

how many photons must be absorbed, to make the water boil? The specific heat capacity of water is 4.18 J/°C · g, and assume only the water absorbs the energy of the microwaves.

139. Mars is roughly 60 million km from Earth. How long does it take for a radio signal originating from Earth to reach Mars?

140. Consider the following approximate visible light spectrum:

Wavelength 7×10^{-5} 6×10^{-5} 5×10^{-5} 4×10^{-5} cm

| Infrared | Red | Orange | Yellow | Green | Blue | Violet | Ultraviolet |

Barium emits light in the visible region of the spectrum. If each photon of light emitted from barium has an energy of 3.59×10^{-19} J, what color of visible light is emitted?

141. One of the visible lines in the hydrogen emission spectrum corresponds to the $n = 6$ to $n = 2$ electronic transition. What color light is this transition? See Exercise 140.

142. Using Fig. 7.29, list the elements (ignore the lanthanides and actinides) that have ground-state electron configurations that differ from those we would expect from their positions in the periodic table.

143. Are the following statements true for the hydrogen atom only, true for all atoms, or not true for any atoms?
   a. The principal quantum number completely determines the energy of a given electron.
   b. The angular momentum quantum number, ℓ, determines the shapes of the atomic orbitals.
   c. The magnetic quantum number, m_ℓ, determines the direction that the atomic orbitals point in space.

144. Although no currently known elements contain electrons in g orbitals in the ground state, it is possible that these elements will be found or that electrons in excited states of known elements could be in g orbitals. For g orbitals, the value of ℓ is 4. What is the lowest value of n for which g orbitals could exist? What are the possible values of m_ℓ? How many electrons could a set of g orbitals hold?

145. Which of the following orbital designations are incorrect: 1s, 1p, 7d, 9s, 3f, 4f, 2d?

146. The four most abundant elements by mass in the human body are oxygen, carbon, hydrogen, and nitrogen. These four elements make up about 96% of the human body. The next four most abundant elements are calcium, phosphorus, magnesium, and potassium. Write the expected ground-state electron configurations for these eight most abundant elements in the human body.

147. Consider the eight most abundant elements in the human body, as outlined in Exercise 146. Excluding hydrogen, which of these elements would have the smallest size? largest size? smallest first ionization energy? largest first ionization energy?

148. An ion having a 4+ charge and a mass of 49.9 amu has 2 electrons with principal quantum number $n = 1$, 8 electrons with $n = 2$, and 10 electrons with $n = 3$. Supply as many of the properties for the ion as possible from the information given. (*Hint:* In forming ions for this species, the 4s electrons are lost before the 3d electrons.)
   a. the atomic number
   b. total number of s electrons

   c. total number of p electrons
   d. total number of d electrons
   e. the number of neutrons in the nucleus
   f. the ground-state electron configuration of the neutral atom

149. The successive ionization energies for an unknown element are
   $I_1 = 896$ kJ/mol
   $I_2 = 1752$ kJ/mol
   $I_3 = 14,807$ kJ/mol
   $I_4 = 17,948$ kJ/mol
   To which family in the periodic table does the unknown element most likely belong?

150. An unknown element is a nonmetal and has a valence electron configuration of ns^2np^4.
   a. How many valence electrons does this element have?
   b. What are some possible identities for this element?
   c. What is the formula of the compound this element would form with potassium?
   d. Would this element have a larger or smaller radius than barium?
   e. Would this element have a greater or smaller ionization energy than fluorine?

151. Using data from this chapter, calculate the change in energy expected for each of the following processes.
   a. $Na(g) + Cl(g) \rightarrow Na^+(g) + Cl^-(g)$
   b. $Mg(g) + F(g) \rightarrow Mg^+(g) + F^-(g)$
   c. $Mg^+(g) + F(g) \rightarrow Mg^{2+}(g) + F^-(g)$
   d. $Mg(g) + 2F(g) \rightarrow Mg^{2+}(g) + 2F^-(g)$

152. How many unpaired electrons are present in each of the first-row transition metals in the ground state?

ChemWork Problems

These multiconcept problems (and additional ones) are found interactively online with the same type of assistance a student would get from an instructor.

153. It takes 476 kJ to remove 1 mole of electrons from the atoms at the surface of a solid metal. How much energy (in kJ) does it take to remove a single electron from an atom at the surface of this solid metal?

154. Calculate, to four significant figures, the longest and shortest wavelengths of light emitted by electrons in the hydrogen atom that begin in the $n = 5$ state and then fall to states with smaller values of n.

155. Assume that a hydrogen atom's electron has been excited to the $n = 6$ level. How many different wavelengths of light can be emitted as this excited atom loses energy?

156. Determine the maximum number of electrons that can have each of the following designations: $2f$, $2d_{xy}$, $3p$, $5d_{yz}$, and $4p$.

157. Consider the ground state of arsenic, As. How many electrons have $\ell = 1$ as one of their quantum numbers? How many electrons have $m_\ell = 0$? How many electrons have $m_\ell = +1$?

158. Which of the following statements is(are) *true*?
   a. The 2s orbital in the hydrogen atom is larger than the 3s orbital also in the hydrogen atom.

b. The Bohr model of the hydrogen atom has been found to be incorrect.

c. The hydrogen atom has quantized energy levels.

d. An orbital is the same as a Bohr orbit.

e. The third energy level has three sublevels, the *s*, *p*, and *d* sublevels.

159. Identify the following three elements.

    a. The ground-state electron configuration is $[Kr]5s^24d^{10}5p^4$.

    b. The ground-state electron configuration is $[Ar]4s^23d^{10}4p^2$.

    c. An excited state of this element has the electron configuration $1s^22s^22p^43s^1$.

160. For each of the following pairs of elements, choose the one that correctly completes the following table.

	K and Cs	Te and Br	Ge and Se
The more favorable (exothermic) electron affinity	——	——	——
The higher ionization energy	——	——	——
The larger size (atomic radius)	——	——	——

161. Which of the following statements is(are) *true*?

    a. F has a larger first ionization energy than does Li.

    b. Cations are larger than their parent atoms.

    c. The removal of the first electron from a lithium atom (electron configuration is $1s^22s^1$) is exothermic—that is, removing this electron gives off energy.

    d. The He atom is larger than the H^+ ion.

    e. The Al atom is smaller than the Li atom.

162. Three elements have the electron configurations $1s^22s^22p^63s^2$, $1s^22s^22p^63s^23p^4$, and $1s^22s^22p^63s^23p^64s^2$. The first ionization energies of these elements (not in the same order) are 0.590, 0.999, and 0.738 MJ/mol. The atomic radii are 104, 160, and 197 pm. Identify the three elements, and match the appropriate values of ionization energy and atomic radius to each configuration. Complete the following table with the correct information.

Electron Configuration	Element Symbol	First Ionization Energy (MJ/mol)	Atomic Radius (pm)
$1s^22s^22p^63s^2$	——	——	——
$1s^22s^22p^63s^23p^4$	——	——	——
$1s^22s^22p^63s^23p^64s^2$	——	——	——

Challenge Problems

163. An atom moving at its root mean square velocity at 100.°C has a wavelength of 2.31×10^{-11} m. Which atom is it?

164. One of the emission spectral lines for Be^{3+} has a wavelength of 253.4 nm for an electronic transition that begins in the state with $n = 5$. What is the principal quantum number of the lower-energy state corresponding to this emission? (*Hint:* The Bohr model can be applied to one-electron ions. Don't forget the Z factor: Z = nuclear charge = atomic number.)

165. The figure below represents part of the emission spectrum for a one-electron ion in the gas phase. All the lines result from electronic transitions from excited states to the $n = 3$ state. (See Exercise 164.)

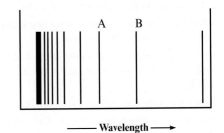

⟵ Wavelength ⟶

    a. What electronic transitions correspond to lines *A* and *B*?

    b. If the wavelength of line *B* is 142.5 nm, calculate the wavelength of line *A*.

166. When the excited electron in a hydrogen atom falls from $n = 5$ to $n = 2$, a photon of blue light is emitted. If an excited electron in He^+ falls from $n = 4$, to which energy level must it fall so that a similar blue light (as with the hydrogen) is emitted? Prove it. (See Exercise 164.)

167. The ground state ionization energy for the one electron ion X^{m+} is 4.72×10^4 kJ/mol. Identify X and m. (See Exercise 164.)

168. For hydrogen atoms, the wave function for the state $n = 3$, $\ell = 0$, $m_\ell = 0$ is

$$\psi_{300} = \frac{1}{81\sqrt{3\pi}}\left(\frac{1}{a_0}\right)^{3/2}(27 - 18\sigma + 2\sigma^2)e^{-\sigma/3}$$

where $\sigma = r/a_0$ and a_0 is the Bohr radius (5.29×10^{-11} m). Calculate the position of the nodes for this wave function.

169. The wave function for the $2p_z$ orbital in the hydrogen atom is

$$\psi_{2p_z} = \frac{1}{4\sqrt{2\pi}}\left(\frac{Z}{a_0}\right)^{3/2}\sigma e^{-\sigma/2}\cos\theta$$

where a_0 is the value for the radius of the first Bohr orbit in meters (5.29×10^{-11}), σ is $Z(r/a_0)$, r is the value for the distance from the nucleus in meters, and θ is an angle. Calculate the value of $\psi_{2p_z}^2$ at $r = a_0$ for $\theta = 0°$ (z axis) and for $\theta = 90°$ (xy plane).

170. Answer the following questions assuming that m_s could have three values rather than two and that the rules for *n*, ℓ, and m_ℓ are the normal ones.

    a. How many electrons would an orbital be able to hold?

    b. How many elements would the first and second periods in the periodic table contain?

    c. How many elements would be contained in the first transition metal series?

    d. How many electrons would the set of 4*f* orbitals be able to hold?

171. Assume that we are in another universe with different physical laws. Electrons in this universe are described by four quantum numbers with meanings similar to those we use. We will call these quantum numbers *p*, *q*, *r*, and *s*. The rules for these quantum numbers are as follows:

$p = 1, 2, 3, 4, 5, \ldots$.

q takes on positive odd integers and $q \leq p$.

r takes on all even integer values from $-q$ to $+q$. (Zero is considered an even number.)

$s = +\frac{1}{2}$ or $-\frac{1}{2}$

  a. Sketch what the first four periods of the periodic table will look like in this universe.

  b. What are the atomic numbers of the first four elements you would expect to be least reactive?

  c. Give an example, using elements in the first four rows, of ionic compounds with the formulas XY, XY_2, X_2Y, XY_3, and X_2Y_3.

  d. How many electrons can have $p = 4$, $q = 3$?

  e. How many electrons can have $p = 3$, $q = 0$, $r = 0$?

  f. How many electrons can have $p = 6$?

172. Without looking at data in the text, sketch a qualitative graph of the third ionization energy versus atomic number for the elements Na through Ar, and explain your graph.

173. The following numbers are the ratios of second ionization energy to first ionization energy:

Na:	9.2	P:	1.8
Mg:	2.0	S:	2.3
Al:	3.1	Cl:	1.8
Si:	2.0	Ar:	1.8

Explain these relative numbers.

174. We expect the atomic radius to increase going down a group in the periodic table. Can you suggest why the atomic radius of hafnium breaks this rule? (See data below.)

Atomic Radii (in pm)			
Sc	157	Ti	147.7
Y	169.3	Zr	159.3
La	191.5	Hf	147.6

175. The ionization energy for a $1s$ electron in a silver atom is 2.462×10^6 kJ/mol.

  a. Determine an approximate value for Z_{eff} for the Ag $1s$ electron. Assume the Bohr model applies to the $1s$ electron. Z_{eff} is the apparent nuclear charge experienced by the electrons.

  b. How does Z_{eff} from part a compare to Z for Ag? Rationalize the relative numbers.

176. While Mendeleev predicted the existence of several undiscovered elements, he did not predict the existence of the noble gases, the lanthanides, or the actinides. Propose reasons why Mendeleev was not able to predict the existence of the noble gases.

177. An atom of a particular element is traveling at 1.00% of the speed of light. The de Broglie wavelength is found to be 3.31×10^{-3} pm. Which element is this? Prove it.

Integrative Problems

These problems require the integration of multiple concepts to find the solutions.

178. As the weapons officer aboard the *Starship Chemistry,* it is your duty to configure a photon torpedo to remove an electron from the outer hull of an enemy vessel. You know that the work function (the binding energy of the electron) of the hull of the enemy ship is 7.52×10^{-19} J.

  a. What wavelength does your photon torpedo need to be to eject an electron?

  b. You find an extra photon torpedo with a wavelength of 259 nm and fire it at the enemy vessel. Does this photon torpedo do any damage to the ship (does it eject an electron)?

  c. If the hull of the enemy vessel is made of the element with an electron configuration of $[Ar]4s^13d^{10}$, what metal is this?

179. Francium, Fr, is a radioactive element found in some uranium minerals and is formed as a result of the decay of actinium.

  a. What are the electron configurations of francium and its predicted most common ion?

  b. It has been estimated that at any one time, there is only one (1.0) ounce of francium on earth. Assuming this is true, what number of francium atoms exist on earth?

  c. The longest-lived isotope of francium is ^{223}Fr. What is the total mass in grams of the neutrons in one atom of this isotope?

180. Answer the following questions based on the given electron configurations, and identify the elements.

  a. Arrange these atoms in order of increasing size: $[Kr]5s^24d^{10}5p^6$; $[Kr]5s^24d^{10}5p^1$; $[Kr]5s^24d^{10}5p^3$.

  b. Arrange these atoms in order of decreasing first ionization energy: $[Ne]3s^23p^5$; $[Ar]4s^23d^{10}4p^3$; $[Ar]4s^23d^{10}4p^5$.

Chapter 8

Bonding: General Concepts

The nudibranch uses a particular molecule (called an allomone) to defend itself against predators. The structure of the molecule is specific to its purpose. (Norbert Wu/Minden Pictures)

As we examine the world around us, we find it to be composed almost entirely of compounds and mixtures of compounds: Rocks, coal, soil, petroleum, trees, and human bodies are all complex mixtures of chemical compounds in which different kinds of atoms are bound together. Substances composed of unbound atoms do exist in nature, but they are very rare. Examples are the argon in the atmosphere and the helium mixed with natural gas reserves.

The manner in which atoms are bound together has a profound effect on chemical and physical properties. For example, graphite is a soft, slippery material used as a lubricant in locks, and diamond is one of the hardest materials known, valuable both as a gemstone and in industrial cutting tools. Why do these materials, both composed solely of carbon atoms, have such different properties? The answer, as we will see, lies in the bonding in these substances.

Silicon and carbon are next to each other in Group 4A of the periodic table. From our knowledge of periodic trends, we might expect SiO_2 and CO_2 to be very similar. But SiO_2 is the empirical formula of silica, which is found in sand and quartz, and carbon dioxide is a gas, a product of respiration. Why are they so different? We will be able to answer this question after we have developed models for bonding.

Molecular bonding and structure play the central role in determining the course of all chemical reactions, many of which are vital to our survival. Later in this book we will demonstrate their importance by showing how enzymes facilitate complex chemical reactions, how genetic characteristics are transferred, and how hemoglobin in the blood carries oxygen throughout the body. All of these fundamental biological reactions hinge on the geometric structures of molecules, sometimes depending on very subtle differences in molecular shape to channel the chemical reaction one way rather than another.

Many of the world's current problems require fundamentally chemical answers: disease and pollution control, the search for new energy sources, the development of new fertilizers to increase crop yields, the improvement of the protein content in various staple grains, and many more. To understand the behavior of natural materials, we must understand the nature of chemical bonding and the factors that control the structures of compounds. In this chapter we will present various classes of compounds that illustrate the different types of bonds and then develop models to describe the structure and bonding that characterize materials found in nature. Later these models will be useful in understanding chemical reactions.

Quartz grows in beautiful, regular crystals.

8.1 | Types of Chemical Bonds

What is a chemical bond? There is no simple and yet complete answer to this question. In Chapter 2 we defined bonds as forces that hold groups of atoms together and make them function as a unit.

There are many types of experiments we can perform to determine the fundamental nature of materials. For example, we can study physical properties such as melting point, hardness, and electrical and thermal conductivity. We can also study solubility characteristics and the properties of the resulting solutions. To determine the charge distribution in a molecule, we can study its behavior in an electric field. We can obtain information about the strength of a bonding interaction by measuring the **bond energy**, which is the energy required to break the bond.

There are several ways in which atoms can interact with one another to form aggregates. We will consider several specific examples to illustrate the various types of chemical bonds.

Earlier, we saw that when solid sodium chloride is dissolved in water, the resulting solution conducts electricity, a fact that helps to convince us that sodium chloride is composed of Na^+ and Cl^- ions. Therefore, when sodium and chlorine react to form sodium chloride, electrons are transferred from the sodium atoms to the chlorine atoms to form Na^+ and Cl^- ions, which then aggregate to form solid sodium chloride. Why does this happen? The best simple answer is that *the system can achieve the lowest possible energy by behaving in this way*. The attraction of a chlorine atom for the extra electron and the very strong mutual attractions of the oppositely charged ions provide the driving forces for the process. The resulting solid sodium chloride is a very sturdy material; it has a melting point of approximately 800°C. The bonding forces that produce this great thermal stability result from the electrostatic attractions of the closely packed, oppositely charged ions. This is an example of **ionic bonding**. Ionic substances are formed when an atom that loses electrons relatively easily reacts with an atom that has a high affinity for electrons. That is, an **ionic compound** results when a metal reacts with a nonmetal.

The energy of interaction between a pair of ions can be calculated using **Coulomb's law** in the form

$$E = (2.31 \times 10^{-19} \, \text{J} \cdot \text{nm}) \left(\frac{Q_1 Q_2}{r} \right)$$

0.276 nm

Na^+ Cl^-

where E has units of joules, r is the distance between the ion centers in nanometers, and Q_1 and Q_2 are the numerical ion charges.

For example, in solid sodium chloride the distance between the centers of the Na^+ and Cl^- ions is 2.76 Å (0.276 nm), and the ionic energy per pair of ions is

$$E = (2.31 \times 10^{-19} \, \text{J} \cdot \text{nm}) \left[\frac{(+1)(-1)}{0.276 \, \text{nm}} \right] = -8.37 \times 10^{-19} \, \text{J}$$

where the negative sign indicates an attractive force. That is, the *ion pair has lower energy than the separated ions*.

Coulomb's law also can be used to calculate the repulsive energy when two like-charged ions are brought together. In this case the calculated value of the energy will have a positive sign.

We have seen that a bonding force develops when two different types of atoms react to form oppositely charged ions. But how does a bonding force develop between two identical atoms? Let's explore this situation from a very simple point of view by considering the energy terms that result when two hydrogen atoms are brought close together, as shown in Fig. 8.1(a). When hydrogen atoms are brought close together, there are two unfavorable potential energy terms, proton–proton repulsion and electron–electron repulsion, and one favorable term, proton–electron attraction. Under what conditions will the H_2 molecule be favored over the separated hydrogen atoms? That is, what conditions will favor bond formation? The answer lies in the strong tendency in nature for any system to achieve the lowest possible energy. A bond will form (that is, the two hydrogen atoms will exist as a molecular unit) if the system can lower its total energy in the process.

In this case, then, the hydrogen atoms will position themselves so that the system will achieve the lowest possible energy; the system will act to minimize the sum of the positive (repulsive) energy terms and the negative (attractive) energy term. The distance where the energy is minimal is called the **bond length**. The total energy of this system as a function of distance between the hydrogen nuclei is shown in Fig. 8.1(b). Note several important features of this diagram:

The energy terms involved are the net potential energy that results from the attractions and repulsions among the charged particles and the kinetic energy due to the motions of the electrons.

The zero point of energy is defined with the atoms at infinite separation.

A bond will form if the energy of the aggregate is lower than that of the separated atoms.

Potential energy was discussed in Chapter 6.

Chemical connections
No Lead Pencils

Did you ever wonder why the part of a pencil that makes the mark is called the "lead"? Pencils have no lead in them now—and they never have. Apparently the association between writing and the element lead arose during the Roman Empire, when lead rods were used as writing utensils because they leave a gray mark on paper. Many centuries later, in 1564, a deposit of a black substance found to be very useful for writing was discovered in Borrowdale, England. This substance, originally called "black lead," was shown in 1879 by Swedish chemist Carl Scheele to be a form of carbon and was subsequently named graphite (after the Greek *graphein,* meaning "to write").

Originally, chunks of graphite from Borrowdale, called marking stones, were used as writing instruments. Later, sticks of graphite were used. Because graphite is brittle, the sticks needed reinforcement. At first they were wrapped in string, which was unwound as the core wore down. Eventually, graphite rods were tied between two wooden slats or inserted into hollowed-out wooden sticks to form the first crude pencils.

Although Borrowdale graphite was pure enough to use directly, most graphite must be mixed with other materials to be useful for writing instruments. In 1795, the French chemist Nicolas-Jaques Conté

invented a process in which graphite is mixed with clay and water to produce pencil "lead," a recipe that is still used today. In modern pencil manufacture, graphite and clay are mixed and crushed into a fine powder to which water is added. After the gray sludge is blended for several days, it is dried, ground up again, and mixed with more water to give a gray paste. The paste is extruded through a metal tube to form thin rods, which are then cut into pencil-length pieces called "leads." These leads are heated in an oven to 1000°C until they are smooth and hard. The ratio of clay to graphite is adjusted to vary the hardness of the lead—the more clay in the mix, the

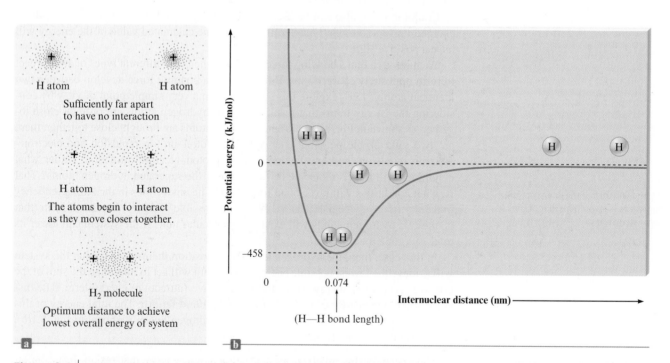

Figure 8.1 | (a) The interaction of two hydrogen atoms. (b) Energy profile as a function of the distance between the nuclei of the hydrogen atoms. As the atoms approach each other (right side of graph), the energy decreases until the distance reaches 0.074 nm (74 pm) and then begins to increase again due to repulsions.

harder the lead and the lighter the line it makes.

Pencils are made from a slat of wood with several grooves cut in it to hold the leads. A similar grooved slat is then placed on top and glued to form a "sandwich" from which individual pencils are cut, sanded smooth, and painted. Although many types of wood have been used over the years to make pencils, the current favorite is incense cedar from the Sierra Nevada Mountains of California.

Modern pencils are simple but amazing instruments. The average pencil can write approximately 45,000 words, which is equivalent to a line 35 miles long. The graphite in a pencil is easily transferred to paper because graphite contains layers of carbon atoms bound together in a "chicken-wire" structure. Although the bonding *within* each layer is very strong, the bonding *between* layers is weak, giving graphite its slippery, soft nature. In this way, graphite is much

different from diamond, the other common elemental form of carbon. In diamond the carbon atoms are bound tightly in all three dimensions, making it extremely hard—the hardest natural substance.

Pencils are very useful—especially for doing chemistry problems—because we can erase our mistakes. Most pencils used in the United States have erasers (first attached to pencils in 1858), although most European pencils do not. Laid end to end, the number of pencils made in the United States each year would circle the earth about 15 times. Pencils illustrate how useful a simple substance like graphite can be.

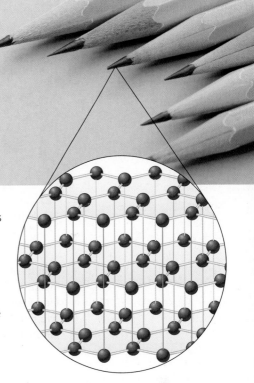

Graphite

At very short distances the energy rises steeply because of the importance of the repulsive forces when the atoms are very close together.

The bond length is the distance at which the system has minimum energy.

In the H_2 molecule, the electrons reside primarily in the space between the two nuclei, where they are attracted simultaneously by both protons. This positioning is precisely what leads to the stability of the H_2 molecule compared with two separated hydrogen atoms. The potential energy of each electron is lowered because of the increased attractive forces in this area. When we say that a bond is formed between the hydrogen atoms, we mean that the H_2 molecule is more stable than two separated hydrogen atoms by a certain quantity of energy (the bond energy).

We can also think of a bond in terms of forces. The simultaneous attraction of each electron by the protons generates a force that pulls the protons toward each other and that just balances the proton–proton and electron–electron repulsive forces at the distance corresponding to the bond length.

The type of bonding we encounter in the hydrogen molecule and in many other molecules in which *electrons are shared by nuclei* is called **covalent bonding**.

So far we have considered two extreme types of bonding. In ionic bonding the participating atoms are so different that one or more electrons are transferred to form oppositely charged ions, which then attract each other. In covalent bonding two identical atoms share electrons equally. The bonding results from the mutual attraction of the two nuclei for the shared electrons. Between these extremes are intermediate cases in which the atoms are not so different that electrons are completely transferred but are different enough that unequal sharing results, forming what is called a **polar covalent**

Ionic and covalent bonds are the extreme bond types.

Figure 8.2 | The effect of an electric field on hydrogen fluoride molecules. (a) When no electric field is present, the molecules are randomly oriented. (b) When the field is turned on, the molecules tend to line up with their negative ends toward the positive pole and their positive ends toward the negative pole.

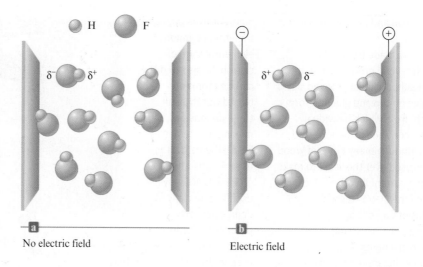

No electric field

Electric field

bond. An example of this type of bond occurs in the hydrogen fluoride (HF) molecule. When a sample of hydrogen fluoride gas is placed in an electric field, the molecules tend to orient themselves as shown in Fig. 8.2, with the fluoride end closest to the positive pole and the hydrogen end closest to the negative pole. This result implies that the HF molecule has the following charge distribution:

$$\underset{\delta^+ \quad \delta^-}{H—F}$$

where δ (lowercase delta) is used to indicate a fractional charge. This same effect was noted in Chapter 4, where many of water's unusual properties were attributed to the polar O—H bonds in the H_2O molecule.

   The most logical explanation for the development of the partial positive and negative charges on the atoms (bond polarity) in such molecules as HF and H_2O is that the electrons in the bonds are not shared equally. For example, we can account for the polarity of the HF molecule by assuming that the fluorine atom has a stronger attraction for the shared electrons than the hydrogen atom. Likewise, in the H_2O molecule the oxygen atom appears to attract the shared electrons more strongly than the hydrogen atoms do. Because bond polarity has important chemical implications, we find it useful to quantify the ability of an atom to attract shared electrons. In the next section we show how this is done.

8.2 | Electronegativity

The different affinities of atoms for the electrons in a bond are described by a property called **electronegativity**: *the ability of an atom in a molecule to attract shared electrons to itself.*

   The most widely accepted method for determining values of electronegativity is that of Linus Pauling (1901–1995), an American scientist who won the Nobel Prizes for both chemistry and peace. To understand Pauling's model, consider a hypothetical molecule HX. The relative electronegativities of the H and X atoms are determined by comparing the measured H—X bond energy with the "expected" H—X bond energy, which is an average of the H—H and X—X bond energies:

$$\text{Expected H—X bond energy} = \frac{\text{H—H bond energy} + \text{X—X bond energy}}{2}$$

The difference (Δ) between the actual (measured) and expected bond energies is

$$\Delta = (\text{H—X})_{\text{act}} - (\text{H—X})_{\text{exp}}$$

Increasing electronegativity

Decreasing electronegativity

						H 2.1											
Li 1.0	Be 1.5											B 2.0	C 2.5	N 3.0	O 3.5	F 4.0	
Na 0.9	Mg 1.2											Al 1.5	Si 1.8	P 2.1	S 2.5	Cl 3.0	
K 0.8	Ca 1.0	Sc 1.3	Ti 1.5	V 1.6	Cr 1.6	Mn 1.5	Fe 1.8	Co 1.9	Ni 1.9	Cu 1.9	Zn 1.6	Ga 1.6	Ge 1.8	As 2.0	Se 2.4	Br 2.8	
Rb 0.8	Sr 1.0	Y 1.2	Zr 1.4	Nb 1.6	Mo 1.8	Tc 1.9	Ru 2.2	Rh 2.2	Pd 2.2	Ag 1.9	Cd 1.7	In 1.7	Sn 1.8	Sb 1.9	Te 2.1	I 2.5	
Cs 0.7	Ba 0.9	La–Lu 1.0–1.2	Hf 1.3	Ta 1.5	W 1.7	Re 1.9	Os 2.2	Ir 2.2	Pt 2.2	Au 2.4	Hg 1.9	Tl 1.8	Pb 1.9	Bi 1.9	Po 2.0	At 2.2	
Fr 0.7	Ra 0.9	Ac 1.1	Th 1.3	Pa 1.4	U 1.4	Np–No 1.4–1.3											

Figure 8.3 | The Pauling electronegativity values. Electronegativity generally increases across a period and decreases down a group.

If H and X have identical electronegativities, $(H—X)_{act}$ and $(H—X)_{exp}$ are the same, and Δ is 0. On the other hand, if X has a greater electronegativity than H, the shared electron(s) will tend to be closer to the X atom. The molecule will be polar, with the following charge distribution:

$$H—X$$
$$\delta^+ \quad \delta^-$$

Note that this bond can be viewed as having an ionic as well as a covalent component. The attraction between the partially (and oppositely) charged H and X atoms will lead to a greater bond strength. Thus $(H—X)_{act}$ will be larger than $(H—X)_{exp}$. The greater the difference in the electronegativities of the atoms, the greater is the ionic component of the bond and the greater is the value of Δ. Thus the relative electronegativities of H and X can be assigned from the Δ values.

Electronegativity values have been determined by this process for virtually all the elements; the results are given in Fig. 8.3. Note that electronegativity generally increases going from left to right across a period and decreases going down a group for the representative elements. The range of electronegativity values is from 4.0 for fluorine to 0.7 for cesium.

The relationship between electronegativity and bond type is shown in Table 8.1. For identical atoms (an electronegativity difference of zero), the electrons in the bond are

Table 8.1 | The Relationship Between Electronegativity and Bond Type

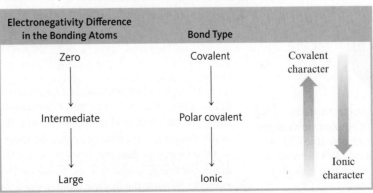

shared equally, and no polarity develops. When two atoms with very different electro-negativities interact, electron transfer can occur to form the ions that make up an ionic substance. Intermediate cases give polar covalent bonds with unequal electron sharing.

Interactive Example 8.1

Sign in at http://login.cengagebrain.com to try this Interactive Example in OWL.

Relative Bond Polarities

Order the following bonds according to polarity: H—H, O—H, Cl—H, S—H, and F—H.

Solution

The polarity of the bond increases as the difference in electronegativity increases. From the electronegativity values in Fig. 8.3, the following variation in bond polarity is expected (the electronegativity value appears in parentheses below each element):

$$H—H < S—H < Cl—H < O—H < F—H$$
$$(2.1)(2.1) \quad (2.5)(2.1) \quad (3.0)(2.1) \quad (3.5)(2.1) \quad (4.0)(2.1)$$

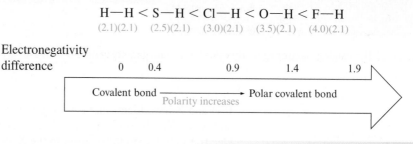

Electronegativity difference

See Exercises 8.37 and 8.38

Critical Thinking

We use differences in electronegativity to account for certain properties of bonds. What if all atoms had the same electronegativity values? How would bonding between atoms be affected? What are some differences we would notice?

8.3 | Bond Polarity and Dipole Moments

We have seen that when hydrogen fluoride is placed in an electric field, the molecules have a preferential orientation (see Fig. 8.2). This follows from the charge distribution in the HF molecule, which has a positive end and a negative end. A molecule such as HF that has a center of positive charge and a center of negative charge is said to be **dipolar**, or to have a **dipole moment**. The dipolar character of a molecule is often represented by an arrow pointing to the negative charge center with the tail of the arrow indicating the positive center of charge:

Figure 8.4 | An electrostatic potential map of HF. Red indicates the most electron-rich area (the fluorine atom), and blue indicates the most electron-poor region (the hydrogen atom).

Another way to represent the charge distribution in HF is by an electrostatic potential diagram (Fig. 8.4). For this representation the colors of visible light are used to show the variation in charge distribution. Red indicates the most electron-rich region of the molecule, and blue indicates the most electron-poor region.

Of course, any diatomic (two-atom) molecule that has a polar bond also will show a molecular dipole moment. Polyatomic molecules also can exhibit dipolar behavior. For example, because the oxygen atom in the water molecule has a greater electronegativity

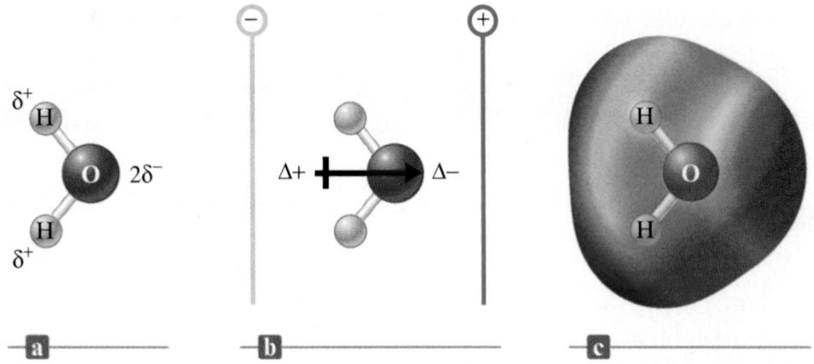

Figure 8.5 | (a) The charge distribution in the water molecule. (b) The water molecule in an electric field. (c) The electrostatic potential diagram of the water molecule.

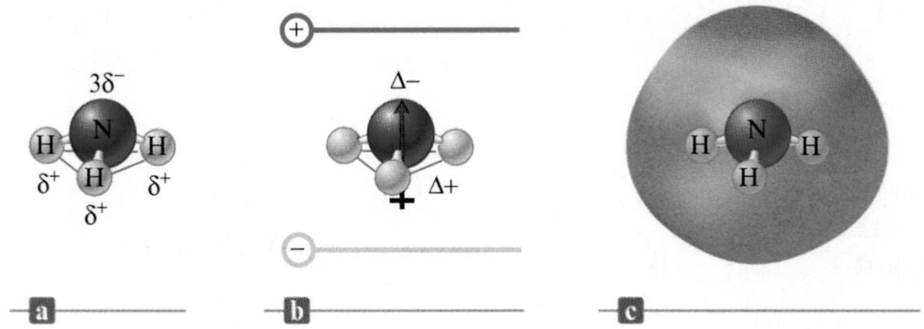

Figure 8.6 | (a) The structure and charge distribution of the ammonia molecule. The polarity of the N—H bonds occurs because nitrogen has a greater electronegativity than hydrogen. (b) The dipole moment of the ammonia molecule oriented in an electric field. (c) The electrostatic potential diagram for ammonia.

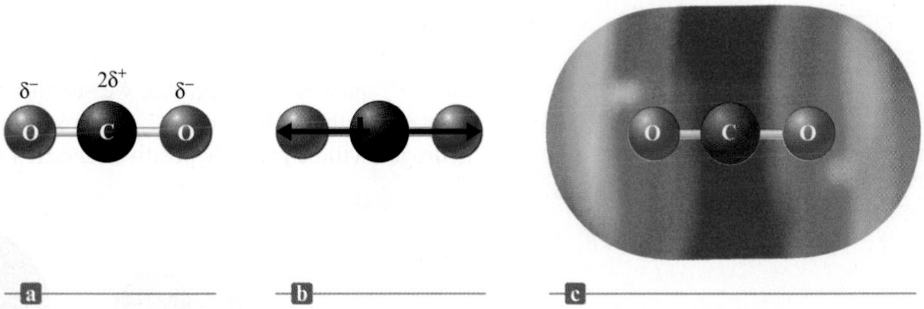

Figure 8.7 | (a) The carbon dioxide molecule. (b) The opposed bond polarities cancel out, and the carbon dioxide molecule has no dipole moment. (c) The electrostatic potential diagram for carbon dioxide.

than the hydrogen atoms, the molecular charge distribution is that shown in Fig. 8.5(a). Because of this charge distribution, the water molecule behaves in an electric field as if it had two centers of charge—one positive and one negative—as shown in Fig. 8.5(b). The water molecule has a dipole moment. The same type of behavior is observed for the NH_3 molecule (Fig. 8.6). Some molecules have polar bonds but do not have a dipole moment. This occurs when the individual bond polarities are arranged in such a way that they cancel each other out. An example is the CO_2 molecule, which is a linear molecule that has the charge distribution shown in Fig. 8.7. In this case the opposing bond polarities cancel out, and the carbon dioxide molecule does not have a dipole moment. There is no preferential way for this molecule to line up in an electric field. (Try to find a preferred orientation to make sure you understand this concept.)

There are many cases besides that of carbon dioxide where the bond polarities oppose and exactly cancel each other. Some common types of molecules with polar bonds but no dipole moment are shown in Table 8.2.

| Example 8.2 | Bond Polarity and Dipole Moment |

For each of the following molecules, show the direction of the bond polarities and indicate which ones have a dipole moment: HCl, Cl_2, SO_3 (a planar molecule with the oxygen atoms spaced evenly around the central sulfur atom), CH_4 [tetrahedral (see Table 8.2) with the carbon atom at the center], and H_2S (V-shaped with the sulfur atom at the point).

Table 8.2 | Types of Molecules with Polar Bonds but No Resulting Dipole Moment

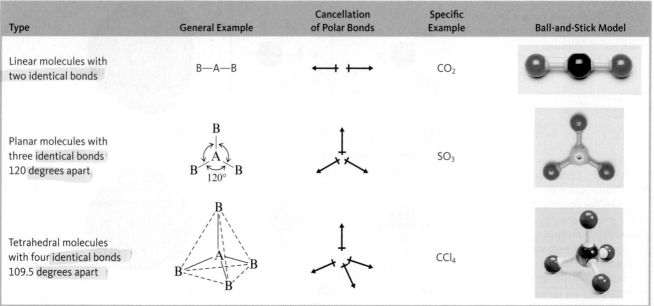

Type	General Example	Cancellation of Polar Bonds	Specific Example	Ball-and-Stick Model
Linear molecules with two identical bonds	B—A—B		CO_2	
Planar molecules with three identical bonds 120 degrees apart			SO_3	
Tetrahedral molecules with four identical bonds 109.5 degrees apart			CCl_4	

Photos: Ken O'Donoghue © Cengage Learning

Solution

The HCl molecule: In Fig. 8.3, we see that the electronegativity of chlorine (3.0) is greater than that of hydrogen (2.1). Thus the chlorine will be partially negative, and the hydrogen will be partially positive. The HCl molecule has a dipole moment:

Recall that a blue color is an electron-poor region and red is an electron-rich region.

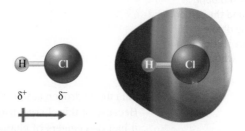

The Cl₂ molecule: The two chlorine atoms share the electrons equally. No bond polarity occurs, and the Cl_2 molecule has no dipole moment.

The SO₃ molecule: The electronegativity of oxygen (3.5) is greater than that of sulfur (2.5). This means that each oxygen will have a partial negative charge, and the sulfur will have a partial positive charge:

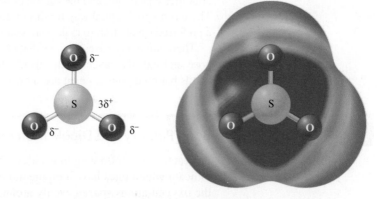

The presence of polar bonds does not always yield a polar molecule.

The bond polarities arranged symmetrically as shown cancel, and the molecule has no dipole moment. This molecule is the second type shown in Table 8.2.

The CH₄ molecule: Carbon has a slightly higher electronegativity (2.5) than does hydrogen (2.1). This leads to small partial positive charges on the hydrogen atoms and a small partial negative charge on the carbon:

This case is similar to the third type in Table 8.2, and the bond polarities cancel. The molecule has no dipole moment.

The H₂S molecule: Since the electronegativity of sulfur (2.5) is slightly greater than that of hydrogen (2.1), the sulfur will have a partial negative charge, and the hydrogen atoms will have a partial positive charge, which can be represented as

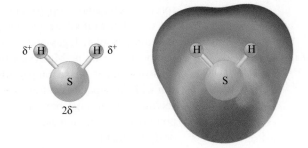

This case is analogous to the water molecule, and the polar bonds result in a dipole moment oriented as shown:

See Exercise 8.136

8.4 | Ions: Electron Configurations and Sizes

The description of the electron arrangements in atoms that emerged from the quantum mechanical model has helped a great deal in our understanding of what constitutes a stable compound. In virtually every case, the atoms in a stable compound have a noble gas arrangement of electrons. Nonmetallic elements achieve a noble gas electron configuration either by sharing electrons with other nonmetals to form covalent bonds or by taking electrons from metals to form ions. In the second case, the nonmetals form anions, and the metals form cations. The generalizations that apply to electron configurations in stable compounds are as follows:

Atoms in stable compounds usually have a noble gas electron configuration.

Electron Configuration of Compounds

❯ When *two nonmetals* react to form a covalent bond, they share electrons in a way that completes the valence electron configurations of both atoms. That is, both nonmetals attain noble gas electron configurations.

❯ When *a nonmetal and a representative-group metal* react to form a binary ionic compound, the ions form so that the valence electron configuration of the nonmetal achieves the electron configuration of the next noble gas atom and the valence orbitals of the metal are emptied. In this way both ions achieve noble gas electron configurations.

These generalizations apply to the vast majority of compounds and are important to remember. We will deal with covalent bonds more thoroughly later, but now we will consider what implications these rules hold for ionic compounds.

Predicting Formulas of Ionic Compounds

In the solid state of an ionic compound, the ions are relatively close together and many ions are simultaneously interacting:

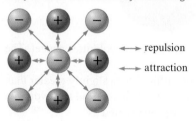

← → repulsion

← → attraction

In the gas phase of an ionic substance, the ions would be relatively far apart and would not contain large groups of ions:

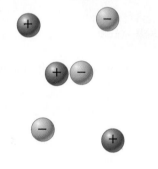

At the beginning of this discussion, it should be emphasized that when chemists use the term *ionic compound,* they are usually referring to the solid state of that compound. In the solid state, the ions are close together. That is, solid ionic compounds contain a large collection of positive and negative ions packed together in a way that minimizes the $\ominus\cdots\ominus$ and $\oplus\cdots\oplus$ repulsions and maximizes the $\oplus\cdots\ominus$ attractions. This situation stands in contrast to the gas phase of an ionic substance, where the ions are quite far apart on average. In the gas phase, a pair of ions may get close enough to interact, but large collections of ions do not exist. Thus, when we speak in this text of the stability of an ionic compound, we are referring to the solid state, where the large attractive forces present among oppositely charged ions tend to stabilize (favor the formation of) the ions. For example, as we mentioned in the preceding chapter, the O^{2-} ion is not stable as an isolated, gas-phase species but, of course, is very stable in many solid ionic compounds. That is, $MgO(s)$, which contains Mg^{2+} and O^{2-} ions, is very stable, but the isolated, gas-phase ion pair $Mg^{2+}\cdots O^{2-}$ is not energetically favorable in comparison with the separate neutral gaseous atoms. Thus you should keep in mind that in this section, and in most other cases where we are describing the nature of ionic compounds, the discussion usually refers to the solid state, where many ions are simultaneously interacting.

To illustrate the principles of electron configurations in stable, solid ionic compounds, we will consider the formation of an ionic compound from calcium and oxygen. We can predict what compound will form by considering the valence electron configurations of the two atoms:

$$Ca \quad [Ar]4s^2$$
$$O \quad [He]2s^22p^4$$

From Fig. 8.3 we see that the electronegativity of oxygen (3.5) is much greater than that of calcium (1.0). Because of this large difference, electrons will be transferred from calcium to oxygen to form oxygen anions and calcium cations in the compound. How many electrons are transferred? We can base our prediction on the observation that noble gas configurations are generally the most stable. Note that oxygen needs two electrons to fill its $2s$ and $2p$ valence orbitals and to achieve the configuration of neon ($1s^22s^22p^6$). And by losing two electrons, calcium can achieve the configuration of argon. Two electrons therefore transferred:

$$Ca + O \longrightarrow Ca^{2+} + O^{2-}$$
$$\underbrace{\qquad}_{2e^-}$$

To predict the formula of the ionic compound, we simply recognize that chemical compounds are always electrically neutral—they have the same quantities of positive and negative charges. In this case we have equal numbers of Ca^{2+} and O^{2-} ions, and the empirical formula of the compound is CaO.

The same principles can be applied to many other cases. For example, consider the compound formed between aluminum and oxygen. Because aluminum has the configuration $[Ne]3s^23p^1$, it loses three electrons to form the Al^{3+} ion and thus achieves the neon configuration. Therefore, the Al^{3+} and O^{2-} ions form in this case. Since the compound must be electrically neutral, there must be three O^{2-} ions for every two Al^{3+} ions, and the compound has the empirical formula Al_2O_3.

Table 8.3 shows common elements that form ions with noble gas electron configurations in ionic compounds. In losing electrons to form cations, metals in Group 1A

A bauxite mine. Bauxite contains Al_2O_3, the main source of aluminum.

Courtesy, Aluminum Company of America

Table 8.3 | Common Ions with Noble Gas Configurations in Ionic Compounds

Group 1A	Group 2A	Group 3A	Group 6A	Group 7A	Electron Configuration
H^-, Li^+	Be^{2+}				[He]
Na^+	Mg^{2+}	Al^{3+}	O^{2-}	F^-	[Ne]
K^+	Ca^{2+}		S^{2-}	Cl^-	[Ar]
Rb^+	Sr^{2+}		Se^{2-}	Br^-	[Kr]
Cs^+	Ba^{2+}		Te^{2-}	I^-	[Xe]

lose one electron, those in Group 2A lose two electrons, and those in Group 3A lose three electrons. In gaining electrons to form anions, nonmetals in Group 7A (the halogens) gain one electron, and those in Group 6A gain two electrons. Hydrogen typically behaves as a nonmetal and can gain one electron to form the hydride ion (H^-), which has the electron configuration of helium.

There are some important exceptions to the rules discussed here. For example, tin forms both Sn^{2+} and Sn^{4+} ions, and lead forms both Pb^{2+} and Pb^{4+} ions. Also, bismuth forms Bi^{3+} and Bi^{5+} ions, and thallium forms Tl^+ and Tl^{3+} ions. There are no simple explanations for the behavior of these ions. For now, just note them as exceptions to the very useful rule that ions generally adopt noble gas electron configurations in ionic compounds. Our discussion here refers to representative metals. The transition metals exhibit more complicated behavior, forming a variety of ions that will be considered in Chapter 21.

Sizes of Ions

Ion size plays an important role in determining the structure and stability of ionic solids, the properties of ions in aqueous solution, and the biologic effects of ions. As with atoms, it is impossible to define precisely the sizes of ions. Most often, ionic radii are determined from the measured distances between ion centers in ionic compounds. This method, of course, involves an assumption about how the distance should be divided up between the two ions. Thus you will note considerable disagreement among ionic sizes given in various sources. Here we are mainly interested in trends and will be less concerned with absolute ion sizes.

Various factors influence ionic size. We will first consider the relative sizes of an ion and its parent atom. Since a positive ion is formed by removing one or more electrons from a neutral atom, the resulting cation is smaller than its parent atom. The opposite is true for negative ions; the addition of electrons to a neutral atom produces an anion significantly larger than its parent atom.

It is also important to know how the sizes of ions vary depending on the positions of the parent elements in the periodic table. Figure 8.8 shows the sizes of the most important ions (each with a noble gas configuration) and their position in the periodic table. Note that ion size increases down a group. The changes that occur horizontally are complicated because of the change from predominantly metals on the left-hand side of the periodic table to nonmetals on the right-hand side. A given period thus contains both elements that give up electrons to form cations and ones that accept electrons to form anions.

One trend worth noting involves the relative sizes of a set of **isoelectronic ions**—*ions containing the same number of electrons.* Consider the ions O^{2-}, F^-, Na^+, Mg^{2+}, and Al^{3+}. Each of these ions has the neon electron configuration. How do the sizes of these ions vary? In general, there are two important facts to consider in predicting the relative sizes of ions: the number of electrons and the number of protons. Since these ions are isoelectronic, the number of electrons is 10 in each case. Electron repulsions

Figure 8.8 | Sizes of ions related to positions of the elements on the periodic table. Note that size generally increases down a group. Also note that in a series of isoelectronic ions, size decreases with increasing atomic number. The ionic radii are given in units of picometers.

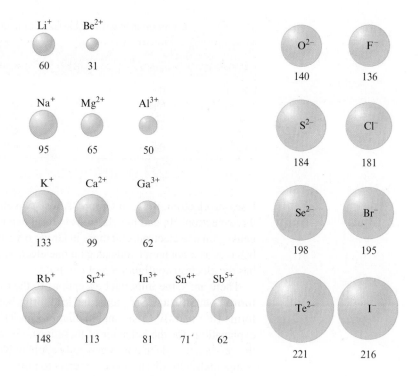

therefore should be about the same in all cases. However, the number of protons increases from 8 to 13 as we go from the O^{2-} ion to the Al^{3+} ion. Thus, in going from O^{2-} to Al^{3+}, the 10 electrons experience a greater attraction as the positive charge on the nucleus increases. This causes the ions to become smaller. You can confirm this by looking at Fig. 8.8. In general, for a series of isoelectronic ions, the size decreases as the nuclear charge Z increases.

For isoelectronic ions, size decreases as Z increases.

Critical Thinking

Ions have different radii than their parent atoms. What if ions stayed the same size as their parent atoms? How would this affect ionic bonding in compounds?

Interactive Example 8.3

Sign in at http://login.cengagebrain .com to try this Interactive Example in **OWL**.

Relative Ion Size I

Arrange the ions Se^{2-}, Br^-, Rb^+, and Sr^{2+} in order of decreasing size.

Solution

This is an isoelectronic series of ions with the krypton electron configuration. Since these ions all have the same number of electrons, their sizes will depend on the nuclear charge. The Z values are 34 for Se^{2-}, 35 for Br^-, 37 for Rb^+, and 38 for Sr^{2+}. Since the nuclear charge is greatest for Sr^{2+}, it is the smallest of these ions. The Se^{2-} ion is largest:

$$Se^{2-} > Br^- > Rb^+ > Sr^{2+}$$
$$\uparrow \qquad\qquad\qquad \uparrow$$
Largest   Smallest

See Exercises 8.51 and 8.52

Relative Ion Size II

Choose the largest ion in each of the following groups.

a. $Li^+, Na^+, K^+, Rb^+, Cs^+$

b. $Ba^{2+}, Cs^+, I^-, Te^{2-}$

Solution

a. The ions are all from Group 1A elements. Since size increases down a group (the ion with the greatest number of electrons is largest), Cs^+ is the largest ion.

b. This is an isoelectronic series of ions, all of which have the xenon electron configuration. The ion with the smallest nuclear charge is largest:

$$Te^{2-} > I^- > Cs^+ > Ba^{2+}$$
$$Z = 52 \quad Z = 53 \quad Z = 55 \quad Z = 56$$

See Exercises 8.53 and 8.54

8.5 | Energy Effects in Binary Ionic Compounds

In this section we will introduce the factors that influence the stability and the structures of solid binary ionic compounds. We know that metals and nonmetals react by transferring electrons to form cations and anions that are mutually attractive. The resulting ionic solid forms because the aggregated oppositely charged ions have a lower energy than the original elements. Just how strongly the ions attract each other in the solid state is indicated by the **lattice energy**—*the change in energy that takes place when separated gaseous ions are packed together to form an ionic solid:*

$$M^+(g) + X^-(g) \longrightarrow MX(s)$$

The structures of ionic solids will be discussed in detail in Chapter 10.

The lattice energy is often defined as the energy *released* when an ionic solid forms from its ions. However, in this book the sign of an energy term is always determined from the system's point of view: negative if the process is exothermic, positive if endothermic. Thus, using this convention, the lattice energy has a negative sign.

We can illustrate the energy changes involved in the formation of an ionic solid by considering the formation of solid lithium fluoride from its elements:

$$Li(s) + \tfrac{1}{2}F_2(g) \longrightarrow LiF(s)$$

To see the energy terms associated with this process, we take advantage of the fact that energy is a state function and break this reaction into steps, the sum of which gives the overall reaction.

1. Sublimation of solid lithium. Sublimation involves taking a substance from the solid state to the gaseous state:

$$Li(s) \longrightarrow Li(g)$$

   The enthalpy of sublimation for $Li(s)$ is 161 kJ/mol.

2. Ionization of lithium atoms to form Li^+ ions in the gas phase:

$$Li(g) \longrightarrow Li^+(g) + e^-$$

   This process corresponds to the first ionization energy for lithium, which is 520 kJ/mol.

Lithium fluoride.

3. Dissociation of fluorine molecules. We need to form a mole of fluorine atoms by breaking the F—F bonds in a half mole of F_2 molecules:

$$\tfrac{1}{2}F_2(g) \longrightarrow F(g)$$

The energy required to break this bond is 154 kJ/mol. In this case we are breaking the bonds in a half mole of fluorine, so the energy required for this step is (154 kJ)/2, or 77 kJ.

4. Formation of F^- ions from fluorine atoms in the gas phase:

$$F(g) + e^- \longrightarrow F^-(g)$$

The energy change for this process corresponds to the electron affinity of fluorine, which is −328 kJ/mol.

5. Formation of solid lithium fluoride from the gaseous Li^+ and F^- ions:

$$Li^+(g) + F^-(g) \longrightarrow LiF(s)$$

This corresponds to the lattice energy for LiF, which is −1047 kJ/mol.

Since the sum of these five processes yields the desired overall reaction, the sum of the individual energy changes gives the overall energy change:

Process	Energy Change (kJ)
$Li(s) \rightarrow Li(g)$	161
$Li(g) \rightarrow Li^+(g) + e^-$	520
$\tfrac{1}{2}F_2(g) \rightarrow F(g)$	77
$F(g) + e^- \rightarrow F^-(g)$	−328
$Li^+(g) + F^-(g) \rightarrow LiF(s)$	−1047
Overall: $Li(s) + \tfrac{1}{2}F_2(g) \rightarrow LiF(s)$	−617 kJ (per mole of LiF)

In doing this calculation, we have ignored the small difference between ΔH_{sub} and ΔE_{sub}.

This process is summarized by the energy diagram in Fig. 8.9. Note that the formation of solid lithium fluoride from its elements is highly exothermic, mainly because of the very large negative lattice energy. A great deal of energy is released when the ions combine to form the solid. In fact, note that the energy released when an electron is added to a fluorine atom to form the F^- ion (328 kJ/mol) is not enough to remove an electron from lithium (520 kJ/mol). That is, when a metallic lithium atom reacts with a nonmetallic fluorine atom to form *separated* ions,

$$Li(g) + F(g) \longrightarrow Li^+(g) + F^-(g)$$

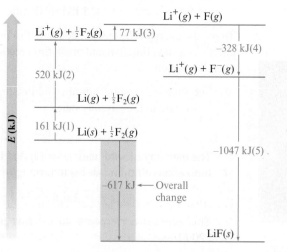

Figure 8.9 | The energy changes involved in the formation of solid lithium fluoride from its elements. The numbers in parentheses refer to the reaction steps discussed in the text.

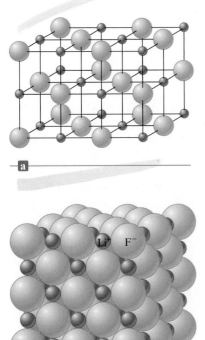

the process is endothermic and thus unfavorable. Clearly, then, the main impetus for the formation of an ionic compound rather than a covalent compound results from the strong mutual attractions among the Li$^+$ and F$^-$ ions in the solid. The lattice energy is the dominant energy term.

The structure of solid lithium fluoride is represented in Fig. 8.10. Note the alternating arrangement of the Li$^+$ and F$^-$ ions. Also note that each Li$^+$ is surrounded by six F$^-$ ions, and each F$^-$ ion is surrounded by six Li$^+$ ions. This structure can be rationalized by assuming that the ions behave as hard spheres that pack together in a way that both maximizes the attractions among the oppositely charged ions and minimizes the repulsions among the identically charged ions.

All the binary ionic compounds formed by an alkali metal and a halogen have the structure shown in Fig. 8.10, except for the cesium salts. The arrangement of ions shown in Fig. 8.10 is often called the *sodium chloride structure,* after the most common substance that possesses it.

Lattice Energy Calculations

In discussing the energetics of the formation of solid lithium fluoride, we emphasized the importance of lattice energy in contributing to the stability of the ionic solid. Lattice energy can be represented by a modified form of Coulomb's law:

$$\text{Lattice energy} = k\left(\frac{Q_1 Q_2}{r}\right)$$

where k is a proportionality constant that depends on the structure of the solid and the electron configurations of the ions, Q_1 and Q_2 are the charges on the ions, and r is the shortest distance between the centers of the cations and anions. Note that the lattice energy has a negative sign when Q_1 and Q_2 have opposite signs. This result is expected, since bringing cations and anions together is an exothermic process. Also note that the process becomes more exothermic as the ionic charges increase and as the distances between the ions in the solid decrease.

The importance of the charges in ionic solids can be illustrated by comparing the energies involved in the formation of NaF(s) and MgO(s). These solids contain the isoelectronic ions Na$^+$, F$^-$, Mg^{2+}, and O^{2-}. The energy diagram for the formation of the two solids is given in Fig. 8.11. Note several important features:

The energy released when the gaseous Mg^{2+} and O^{2-} ions combine to form solid MgO is much greater (more than four times greater) than that released when the gaseous Na$^+$ and F$^-$ ions combine to form solid NaF.

The energy required to remove two electrons from the magnesium atom (735 kJ/mol for the first and 1445 kJ/mol for the second, yielding a total of 2180 kJ/mol) is much greater than the energy required to remove one electron from a sodium atom (495 kJ/mol).

Energy (737 kJ/mol) is required to add two electrons to the oxygen atom in the gas phase. Addition of the first electron is exothermic (−141 kJ/mol), but addition of the second electron is quite endothermic (878 kJ/mol). This latter energy must be obtained indirectly, since the O^{2-}(g) ion is not stable.

In view of the facts that twice as much energy is required to remove the second electron from magnesium as to remove the first and that addition of an electron to the gaseous O$^-$ ion is quite endothermic, it seems puzzling that magnesium oxide contains Mg^{2+} and O^{2-} ions rather than Mg$^+$ and O$^-$ ions. The answer lies in the lattice energy. Note that the lattice energy for combining gaseous Mg^{2+} and O^{2-} ions to form MgO(s) is 3000 kJ/mol more negative than that for combining gaseous Na$^+$ and F$^-$ ions to form NaF(s). Thus the energy released in forming a solid containing Mg^{2+} and O^{2-}

Since the equation for lattice energy contains the product Q_1Q_2, the lattice energy for a solid with 2+ and 2− ions should be four times that for a solid with 1+ and 1− ions. That is,

$$\frac{(+2)(-2)}{(+1)(-1)} = 4$$

For MgO and NaF, the observed ratio of lattice energies (see Fig. 8.11) is

$$\frac{-3916 \text{ kJ}}{-923 \text{ kJ}} = 4.24$$

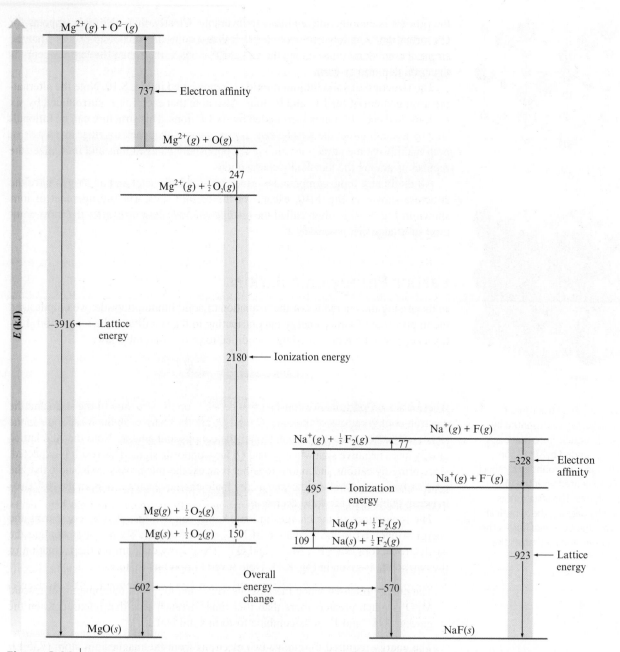

Figure 8.11 | Comparison of the energy changes involved in the formation of solid sodium fluoride and solid magnesium oxide. Note the large lattice energy for magnesium oxide (where doubly charged ions are combining) compared with that for sodium fluoride (where singly charged ions are combining).

ions rather than Mg^+ and O^- ions more than compensates for the energies required for the processes that produce the Mg^{2+} and O^{2-} ions.

If there is so much lattice energy to be gained in going from singly charged to doubly charged ions in the case of magnesium oxide, why then does solid sodium fluoride contain Na^+ and F^- ions rather than Na^{2+} and F^{2-} ions? We can answer this question by recognizing that both Na^+ and F^- ions have the neon electron configuration. Removal of an electron from Na^+ requires an extremely large quantity of energy (4560 kJ/mol) because a $2p$ electron must be removed. Conversely, the addition of an electron to F^- would require use of the relatively high-energy $3s$ orbital, which is also an unfavorable process. Thus we can say that for sodium fluoride the extra energy required to form the doubly charged ions is greater than the gain in lattice energy that would result.

Figure 8.12 | The three possible types of bonds: (a) a covalent bond formed between identical F atoms; (b) the polar covalent bond of HF, with both ionic and covalent components; and (c) an ionic bond with no electron sharing.

This discussion of the energies involved in the formation of solid ionic compounds illustrates that a variety of factors operate to determine the composition and structure of these compounds. The most important of these factors involve the balancing of the energies required to form highly charged ions and the energy released when highly charged ions combine to form the solid.

8.6 | Partial Ionic Character of Covalent Bonds

Recall that when atoms with different electronegativities react to form molecules, the electrons are not shared equally. The possible result is a polar covalent bond or, in the case of a large electronegativity difference, a complete transfer of one or more electrons to form ions. The cases are summarized in Fig. 8.12.

How well can we tell the difference between an ionic bond and a polar covalent bond? The only honest answer to this question is that there are probably no totally ionic bonds between *discrete pairs of atoms*. The evidence for this statement comes from calculations of the percent ionic character for the bonds of various binary compounds in the gas phase. These calculations are based on comparisons of the measured dipole moments for molecules of the type X—Y with the calculated dipole moments for the completely ionic case, X^+Y^-. The percent ionic character of a bond can be defined as

$$\text{Percent ionic character of a bond} = \left(\frac{\text{measured dipole moment of X—Y}}{\text{calculated dipole moment of }X^+Y^-}\right) \times 100\%$$

Application of this definition to various compounds (in the gas phase) gives the results shown in Fig. 8.13, where percent ionic character is plotted versus the difference in the electronegativity values of X and Y. Note from this plot that ionic character increases with electronegativity difference, as expected. However, none of the bonds reaches 100% ionic character, even though compounds with the maximum possible electronegativity differences are considered. Thus, according to this definition, no individual bonds are completely ionic. This conclusion is in contrast to the usual classification of many of these compounds (as ionic solids). All the compounds shown in Fig. 8.13 with more than 50% ionic character are normally considered to be ionic solids. Recall, however, the results in Fig. 8.13 are for the gas phase, where individual XY molecules exist. These results cannot necessarily be assumed to apply to the solid state, where the existence of ions is favored by the multiple ion interactions.

Another complication in identifying ionic compounds is that many substances contain polyatomic ions. For example, NH_4Cl contains NH_4^+ and Cl^- ions, and Na_2SO_4

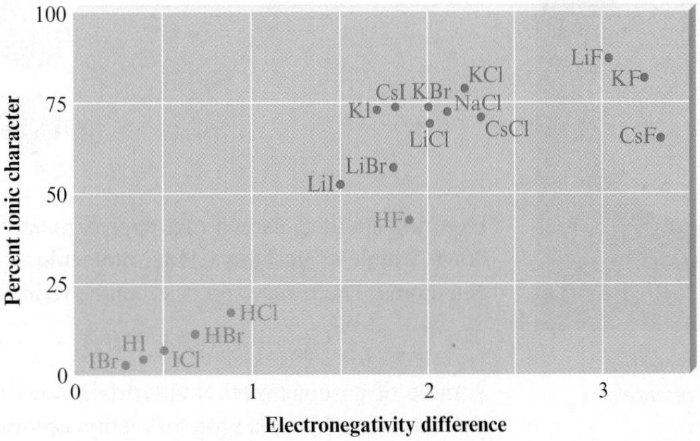

Figure 8.13 | The relationship between the ionic character of a covalent bond and the electronegativity difference of the bonded atoms. Note that the compounds with ionic character greater than 50% (red) are normally considered to be ionic compounds.

contains Na^+ and SO_4^{2-} ions. The bonds within the ammonium and sulfate ions are covalent bonds.

We will avoid these problems by adopting an operational definition of ionic compounds: *Any compound that conducts an electric current when melted will be classified as ionic.*

8.7 | The Covalent Chemical Bond: A Model

Before we develop specific models for covalent chemical bonding, it will be helpful to summarize some of the concepts introduced in this chapter.

What is a chemical bond? Chemical bonds can be viewed as forces that cause a group of atoms to behave as a unit.

Why do chemical bonds occur? There is no principle of nature that states that bonds are favored or disfavored. Bonds are neither inherently "good" nor inherently "bad" as far as nature is concerned; *bonds result from the tendency of a system to seek its lowest possible energy.* From a simplistic point of view, bonds occur when collections of atoms are more stable (lower in energy) than the separate atoms. For example, approximately 1652 kJ of energy is required to break a mole of methane (CH_4) molecules into separate C and H atoms. Or, taking the opposite view, 1652 kJ of energy is released when 1 mole of methane is formed from 1 mole of gaseous C atoms and 4 moles of gaseous H atoms. Thus we can say that 1 mole of CH_4 molecules in the gas phase is 1652 kJ lower in energy than 1 mole of carbon atoms plus 4 moles of hydrogen atoms. Methane is therefore a stable molecule relative to its separated atoms.

We find it useful to interpret molecular stability in terms of a model called a *chemical bond.* To understand why this model was invented, let's continue with methane, which consists of four hydrogen atoms arranged at the corners of a tetrahedron around a carbon atom:

A tetrahedron has four equal triangular faces.

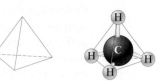

Given this structure, it is natural to envision four individual C—H interactions (we call them *bonds*). The energy of stabilization of CH_4 is divided equally among the four bonds to give an average C—H bond energy per mole of C—H bonds:

$$\frac{1652 \text{ kJ/mol}}{4} = 413 \text{ kJ/mol}$$

Next, consider methyl chloride, which consists of CH_3Cl molecules having the structure

Experiments have shown that approximately 1578 kJ of energy is required to break down 1 mole of gaseous CH_3Cl molecules into gaseous carbon, chlorine, and hydrogen atoms. The reverse process can be represented as

$$C(g) + Cl(g) + 3H(g) \longrightarrow CH_3Cl(g) + 1578 \text{ kJ/mol}$$

A mole of gaseous methyl chloride is lower in energy by 1578 kJ than its separate gaseous atoms. Thus a mole of methyl chloride is held together by 1578 kJ of energy.

Molten NaCl conducts an electric current, indicating the presence of mobile Na^+ and Cl^- ions.

Ken O'Donoghue © Cengage Learning

Again, it is very useful to divide this energy into individual bonds. Methyl chloride can be visualized as containing one C—Cl bond and three C—H bonds. If we assume arbitrarily that a C—H interaction represents the same quantity of energy in any situation (that is, that the strength of a C—H bond is independent of its molecular environment), we can do the following bookkeeping:

$$1 \text{ mol C—Cl bonds plus 3 mol C—H bonds} = 1578 \text{ kJ}$$
$$\text{C—Cl bond energy} + 3(\text{average C—H bond energy}) = 1578 \text{ kJ}$$
$$\text{C—Cl bond energy} + 3(413 \text{ kJ/mol}) = 1578 \text{ kJ}$$
$$\text{C—Cl bond energy} = 1578 - 1239 = 339 \text{ kJ/mol}$$

These assumptions allow us to associate given quantities of energy with C—H and C—Cl bonds.

It is important to note that the bond concept is a human invention. Bonds provide a method for dividing up the energy evolved when a stable molecule is formed from its component atoms. Thus in this context a bond represents a quantity of energy obtained from the overall molecular energy of stabilization in a rather arbitrary way. This is not to say that the concept of individual bonds is a bad idea. In fact, the modern concept of the chemical bond, conceived by the American chemists G. N. Lewis and Linus Pauling, is one of the most useful ideas chemists have ever developed.

Models: An Overview

The framework of chemistry, like that of any science, consists of *models*—attempts to explain how nature operates on the microscopic level based on experiences in the macroscopic world. To understand chemistry, one must understand its models and how they are used. We will use the concept of bonding to reemphasize the important characteristics of models, including their origin, structure, and uses.

Models originate from our observations of the properties of nature. For example, the concept of bonds arose from the observations that most chemical processes involve collections of atoms and that chemical reactions involve rearrangements of the ways the atoms are grouped. Therefore, to understand reactions, we must understand the forces that bind atoms together.

Bonding is a model proposed to explain molecular stability.

In natural processes there is a tendency toward lower energy. Collections of atoms therefore occur because the aggregated state has lower energy than the separated atoms. Why? As we saw earlier in this chapter, the best explanations for the energy change involve atoms sharing electrons or atoms transferring electrons to become ions. In the case of electron sharing, we find it convenient to assume that individual bonds occur between pairs of atoms. Let's explore the validity of this assumption and see how it is useful.

In a diatomic molecule such as H_2, it is natural to assume that a bond exists between the atoms, holding them together. It is also useful to assume that individual bonds are present in polyatomic molecules such as CH_4. Therefore, instead of thinking of CH_4 as a unit with a stabilization energy of 1652 kJ per mole, we choose to think of CH_4 as containing four C—H bonds, each worth 413 kJ of energy per mole of bonds. Without this concept of individual bonds in molecules, chemistry would be hopelessly complicated. There are millions of different chemical compounds, and if each of these compounds had to be considered as an entirely new entity, the task of understanding chemical behavior would be overwhelming.

The bonding model provides a framework to systematize chemical behavior by enabling us to think of molecules as collections of common fundamental components. For example, a typical biomolecule, such as a protein, contains hundreds of atoms and might seem discouragingly complex. However, if we think of a protein as constructed of individual bonds, C—C, C—H, C—N, C—O, N—H, and so on, it helps tremendously in predicting and understanding the protein's behavior. The essential idea is that we expect a given bond to behave about the same in any molecular environment.

bond strength assignment

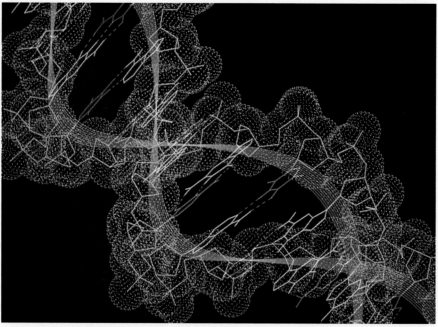

The concept of individual bonds makes it much easier to deal with complex molecules such as DNA. A small segment of a DNA molecule is shown here.

Used in this way, the model of the chemical bond has helped chemists to systematize the reactions of the millions of existing compounds.

In addition to being useful, the bonding model is physically sensible. It makes sense that atoms can form stable groups by sharing electrons; shared electrons give a lower energy state because they are simultaneously attracted by two nuclei.

Also, as we will see in the next section, bond energy data support the existence of discrete bonds that are relatively independent of the molecular environment. It is very important to remember, however, that the chemical bond is only a model. Although our concept of discrete bonds in molecules agrees with many of our observations, some molecular properties require that we think of a molecule as a whole, with the electrons free to move through the entire molecule. This is called *delocalization* of the electrons, a concept that will be discussed more completely in the next chapter.

Let's Review | *Fundamental Properties of Models*

❯ Models are human inventions, always based on an incomplete understanding of how nature works. *A model does not equal reality.*

❯ Models are often wrong. This property derives from the first property. Models are based on speculation and are always oversimplifications.

❯ Models tend to become more complicated as they age. As flaws are discovered in our models, we "patch" them and thus add more detail.

❯ It is very important to understand the assumptions inherent in a particular model before you use it to interpret observations or to make predictions. Simple models usually involve very restrictive assumptions and can be expected to yield only qualitative information. Asking for a sophisticated explanation from a simple model is like expecting to get an accurate mass for a diamond using a bathroom scale.

For a model to be used effectively, we must understand its strengths and weaknesses and ask only appropriate questions. An illustration of this point is the simple aufbau principle used to account for the electron configurations of the elements. Although this model correctly predicts the configuration for most atoms, chromium and copper, for example, do not agree with the predictions.

> Detailed studies show that the configurations of chromium and copper result from complex electron interactions that are not taken into account in the simple model. However, this does not mean that we should discard the simple model that is so useful for most atoms. Instead, we must apply it with caution and not expect it to be correct in every case.

> ❯ When a model is wrong, we often learn much more than when it is right. If a model makes a wrong prediction, it usually means we do not understand some fundamental characteristics of nature. We often learn by making mistakes. (Try to remember this when you get back your next chemistry test.)

8.8 | Covalent Bond Energies and Chemical Reactions

In this section we will consider the energies associated with various types of bonds and see how the bonding concept is useful in dealing with the energies of chemical reactions. One important consideration is to establish the sensitivity of a particular type of bond to its molecular environment. For example, consider the stepwise decomposition of methane:

Process	Energy Required (kJ/mol)
$CH_4(g) \rightarrow CH_3(g) + H(g)$	435
$CH_3(g) \rightarrow CH_2(g) + H(g)$	453
$CH_2(g) \rightarrow CH(g) + H(g)$	425
$CH(g) \rightarrow C(g) + H(g)$	339
	Total $= 1652$
	Average $= \dfrac{1652}{4} = 413$

Although a C—H bond is broken in each case, the energy required varies in a non-systematic way. This example shows that the C—H bond is somewhat sensitive to its environment. We use the *average* of these individual bond dissociation energies even though this quantity only approximates the energy associated with a C—H bond in a particular molecule. The degree of sensitivity of a bond to its environment also can be seen from experimental measurements of the energy required to break the C—H bond in the following molecules:

Molecule	Measured C—H Bond Energy (kJ/mol)
$HCBr_3$	380
$HCCl_3$	380
HCF_3	430
C_2H_6	410

These data show that the C—H bond strength varies significantly with its environment, but the concept of an average C—H bond strength remains useful to chemists. The average values of bond energies for various types of bonds are listed in Table 8.4.

So far we have discussed bonds in which one pair of electrons is shared. This type of bond is called a **single bond**. As we will see in more detail later, atoms sometimes share two pairs of electrons, forming a **double bond**, or share three pairs of electrons, forming a **triple bond**. The bond energies for these *multiple bonds* are also given in Table 8.4.

Table 8.4 | Average Bond Energies (kJ/mol)

Single Bonds						Multiple Bonds	
H—H	432	N—H	391	I—I	149	C=C	614
H—F	565	N—N	160	I—Cl	208	C≡C	839
H—Cl	427	N—F	272	I—Br	175	O=O	495
H—Br	363	N—Cl	200			C=O*	745
H—I	295	N—Br	243	S—H	347	C≡O	1072
		N—O	201	S—F	327	N=O	607
C—H	413	O—H	467	S—Cl	253	N=N	418
C—C	347	O—O	146	S—Br	218	N≡N	941
C—N	305	O—F	190	S—S	266	C≡N	891
C—O	358	O—Cl	203			C=N	615
C—F	485	O—I	234				
C—Cl	339			Si—Si	340		
C—Br	276	F—F	154	Si—H	393		
C—I	240	F—Cl	253	Si—C	360		
C—S	259	F—Br	237	Si—O	452		
		Cl—Cl	239				
		Cl—Br	218				
		Br—Br	193				

*C=O(CO_2) = 799

A relationship also exists between the number of shared electron pairs and the bond length. As the number of shared electrons increases, the bond length shortens. This relationship is shown for selected bonds in Table 8.5.

Bond Energy and Enthalpy

In this discussion we are ignoring the small difference between enthalpy and energy.

Bond energy values can be used to calculate approximate energies for reactions. To illustrate how this is done, we will calculate the change in energy that accompanies the following reaction:

$$H_2(g) + F_2(g) \longrightarrow 2HF(g)$$

This reaction involves breaking one H—H and one F—F bond and forming two H—F bonds. For bonds to be broken, energy must be *added* to the system—an endothermic process. Consequently, the energy terms associated with bond breaking have *positive* signs. The formation of a bond *releases* energy, an exothermic process, so the

Table 8.5 | Bond Lengths and Bond Energies for Selected Bonds

Bond	Bond Type	Bond Length (pm)	Bond Energy (kJ/mol)
C—C	Single	154	347
C=C	Double	134	614
C≡C	Triple	120	839
C—O	Single	143	358
C=O	Double	123	745
C—N	Single	143	305
C=N	Double	138	615
C≡N	Triple	116	891

energy terms associated with bond making carry a *negative* sign. We can write the enthalpy change for a reaction as follows:

ΔH = sum of the energies required to break old bonds (positive signs) plus the sum of the energies released in the formation of new bonds (negative signs)

This leads to the expression

$$\Delta H = \underbrace{\Sigma n \times D \text{ (bonds broken)}}_{\text{Energy required}} - \underbrace{\Sigma n \times D \text{ (bonds formed)}}_{\text{Energy released}}$$

where Σ represents the sum of terms, D represents the bond energy per mole of bonds (D *always* has a positive sign), and n represents the moles of a particular type of bond.

In the case of the formation of HF,

$$\Delta H = D_{H-H} + D_{F-F} - 2D_{H-F}$$
$$= 1 \text{ mol} \times \frac{432 \text{ kJ}}{\text{mol}} + 1 \text{ mol} \times \frac{154 \text{ kJ}}{\text{mol}} - 2 \text{ mol} \times \frac{565 \text{ kJ}}{\text{mol}}$$
$$= -544 \text{ kJ}$$

Thus, when 1 mole of $H_2(g)$ and 1 mole of $F_2(g)$ react to form 2 moles of $HF(g)$, 544 kJ of energy should be released.

This result can be compared with the calculation of ΔH for this reaction from the standard enthalpy of formation for HF (-271 kJ/mol):

$$\Delta H^\circ = 2 \text{ mol} \times (-271 \text{ kJ/mol}) = -542 \text{ kJ}$$

Thus the use of bond energies to calculate ΔH works quite well in this case.

Interactive Example 8.5

Sign in at http://login.cengagebrain.com to try this Interactive Example in **OWL**.

ΔH from Bond Energies

Using the bond energies listed in Table 8.4, calculate ΔH for the reaction of methane with chlorine and fluorine to give Freon-12 (CF_2Cl_2).

$$CH_4(g) + 2Cl_2(g) + 2F_2(g) \longrightarrow CF_2Cl_2(g) + 2HF(g) + 2HCl(g)$$

Solution

The idea here is to break the bonds in the gaseous reactants to give individual atoms and then assemble these atoms into the gaseous products by forming new bonds:

$$\text{Reactants} \xrightarrow[\text{required}]{\text{Energy}} \text{atoms} \xrightarrow[\text{released}]{\text{Energy}} \text{products}$$

We then combine the energy changes to calculate ΔH:

$$\Delta H = \text{energy required to break bonds} - \text{energy released when bonds form}$$

where the minus sign gives the correct sign to the energy terms for the exothermic processes.

Reactant Bonds Broken:

CH_4: 4 mol C—H $4 \text{ mol} \times \dfrac{413 \text{ kJ}}{\text{mol}} = 1652 \text{ kJ}$

$2Cl_2$: 2 mol Cl—Cl $2 \text{ mol} \times \dfrac{239 \text{ kJ}}{\text{mol}} = 478 \text{ kJ}$

$2F_2$: 2 mol F—F $2 \text{ mol} \times \dfrac{154 \text{ kJ}}{\text{mol}} = 308 \text{ kJ}$

Total energy required = 2438 kJ

Product Bonds Formed:

CF_2Cl_2:   2 mol C—F $2 \text{ mol} \times \dfrac{485 \text{ kJ}}{\text{mol}} = $ 970 kJ

and

2 mol C—Cl $2 \text{ mol} \times \dfrac{339 \text{ kJ}}{\text{mol}} = $ 678 kJ

2HF: 2 mol H—F $2 \text{ mol} \times \dfrac{565 \text{ kJ}}{\text{mol}} = $ 1130 kJ

2HCl: 2 mol H—Cl $2 \text{ mol} \times \dfrac{427 \text{ kJ}}{\text{mol}} = $ 854 kJ

Total energy released = 3632 kJ

We now can calculate ΔH:

ΔH = energy required to break bonds − energy released when bonds form

= 2438 kJ − 3632 kJ

= −1194 kJ

Since the sign of the value for the enthalpy change is negative, this means that 1194 kJ of energy is released per mole of CF_2Cl_2 formed.

See Exercises 8.65 through 8.72

8.9 | The Localized Electron Bonding Model

So far we have discussed the general characteristics of the chemical bonding model and have seen that properties such as bond strength and polarity can be assigned to individual bonds. In this section we introduce a specific model used to describe covalent bonds. We need a simple model that can be applied easily even to very complicated molecules and that can be used routinely by chemists to interpret and organize the wide variety of chemical phenomena. The model that serves this purpose is the **localized electron (LE) model**, which assumes that *a molecule is composed of atoms that are bound together by sharing pairs of electrons using the atomic orbitals of the bound atoms.* Electron pairs in the molecule are assumed to be localized on a particular atom or in the space between two atoms. Those pairs of electrons localized on an atom are called **lone pairs**, and those found in the space between the atoms are called **bonding pairs**.

As we will apply it, the LE model has three parts:

1. Description of the valence electron arrangement in the molecule using Lewis structures (will be discussed in the next section).
2. Prediction of the geometry of the molecule using the valence shell electron-pair repulsion (VSEPR) model (will be discussed in Section 8.13).
3. Description of the type of atomic orbitals used by the atoms to share electrons or hold lone pairs (will be discussed in Chapter 9).

8.10 | Lewis Structures

The **Lewis structure** of a molecule shows how the valence electrons are arranged among the atoms in the molecule. These representations are named after G. N. Lewis (Fig. 8.14). The rules for writing Lewis structures are based on observations of thousands of molecules. From experiment, chemists have learned that the *most important requirement for the formation of a stable compound is that the atoms achieve noble gas electron configurations.*

Lewis structures show only valence electrons.

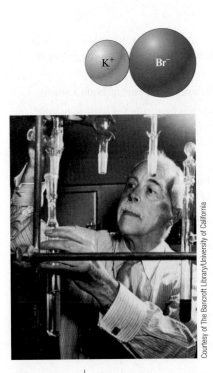

Courtesy of The Bancroft Library/University of California

Figure 8.14 | G. N. Lewis (1875–1946).

Carbon, nitrogen, oxygen, and fluorine always obey the octet rule in stable molecules.

F_2

We have already seen that when metals and nonmetals react to form binary ionic compounds, electrons are transferred and the resulting ions typically have noble gas electron configurations. An example is the formation of KBr, where the K^+ ion has the [Ar] electron configuration and the Br^- ion has the [Kr] electron configuration. In writing Lewis structures, the rule is that *only the valence electrons are included.* Using dots to represent electrons, the Lewis structure for KBr is

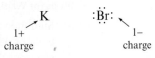

No dots are shown on the K^+ ion because it has no valence electrons. The Br^- ion is shown with eight electrons because it has a filled valence shell.

Next we will consider Lewis structures for molecules with covalent bonds, involving elements in the first and second periods. The principle of achieving a noble gas electron configuration applies to these elements as follows:

- Hydrogen forms stable molecules where it shares two electrons. That is, it follows a **duet rule**. For example, when two hydrogen atoms, each with one electron, combine to form the H_2 molecule, we have

$$H \cdot \quad \longrightarrow \quad \cdot H$$
$$H : H$$

By sharing electrons, each hydrogen in H_2, in effect, has two electrons; that is, each hydrogen has a filled valence shell.

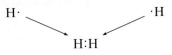

- Helium does not form bonds because its valence orbital is already filled; it is a noble gas. Helium has the electron configuration $1s^2$ and can be represented by the Lewis structure

$$He :$$

- The second-row nonmetals carbon through fluorine form stable molecules when they are surrounded by enough electrons to fill the valence orbitals, that is, the $2s$ and the three $2p$ orbitals. Since eight electrons are required to fill these orbitals, these elements typically obey the **octet rule**; they are surrounded by eight electrons. An example is the F_2 molecule, which has the following Lewis structure:

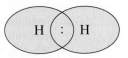

:F·	→	:F:F:	←	·F:
F atom with seven valence electrons		F_2 molecule		F atom with seven valence electrons

Note that each fluorine atom in F_2 is, in effect, surrounded by eight electrons, two of which are shared with the other atom. This is a *bonding pair* of electrons, as discussed earlier. Each fluorine atom also has three pairs of electrons not involved in bonding. These are the *lone pairs*.

- Neon does not form bonds because it already has an octet of valence electrons (it is a noble gas). The Lewis structure is

$$:Ne:$$

Note that only the valence electrons of the neon atom ($2s^2 2p^6$) are represented by the Lewis structure. The $1s^2$ electrons are core electrons and are not shown.

From the preceding discussion we can formulate the following rules for writing the Lewis structures of molecules containing atoms from the first two periods.

Problem-Solving Strategy

Steps for Writing Lewis Structures

1. Sum the valence electrons from all the atoms. Do not worry about keeping track of which electrons come from which atoms. It is the *total* number of electrons that is important.

2. Use a pair of electrons to form a bond between each pair of bound atoms.

3. Arrange the remaining electrons to satisfy the duet rule for hydrogen and the octet rule for the second-row elements.

To see how these steps are applied, we will draw the Lewis structures of a few molecules. We will first consider the water molecule and follow the previous steps.

1. We sum the *valence* electrons for H_2O as shown:

$$1 + 1 + 6 = 8 \text{ valence electrons}$$
$$\nearrow \quad \nearrow \quad \nearrow$$
$$H \quad H \quad O$$

2. Using a pair of electrons per bond, we draw in the two O—H single bonds:

$$H—O—H$$

Note that *a line instead of a pair of dots is used to indicate each pair of bonding electrons.* This is the standard notation.

3. We distribute the remaining electrons to achieve a noble gas electron configuration for each atom. Since four electrons have been used in forming the two bonds, four electrons $(8 - 4)$ remain to be distributed. Hydrogen is satisfied with two electrons (duet rule), but oxygen needs eight electrons to have a noble gas configuration. Thus the remaining four electrons are added to oxygen as two lone pairs. Dots are used to represent the lone pairs:

H—O—H represents H:O:H

H—Ö—H   Lone pairs

This is the correct Lewis structure for the water molecule. Each hydrogen has two electrons and the oxygen has eight, as shown below:

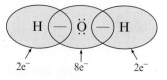

H_2O

As a second example, let's write the Lewis structure for carbon dioxide.

1. Summing the valence electrons gives

$$4 + 6 + 6 = 16$$
$$\nearrow \quad \nearrow \quad \nearrow$$
$$C \quad O \quad O$$

2. Parts of electrons are used to form a bond between the carbon and each oxygen,

$$O—C—O$$

the remaining electrons are distributed to achieve noble gas configurations on each atom. In this case we have 12 electrons (16 − 4) remaining after the bonds are drawn. The distribution of these electrons is determined by a trial-and-error process. We have 6 pairs of electrons to distribute. Suppose we try 3 pairs on each oxygen to give

3. Is this correct? To answer this question, we need to check two things:

1. The total number of electrons. There are 16 valence electrons in this structure, which is the correct number.
2. The octet rule for each atom. Each oxygen has 8 electrons, but the carbon has only 4. This cannot be the correct Lewis structure.

How can we arrange the 16 available electrons to achieve an octet for each atom? Suppose there are 2 shared pairs between the carbon and each oxygen:

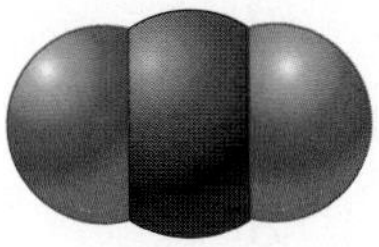

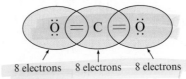

8 electrons 8 electrons 8 electrons

Now each atom is surrounded by 8 electrons, and the total number of electrons is 16, as required. This is the correct Lewis structure for carbon dioxide, which has two double bonds and four lone pairs.

Finally, let's consider the Lewis structure of the CN^- (cyanide) ion. Summing the valence electrons, we have

Note that the negative charge means an extra electron is present. After drawing a single bond (C—N), we distribute the remaining electrons to achieve a noble gas configuration for each atom. Eight electrons remain to be distributed. We can try various possibilities, for example:

This structure is incorrect because C and N have only six electrons each instead of eight. The correct arrangement is

$$[:C{\equiv}N:]^-$$

(Satisfy yourself that both carbon and nitrogen have eight electrons.)

CO₂ figure caption:

CO₂

$\ddot{O}{=}C{=}\ddot{O}$ represents $\ddot{O}:\,:C:\,:\ddot{O}$

Interactive Example 8.6

Sign in at http://login.cengagebrain .com to try this Interactive Example in OWL.

Writing Lewis Structures

Give the Lewis structure for each of the following.

a. HF   d. CH_4

b. N_2 e. CF_4

c. NH_3 f. NO^+

Solution

In each case we apply the three steps for writing Lewis structures. Recall that lines are used to indicate shared electron pairs and that dots are used to indicate nonbonding pairs (lone pairs). We have the following tabulated results:

	Total Valence Electrons	Draw Single Bonds	Calculate Number of Electrons Remaining	Use Remaining Electrons to Achieve Noble Gas Configurations	Check Number of Electrons
a. HF	$1 + 7 = 8$	H—F	6	H—$\ddot{\underset{\displaystyle\cdot}{F}}$:	H, 2 F, 8
b. N_2	$5 + 5 = 10$	N—N	8	:N≡N:	N, 8

c. NH_3	$5 + 3(1) = 8$	H—N—H
		      |
		      H
			2	H—$\ddot{N}$—H
				      |
				      H
					H, 2
					 N, 8

d. CH_4	$4 + 4(1) = 8$	H
		     |
		 H—C—H
		     |
		     H
			0	H
				     |
				 H—C—H
				     |
				     H
					H, 2
					 C, 8

e. CF_4	$4 + 4(7) = 32$	F
		     |
		 F—C—F
		     |
		     F
			24	:$\ddot{F}$:
				     |
				 :$\ddot{F}$—C—$\ddot{F}$:
				     |
				    :$\ddot{F}$:
					F, 8
					 C, 8

f. NO^+	$5 + 6 - 1 = 10$	N—O	8	$[:N≡O:]^+$	N, 8 O, 8

See Exercises 8.81 Through 8.84

When writing Lewis structures, do not worry about which electrons come from which atoms in a molecule. The best way to look at a molecule is to regard it as a new entity that uses all the available valence electrons of the atoms to achieve the lowest possible energy.* The valence electrons belong to the molecule, rather than to the individual atoms. Simply distribute all valence electrons so that the various rules are satisfied, without regard for the origin of each particular electron.

8.11 | Exceptions to the Octet Rule

The localized electron model is a simple but very successful model, and the rules we have used for Lewis structures apply to most molecules. However, with such a simple model, some exceptions are inevitable. Boron, for example, tends to form compounds in which the boron atom has fewer than eight electrons around it—it does not have a complete octet. Boron trifluoride (BF_3), a gas at normal temperatures and pressures, reacts very energetically with molecules such as water and ammonia that have available electron pairs (lone pairs). The violent reactivity of BF_3 with electron-rich mole-

---

*In a sense this approach corrects for the fact that the localized electron model overemphasizes that a molecule is simply a sum of its parts—that is, that the atoms retain their individual identities in the molecule.

Chemical connections

Nitrogen Under Pressure

The element nitrogen exists at normal temperatures and pressures as a gas containing N_2, a molecule with a very strong triple bond. In the gas phase, the diatomic molecules move around independently with almost no tendency to associate with each other. Under intense pressure, however, nitrogen changes to a dramatically different form. This conclusion was reached at the Carnegie Institution in Washington, D.C., by Mikhail Erements and his colleagues, who subjected nitrogen to a pressure of 2.4 million atmospheres in a special diamond anvil press. Under this tremendous pressure, the bonds of the N_2 molecules break and a substance containing an aggregate of nitrogen atoms forms. In other words, under great pressure elemental nitrogen changes from a substance containing diatomic molecules to one containing many nitrogen atoms bonded to each other. Interestingly, this substance remains intact even after the pressure is released—as long as the temperature remains at 100 K. This new form of nitrogen has a very high potential energy relative to N_2. Thus this substance would be an extraordinarily powerful propellant or explosive if enough of it could be made. This new form of nitrogen is also a semiconductor for electricity; normal nitrogen gas is an insulator.

The newly discovered form of nitrogen is significant for several reasons. For one thing, it may help us understand the nature of the interiors of the giant gas planets such as Jupiter. Also, their success in changing nitrogen to an atomic solid encourages high-pressure scientists who are trying to accomplish the same goal with hydrogen. It is surprising that nitrogen, which has diatomic molecules containing bonds more than twice as strong as those in hydrogen, will form an atomic solid at these pressures but hydrogen does not. Hydrogen remains a molecular solid at far greater pressures than nitrogen can endure.

A diamond anvil cell used to study materials at very high pressures.

Steven D. Jacobsen/Northwestern University

---

cules arises because the boron atom is electron-deficient. Boron trifluoride has 24 valence electrons. The Lewis structure often drawn for BF_3 is

BF₃

Note that in this structure boron has only 6 electrons around it. The octet rule for boron can be satisfied by drawing a structure with a double bond, such as

Recent studies indicate that double bonding may be important in BF_3. However, the boron atom in BF_3 certainly behaves as if it is electron-deficient, as indicated by the

reactivity of BF_3 toward electron-rich molecules, for example, toward NH_3 to form H_3NBF_3:

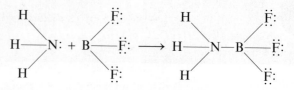

In this stable compound, boron has an octet of electrons.

It is characteristic of boron to form molecules in which the boron atom is electron-deficient. On the other hand, carbon, nitrogen, oxygen, and fluorine can be counted on to obey the octet rule.

Some atoms exceed the octet rule. This behavior is observed only for those elements in Period 3 of the periodic table and beyond. To see how this arises, we will consider the Lewis structure for sulfur hexafluoride (SF_6), a well-known and very stable molecule. The sum of the valence electrons is

$$6 + 6(7) = 48 \text{ electrons}$$

Indicating the single bonds gives the structure on the left below:

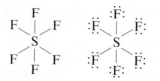

SF_6

We have used 12 electrons to form the S—F bonds, which leaves 36 electrons. Since fluorine always follows the octet rule, we complete the six fluorine octets to give the structure on the right above. This structure uses all 48 valence electrons for SF_6, but sulfur has 12 electrons around it; that is, sulfur *exceeds* the octet rule. How can this happen?

To answer this question, we need to consider the different types of valence orbitals characteristic of second- and third-period elements. The second-row elements have $2s$ and $2p$ valence orbitals, and the third-row elements have $3s$, $3p$, and $3d$ orbitals. The $3s$ and $3p$ orbitals fill with electrons in going from sodium to argon, but the $3d$ orbitals remain empty. For example, the valence orbital diagram for a sulfur atom is

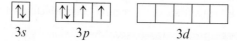

The localized electron model assumes that the empty $3d$ orbitals can be used to accommodate extra electrons. Thus the sulfur atom in SF_6 can have 12 electrons around it by using the $3s$ and $3p$ orbitals to hold 8 electrons, with the extra 4 electrons placed in the formerly empty $3d$ orbitals.

Let's Review | *Lewis Structures: Comments About the Octet Rule*

> The second-row elements C, N, O, and F should always be assumed to obey the octet rule.

> The second-row elements B and Be often have fewer than eight electrons around them in their compounds. These electron-deficient compounds are very reactive.

> The second-row elements never exceed the octet rule, since their valence orbitals ($2s$ and $2p$) can accommodate only eight electrons.

> Third-row and heavier elements often satisfy the octet rule but can exceed the octet rule by using their empty valence *d* orbitals.

> When writing the Lewis structure for a molecule, satisfy the octet rule for the atoms first. If electrons remain after the octet rule has been satisfied, then place them on the elements having available *d* orbitals (elements in Period 3 or beyond).

**Interactive
Example 8.7**

Sign in at http://login.cengagebrain
.com to try this Interactive Example
in OWL.

Lewis Structures for Molecules That Violate the Octet Rule I

Write the Lewis structure for PCl_5.

Solution

We can follow the same stepwise procedure we used above for sulfur hexafluoride.

1. Sum the valence electrons.

$$5 + 5(7) = 40 \text{ electrons}$$
$$\uparrow \quad\quad \uparrow$$
$$P \quad\quad Cl$$

2. Indicate single bonds between bound atoms.

3. Distribute the remaining electrons. In this case, 30 electrons $(40 - 10)$ remain. These are used to satisfy the octet rule for each chlorine atom. The final Lewis structure is

Note that phosphorus, which is a third-row element, has exceeded the octet rule by two electrons.

See Exercises 8.87 and 8.88

PCl_5

In the PCl_5 and SF_6 molecules, the central atoms (P and S, respectively) must have the extra electrons. However, in molecules having more than one atom that can exceed the octet rule, it is not always clear which atom should have the extra electrons. Consider the Lewis structure for the triiodide ion (I_3^-), which has

$$3(7) + 1 = 22 \text{ valence electrons}$$
$$\uparrow \quad\quad \uparrow$$
$$I \quad -1 \text{ charge}$$

Indicating the single bonds gives I—I—I. At this point, 18 electrons $(22 - 4)$ remain. Trial and error will convince you that one of the iodine atoms must exceed the octet rule, but *which* one?

The rule we will follow is that *when it is necessary to exceed the octet rule for one of several third-row (or higher) elements, assume that the extra electrons should be placed on the central atom.*

Thus for I_3^- the Lewis structure is

$$\left[\ddot{\text{I}} - \ddot{\text{I}} - \ddot{\text{I}} \right]^-$$

where the central iodine exceeds the octet rule. This structure agrees with known properties of I_3^-.

Interactive
Example 8.8

Sign in at http://login.cengagebrain
.com to try this Interactive Example
in **OWL**.

Lewis Structures for Molecules That Violate the Octet Rule II

Write the Lewis structure for each molecule or ion.

a. ClF_3 **b.** XeO_3 **c.** $RnCl_2$ **d.** $BeCl_2$ **e.** ICl_4^-

Solution

a. The chlorine atom (third row) accepts the extra electrons.

b. All atoms obey the octet rule.

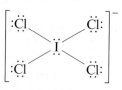

c. Radon, a noble gas in Period 6, accepts the extra electrons.

$$:\ddot{C}l-\ddot{R}n-\ddot{C}l:$$

d. Beryllium is electron-deficient.

$$:\ddot{C}l-Be-\ddot{C}l:$$

e. Iodine exceeds the octet rule.

See Exercises 8.85, 8.87, and 8.88

8.12 | Resonance

Sometimes more than one valid Lewis structure (one that obeys the rules we have outlined) is possible for a given molecule. Consider the Lewis structure for the nitrate ion (NO_3^-), which has 24 valence electrons. To achieve an octet of electrons around each atom, a structure like this is required:

NO_3^-

If this structure accurately represents the bonding in NO_3^-, there should be two types of N—O bonds observed in the molecule: one shorter bond (the double bond) and two identical longer ones (the two single bonds). However, experiments clearly show that

NO_3^- exhibits only *one* type of N—O bond with a length and strength *between* those expected for a single bond and a double bond. Thus, although the structure we have shown above is a valid Lewis structure, it does *not* correctly represent the bonding in NO_3^-. This is a serious problem, and it means that the model must be modified.

Look again at the proposed Lewis structure for NO_3^-. There is no reason for choosing a particular oxygen atom to have the double bond. There are really three valid Lewis structures:

Is any of these structures a correct description of the bonding in NO_3^-? No, because NO_3^- does not have one double and two single bonds—it has three equivalent bonds. We can solve this problem by making the following assumption: The correct description of NO_3^- is *not given by any one* of the three Lewis structures but is given only by the *superposition of all three.*

The nitrate ion does not exist as any of the three extreme structures but exists as an average of all three. **Resonance is invoked when more than one valid Lewis structure can be written for a particular molecule. The resulting electron structure of the molecule is given by the average of these resonance structures.** This situation is usually represented by double-headed arrows as follows:

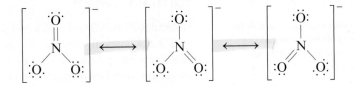

Note that in all these resonance structures the arrangement of the nuclei is the same. Only the placement of the electrons differs. The arrows do not mean that the molecule "flips" from one resonance to another. They simply show that the *actual structure is an average of the three resonance structures.*

The concept of resonance is necessary because the localized electron model postulates that electrons are localized between a given pair of atoms. However, nature does not really operate this way. Electrons are really delocalized—they can move around the entire molecule. The valence electrons in the NO_3^- molecule distribute themselves to provide equivalent N—O bonds. Resonance is necessary to compensate for the defective assumption of the localized electron model. However, this model is so useful that we retain the concept of localized electrons and add resonance to allow the model to treat species such as NO_3^-.

Example 8.9

Resonance Structures

Describe the electron arrangement in the nitrite anion (NO_2^-) using the localized electron model.

Solution

We will follow the usual procedure for obtaining the Lewis structure for the NO_2^- ion. In NO_2^- there are $5 + 2(6) + 1 = 18$ valence electrons. Indicating the single bonds gives the structure

O—N—O

NO₂⁻

The remaining 14 electrons (18 − 4) can be distributed to produce these structures:

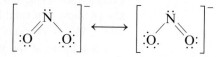

This is a resonance situation. Two equivalent Lewis structures can be drawn. *The electronic structure of the molecule is correctly represented not by either resonance structure but by the average of the two.* There are two equivalent N—O bonds, each one intermediate between a single and a double bond.

See Exercises 8.89 through 8.94

Odd-Electron Molecules

Relatively few molecules formed from nonmetals contain odd numbers of electrons. One common example is nitric oxide (NO), which is formed when nitrogen and oxygen gases react at the high temperatures in automobile engines. Nitric oxide is emitted into the air, where it immediately reacts with oxygen to form gaseous nitrogen dioxide (NO₂), another odd-electron molecule.

Since the localized electron model is based on pairs of electrons, it does not handle odd-electron cases in a natural way. To treat odd-electron molecules, a more sophisticated model is needed.

Formal Charge

Molecules or polyatomic ions containing atoms that can exceed the octet rule often have many nonequivalent Lewis structures (see margin note), all of which obey the rules for writing Lewis structures. For example, as we will see in detail below, the sulfate ion has a Lewis structure with all single bonds and several Lewis structures that contain double bonds. How do we decide which of the many possible Lewis structures best describes the actual bonding in sulfate? One method is to estimate the charge on each atom in the various possible Lewis structures and use these charges to select the most appropriate structure(s). We will see below how this is done, but first we must decide on a method to assign atomic charges in molecules.

In Chapter 4 we discussed one system for obtaining charges, called *oxidation states.* However, in assigning oxidation states, we always count *both* the shared electrons as belonging to the more electronegative atom in a bond. This practice leads to highly exaggerated estimates of charge. In other words, although oxidation states are useful for bookkeeping electrons in redox reactions, they are not realistic estimates of the actual charges on individual atoms in a molecule, so they are not suitable for judging the appropriateness of Lewis structures. Another definition of the charge on an atom in a molecule, the formal charge, can, however, be used to evaluate Lewis structures. As we will see below, the **formal charge** of an atom in a molecule is *the difference between the number of valence electrons on the free atom and the number of valence electrons assigned to the atom in the molecule.*

Therefore, to determine the formal charge of a given atom in a molecule, we need to know two things:

1. The number of valence electrons on the free neutral atom (which has zero net charge because the number of electrons equals the number of protons)
2. The number of valence electrons "belonging" to the atom in a molecule

We then compare these numbers. If in the molecule the atom has the same number of valence electrons as it does in the free state, the positive and negative charges just balance, and it has a formal charge of zero. If the atom has one more valence electron in

Equivalent Lewis structures contain the same numbers of single and multiple bonds. For example, the resonance structures for O₃

are equivalent Lewis structures. These are equally important in describing the bonding in O₃. Nonequivalent Lewis structures contain different numbers of single and multiple bonds.

a molecule than it has as a free atom, it has a formal charge of -1, and so on. Thus the formal charge on an atom in a molecule is defined as

Formal charge = (number of valence electrons on free atom)
        $-$ (number of valence electrons assigned to the atom in the molecule)

To compute the formal charge of an atom in a molecule, we assign the valence electrons in the molecule to the various atoms, making the following assumptions:

1. Lone pair electrons belong entirely to the atom in question.
2. Shared electrons are *divided equally* between the two sharing atoms.

Thus the number of valence electrons assigned to a given atom is calculated as follows:

(Valence electrons)$_{assigned}$ = (number of lone pair electrons)
        $+ \frac{1}{2}$ (number of shared electrons)

We will illustrate the procedure for calculating formal charges by considering two of the possible Lewis structures for the sulfate ion, which has 32 valence electrons. For the Lewis structure

SO_4^{2-}

each oxygen atom has 6 lone pair electrons and shares 2 electrons with the sulfur atom. Thus, using the preceding assumptions, each oxygen is assigned 7 valence electrons.

Valence electrons assigned to each oxygen = 6 *plus* $\frac{1}{2}$(2) = 7

        Lone   Shared
        pair   electrons
        electrons

Formal charge on oxygen = 6 *minus* 7 = -1

    Valence electrons
    on a free O atom
                Valence electrons
                assigned to each O
                in SO_4^{2-}

The formal charge on each oxygen is -1.

For the sulfur atom there are no lone pair electrons, and eight electrons are shared with the oxygen atoms. Thus, for sulfur,

Valence electrons assigned to sulfur = 0 *plus* $\frac{1}{2}$(8) = 4

        Lone   Shared
        pair   electrons
        electrons

Formal charge on sulfur = 6 *minus* 4 = 2

    Valence electrons
    on a free S atom
                Valence
                electrons
                assigned to S
                in SO_4^{2-}

A second possible Lewis structure is

$$\left[\ \ddot{O} \atop {\overset{\displaystyle\|}{:\ddot{O}-S-\ddot{O}:}} \atop \ddot{O}.\ \right]^{2-}$$

In this case the formal charges are as follows:

For oxygen atoms with single bonds:

$$\text{Valence electrons assigned} = 6 + \tfrac{1}{2}(2) = 7$$
$$\text{Formal charge} = 6 - 7 = -1$$

For oxygen atoms with double bonds:

$$\text{Valence electrons assigned} = 4 + \tfrac{1}{2}(4) = 6$$

Each double
bond has 4
electrons

$$\text{Formal charge} = 6 - 6 = 0$$

For the sulfur atom:

$$\text{Valence electrons assigned} = 0 + \tfrac{1}{2}(12) = 6$$
$$\text{Formal charge} = 6 - 6 = 0$$

We will use two fundamental assumptions about formal charges to evaluate Lewis structures:

1. Atoms in molecules try to achieve formal charges as close to zero as possible.
2. Any negative formal charges are expected to reside on the most electronegative atoms.

We can use these principles to evaluate the two nonequivalent Lewis structures for sulfate given previously. Notice that in the structure with only single bonds, each oxygen has a formal charge of -1, while the sulfur has a formal charge of $+2$. In contrast, in the structure with two double bonds and two single bonds, the sulfur and two oxygen atoms have a formal charge of 0, while two oxygens have a formal charge of -1. Based on the assumptions given above, the structure with two double bonds is preferred—it has lower formal charges and the -1 formal charges are on electronegative oxygen atoms. Thus, for the sulfate ion, we might expect resonance structures such as

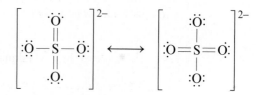

to more closely describe the bonding than the Lewis structure with only single bonds.

> ## Rules Governing Formal Charge
>
> ❯ To calculate the formal charge on an atom:
>   1. Take the sum of the lone pair electrons and one-half the shared electrons. This is the number of valence electrons assigned to the atom in the molecule.
>   2. Subtract the number of assigned electrons from the number of valence electrons on the free, neutral atom to obtain the formal charge.
> ❯ The sum of the formal charges of all atoms in a given molecule or ion must equal the overall charge on that species.
> ❯ If nonequivalent Lewis structures exist for a species, those with formal charges closest to zero and with any negative formal charges on the most electronegative atoms are considered to best describe the bonding in the molecule or ion.

Example 8.10

Formal Charges

Give possible Lewis structures for XeO_3, an explosive compound of xenon. Which Lewis structure or structures are most appropriate according to the formal charges?

Solution

For XeO_3 (26 valence electrons) we can draw the following possible Lewis structures (formal charges are indicated in parentheses):

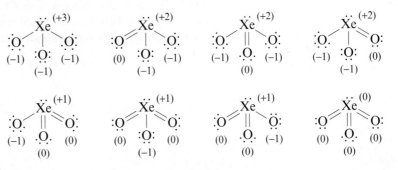

Based on the ideas of formal charge, we would predict that the Lewis structures with the lower values of formal charge would be most appropriate for describing the bonding in XeO_3.

See Exercises 8.101 and 8.102

As a final note, there are a couple of cautions about formal charge to keep in mind. First, although formal charges are closer to actual atomic charges in molecules than are oxidation states, formal charges still provide only *estimates* of charge—they should not be taken as actual atomic charges. Second, the evaluation of Lewis structures using formal charge ideas can lead to erroneous predictions. Tests based on experiments must be used to make the final decisions on the correct description of the bonding in a molecule or polyatomic ion.

8.13 | Molecular Structure: The VSEPR Model

The structures of molecules play a very important role in determining their chemical properties. As we will see later, this is particularly important for biological molecules; a slight change in the structure of a large biomolecule can completely destroy its usefulness to a cell or may even change the cell from a normal one to a cancerous one.

Many accurate methods now exist for determining **molecular structure**, the three-dimensional arrangement of the atoms in a molecule. These methods must be used if precise information about structure is required. However, it is often useful to be able to predict the approximate molecular structure of a molecule. In this section we consider a simple model that allows us to do this. This model, called the **valence shell electron-pair repulsion (VSEPR) model**, is useful in predicting the geometries of molecules formed from nonmetals. The main postulate of this model is that *the structure around a given atom is determined principally by minimizing electron-pair repulsions.* The idea here is that the bonding and nonbonding pairs around a given atom will be positioned as far apart as possible. To see how this model works, we will first consider the molecule $BeCl_2$, which has the Lewis structure

$$:\ddot{C}l-Be-\ddot{C}l:$$

BeCl₂ has only four electrons around Be and is expected to be very reactive with electron-pair donors.

Note that there are two pairs of electrons around the beryllium atom. What arrangement of these electron pairs allows them to be as far apart as possible to minimize the repulsions? Clearly, the best arrangement places the pairs on opposite sides of the beryllium atom at 180 degrees from each other:

This is the maximum possible separation for two electron pairs. Once we have determined the optimal arrangement of the electron pairs around the central atom, we can specify the molecular structure of $BeCl_2$, that is, the positions of the atoms. Since each electron pair on beryllium is shared with a chlorine atom, the molecule has a **linear structure** with a 180-degree bond angle:

Next, let's consider BF_3, which has the Lewis structure

$$:\ddot{F}-B-\ddot{F}:$$
with an $:\ddot{F}:$ above B.

Here the boron atom is surrounded by three pairs of electrons. What arrangement will minimize the repulsions? The electron pairs are farthest apart at angles of 120 degrees:

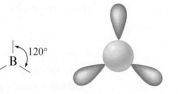

Since each of the electron pairs is shared with a fluorine atom, the molecular structure will be

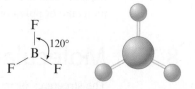

This is a planar (flat) and triangular molecule, which is commonly described as a **trigonal planar structure.**

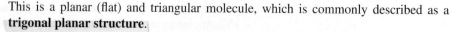

Next, let's consider the methane molecule, which has the Lewis structure

There are four pairs of electrons around the central carbon atom. What arrangement of these electron pairs best minimizes the repulsions? First, let's try a square planar arrangement:

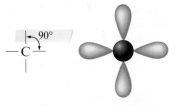

The carbon atom and the electron pairs are centered in the plane of the paper, and the angles between the pairs are all 90 degrees.

Is there another arrangement with angles greater than 90 degrees that would put the electron pairs even farther away from each other? The answer is yes. The **tetrahedral structure** has angles of 109.5 degrees:

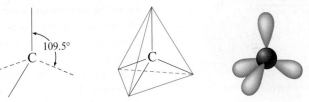

It can be shown that this is the maximum possible separation of four pairs around a given atom. This means that *whenever four pairs of electrons are present around an atom, they should always be arranged tetrahedrally.*

Now that we have the electron-pair arrangement that gives the least repulsion, we can determine the positions of the atoms and thus the molecular structure of CH_4. In methane, each of the four electron pairs is shared between the carbon atom and a hydrogen atom. Thus the hydrogen atoms are placed as in Fig. 8.15, and the molecule has a tetrahedral structure with the carbon atom at the center.

Recall that the main idea of the VSEPR model is to find the arrangement of electron pairs around the central atom that minimizes the repulsions. Then we can determine the molecular structure from knowing how the electron pairs are shared with the peripheral atoms. Use the following steps to predict the structure of a molecule using the VSEPR model.

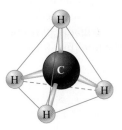

Figure 8.15 | The molecular structure of methane. The tetrahedral arrangement of electron pairs produces a tetrahedral arrangement of hydrogen atoms.

Problem-Solving Strategy

Steps to Apply the VSEPR Model

1. Draw the Lewis structure for the molecule.
2. Count the electron pairs and arrange them in the way that minimizes repulsion (that is, put the pairs as far apart as possible).
3. Determine the positions of the atoms from the *way the electron pairs are shared*.
4. Determine the name of the molecular structure from the *positions of the atoms*.

Figure 8.16 | (a) The tetrahedral arrangement of electron pairs around the nitrogen atom in the ammonia molecule. (b) Three of the electron pairs around nitrogen are shared with hydrogen atoms as shown and one is a lone pair. Although the arrangement of *electron pairs* is tetrahedral, as in the methane molecule, the hydrogen atoms in the ammonia molecule occupy only three corners of the tetrahedron. A lone pair occupies the fourth corner. (c) Note that molecular geometry is trigonal pyramidal, not tetrahedral.

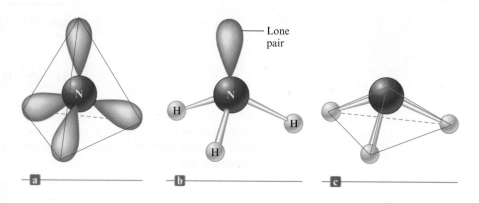

When four uniform balloons are tied together, they naturally form a tetrahedral shape.

We will predict the structure of ammonia (NH_3) using this stepwise approach.

1. Draw the Lewis structure:

$$H - \overset{\cdot\cdot}{N} - H$$
$$|$$
$$H$$

2. Count the pairs of electrons and arrange them to minimize repulsions. The NH_3 molecule has four pairs of electrons: three bonding pairs and one nonbonding pair. From the discussion of the methane molecule, we know that the best arrangement of four electron pairs is a tetrahedral array [Fig. 8.16(a)].
3. Determine the positions of the atoms. The three H atoms share electron pairs [Fig. 8.16(b)].
4. Name the molecular structure. It is very important to recognize that the *name* of the molecular structure is always based on the *positions of the atoms*. The placement of the electron pairs determines the structure, but the name is based on the positions of the atoms. Thus it is incorrect to say that the NH_3 molecule is tetrahedral. It has a tetrahedral arrangement of electron pairs but not a tetrahedral arrangement of atoms. The molecular structure of ammonia is a **trigonal pyramid** (one side is different from the other three) rather than a tetrahedron [Fig. 8.16(c)].

Example 8.11

Prediction of Molecular Structure I

Describe the molecular structure of the water molecule.

Solution

The Lewis structure for water is

$$H - \overset{\cdot\cdot}{\underset{\cdot\cdot}{O}} - H$$

There are four pairs of electrons: two bonding pairs and two nonbonding pairs. To minimize repulsions, these are best arranged in a tetrahedral array [Fig. 8.17(a)]. Although H_2O has a tetrahedral arrangement of electron pairs, it is not a tetrahedral molecule. The atoms in the H_2O molecule form a V shape [Fig. 8.17(b) and (c)].

See Exercises 8.113 and 8.114

From Example 8.11 we see that the H_2O molecule is V-shaped, or bent, because of the presence of the lone pairs. If no lone pairs were present, the molecule would be linear, the polar bonds would cancel, and the molecule would have no dipole moment. This would make water very different from the polar substance so familiar to us.

Figure 8.17 | (a) The tetrahedral arrangement of the four electron pairs around oxygen in the water molecule. (b) Two of the electron pairs are shared between oxygen and the hydrogen atoms and two are lone pairs. (c) The V-shaped molecular structure of the water molecule.

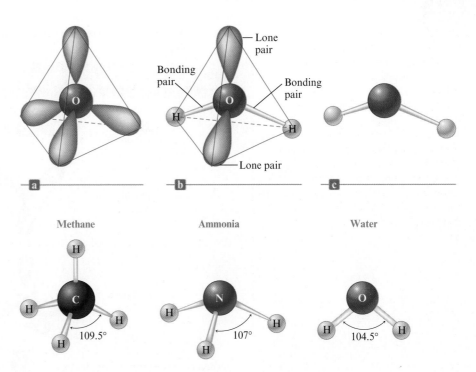

Figure 8.18 | The bond angles in the CH$_4$, NH$_3$, and H$_2$O molecules. Note that the bond angle between bonding pairs decreases as the number of lone pairs increases. Note that all of the angles in CH$_4$ are 109.5 degrees and all of the angles in NH$_3$ are 107 degrees.

From the previous discussion, we would predict that the H—X—H bond angle (where X is the central atom) in CH$_4$, NH$_3$, and H$_2$O should be the tetrahedral angle of 109.5 degrees. Experimental studies, however, show that the actual bond angles are those given in Fig. 8.18. What significance do these results have for the VSEPR model? One possible point of view is that we should be pleased to have the observed angles so close to the tetrahedral angle. The opposite view is that the deviations are significant enough to require modification of the simple model so that it can more accurately handle similar cases. We will take the latter view.

Let us examine the following data:

	CH$_4$	NH$_3$	H$_2$O
Number of lone pairs	0	1	2
Bond angle	109.5°	107°	104.5°

One interpretation of the trend observed here is that lone pairs require more space than bonding pairs; in other words, as the number of lone pairs increases, the bonding pairs are increasingly squeezed together.

This interpretation seems to make physical sense if we think in the following terms. A bonding pair is shared between two nuclei, and the electrons can be close to either nucleus. They are relatively confined between the two nuclei. A lone pair is localized on only one nucleus, and both electrons will be close only to that nucleus, as shown schematically in Fig. 8.19. These pictures help us understand why a lone pair may require more space near an atom than a bonding pair.

As a result of these observations, we make the following addition to the original postulate of the VSEPR model: Lone pairs require more room than bonding pairs and tend to compress the angles between the bonding pairs.

So far we have considered cases with two, three, and four electron pairs around the central atom. These are summarized in Table 8.6.

Table 8.7 summarizes the structures possible for molecules in which there are four electron pairs around the central atom with various numbers of atoms bonded to it. Note that molecules with four pairs of electrons around the central atom can be tetrahedral (AB$_4$), trigonal pyramidal (AB$_3$), and V-shaped (AB$_2$).

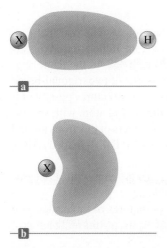

Figure 8.19 | (a) In a bonding pair of electrons, the electrons are shared by two nuclei. (b) In a lone pair, both electrons must be close to a single nucleus and tend to take up more of the space around that atom.

Table 8.6 | Arrangements of Electron Pairs Around an Atom Yielding Minimum Repulsion

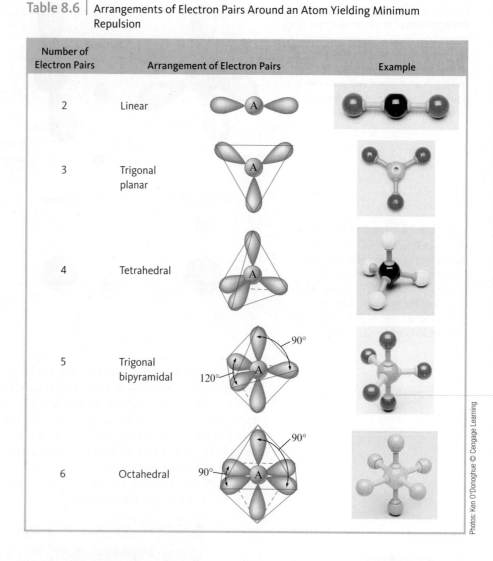

Number of Electron Pairs	Arrangement of Electron Pairs	Example
2	Linear	
3	Trigonal planar	
4	Tetrahedral	
5	Trigonal bipyramidal	
6	Octahedral	

Photos: Ken O'Donoghue © Cengage Learning

For five pairs of electrons, there are several possible choices. The one that produces minimum repulsion is a **trigonal bipyramid**. Note from Table 8.6 that this arrangement has two different angles, 90 degrees and 120 degrees. As the name suggests, the structure formed by this arrangement of pairs consists of two trigonal-based pyramids that share a common base. Table 8.8 summarizes the structures possible for molecules in which there are five electron pairs around the central atom with various numbers of atoms bonded to it. Note that molecules with five pairs of electrons around the central atom can be trigonal bipyramidal (AB_5), see-saw (AB_4), T-shaped (AB_3), and linear (AB_2).

Six pairs of electrons can best be arranged around a given atom with 90-degree angles to form an **octahedral structure**, as shown in Table 8.6.

To use the VSEPR model to determine the geometric structures of molecules, you should memorize the relationships between the number of electron pairs and their best arrangement.

Critical Thinking

You and a friend are studying for a chemistry exam. What if your friend tells you that all molecules with polar bonds are polar molecules? How would you explain to your friend that this is not correct? Provide two examples to support your answer.

Table 8.7 | Structures of Molecules with Four Electron Pairs Around the Central Atom

Table 8.8 | Structures of Molecules with Five Electron Pairs Around the Central Atom

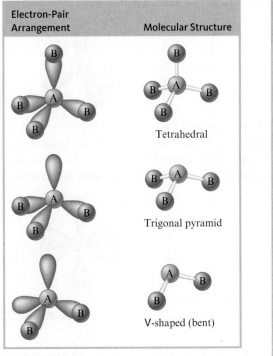

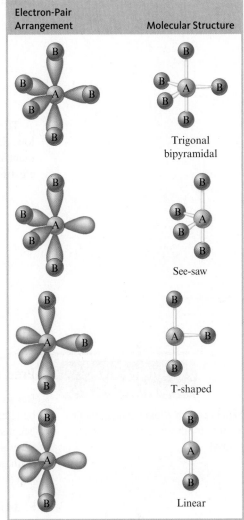

Interactive Example 8.12

Sign in at http://login.cengagebrain.com to try this Interactive Example in **OWL**.

Prediction of Molecular Structure II

When phosphorus reacts with excess chlorine gas, the compound phosphorus penta-chloride (PCl_5) is formed. In the gaseous and liquid states, this substance consists of PCl_5 molecules, but in the solid state it consists of a 1:1 mixture of PCl_4^+ and PCl_6^- ions. Predict the geometric structures of PCl_5, PCl_4^+, and PCl_6^-.

Solution

The Lewis structure for PCl_5 is shown. Five pairs of electrons around the phosphorus atom require a trigonal bipyramidal arrangement (see Table 8.6). When the chlorine atoms are included, a trigonal bipyramidal molecule results:

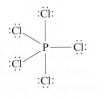

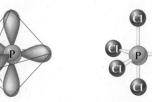

The Lewis structure for the PCl_4^+ ion [$5 + 4(7) - 1 = 32$ valence electrons] is shown below. There are four pairs of electrons surrounding the phosphorus atom in the PCl_4^+ ion, which requires a tetrahedral arrangement of the pairs. Since each pair is shared with a chlorine atom, a tetrahedral PCl_4^+ cation results.

The Lewis structure for PCl_6^- [$5 + 6(7) + 1 = 48$ valence electrons] is shown below. Since phosphorus is surrounded by six pairs of electrons, an octahedral arrangement is required to minimize repulsions, as shown below in the center. Since each electron pair is shared with a chlorine atom, an octahedral PCl_6^- anion is predicted.

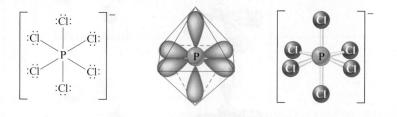

See Exercises 8.111, 8.112, 8.115, and 8.116

Interactive Example 8.13

Sign in at http://login.cengagebrain .com to try this Interactive Example in **OWL**.

Prediction of Molecular Structure III

Because the noble gases have filled s and p valence orbitals, they were not expected to be chemically reactive. In fact, for many years these elements were called *inert gases* because of this supposed inability to form any compounds. However, in the early 1960s several compounds of krypton, xenon, and radon were synthesized. For example, a team at the Argonne National Laboratory produced the stable colorless compound xenon tetrafluoride (XeF_4). Predict its structure and whether it has a dipole moment.

Solution

The Lewis structure for XeF_4 is

The xenon atom in this molecule is surrounded by six pairs of electrons, which means an octahedral arrangement.

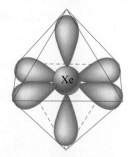

The structure predicted for this molecule will depend on how the lone pairs and bonding pairs are arranged. Consider the two possibilities shown in Fig. 8.20. The bonding pairs are indicated by the presence of the fluorine atoms. Since the structure

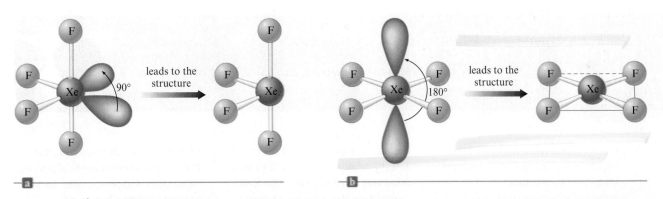

Figure 8.20 | Possible electron-pair arrangements for XeF₄. Since arrangement (a) has lone pairs at 90 degrees from each other, it is less favorable than arrangement (b), where the lone pairs are at 180 degrees.

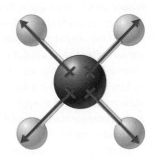

predicted differs in the two cases, we must decide which of these arrangements is preferable. The key is to look at the lone pairs. In the structure in part (a), the lone pair–lone pair angle is 90 degrees; in the structure in part (b), the lone pairs are separated by 180 degrees. Since lone pairs require more room than bonding pairs, a structure with two lone pairs at 90 degrees is unfavorable. Thus the arrangement in Fig. 8.20(b) is preferred, and the molecular structure is predicted to be square planar. Note that this molecule is *not* described as being octahedral. There is an *octahedral arrangement of electron pairs,* but the *atoms* form a **square planar structure**.

Although each Xe—F bond is polar (fluorine has a greater electronegativity than xenon), the square planar arrangement of these bonds causes the polarities to cancel. Thus XeF_4 has no dipole moment, as shown in the margin.

See Exercises 8.117 through 8.120

We can further illustrate the use of the VSEPR model for molecules or ions with lone pairs by considering the triiodide ion (I_3^-).

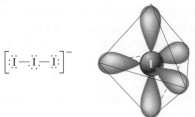

$$\left[:\ddot{I}-\ddot{I}-\ddot{I}:\right]^-$$

The central iodine atom has five pairs around it, which requires a trigonal bipyramidal arrangement. Several possible arrangements of lone pairs are shown in Fig. 8.21. Note that structures (a) and (b) have lone pairs at 90 degrees, whereas in (c) all lone pairs are at 120 degrees. Thus structure (c) is preferred. The resulting molecular structure for I_3^- is linear:

$$[I—I—I]^-$$

Figure 8.21 | Three possible arrangements of the electron pairs in the I_3^- ion. Arrangement (c) is preferred because there are no 90-degree lone pair–lone pair interactions.

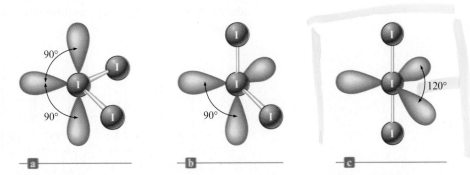

Chemical connections
Chemical Structure and Communication: Semiochemicals

In this chapter we have stressed the importance of being able to predict the three-dimensional structure of a molecule. Molecular structure is important because of its effect on chemical reactivity. This is especially true in biological systems, where reactions must be efficient and highly specific. Among the hundreds of types of molecules in the fluids of a typical biological system, the appropriate reactants must find and react only with each other—they must be very discriminating. This specificity depends largely on structure. The molecules are constructed so that only the appropriate partners can approach each other in a way that allows reaction.

Another area where molecular structure is central is in the use of molecules as a means of communication. Examples of a chemical communication occur in humans in the conduction of nerve impulses across synapses, the control of the manufacture and storage of key chemicals in cells, and the senses of smell and taste. Plants and animals also use chemical communication. For example, ants lay down a chemical trail so that other ants can find a particular food supply. Ants also warn their fellow workers of approaching danger by emitting certain chemicals.

Molecules convey messages by fitting into appropriate receptor sites in a very specific way, which is determined by their structure. When a molecule occupies a receptor site, chemical processes are stimulated that produce the appropriate response. Sometimes receptors can be fooled, as in the use of artificial sweeteners— molecules fit the sites on the taste buds that stimulate a "sweet" response in the brain, but they are not metabolized in the same way as natural sugars. Similar deception is useful in insect control. If an area is sprayed with synthetic female sex attractant molecules, the males of that species become so confused that mating does not occur.

A *semiochemical* is a molecule that delivers a message between members of the same or different species of plant or animal. There are three groups of these chemical messengers: allomones, kairomones, and pheromones. Each is of great ecological importance.

An *allomone* is defined as a chemical that somehow gives adaptive advantage to the producer. For example, leaves of the black walnut tree contain an herbicide, juglone, that appears after the leaves fall to the ground. Juglone is not toxic to grass or certain grains, but it is effective against plants such as apple trees that would compete for the available water and food supplies.

Antibiotics are also allomones, since the microorganisms produce them to inhibit other species from growing near them.

Many plants produce bad-tasting chemicals to protect themselves from plant-eating insects and animals. The familiar compound nicotine deters animals from eating the tobacco plant. The millipede sends an unmistakable "back off" message by squirting a predator with benzaldehyde and hydrogen cyanide.

The VSEPR Model and Multiple Bonds

So far in our treatment of the VSEPR model we have not considered any molecules with multiple bonds. To see how these molecules are handled by this model, let's consider the NO_3^- ion, which requires three resonance structures to describe its electronic structure:

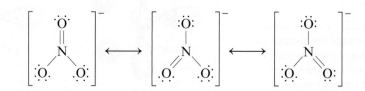

The queen bee secretes a chemical that prevents the worker bees from raising a competitive sovereign.

Kenneth Lorenzen

Defense is not the only use of allomones, however. Flowers use scent as a way to attract pollinating insects. Honeybees, for instance, are guided to alfalfa and flowers by a series of sweet-scented compounds.

Kairomones are chemical messengers that bring advantageous news to the receiver, and the floral scents are kairomones from the honeybees' viewpoint. Many predators are guided by kairomones emitted by their food. For example, apple skins exude a chemical that attracts the codling moth larva. In some cases kairomones help the underdog. Certain marine mollusks can pick up the "scent" of their predators, the sea stars, and make their escape.

Pheromones are chemicals that affect receptors of the same species as the donor. That is, they are specific within a species. *Releaser pheromones* cause an immediate reaction in the receptor, and *primer pheromones* cause long-term effects. Examples of releaser pheromones are sex attractants of insects, generated in some species by the males and in others by the females. Sex pheromones also have been found in plants and mammals.

Alarm pheromones are highly volatile compounds (ones easily changed to a gas) released to warn of danger. Honeybees produce isoamyl acetate ($C_7H_{14}O_2$) in their sting glands. Because of its high volatility, this compound does not linger after the state of alert is over. Social behavior in insects is characterized by the use of *trail pheromones*, which are used to indicate a food source. Social insects such as bees, ants, wasps, and termites use these substances. Since trail pheromones are less volatile compounds, the indicators persist for some time.

Primer pheromones, which cause long-term behavioral changes, are harder to isolate and identify. One example, however, is the "queen substance" produced by queen honeybees. All the eggs in a colony are laid by one queen bee. If she is removed from the hive or dies, the worker bees are activated by the absence of the queen substance and begin to feed royal jelly to bee larvae so as to raise a new queen. The queen substance also prevents the development of the workers' ovaries so that only the queen herself can produce eggs.

Many studies of insect pheromones are now under way in the hope that they will provide a method of controlling insects that is more efficient and safer than the current chemical pesticides.

The NO_3^- ion is known to be planar with 120-degree bond angles:

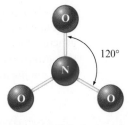

This planar structure is the one expected for three pairs of electrons around a central atom, which means that *a double bond should be counted as one effective pair* in using the VSEPR model. This makes sense because the two pairs of electrons involved in the double bond are *not* independent pairs. Both the electron pairs must be in the space between the nuclei of the two atoms to form the double bond. In other words, the double bond acts as one center of electron density to repel the other pairs of electrons. The same holds true for triple bonds. This leads us to another general rule: For the VSEPR model, multiple bonds count as one effective electron pair.

The molecular structure of nitrate also shows us one more important point: When a molecule exhibits resonance, any one of the resonance structures can be used to predict the molecular structure using the VSEPR model. These rules are illustrated in Example 8.14.

Interactive Example 8.14

Sign in at http://login.cengagebrain.com to try this Interactive Example in **OWL**.

SO_2

Structures of Molecules with Multiple Bonds

Predict the molecular structure of the sulfur dioxide molecule. Is this molecule expected to have a dipole moment?

Solution

First, we must determine the Lewis structure for the SO_2 molecule, which has 18 valence electrons. The expected resonance structures are

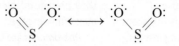

To determine the molecular structure, we must count the electron pairs around the sulfur atom. In each resonance structure the sulfur has one lone pair, one pair in a single bond, and one double bond. Counting the double bond as one pair yields three effective pairs around the sulfur. According to Table 8.6, a trigonal planar arrangement is required, which yields a V-shaped molecule:

$$O \overset{120°}{\underset{S}{\diagdown\diagup}} O$$

Thus the structure of the SO_2 molecule is expected to be V-shaped, with a 120-degree bond angle. The molecule has a dipole moment directed as shown:

$$O \overset{\uparrow}{\underset{S}{\diagdown\diagup}} O$$

Since the molecule is V-shaped, the polar bonds do not cancel.

See Exercises 8.121 and 8.122

It should be noted at this point that lone pairs that are oriented at least 120 degrees from other pairs do not produce significant distortions of bond angles. For example, the angle in the SO_2 molecule is actually quite close to 120 degrees. We will follow the general principle that *a 120-degree angle provides lone pairs with enough space so that distortions do not occur. Angles less than 120 degrees are distorted when lone pairs are present.*

Molecules Containing No Single Central Atom

So far we have considered molecules consisting of one central atom surrounded by other atoms. The VSEPR model can be readily extended to more complicated molecules, such as methanol (CH_3OH). This molecule is represented by the following Lewis structure:

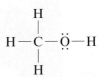

The molecular structure can be predicted from the arrangement of pairs around the carbon and oxygen atoms. Note that there are four pairs of electrons around the carbon, which requires a tetrahedral arrangement [Fig. 8.22(a)]. The oxygen also has four pairs, which requires a tetrahedral arrangement. However, in this case the tetrahedron will be slightly distorted by the space requirements of the lone pairs [Fig. 8.22(b)]. The overall geometric arrangement for the molecule is shown in Fig. 8.22(c).

> ### Let's Review │ *Summary of the VSEPR Model*
>
> The rules for using the VSEPR model to predict molecular structure are as follows:
> ❯ Determine the Lewis structure(s) for the molecule.
> ❯ For molecules with resonance structures, use any of the structures to predict the molecular structure.
> ❯ Sum the electron pairs around the central atom.
> ❯ In counting pairs, count each multiple bond as a single effective pair.
> ❯ The arrangement of the pairs is determined by minimizing electron-pair repulsions. These arrangements are shown in Table 8.6.
> ❯ Lone pairs require more space than bonding pairs do. Choose an arrangement that gives the lone pairs as much room as possible. Recognize that the lone pairs may produce a slight distortion of the structure at angles less than 120 degrees.

The VSEPR Model—How Well Does It Work?

The VSEPR model is very simple. There are only a few rules to remember, yet the model correctly predicts the molecular structures of most molecules formed from nonmetallic elements. Molecules of any size can be treated by applying the VSEPR model to each appropriate atom (those bonded to at least two other atoms) in the molecule. Thus we can use this model to predict the structures of molecules with hundreds of atoms. It does, however, fail in a few instances. For example, phosphine (PH_3), which has a Lewis structure analogous to that of ammonia,

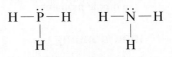

would be predicted to have a molecular structure similar to that for NH_3, with bond angles of approximately 107 degrees. However, the bond angles of phosphine are actually 94 degrees. There are ways of explaining this structure, but more rules have to be added to the model.

This again illustrates the point that simple models are bound to have exceptions. In introductory chemistry we want to use simple models that fit the majority of cases; we are willing to accept a few failures rather than complicate the model. The amazing thing about the VSEPR model is that such a simple model predicts correctly the structures of so many molecules.

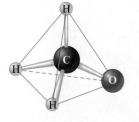

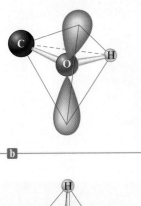

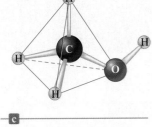

Figure 8.22 │ The molecular structure of methanol. (a) The arrangement of electron pairs and atoms around the carbon atom. (b) The arrangement of bonding and lone pairs around the oxygen atom. (c) The molecular structure.

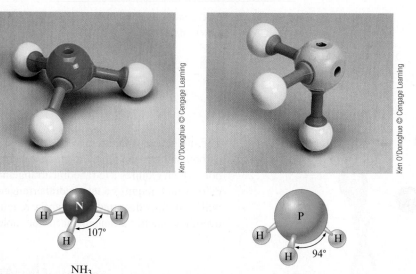

NH₃

PH₃

For review

Key terms

Chemical bonds

› Hold groups of atoms together
› Occur when a group of atoms can lower its total energy by aggregating
› Types of chemical bonds
  › Ionic: electrons are transferred to form ions
  › Covalent: equal sharing of electrons
  › Polar covalent: unequal electron sharing
› Percent ionic character of a bond X—Y

$$\frac{\text{Measured dipole moment of X—Y}}{\text{Calculated dipole moment for X}^+\text{ Y}^-} \times 100\%$$

› Electronegativity: the relative ability of an atom to attract shared electrons
  › The polarity of a bond depends on the electronegativity difference of the bonded atoms
› The spatial arrangement of polar bonds in a molecule determines whether the molecule has a dipole moment

Ionic bonding

› An ion has a different size than its parent atom
  › An anion is larger than its parent ion
  › A cation is smaller than its parent atom
› Lattice energy: the change in energy when ions are packed together to form an ionic solid

Bond energy

› The energy necessary to break a covalent bond
› Increases as the number of shared pairs increases
› Can be used to estimate the enthalpy change for a chemical reaction

Key terms

Section 8.10
Lewis structure
duet rule
octet rule

Section 8.12
resonance
resonance structure
formal charge

Section 8.13
molecular structure
valence shell electron-pair
    repulsion (VSEPR) model
linear structure
trigonal planar structure
tetrahedral structure
trigonal pyramid
trigonal bipyramid
octahedral structure
square planar structure

Lewis structures

> Show how the valence electron pairs are arranged among the atoms in a molecule or poly-atomic ion
> Stable molecules usually contain atoms that have their valence orbitals filled
  > Leads to a duet rule for hydrogen
  > Leads to an octet rule for second-row elements
  > The atoms of elements in the third row and beyond can exceed the octet rule
> Several equivalent Lewis structures can be drawn for some molecules, a concept called resonance
> When several nonequivalent Lewis structures can be drawn for a molecule, formal charge is often used to choose the most appropriate structure(s)

VSEPR model

> Based on the idea that electron pairs will be arranged around a central atom in a way that minimizes the electron repulsions
> Can be used to predict the geometric structure of most molecules

Review questions   *Answers to the Review Questions can be found on the Student website (accessible from* **www.cengagebrain.com***).*

1. Distinguish between the terms *electronegativity* versus *electron affinity, covalent bond* versus *ionic bond,* and *pure covalent bond* versus *polar covalent bond.* Characterize the types of bonds in terms of electronegativity difference. Energetically, why do ionic and covalent bonds form?

2. When an element forms an anion, what happens to the radius? When an element forms a cation, what happens to the radius? Why? Define the term *isoelectronic.* When comparing sizes of ions, which ion has the largest radius and which ion has the smallest radius in an isoelectronic series? Why?

3. Define the term *lattice energy.* Why, energetically, do ionic compounds form? Figure 8.11 illustrates the energy changes involved in the formation of $MgO(s)$ and $NaF(s)$. Why is the lattice energy of $MgO(s)$ so different from that of $NaF(s)$? Magnesium oxide is composed of Mg^{2+} and O^{2-} ions. Energetically, why does $Mg^{2+}O^{2-}$ form and not Mg^+O^-? Why doesn't $Mg^{3+}O^{3-}$ form?

4. Explain how bond energies can be used to estimate ΔH for a reaction. Why is this an estimate of ΔH? How do the product bond strengths compare to the reactant bond strengths for an exothermic reaction? For an endothermic reaction? What is the relationship between the number of bonds between two atoms and bond strength? Bond length?

5. Give a rationale for the octet rule and the duet rule for H in terms of orbitals. Give the steps for drawing a Lewis structure for a molecule or ion. In general, molecules and ions always follow the octet rule unless it is impossible. The three types of exceptions are molecules/ions with too few electrons, molecules/ions with an odd number of electrons, and molecules/ions with too many electrons. Which atoms sometimes have fewer than 8 electrons around them? Give an example. Which atoms sometimes have more than 8 electrons around them? Give some examples. Why are odd-electron species generally very reactive and uncommon? Give an example of an odd-electron molecule.

6. Explain the terms *resonance* and *delocalized electrons.* When a substance exhibits resonance, we say that none of the individual Lewis structures accurately portrays the bonding in the substance. Why do we draw resonance structures?

7. Define *formal charge* and explain how to calculate it. What is the purpose of the formal charge? Organic compounds are composed mostly of carbon and hydrogen, but also may have oxygen, nitrogen, and/or halogens in the formula. Formal charge arguments work very well for organic compounds when drawing the best Lewis structure. How do C, H, N, O, and Cl satisfy the octet rule in organic compounds so as to have a formula charge of zero?

8. Explain the main postulate of the VSEPR model. List the five base geometries (along with bond angles) that most molecules or ions adopt to minimize electron-pair repulsions. Why are bond angles sometimes slightly less than predicted in actual molecules as compared to what is predicted by the VSEPR model?

9. Give two requirements that should be satisfied for a molecule to be polar. Explain why CF_4 and XeF_4 are nonpolar compounds (have no net dipole moments) while SF_4 is polar (has a net dipole moment). Is CO_2 polar? What about COS? Explain.

10. Consider the following compounds: CO_2, SO_2, KrF_2, SO_3, NF_3, IF_3, CF_4, SF_4, XeF_4, PF_5, IF_5, and SCl_6. These 12 compounds are all examples of different molecular structures. Draw the Lewis structures for each and predict the molecular structure. Predict the bond angles and the polarity of each. (A polar molecule has a net dipole moment, while a nonpolar molecule does not.) See Exercises 111 and 112 for the molecular structures based on the trigonal bipyramid and the octahedral geometries.

Active Learning Questions

These questions are designed to be used by groups of students in class.

1. Explain the electronegativity trends across a row and down a column of the periodic table. Compare these trends with those of ionization energies and atomic radii. How are they related?

2. The ionic compound AB is formed. The charges on the ions may be $+1$, -1; $+2$, -2; $+3$, -3; or even larger. What are the factors that determine the charge for an ion in an ionic compound?

3. Using only the periodic table, predict the most stable ion for Na, Mg, Al, S, Cl, K, Ca, and Ga. Arrange these from largest to smallest radius, and explain why the radius varies as it does. Compare your predictions with Fig. 8.8.

4. The bond energy for a C—H bond is about 413 kJ/mol in CH_4 but 380 kJ/mol in $CHBr_3$. Although these values are relatively close in magnitude, they are different. Explain why they are different. Does the fact that the bond energy is lower in $CHBr_3$ make any sense? Why?

5. Consider the following statement: "Because oxygen wants to have a negative-two charge, the second electron affinity is more negative than the first." Indicate everything that is correct in this statement. Indicate everything that is incorrect. Correct the incorrect statements and explain.

6. Which has the greater bond lengths: NO_2^- or NO_3^-? Explain.

7. The following ions are best described with resonance structures. Draw the resonance structures, and using formal charge arguments, predict the best Lewis structure for each ion.
   a. NCO^-
   b. CNO^-

8. Would you expect the electronegativity of titanium to be the same in the species Ti, Ti^{2+}, Ti^{3+}, and Ti^{4+}? Explain.

9. The second electron affinity values for both oxygen and sulfur are unfavorable (endothermic). Explain.

10. What is meant by a chemical bond? Why do atoms form bonds with each other? Why do some elements exist as molecules in nature instead of as free atoms?

11. Why are some bonds ionic and some covalent?

12. How does a bond between Na and Cl differ from a bond between C and O? What about a bond between N and N?

13. Arrange the following molecules from most to least polar and explain your order: CH_4, CF_2Cl_2, CF_2H_2, CCl_4, and CCl_2H_2.

14. Does a Lewis structure tell which electrons come from which atoms? Explain.

A blue question or exercise number indicates that the answer to that question or exercise appears at the back of this book and a solution appears in the *Solutions Guide*, as found on PowerLecture.

Questions

15. The following electrostatic potential diagrams represent H_2, HCl, or NaCl. Label each and explain your choices.

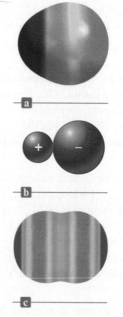

16. Describe the type of bonding that exists in the $F_2(g)$ molecule. How does this type of bonding differ from that found in the HF(g) molecule? How is it similar?

17. Some plant fertilizer compounds are $(NH_4)_2SO_4$, $Ca_3(PO_4)_2$, K_2O, P_2O_5, and KCl. Which of these compounds contain both ionic and covalent bonds?

18. Some of the important properties of ionic compounds are as follows:

    i. low electrical conductivity as solids and high conductivity in solution or when molten

    ii. relatively high melting and boiling points

    iii. brittleness

    iv. solubility in polar solvents

    How does the concept of ionic bonding discussed in this chapter account for these properties?

19. What is the electronegativity trend? Where does hydrogen fit into the electronegativity trend for the other elements in the periodic table?

20. Give one example of a compound having a linear molecular structure that has an overall dipole moment (is polar) and one example that does not have an overall dipole moment (is nonpolar). Do the same for molecules that have trigonal planar and tetrahedral molecular structures.

21. When comparing the size of different ions, the general radii trend discussed in Chapter 7 is usually not very useful. What do you concentrate on when comparing sizes of ions to each other or when comparing the size of an ion to its neutral atom?

22. In general, the higher the charge on the ions in an ionic compound, the more favorable the lattice energy. Why do some stable ionic compounds have $+1$ charged ions even though $+4$, $+5$, and $+6$ charged ions would have a more favorable lattice energy?

23. Combustion reactions of fossil fuels provide most of the energy needs of the world. Why are combustion reactions of fossil fuels so exothermic?

24. Which of the following statements is(are) *true*? Correct the false statements.

    a. It is impossible to satisfy the octet rule for all atoms in XeF_2.

    b. Because SF_4 exists, OF_4 should also exist because oxygen is in the same family as sulfur.

    c. The bond in NO^+ should be stronger than the bond in NO^-.

    d. As predicted from the two Lewis structures for ozone, one oxygen–oxygen bond is stronger than the other oxygen–oxygen bond.

25. Three resonance structures can be drawn for CO_2. Which resonance structure is best from a formal charge standpoint?

26. Which of the following statements is(are) *true*? Correct the false statements.

    a. The molecules SeS_3, SeS_2, PCl_5, $TeCl_4$, ICl_3, and $XeCl_2$ all exhibit at least one bond angle, which is approximately $120°$.

    b. The bond angle in SO_2 should be similar to the bond angle in CS_2 or SCl_2.

    c. Of the compounds CF_4, KrF_4, and SeF_4, only SeF_4 exhibits an overall dipole moment (is polar).

    d. Central atoms in a molecule adopt a geometry of the bonded atoms and lone pairs about the central atom in order to maximize electron repulsions.

Exercises

In this section similar exercises are paired.

Chemical Bonds and Electronegativity

27. Without using Fig. 8.3, predict the order of increasing electronegativity in each of the following groups of elements.
    a. C, N, O   **c.** Si, Ge, Sn
    b. S, Se, Cl   **d.** Tl, S, Ge

28. Without using Fig. 8.3, predict the order of increasing electronegativity in each of the following groups of elements.
    a. Na, K, Rb   **c.** F, Cl, Br
    b. B, O, Ga   **d.** S, O, F

29. Without using Fig. 8.3, predict which bond in each of the following groups will be the most polar.
    a. C—F, Si—F, Ge—F
    b. P—Cl or S—Cl
    c. S—F, S—Cl, S—Br
    d. Ti—Cl, Si—Cl, Ge—Cl

30. Without using Fig. 8.3, predict which bond in each of the following groups will be the most polar.
    a. C—H, Si—H, Sn—H
    b. Al—Br, Ga—Br, In—Br, Tl—Br
    c. C—O or Si—O
    d. O—F or O—Cl

31. Repeat Exercises 27 and 29, this time using the values for the electronegativities of the elements given in Fig. 8.3. Are there differences in your answers?

32. Repeat Exercises 28 and 30, this time using the values for the electronegativities of the elements given in Fig. 8.3. Are there differences in your answers?

33. Which of the following incorrectly shows the bond polarity? Show the correct bond polarity for those that are incorrect.
    a. $^{\delta+}H—F^{\delta-}$ **d.** $^{\delta+}Br—Br^{\delta-}$
    b. $^{\delta+}Cl—I^{\delta-}$ **e.** $^{\delta+}O—P^{\delta-}$
    c. $^{\delta+}Si—S^{\delta-}$

34. Indicate the bond polarity (show the partial positive and partial negative ends) in the following bonds.
    a. C—O   **d.** Br—Te
    b. P—H   **e.** Se—S
    c. H—Cl

35. Predict the type of bond (ionic, covalent, or polar covalent) one would expect to form between the following pairs of elements.
    a. Rb and Cl   **d.** Ba and S
    b. S and S   **e.** N and P
    c. C and F   **f.** B and H

36. List all the possible bonds that can occur between the elements P, Cs, O, and H. Predict the type of bond (ionic, covalent, or polar covalent) one would expect to form for each bond.

37. Hydrogen has an electronegativity value between boron and carbon and identical to phosphorus. With this in mind, rank the following bonds in order of decreasing polarity: P—H, O—H, N—H, F—H, C—H.

38. Rank the following bonds in order of increasing ionic character: N—O, Ca—O, C—F, Br—Br, K—F.

39. State whether or not each of the following has a permanent dipole moment.

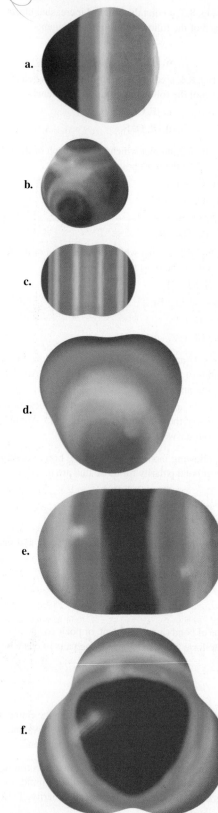

a.

b.

c.

d.

e.

f.

40. The following electrostatic potential diagrams represent CH_4, NH_3, or H_2O. Label each and explain your choices.

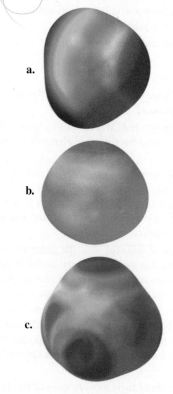

a.

b.

c.

Ions and Ionic Compounds

41. Write electron configurations for the most stable ion formed by each of the elements Al, Ba, Se, and I (when in stable ionic compounds).

42. Write electron configurations for the most stable ion formed by each of the elements Te, Cl, Sr, and Li (when in stable ionic compounds).

43. Predict the empirical formulas of the ionic compounds formed from the following pairs of elements. Name each compound.

    a. Li and N   **c.** Rb and Cl

    b. Ga and O   **d.** Ba and S

44. Predict the empirical formulas of the ionic compounds formed from the following pairs of elements. Name each compound.

    a. Al and Cl   **c.** Sr and F

    b. Na and O   **d.** Ca and Se

45. Write electron configurations for

    a. the cations Mg^{2+}, K^+, and Al^{3+}.

    b. the anions N^{3-}, O^{2-}, F^-, and Te^{2-}.

46. Write electron configurations for

    a. the cations Sr^{2+}, Cs^+, In^+, and Pb^{2+}.

    b. the anions P^{3-}, S^{2-}, and Br^-.

47. Which of the following ions have noble gas electron configurations?

    a. Fe^{2+}, Fe^{3+}, Sc^{3+}, Co^{3+}

    b. Tl^+, Te^{2-}, Cr^{3+}

    c. Pu^{4+}, Ce^{4+}, Ti^{4+}

    d. Ba^{2+}, Pt^{2+}, Mn^{2+}

48. What noble gas has the same election configuration as each of the ions in the following compounds?

   a. cesium sulfide

   b. strontium fluoride

   c. calcium nitride

   d. aluminum bromide

49. Give the formula of a *negative* ion that would have the same number of electrons as each of the following *positive* ions.

   a. Na^+ **c.** Al^{3+}

   b. Ca^{2+} **d.** Rb^+

50. Give an example of an ionic compound where both the anion and the cation are isoelectronic with each of the following noble gases.

   a. Ne   **c.** Kr

   b. Ar   **d.** Xe

51. Give three ions that are isoelectronic with neon. Place these ions in order of increasing size.

52. Consider the ions Sc^{3+}, Cl^-, K^+, Ca^{2+}, and S^{2-}. Match these ions to the following pictures that represent the relative sizes of the ions.

53. For each of the following groups, place the atoms and/or ions in order of decreasing size.

   a. Cu, Cu^+, Cu^{2+}

   b. Ni^{2+}, Pd^{2+}, Pt^{2+}

   c. O, O^-, O^{2-}

   d. La^{3+}, Eu^{3+}, Gd^{3+}, Yb^{3+}

   e. Te^{2-}, I^-, Cs^+, Ba^{2+}, La^{3+}

54. For each of the following groups, place the atoms and/or ions in order of decreasing size.

   a. V, V^{2+}, V^{3+}, V^{5+}

   b. Na^+, K^+, Rb^+, Cs^+

   c. Te^{2-}, I^-, Cs^+, Ba^{2+}

   d. P, P^-, P^{2-}, P^{3-}

   e. O^{2-}, S^{2-}, Se^{2-}, Te^{2-}

55. Which compound in each of the following pairs of ionic substances has the most exothermic lattice energy? Justify your answers.

   a. NaCl, KCl

   b. LiF, LiCl

   c. $Mg(OH)_2$, MgO

   d. $Fe(OH)_2$, $Fe(OH)_3$

   e. NaCl, Na_2O

   f. MgO, BaS

56. Which compound in each of the following pairs of ionic substances has the most exothermic lattice energy? Justify your answers.

   a. LiF, CsF

   b. NaBr, NaI

   c. $BaCl_2$, BaO

   d. Na_2SO_4, $CaSO_4$

   e. KF, K_2O

   f. Li_2O, Na_2S

57. Use the following data to estimate ΔH_f° for potassium chloride.

$$K(s) + \tfrac{1}{2}Cl_2(g) \longrightarrow KCl(s)$$

Lattice energy	−690. kJ/mol
Ionization energy for K	419 kJ/mol
Electron affinity of Cl	−349 kJ/mol
Bond energy of Cl_2	239 kJ/mol
Enthalpy of sublimation for K	90. kJ/mol

58. Use the following data to estimate ΔH_f° for magnesium fluoride.

$$Mg(s) + F_2(g) \longrightarrow MgF_2(s)$$

Lattice energy	−2913 kJ/mol
First ionization energy of Mg	735 kJ/mol
Second ionization energy of Mg	1445 kJ/mol
Electron affinity of F	−328 kJ/mol
Bond energy of F_2	154 kJ/mol
Enthalpy of sublimation for Mg	150. kJ/mol

59. Consider the following energy changes:

	ΔH (kJ/mol)
$Mg(g) \rightarrow Mg^+(g) + e^-$	735
$Mg^+(g) \rightarrow Mg^{2+}(g) + e^-$	1445
$O(g) + e^- \rightarrow O^-(g)$	−141
$O^-(g) + e^- \rightarrow O^{2-}(g)$	878

Magnesium oxide exists as $Mg^{2+}O^{2-}$ and not as Mg^+O^-. Explain.

60. Compare the electron affinity of fluorine to the ionization energy of sodium. Is the process of an electron being "pulled" from the sodium atom to the fluorine atom exothermic or endothermic? Why is NaF a stable compound? Is the overall formation of NaF endothermic or exothermic? How can this be?

61. Consider the following: $Li(s) + \tfrac{1}{2} I_2(g) \rightarrow LiI(s)$ $\Delta H = -292$ kJ. $LiI(s)$ has a lattice energy of -753 kJ/mol. The ionization energy of $Li(g)$ is 520. kJ/mol, the bond energy of $I_2(g)$ is 151 kJ/mol, and the electron affinity of $I(g)$ is -295 kJ/mol. Use these data to determine the heat of sublimation of $Li(s)$.

62. Use the following data (in kJ/mol) to estimate ΔH for the reaction $S^-(g) + e^- \rightarrow S^{2-}(g)$. Include an estimate of uncertainty.

	ΔH_f°	Lattice Energy	Ionization Energy of M	ΔH_{sub} of M
Na_2S	−365	−2203	495	109
K_2S	−381	−2052	419	90
Rb_2S	−361	−1949	409	82
Cs_2S	−360	−1850	382	78

$$S(s) \longrightarrow S(g) \qquad \Delta H = 277 \text{ kJ/mol}$$
$$S(g) + e^- \longrightarrow S^-(g) \qquad \Delta H = -200 \text{ kJ/mol}$$

Assume that all values are known to ± 1 kJ/mol.

63. Rationalize the following lattice energy values:

Compound	Lattice Energy (kJ/mol)
CaSe	−2862
Na_2Se	−2130
CaTe	−2721
Na_2Te	−2095

64. The lattice energies of $FeCl_3$, $FeCl_2$, and Fe_2O_3 are (in no particular order) −2631, −5359, and −14,774 kJ/mol. Match the appropriate formula to each lattice energy. Explain.

Bond Energies

65. Use bond energy values (Table 8.4) to estimate ΔH for each of the following reactions in the gas phase.
   a. $H_2 + Cl_2 \rightarrow 2HCl$
   b. $N{\equiv}N + 3H_2 \rightarrow 2NH_3$

66. Use bond energy values (Table 8.4) to estimate ΔH for each of the following reactions.

   a. $H-C{\equiv}N(g) + 2H_2(g) \longrightarrow$ H—C—N(g) (with H, H above and H, H below the C—N)

   b.
   H₂N—NH₂ $(g) + 2F_2(g) \longrightarrow N{\equiv}N(g) + 4HF(g)$

67. Use bond energies (Table 8.4) to predict ΔH for the isomerization of methyl isocyanide to acetonitrile:
   $$CH_3N{\equiv}C(g) \longrightarrow CH_3C{\equiv}N(g)$$

68. Acetic acid is responsible for the sour taste of vinegar. It can be manufactured using the following reaction:
   $$CH_3OH(g) + C{\equiv}O(g) \longrightarrow CH_3\overset{\displaystyle O}{\overset{\|}{C}}-OH(l)$$
   Use tabulated values of bond energies (Table 8.4) to estimate ΔH for this reaction.

69. Use bond energies to predict ΔH for the following reaction:
   $$H_2S(g) + 3F_2(g) \longrightarrow SF_4(g) + 2HF(g)$$

70. The major industrial source of hydrogen gas is by the following reaction:
   $$CH_4(g) + H_2O(g) \longrightarrow CO(g) + 3H_2(g)$$
   Use bond energies to predict ΔH for this reaction.

71. Use bond energies to estimate ΔH for the combustion of one mole of acetylene:
   $$C_2H_2(g) + \tfrac{5}{2}O_2(g) \longrightarrow 2CO_2(g) + H_2O(g)$$

72. Use data from Table 8.4 to estimate ΔH for the combustion of methane (CH_4), as shown below:

73. Consider the following reaction:

H₂C=CH₂ $(g) + F_2(g) \longrightarrow$

F₂HC—CHF₂ ... H—C—C—H(g) $\Delta H = -549$ kJ

Estimate the carbon–fluorine bond energy given that the C—C bond energy is 347 kJ/mol, the C=C bond energy is 614 kJ/mol, and the F—F bond energy is 154 kJ/mol.

74. Consider the following reaction:
   $$A_2 + B_2 \longrightarrow 2AB \qquad \Delta H = -285 \text{ kJ}$$
   The bond energy for A_2 is one-half the amount of the AB bond energy. The bond energy of $B_2 = 432$ kJ/mol. What is the bond energy of A_2?

75. Compare your answers from parts a and b of Exercise 65 with ΔH values calculated for each reaction using standard enthalpies of formation in Appendix 4. Do enthalpy changes calculated from bond energies give a reasonable estimate of the actual values?

76. Compare your answer from Exercise 68 to the ΔH value calculated from standard enthalpies of formation in Appendix 4. Explain any discrepancies.

77. The standard enthalpies of formation for $S(g)$, $F(g)$, $SF_4(g)$, and $SF_6(g)$ are +278.8, +79.0, −775, and −1209 kJ/mol, respectively.
   a. Use these data to estimate the energy of an S—F bond.
   b. Compare your calculated value to the value given in Table 8.4. What conclusions can you draw?
   c. Why are the ΔH_f° values for $S(g)$ and $F(g)$ not equal to zero, since sulfur and fluorine are elements?

78. Use the following standard enthalpies of formation to estimate the N—H bond energy in ammonia: $N(g)$, 472.7 kJ/mol; $H(g)$, 216.0 kJ/mol; $NH_3(g)$, −46.1 kJ/mol. Compare your value to the one in Table 8.4.

79. The standard enthalpy of formation for $N_2H_4(g)$ is 95.4 kJ/mol. Use this and the data in Exercise 78 to estimate the N—N single bond energy. Compare this with the value in Table 8.4.

80. The standard enthalpy of formation for $NO(g)$ is 90. kJ/mol. Use this and the values for the O=O and N≡N bond energies to estimate the bond strength in NO.

Lewis Structures and Resonance

81. Write Lewis structures that obey the octet rule (duet rule for H) for each of the following molecules. Carbon is the central atom in CH_4, nitrogen is the central atom in NH_3, and oxygen is the central atom in H_2O.
   a. F_2 **e.** NH_3
   b. O_2 **f.** H_2O
   c. CO   **g.** HF
   d. CH_4

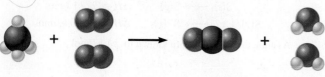

82. Write Lewis structures that obey the octet rule (duet rule for H) for each of the following molecules.

   a. H_2CO

   b. CO_2

   c. HCN

   Carbon is the central atom in all of these molecules.

83. Write Lewis structures that obey the octet rule for each of the following molecules.

   a. CCl_4 c. $SeCl_2$

   b. NCl_3 d. ICl

   In each case, the atom listed first is the central atom.

84. Write Lewis structures that obey the octet rule for each of the following molecules and ions. (In each case the first atom listed is the central atom.)

   a. $POCl_3$, SO_4^{2-}, XeO_4, PO_4^{3-}, ClO_4^-

   b. NF_3, SO_3^{2-}, PO_3^{3-}, ClO_3^-

   c. ClO_2^-, SCl_2, PCl_2^-

   d. Considering your answers to parts a, b, and c, what conclusions can you draw concerning the structures of species containing the same number of atoms and the same number of valence electrons?

85. One type of exception to the octet rule are compounds with central atoms having fewer than eight electrons around them. BeH_2 and BH_3 are examples of this type of exception. Draw the Lewis structures for BeH_2 and BH_3.

86. Lewis structures can be used to understand why some molecules react in certain ways. Write the Lewis structures for the reactants and products in the reactions described below.

   a. Nitrogen dioxide dimerizes to produce dinitrogen tetroxide.

   b. Boron trihydride accepts a pair of electrons from ammonia, forming BH_3NH_3.

   Give a possible explanation for why these two reactions occur.

87. The most common type of exception to the octet rule are compounds or ions with central atoms having more than eight electrons around them. PF_5, SF_4, ClF_3 and Br_3^- are examples of this type of exception. Draw the Lewis structure for these compounds or ions. Which elements, when they have to, can have more than eight electrons around them? How is this rationalized?

88. SF_6, ClF_5, and XeF_4 are three compounds whose central atoms do not follow the octet rule. Draw Lewis structures for these compounds.

89. Write Lewis structures for the following. Show all resonance structures where applicable.

   a. NO_2^-, NO_3^-, N_2O_4 (N_2O_4 exists as O_2N—NO_2.)

   b. OCN^-, SCN^-, N_3^- (Carbon is the central atom in OCN^- and SCN^-.)

90. Some of the important pollutants in the atmosphere are ozone (O_3), sulfur dioxide, and sulfur trioxide. Write Lewis structures for these three molecules. Show all resonance structures where applicable.

91. Benzene (C_6H_6) consists of a six-membered ring of carbon atoms with one hydrogen bonded to each carbon. Write Lewis structures for benzene, including resonance structures.

92. Borazine ($B_3N_3H_6$) has often been called "inorganic" benzene. Write Lewis structures for borazine. Borazine contains a six-membered ring of alternating boron and nitrogen atoms with one hydrogen bonded to each boron and nitrogen.

93. An important observation supporting the concept of resonance in the localized electron model was that there are only three different structures of dichlorobenzene ($C_6H_4Cl_2$). How does this fact support the concept of resonance? (See Exercise 91.)

94. Consider the following bond lengths:

   C—O 143 pm   C=O 123 pm   C≡O 109 pm

   In the CO_3^{2-} ion, all three C—O bonds have identical bond lengths of 136 pm. Why?

95. A toxic cloud covered Bhopal, India, in December 1984 when water leaked into a tank of methyl isocyanate, and the product escaped into the atmosphere. Methyl isocyanate is used in the production of many pesticides. Draw the Lewis structures for methyl isocyanate, CH_3NCO, including resonance forms. The skeletal structure is

$$\begin{array}{c} H \\ | \\ H-C-N-C-O \\ | \\ H \end{array}$$

96. Peroxyacetyl nitrate, or PAN, is present in photochemical smog. Draw Lewis structures (including resonance forms) for PAN. The skeletal structure is

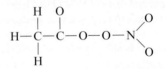

97. Order the following species with respect to carbon–oxygen bond length (longest to shortest).

      CO, CO_2, CO_3^{2-}, CH_3OH

   What is the order from the weakest to the strongest carbon–oxygen bond? (CH_3OH exists as H_3C—OH.)

98. Place the species below in order of the shortest to the longest nitrogen–oxygen bond.

     H_2NOH, N_2O, NO^+, NO_2^-, NO_3^-

   (H_2NOH exists as H_2N—OH.)

Formal Charge

99. Use the formal charge arguments to rationalize why BF_3 would not follow the octet rule.

100. Use formal charge arguments to explain why CO has a much smaller dipole moment than would be expected on the basis of electronegativity.

101. Write Lewis structures that obey the octet rule for the following species. Assign the formal charge for each central atom.

   a. $POCl_3$ e. SO_2Cl_2

   b. SO_4^{2-} f. XeO_4

   c. ClO_4^- g. ClO_3^-

   d. PO_4^{3-} h. NO_4^{3-}

102. Write Lewis structures for the species in Exercise 101 that involve minimum formal charges.

103. Write the Lewis structure for O_2F_2 (O_2F_2 exists as F—O—O—F). Assign oxidation states and formal charges to the atoms in O_2F_2. This compound is a vigorous and potent oxidizing and fluorinating agent. Are oxidation states or formal charges more useful in accounting for these properties of O_2F_2?

104. Oxidation of the cyanide ion produces the stable cyanate ion, OCN^-. The fulminate ion, CNO^-, on the other hand, is very unstable. Fulminate salts explode when struck; $Hg(CNO)_2$ is used in blasting caps. Write the Lewis structures and assign formal charges for the cyanate and fulminate ions. Why is the fulminate ion so unstable? (C is the central atom in OCN^- and N is the central atom in CNO^-.)

105. When molten sulfur reacts with chlorine gas, a vile-smelling orange liquid forms that has an empirical formula of SCl. The structure of this compound has a formal charge of zero on all elements in the compound. Draw the Lewis structure for the vile-smelling orange liquid.

106. Nitrous oxide (N_2O) has three possible Lewis structures:

$$:\!N\!=\!N\!=\!\ddot{O}: \longleftrightarrow :N\!\equiv\!N\!-\!\ddot{O}: \longleftrightarrow :\!\ddot{N}\!-\!N\!\equiv\!O:$$

Given the following bond lengths,

N—N	167 pm	N=O	115 pm
N=N	120 pm	N—O	147 pm
N≡N	110 pm		

rationalize the observations that the N—N bond length in N_2O is 112 pm and that the N—O bond length is 119 pm. Assign formal charges to the resonance structures for N_2O. Can you eliminate any of the resonance structures on the basis of formal charges? Is this consistent with observation?

107. A common trait of simple organic compounds is to have Lewis structures where all atoms have a formal charge of zero. Consider the following incomplete Lewis structure for an organic compound called *methyl cyanoacrylate*, the main ingredient in Super Glue.

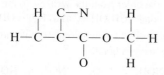

Draw a complete Lewis structure for methyl cyanoacrylate in which all atoms have a formal charge of zero.

108. Benzoic acid is a food preservative. The space-filling model for benzoic acid is shown below.

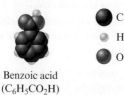

Benzoic acid
($C_6H_5CO_2H$)

Draw the Lewis structure for benzoic acid, including all resonance structures in which all atoms have a formal charge of zero.

Molecular Structure and Polarity

109. Predict the molecular structure and bond angles for each molecule or ion in Exercises 83 and 89.

110. Predict the molecular structure and bond angles for each molecule or ion in Exercises 84 and 90.

111. There are several molecular structures based on the trigonal bipyramid geometry (see Table 8.8). Three such structures are

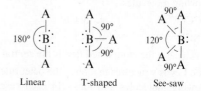

| Linear | T-shaped | See-saw |

Which of the compounds in Exercises 87 and 88 have these molecular structures?

112. Two variations of the octahedral geometry (see Table 8.6) are illustrated below.

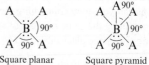

| Square planar | Square pyramid |

Which of the compounds in Exercises 87 and 88 have these molecular structures?

113. Predict the molecular structure (including bond angles) for each of the following.
  a. SeO_3
  b. SeO_2

114. Predict the molecular structure (including bond angles) for each of the following.
  a. PCl_3
  b. SCl_2
  c. SiF_4

115. Predict the molecular structure (including bond angles) for each of the following. (See Exercises 111 and 112.)
  a. $XeCl_2$
  b. ICl_3
  c. TeF_4
  d. PCl_5

116. Predict the molecular structure (including bond angles) for each of the following. (See Exercises 111 and 112.)
  a. ICl_5
  b. $XeCl_4$
  c. $SeCl_6$

117. Which of the molecules in Exercise 113 have net dipole moments (are polar)?

118. Which of the molecules in Exercise 114 have net dipole moments (are polar)?

119. Which of the molecules in Exercise 115 have net dipole moments (are polar)?

120. Which of the molecules in Exercise 116 have net dipole moments (are polar)?

121. Write Lewis structures and predict the molecular structures of the following. (See Exercises 111 and 112.)

  a. OCl_2, KrF_2, BeH_2, SO_2

  b. SO_3, NF_3, IF_3

  c. CF_4, SeF_4, KrF_4

  d. IF_5, AsF_5

  Which of these compounds are polar?

122. Write Lewis structures and predict whether each of the following is polar or nonpolar.

  a. HOCN (exists as HO—CN)

  b. COS

  c. XeF_2

  d. CF_2Cl_2

  e. SeF_6

  f. H_2CO (C is the central atom)

123. Consider the following Lewis structure where E is an unknown element:

  What are some possible identities for element E? Predict the molecular structure (including bond angles) for this ion.

124. Consider the following Lewis structure where E is an unknown element:

  What are some possible identities for element E? Predict the molecular structure (including bond angles) for this ion. (See Exercises 111 and 112.)

125. The molecules BF_3, CF_4, CO_2, PF_5, and SF_6 are all nonpolar, even though they all contain polar bonds. Why?

126. Two different compounds have the formula XeF_2Cl_2. Write Lewis structures for these two compounds, and describe how measurement of dipole moments might be used to distinguish between them.

Additional Exercises

127. Arrange the following in order of increasing radius and increasing ionization energy.

  a. N^+, N, N^-

  b. Se, Se^-, Cl, Cl^+

  c. Br^-, Rb^+, Sr^{2+}

128. For each of the following, write an equation that corresponds to the energy given.

  a. lattice energy of NaCl

  b. lattice energy of NH_4Br

  c. lattice energy of MgS

  d. O=O double bond energy beginning with $O_2(g)$ as a reactant

129. Use bond energies (Table 8.4), values of electron affinities (Table 7.7), and the ionization energy of hydrogen (1312 kJ/mol) to estimate ΔH for each of the following reactions.

  a. $HF(g) \rightarrow H^+(g) + F^-(g)$

  b. $HCl(g) \rightarrow H^+(g) + Cl^-(g)$

  c. $HI(g) \rightarrow H^+(g) + I^-(g)$

  d. $H_2O(g) \rightarrow H^+(g) + OH^-(g)$

  (Electron affinity of $OH(g) = -180.$ kJ/mol.)

130. Write Lewis structures for CO_3^{2-}, HCO_3^-, and H_2CO_3. When acid is added to an aqueous solution containing carbonate or bicarbonate ions, carbon dioxide gas is formed. We generally say that carbonic acid (H_2CO_3) is unstable. Use bond energies to estimate ΔH for the reaction (in the gas phase)

$$H_2CO_3 \longrightarrow CO_2 + H_2O$$

  Specify a possible cause for the instability of carbonic acid.

131. Which member of the following pairs would you expect to be more energetically stable? Justify each choice.

  a. NaBr or $NaBr_2$

  b. ClO_4 or ClO_4^-

  c. SO_4 or XeO_4

  d. OF_4 or SeF_4

132. What do each of the following sets of compounds/ions have in common with each other?

  a. SO_3, NO_3^-, CO_3^{2-}

  b. O_3, SO_2, NO_2^-

133. What do each of the following sets of compounds/ions have in common with each other? See your Lewis structures for Exercises 113 through 116.

  a. $XeCl_4$, $XeCl_2$

  b. ICl_5, TeF_4, ICl_3, PCl_3, SCl_2, SeO_2

134. Although both Br_3^- and I_3^- ions are known, the F_3^- ion has not been observed. Explain.

135. Refer back to Exercises 101 and 102. Would you make the same prediction for the molecular structure for each case using the Lewis structure obtained in Exercise 101 as compared with the one obtained in Exercise 102?

136. Which of the following molecules have net dipole moments? For the molecules that are polar, indicate the polarity of each bond and the direction of the net dipole moment of the molecule.

  a. CH_2Cl_2, $CHCl_3$, CCl_4

  b. CO_2, N_2O

  c. PH_3, NH_3

137. The structure of TeF_5^- is

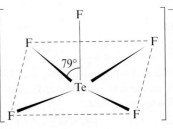

  Draw a complete Lewis structure for TeF_5^-, and explain the distortion from the ideal square pyramidal structure. (See Exercise 112.)

138. Look up the energies for the bonds in CO and N_2. Although the bond in CO is stronger, CO is considerably more reactive than N_2. Give a possible explanation.

ChemWork Problems

These multiconcept problems (and additional ones) are found interactively online with the same type of assistance a student would get from an instructor.

139. Classify the bonding in each of the following molecules as ionic, polar covalent, or nonpolar covalent.

a. H_2 e. HF

b. K_3P f. CCl_4

c. NaI   g. CF_4

d. SO_2 h. K_2S

140. List the bonds P—Cl, P—F, O—F, and Si—F from least polar to most polar.

141. Arrange the atoms and/or ions in the following groups in order of decreasing size.

a. O, O^-, O^{2-}

b. Fe^{2+}, Ni^{2+}, Zn^{2+}

c. Ca^{2+}, K^+, Cl^-

142. Use the following data to estimate ΔH_f° for barium bromide.

$$Ba(s) + Br_2(g) \longrightarrow BaBr_2(s)$$

Lattice energy	-1985 kJ/mol
First ionization energy of Ba	503 kJ/mol
Second ionization energy of Ba	965 kJ/mol
Electron affinity of Br	-325 kJ/mol
Bond energy of Br_2	193 kJ/mol
Enthalpy of sublimation of Ba	178 kJ/mol

143. Use bond energy values to estimate ΔH for the following gas phase reaction:

$$C_2H_4 + H_2O_2 \longrightarrow CH_2OHCH_2OH$$

144. Which of the following compounds or ions exhibit resonance?

a. O_3 d. CO_3^{2-}

b. CNO^- d. AsF_3

c. AsI_3

145. The formulas of several chemical substances are given in the table below. For each substance in the table, give its chemical name and predict its molecular structure.

Formula	Compound Name	Molecular Structure
CO_2		
NH_3		
SO_3		
H_2O		
ClO_4^-		

146. Predict the molecular structure, bond angles, and polarity (has a net dipole moment or has no net dipole moment) for each of the following compounds.

a. $SeCl_4$ d. CBr_4

b. SO_2 e. IF_3

c. KrF_4 f. ClF_5

Challenge Problems

147. Use Coulomb's law,

$$V = \frac{Q_1 Q_2}{4\pi\epsilon_0 r} = 2.31 \times 10^{-19}\,\text{J} \cdot \text{nm}\left(\frac{Q_1 Q_2}{r}\right)$$

to calculate the energy of interaction for the following two arrangements of charges, each having a magnitude equal to the electron charge.

a.

b.

148. An alternative definition of electronegativity is

$$\text{Electronegativity} = \text{constant (I.E.} - \text{E.A.)}$$

where I.E. is the ionization energy and E.A. is the electron affinity using the sign conventions of this book. Use data in Chapter 7 to calculate the (I.E. $-$ E.A.) term for F, Cl, Br, and I. Do these values show the same trend as the electronegativity values given in this chapter? The first ionization energies of the halogens are 1678, 1255, 1138, and 1007 kJ/mol, respectively. (*Hint:* Choose a constant so that the electronegativity of fluorine equals 4.0. Using this constant, calculate relative electronegativities for the other halogens and compare to values given in the text.)

149. Calculate the standard heat of formation of the compound $ICl(g)$ at 25°C. (*Hint:* Use Table 8.4 and Appendix 4 data.)

150. Given the following information:

Heat of sublimation of $Li(s) = 166$ kJ/mol
Bond energy of $HCl = 427$ kJ/mol
Ionization energy of $Li(g) = 520.$ kJ/mol
Electron affinity of $Cl(g) = -349$ kJ/mol
Lattice energy of $LiCl(s) = -829$ kJ/mol
Bond energy of $H_2 = 432$ kJ/mol

Calculate the net change in energy for the following reaction:

$$2Li(s) + 2HCl(g) \longrightarrow 2LiCl(s) + H_2(g)$$

151. Use data in this chapter (and Chapter 7) to discuss why MgO is an ionic compound but CO is not an ionic compound.

152. Think of forming an ionic compound as three steps (this is a simplification, as with all models): (1) removing an electron from the metal; (2) adding an electron to the nonmetal; and (3) allowing the metal cation and nonmetal anion to come together.

a. What is the sign of the energy change for each of these three processes?

b. In general, what is the sign of the sum of the first two processes? Use examples to support your answer.

c. What must be the sign of the sum of the three processes?

d. Given your answer to part c, why do ionic bonds occur?

e. Given your above explanations, why is NaCl stable but not Na_2Cl? $NaCl_2$? What about MgO compared to MgO_2? Mg_2O?

153. The compound NF_3 is quite stable, but NCl_3 is very unstable (NCl_3 was first synthesized in 1811 by P. L. Dulong, who lost three fingers and an eye studying its properties). The compounds NBr_3 and NI_3 are unknown, although the explosive compound $NI_3 \cdot NH_3$ is known. Account for the instability of these halides of nitrogen.

154. Three processes that have been used for the industrial manufacture of acrylonitrile (CH_2CHCN), an important chemical used in the manufacture of plastics, synthetic rubber, and fibers, are shown below. Use bond energy values (Table 8.4) to estimate ΔH for each of the reactions.

a.

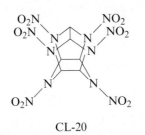

$$HOCH_2CH_2CN \longrightarrow$$

b. $4CH_2{=}CHCH_3 + 6NO \xrightarrow[Ag]{700°C}$
$$4CH_2{=}CHCN + 6H_2O + N_2$$

The nitrogen–oxygen bond energy in nitric oxide (NO) is 630. kJ/mol.

c. $2CH_2{=}CHCH_3 + 2NH_3 + 3O_2 \xrightarrow[425–510°C]{Catalyst}$
$$2CH_2{=}CHCN + 6H_2O$$

d. Is the elevated temperature noted in parts b and c needed to provide energy to endothermic reactions?

155. The compound hexaazaisowurtzitane is one of the highest-energy explosives known (*C & E News,* Jan. 17, 1994, p. 26). The compound, also known as CL-20, was first synthesized in 1987. The method of synthesis and detailed performance data are still classified because of CL-20's potential military application in rocket boosters and in warheads of "smart" weapons. The structure of CL-20 is

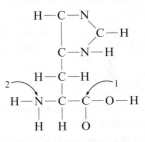

CL-20

In such shorthand structures, each point where lines meet represents a carbon atom. In addition, the hydrogens attached to the carbon atoms are omitted; each of the six carbon atoms has one hydrogen atom attached. Finally, assume that the two O atoms in the NO_2 groups are attached to N with one single bond and one double bond.

Three possible reactions for the explosive decomposition of CL-20 are

i. $C_6H_6N_{12}O_{12}(s) \rightarrow 6CO(g) + 6N_2(g) + 3H_2O(g) + \frac{3}{2}O_2(g)$
ii. $C_6H_6N_{12}O_{12}(s) \rightarrow 3CO(g) + 3CO_2(g) + 6N_2(g) + 3H_2O(g)$
iii. $C_6H_6N_{12}O_{12}(s) \rightarrow 6CO_2(g) + 6N_2(g) + 3H_2(g)$

a. Use bond energies to estimate ΔH for these three reactions.

b. Which of the above reactions releases the largest amount of energy per kilogram of CL-20?

156. Many times extra stability is characteristic of a molecule or ion in which resonance is possible. How could this be used to explain the acidities of the following compounds? (The acidic hydrogen is marked by an asterisk.) Part c shows resonance in the C_6H_5 ring.

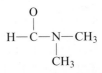

157. The study of carbon-containing compounds and their properties is called *organic chemistry.* Besides carbon atoms, organic compounds also can contain hydrogen, oxygen, and nitrogen atoms (as well as other types of atoms). A common trait of simple organic compounds is to have Lewis structures where all atoms have a formal charge of zero. Consider the following incomplete Lewis structure for an organic compound called *histidine* (an amino acid), which is one of the building blocks of proteins found in our bodies:

Draw a complete Lewis structure for histidine in which all atoms have a formal charge of zero. What should be the approximate bond angles about the carbon atom labeled 1 and the nitrogen atom labeled 2?

158. Draw a Lewis structure for the *N,N*-dimethylformamide molecule. The skeletal structure is

Various types of evidence lead to the conclusion that there is some double bond character to the C—N bond. Draw one or more resonance structures that support this observation.

159. Predict the molecular structure for each of the following. (See Exercises 111 and 112.)

a. $BrFI_2$
b. XeO_2F_2
c. $TeF_2Cl_3^-$

For each formula there are at least two different structures that can be drawn using the same central atom. Draw all possible structures for each formula.

160. Consider the following computer-generated model of caffeine.

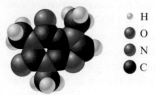

H
O
N
C

Draw a Lewis structure for caffeine in which all atoms have a formal charge of zero.

Integrative Problems

These problems require the integration of multiple concepts to find the solutions.

161. A compound, XF_5, is 42.81% fluorine by mass. Identify the element X. What is the molecular structure of XF_5?

162. A polyatomic ion is composed of C, N, and an unknown element X. The skeletal Lewis structure of this polyatomic ion is $[X—C—N]^-$. The ion X^{2-} has an electron configuration of $[Ar]4s^23d^{10}4p^6$. What is element X? Knowing the identity of X, complete the Lewis structure of the polyatomic ion, including all important resonance structures.

163. Identify the following elements based on their electron configurations and rank them in order of increasing electronegativity: $[Ar]4s^13d^5$; $[Ne]3s^23p^3$; $[Ar]4s^23d^{10}4p^3$; $[Ne]3s^23p^5$.

Marathon Problem

This problem is designed to incorporate several concepts and techniques into one situation.

164. Identify the five compounds of H, N, and O described as follows. For each compound, write a Lewis structure that is consistent with the information given.

a. All the compounds are electrolytes, although not all of them are strong electrolytes. Compounds C and D are ionic and compound B is covalent.

b. Nitrogen occurs in its highest possible oxidation state in compounds A and C; nitrogen occurs in its lowest possible oxidation state in compounds C, D, and E. The formal charge on both nitrogens in compound C is +1; the formal charge on the only nitrogen in compound B is 0.

c. Compounds A and E exist in solution. Both solutions give off gases. Commercially available concentrated solutions of compound A are normally 16 *M*. The commercial, concentrated solution of compound E is 15 *M*.

d. Commercial solutions of compound E are labeled with a misnomer that implies that a binary, gaseous compound of nitrogen and hydrogen has reacted with water to produce ammonium ions and hydroxide ions. Actually, this reaction occurs to only a slight extent.

e. Compound D is 43.7% N and 50.0% O by mass. If compound D were a gas at STP, it would have a density of 2.86 g/L.

f. A formula unit of compound C has one more oxygen than a formula unit of compound D. Compounds C and A have one ion in common when compound A is acting as a strong electrolyte.

g. Solutions of compound C are weakly acidic; solutions of compound A are strongly acidic; solutions of compounds B and E are basic. The titration of 0.726 g compound B requires 21.98 mL of 1.000 *M* HCl for complete neutralization.

Chapter 9

Covalent Bonding: Orbitals

Chili peppers taste hot because of capsaicin, a complex molecule containing atoms with many different hybridizations. (Image Source)

In Chapter 8 we discussed the fundamental concepts of bonding and introduced the most widely used simple model for covalent bonding: the localized electron model. We saw the usefulness of a bonding model as a means for systematizing chemistry by allowing us to look at molecules in terms of individual bonds. We also saw that molecular structure can be predicted by minimizing electron-pair repulsions. In this chapter we will examine bonding models in more detail, particularly focusing on the role of orbitals.

9.1 | Hybridization and the Localized Electron Model

As we saw in Chapter 8, the localized electron model views a molecule as a collection of atoms bound together by sharing electrons between their atomic orbitals. The arrangement of valence electrons is represented by the Lewis structure (or structures, where resonance occurs), and the molecular geometry can be predicted from the VSEPR model. In this section we will describe the atomic orbitals used to share electrons and hence to form the bonds.

*sp*³ Hybridization

Let us reconsider the bonding in methane, which has the Lewis structure and molecular geometry shown in Fig. 9.1. In general, we assume that bonding involves only the valence orbitals. This means that the hydrogen atoms in methane use $1s$ orbitals. The valence orbitals of a carbon atom are the $2s$ and $2p$ orbitals shown in Fig. 9.2. In thinking about how carbon can use these orbitals to bond to the hydrogen atoms, we can see two related problems:

> The valence orbitals are the orbitals associated with the highest principal quantum level that contains electrons on a given atom.

1. Using the $2p$ and $2s$ atomic orbitals will lead to two different types of C—H bonds: (a) those from the overlap of a $2p$ orbital of carbon and a $1s$ orbital of hydrogen (there will be three of these) and (b) those from the overlap of a $2s$ orbital

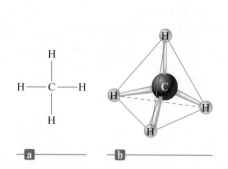

Figure 9.1 | (a) The Lewis structure of the methane molecule. (b) The tetrahedral molecular geometry of the methane molecule.

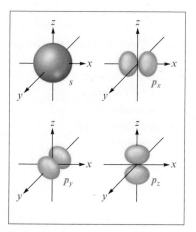

Figure 9.2 | The valence orbitals on a free carbon atom: $2s$, $2p_x$, $2p_y$, and $2p_z$.

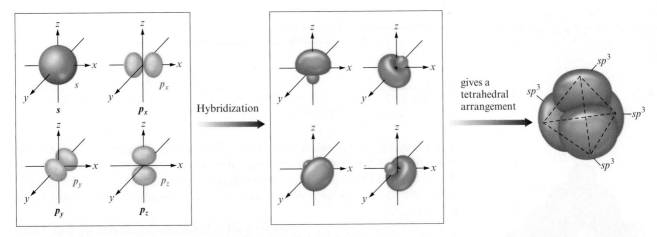

Figure 9.3 | The "native" 2s and three 2p atomic orbitals characteristic of a free carbon atom are combined to form a new set of four sp^3 orbitals. The small lobes of the orbitals are usually omitted from diagrams for clarity.

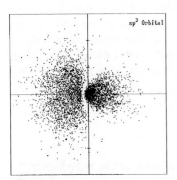

Figure 9.4 | Cross section of an sp^3 orbital. This shows a "slice" of the electron density of the sp^3 orbitals illustrated in the center diagram of Fig. 9.3.

(Generated from a program by Robert Allendoerfer on Project SERAPHIM disk PC2402; printed with permission.)

Hybridization is a modification of the localized electron model to account for the observation that atoms often seem to use special atomic orbitals in forming molecules.

sp^3 hybridization gives a tetrahedral set of orbitals.

Note that the shapes of the hybrid orbitals shown in Figure 9.3 are "mushroom-like," calculated from wave functions.

of carbon and a 1s orbital of hydrogen (there will be one of these). This is a problem because methane is known to have four identical C—H bonds.

2. Since the carbon 2p orbitals are mutually perpendicular, we might expect the three C—H bonds formed with these orbitals to be oriented at 90-degree angles:

However, the methane molecule is known by experiment to be tetrahedral with bond angles of 109.5 degrees.

This analysis leads to one of two conclusions: Either the simple localized electron model is wrong or carbon adopts a set of atomic orbitals other than its "native" 2s and 2p orbitals to bond to the hydrogen atoms in forming the methane molecule. The second conclusion seems more reasonable. The 2s and 2p orbitals present on an *isolated* carbon atom may not be the best set of orbitals for bonding; a new set of atomic orbitals might better serve the carbon atom in forming molecules. To account for the known structure of methane, it makes sense to assume that the carbon atom has four equivalent atomic orbitals, arranged tetrahedrally. In fact, such a set of orbitals can be obtained quite readily by combining the carbon 2s and 2p orbitals, as shown schematically in Fig. 9.3. This mixing of the native atomic orbitals to form special orbitals for bonding is called **hybridization**. The four new orbitals are called sp^3 orbitals because they are formed from one 2s and three 2p orbitals. We say that the carbon atom undergoes sp^3 **hybridization** or is sp^3 **hybridized**. The four sp^3 orbitals are identical in shape, each one having a large lobe and a small lobe (Fig. 9.4). The four orbitals are oriented in space so that the large lobes form a tetrahedral arrangement, as shown in Fig. 9.3.

The hybridization of the carbon 2s and 2p orbitals also can be represented by an orbital energy-level diagram (Fig. 9.5). Note that electrons have been omitted because we do not need to be concerned with the electron arrangements on the individual atoms—it is the total number of electrons and the arrangement of these electrons in the *molecule* that are important. We are assuming that carbon's atomic orbitals are rearranged to accommodate the best electron arrangement for the molecule as a whole. The new sp^3 atomic orbitals on carbon are used to share electron pairs with the 1s orbitals from the four hydrogen atoms (Fig. 9.6).

At this point let's summarize the bonding in the methane molecule. The experimentally known structure of this molecule can be explained if we assume that the carbon atom

Figure 9.5 | An energy-level diagram showing the formation of four sp^3 orbitals.

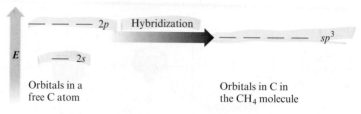

adopts a special set of atomic orbitals. These new orbitals are obtained by combining the $2s$ and the three $2p$ orbitals of the carbon atom to produce four identically shaped orbitals that are oriented toward the corners of a tetrahedron and are used to bond to the hydrogen atoms. Thus the four sp^3 orbitals on carbon in methane are postulated to account for its known structure.

Remember this principle: Whenever a set of equivalent tetrahedral atomic orbitals is required by an atom, this model assumes that the atom adopts a set of sp^3 orbitals; the atom becomes sp^3 hybridized.

It is really not surprising that an atom in a molecule might adopt a different set of atomic orbitals (called **hybrid orbitals**) from those it has in the free state. It does not seem unreasonable that to achieve minimum energy, an atom uses one set of atomic orbitals in the free state and a different set in a molecule. This is consistent with the idea that a molecule is more than simply a sum of its parts. What the atoms in a molecule were like before the molecule was formed is not as important as how the electrons are best arranged in the molecule. Therefore, this model assumes that the individual atoms respond as needed to achieve the minimum energy for the molecule.

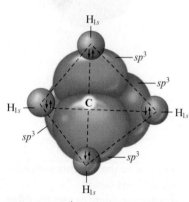

Figure 9.6 | The tetrahedral set of four sp^3 orbitals of the carbon atom are used to share electron pairs with the four 1s orbitals of the hydrogen atoms to form the four equivalent C—H bonds. This accounts for the known tetrahedral structure of the CH₄ molecule.

Critical Thinking

What if the sp^3 hybrid orbitals were higher in energy than the p orbitals in the free atom? How would this affect our model of bonding?

Example 9.1

The Localized Electron Model I

Describe the bonding in the ammonia molecule using the localized electron model.

Solution

A complete description of the bonding involves three steps:

1. writing the Lewis structure
2. determining the arrangement of electron pairs using the VSEPR model
3. determining the hybrid atomic orbitals needed to describe the bonding in the molecule

The Lewis structure for NH₃ is

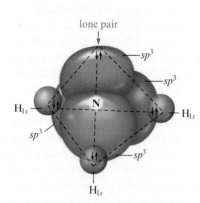

Figure 9.7 | The nitrogen atom in ammonia is sp^3 hybridized.

The four electron pairs around the nitrogen atom require a tetrahedral arrangement to minimize repulsions. We have seen that a tetrahedral set of sp^3 hybrid orbitals is obtained by combining the $2s$ and three $2p$ orbitals. In the NH₃ molecule, three of the sp^3 orbitals are used to form bonds to the three hydrogen atoms, and the fourth sp^3 orbital holds the lone pair (Fig. 9.7).

See Exercises 9.17 and 9.18

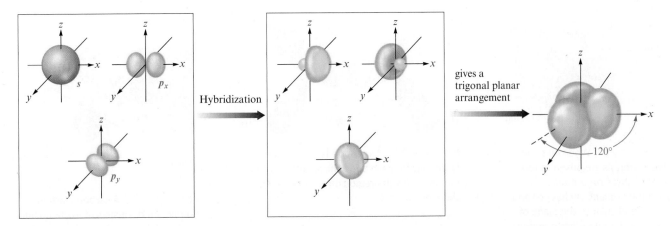

Figure 9.8 | The hybridization of the s, p_x, and p_y atomic orbitals results in the formation of three sp^2 orbitals centered in the xy plane. The large lobes of the orbitals lie in the plane at angles of 120 degrees and point toward the corners of a triangle.

sp^2 Hybridization

Ethylene (C_2H_4) is an important starting material in the manufacture of plastics. The C_2H_4 molecule has 12 valence electrons and the following Lewis structure:

A double bond acts as one effective electron pair.

We saw in Chapter 8 that a double bond acts as one effective pair, so in the ethylene molecule each carbon is surrounded by three effective pairs. This requires a trigonal planar arrangement with bond angles of 120 degrees. What orbitals do the carbon atoms in this molecule employ? The molecular geometry requires a set of orbitals in one plane at angles of 120 degrees. Since the 2s and 2p valence orbitals of carbon do not have the required arrangement, we need a set of hybrid orbitals.

sp^2 hybridization gives a trigonal planar arrangement of atomic orbitals.

The sp^3 orbitals we have just considered will not work because they are at angles of 109.5 degrees rather than the required 120 degrees. In ethylene the carbon atom must hybridize in a different manner. A set of three orbitals arranged at 120-degree angles in the same plane can be obtained by combining one s orbital and two p orbitals, as shown in Fig. 9.8. The orbital energy-level diagram for this arrangement is shown in Fig. 9.9. Since one 2s and two 2p orbitals are used to form these hybrid orbitals, this is called **sp^2 hybridization**. Note from Fig. 9.8 that the plane of the sp^2 hybridized orbitals is determined by which p orbitals are used. Since in this case we have arbitrarily decided to use the p_x and p_y orbitals, the hybrid orbitals are centered in the xy plane.

Note in Fig. 9.10 and the figures that follow that the orbital lobes are artificially narrowed to more clearly show their relative orientations.

In forming the sp^2 orbitals, one 2p orbital on carbon has not been used. This remaining p orbital (p_z) is oriented perpendicular to the plane of the sp^2 orbitals, as shown in Fig. 9.10.

Now we will see how these orbitals can be used to account for the bonds in ethylene. The three sp^2 orbitals on each carbon can be used to share electrons (Fig. 9.11). In

Figure 9.9 | An orbital energy-level diagram for sp^2 hybridization. Note that one p orbital remains unchanged.

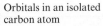
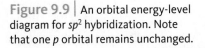

Orbitals in an isolated carbon atom

Carbon orbitals in ethylene

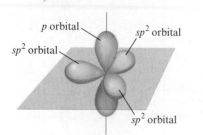

Figure 9.10 │ When an *s* and two *p* orbitals are mixed to form a set of three *sp²* orbitals, one *p* orbital remains unchanged and is perpendicular to the plane of the hybrid orbitals. Note that in this figure and those that follow, the orbitals are drawn with narrowed lobes to show their orientations more clearly.

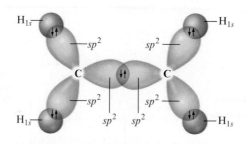

Figure 9.11 │ The σ bonds in ethylene. Note that for each bond the shared electron pair occupies the region directly between the atoms.

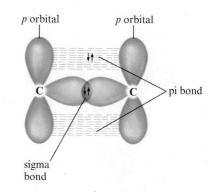

Figure 9.12 │ A carbon–carbon double bond consists of a σ bond and a π bond. In the σ bond the shared electrons occupy the space directly between the atoms. The π bond is formed from the unhybridized *p* orbitals on the two carbon atoms. In a π bond the shared electron pair occupies the space above and below a line joining the atoms.

each of these bonds, the electron pair is shared in an area centered on a line running between the atoms. This type of covalent bond is called a **sigma (σ) bond**. In the ethylene molecule, the σ bonds are formed using *sp²* orbitals on each carbon atom and the 1*s* orbital on each hydrogen atom.

How can we explain the double bond between the carbon atoms? In the σ bond the electron pair occupies the space between the carbon atoms. The second bond must therefore result from sharing an electron pair in the space *above and below* the σ bond. This type of bond can be formed using the 2*p* orbital perpendicular to the *sp²* hybrid orbitals on each carbon atom (see Fig. 9.10). These parallel *p* orbitals can share an electron pair, which occupies the space above and below a line joining the atoms, to form a **pi (π) bond** (Fig. 9.12).

Note that σ bonds are formed from orbitals whose lobes point toward each other, but π bonds result from parallel orbitals. A *double bond always consists of one σ bond,* where the electron pair is located directly between the atoms, *and one π bond,* where the shared pair occupies the space above and below the σ bond.

We can now completely specify the orbitals that this model assumes are used to form the bonds in the ethylene molecule. As shown in Fig. 9.13, the carbon atoms use *sp²* hybrid orbitals to form the σ bonds to the hydrogen atoms and to each other, and they use *p* orbitals to form the π bond with each other. Note that we have accounted fully for the Lewis structure of ethylene with its carbon–carbon double bond and carbon–hydrogen single bonds.

This example illustrates an important general principle of this model: Whenever an atom is surrounded by three effective pairs, a set of *sp²* hybrid orbitals is required.

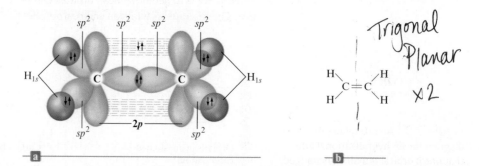

Figure 9.13 │ (a) The orbitals used to form the bonds in ethylene. (b) The Lewis structure for ethylene.

Trigonal Planar ×2

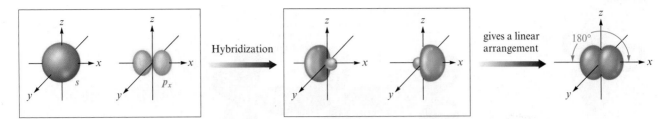

Figure 9.14 | When one *s* orbital and one *p* orbital are hybridized, a set of two *sp* orbitals oriented at 180 degrees results.

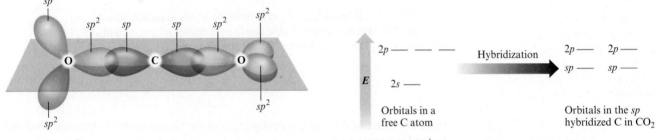

Figure 9.15 | The hybrid orbitals in the CO_2 molecule.

Figure 9.16 | The orbital energy-level diagram for the formation of *sp* hybrid orbitals on carbon.

sp Hybridization

Another type of hybridization occurs in carbon dioxide, which has the following Lewis structure:

$$\ddot{\text{O}}\!=\!\text{C}\!=\!\ddot{\text{O}}$$

In the CO_2 molecule, the carbon atom has two effective pairs that will be arranged at an angle of 180 degrees. We therefore need a pair of atomic orbitals oriented in opposite directions. This requires a new type of hybridization, since neither sp^3 nor sp^2 hybrid orbitals will fit this case. To obtain two hybrid orbitals arranged at 180 degrees requires **sp hybridization,** involving one *s* orbital and one *p* orbital (Fig. 9.14).

In terms of this model, two effective pairs around an atom will always require *sp* hybridization of that atom. The *sp* orbitals of carbon in carbon dioxide can be seen in Fig. 9.15, and the corresponding orbital energy-level diagram for their formation is given in Fig. 9.16. These *sp* hybrid orbitals are used to form the σ bonds between the carbon and the oxygen atoms. Note that two $2p$ orbitals remain unchanged on the *sp* hybridized carbon. These are used to form the π bonds with the oxygen atoms.

In the CO_2 molecule, each oxygen atom* has three effective pairs around it, requiring a trigonal planar arrangement of the pairs. Since a trigonal set of hybrid orbitals requires sp^2 hybridization, each oxygen atom is sp^2 hybridized. One *p* orbital on each oxygen is unchanged and is used for the π bond with the carbon atom.

Now we are ready to use our model to describe the bonding in carbon dioxide. The *sp* orbitals on carbon form σ bonds with the sp^2 orbitals on the two oxygen atoms (see Fig. 9.15). The remaining sp^2 orbitals on the oxygen atoms hold lone pairs. The π bonds between the carbon atom and each oxygen atom are formed by the overlap of parallel $2p$ orbitals. The *sp* hybridized carbon atom has two unhybridized *p* orbitals (Fig. 9.17). Each of these *p* orbitals is used to form a π bond with an oxygen atom (Fig. 9.18). The total bonding picture for the CO_2 molecule is shown in Fig. 9.19. Note

More rigorous theoretical models of CO_2 indicate that each of the oxygen atoms uses two *p* orbitals simultaneously to form the pi bonds to the carbon atom, thus leading to unusually strong C=O bonds.

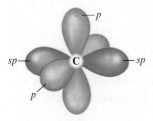

Figure 9.17 | The orbitals of an *sp* hybridized carbon atom.

*We will assume that minimizing electron repulsions also is important for the peripheral atoms in a molecule and apply the VSEPR model to these atoms as well.

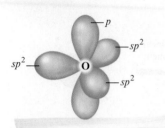

Figure 9.18 | The orbital arrangement for an sp^2 hybridized oxygen atom.

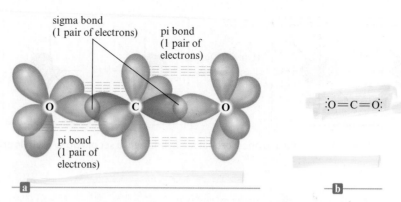

Figure 9.19 | (a) The orbitals used to form the bonds in carbon dioxide. Note that the carbon–oxygen double bonds each consist of one σ bond and one π bond. (b) The Lewis structure for carbon dioxide.

that this picture of the bonding neatly explains the arrangement of electrons predicted by the Lewis structure.

Another molecule whose bonding can be described by sp hybridization is acetylene (C_2H_2), which has the systematic name ethyne. The Lewis structure for acetylene is

$$H-C\equiv C-H$$

Because the triple bond counts as one effective repulsive unit, each carbon has two effective pairs, which requires a linear arrangement. Thus each carbon atom requires sp hybridization, leaving two unchanged p orbitals (see Fig. 9.16). One of the oppositely oriented (see Fig. 9.14) sp orbitals is used to form a bond to the hydrogen atom; the other sp orbital overlaps with the similar sp orbital on the other carbon to form the sigma bond. The two pi bonds are formed from the overlap of the two p orbitals on each carbon. This accounts for the triple bond (one sigma and two pi bonds) in acetylene.

Example 9.2

The Localized Electron Model II

Describe the bonding in the N_2 molecule.

Solution

The Lewis structure for the nitrogen molecule is

$$:N\equiv N:$$

where each nitrogen atom is surrounded by two effective pairs. (Remember that a multiple bond counts as one effective pair.) This gives a linear arrangement (180 degrees) requiring a pair of oppositely directed orbitals. This situation requires sp hybridization. Each nitrogen atom in the nitrogen molecule has two sp hybrid orbitals and two unchanged p orbitals [Fig. 9.20(a)]. The sp orbitals are used to form the σ bond between the nitrogen atoms and to hold lone pairs [Fig. 9.20(b)]. The p orbitals are used to form the two π bonds [Fig. 9.20(c)]; each pair of overlapping parallel p orbitals holds one electron pair. Such bonding accounts for the electron arrangement given by the Lewis structure. The triple bond consists of a σ bond (overlap of two sp orbitals) and two π bonds (each one from an overlap of two p orbitals). In addition, a lone pair occupies an sp orbital on each nitrogen atom.

See Exercises 9.19 and 9.20

Figure 9.20 | (a) An *sp* hybridized nitrogen atom. There are two *sp* hybrid orbitals and two unhybridized *p* orbitals. (b) The σ bond in the N₂ molecule. (c) The two π bonds in N₂ are formed when electron pairs are shared between two sets of parallel *p* orbitals. (d) The total bonding picture for N₂.

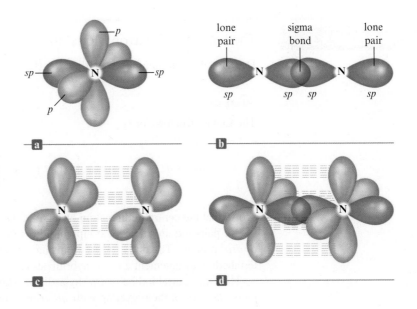

*dsp*³ Hybridization

To illustrate the treatment of a molecule in which the central atom exceeds the octet rule, consider the bonding in the phosphorus pentachloride molecule (PCl_5). The Lewis structure

shows that the phosphorus atom is surrounded by five electron pairs. Since five pairs require a trigonal bipyramidal arrangement, we need a trigonal bipyramidal set of atomic orbitals on phosphorus. Such a set of orbitals is formed by ***dsp*³ hybridization** of one *d* orbital, one *s* orbital, and three *p* orbitals (Fig. 9.21).

The *dsp*³ hybridized phosphorus atom in the PCl_5 molecule uses its five *dsp*³ orbitals to share electrons with the five chlorine atoms. Note that a set of five effective pairs around a given atom always requires a trigonal bipyramidal arrangement, which in turn requires *dsp*³ hybridization of that atom.

The Lewis structure for PCl_5 shows that each chlorine atom is surrounded by four electron pairs. This requires a tetrahedral arrangement, which in turn requires a set of four *sp*³ orbitals on each chlorine atom.

Now we can describe the bonding in the PCl_5 molecule shown in Fig. 9.22. The five P—Cl σ bonds are formed by sharing electrons between a *dsp*³ orbital* on the phosphorus atom and an orbital on each chlorine.

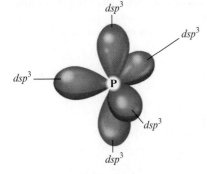

Figure 9.21 | A set of *dsp*³ hybrid orbitals on a phosphorus atom. Note that the set of five *dsp*³ orbitals has a trigonal bipyramidal arrangement. (Each *dsp*³ orbital also has a small lobe that is not shown in this diagram.)

Figure 9.22 | The structure of the PCl_5 molecule.

---

*There is considerable controversy about whether the *d* orbitals are as heavily involved in the bonding in these molecules as this model predicts. However, this matter is beyond the scope of this text.

| Example 9.3 | The Localized Electron Model III |

Describe the bonding in the triiodide ion (I_3^-).

Solution

The Lewis structure for I_3^-

$$\left[\; :\ddot{I}-\ddot{I}-\ddot{I}: \; \right]^-$$

shows that the central iodine atom has five pairs of electrons (see Section 8.11). A set of five pairs requires a trigonal bipyramidal arrangement, which in turn requires a set of dsp^3 orbitals. The outer iodine atoms have four pairs of electrons, which calls for a tetrahedral arrangement and sp^3 hybridization.

Thus the central iodine is dsp^3 hybridized. Three of these hybrid orbitals hold lone pairs, and two of them overlap with sp^3 orbitals of the other two iodine atoms to form σ bonds.

See Exercise 9.25

d^2sp^3 hybridization gives six orbitals arranged octahedrally.

d^2sp^3 Hybridization

Some molecules have six pairs of electrons around a central atom; an example is sulfur hexafluoride (SF_6), which has the Lewis structure

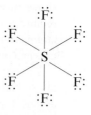

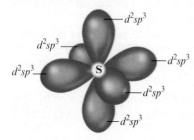

Figure 9.23 | An octahedral set of d^2sp^3 orbitals on a sulfur atom. The small lobe of each hybrid orbital has been omitted for clarity.

This requires an octahedral arrangement of pairs and in turn an octahedral set of six hybrid orbitals, or **d^2sp^3 hybridization**, in which two d orbitals, one s orbital, and three p orbitals are combined (Fig. 9.23). Six electron pairs around an atom are always arranged octahedrally and require d^2sp^3 hybridization of the atom. Each of the d^2sp^3 orbitals on the sulfur atom is used to bond to a fluorine atom. Since there are four pairs on each fluorine atom, the fluorine atoms are assumed to be sp^3 hybridized.

| Interactive Example 9.4 | The Localized Electron Model IV |

How is the xenon atom in XeF_4 hybridized?

Solution

As seen in Example 8.13, XeF_4 has six pairs of electrons around xenon that are arranged octahedrally to minimize repulsions. An octahedral set of six atomic orbitals is required to hold these electrons, and the xenon atom is d^2sp^3 hybridized.

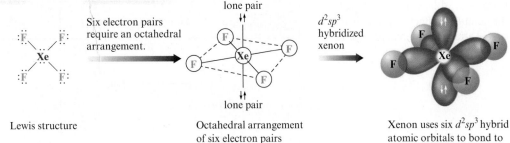

Lewis structure

Six electron pairs require an octahedral arrangement.

Octahedral arrangement of six electron pairs

d^2sp^3 hybridized xenon

Xenon uses six d^2sp^3 hybrid atomic orbitals to bond to the four fluorine atoms and to hold the two lone pairs.

See Exercise 9.26

The Localized Electron Model: A Summary

The description of a molecule using the localized electron model involves three distinct steps.

Problem-Solving Strategy

Using the Localized Electron Model

1. Draw the Lewis structure(s).
2. Determine the arrangement of electron pairs using the VSEPR model.
3. Specify the hybrid orbitals needed to accommodate the electron pairs.

It is important to do the steps in this order. For a model to be successful, it must follow nature's priorities. In the case of bonding, it seems clear that the tendency for a molecule to minimize its energy is more important than the maintenance of the characteristics of atoms as they exist in the free state. The atoms adjust to meet the "needs" of the molecule. When considering the bonding in a particular molecule, therefore, we always start with the molecule rather than the component atoms. In the molecule the electrons will be arranged to give each atom a noble gas configuration, where possible, and to minimize electron-pair repulsions. We then assume that the atoms adjust their orbitals by hybridization to allow the molecule to adopt the structure that gives the minimum energy.

In applying the localized electron model, we must remember not to overemphasize the characteristics of the separate atoms. It is not where the valence electrons originate that is important; it is where they are needed in the molecule to achieve stability. In the same vein, it is not the orbitals in the isolated atom that matter, but which orbitals the molecule requires for minimum energy.

The requirements for the various types of hybridization are summarized in Fig. 9.24 on the following page.

Interactive Example 9.5

Sign in at http://login.cengagebrain .com to try this Interactive Example in **OWL**.

Unless otherwise noted, all art on this page is © Cengage Learning 2014.

The Localized Electron Model V

For each of the following molecules or ions, predict the hybridization of each atom, and describe the molecular structure.

a. CO   **b.** BF_4^- **c.** XeF_2

Figure 9.24 | The relationship of the number of effective pairs, their spatial arrangement, and the hybrid orbital set required.

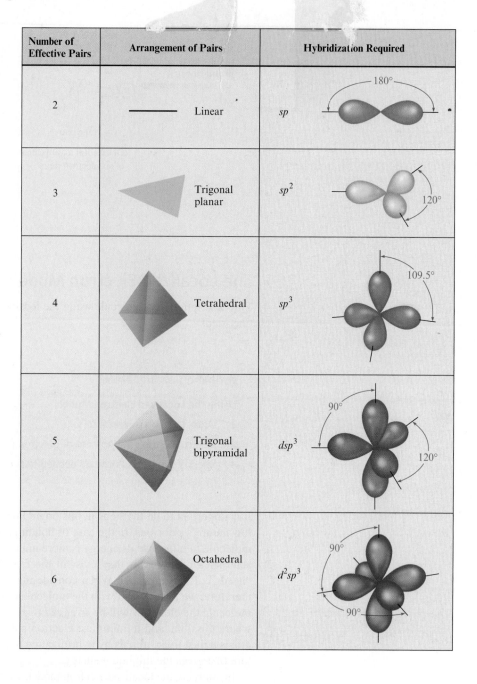

Number of Effective Pairs	Arrangement of Pairs	Hybridization Required
2	Linear	sp
3	Trigonal planar	sp^2
4	Tetrahedral	sp^3
5	Trigonal bipyramidal	dsp^3
6	Octahedral	d^2sp^3

Solution

a. The CO molecule has 10 valence electrons, and its Lewis structure is

$$: C \equiv O :$$

Each atom has two effective pairs, which means that both are sp hybridized. The triple bond consists of a σ bond produced by overlap of an sp orbital from each atom and two π bonds produced by overlap of $2p$ orbitals from each atom. The lone pairs are in sp orbitals. Since the CO molecule has only two atoms, it must be linear.

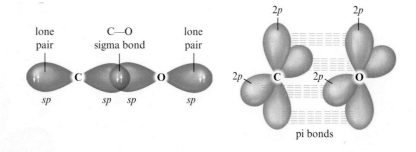

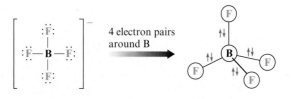

b. The BF_4^- ion has 32 valence electrons. The Lewis structure shows four pairs of electrons around the boron atom, which means a tetrahedral arrangement:

This requires sp^3 hybridization of the boron atom. Each fluorine atom also has four electron pairs and can be assumed to be sp^3 hybridized (only one sp^3 orbital is shown for each fluorine atom). The BF_4^- ion's molecular structure is tetrahedral.

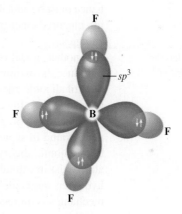

c. The XeF_2 molecule has 22 valence electrons. The Lewis structure shows five electron pairs on the xenon atom, which requires a trigonal bipyramidal arrangement:

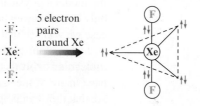

Note that the lone pairs are placed in the plane where they are 120 degrees apart. To accommodate five pairs at the vertices of a trigonal bipyramid requires that the xenon atom adopt a set of five dsp^3 orbitals. Each fluorine atom has four electron pairs and can be assumed to be sp^3 hybridized. The XeF_2 molecule has a linear arrangement of atoms.

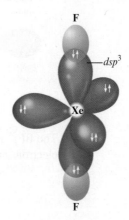

See Exercises 9.29 and 9.30

9.2 | The Molecular Orbital Model

Molecular orbital theory parallels the atomic theory discussed in Chapter 7.

We have seen that the localized electron model is of great value in interpreting the structure and bonding of molecules. However, there are some problems with this model. For example, it incorrectly assumes that electrons are localized, and so the concept of resonance must be added. Also, the model does not deal effectively with molecules containing unpaired electrons. And finally, the model gives no direct information about bond energies.

Another model often used to describe bonding is the **molecular orbital model**. To introduce the assumptions, methods, and results of this model, we will consider the simplest of all molecules, H_2, which consists of two protons and two electrons. A very stable molecule, H_2 is lower in energy than the separated hydrogen atoms by 432 kJ/mol.

Since the hydrogen molecule consists of protons and electrons, the same components found in separated hydrogen atoms, it seems reasonable to use a theory similar to the atomic theory discussed in Chapter 7, which assumes that the electrons in an atom exist in orbitals of a given energy. Can we apply this same type of model to the hydrogen molecule? Yes. In fact, describing the H_2 molecule in terms of quantum mechanics is quite straightforward.

However, even though it is formulated rather easily, this problem cannot be solved exactly. The difficulty is the same as that in dealing with polyelectronic atoms—the electron correlation problem. Since we do not know the details of the electron movements, we cannot deal with the electron–electron interactions in a specific way. We need to make approximations that allow a solution of the problem but do not destroy the model's physical integrity. The success of these approximations can be measured only by comparing predictions based on theory with experimental observations. In this case we will see that the simplified model works very well.

Just as atomic orbitals are solutions to the quantum mechanical treatment of atoms, **molecular orbitals (MOs)** are solutions to the molecular problem. Molecular orbitals have many of the same characteristics as atomic orbitals. Two of the most important are that they can hold two electrons with opposite spins and that the square of the molecular orbital wave function indicates electron probability.

We will now describe the bonding in the hydrogen molecule using this model. The first step is to obtain the hydrogen molecule's orbitals, a process that is greatly simplified if we assume that the molecular orbitals can be constructed from the hydrogen $1s$ atomic orbitals.

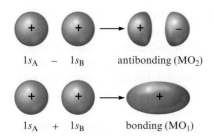

Figure 9.25 | The combination of hydrogen 1s atomic orbitals to form MOs. The phases of the orbitals are shown by signs inside the boundary surfaces. When the orbitals are added, the matching phases produce constructive interference, which gives enhanced electron probability between the nuclei. This results in a bonding molecular orbital. When one orbital is subtracted from the other, destructive interference occurs between the opposite phases, leading to a node between the nuclei. This is an antibonding MO.

When the quantum mechanical equations for the hydrogen molecule are solved, two molecular orbitals result, which can be represented as

$$MO_1 = 1s_A + 1s_B$$
$$MO_2 = 1s_A - 1s_B$$

where $1s_A$ and $1s_B$ represent the $1s$ orbitals from the two separated hydrogen atoms. This process is shown schematically in Fig. 9.25.

The orbital properties of most interest are size, shape (described by the electron probability distribution), and energy. These properties for the hydrogen molecular orbitals are represented in Fig. 9.26. From Fig. 9.26 we can note several important points:

1. The electron probability of both molecular orbitals is centered along the line passing through the two nuclei. For MO_1 the greatest electron probability is *between* the nuclei, and for MO_2 it is on *either side* of the nuclei. This type of electron distribution is described as *sigma* (σ), as in the localized electron model. Accordingly, we refer to MO_1 and MO_2 as **sigma (σ) molecular orbitals**.

2. In the molecule only the molecular orbitals are available for occupation by electrons. The $1s$ atomic orbitals of the hydrogen atoms no longer exist, because the H_2 molecule—a new entity—has its own set of new orbitals.

3. MO_1 is lower in energy than the $1s$ orbitals of free hydrogen atoms, while MO_2 is higher in energy than the $1s$ orbitals. This fact has very important implications for the stability of the H_2 molecule, since if the two electrons (one from each hydrogen atom) occupy the lower-energy MO_1, they will have lower energy than they do in the two separate hydrogen atoms. This situation favors molecule formation, because nature tends to seek the lowest energy state. That is, the driving force for molecule formation is that the molecular orbital available to the two electrons has lower energy than the atomic orbitals these electrons occupy in the separated atoms. This situation is favorable to bonding, or *probonding*.

On the other hand, if the two electrons were forced to occupy the higher-energy MO_2, they would be definitely *antibonding*. In this case, these electrons would have lower energy in the separated atoms than in the molecule, and the separated state would be favored. Of course, since the lower-energy MO_1 *is* available, the two electrons occupy that MO and the molecule is stable.

We have seen that the molecular orbitals of the hydrogen molecule fall into two classes: bonding and antibonding. A **bonding molecular orbital** is *lower in energy than the atomic orbitals of which it is composed.* Electrons in this type of orbital will favor the molecule; that is, they will favor bonding. An **antibonding molecular orbital** is *higher in energy than the atomic orbitals of which it is composed.* Electrons in this type of orbital will favor the separated atoms (they are antibonding). Figure 9.27 illustrates these ideas.

4. Figure 9.26 shows that for the bonding molecular orbital in the H_2 molecule the electrons have the greatest probability of being between the nuclei. This is exactly what we would expect, since the electrons can lower their energies by being

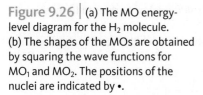

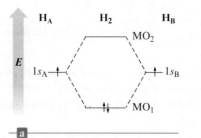

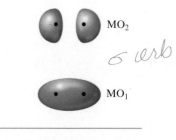

Figure 9.26 | (a) The MO energy-level diagram for the H_2 molecule. (b) The shapes of the MOs are obtained by squaring the wave functions for MO_1 and MO_2. The positions of the nuclei are indicated by •.

Energy diagram

Electron probability distribution

simultaneously attracted by both nuclei. On the other hand, the electron distribution for the antibonding molecular orbital is such that the electrons are mainly outside the space between the nuclei. This type of distribution is not expected to provide any bonding force. In fact, it causes the electrons to be higher in energy than in the separated atoms. Thus the molecular orbital model produces electron distributions and energies that agree with our basic ideas of bonding. This fact reassures us that the model is physically reasonable.

5. The labels on molecular orbitals indicate their symmetry (shape), the parent atomic orbitals, and whether they are bonding or antibonding. Antibonding character is indicated by an asterisk. For the H_2 molecule, both MOs have σ symmetry, and both are constructed from hydrogen $1s$ atomic orbitals. The molecular orbitals for H_2 are therefore labeled as follows:

$$MO_1 = \sigma_{1s}$$
$$MO_2 = \sigma_{1s}{}^*$$

6. Molecular electron configurations can be written in much the same way as atomic (electron) configurations. Since the H_2 molecule has two electrons in the σ_{1s} molecular orbital, the electron configuration is $\sigma_{1s}{}^2$.

7. Each molecular orbital can hold two electrons, but the spins must be opposite.

8. Orbitals are conserved. The number of molecular orbitals will always be the same as the number of atomic orbitals used to construct them.

Many of the above points are summarized in Fig. 9.28.

Now suppose we could form the $H_2{}^-$ ion from a hydride ion (H^-) and a hydrogen atom. Would this species be stable? Since the H^- ion has the configuration $1s^2$ and the H atom has a $1s^1$ configuration, we will use $1s$ atomic orbitals to construct the MO diagram for the $H_2{}^-$ ion (Fig. 9.29). The electron configuration for $H_2{}^-$ is $(\sigma_{1s})^2(\sigma_{1s}{}^*)^1$.

The key idea is that the $H_2{}^-$ ion will be stable if it has a lower energy than its separated parts. From Fig. 9.29 we see that in going from the separated H^- ion and H atom to the $H_2{}^-$ ion, the model predicts that two electrons are lowered in energy and one electron is raised in energy. In other words, two electrons are bonding and one electron is antibonding. Since more electrons favor bonding, $H_2{}^-$ is predicted to be a stable entity—a bond has formed. But how would we expect the bond strengths in the molecules of H_2 and $H_2{}^-$ to compare?

In the formation of the H_2 molecule, two electrons are lowered in energy and no electrons are raised in energy compared with the parent atoms. When $H_2{}^-$ is formed, two electrons are lowered in energy and one is raised, producing *a net lowering of the energy of only one electron*. Thus the model predicts that H_2 is *twice as stable* as $H_2{}^-$ with respect to their separated components. In other words, the bond in the H_2 molecule is predicted to be about twice as strong as the bond in the $H_2{}^-$ ion.

Bond Order

To indicate bond strength, we use the concept of bond order. **Bond order** is *the difference between the number of bonding electrons and the number of antibonding electrons divided by 2.*

$$\text{Bond order} = \frac{\text{number of bonding electrons} - \text{number of antibonding electrons}}{2}$$

We divide by 2 because, from the localized electron model, we are used to thinking of bonds in terms of *pairs* of electrons.

Since the H_2 molecule has two bonding electrons and no antibonding electrons, the bond order is

$$\text{Bond order} = \frac{2 - 0}{2} = 1$$

Bonding will result if the molecule has lower energy than the separated atoms.

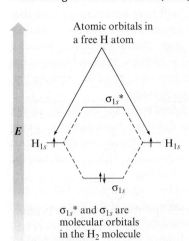

Figure 9.27 | Bonding and antibonding molecular orbitals (MOs).

$\sigma_{1s}{}^*$ and σ_{1s} are molecular orbitals in the H_2 molecule

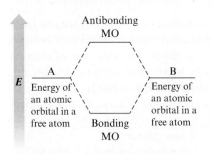

Figure 9.28 | A molecular orbital energy-level diagram for the H_2 molecule.

Although the model predicts that $H_2{}^-$ should be stable, this ion has never been observed, again emphasizing the perils of simple models.

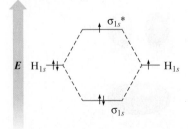

Figure 9.29 | The molecular orbital energy-level diagram for the $H_2{}^-$ ion.

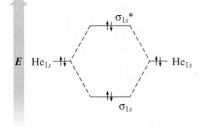

Figure 9.30 | The molecular orbital energy-level diagram for the He₂ molecule.

The H₂⁻ ion has two bonding electrons and one antibonding electron; the bond order is

$$\text{Bond order} = \frac{2-1}{2} = \frac{1}{2}$$

Bond order is an indication of bond strength because it reflects the difference between the number of bonding electrons and the number of antibonding electrons. *Larger bond order means greater bond strength.*

We will now apply the molecular orbital model to the helium molecule (He₂). Does this model predict that this molecule will be stable? Since the He atom has a $1s^2$ configuration, $1s$ orbitals are used to construct the molecular orbitals, and the molecule will have four electrons. From the diagram shown in Fig. 9.30, it is apparent that two electrons are raised in energy and two are lowered in energy. Thus the bond order is zero:

$$\frac{2-2}{2} = 0$$

This implies that the He₂ molecule is *not* stable with respect to the two free He atoms, which agrees with the observation that helium gas consists of individual He atoms.

9.3 | Bonding in Homonuclear Diatomic Molecules

In this section we consider *homonuclear diatomic molecules* (those composed of two identical atoms) of elements in Period 2 of the periodic table. Since the lithium atom has a $1s^2 2s^1$ electron configuration, it would seem that we should use the Li $1s$ and $2s$ orbitals to form the molecular orbitals of the Li₂ molecule. However, the $1s$ orbitals on the lithium atoms are much smaller than the $2s$ orbitals and therefore do not overlap in space to any appreciable extent (Fig. 9.31). Thus the two electrons in each $1s$ orbital can be assumed to be localized and not to participate in the bonding. *To participate in molecular orbitals, atomic orbitals must overlap in space.* This means that only the valence orbitals of the atoms contribute significantly to the molecular orbitals of a particular molecule.

The molecular orbital diagram of the Li₂ molecule and the shapes of its bonding and antibonding MOs are shown in Fig. 9.32. The electron configuration for Li₂ (valence electrons only) is σ_{2s}^2, and the bond order is

$$\frac{2-0}{2} = 1$$

Li₂ is a stable molecule (has lower energy than two separated lithium atoms). However, this does not mean that Li₂ is the most stable form of elemental lithium. In fact, at normal temperature and pressure, lithium exists as a solid containing many lithium atoms bound to each other.

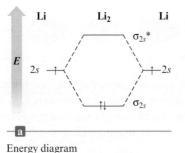

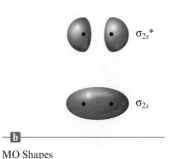

Figure 9.32 | The molecular orbital energy-level diagram for the Li₂ molecule.

Figure 9.31 | The relative sizes of the lithium $1s$ and $2s$ atomic orbitals in Li₂.

Figure 9.33 | (a) The three mutually perpendicular 2p orbitals on two adjacent boron atoms. The signs indicate the orbital phases. Two pairs of parallel p orbitals can overlap, as shown in (b) and (c), and the third pair can overlap head-on, as shown in (d).

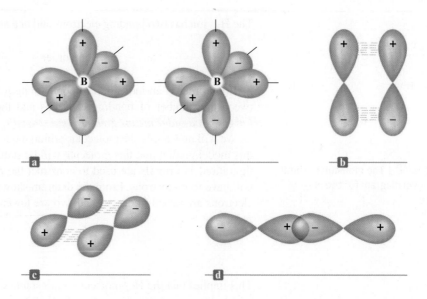

Beryllium metal.

For the beryllium molecule (Be_2), the bonding and antibonding orbitals both contain two electrons. In this case the bond order is $(2 - 2)/2 = 0$, and since Be_2 is not more stable than two separated Be atoms, no molecule forms. However, beryllium metal contains many beryllium atoms bonded to each other and is stable for reasons we will discuss in Chapter 10.

Since the boron atom has a $1s^2 2s^2 2p^1$ configuration, we describe the B_2 molecule by considering how p atomic orbitals combine to form molecular orbitals. Recall that p orbitals have two lobes and that they occur in sets of three mutually perpendicular orbitals [Fig. 9.33(a)]. When two B atoms approach each other, two pairs of p orbitals can overlap in a parallel fashion [Fig. 9.33(b) and (c)] and one pair can overlap head-on [Fig. 9.33(d)].

First, let's consider the molecular orbitals from the head-on overlap. The bonding orbital is formed by reversing the sign of the right orbital so the positive phases of both orbitals match between the nuclei to produce constructive interference. This leads to enhanced electron probability between the nuclei. The antibonding orbital is formed by the direct combination of the orbitals, which gives destructive interference of the positive phase of one orbital with the negative phase of the second orbital. This produces a node between the nuclei, which gives decreased electron probability. Note that the electrons in the bonding MO are, as expected, concentrated between the nuclei, and the electrons in the antibonding MO are concentrated outside the area between the two nuclei. Also, both these MOs are σ molecular orbitals.

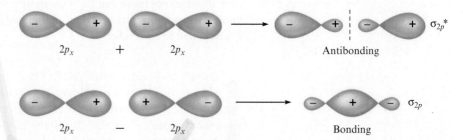

When the parallel p orbitals are combined with the positive and negative phases matched, constructive interference occurs, giving a bonding π orbital. When the orbitals have opposite phases (the signs of one orbital are reversed), destructive interference occurs, resulting in an antibonding π orbital.

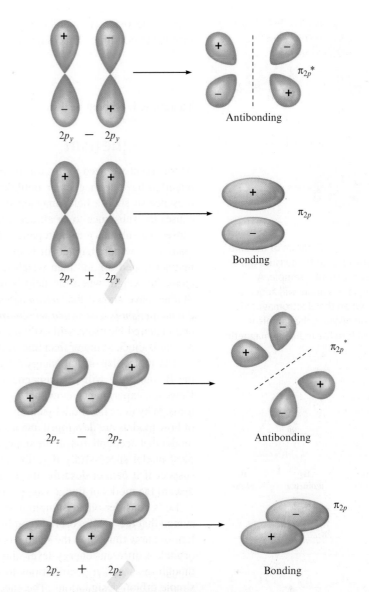

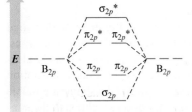

Figure 9.34 | The *expected* molecular orbital energy-level diagram resulting from the combination of the $2p$ orbitals on two boron atoms.

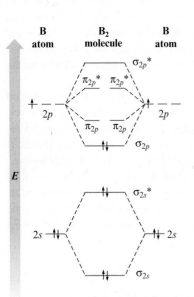

Figure 9.35 | The *expected* molecular orbital energy-level diagram for the B_2 molecule.

Since the electron probability lies above and below the line between the nuclei, both the orbitals are **pi (π) molecular orbitals**. They are designated as π_{2p} for the bonding MO and π_{2p}^* for the antibonding MO. A similar set of π molecular orbitals is formed from overlap of the parallel p_z atomic orbitals.

Let's try to make an educated guess about the relative energies of the σ and π molecular orbitals formed from the $2p$ atomic orbitals. Would we expect the electrons to prefer the σ bonding orbital (where the electron probability is concentrated in the area between the nuclei) or the π bonding orbital? The σ orbital would seem to have the lower energy, since the electrons are closest to the two nuclei. This agrees with the observation that σ interactions are stronger than π interactions.

Figure 9.34 gives the molecular orbital energy-level diagram *expected* when the two sets of $2p$ orbitals on the boron atoms combine to form molecular orbitals. Note that there are two π bonding orbitals at the same energy (degenerate orbitals) formed from the two pairs of parallel p orbitals, and there are two degenerate π antibonding orbitals. The energy of the π_{2p} orbitals is expected to be higher than that of the σ_{2p} orbital because σ interactions are generally stronger than π interactions.

To construct the total molecular orbital diagram for the B_2 molecule, we make the assumption that the $2s$ and $2p$ orbitals combine separately (in other words, there is no $2s$–$2p$ mixing). The resulting diagram is shown in Fig. 9.35. Note that B_2 has six

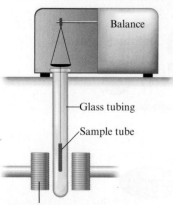

Figure 9.36 | Diagram of the kind of apparatus used to measure the paramagnetism of a sample. A paramagnetic sample will appear heavier when the electromagnet is turned on because the sample is attracted into the inducing magnetic field.

valence electrons. (Remember the 1s orbitals and electrons are assumed not to participate in the bonding.) This diagram predicts the bond order:

$$\frac{4-2}{2} = 1$$

Therefore, B_2 should be a stable molecule.

Paramagnetism

At this point we need to discuss an additional molecular property—magnetism. Most materials have no magnetism until they are placed in a magnetic field. However, in the presence of such a field, magnetism of two types can be induced. **Paramagnetism** causes the substance to be attracted into the inducing magnetic field. **Diamagnetism** causes the substance to be repelled from the inducing magnetic field. Figure 9.36 illustrates how paramagnetism is measured. The sample is weighed with the electromagnet turned off and then weighed again with the electromagnet turned on. An increase in weight when the field is turned on indicates the sample is paramagnetic. Studies have shown that *paramagnetism is associated with unpaired electrons and diamagnetism is associated with paired electrons.* Any substance that has both paired and unpaired electrons will exhibit a net paramagnetism, since the effect of paramagnetism is much stronger than that of diamagnetism.

The molecular orbital energy-level diagram represented in Fig. 9.35 predicts that the B_2 molecule will be diamagnetic, since the MOs contain only paired electrons. However, experiments show that B_2 is actually paramagnetic with two unpaired electrons. Why does the model yield the wrong prediction? This is yet another illustration of how models are developed and used. In general, we try to use the simplest possible model that accounts for all the important observations. In this case, although the simplest model successfully describes the diatomic molecules up to B_2, it certainly is suspect if it cannot describe the B_2 molecule correctly. This means we must either discard the model or find a way to modify it.

Let's consider one assumption that we made. In our treatment of B_2, we have assumed that the s and p orbitals combine separately to form molecular orbitals. Calculations show that when the s and p orbitals are allowed to mix in the same molecular orbital, a different energy-level diagram results for B_2 (Fig. 9.37). Note that even though the s and p contributions to the MOs are no longer separate, we retain the simple orbital designations. The energies of π_{2p} and σ_{2p} orbitals are reversed by p–s mixing, and the σ_{2s} and the σ_{2s}* orbitals are no longer equally spaced relative to the energy of the free 2s orbital.

When the six valence electrons for the B_2 molecule are placed in the modified energy-level diagram, each of the last two electrons goes into one of the degenerate π_{2p} orbitals. This produces a paramagnetic molecule in agreement with experimental results. Thus when the model is extended to allow p–s mixing in molecular orbitals, it predicts the correct magnetism. Note that the bond order is $(4-2)/2 = 1$, as before.

The remaining diatomic molecules of the elements in Period 2 can be described using similar ideas. For example, the C_2 and N_2 molecules use the same set of orbitals as for B_2 (see Fig. 9.37). Because the importance of 2s–2p mixing decreases across the period, the σ_{2p} and π_{2p} orbitals revert to the order expected in the absence of 2s–2p mixing for the molecules O_2 and F_2 (Fig. 9.38).

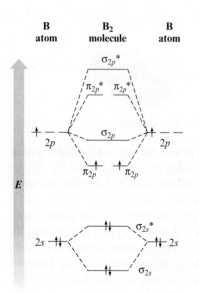

Figure 9.37 | The *correct* molecular orbital energy-level diagram for the B_2 molecule. When p–s mixing is allowed, the energies of the σ_{2p} and π_{2p} orbitals are reversed. The two electrons from the B 2p orbitals now occupy separate, degenerate π_{2p} molecular orbitals and thus have parallel spins. Therefore, this diagram explains the observed paramagnetism of B_2.

Critical Thinking

What if π_{2p} orbitals were lower in energy than σ_{2p} orbitals? What would you *expect* the B_2 molecular orbital energy-level diagram to look like (without considering p–s mixing)? Compare your expected diagram to Figs. 9.34 and 9.35, and state the differences from each.

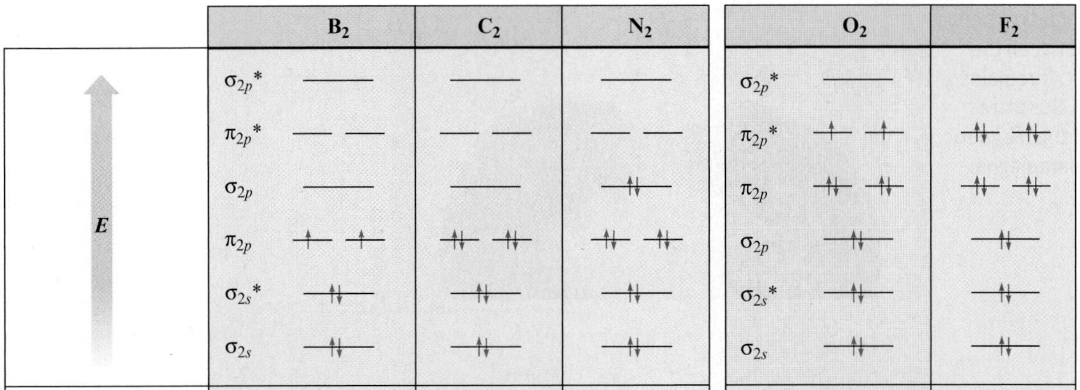

		B_2	C_2	N_2		O_2	F_2
	σ_{2p}^*	──	──	──	σ_{2p}^* ──	──
	π_{2p}^*	── ──	── ──	── ──	π_{2p}^* ↑ ↑	↑↓ ↑↓
	σ_{2p}	──	──	↑↓	π_{2p} ↑↓ ↑↓	↑↓ ↑↓
E	π_{2p}	↑ ↑	↑↓ ↑↓	↑↓ ↑↓	σ_{2p} ↑↓	↑↓
	σ_{2s}^*	↑↓	↑↓	↑↓	σ_{2s}^* ↑↓	↑↓
	σ_{2s}	↑↓	↑↓	↑↓	σ_{2s} ↑↓	↑↓
Magnetism		Paramagnetic	Diamagnetic	Diamagnetic	Paramagnetic	Diamagnetic
Bond order		1	2	3	2	1
Observed bond dissociation energy (kJ/mol)		290	620	942	495	154
Observed bond length (pm)		159	131	110	121	143

Figure 9.38 | The molecular orbital energy-level diagrams, bond orders, bond energies, and bond lengths for the diatomic molecules B_2 through F_2. Note that for O_2 and F_2 the σ_{2p} orbital is lower in energy than the π_{2p} orbitals.

Several significant points arise from the orbital diagrams, bond strengths, and bond lengths summarized in Fig. 9.38 for the Period 2 diatomics:

1. There are definite correlations between bond order, bond energy, and bond length. As the bond order predicted by the molecular orbital model increases, the bond energy increases and the bond length decreases. This is a clear indication that the bond order predicted by the model accurately reflects bond strength, and it strongly supports the reasonableness of the MO model.
2. Comparison of the bond energies of the B_2 and F_2 molecules indicates that bond order cannot automatically be associated with a particular bond energy. Although both molecules have a bond order of 1, the bond in B_2 appears to be about twice as strong as the bond in F_2. As we will see in our later discussion of the halogens, F_2 has an unusually weak single bond due to larger than usual electron–electron repulsions (there are 14 valence electrons on the small F_2 molecule).
3. Note the very large bond energy associated with the N_2 molecule, which the molecular orbital model predicts will have a bond order of 3, a triple bond. The very strong bond in N_2 is the principal reason that many nitrogen-containing compounds are used as high explosives. The reactions involving these explosives give the very stable N_2 molecule as a product, thus releasing large quantities of energy.
4. The O_2 molecule is known to be paramagnetic. This can be very convincingly demonstrated by pouring liquid oxygen between the poles of a strong magnet (Fig. 9.39). The oxygen remains there until it evaporates. Significantly, the molecular orbital model correctly predicts oxygen's paramagnetism, while the localized electron model predicts a diamagnetic molecule.

Figure 9.39 | When liquid oxygen is poured into the space between the poles of a strong magnet, it remains there until it boils away. This attraction of liquid oxygen for the magnetic field demonstrates the paramagnetism of the O_2 molecule.

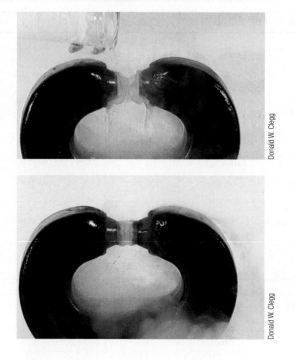

Donald W. Clegg

Donald W. Clegg

The Molecular Orbital Model I

For the species O_2, O_2^+, and O_2^-, give the electron configuration and the bond order for each. Which has the strongest bond?

Solution

The O_2 molecule has 12 valence electrons (6 + 6); O_2^+ has 11 valence electrons (6 + 6 − 1); and O_2^- has 13 valence electrons (6 + 6 + 1). We will assume that the ions can be treated using the same molecular orbital diagram as for the neutral diatomic molecule:

	O_2	O_2^+	O_2^-
σ_{2p}^*	—	—	—
π_{2p}^*	↑ ↑	↑ —	↑↓ ↑
π_{2p}	↑↓ ↑↓	↑↓ ↑↓	↑↓ ↑↓
σ_{2p}	↑↓	↑↓	↑↓
σ_{2s}^*	↑↓	↑↓	↑↓
σ_{2s}	↑↓	↑↓	↑↓

The electron configuration for each species can then be taken from the diagram:

$$O_2: \quad (\sigma_{2s})^2(\sigma_{2s}^*)^2(\sigma_{2p})^2(\pi_{2p})^4(\pi_{2p}^*)^2$$
$$O_2^+: \quad (\sigma_{2s})^2(\sigma_{2s}^*)^2(\sigma_{2p})^2(\pi_{2p})^4(\pi_{2p}^*)^1$$
$$O_2^-: \quad (\sigma_{2s})^2(\sigma_{2s}^*)^2(\sigma_{2p})^2(\pi_{2p})^4(\pi_{2p}^*)^3$$

The bond orders are:

$$\text{For } O_2: \quad \frac{8-4}{2} = 2$$

$$\text{For } O_2^+: \quad \frac{8-3}{2} = 2.5$$

$$\text{For } O_2^-: \quad \frac{8-5}{2} = 1.5$$

Thus O_2^+ is expected to have the strongest bond of the three species.

See Exercises 9.47 and 9.48

Interactive Example 9.7

Sign in at http://login.cengagebrain.com to try this Interactive Example in **OWL**.

The Molecular Orbital Model II

Use the molecular orbital model to predict the bond order and magnetism of each of the following molecules.

a. Ne_2 **b.** P_2

Solution

a. The valence orbitals for Ne are $2s$ and $2p$. Thus we can use the molecular orbitals we have already constructed for the diatomic molecules of the Period 2 elements. The Ne_2 molecule has 16 valence electrons (8 from each atom). Placing these electrons in the appropriate molecular orbitals produces the following diagram:

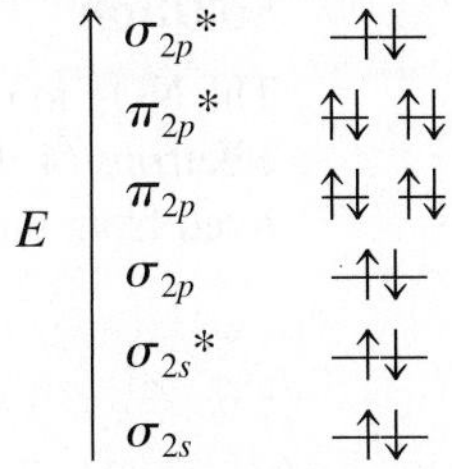

The bond order is $(8-8)/2 = 0$, and Ne_2 does not exist.

b. The P_2 molecule contains phosphorus atoms from the third row of the periodic table. We will assume that the diatomic molecules of the Period 3 elements can be treated in a way very similar to that which we have used so far. Thus we will draw the MO diagram for P_2 analogous to that for N_2. The only change will be that the molecular orbitals will be formed from $3s$ and $3p$ atomic orbitals. The P_2 model has 10 valence electrons (5 from each phosphorus atom). The resulting molecular orbital diagram is

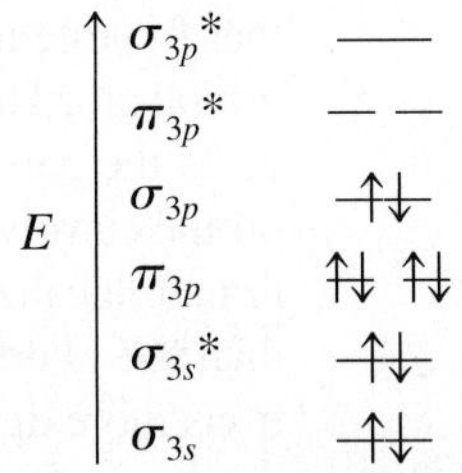

The molecule has a bond order of 3 and is expected to be diamagnetic.

See Exercises 9.45 through 9.48, and 9.50

9.4 | Bonding in Heteronuclear Diatomic Molecules

In this section we will deal with selected examples of **heteronuclear** (different atoms) **diatomic molecules**. A special case involves molecules containing atoms adjacent to each other in the periodic table. Since the atoms involved in such a molecule are so similar, we can use the molecular orbital diagram for homonuclear molecules. For example, we can predict the bond order and magnetism of nitric oxide (NO) by placing its 11 valence electrons (5 from nitrogen and 6 from oxygen) in the molecular orbital energy-level diagram shown in Fig. 9.40. The molecule should be paramagnetic and has a bond order of

$$\frac{8 - 3}{2} = 2.5$$

Experimentally, nitric oxide is indeed found to be paramagnetic. Notice that this odd-electron molecule is described very naturally by the MO model. In contrast, the localized electron model, in the simple form used in this text, cannot be used readily to treat such molecules.

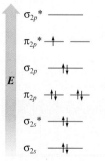

Figure 9.40 | The molecular orbital energy-level diagram for the NO molecule. We assume that orbital order is the same as that for N_2. The bond order is 2.5.

The Molecular Orbital Model III

Use the molecular orbital model to predict the magnetism and bond order of the NO^+ and CN^- ions.

Solution

The NO^+ ion has 10 valence electrons ($5 + 6 - 1$). The CN^- ion also has 10 valence electrons ($4 + 5 + 1$). Both ions are therefore diamagnetic and have a bond order derived from the equation

$$\frac{8 - 2}{2} = 3$$

The molecular orbital diagram for these two ions is the same (Fig. 9.41).

See Exercises 9.51 and 9.52

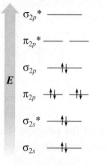

Figure 9.41 | The molecular orbital energy-level diagram for both the NO^+ and CN^- ions.

When the two atoms of a diatomic molecule are very different, the energy-level diagram for homonuclear molecules can no longer be used. A new diagram must be devised for each molecule. We will illustrate this case by considering the hydrogen fluoride (HF) molecule. The electron configurations of the hydrogen and fluorine atoms are $1s^1$ and $1s^2 2s^2 2p^5$, respectively. To keep things as simple as possible, we will assume that fluorine uses only one of its $2p$ orbitals to bond to hydrogen. Thus the molecular orbitals for HF will be composed of fluorine $2p$ and hydrogen $1s$ orbitals. Figure 9.42 gives the partial molecular orbital energy-level diagram for HF, focusing only on the orbitals involved in the bonding. We are assuming that fluorine's other valence electrons remain localized on the fluorine. The $2p$ orbital of fluorine is shown at a lower energy than the $1s$ orbital of hydrogen on the diagram because fluorine binds its valence electrons more tightly. Thus the $2p$ electron on a free fluorine atom is at lower energy than the $1s$ electron on a free hydrogen atom. The diagram predicts that the HF molecule should be stable because both electrons are lowered in energy relative to their energy in the free hydrogen and fluorine atoms, which is the driving force for bond formation.

Because the fluorine $2p$ orbital is lower in energy than the hydrogen $1s$ orbital, the electrons prefer to be closer to the fluorine atom. That is, the σ molecular orbital containing the bonding electron pair shows greater electron probability close to the

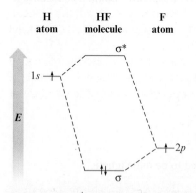

Figure 9.42 | A partial molecular orbital energy-level diagram for the HF molecule.

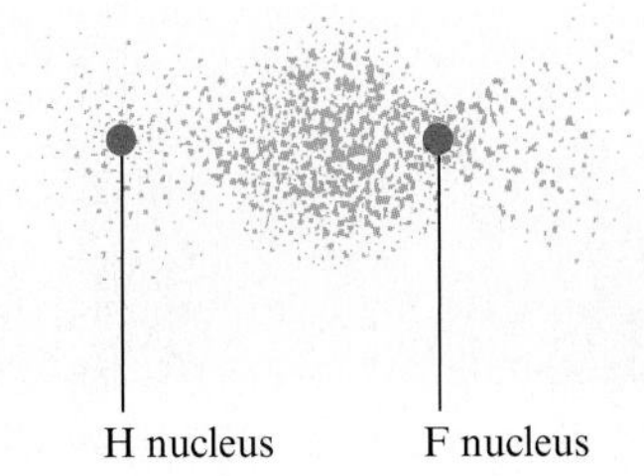

Figure 9.43 | The electron probability distribution in the bonding molecular orbital of the HF molecule. Note the greater electron density close to the fluorine atom.

H nucleus   F nucleus

fluorine (Fig. 9.43). The electron pair is not shared equally. This causes the fluorine atom to have a slight excess of negative charge and leaves the hydrogen atom partially positive. This is *exactly* the bond polarity observed for HF. Thus the molecular orbital model accounts in a straightforward way for the different electronegativities of hydrogen and fluorine and the resulting unequal charge distribution.

9.5 | Combining the Localized Electron and Molecular Orbital Models

One of the main difficulties with the localized electron model is its assumption that electrons are localized. This problem is most apparent with molecules for which several valid Lewis structures can be drawn. It is clear that none of these structures taken alone adequately describes the electronic structure of the molecule. The concept of resonance was invented to solve this problem. However, even with resonance included, the localized electron model does not describe molecules and ions such as O_3 and NO_3^- in a very satisfying way.

It would seem that the ideal bonding model would be one with the simplicity of the localized electron model but with the delocalization characteristic of the molecular orbital model. We can achieve this by combining the two models to describe molecules that require resonance. Note that for species such as O_3 and NO_3^- the double bond changes position in the resonance structures (Fig. 9.44). Since a double bond involves one σ and one π bond, there is a σ bond between all bound atoms in each resonance structure. It is really the π bond that has different locations in the various resonance structures.

Therefore, we conclude that the σ bonds in a molecule can be described as being localized with no apparent problems. It is the π bonding that must be treated as being delocalized. Thus for molecules that require resonance, we will use the localized electron model to describe the σ bonding and the molecular orbital model to describe the π bonding. This allows us to keep the bonding model as simple as possible and yet give a more physically accurate description of such molecules.

We will illustrate the general method by considering the bonding in benzene, an important industrial chemical that must be handled carefully because it is a known carcinogen. The benzene molecule (C_6H_6) consists of a planar hexagon of carbon atoms with one hydrogen atom bound to each carbon atom [Fig. 9.45(a)]. In the molecule all six C—C bonds are known to be equivalent. To explain this fact, the localized electron model must invoke resonance [Fig. 9.45(b)].

A better description of the bonding in benzene results when we use a combination of the models, as described above. In this description it is assumed that the σ bonds of carbon involve sp^2 orbitals (Fig. 9.46). These σ bonds are all centered in the plane of the molecule.

In molecules that require resonance, it is the π bonding that is most clearly delocalized.

Figure 9.44 | The resonance structures for O_3 and NO_3^-. Note that it is the double bond that occupies various positions in the resonance structures.

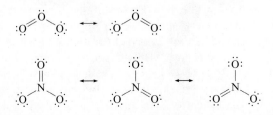

Chemical connections
What's Hot?

One of the best things about New Mexico is the food. Authentic New Mexican cuisine uses liberal amounts of green and red chilies—often called *chili peppers*. Chilies apparently originated in parts of South America and were spread north by birds. When Columbus came to North America, which he originally thought was India, he observed the natives using chilies for spicing foods. When he took chilies back to Europe, Columbus mistakenly called them peppers and the name stuck.

The spicy payload of chilies is delivered mainly by the chemical capsaicin, which has the following structure:

Capsaicin was isolated as a pure substance by L. T. Thresh in 1846. Since then substituted capsaicins have also been found in chilies. The spicy power of chilies derives mostly from capsaicin and dihydrocapsaicin.

The man best known for explaining the "heat" of chilies is Wilbur Scoville, who defined the Scoville unit for measuring chili power. He arbitrarily established the hotness of pure capsaicin as 16 million. On this scale a typical green or red chili has a rating of

about 2500 Scoville units. You may have had an encounter with habanero chilies that left you looking for a firehose to put out the blaze in your mouth—habaneros have a Scoville rating of about 500,000!

Capsaicin has found many uses outside of cooking. It is used in pepper sprays and repellant sprays for many garden pests, although birds are unaffected by capsaicin. Capsaicin also stimulates the body's circulation and causes pain receptors to release endorphins, similar to the effect produced by intense exercise. Instead of jogging you may want to sit on the couch eating chilies. Either way you are going to sweat.

Figure 9.45 | (a) The benzene molecule consists of a ring of six carbon atoms with one hydrogen atom bound to each carbon; all atoms are in the same plane. All the C—C bonds are known to be equivalent. (b) Two of the resonance structures for the benzene molecule. The localized electron model must invoke resonance to account for the six equal C—C bonds.

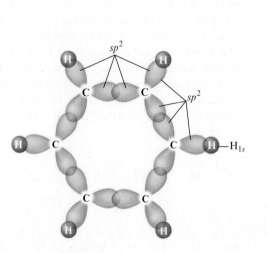

Figure 9.46 | The σ bonding system in the benzene molecule.

Figure 9.47 | (a) The π molecular orbital system in benzene is formed by combining the six p orbitals from the six sp^2 hybridized carbon atoms. (b) The electrons in the resulting π molecular orbitals are delocalized over the entire ring of carbon atoms, giving six equivalent bonds. A composite of these orbitals is represented here.

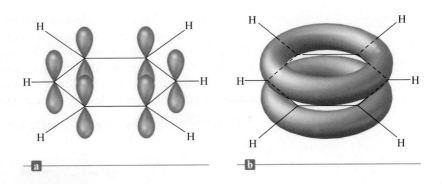

Figure 9.48 | (a) The p orbitals used to form the π bonding system in the NO_3^- ion. (b) A representation of the delocalization of the electrons in the π molecular orbital system of the NO_3^- ion.

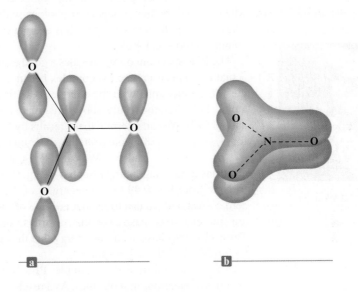

Since each carbon atom is sp^2 hybridized, a p orbital perpendicular to the plane of the ring remains on each carbon atom. These six p orbitals can be used to form π molecular orbitals [Fig. 9.47(a)]. The electrons in the resulting π molecular orbitals are delocalized above and below the plane of the ring [Fig. 9.47(b)]. This gives six equivalent C—C bonds, as required by the known structure of the benzene molecule. The benzene structure is often written as

to indicate the **delocalized π bonding** in the molecule.

Very similar treatments can be applied to other planar molecules for which resonance is required by the localized electron model. For example, the NO_3^- ion can be described using the π molecular orbital system shown in Fig. 9.48. In this molecule each atom is assumed to be sp^2 hybridized, which leaves one p orbital on each atom perpendicular to the plane of the ion. These p orbitals can combine to form the π molecular orbital system.

9.6 | Photoelectron Spectroscopy (PES)

Recall from our discussion of the photoelectric effect (see Section 7.2) that electrons can be ejected from a metal by shining light of sufficient energy on its surface. By gradually increasing the energy of the light used we can determine the exact energy of the photon required to eject an electron and thus determine the binding energy of the electrons in the solid metal. A similar technique, called photoelectron spectroscopy (PES), can be used to determine the relative energies of electrons in individual atoms as well as in molecules. In this technique, high-energy photons are directed at the sample, and the kinetic energies of the ejected electrons are measured. From this information we can determine the energy of the electron in the atom or molecule since:

Energy of electron = energy of photons used − kinetic energy of the electron

or

$$E_{electron} = h\nu - \text{KE}$$

Helium gas usually is used as the source of high-energy photons in PES because, when excited, helium atoms emit photons with a wavelength of 58.4 nm (21.2 eV). These very-high-energy photons are energetic enough to eject electrons from many common atoms and molecules.

The PES spectrum of N_2 provides a good example of how the technique works. The actual PES spectrum itself is complex because of the coupling of the ionization energies of the electrons with the energies resulting from vibrations of the two nitrogen atoms. However, the results show three main areas of energy absorption at 15.6, 16.7, and 18.6 eV. We can interpret these results by looking at the molecular orbital diagram for N_2 (see Fig. 9.38). Note from this diagram that N_2 has five filled molecular orbitals: σ_{2s}, σ_{2s}^*, two π_{2p}, and σ_{2p}. We can represent the energies required to remove these electrons in the diagram represented in Fig. 9.49.

Note from Fig. 9.49 that the energies of the bound electrons are negative and that zero corresponds to electrons that are free of the influence of the two N nuclei—they are free electrons. Also note that no transition is represented for the σ_{2s} electrons. These electrons have such low energy that they require more than the 21.2 eV that can be furnished by the helium source.

As we can see from this example, PES furnishes valuable information about the energies of electrons in molecules. As a result, it is a very useful tool for characterizing and testing our theories of bonding in molecules.

PES is also useful for studying the electron energy levels of atoms. Usually PES is used to analyze for the presence of specific elements in samples by identifying known binding energies. For example, the O_{1s} electrons occur in the PES spectrum at 530 eV and C_{1s} appears at 285 eV. However, we can imagine an idealized schematic PES spectrum of an element such as phosphorus shown in Fig. 9.50. Because the actual spectrum is very complex and hard for a non-expert to interpret, we have represented the data from the PES spectrum in a very simple manner. Note that the area of each peak is proportional to the number of electrons that have that particular energy.

A researcher using a photoelectron spectrometer.

Andrew Brookes, National Physical Laboratory, SPL/Photo Researchers, Inc.

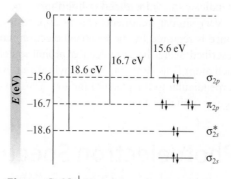

Figure 9.49 | Representation of the removal of various electrons from N_2. The diagram is not drawn to scale.

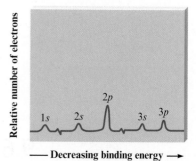

Figure 9.50 | The idealized PES spectrum of phosphorus.

For review

Key terms

Two widely used bonding models

> Localized electron model
> Molecular orbital model

Localized electron model

> Molecule is pictured as a group of atoms sharing electron pairs between atomic orbitals
> Hybrid orbitals, which are combinations of the "native" atomic orbitals, are often required to account for the molecular structure
>> Six electron pairs (octahedral arrangement) require d^2sp^3 orbitals
>> Five electron pairs (trigonal bipyramidal arrangement) require dsp^3 orbitals
>> Four electron pairs (tetrahedral arrangement) require sp^3 orbitals
>> Three electron pairs (trigonal planar arrangement) require sp^2 orbitals
>> Two electron pairs (linear arrangement) require sp orbitals

Two types of bonds

> Sigma (σ): electrons are shared in the area centered on a line joining the atoms
> Pi (π): a shared electron pair occupies the space above and below the line joining the atoms

Molecular orbital model

> A molecule is assumed to be a new entity consisting of positively charged nuclei and electrons
> The electrons in the molecule are contained in molecular orbitals, which in the simplest form of the model are constructed from the atomic orbitals of the constituent atoms
> The model correctly predicts relative bond strength, magnetism, and bond polarity
> It correctly portrays electrons as being delocalized in polyatomic molecules
> The main disadvantage of the model is that it is difficult to apply qualitatively to polyatomic molecules

Molecular orbitals are classified in two ways: energy and shape

> Energy
>> A bonding MO is lower in energy than the atomic orbitals from which it is constructed. Electrons in this type of MO are lower in energy in the molecule than in the separated atoms and thus favor molecule formation.
>> An antibonding MO is higher in energy than the atomic orbitals from which it is constructed. Electrons in this type of MO are higher in energy in the molecule than in the separated atoms and thus do not favor molecule formation.
> Shape (symmetry)
>> Sigma (σ) MOs have their electron probability centered on a line passing through the nuclei
>> Pi (π) MOs have their electron probability concentrated above and below the line connecting the nuclei

Bond order is an index of bond strength

$$\text{Bond order} = \frac{\text{number of bonding electrons} - \text{number of antibonding electrons}}{2}$$

Molecules that require the concept of resonance in the localized electron model can be more accurately described by combining the localized electron and molecular orbital models

> The σ bonds are localized
> The π bonds are delocalized

Review questions
Answers to the Review Questions can be found on the Student website (accessible from **www.cengagebrain.com**).

1. Why do we hybridize atomic orbitals to explain the bonding in covalent compounds? What type of bonds form from hybrid orbitals, σ or π? Explain.

2. What hybridization is required for central atoms that have a tetrahedral arrangement of electron pairs? A trigonal planar arrangement of electron pairs? A linear arrangement of electron pairs? How many unhybridized p atomic orbitals are present when a central atom exhibits tetrahedral geometry? Trigonal planar geometry? Linear geometry? What are the unhybridized p atomic orbitals used for?

3. Describe the bonding in H_2S, CH_4, H_2CO, and HCN using the localized electron model.

4. What hybridization is required for central atoms exhibiting trigonal bipyramidal geometry? Octahedral geometry? Describe the bonding of PF_5, SF_4, SF_6, and IF_5 using the localized electron model.

5. Electrons in σ bonding molecular orbitals are most likely to be found in the region between the two bonded atoms. Why does this arrangement favor bonding? In a σ antibonding orbital, where are the electrons most likely to be found in relation to the nuclei in a bond?

6. Show how 2s orbitals combine to form σ bonding and σ antibonding molecular orbitals. Show how 2p orbitals overlap to form σ bonding, π bonding, π antibonding, and σ antibonding molecular orbitals.

7. What are the relationships among bond order, bond energy, and bond length? Which of these can be measured? Distinguish between the terms *paramagnetic* and *diamagnetic*. What type of experiment can be done to determine if a material is paramagnetic?

8. How does molecular orbital theory explain the following observations?
    a. H_2 is stable, while He_2 is unstable.
    b. B_2 and O_2 are paramagnetic, while C_2, N_2, and F_2 are diamagnetic.
    c. N_2 has a very large bond energy associated with it.
    d. NO^+ is more stable than NO^-.

9. Consider the heteronuclear diatomic molecule HF. Explain in detail how molecular orbital theory is applied to describe the bonding in HF.

10. What is delocalized π bonding, and what does it explain? Explain the delocalized π bonding system in C_6H_6 (benzene) and O_3 (ozone).

Active Learning Questions

These questions are designed to be used by groups of students in class.

1. What are molecular orbitals? How do they compare with atomic orbitals? Can you tell by the shape of the bonding and antibonding orbitals which is lower in energy? Explain.

2. Explain the difference between the σ and π MOs for homonuclear diatomic molecules. How are bonding and antibonding orbitals different? Why are there two π MOs and one σ MO? Why are the π MOs degenerate?

3. Compare Figs. 9.35 and 9.37. Why are they different? Because B_2 is known to be paramagnetic, the π_{2p} and σ_{2p} molecular orbitals must be switched from the first prediction. What is the rationale for this? Why might one expect the σ_{2p} to be lower in energy than the π_{2p}? Why can't we use diatomic oxygen to help us decide whether the σ_{2p} or π_{2p} is lower in energy?

4. Which of the following would you expect to be more favorable energetically? Explain.
    a. an H_2 molecule in which enough energy is added to excite one electron from the bonding to the antibonding MO
    b. two separate H atoms

5. Draw the Lewis structure for HCN. Indicate the hybrid orbitals, and draw a picture showing all the bonds between the atoms, labeling each bond as σ or π.

6. Which is the more correct statement: "The methane molecule (CH_4) is a tetrahedral molecule because it is sp^3 hybridized" or "The methane molecule (CH_4) is sp^3 hybridized because it is a tetrahedral molecule"? What, if anything, is the difference between these two statements?

7. Compare and contrast the MO model with the LE model. When is each useful?

8. What are the relationships among *bond order, bond energy,* and *bond length*? Which of these quantities can be measured?

A blue question or exercise number indicates that the answer to that question or exercise appears at the back of this book and a solution appears in the *Solutions Guide,* as found on PowerLecture.

Questions

9. In the hybrid orbital model, compare and contrast σ bonds with π bonds. What orbitals form the σ bonds and what orbitals form the π bonds? Assume the z-axis is the internuclear axis.

10. In the molecular orbital model, compare and contrast σ bonds with π bonds. What orbitals form the σ bonds and what orbitals form the π bonds? Assume the z-axis is the internuclear axis.

11. Why are d orbitals sometimes used to form hybrid orbitals? Which period of elements does not use d orbitals for hybridization? If necessary, which d orbitals ($3d$, $4d$, $5d$, or $6d$) would sulfur use to form hybrid orbitals requiring d atomic orbitals? Answer the same question for arsenic and for iodine.

12. The atoms in a single bond can rotate about the internuclear axis without breaking the bond. The atoms in a double and triple bond cannot rotate about the internuclear axis unless the bond is broken. Why?

13. Compare and contrast bonding molecular orbitals with antibonding molecular orbitals.

14. What modification to the molecular orbital model was made from the experimental evidence that B_2 is paramagnetic?

15. Why does the molecular orbital model do a better job in explaining the bonding in NO^- and NO than the hybrid orbital model?

16. The three NO bonds in NO_3^- are all equivalent in length and strength. How is this explained even though any valid Lewis structure for NO_3^- has one double bond and two single bonds to nitrogen?

Exercises

In this section similar exercises are paired.

The Localized Electron Model and Hybrid Orbitals

17. Use the localized electron model to describe the bonding in H_2O.

18. Use the localized electron model to describe the bonding in CCl_4.

19. Use the localized electron model to describe the bonding in H_2CO (carbon is the central atom).

20. Use the localized electron model to describe the bonding in C_2H_2 (exists as HCCH).

21. The space-filling models of ethane and ethanol are shown below.

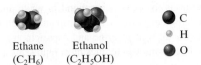

Ethane   Ethanol
(C_2H_6) (C_2H_5OH)

● C
● H
● O

Use the localized electron model to describe the bonding in ethane and ethanol.

22. The space-filling models of hydrogen cyanide and phosgene are shown below.

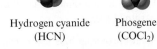

Hydrogen cyanide   Phosgene
(HCN) $(COCl_2)$

● C
● H
● O
● N
● Cl

Use the localized electron model to describe the bonding in hydrogen cyanide and phosgene.

23. Give the expected hybridization of the central atom for the molecules or ions in Exercises 83 and 89 from Chapter 8.

24. Give the expected hybridization of the central atom for the molecules or ions in Exercises 84 and 90 from Chapter 8.

25. Give the expected hybridization of the central atom for the molecules or ions in Exercise 87 from Chapter 8.

26. Give the expected hybridization of the central atom for the molecules in Exercise 88 from Chapter 8.

27. Give the expected hybridization of the central atom for the molecules in Exercises 113 and 114 from Chapter 8.

28. Give the expected hybridization of the central atom for the molecules in Exercises 115 and 116 from Chapter 8.

29. For each of the following molecules, write the Lewis structure(s), predict the molecular structure (including bond angles), give the expected hybrid orbitals on the central atom, and predict the overall polarity.

   a. CF_4 e. BeH_2 i. KrF_4
   b. NF_3 f. TeF_4 j. SeF_6
   c. OF_2 g. AsF_5 k. IF_5
   d. BF_3 h. KrF_2 l. IF_3

30. For each of the following molecules or ions that contain sulfur, write the Lewis structure(s), predict the molecular structure (including bond angles), and give the expected hybrid orbitals for sulfur.

   a. SO_2
   b. SO_3
   c.

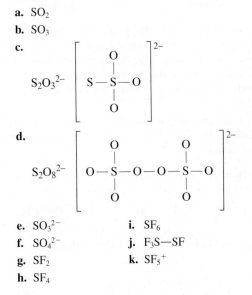

$S_2O_3^{2-}$

$S_2O_8^{2-}$

   e. SO_3^{2-} i. SF_6
   f. SO_4^{2-} j. $F_3S\!-\!SF$
   g. SF_2 k. SF_5^+
   h. SF_4

31. Why must all six atoms in C_2H_4 lie in the same plane?

32. The allene molecule has the following Lewis structure:

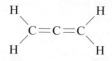

Must all hydrogen atoms lie the same plane? If not, what is their spatial relationship? Explain.

33. Indigo is the dye used in coloring blue jeans. The term *navy blue* is derived from the use of indigo to dye British naval uniforms in the eighteenth century. The structure of the indigo molecule is

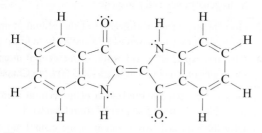

a. How many σ bonds and π bonds exist in the molecule?

b. What hybrid orbitals are used by the carbon atoms in the indigo molecule?

34. Urea, a compound formed in the liver, is one of the ways humans excrete nitrogen. The Lewis structure for urea is

$$H-\underset{\displaystyle \cdot\cdot}{\overset{\displaystyle H}{N}}-\underset{}{\overset{\displaystyle :O:}{\overset{\displaystyle \|}{C}}}-\underset{\displaystyle \cdot\cdot}{\overset{\displaystyle H}{N}}-H$$

Using hybrid orbitals for carbon, nitrogen, and oxygen, determine which orbitals overlap to form the various bonds in urea.

35. Biacetyl and acetoin are added to margarine to make it taste more like butter.

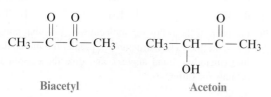

Biacetyl   **Acetoin**

Complete the Lewis structures, predict values for all C—C—O bond angles, and give the hybridization of the carbon atoms in these two compounds. Must the four carbon atoms and two oxygen atoms in biacetyl lie the same plane? How many σ bonds and how many π bonds are there in biacetyl and acetoin?

36. Many important compounds in the chemical industry are derivatives of ethylene (C_2H_4). Two of them are acrylonitrile and methyl methacrylate.

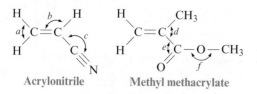

Acrylonitrile   **Methyl methacrylate**

Complete the Lewis structures, showing all lone pairs. Give approximate values for bond angles *a* through *f*. Give the hybridization of all carbon atoms. In acrylonitrile, how many of the atoms in the molecule must lie in the same plane? How many σ bonds and how many π bonds are there in methyl methacrylate and acrylonitrile?

37. Two molecules used in the polymer industry are azodicarbonamide and methyl cyanoacrylate. Their structures are

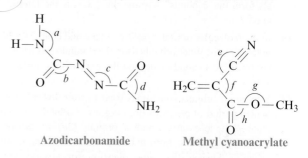

Azodicarbonamide   **Methyl cyanoacrylate**

Azodicarbonamide is used in forming polystyrene. When added to the molten plastic, it decomposes to nitrogen, carbon monoxide, and ammonia gases, which are captured as bubbles in the molten polymer. Methyl cyanoacrylate is the main ingredient in super glue. As the glue sets, methyl cyanoacrylate polymerizes across the carbon–carbon double bond. (See Chapter 22.)

a. Complete the Lewis structures showing all lone pairs of electrons.

b. Which hybrid orbitals are used by the carbon atoms in each molecule and the nitrogen atom in azodicarbonamide?

c. How many π bonds are present in each molecule?

d. Give approximate values for the bond angles marked *a* through *h* in the above structures.

38. Hot and spicy foods contain molecules that stimulate pain-detecting nerve endings. Two such molecules are piperine and capsaicin:

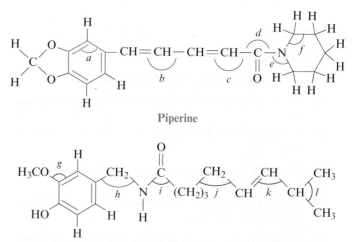

Piperine

Capsaicin

Piperine is the active compound in white and black pepper, and capsaicin is the active compound in chili peppers. The ring structures in piperine and capsaicin are shorthand notation. Each point where lines meet represents a carbon atom.

a. Complete the Lewis structure for piperine and capsaicin showing all lone pairs of electrons.

b. How many carbon atoms are sp, sp^2, and sp^3 hybridized in each molecule?

c. Which hybrid orbitals are used by the nitrogen atoms in each molecule?

d. Give approximate values for the bond angles marked a through l in the above structures.

39. One of the first drugs to be approved for use in treatment of acquired immune deficiency syndrome (AIDS) was azidothymidine (AZT). Complete the Lewis structure for AZT.

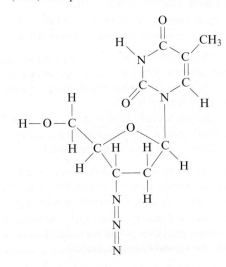

a. How many carbon atoms are sp^3 hybridized?

b. How many carbon atoms are sp^2 hybridized?

c. Which atom is sp hybridized?

d. How many σ bonds are in the molecule?

e. How many π bonds are in the molecule?

f. What is the N=N=N bond angle in the azide ($-N_3$) group?

g. What is the H—O—C bond angle in the side group attached to the five-membered ring?

h. What is the hybridization of the oxygen atom in the $-CH_2OH$ group?

40. The antibiotic thiarubin-A was discovered by studying the feeding habits of wild chimpanzees in Tanzania. The structure for thiarubin-A is

$$H_3C-C\equiv C-C \underset{S-S}{\overset{\underset{C-C}{\overset{H\quad H}{|\quad|}}}{\diagup\diagdown}} C-C\equiv C-C\equiv C-CH=CH_2$$

a. Complete the Lewis structure showing all lone pairs of electrons.

b. Indicate the hybrid orbitals used by the carbon and sulfur atoms in thiarubin-A.

c. How many σ and π bonds are present in this molecule?

The Molecular Orbital Model

41. Consider the following molecular orbitals formed from the combination of two hydrogen $1s$ orbitals:

a. Which is the bonding molecular orbital and which is the antibonding molecular orbital? Explain how you can tell by looking at their shapes.

b. Which of the two molecular orbitals is lower in energy? Why is this true?

42. Sketch the molecular orbital and label its type (σ or π; bonding or antibonding) that would be formed when the following atomic orbitals overlap. Explain your labels.

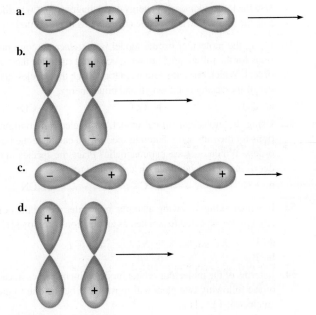

43. Which of the following are predicted by the molecular orbital model to be stable diatomic species?

a. H_2^+, H_2, H_2^-, H_2^{2-}

b. He_2^{2+}, He_2^+, He_2

44. Which of the following are predicted by the molecular orbital model to be stable diatomic species?

a. N_2^{2-}, O_2^{2-}, F_2^{2-} **b.** Be_2, B_2, Ne_2

45. Using the molecular orbital model, write electron configurations for the following diatomic species and calculate the bond orders. Which ones are paramagnetic?

a. Li_2 **b.** C_2 **c.** S_2

46. Consider the following electron configuration:

$$(\sigma_{3s})^2(\sigma_{3s}{}^*)^2(\sigma_{3p})^2(\pi_{3p})^4(\pi_{3p}{}^*)^4$$

Give four species that, in theory, would have this electron configuration.

47. Using molecular orbital theory, explain why the removal of one electron in O_2 strengthens bonding, while the removal of one electron in N_2 weakens bonding.

48. Using the molecular orbital model to describe the bonding in F_2^+, F_2, and F_2^-, predict the bond orders and the relative bond lengths for these three species. How many unpaired electrons are present in each species?

49. The transport of O_2 in the blood is carried out by hemoglobin. Carbon monoxide can interfere with oxygen transport because hemoglobin has a stronger affinity for CO than for O_2. If CO is present, normal uptake of O_2 is prevented, depriving the body of needed oxygen. Using the molecular orbital model, write the electron configurations for CO and for O_2. From your configurations, give two property differences between CO and O_2.

50. A Lewis structure obeying the octet rule can be drawn for O_2 as follows:

$$:\ddot{O}=\ddot{O}:$$

Use the molecular orbital energy-level diagram for O_2 to show that the above Lewis structure corresponds to an excited state.

51. Using the molecular orbital model, write electron configurations for the following diatomic species and calculate the bond orders. Which ones are paramagnetic? Place the species in order of increasing bond length and bond energy.

   a. CO   **b.** CO^+ **c.** CO^{2+}

52. Using the molecular orbital model, write electron configurations for the following diatomic species and calculate the bond orders. Which ones are paramagnetic? Place the species in order of increasing bond length and bond energy.

   a. CN^+ **b.** CN   **c.** CN^-

53. In which of the following diatomic molecules would the bond strength be expected to weaken as an electron is removed?

   a. H_2 **c.** C_2^{2-}
   b. B_2 **d.** OF

54. In terms of the molecular orbital model, which species in each of the following two pairs will most likely be the one to gain an electron? Explain.

   a. CN or NO
   b. O_2^{2+} or N_2^{2+}

55. Show how two $2p$ atomic orbitals can combine to form a σ or a π molecular orbital.

56. Show how a hydrogen $1s$ atomic orbital and a fluorine $2p$ atomic orbital overlap to form bonding and antibonding molecular orbitals in the hydrogen fluoride molecule. Are these molecular orbitals σ or π molecular orbitals?

57. Use Figs. 9.42 and 9.43 to answer the following questions.

   a. Would the bonding molecular orbital in HF place greater electron density near the H or the F atom? Why?

   b. Would the bonding molecular orbital have greater fluorine $2p$ character, greater hydrogen $1s$ character, or an equal contribution from both? Why?

   c. Answer the previous two questions for the antibonding molecular orbital in HF.

58. The diatomic molecule OH exists in the gas phase. The bond length and bond energy have been measured to be 97.06 pm and 424.7 kJ/mol, respectively. Assume that the OH molecule is analogous to the HF molecule discussed in the chapter and

that molecular orbitals result from the overlap of a lower-energy p_z orbital from oxygen with the higher-energy $1s$ orbital of hydrogen (the O—H bond lies along the z-axis).

   a. Which of the two molecular orbitals will have the greater hydrogen $1s$ character?

   b. Can the $2p_x$ orbital of oxygen form molecular orbitals with the $1s$ orbital of hydrogen? Explain.

   c. Knowing that only the $2p$ orbitals of oxygen will interact significantly with the $1s$ orbital of hydrogen, complete the molecular orbital energy-level diagram for OH. Place the correct number of electrons in the energy levels.

   d. Estimate the bond order for OH.

   e. Predict whether the bond order of OH^+ will be greater than, less than, or the same as that of OH. Explain.

59. Acetylene (C_2H_2) can be produced from the reaction of calcium carbide (CaC_2) with water. Use both the localized electron and molecular orbital models to describe the bonding in the acetylide anion (C_2^{2-}).

60. Describe the bonding in NO^+, NO^-, and NO using both the localized electron and molecular orbital models. Account for any discrepancies between the two models.

61. Describe the bonding in the O_3 molecule and the NO_2^- ion using the localized electron model. How would the molecular orbital model describe the π bonding in these two species?

62. Describe the bonding in the CO_3^{2-} ion using the localized electron model. How would the molecular orbital model describe the π bonding in this species?

Additional Exercises

63. Draw the Lewis structures, predict the molecular structures, and describe the bonding (in terms of the hybrid orbitals for the central atom) for the following.

   a. XeO_3 **d.** $XeOF_2$
   b. XeO_4 **e.** XeO_3F_2
   c. $XeOF_4$

64. $FClO_2$ and F_3ClO can both gain a fluoride ion to form stable anions. F_3ClO and F_3ClO_2 will both lose a fluoride ion to form stable cations. Draw the Lewis structures and describe the hybrid orbitals used by chlorine in these ions.

65. Two structures can be drawn for cyanuric acid:

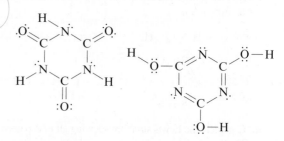

   a. Are these two structures the same molecule? Explain.

   b. Give the hybridization of the carbon and nitrogen atoms in each structure.

   c. Use bond energies (Table 8.4) to predict which form is more stable; that is, which contains the strongest bonds?

66. Give the expected hybridization for the molecular structures illustrated below.

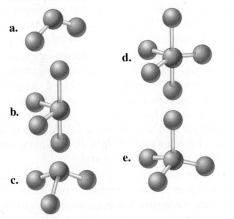

a.

b.

c.

d.

e.

67. Vitamin B₆ is an organic compound whose deficiency in the human body can cause apathy, irritability, and an increased susceptibility to infections. An incomplete Lewis structure for vitamin B₆ is shown below. Complete the Lewis structure and answer the following questions. *Hint:* Vitamin B₆ can be classified as an organic compound (a compound based on carbon atoms). The majority of Lewis structures for simple organic compounds have all atoms with a formal charge of zero. Therefore, add lone pairs and multiple bonds to the structure below to give each atom a formal charge of zero.

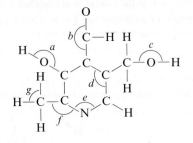

a. How many σ bonds and π bonds exist in vitamin B₆?
b. Give approximate values for the bond angles marked *a* through *g* in the structure.
c. How many carbon atoms are sp^2 hybridized?
d. How many carbon, oxygen, and nitrogen atoms are sp^3 hybridized?
e. Does vitamin B₆ exhibit delocalized π bonding? Explain.

68. Aspartame is an artificial sweetener marketed under the name NutraSweet. A partial Lewis structure for aspartame is shown below.

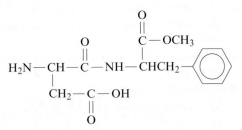

Aspartame can be classified as an organic compound (a compound based on carbon atoms). The majority of Lewis structures for simple organic compounds have all atoms with a

formal charge of zero. Therefore, add lone pairs and multiple bonds to the structure above to give each atom a formal charge of zero when drawing the Lewis structure. Also note that the six-sided ring is shorthand notation for a benzene ring ($-C_6H_5$). Benzene is discussed in Section 9.5. Complete the Lewis structure for aspartame. How many C and N atoms exhibit sp^2 hybridization? How many C and O atoms exhibit sp^3 hybridization? How many σ and π bonds are in aspartame?

69. Using bond energies from Table 8.4, estimate the barrier to rotation about a carbon–carbon double bond. To do this, consider what must happen to go from

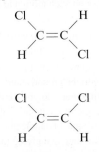

to

in terms of making and breaking chemical bonds; that is, what must happen in terms of the π bond?

70. The three most stable oxides of carbon are carbon monoxide (CO), carbon dioxide (CO_2), and carbon suboxide (C_3O_2). The space-filling models for these three compounds are

For each oxide, draw the Lewis structure, predict the molecular structure, and describe the bonding (in terms of the hybrid orbitals for the carbon atoms).

71. Complete the Lewis structures of the following molecules. Predict the molecular structure, polarity, bond angles, and hybrid orbitals used by the atoms marked by asterisks for each molecule.
a. BH_3

b. N_2F_2

F—N*—N*—F

c. C_4H_6

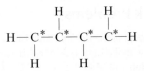

72. Complete the following resonance structures for $POCl_3$.

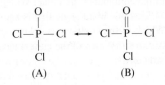

a. Would you predict the same molecular structure from each resonance structure?

b. What is the hybridization of P in each structure?

c. What orbitals can the P atom use to form the π bond in structure B?

d. Which resonance structure would be favored on the basis of formal charges?

73. The N_2O molecule is linear and polar.

   a. On the basis of this experimental evidence, which arrangement, NNO or NON, is correct? Explain your answer.

   b. On the basis of your answer to part a, write the Lewis structure of N_2O (including resonance forms). Give the formal charge on each atom and the hybridization of the central atom.

   c. How would the multiple bonding in :N≡N—Ö: be described in terms of orbitals?

74. Describe the bonding in the first excited state of N_2 (the one closest in energy to the ground state) using the molecular orbital model. What differences do you expect in the properties of the molecule in the ground state as compared to the first excited state? (An excited state of a molecule corresponds to an electron arrangement other than that giving the lowest possible energy.)

75. Using an MO energy-level diagram, would you expect F_2 to have a lower or higher first ionization energy than atomic fluorine? Why?

76. Show how a d_{xz} atomic orbital and a p_z atomic orbital combine to form a bonding molecular orbital. Assume the x-axis is the internuclear axis. Is a σ or a π molecular orbital formed? Explain.

77. What type of molecular orbital would result from the in-phase combination of two d_{xz} atomic orbitals shown below? Assume the x-axis is the internuclear axis.

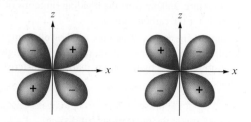

78. Consider three molecules: A, B, and C. Molecule A has a hybridization of sp^3. Molecule B has two more effective pairs (electron pairs around the central atom) than molecule A. Molecule C consists of two σ bonds and two π bonds. Give the molecular structure, hybridization, bond angles, and an example for each molecule.

ChemWork Problems

These multiconcept problems (and additional ones) are found interactively online with the same type of assistance a student would get from an instructor.

79. Draw the Lewis structures for SO_2, PCl_3, NNO, COS, and PF_3. Which of the compounds are polar? Which of the compounds exhibit at least one bond angle that is approximately 120 degrees? Which of the compounds exhibit sp^3 hybridization by the central atom? Which of the compounds have a linear molecular structure?

80. Draw the Lewis structures for $TeCl_4$, ICl_5, PCl_5, $KrCl_4$, and $XeCl_2$. Which of the compounds exhibit at least one bond angle that is approximately 120 degrees? Which of the compounds exhibit d^2sp^3 hybridization? Which of the compounds have a square planar molecular structure? Which of the compounds are polar?

81. A variety of chlorine oxide fluorides and related cations and anions are known. They tend to be powerful oxidizing and fluorinating agents. $FClO_3$ is the most stable of this group of compounds and has been studied as an oxidizing component in rocket propellants. Draw a Lewis structure for F_3ClO, $F_2ClO_2^+$, and F_3ClO_2. What is the molecular structure for each species, and what is the expected hybridization of the central chlorine atom in each compound or ion?

82. Pelargondin is the molecule responsible for the red color of the geranium flower. It also contributes to the color of ripe strawberries and raspberries. The structure of pelargondin is:

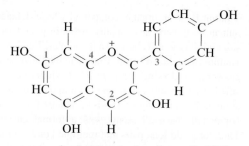

How many σ and π bonds exist in pelargondin? What is the hybridization of the carbon atoms marked 1–4?

83. Complete a Lewis structure for the compound shown below, then answer the following questions. How many carbon atoms are sp^2 hybridized? How many C—N bonds are formed by the overlap of an sp^3 hybridized carbon with an sp^3 hybridized nitrogen? How many lone pairs of electrons are in the Lewis structure of your molecule? How many π bonds are present?

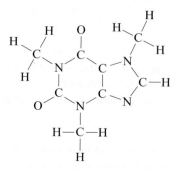

84. Which of the following statements concerning SO_2 is(are) *true*?

   a. The central sulfur atom is sp^2 hybridized.

   b. One of the sulfur–oxygen bonds is longer than the other(s).

   c. The bond angles about the central sulfur atom are about 120 degrees.

   d. There are two σ bonds in SO_2.

   e. There are no resonance structures for SO_2.

85. Consider the molecular orbital electron configurations for N_2, N_2^+, and N_2^-. For each compound or ion, fill in the table below with the correct number of electrons in each molecular orbital.

MO	N_2	N_2^+	N_2^-
σ_{2p}^{*}			
π_{2p}^{*}			
σ_{2p}			
π_{2p}			
σ_{2s}^{*}			
σ_{2s}			

86. Place the species B_2^+, B_2, and B_2^- in order of increasing bond length and increasing bond energy.

Challenge Problems

87. Consider the following computer-generated model of caffeine:

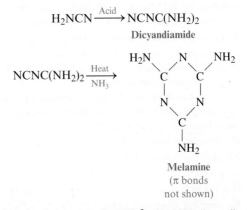

H
O
N
C

Complete a Lewis structure for caffeine in which all atoms have a formal charge of zero (as is typical with most organic compounds). How many C and N atoms are sp^2 hybridized? How many C and N atoms are sp^3 hybridized? sp hybridized? How many σ and π bonds are there?

88. Cholesterol ($C_{27}H_{46}O$) has the following structure:

In such shorthand structures, each point where lines meet represents a carbon atom and most H atoms are not shown. Draw the complete structure showing all carbon and hydrogen atoms. (There will be four bonds to each carbon atom.) Indicate which carbon atoms use sp^2 or sp^3 hybrid orbitals. Are all carbon atoms in the same plane, as implied by the structure?

89. Cyanamide (H_2NCN), an important industrial chemical, is produced by the following steps:

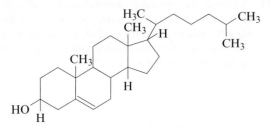

$$CaC_2 + N_2 \longrightarrow CaNCN + C$$

$$CaNCN \xrightarrow{\text{Acid}} H_2NCN$$
Cyanamide

Calcium cyanamide (CaNCN) is used as a direct-application fertilizer, weed killer, and cotton defoliant. It is also used to make cyanamide, dicyandiamide, and melamine plastics:

$$H_2NCN \xrightarrow{\text{Acid}} NCNC(NH_2)_2$$
Dicyandiamide

$$NCNC(NH_2)_2 \xrightarrow[\text{NH}_3]{\text{Heat}}$$

Melamine
(π bonds not shown)

a. Write Lewis structures for NCN^{2-}, H_2NCN, dicyandiamide, and melamine, including resonance structures where appropriate.

b. Give the hybridization of the C and N atoms in each species.

c. How many σ bonds and how many π bonds are in each species?

d. Is the ring in melamine planar?

e. There are three different C—N bond distances in dicyandiamide, $NCNC(NH_2)_2$, and the molecule is nonlinear. Of all the resonance structures you drew for this molecule, predict which should be the most important.

90. In Exercise 91 in Chapter 8, the Lewis structures for benzene (C_6H_6) were drawn. Using one of the Lewis structures, estimate ΔH_f° for $C_6H_6(g)$ using bond energies and given that the standard enthalpy of formation of $C(g)$ is 717 kJ/mol. The experimental ΔH_f° value of $C_6H_6(g)$ is 83 kJ/mol. Explain the discrepancy between the experimental value and the calculated ΔH_f° value for $C_6H_6(g)$.

91. A flask containing gaseous N_2 is irradiated with 25-nm light.

a. Using the following information, indicate what species can form in the flask during irradiation.

$$N_2(g) \longrightarrow 2N(g) \qquad \Delta H = 941 \text{ kJ/mol}$$
$$N_2(g) \longrightarrow N_2^+(g) + e^- \qquad \Delta H = 1501 \text{ kJ/mol}$$
$$N(g) \longrightarrow N^+(g) + e^- \qquad \Delta H = 1402 \text{ kJ/mol}$$

b. What range of wavelengths will produce atomic nitrogen in the flask but will not produce any ions?

c. Explain why the first ionization energy of N_2 (1501 kJ/mol) is greater than the first ionization energy of atomic nitrogen (1402 kJ/mol).

92. As compared with CO and O_2, CS and S_2 are very unstable molecules. Give an explanation based on the relative abilities of the sulfur and oxygen atoms to form π bonds.

93. Values of measured bond energies may vary greatly depending on the molecule studied. Consider the following reactions:

$$NCl_3(g) \longrightarrow NCl_2(g) + Cl(g) \qquad \Delta H = 375 \text{ kJ/mol}$$
$$ONCl(g) \longrightarrow NO(g) + Cl(g) \qquad \Delta H = 158 \text{ kJ/mol}$$

Rationalize the difference in the values of ΔH for these reactions, even though each reaction appears to involve only the breaking of one N—Cl bond. (*Hint:* Consider the bond order of the NO bond in ONCl and in NO.)

94. Use the MO model to explain the bonding in BeH_2. When constructing the MO energy-level diagram, assume that the Be's $1s$ electrons are not involved in bond formation.

95. *Bond energy* has been defined in the text as the amount of energy required to break a chemical bond, so we have come to think of the addition of energy as breaking bonds. However, in some cases the addition of energy can cause the formation of bonds. For example, in a sample of helium gas subjected to a high-energy source, some He_2 molecules exist momentarily and then dissociate. Use MO theory (and diagrams) to explain why He_2 molecules can come to exist and why they dissociate.

96. Arrange the following from lowest to highest ionization energy: O, O_2, O_2^-, O_2^+. Explain your answer.

97. Use the MO model to determine which of the following has the smallest ionization energy: N_2, O_2, N_2^{2-}, N_2^-, O_2^+. Explain your answer.

98. Given that the ionization energy of F_2^- is 290 kJ/mol, do the following:

    a. Calculate the bond energy of F_2^-. You will need to look up the bond energy of F_2 and ionization energy of F^-.

    b. Explain the difference in bond energy between F_2^- and F_2 using MO theory.

99. Carbon monoxide (CO) forms bonds to a variety of metals and metal ions. Its ability to bond to iron in hemoglobin is the reason that CO is so toxic. The bond carbon monoxide forms to metals is through the carbon atom:

$$M—C≡O$$

    a. On the basis of electronegativities, would you expect the carbon atom or the oxygen atom to form bonds to metals?

    b. Assign formal charges to the atoms in CO. Which atom would you expect to bond to a metal on this basis?

    c. In the MO model, bonding MOs place more electron density near the more electronegative atom. (See the HF molecule in Figs. 9.42 and 9.43.) Antibonding MOs place more electron density near the less electronegative atom in the diatomic molecule. Use the MO model to predict which atom of carbon monoxide should form bonds to metals.

100. The space-filling model for benzoic acid, a food preservative, is shown below.

● C
● H
● O

Benzoic acid
($C_6H_5CO_2H$)

Describe the bonding in benzoic acid using the localized electron model combined with the molecular orbital model.

Integrative Problems

These problems require the integration of multiple concepts to find the solutions.

101. As the head engineer of your starship in charge of the warp drive, you notice that the supply of dilithium is critically low. While searching for a replacement fuel, you discover some diboron, B_2.

    a. What is the bond order in Li_2 and B_2?

    b. How many electrons must be removed from B_2 to make it isoelectronic with Li_2 so that it might be used in the warp drive?

    c. The reaction to make B_2 isoelectronic with Li_2 is generalized (where n = number of electrons determined in part b) as follows:

$$B_2 \longrightarrow B_2^{n+} + ne^- \qquad \Delta H = 6455 \text{ kJ/mol}$$

How much energy is needed to ionize 1.5 kg B_2 to the desired isoelectronic species?

102. An unusual category of acids known as superacids, which are defined as any acid stronger than 100% sulfuric acid, can be prepared by seemingly simple reactions similar to the one below. In this example, the reaction of anhydrous HF with SbF_5 produces the superacid $[H_2F]^+[SbF_6]^-$:

$$2HF(l) + SbF_5(l) \longrightarrow [H_2F]^+[SbF_6]^-(l)$$

    a. What are the molecular structures of all species in this reaction? What are the hybridizations of the central atoms in each species?

    b. What mass of $[H_2F]^+[SbF_6]^-$ can be prepared when 2.93 mL anhydrous HF (density = 0.975 g/mL) and 10.0 mL SbF_5 (density = 3.10 g/mL) are allowed to react?

103. Determine the molecular structure and hybridization of the central atom X in the polyatomic ion XY_3^+ given the following information: A neutral atom of X contains 36 electrons, and the element Y makes an anion with a 1– charge, which has the electron configuration $1s^2 2s^2 2p^6$.

Liquids and Solids

Huge crystals of calcium sulfate in The Cave of the Crystals in Mexico. (Carsten Peter/Speleoresearch Films/ National Geographic Stock)

You have only to think about water to appreciate how different the three states of matter are. Flying, swimming, and ice skating are all done in contact with water in its various forms. Clearly, the arrangements of the water molecules must be significantly different in its gas, liquid, and solid forms.

In Chapter 5 we saw that a gas can be pictured as a substance whose component particles are far apart, are in rapid random motion, and exert relatively small forces on each other. The kinetic molecular model was constructed to account for the ideal behavior that most gases approach at high temperatures and low pressures.

Solids are obviously very different from gases. A gas has low density and high compressibility and completely fills its container. Solids have much greater densities, are compressible only to a very slight extent, and are rigid—a solid maintains its shape irrespective of its container. These properties indicate that the components of a solid are close together and exert large attractive forces on each other.

The properties of liquids lie somewhere between those of solids and gases but not midway between, as can be seen from some of the properties of the three states of water. For example, compare the enthalpy change for the melting of ice at 0°C (the heat of fusion) with that for vaporizing liquid water at 100°C (the heat of vaporization):

$$H_2O(s) \longrightarrow H_2O(l) \qquad \Delta H°_{fus} = 6.02 \text{ kJ/mol}$$
$$H_2O(l) \longrightarrow H_2O(g) \qquad \Delta H°_{vap} = 40.7 \text{ kJ/mol}$$

These values show a much greater change in structure in going from the liquid to the gaseous state than in going from the solid to the liquid state. This suggests that there are extensive attractive forces among the molecules in liquid water, similar to but not as strong as those in the solid state.

The relative similarity of the liquid and solid states also can be seen in the densities of the three states of water. As shown in Table 10.1, the densities for liquid and solid water are quite close.* Compressibilities also can be used to explore the relationship among water's states. At 25°C, the density of liquid water changes from 0.9971 g/cm^3 at a pressure of 1 atm to 1.046 g/cm^3 at 1065 atm. Given the large change in pressure, this is a very small variation in the density. Ice also shows little variation in density with increased pressure. On the other hand, at 400°C, the density of gaseous water changes from 3.26×10^{-4} g/cm^3 at 1 atm pressure to 0.157 g/cm^3 at 242 atm—a huge variation.

The conclusion is clear. The liquid and solid states show many similarities and are strikingly different from the gaseous state, as shown schematically in Fig. 10.1. We must bear this in mind as we develop models for the structures of solids and liquids.

Table 10.1 | Densities of the Three States of Water

State	Density (g/cm^3)
Solid (0°C, 1 atm)	0.9168
Liquid (25°C, 1 atm)	0.9971
Gas (400°C, 1 atm)	3.26×10^{-4}

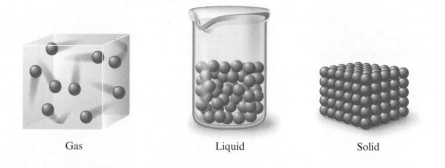

Figure 10.1 | Schematic representations of the three states of matter.

Gas Liquid Solid

*Although the densities of solid and liquid water are quite similar, as is typical for most substances, water is quite unusual in that the density of its solid state is slightly less than that of its liquid state. For most substances, the density of the solid state is slightly greater than that of the liquid state.

We will proceed in our study of liquids and solids by first considering the properties and structures of liquids and solids. Then we will consider the changes in state that occur between solid and liquid, liquid and gas, and solid and gas.

10.1 | Intermolecular Forces

In Chapters 8 and 9 we saw that atoms can form stable units called *molecules* by sharing electrons. This is called *intramolecular* (within the molecule) *bonding*. In this chapter we consider the properties of the **condensed states** of matter (liquids and solids) and the forces that cause the aggregation of the components of a substance to form a liquid or a solid. These forces may involve covalent or ionic bonding, or they may involve weaker interactions usually called **intermolecular forces** (because they occur between, rather than within, molecules).

It is important to recognize that when a substance such as water changes from solid to liquid to gas, *the molecules remain intact.* The changes in states are due to changes in the forces *among* the molecules rather than in those *within* the molecules. In ice, as we will see later in this chapter, the molecules are virtually locked in place, although they can vibrate about their positions. If energy is added, the motions of the molecules increase, and they eventually achieve the greater movement and disorder characteristic of liquid water. The ice has melted. As more energy is added, the gaseous state is eventually reached, with the individual molecules far apart and interacting relatively little. However, the gas still consists of water molecules. It would take much energy to overcome the covalent bonds and decompose the water molecules into their component atoms. This can be seen by comparing the energy needed to vaporize 1 mole of liquid water (40.7 kJ) with that needed to break the O—H bonds in 1 mole of water molecules (934 kJ).

Intermolecular forces were introduced in Chapter 5 to explain nonideal gas behavior.

Remember that temperature is a measure of the random motions of the particles in a substance.

Dipole–Dipole Forces

As we saw in Section 8.3, molecules with polar bonds often behave in an electric field as if they had a center of positive charge and a center of negative charge. That is, they exhibit a dipole moment. Molecules with dipole moments can attract each other electrostatically by lining up so that the positive and negative ends are close to each other [Fig. 10.2(a)]. This is called a **dipole–dipole attraction**. In a condensed state such as a liquid, where many molecules are in close proximity, the dipoles find the best compromise between attraction and repulsion. That is, the molecules orient themselves to maximize the $\oplus$---$\ominus$ interactions and to minimize $\oplus$---$\oplus$ and $\ominus$---$\ominus$ interactions [Fig. 10.2(b)].

Dipole–dipole forces are typically only about 1% as strong as covalent or ionic bonds, and they rapidly become weaker as the distance between the dipoles increases. At low pressures in the gas phase, where the molecules are far apart, these forces are relatively unimportant.

Dipole–dipole forces are forces that act between polar molecules.

Particularly strong dipole–dipole forces, however, are seen among molecules in which hydrogen is bound to a highly electronegative atom, such as nitrogen, oxygen, or fluorine. Two factors account for the strengths of these interactions: the great polarity of the bond and the close approach of the dipoles, allowed by the very small size of the hydrogen atom. Because dipole–dipole attractions of this type are so unusually strong, they are given a special name—**hydrogen bonding**. Figure 10.3 shows hydrogen bonding among water molecules, which occurs between the partially positive H atoms and the lone pairs on adjacent water molecules.

Hydrogen bonding has a very important effect on physical properties. For example, the boiling points for the covalent hydrides of the elements in Groups 4A, 5A, 6A, and 7A are given in Fig. 10.4. Note that the nonpolar tetrahedral hydrides of Group 4A show a steady increase in boiling point with molar mass (that is, in going down the group), whereas, for the other groups, the lightest member has an unexpectedly high

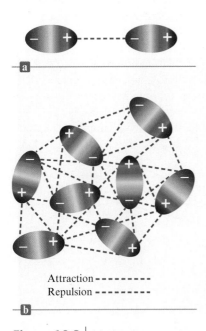

Attraction - - - - - - - -
Repulsion - - - - - - - -

b

Figure 10.2 | (a) The electrostatic interaction of two polar molecules. (b) The interaction of many dipoles in a condensed state.

Figure 10.3 | (a) The polar water molecule. (b) Hydrogen bonding (•••) among water molecules. Note that the small size of the hydrogen atom allows for close interactions.

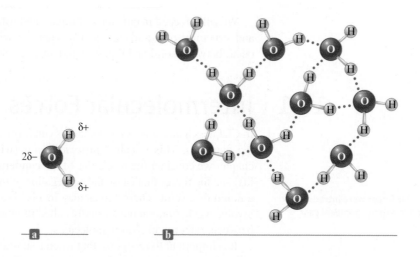

Boiling point will be defined precisely in Section 10.8.

boiling point. Why? The answer lies in the especially large hydrogen bonding interactions that exist among the smallest molecules with the most polar X—H bonds. These unusually strong hydrogen bonding forces are due primarily to two factors. One factor is the relatively large electronegativity values of the lightest elements in each group, which leads to especially polar X—H bonds. The second factor is the small size of the first element of each group, which allows for the close approach of the dipoles, further strengthening the intermolecular forces. Because the interactions among the molecules containing the lightest elements in Groups 5A and 6A are so strong, an unusually large quantity of energy must be supplied to overcome these interactions and separate the molecules to produce the gaseous state. These molecules will remain together in the liquid state even at high temperatures—hence the very high boiling points.

Hydrogen bonding is also important in organic molecules (molecules with a carbon chain backbone). For example, the alcohols methanol (CH_3OH) and ethanol (CH_3CH_2OH) have much higher boiling points than would be expected from their molar masses because of the polar O—H bonds in these molecules, which produce hydrogen bonding.

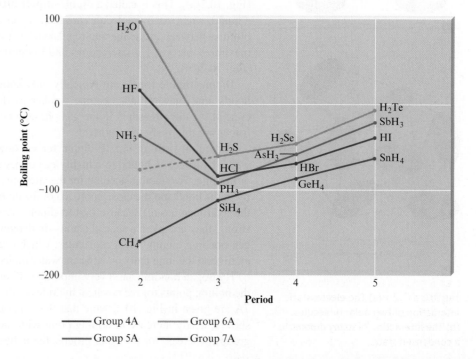

Figure 10.4 | The boiling points of the covalent hydrides of the elements in Groups 4A, 5A, 6A, and 7A. The dashed line shows the expected boiling point of water if it had no hydrogen bonding.

London Dispersion Forces

Even molecules without dipole moments must exert forces on each other. We know this because all substances—even the noble gases—exist in the liquid and solid states under certain conditions. The forces that exist among noble gas atoms and nonpolar molecules are called **London dispersion forces**. To understand the origin of these forces, let's consider a pair of noble gas atoms. Although we usually assume that the electrons of an atom are uniformly distributed about the nucleus, this is apparently not true at every instant. As the electrons move about the nucleus, a momentary nonsymmetrical electron distribution can develop that produces a temporary dipolar arrangement of charge. The formation of this temporary dipole can, in turn, affect the electron distribution of a neighboring atom. That is, this *instantaneous dipole* that occurs accidentally in a given atom can then *induce* a similar dipole in a neighboring atom [Fig. 10.5(a)]. This phenomenon leads to an interatomic attraction that is relatively weak and short-lived but that can be very significant especially for large atoms (see below). For these interactions to become strong enough to produce a solid, the motions of the atoms must be greatly slowed down. This explains, for instance, why the noble gas elements have such low freezing points (Table 10.2).

Note from Table 10.2 that the freezing point rises going down the group. The principal cause for this trend is that as the atomic number increases, the number of electrons increases, and there is an increased chance of the occurrence of momentary dipole interactions. We describe this phenomenon using the term *polarizability,* which indicates the ease with which the electron "cloud" of an atom can be distorted to give a dipolar charge distribution. Thus we say that large atoms with many electrons exhibit a higher polarizability than small atoms. This means that the importance of London dispersion forces increases greatly as the size of the atom increases.

These same ideas also apply to nonpolar molecules such as H_2, CH_4, CCl_4, and CO_2 [Fig. 10.5(b)]. Since none of these molecules has a permanent dipole moment, their principal means of attracting each other is through London dispersion forces.

Table 10.2 | The Freezing Points of the Group 8A Elements

Element	Freezing Point (°C)
Helium*	−269.7
Neon	−248.6
Argon	−189.4
Krypton	−157.3
Xenon	−111.9

*Helium is the only element that will not freeze by lowering its temperature at 1 atm. Pressure must be applied to freeze helium.

The dispersion forces in molecules with large atoms are quite significant and are often actually more important than dipole–dipole forces.

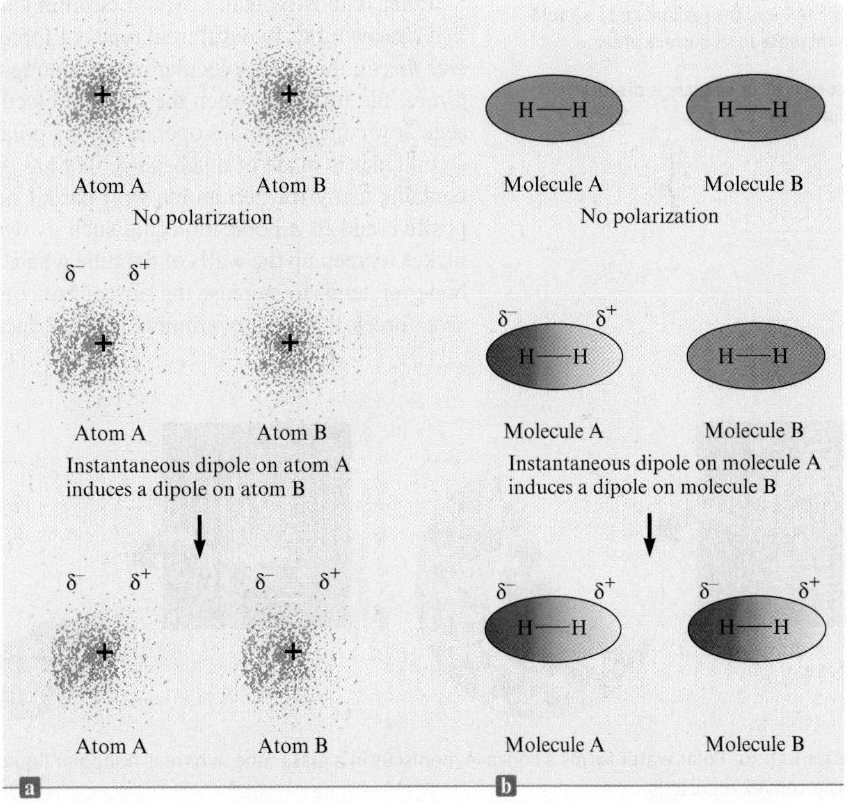

Figure 10.5 | (a) An instantaneous polarization can occur on atom A, creating an instantaneous dipole. This dipole creates an induced dipole on neighboring atom B. (b) Nonpolar molecules such as H_2 also can develop instantaneous and induced dipoles.

10.2 | The Liquid State

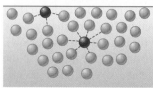

Figure 10.6 | A molecule in the interior of a liquid is attracted by the molecules surrounding it, whereas a molecule at the surface of a liquid is attracted only by molecules below it and on each side. Remember that the molecules in a liquid are in constant motion.

For a given volume, a sphere has a smaller surface area than any other shape.

Surface tension: the resistance of a liquid to an increase in its surface area.

The composition of glass is discussed in Section 10.5.

Liquids and liquid solutions are vital to our lives. Of course, water is the most important liquid. Besides being essential to life, water provides a medium for food preparation, for transportation, for cooling in many types of machines and industrial processes, for recreation, for cleaning, and for a myriad of other uses.

Liquids exhibit many characteristics that help us understand their nature. We have already mentioned their low compressibility, lack of rigidity, and high density compared with gases. Many of the properties of liquids give us direct information about the forces that exist among the particles. For example, when a liquid is poured onto a solid surface, it tends to bead as droplets, a phenomenon that depends on the intermolecular forces. Although molecules in the interior of the liquid are completely surrounded by other molecules, those at the liquid surface are subject to attractions only from the side and from below (Fig. 10.6). The effect of this uneven pull on the surface molecules tends to draw them into the body of the liquid and causes a droplet of liquid to assume the shape that has the minimum surface area—a sphere.

To increase a liquid's surface area, molecules must move from the interior of the liquid to the surface. This requires energy, since some intermolecular forces must be overcome. The resistance of a liquid to an increase in its surface area is called the **surface tension** of the liquid. As we would expect, liquids with relatively large intermolecular forces, such as those with polar molecules, tend to have relatively high surface tensions.

Polar liquids typically exhibit **capillary action**, the spontaneous rising of a liquid in a narrow tube. Two different types of forces are responsible for this property: *cohesive forces,* the intermolecular forces among the molecules of the liquid, and *adhesive forces,* the forces between the liquid molecules and their container. We have already seen how cohesive forces operate among polar molecules. Adhesive forces occur when a container is made of a substance that has polar bonds. For example, a glass surface contains many oxygen atoms with partial negative charges that are attractive to the positive end of a polar molecule such as water. This ability of water to "wet" glass makes it creep up the walls of the tube where the water surface touches the glass. This, however, tends to increase the surface area of the water, which is opposed by the cohesive forces that try to minimize the surface. Thus because water has both strong

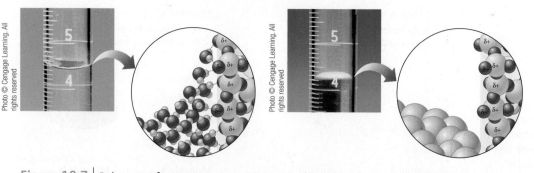

Figure 10.7 | Polar water forms a concave meniscus in a glass tube, whereas nonpolar liquid mercury forms a convex meniscus.

cohesive (intermolecular) forces and strong adhesive forces to glass, it "pulls itself" up a glass capillary tube (a tube with a small diameter) to a height where the weight of the column of water just balances the water's tendency to be attracted to the glass surface. The concave shape of the meniscus (Fig. 10.7) shows that water's adhesive forces toward the glass are stronger than its cohesive forces. A nonpolar liquid such as mercury (see Fig. 10.7) shows a convex meniscus. This behavior is characteristic of a liquid in which the cohesive forces are stronger than the adhesive forces toward glass.

Another property of liquids strongly dependent on intermolecular forces is **viscosity**, a measure of a liquid's resistance to flow. As might be expected, liquids with large intermolecular forces tend to be highly viscous. For example, glycerol, whose structure is

Viscosity: a measure of a liquid's resistance to flow.

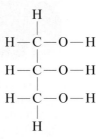

has an unusually high viscosity due mainly to its high capacity to form hydrogen bonds using its O—H groups (see margin).

Molecular complexity also leads to higher viscosity because very large molecules can become entangled with each other. For example, gasoline, a nonviscous liquid, contains hydrocarbon molecules of the type $CH_3—(CH_2)_n—CH_3$, where n varies from about 3 to 8. However, grease, which is very viscous, contains much larger hydrocarbon molecules in which n varies from 20 to 25.

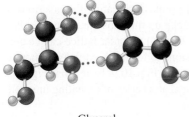

Glycerol

Structural Model for Liquids

In many respects, the development of a structural model for liquids presents greater challenges than the development of such a model for the other two states of matter. In the gaseous state, the particles are so far apart and are moving so rapidly that intermolecular forces are negligible under most circumstances. This means that we can use a relatively simple model for gases. In the solid state, although the intermolecular forces are large, the molecular motions are minimal, and fairly simple models are again possible. The liquid state, however, has both strong intermolecular forces *and* significant molecular motions. Such a situation precludes the use of really simple models for liquids. Recent advances in *spectroscopy,* the study of the manner in which substances interact with electromagnetic radiation, make it possible to follow the very rapid changes that occur in liquids. As a result, our models of liquids are becoming more accurate. As a starting point, a typical liquid might best be viewed as containing a large number of regions where the arrangements of the components are similar to those found in the solid, but with more disorder, and a smaller number of regions where holes are present. The situation is highly dynamic, with rapid fluctuations occurring in both types of regions.

10.3 | An Introduction to Structures and Types of Solids

There are many ways to classify solids, but the broadest categories are **crystalline solids**, those with a highly regular arrangement of their components, and **amorphous solids**, those with considerable disorder in their structures.

Figure 10.8 | Crystalline solid.

The regular arrangement of the components of a crystalline solid at the microscopic level produces the beautiful, characteristic shapes of crystals, such as those shown in Fig. 10.8. The positions of the components in a crystalline solid are usually represented by a **lattice**, a three-dimensional system of points designating the positions of the components (atoms, ions, or molecules) that make up the substance. The *smallest repeating unit* of the lattice is called the **unit cell**. Thus a particular lattice can be generated by repeating the unit cell in all three dimensions to form the extended structure. Three common unit cells and their lattices are shown in Fig. 10.9. Note from Fig. 10.9 that the extended structure in each case can be viewed as a series of repeating unit cells that share common faces in the interior of the solid.

Although we will concentrate on crystalline solids in this book, there are many important noncrystalline (amorphous) materials. An example is common glass, which is best pictured as a solution in which the components are "frozen in place" before they can achieve an ordered arrangement. Although glass is a solid (it has a rigid shape), a great deal of disorder exists in its structure.

X-Ray Analysis of Solids

The structures of crystalline solids are most commonly determined by **X-ray diffraction**. Diffraction occurs when beams of light are scattered from a regular array of points in which the spacings between the components are comparable with the wavelength of the light. Diffraction is due to constructive interference when the waves of parallel beams are in phase and to destructive interference when the waves are out of phase.

When X rays of a single wavelength are directed at a crystal, a diffraction pattern is obtained, as we saw in Fig. 7.5. The light and dark areas on the photographic plate occur because the waves scattered from various atoms may reinforce or cancel each other (Fig. 10.10). The key to whether the waves reinforce or cancel is the difference in distance traveled by the waves after they strike the atoms. The waves are in phase before they are reflected, so if the difference in distance traveled is an *integral number of wavelengths,* the waves will still be in phase.

Since the distance traveled depends on the distance between the atoms, the diffraction pattern can be used to determine the interatomic spacings. The exact relationship can be worked out using the diagram in Fig. 10.11, which shows two in-phase waves being reflected by atoms in two different layers in a crystal. The extra distance traveled by the lower wave is the sum of the distances xy and yz, and the waves will be in phase after reflection if

$$xy + yz = n\lambda \tag{10.1}$$

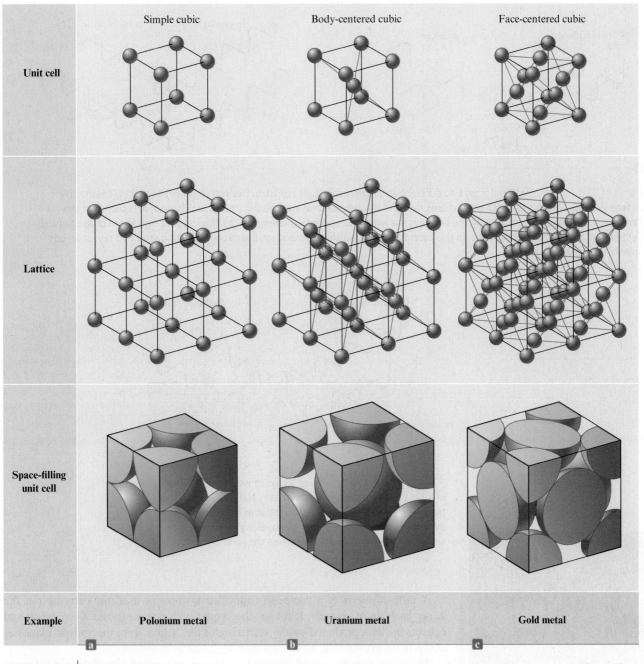

Figure 10.9 | Three cubic unit cells and the corresponding lattices. Note that only parts of spheres on the corners and faces of the unit cells reside inside the unit cell, as shown by the "cutoff" versions.

where n is an integer and λ is the wavelength of the X rays. Using trigonometry (see Fig. 10.11), we can show that

$$xy + yz = 2d \sin \theta \qquad (10.2)$$

where d is the distance between the atoms and θ is the angle of incidence and reflection. Combining Equation (10.1) and Equation (10.2) gives

$$n\lambda = 2d \sin \theta \qquad (10.3)$$

Equation (10.3) is called the *Bragg equation* after William Henry Bragg (1862–1942) and his son William Lawrence Bragg (1890–1972), who shared the Nobel Prize in physics in 1915 for their pioneering work in X-ray crystallography.

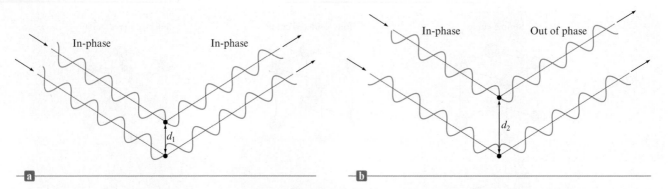

Figure 10.10 | X rays scattered from two different atoms may reinforce (constructive interference) or cancel (destructive interference) one another. (a) Both the incident rays and the reflected rays are also in phase. In this case, d_1 is such that the difference in the distances traveled by the two rays is a whole number of wavelengths. (b) The incident rays are in phase but the reflected rays are exactly out of phase. In this case d_2 is such that the difference in distances traveled by the two rays is an odd number of half wavelengths.

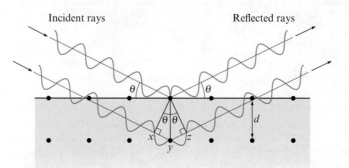

Figure 10.11 | Reflection of X rays of wavelength λ from a pair of atoms in two different layers of a crystal. The lower wave travels an extra distance equal to the sum of xy and yz. If this distance is an integral number of wavelengths ($n = 1, 2, 3, \ldots$), the waves will reinforce each other when they exit the crystal.

Student operating an X-ray diffractometer at Colgate University.

A diffractometer is a computer-controlled instrument used for carrying out the X-ray analysis of crystals. It rotates the crystal with respect to the X-ray beam and collects the data produced by the scattering of the X rays from the various planes of atoms in the crystal. The results are then analyzed by computer.

The techniques for crystal structure analysis have reached a level of sophistication that allows the determination of very complex structures, such as those important in biological systems. For example, the structures of several enzymes have been determined, thus enabling biochemists to understand how they perform their functions. We will explore this topic further in Chapter 12. Using X-ray diffraction, we can gather data on bond lengths and angles and in so doing can test the predictions of our models of molecular geometry.

Interactive Example 10.1

Sign in at http://login.cengagebrain.com to try this Interactive Example in OWL.

Using the Bragg Equation

X rays of wavelength 1.54 Å were used to analyze an aluminum crystal. A reflection was produced at $\theta = 19.3$ degrees. Assuming $n = 1$, calculate the distance d between the planes of atoms producing this reflection.

Chemical connections
Smart Fluids

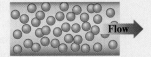

Magnetic field off
Magnetic particles
flow randomly

Magnetic field on
Applied field (H)
creates structure that
increases viscosity

Matter seems to be getting smarter these days. Increasingly, we have discovered materials that can remember their initial shape after being deformed or can sense and respond to their environment. In particular, valuable new materials have been formulated whose properties can be changed instantly by applying a magnetic or electric field.

One example of such a substance is a fluid whose flow characteristics (rheology) can be changed from free flowing to almost solid in about 0.01 second by the application of an electromagnetic field. This "magneto-rheological" (MR) fluid was developed by Lord Corporation. Working in collaboration with Delphi Corporation, the company is applying the fluid in suspension control of General Motors automobiles such as Cadillacs and Corvettes. The so-called Magneride system has sensors that monitor the road surface and provide information about what suspension damping is needed. In response, a message is instantly sent to an electromagnetic coil in the shock absorbers, which adjusts the viscosity of the MR fluid to provide continuously variable damping. The result: an amazingly smooth ride and unerring road-holding ability.

The MR fluid is composed of a synthetic oil in which particles of an iron-containing compound are suspended. When the magnetic field is turned off, these particles flow freely in all directions (see the figure above). When the field is turned on, the particles aggregate into chains that line up perpendicular to the flow of the fluid, thereby increasing its viscosity in proportion to the strength of the applied field.

Many other applications of MR fluids besides auto suspensions are under development. One very large-scale application is in Japan's National Museum of Emerging Science and Innovation, where an MR fluid is being used in dampers to protect the building against earthquake damage. Large MR-fluid dampers are also being used for stabilizing bridges such as the Dong Ting Lake Bridge in China's Hunan province to steady it in high winds.

This Corvette ZR1 uses magnetorheological fluid in its suspension system.

Frederic J. Brown/AFP/Getty Images/Newscom.com

Solution
To determine the distance between the planes, we use Equation (10.3) with $n = 1$, $\lambda = 1.54$ Å, and $\theta = 19.3$ degrees. Since $2d \sin \theta = n\lambda$,

$$d = \frac{n\lambda}{2 \sin \theta} = \frac{(1)(1.54 \text{ Å})}{(2)(0.3305)} = 2.33 \text{ Å} = 233 \text{ pm}$$

See Exercises 10.47 through 10.50

Types of Crystalline Solids

There are many different types of crystalline solids. For example, although both sugar and salt dissolve readily in water, the properties of the resulting solutions are quite different. The salt solution readily conducts an electric current, whereas the sugar solution does not. This behavior arises from the nature of the components in these two solids. Common salt (NaCl) is an ionic solid; it contains Na^+ and Cl^- ions. When solid sodium chloride dissolves in the polar water, sodium and chloride ions are distributed throughout the resulting solution and are free to conduct electric current. Table sugar (sucrose), on the other hand, is composed of neutral molecules that are dispersed throughout the water when the solid dissolves. No ions are present, and the resulting solution does not conduct electricity. These examples illustrate two important types of solids: **ionic solids**, represented by sodium chloride, and **molecular solids**, represented by sucrose. Ionic solids have ions at the points of the lattice that describes the structure of the solid. A molecular solid, on the other hand, has discrete covalently bonded molecules at each of its lattice points. Ice is a molecular solid that has an H_2O molecule at each point (Fig. 10.12).

A third type of solid is represented by elements such as carbon (which exists in the forms graphite, diamond, and the fullerenes), boron, silicon, and all metals. These substances all have atoms at the lattice points that describe the structure of the solid. Therefore, we call solids of this type **atomic solids**. Examples of these three types of solids are shown in Fig. 10.12.

Buckminsterfullerene, C_{60}, is a particular member of the fullerene family.

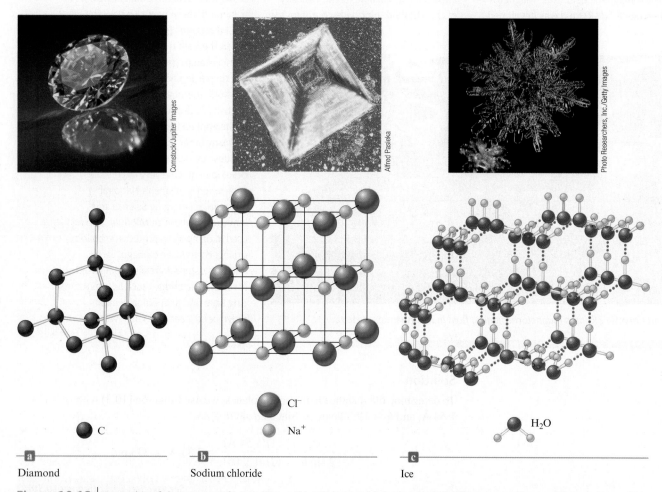

Diamond   Sodium chloride   Ice

Figure 10.12 | Examples of three types of crystalline solids. Only part of the structure is shown in each case. (a) An atomic solid. (b) An ionic solid. (c) A molecular solid. The dotted lines show the hydrogen bonding interactions among the polar water molecules.

Table 10.3 | Classification of Solids

| | Atomic Solids | | | Molecular Solids | Ionic Solids |
	Metallic	Network	Group 8A		
Components That Occupy the Lattice Points	Metal atoms	Nonmetal atoms	Group 8A atoms	Discrete molecules	Ions
Bonding	Delocalized covalent	Directional covalent (leading to giant molecules)	London dispersion forces	Dipole–dipole and/or London dispersion forces	Ionic

To summarize, we find it convenient to classify solids according to what type of component occupies the lattice points. This leads to the classifications *atomic solids* (atoms at the lattice points), *molecular solids* (discrete, relatively small molecules at the lattice points), and *ionic solids* (ions at the lattice points). In addition, atomic solids are placed into the following subgroups based on the bonding that exists among the atoms in the solid: *metallic solids, network solids,* and *Group 8A solids.* In metallic solids, a special type of delocalized nondirectional covalent bonding occurs. In network solids, the atoms bond to each other with strong directional covalent bonds that lead to giant molecules, or networks, of atoms. In the Group 8A solids, the noble gas elements are attracted to each other with London dispersion forces. The classification of solids is summarized in Table 10.3.

The internal forces in a solid determine the properties of the solid.

The markedly different bonding present in the various atomic solids leads to dramatically different properties for the resulting solids. For example, although argon, copper, and diamond all are atomic solids, they have strikingly different properties. Argon (a Group 8A solid) has a very low melting point ($-189°C$), whereas diamond (a network solid) and copper (a metallic solid) melt at high temperatures (about 3500 and 1083°C, respectively). Copper is an excellent conductor of electricity, whereas argon and diamond are both insulators. Copper can be easily changed in shape; it is both malleable (can be formed into thin sheets) and ductile (can be pulled into a wire). Diamond, on the other hand, is the hardest natural substance known. We will explore the structure and bonding of atomic solids in the next two sections.

10.4 | Structure and Bonding in Metals

Metals are characterized by high thermal and electrical conductivity, malleability, and ductility. As we will see, these properties can be traced to the nondirectional covalent bonding found in metallic crystals.

A metallic crystal can be pictured as containing spherical atoms packed together and bonded to each other equally in all directions. We can model such a structure by packing uniform, hard spheres in a manner that most efficiently uses the available space. Such an arrangement is called **closest packing**. The spheres are packed in layers (Fig. 10.13), in which each sphere is surrounded by six others. In the second layer, the spheres do not lie directly over those in the first layer. Instead, each one occupies an indentation (or dimple) formed by three spheres in the first layer. In the third layer, the spheres can occupy the dimples of the second layer in two possible ways: They can occupy positions so that each sphere in the third layer lies directly over a sphere in the first layer [the *aba* arrangement; Fig. 10.13(a)], or they can occupy positions so that no sphere in the third layer lies over one in the first layer [the *abc* arrangement; Fig. 10.13(b)].

The closest packing model for metallic crystals assumes that metal atoms are uniform, hard spheres.

The *aba* arrangement has the *hexagonal* unit cell shown in Fig. 10.14, and the resulting structure is called the **hexagonal closest packed (hcp) structure**. The *abc* arrangement has a *face-centered cubic* unit cell (Fig. 10.15), and the resulting structure is called the **cubic closest packed (ccp) structure**. Note that in the hcp structure the

Figure 10.13 │ The closest packing arrangement of uniform spheres. In each layer a given sphere is surrounded by six others, creating six dimples, only three of which can be occupied in the next layer. (a) *aba* packing: The second layer is like the first, but it is displaced so that each sphere in the second layer occupies a dimple in the first layer. The spheres in the third layer occupy dimples in the second layer so that the spheres in the third layer lie directly over those in the first layer (*aba*). (b) *abc* packing: The spheres in the third layer occupy dimples in the second layer so that no spheres in the third layer lie above any in the first layer (*abc*). The fourth layer is like the first.

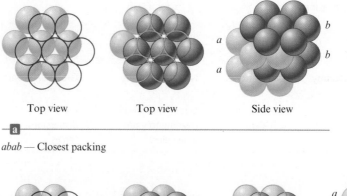

Top view   Top view   Side view

a

abab — Closest packing

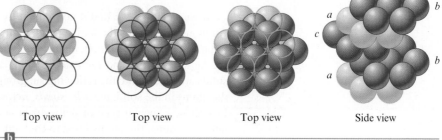

Top view   Top view   Top view   Side view

b

abca — Closest packing

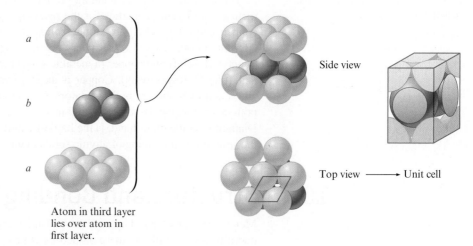

Figure 10.14 │ When spheres are closest packed so that the spheres in the third layer are directly over those in the first layer (*aba*), the unit cell is the hexagonal prism illustrated here in red.

Atom in third layer lies over atom in first layer.

Side view

Top view ⟶ Unit cell

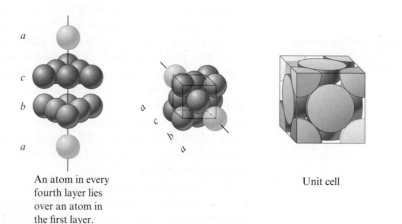

Figure 10.15 │ When spheres are packed in the *abc* arrangement, the unit cell is face-centered cubic. To make the cubic arrangement easier to see, the vertical axis has been tilted as shown.

An atom in every fourth layer lies over an atom in the first layer.

Unit cell

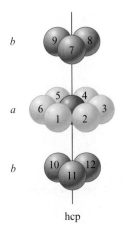

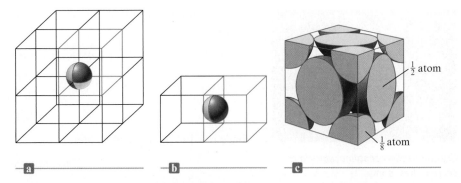

Figure 10.16 | The indicated sphere has 12 nearest neighbors.

Figure 10.17 | The net number of spheres in a face-centered cubic unit cell. (a) Note that the sphere on a corner of the colored cell is shared with 7 other unit cells (a total of 8). Thus $\frac{1}{8}$ of such a sphere lies within a given unit cell. Since there are 8 corners in a cube, there are 8 of these $\frac{1}{8}$ pieces, or 1 net sphere. (b) The sphere on the center of each face is shared by 2 unit cells, and thus each unit cell has $\frac{1}{2}$ of each of these types of spheres. There are 6 of these $\frac{1}{2}$ spheres to give 3 net spheres. (c) Thus the face-centered cubic unit cell contains 4 net spheres (all of the pieces can be assembled to give 4 spheres).

spheres in every other layer occupy the same vertical position ($ababab \cdots$), whereas in the ccp structure the spheres in every fourth layer occupy the same vertical position ($abcabca \cdots$). A characteristic of both structures is that each sphere has 12 equivalent nearest neighbors: 6 in the same layer, 3 in the layer above, and 3 in the layer below (that form the dimples). This is illustrated for the hcp structure in Fig. 10.16.

Knowing the *net* number of spheres (atoms) in a particular unit cell is important for many applications involving solids. To illustrate how to find the net number of spheres in a unit cell, we will consider a face-centered cubic unit cell (Fig. 10.17). Note that this unit cell is defined by the *centers* of the spheres on the cube's corners. Thus 8 cubes share a given sphere, so $\frac{1}{8}$ of this sphere lies inside each unit cell. Since a cube has 8 corners, there are $8 \times \frac{1}{8}$ pieces, or enough to put together 1 whole sphere. The spheres at the center of each face are shared by 2 unit cells, so $\frac{1}{2}$ of each lies inside a particular unit cell. Since the cube has 6 faces, we have $6 \times \frac{1}{2}$ pieces, or enough to construct 3 whole spheres. Thus the net number of spheres in a face-centered cubic unit cell is

$$\left(8 \times \frac{1}{8}\right) + \left(6 \times \frac{1}{2}\right) = 4$$

Calculating the Density of a Closest Packed Solid

Silver crystallizes in a cubic closest packed structure. The radius of a silver atom is 144 pm. Calculate the density of solid silver.

Solution

Density is mass per unit volume. Thus we need to know how many silver atoms occupy a given volume in the crystal. The structure is cubic closest packed, which means the unit cell is face-centered cubic, as shown in the accompanying figure.

We must find the volume of this unit cell for silver and the net number of atoms it contains. Note that in this structure the atoms touch along the diagonals for each face and not along the edges of the cube. Thus the length of the diagonal is $r + 2r + r$,

Crystalline silver contains cubic closest packed silver atoms.

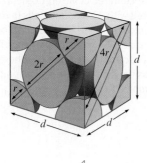

or 4r. We use this fact to find the length of the edge of the cube by the Pythagorean theorem:

$$d^2 + d^2 = (4r)^2$$
$$2d^2 = 16r^2$$
$$d^2 = 8r^2$$
$$d = \sqrt{8r^2} = r\sqrt{8}$$

Since $r = 144$ pm for a silver atom,

$$d = (144 \text{ pm})(\sqrt{8}) = 407 \text{ pm}$$

The volume of the unit cell is d^3, which is $(407 \text{ pm})^3$, or $6.74 \times 10^7 \text{ pm}^3$. We convert this to cubic centimeters as follows:

$$6.74 \times 10^7 \text{ pm}^3 \times \left(\frac{1.00 \times 10^{-10} \text{ cm}}{\text{pm}}\right)^3 = 6.74 \times 10^{-23} \text{ cm}^3$$

Since we know that the net number of atoms in the face-centered cubic unit cell is 4, we have 4 silver atoms contained in a volume of $6.74 \times 10^{-23} \text{ cm}^3$. The density is therefore

$$\text{Density} = \frac{\text{mass}}{\text{volume}} = \frac{(4 \text{ atoms})(107.9 \text{ g/mol})(1 \text{ mol}/6.022 \times 10^{23} \text{ atoms})}{6.74 \times 10^{-23} \text{ cm}^3}$$

$$= 10.6 \text{ g/cm}^3$$

See Exercises 10.51 through 10.54

Examples of metals that form cubic closest packed solids are aluminum, iron, copper, cobalt, and nickel. Magnesium and zinc are hexagonal closest packed. Calcium and certain other metals can crystallize in either of these structures. Some metals, however, assume structures that are not closest packed. For example, the alkali metals have structures characterized by a *body-centered cubic (bcc) unit cell* (see Fig. 10.9), where the spheres touch along the body diagonal of the cube. In this structure, each sphere has 8 nearest neighbors (count the number of atoms around the atom at the center of the unit cell), as compared with 12 in the closest packed structures. Why a particular metal adopts the structure it does is not well understood.

Bonding Models for Metals

Malleable: can be pounded into thin sheets.

Ductile: can be drawn to form a wire.

Any successful bonding model for metals must account for the typical physical properties of metals: malleability, ductility, and the efficient and uniform conduction of heat and electricity in all directions. Although the shapes of most pure metals can be changed relatively easily, most metals are durable and have high melting points. These facts indicate that the bonding in most metals is both *strong* and *nondirectional.* That is, although it is difficult to separate metal atoms, it is relatively easy to move them, provided the atoms stay in contact with each other.

The simplest picture that explains these observations is the *electron sea model,* which envisions a regular array of metal cations in a "sea" of valence electrons (Fig. 10.18). The mobile electrons can conduct heat and electricity, and the metal ions can be easily moved around as the metal is hammered into a sheet or pulled into a wire.

A related model that gives a more detailed view of the electron energies and motions is the **band model**, or **molecular orbital (MO) model**, for metals. In this model, the electrons are assumed to travel around the metal crystal in molecular orbitals formed from the valence atomic orbitals of the metal atoms (Fig. 10.19).

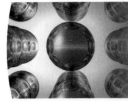

Chemical connections
Closest Packing of M & Ms

Although we usually think of scientists as dealing with esoteric and often toxic materials, sometimes they surprise us. For example, scientists at several prestigious universities have lately shown a lot of interest in M & M candies.

To appreciate the scientists' interest in M & Ms, we must consider the importance of packing atoms, molecules, or microcrystals in understanding the structures of solids. The most efficient use of space is the closest packing of uniform spheres, where 74% of the space is occupied by the spheres and 26% of space is left unoccupied. Although the structures of most pure metals can be explained in terms of closest packing, most other substances—such as many alloys and ceramics—consist of random arrays of microscopic particles. For this reason, it is of interest to study how such objects pack in a random way.

When uniform spheres, such as marbles, are poured into a large

container, the resulting random packing of the spheres results in only 64% of the space being occupied by the spheres. Thus it was very surprising when Princeton University chemist Salvatore Torquato and his colleagues at Cornell and North Carolina Central Universities discovered that, when the ellipsoidal-shaped M & Ms are poured into a large container, the candies occupy 73.5% of the available space. In other words, the randomly packed M & Ms occupy space with almost the same efficiency as closest packed spheres do.

Why do randomly packed ellipsoids occupy space so much more efficiently than randomly packed spheres? The scientists speculate that because the ellipsoids can tip and rotate in ways that spheres cannot, they can pack more closely to their neighbors.

According to Torquato, these results are important because they will help us better understand the properties of disordered materials

ranging from powders to glassy solids. He also says that M & Ms make ideal test objects because they are inexpensive and uniform and "you can eat the experiment afterward."

Princeton University. Photo by Denise Applewhite

Recall that in the MO model for the gaseous Li_2 molecule (Section 9.3), two widely spaced molecular orbital energy levels (bonding and antibonding) result when two identical atomic orbitals interact. However, when many metal atoms interact, as in a metal crystal, the large number of resulting molecular orbitals become more closely spaced and finally form a virtual continuum of levels, called *bands* (see Fig. 10.19).

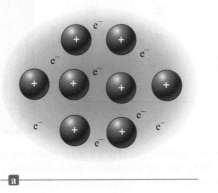
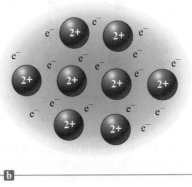

Figure 10.18 | The electron sea model for metals postulates a regular array of cations in a "sea" of valence electrons. (a) Representation of an alkali metal (Group 1A) with one valence electron. (b) Representation of an alkaline earth metal (Group 2A) with two valence electrons.

Number of interacting atomic orbitals

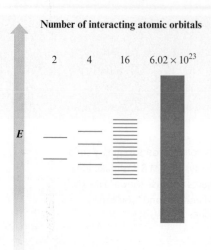

E

2   4   16 6.02×10^{23}

As an illustration, picture a magnesium metal crystal, which has an hcp structure. Since each magnesium atom has one 3*s* and three 3*p* valence atomic orbitals, a crystal with *n* magnesium atoms has available *n*(3*s*) and 3*n*(3*p*) orbitals to form the molecular orbitals (Fig. 10.20). Note that the core electrons are localized, as shown by their presence in the energy "well" around each magnesium atom. However, the valence electrons occupy closely spaced molecular orbitals, which are only partially filled.

The existence of empty molecular orbitals close in energy to filled molecular orbitals explains the thermal and electrical conductivity of metal crystals. Metals conduct electricity and heat very efficiently because of the availability of highly mobile electrons. For example, when an electric potential is placed across a strip of metal, for current to flow, electrons must be free to move. In the band model for metals, the electrons in partially filled bonds are mobile. These conduction electrons are free to travel throughout the metal crystal as dictated by the potential imposed on the metal. The molecular orbitals occupied by these conducting electrons are called *conduction bands*. These mobile electrons also account for the efficiency of the conduction of heat through metals. When one end of a metal rod is heated, the mobile electrons can rapidly transmit the thermal energy to the other end.

Figure 10.19 | The molecular orbital energy levels produced when various numbers of atomic orbitals interact. Note that for two atomic orbitals two rather widely spaced energy levels result. (Recall the description of H_2 in Section 9.2.) As more atomic orbitals are available to form molecular orbitals, the resulting energy levels are more closely spaced, finally producing a band of very closely spaced orbitals.

Metal Alloys

Because of the nature of the structure and bonding of metals, other elements can be introduced into a metallic crystal relatively easily to produce substances called alloys. An **alloy** is best defined as *a substance that contains a mixture of elements and has metallic properties.* Alloys can be conveniently classified into two types.

In a **substitutional alloy** some of the host metal atoms are *replaced* by other metal atoms of similar size. For example, in brass, approximately one-third of the atoms in the host copper metal have been replaced by zinc atoms [Fig. 10.21(a)]. Sterling silver (93% silver and 7% copper), pewter (85% tin, 7% copper, 6% bismuth, and 2% antimony), and plumber's solder (95% tin and 5% antimony) are other examples of substitutional alloys.

An **interstitial alloy** is formed when some of the interstices (holes) in the closest packed metal structure are occupied by small atoms [Fig. 10.21(b)]. Steel, the best-

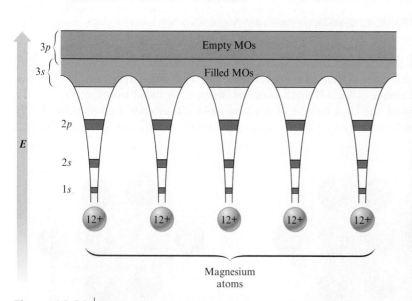

Figure 10.20 | (left) A representation of the energy levels (bands) in a magnesium crystal. The electrons in the 1*s*, 2*s*, and 2*p* orbitals are close to the nuclei and thus are localized on each magnesium atom as shown. However, the 3*s* and 3*p* valence orbitals overlap and mix to form molecular orbitals. Electrons in these energy levels can travel throughout the crystal. (right) Crystals of magnesium grown from a vapor.

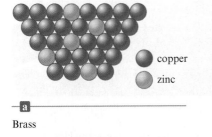

copper

zinc

Brass

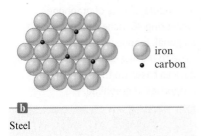

iron

carbon

b

Steel

Figure 10.21 | Two types of alloys.

Table 10.4 | The Composition of the Two Brands of Steel Tubing Commonly Used to Make Lightweight Racing Bicycles

Brand of Tubing	% C	% Si	% Mn	% Mo	% Cr
Reynolds	0.25	0.25	1.3	0.20	—
Columbus	0.25	0.30	0.65	0.20	1.0

known interstitial alloy, contains carbon atoms in the holes of an iron crystal. The presence of the interstitial atoms changes the properties of the host metal. Pure iron is relatively soft, ductile, and malleable due to the absence of directional bonding. The spherical metal atoms can be rather easily moved with respect to each other. However, when carbon, which forms strong directional bonds, is introduced into an iron crystal, the presence of the directional carbon–iron bonds makes the resulting alloy harder, stronger, and less ductile than pure iron. The amount of carbon directly affects the properties of steel. *Mild steels,* containing less than 0.2% carbon, are ductile and malleable and are used for nails, cables, and chains. *Medium steels,* containing 0.2 to 0.6% carbon, are harder than mild steels and are used in rails and structural steel beams. *High-carbon steels,* containing 0.6 to 1.5% carbon, are tough and hard and are used for springs, tools, and cutlery.

Many types of steel also contain elements in addition to iron and carbon. Such steels are often called *alloy steels,* and they can be viewed as being mixed interstitial (carbon) and substitutional (other metals) alloys. Bicycle frames, for example, are constructed from a wide variety of alloy steels. The compositions of the two brands of steel tubing most commonly used in expensive racing bicycles are given in Table 10.4.

10.5 | Carbon and Silicon: Network Atomic Solids

Many atomic solids contain strong directional covalent bonds to form a solid that might best be viewed as a "giant molecule." We call these substances **network solids**. In contrast to metals, these materials are typically brittle and do not efficiently conduct heat or electricity. To illustrate network solids, in this section we will discuss two very important elements, carbon and silicon, and some of their compounds.

The two most common forms of carbon, diamond and graphite, are typical network solids. In diamond, the hardest naturally occurring substance, each carbon atom is surrounded by a tetrahedral arrangement of other carbon atoms to form a huge molecule [Fig. 10.22(a)]. This structure is stabilized by covalent bonds, which, in terms of the

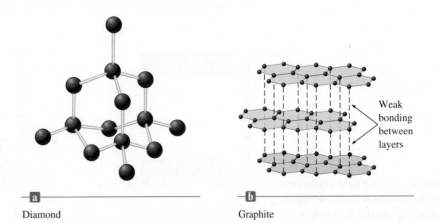

Figure 10.22 | The structures of diamond and graphite. In each case only a small part of the entire structure is shown.

Weak bonding between layers

a

Diamond

b

Graphite

Chemical connections

Graphene—Miracle Substance?

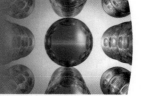

Graphite is a fascinating substance consisting of "chicken-wire" layers of carbon atoms. Due to the strong carbon–carbon bonds in each of its layers, graphite is even more thermodynamically stable than diamond. Because of the unusual stability of graphite, scientists have long wondered whether graphene (the individual chicken-wire layers) could exist independently. This question was answered only recently by Andre Gein and Konstantin Novoselov, two scientists working at the University of Manchester in the United Kingdom, when they were able to pull layers a single atom thick from graphite using Scotch tape. This work was deemed so important that Gein and Novoselov were awarded the Nobel Prize in physics in 2010.

Since the isolation of sheets of graphene by Gein and Novoselov, research in the properties and possible uses of this material has virtually exploded. Graphene has truly amazing properties. For example, it is believed to be the strongest material known, more than 100 times stronger than steel. Also, it is stiffer than diamond, yet it can be stretched like rubber.

Much of the interest in graphene centers on the fact that it is the best conductor of heat and electricity known. Electrons travel through graphene at ultrafast speeds. In fact, electrons in graphene exhibit the fractional quantum Hall effect: The electrons act collectively as if they are particles with only a fraction of the charge of an electron. Because of its exceptional thermal and electrical conductivities, graphene appears to be an ideal candidate for future electronic devices. Scientists at IBM have already built graphene-based transistors that can switch on and off 26 billion times per second, far faster than conventional silicon-based devices.

One problem that has hampered the development of graphene-based devices is the difficulty in producing large sheets of graphene. Byung Hee Hong and his team, from South Korea, have recently reported making rectangular sheets of graphene measuring 30 inches along the diagonal. It appears that we are very close to making commercial electronic devices based on the amazing properties of graphene.

A flexible electrode made of graphene.

Byung Hee Hong. Photo by Ji Hye Hong

localized electron model, are formed by the overlap of *sp*³ hybridized carbon atomic orbitals.

It is also useful to consider the bonding among the carbon atoms in diamond in terms of the molecular orbital model. Energy-level diagrams for diamond and a typical metal are given in Fig. 10.23. Recall that the conductivity of metals can be explained

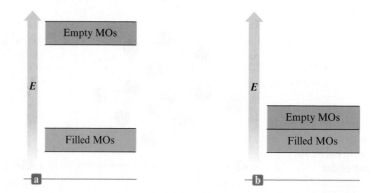

Figure 10.23 | Partial representation of the molecular orbital energies in (a) diamond and (b) a typical metal.

by postulating that electrons are excited from filled levels into the very near empty levels, or conduction bands. However, note that in the energy-level diagram for diamond there is *a large gap between the filled and the empty levels.* This means that electrons cannot be transferred easily to the empty conduction bands. As a result, diamond is not expected to be a good electrical conductor. In fact, this prediction of the model agrees exactly with the observed behavior of diamond, which is known to be an electrical insulator—it does not conduct an electric current.

Graphite is very different from diamond. While diamond is hard, basically colorless, and an insulator, graphite is slippery, black, and a conductor. These differences, of course, arise from the differences in bonding in the two types of solids. In contrast to the tetrahedral arrangement of carbon atoms in diamond, the structure of graphite is based on layers of carbon atoms arranged in fused six-membered rings [Fig. 10.22(b)]. Each carbon atom in a particular layer of graphite is surrounded by the three other carbon atoms in a trigonal planar arrangement with 120-degree bond angles. The localized electron model predicts sp^2 hybridization in this case. The three sp^2 orbitals on each carbon are used to form σ bonds with three other carbon atoms. One $2p$ orbital remains unhybridized on each carbon and is perpendicular to the plane of carbon atoms, as shown in Fig. 10.24. These orbitals combine to form a group of closely spaced π molecular orbitals that are important in two ways. First, they contribute significantly to the stability of the graphite layers because of the π bond formation. Second, the π molecular orbitals with their delocalized electrons account for the electrical conductivity of graphite. These closely spaced orbitals are exactly analogous to the conduction bands found in metal crystals.

Graphite is often used as a lubricant in locks (where oil is undesirable because it collects dirt). The slipperiness that is characteristic of graphite can be explained by noting that graphite has very strong bonding *within* the layers of carbon atoms but little bonding *between* the layers (the valence electrons are all used to form σ and π bonds among carbons within the layers). This arrangement allows the layers to slide past one another quite readily. Graphite's layered structure is shown in Fig. 10.25. This is in contrast to diamond, which has uniform bonding in all directions in the crystal.

Because of their extreme hardness, diamonds are used extensively in industrial cutting implements. Thus it is desirable to convert cheaper graphite to diamond. As we might expect from the higher density of diamond (3.5 g/cm³) compared with that of graphite (2.2 g/cm³), this transformation can be accomplished by applying very high pressures to graphite. The application of 150,000 atm of pressure at 2800°C converts graphite virtually completely to diamond. The high temperature is required to break the strong bonds in graphite so the rearrangement can occur.

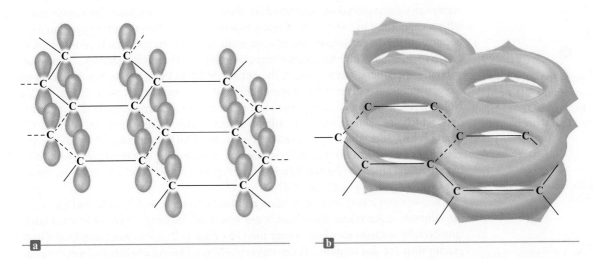

Figure 10.24 | The *p* orbitals (a) perpendicular to the plane of the carbon ring system in graphite can combine to form (b) an extensive π-bonding network.

Figure 10.25 | Graphite consists of layers of carbon atoms.

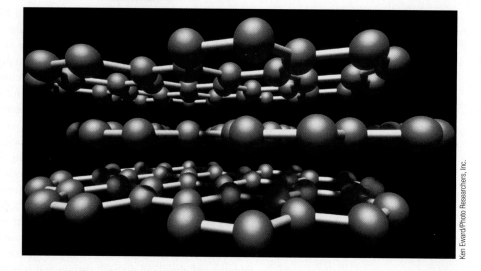

Ken Eward/Photo Researchers, Inc.

Silicon is an important constituent of the compounds that make up the earth's crust. In fact, silicon is to geology as carbon is to biology. Just as carbon compounds are the basis for most biologically significant systems, silicon compounds are fundamental to most of the rocks, sands, and soils found in the earth's crust. However, although carbon and silicon are next to each other in Group 4A of the periodic table, the carbon-based compounds of biology and the silicon-based compounds of geology have markedly different structures. Carbon compounds typically contain long strings of carbon–carbon bonds, whereas the most stable silicon compounds involve chains with silicon–oxygen bonds.

The fundamental silicon–oxygen compound is **silica**, which has the empirical formula SiO_2. Knowing the properties of the similar compound carbon dioxide, one might expect silica to be a gas that contains discrete SiO_2 molecules. In fact, nothing could be further from the truth—quartz and some types of sand are typical of the materials composed of silica. What accounts for this difference? The answer lies in the bonding.

Recall that the Lewis structure of CO_2 is

$$\ddot{O}\!=\!C\!=\!\ddot{O}$$

and that each C=O bond can be viewed as a combination of a σ bond involving a carbon sp hybrid orbital and a π bond involving a carbon $2p$ orbital. On the contrary, silicon cannot use its valence $3p$ orbitals to form strong π bonds with oxygen, mainly because of the larger size of the silicon atom and its orbitals, which results in less effective overlap with the smaller oxygen orbitals. Therefore, instead of forming π bonds, the silicon atom satisfies the octet rule by forming single bonds with four oxygen atoms, as shown in the representation of the structure of quartz in Fig. 10.26. Note that each silicon atom is at the center of a tetrahedral arrangement of oxygen atoms, which are shared with other silicon atoms. Although the empirical formula for quartz is SiO_2, the structure is based on a *network* of SiO_4 tetrahedra with shared oxygen atoms rather than discrete SiO_2 molecules. It is obvious that the differing abilities of carbon and silicon to form π bonds with oxygen have profound effects on the structures and properties of CO_2 and SiO_2.

Compounds closely related to silica and found in most rocks, soils, and clays are the **silicates**. Like silica, the silicates are based on interconnected SiO_4 tetrahedra. However, in contrast to silica, where the O/Si ratio is 2:1, silicates have O/Si ratios greater than 2:1 and contain silicon–oxygen *anions*. This means that to form the neutral solid silicates, cations are needed to balance the excess negative charge. In other

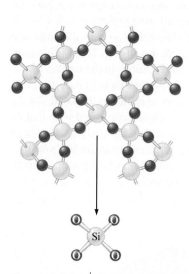

Figure 10.26 | (top) The structure of quartz (empirical formula SiO_2). Quartz contains chains of SiO_4 tetrahedra (bottom) that share oxygen atoms.

The bonding in the CO_2 molecule was described in Section 9.1.

Figure 10.27 | Examples of silicate anions, all of which are based on SiO_4^{4-} tetrahedra.

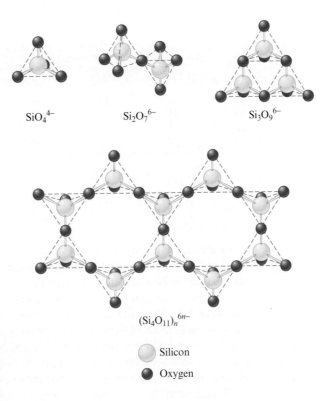

SiO_4^{4-} $Si_2O_7^{6-}$ $Si_3O_9^{6-}$

$(Si_4O_{11})_n^{6n-}$

○ Silicon

● Oxygen

A glassblower in Venice, Italy.

words, silicates are salts containing metal cations and polyatomic silicon–oxygen anions. Examples of important silicate anions are shown in Fig. 10.27.

When silica is heated above its melting point (about 1600°C) and cooled rapidly, an amorphous solid called a **glass** results (Fig. 10.28). Note that a glass contains a good deal of disorder, in contrast to the crystalline nature of quartz. Glass more closely resembles a very viscous solution than it does a crystalline solid. Common glass results when substances such as Na_2CO_3 are added to the silica melt, which is then cooled. The properties of glass can be varied greatly by varying the additives. For example, addition of B_2O_3 produces a glass (called *borosilicate glass*) that expands and contracts little under large temperature changes. Thus it is useful for labware and cooking utensils. The most common brand name for this glass is Pyrex. The addition of K_2O produces an especially hard glass that can be ground to the precise shapes needed for eyeglass and contact lenses. The compositions of several types of glass are shown in Table 10.5.

Figure 10.28 | Two-dimensional representations of (a) a quartz crystal and (b) a quartz glass. Note how the irregular structure of the glass contrasts with the regular structure of the crystal.

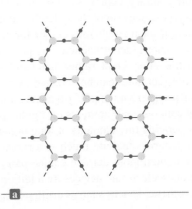

a

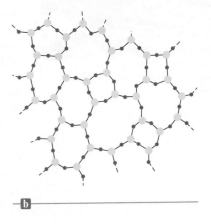

b

Table 10.5 | Compositions of Some Common Types of Glass

Type of Glass	Percentages of Various Components						
	SiO_2	CaO	Na_2O	B_2O_3	Al_2O_3	K_2O	MgO
Window (soda-lime glass)	72	11	13	—	0.3	3.8	—
Cookware (aluminosilicate glass)	55	15	—	—	20	—	10
Heat-resistant (borosilicate glass)	76	3	5	13	2	0.5	—
Optical	69	12	6	0.3	—	12	—

Ceramics

Ceramics are typically made from clays (which contain silicates) and hardened by firing at high temperatures. Ceramics are nonmetallic materials that are strong, brittle, and resistant to heat and attack by chemicals.

Like glass, ceramics are based on silicates, but with that the resemblance ends. Glass can be melted and remelted as often as desired, but once a ceramic has been hardened, it is resistant to extremely high temperatures. This behavior results from the very different structures of glasses and ceramics. A glass is a *homogeneous,* noncrystalline "frozen solution," and a ceramic is *heterogeneous.* A ceramic contains two phases: minute crystals of silicates that are suspended in a glassy cement.

To understand how ceramics harden, it is necessary to know something about the structure of clays. Clays are formed by the weathering action of water and carbon dioxide on the mineral feldspar, which is a mixture of silicates with empirical formulas such as $K_2O \cdot Al_2O_3 \cdot 6SiO_2$ and $Na_2O \cdot Al_2O_3 \cdot 6SiO_2$. Feldspar is really an *aluminosilicate* in which aluminum as well as silicon atoms are part of the oxygen-bridged polyanion. The weathering of feldspar produces kaolinite, consisting of tiny thin platelets with the empirical formula $Al_2Si_2O_5(OH)_4$. When dry, the platelets cling together; when water is present, they can slide over one another, giving clay its plasticity. As clay dries, the platelets begin to interlock again. When the remaining water is driven off during firing, the silicates and cations form a glass that binds the tiny crystals of kaolinite.

Ceramics have a very long history. Rocks, which are natural ceramic materials, served as the earliest tools. Later, clay vessels dried in the sun or baked in fires served as containers for food and water. These early vessels were no doubt crude and quite porous. With the discovery of glazing, which probably occurred about 3000 B.C. in Egypt, pottery became more serviceable as well as more beautiful. Prized porcelain is essentially the same material as crude earthenware, but specially selected clays and glazings are used for porcelain and the clay object is fired at a very high temperature.

Although ceramics have been known since antiquity, they are not obsolete materials. On the contrary, ceramics constitute one of the most important classes of "high-tech" materials. Because of their stability at high temperatures and resistance to corrosion, ceramics seem an obvious choice for constructing jet and automobile engines in which the greatest fuel efficiencies are possible at very high temperatures. But ceramics are brittle—they break rather than bend—which limits their usefulness. However, more flexible ceramics can be obtained by adding small amounts of organic polymers. Taking their cue from natural "organoceramics" such as teeth and shells of sea creatures that contain small amounts of organic polymers, materials scientists have found that incorporating tiny amounts of long organic molecules into ceramics as they form produces materials that are much less subject to fracture. These materials should be useful for lighter, more durable engine parts, as well as for flexible superconducting wire and microelectronic devices. In addition, these organoceramics hold great promise for prosthetic devices such as artificial bones.

An artist paints a ceramic vase before glazing.

Mathias Oppersdorff/Photo Researchers, Inc.

Semiconductors

Elemental silicon has the same structure as diamond, as might be expected from its position in the periodic table (in Group 4A directly under carbon). Recall that in diamond there is a large energy gap between the filled and empty molecular orbitals (see Fig. 10.23). This gap prevents excitation of electrons to the empty molecular orbitals (conduction bands) and makes diamond an insulator. In silicon the situation is similar, but the energy gap is smaller. A few electrons can cross the gap at 25°C, making silicon a **semiconducting element**, or **semiconductor**. In addition, at higher temperatures, where more energy is available to excite electrons into the conduction bands, the conductivity of silicon increases. This is typical behavior for a semiconducting element and is in contrast to that of metals, whose conductivity decreases with increasing temperature.

The small conductivity of silicon can be enhanced at normal temperatures if the silicon crystal is *doped* with certain other elements. For example, when a small fraction of silicon atoms is replaced by arsenic atoms, each having *one more* valence electron than silicon, extra electrons become available for conduction [Fig. 10.29(a)]. This produces an **n-type semiconductor**, a substance whose conductivity is increased by doping it with atoms having more valence electrons than the atoms in the host crystal. These extra electrons lie close in energy to the conduction bands and can be easily excited into these levels, where they can conduct an electric current [Fig. 10.30(a)].

We also can enhance the conductivity of silicon by doping the crystal with an element such as boron, which has only three valence electrons, *one fewer* than silicon. Because boron has one less electron than is required to form the bonds with the surrounding silicon atoms, an electron vacancy, or *hole,* is created [Fig. 10.29(b)]. As an electron fills this hole, it leaves a new hole, and this process can be repeated. Thus the hole advances through the crystal in a direction opposite to the movement of the electrons jumping to fill the hole. Another way of thinking about this phenomenon is that in pure silicon each atom has four valence electrons and the low-energy molecular orbitals are exactly filled. Replacing silicon atoms with boron atoms leaves vacancies in these molecular orbitals [see Fig. 10.30(b)]. This means that there is only one electron in some of the molecular orbitals, and these unpaired electrons can function as conducting electrons. Thus the substance becomes a better conductor. When semiconductors are doped with atoms having fewer valence electrons than the atoms of the host crystal, they are called **p-type semiconductors**, so named because the positive holes can be viewed as the charge carriers.

Most important applications of semiconductors involve connection of a p-type and an n-type to form a **p–n junction**. Figure 10.31(a) shows a typical junction; the red dots represent excess electrons in the n-type semiconductor, and the white circles represent holes (electron vacancies) in the p-type semiconductor. At the junction, a small number of electrons migrate from the n-type region into the p-type region, where there

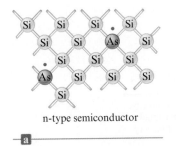

n-type semiconductor

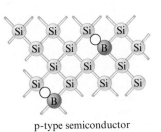

p-type semiconductor

Figure 10.29 | (a) A silicon crystal doped with arsenic, which has one more valence electron than silicon. (b) A silicon crystal doped with boron, which has one less electron than silicon.

Electrons must be in singly occupied molecular orbitals to conduct a current.

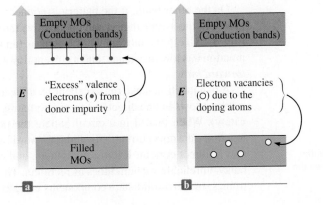

Figure 10.30 | Energy-level diagrams for (a) an n-type semiconductor and (b) a p-type semiconductor.

Figure 10.31 | The p–n junction involves the contact of a p-type and an n-type semiconductor. (a) The charge carriers of the p-type region are holes (◯). In the n-type region, the charge carriers are electrons (●). (b) No current flows (reverse bias). (c) Current readily flows (forward bias). Note that each electron that crosses the boundary leaves a hole behind. Thus the electrons and the holes move in opposite directions.

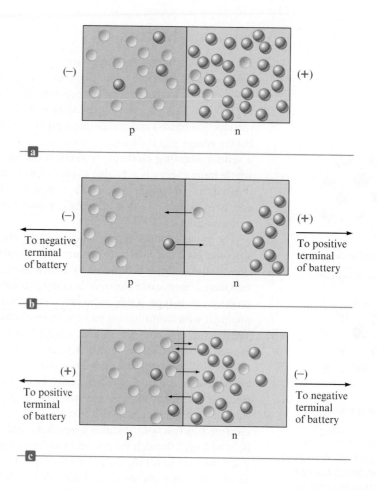

are vacancies in the low-energy molecular orbitals. The effect of these migrations is to place a negative charge on the p-type region (since it now has a surplus of electrons) and a positive charge on the n-type region (since it has lost electrons, leaving holes in its low-energy molecular orbitals). This charge buildup, called the *contact potential,* or *junction potential,* prevents further migration of electrons.

Now suppose an external electric potential is applied by connecting the negative terminal of a battery to the p-type region and the positive terminal to the n-type region. The situation represented in Fig. 10.31(b) results. Electrons are drawn toward the positive terminal, and the resulting holes move toward the negative terminal—exactly opposite to the natural flow of electrons at the p–n junction. The junction resists the imposed current flow in this direction and is said to be under *reverse bias.* No current flows through the system.

On the other hand, if the battery is connected so that the negative terminal is connected to the n-type region and the positive terminal is connected to the p-type region [Fig. 10.31(c)], the movement of electrons (and holes) is in the favored direction. The junction has low resistance, and a current flows easily. The junction is said to be under *forward bias.*

A p–n junction makes an excellent *rectifier,* a device that produces a pulsating direct current (flows in one direction) from alternating current (flows in both directions alternately). When placed in a circuit where the potential is constantly reversing, a p–n junction transmits current only under forward bias, thus converting the alternating current to direct current. Radios, computers, and other electronic devices formerly used bulky, unreliable vacuum tubes as rectifiers. The p–n junction has revolutionized electronics; modern solid-state components contain p–n junctions in printed circuits.

Printed circuits are discussed in the Chemical Connections feature on the student website.

10.6 | Molecular Solids

So far we have considered solids in which atoms occupy the lattice positions. In some of these substances (network solids), the solid can be considered to be one giant molecule. In addition, there are many types of solids that contain discrete molecular units at each lattice position. A common example is ice, where the lattice positions are occupied by water molecules [see Fig. 10.12(c)]. Other examples are dry ice (solid carbon dioxide), some forms of sulfur that contain S_8 molecules [Fig. 10.32(a)], and certain forms of phosphorus that contain P_4 molecules [Fig. 10.32(b)]. These substances are characterized by strong covalent bonding *within* the molecules but relatively weak forces *between* the molecules. For example, it takes only 6 kJ of energy to melt 1 mole of solid water (ice) because only intermolecular (H_2O—H_2O) interactions must be overcome. However, 470 kJ of energy is required to break 1 mole of covalent O—H bonds. The differences between the covalent bonds within the molecules and the forces between the molecules are apparent from the comparison of the interatomic and intermolecular distances in solids shown in Table 10.6.

The forces that exist among the molecules in a molecular solid depend on the nature of the molecules. Many molecules such as CO_2, I_2, P_4, and S_8 have no dipole moment, and the intermolecular forces are London dispersion forces. Because these forces are

Figure 10.32 | (a) Sulfur crystals (yellow) contain S_8 molecules. (b) White phosphorus (containing P_4 molecules) is so reactive with the oxygen in air that it must be stored under water.

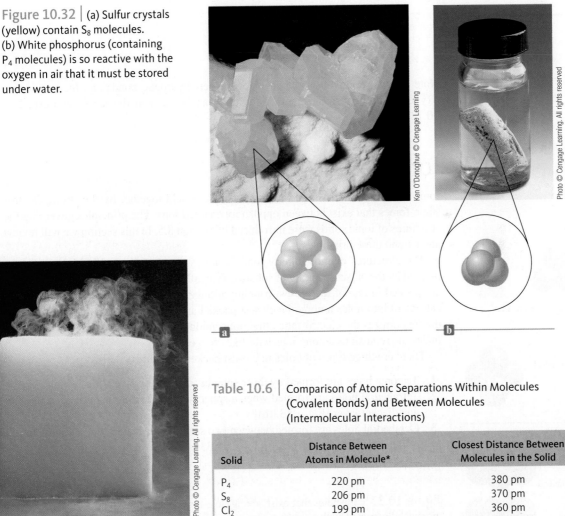

A "steaming" piece of dry ice.

Table 10.6 | Comparison of Atomic Separations Within Molecules (Covalent Bonds) and Between Molecules (Intermolecular Interactions)

Solid	Distance Between Atoms in Molecule*	Closest Distance Between Molecules in the Solid
P_4	220 pm	380 pm
S_8	206 pm	370 pm
Cl_2	199 pm	360 pm

*The shorter distances within the molecules indicate stronger bonding.

often relatively small, we might expect all these substances to be gaseous at 25°C, as is the case for carbon dioxide. However, as the size of the molecules increases, the London forces become quite large, causing many of these substances to be solids at 25°C.

When molecules do have dipole moments, their intermolecular forces are significantly greater, especially when hydrogen bonding is possible. Water molecules are particularly well suited to interact with each other because each molecule has two polar O—H bonds and two lone pairs on the oxygen atom. This can lead to the association of four hydrogen atoms with each oxygen: two by covalent bonds and two by dipole forces:

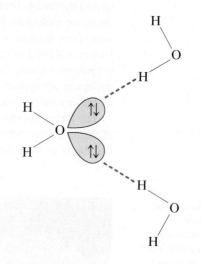

Note the two relatively short covalent oxygen–hydrogen bonds and the two longer oxygen–hydrogen dipole interactions that can be seen in the ice structure in Fig. 10.12(c).

10.7 | Ionic Solids

Ionic solids are stable, high-melting substances held together by the strong electrostatic forces that exist between oppositely charged ions. The principles governing the structures of ionic solids were introduced in Section 8.5. In this section we will review and extend these principles.

The structures of most binary ionic solids, such as sodium chloride, can be explained by the closest packing of spheres. Typically, the larger ions, usually the anions, are packed in one of the closest packing arrangements (hcp or ccp), and the smaller cations fit into holes among the closest packed anions. The packing is done in a way that maximizes the electrostatic attractions among oppositely charged ions and minimizes the repulsions among ions with like charges.

There are three types of holes in closest packed structures:

1. Trigonal holes are formed by three spheres in the same layer [Fig. 10.33(a)].
2. Tetrahedral holes are formed when a sphere sits in the dimple of three spheres in an adjacent layer [Fig. 10.33(b)].
3. Octahedral holes are formed between two sets of three spheres in adjoining layers of the closest packed structures [Fig. 10.33(c)].

Trigonal hole

Tetrahedral hole

Octahedral hole

Figure 10.33 | The holes that exist among closest packed uniform spheres. (a) The trigonal hole formed by three spheres in a given plane. (b) The tetrahedral hole formed when a sphere occupies a dimple formed by three spheres in an adjacent layer. (c) The octahedral hole formed by six spheres in two adjacent layers.

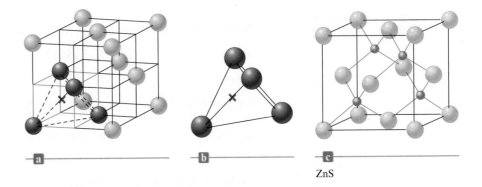

Figure 10.34 | (a) The location (red X) of a tetrahedral hole in the face-centered cubic unit cell. (b) One of the tetrahedral holes. (c) The unit cell for ZnS where the S^{2-} ions (yellow) are closest packed with the Zn^{2+} ions (red) in alternating tetrahedral holes.

ZnS

For spheres of a given diameter, the holes increase in size in the order

trigonal < tetrahedral < octahedral

In fact, trigonal holes are so small that they are never occupied in binary ionic compounds. Whether the tetrahedral or octahedral holes in a given binary ionic solid are occupied depends mainly on the *relative* sizes of the anion and cation. For example, in zinc sulfide the S^{2-} ions (ionic radius = 180 pm) are arranged in a cubic closest packed structure with the smaller Zn^{2+} ions (ionic radius = 70 pm) in the tetrahedral holes. The locations of the tetrahedral holes in the face-centered cubic unit cell of the ccp structure are shown in Fig. 10.34(a). Note from this figure that there are eight tetrahedral holes in the unit cell. Also recall from the discussion in Section 10.4 that there are four net spheres in the face-centered cubic unit cell. Thus there are *twice as many tetrahedral holes as packed anions* in the closest packed structure. Zinc sulfide must have the same number of S^{2-} ions and Zn^{2+} ions to achieve electrical neutrality. Thus in the zinc sulfide structure only *half* the tetrahedral holes contain Zn^{2+} ions, as shown in Fig. 10.34(c).

The structure of sodium chloride can be described in terms of a cubic closest packed array of Cl^- ions with Na^+ ions in all the octahedral holes. The locations of the octahedral holes in the face-centered cubic unit cell are shown in Fig. 10.35(a). The easiest octahedral hole to find in this structure is the one at the center of the cube. Note that this hole is surrounded by six spheres, as is required to form an octahedron. The remaining octahedral holes are shared with other unit cells and are more difficult to visualize. However, it can be shown that the number of octahedral holes in the ccp structure is the *same* as the number of packed anions. Figure 10.35(b) shows the structure for sodium chloride that results from Na^+ ions filling all the octahedral holes in a ccp array of Cl^- ions.

A great variety of ionic solids exists. Our purpose in this section is not to give an exhaustive treatment of ionic solids, but to emphasize the fundamental principles governing their structures. As we have seen, the most useful model for explaining the structures of these solids regards the ions as hard spheres that are packed to maximize attractions and minimize repulsions.

Closest packed structures contain twice as many tetrahedral holes as packed spheres. Closest packed structures contain the same number of octahedral holes as packed spheres.

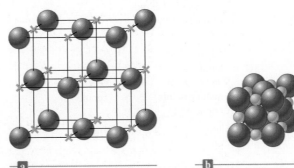

Figure 10.35 | (a) The locations (gray X) of the octahedral holes in the face-centered cubic unit cell. (b) Representation of the unit cell for solid NaCl. The Cl^- ions (green spheres) have a ccp arrangement with Na^+ ions (gray spheres) in all the octahedral holes. Note that this representation shows the idealized closest packed structure of NaCl. In the actual structure, the Cl^- ions do not quite touch.

Example 10.3

Determining the Number of Ions in a Unit Cell

Determine the net number of Na^+ and Cl^- ions in the sodium chloride unit cell.

Solution

Note from Fig. 10.35(b) that the Cl^- ions are cubic closest packed and thus form a face-centered cubic unit cell. There is a Cl^- ion on each corner and one at the center of each face of the cube. Thus the net number of Cl^- ions present in a unit cell is

$$8\left(\tfrac{1}{8}\right) + 6\left(\tfrac{1}{2}\right) = 4$$

The Na^+ ions occupy the octahedral holes located in the center of the cube and midway along each edge. The Na^+ ion in the center of the cube is contained entirely in the unit cell, whereas those on the edges are shared by four unit cells (four cubes share a common edge). Since the number of edges in a cube is 12, the net number of Na^+ ions present is

$$1(1) + 12\left(\tfrac{1}{4}\right) = 4$$

We have shown that the net number of ions in a unit cell is 4 Na^+ ions and 4 Cl^- ions, which agrees with the 1:1 stoichiometry of sodium chloride.

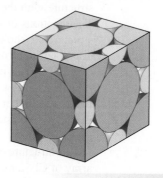

See Exercises 10.69 through 10.76

In this chapter we have considered various types of solids. Table 10.7 summarizes these types of solids and some of their properties.

Table 10.7 | Types and Properties of Solids

Type of Solid	Atomic			Molecular	Ionic
	Network	Metallic	Group 8A		
Structural Unit	Atom	Atom	Atom	Molecule	Ion
Type of Bonding	Directional covalent bonds	Nondirectional covalent bonds involving electrons that are delocalized throughout the crystal	London dispersion forces	Polar molecules: dipole–dipole interactions Nonpolar molecules: London dispersion forces	Ionic
Typical Properties	Hard	Wide range of hardness		Soft	Hard
	High melting point	Wide range of melting points	Very low melting point	Low melting point	High melting point
	Insulator	Conductor		Insulator	Insulator
Examples	Diamond	Silver Iron Brass	Argon(s)	Ice (solid H_2O) Dry ice (solid CO_2)	Sodium chloride Calcium fluoride

Interactive
Example 10.4

Sign in at http://login.cengagebrain
.com to try this Interactive Example
in OWL.

Types of Solids

Using Table 10.7, classify each of the following substances according to the type of solid it forms.

a. Gold

b. Carbon dioxide

c. Lithium fluoride

d. Krypton

Solution

a. Solid gold is an atomic solid with metallic properties.

b. Solid carbon dioxide contains nonpolar carbon dioxide molecules and is a molecular solid.

c. Solid lithium fluoride contains Li^+ and F^- ions and is a binary ionic solid.

d. Solid krypton contains krypton atoms that can interact only through London dispersion forces. It is an atomic solid but has properties characteristic of a molecular solid with nonpolar molecules.

See Exercises 10.81 and 10.82

10.8 | Vapor Pressure and Changes of State

Vapor is the usual term for the gas phase of a substance that exists as a solid or liquid at 25°C and 1 atm.

ΔH_{vap} for water at 100°C is 40.7 kJ/mol.

Figure 10.36 | Behavior of a liquid in a closed container. (a) Initially, net evaporation occurs as molecules are transferred from the liquid to the vapor phase, so the amount of liquid decreases. (b) As the number of vapor molecules increases, the rate of return to the liquid (condensation) increases, until finally the rate of condensation equals the rate of evaporation. The system is at equilibrium, and no further changes occur in the amounts of vapor or liquid.

Now that we have considered the general properties of the three states of matter, we can explore the processes by which matter changes state. One very familiar example of a change in state occurs when a liquid evaporates from an open container. This is clear evidence that the molecules of a liquid can escape the liquid's surface and form a gas, a process called **vaporization**, or **evaporation**. Vaporization is endothermic because energy is required to overcome the relatively strong intermolecular forces in the liquid. The energy required to vaporize 1 mole of a liquid at a pressure of 1 atm is called the **heat of vaporization**, or the **enthalpy of vaporization**, and is usually symbolized as ΔH_{vap}.

The endothermic nature of vaporization has great practical significance; in fact, one of the most important roles that water plays in our world is to act as a coolant. Because of the strong hydrogen bonding among its molecules in the liquid state, water has an unusually large heat of vaporization (40.7 kJ/mol). A significant portion of the sun's energy that reaches earth is spent evaporating water from the oceans, lakes, and rivers rather than warming the earth. The vaporization of water is also crucial to the body's temperature-control system through evaporation of perspiration.

Vapor Pressure

When a liquid is placed in a closed container, the amount of liquid at first decreases but eventually becomes constant. The decrease occurs because there is an initial net transfer of molecules from the liquid to the vapor phase (Fig. 10.36). This evaporation process occurs at a constant rate at a given temperature (Fig. 10.37). However, the reverse process is different. Initially, as the number of vapor molecules increases, so does the rate of return of these molecules to the liquid. The process by which vapor molecules re-form a liquid is called **condensation**. Eventually, enough vapor molecules are present above the liquid so that the rate of condensation equals the rate of evaporation (see Fig. 10.37). *At this point no further net change occurs in the amount of liquid or vapor because the two opposite processes exactly balance each other;* the system is at **equilibrium**. Note that this system is highly *dynamic* on the molecular

Figure 10.37 | The rates of condensation and evaporation over time for a liquid sealed in a closed container. The rate of evaporation remains constant and the rate of condensation increases as the number of molecules in the vapor phase increases, until the two rates become equal. At this point, the equilibrium vapor pressure is attained.

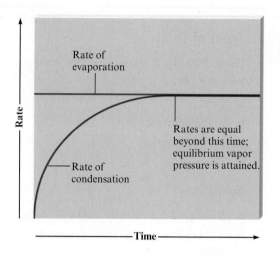

A system at equilibrium is dynamic on the molecular level but shows no macroscopic changes.

level—molecules are constantly escaping from and entering the liquid at a high rate. However, there is no *net* change because the two opposite processes just *balance* each other.

The pressure of the vapor present at equilibrium is called the **equilibrium vapor pressure**, or more commonly, the **vapor pressure** of the liquid. A simple barometer can measure the vapor pressure of a liquid [Fig. 10.38(a)]. The liquid is injected at the bottom of the tube of mercury and floats to the surface because the mercury is so dense. A portion of the liquid evaporates at the top of the column, producing a vapor whose pressure pushes some mercury out of the tube. When the system reaches equilibrium, the vapor pressure can be determined from the change in the height of the mercury column since

$$P_{atmosphere} = P_{vapor} + P_{Hg\ column}$$

Thus

$$P_{vapor} = P_{atmosphere} - P_{Hg\ column}$$

Figure 10.38 | (a) The vapor pressure of a liquid can be measured easily using a simple barometer of the type shown here. (b) The three liquids, water, ethanol (C_2H_5OH), and diethyl ether [$(C_2H_5)_2O$], have quite different vapor pressures. Ether is by far the most volatile of the three. Note that in each case a little liquid remains (floating on the mercury).

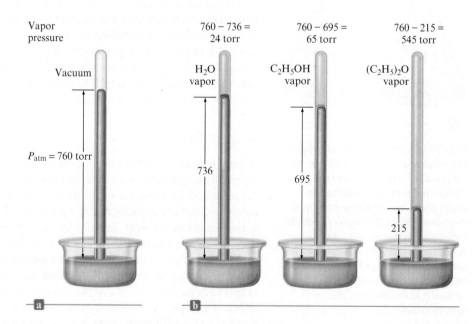

The vapor pressures of liquids vary widely [Fig. 10.38(b)]. Liquids with high vapor pressures are said to be *volatile*—they evaporate rapidly from an open dish. The vapor pressure of a liquid is principally determined by the size of the *intermolecular forces* in the liquid. Liquids in which the intermolecular forces are large have relatively low vapor pressures because the molecules need high energies to escape to the vapor phase. For example, although water has a much lower molar mass than diethyl ether, the strong hydrogen-bonding forces that exist among water molecules in the liquid state cause water's vapor pressure to be much lower than that of diethyl ether [Fig. 10.38(b)]. In general, substances with large molar masses have relatively low vapor pressures, mainly because of the large dispersion forces. The more electrons a substance has, the more polarizable it is, and the greater the dispersion forces are.

Measurements of the vapor pressure for a given liquid at several temperatures show that *vapor pressure increases significantly with temperature*. Figure 10.39 illustrates the distribution of molecular kinetic energy present in a liquid at two different temperatures. To overcome the intermolecular forces in a liquid, a molecule must have sufficient kinetic energy. As the temperature of the liquid is increased, the fraction of molecules having the minimum energy needed to overcome these forces and escape to the vapor phase increases markedly. Thus the vapor pressure of a liquid increases dramatically with temperature. Values for water at several temperatures are given in Table 10.8.

The quantitative nature of the temperature dependence of vapor pressure can be represented graphically. Plots of vapor pressure versus temperature for water, ethanol, and diethyl ether are shown in Fig. 10.40(a). Note the nonlinear increase in vapor pressure for all the liquids as the temperature is increased. We find that a straight line can be obtained by plotting $\ln(P_{vap})$ versus $1/T$, where T is the Kelvin temperature [Fig. 10.40(b)]. We can represent this behavior by the equation

$$\ln(P_{vap}) = -\frac{\Delta H_{vap}}{R}\left(\frac{1}{T}\right) + C \qquad (10.4)$$

where ΔH_{vap} is the enthalpy of vaporization, R is the universal gas constant, and C is a constant characteristic of a given liquid. The symbol *ln* means that the natural logarithm of the vapor pressure is taken.

Table 10.8 | The Vapor Pressure of Water as a Function of Temperature

T (°C)	P (torr)
0.0	4.579
10.0	9.209
20.0	17.535
25.0	23.756
30.0	31.824
40.0	55.324
60.0	149.4
70.0	233.7
90.0	525.8

Figure 10.39 | The number of molecules in a liquid with a given energy versus kinetic energy at two temperatures. Part (a) shows a lower temperature than that in part (b). Note that the proportion of molecules with enough energy to escape the liquid to the vapor phase (indicated by shaded areas) increases dramatically with temperature. This causes vapor pressure to increase markedly with temperature.

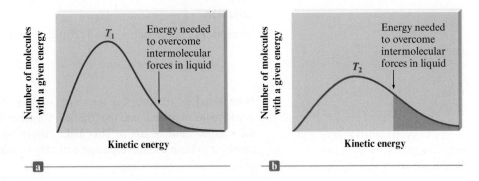

Equation (10.4) is the equation for a straight line of the form $y = mx + b$, where

$$y = \ln(P_{vap})$$

$$x = \frac{1}{T}$$

Natural logarithms are reviewed in
Appendix 1.2.

$$m = \text{slope} = -\frac{\Delta H_{vap}}{R}$$

$$b = \text{intercept} = C$$

Example 10.5	Determining Enthalpies of Vaporization

Using the plots in Fig. 10.40(b), determine whether water or diethyl ether has the larger enthalpy of vaporization.

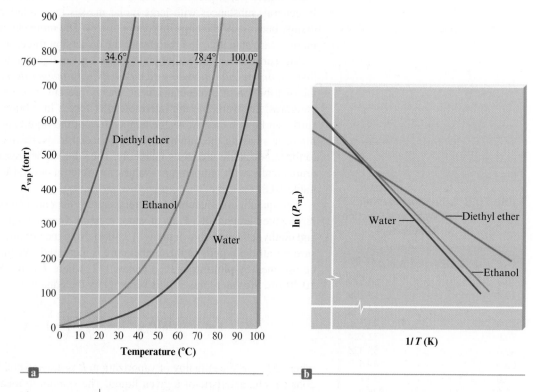

Figure 10.40 | (a) The vapor pressure of water, ethanol, and diethyl ether as a function of temperature. (b) Plots of $\ln(P_{vap})$ versus $1/T$ (Kelvin temperature) for water, ethanol, and diethyl ether.

Solution

When $\ln(P_{vap})$ is plotted versus $1/T$, the slope of the resulting straight line is

$$-\frac{\Delta H_{vap}}{R}$$

Note from Fig. 10.40(b) that the slopes of the lines for water and diethyl ether are both negative, as expected, and that the line for ether has the smaller slope. Thus ether has the smaller value of ΔH_{vap}. This makes sense because the hydrogen bonding in water causes it to have a relatively large enthalpy of vaporization.

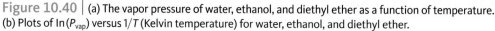

See Exercise 10.89

Equation (10.4) is important for several reasons. For example, we can determine the heat of vaporization for a liquid by measuring P_{vap} at several temperatures and then evaluating the slope of a plot of $\ln(P_{vap})$ versus $1/T$. On the other hand, if we know the values of ΔH_{vap} and P_{vap} at one temperature, we can use Equation (10.4) to calculate P_{vap} at another temperature. This can be done by recognizing that the constant C does not depend on temperature. Thus at two temperatures T_1 and T_2 we can solve Equation (10.4) for C and then write the equality

$$\ln(P_{vap,T_1}) + \frac{\Delta H_{vap}}{RT_1} = C = \ln(P_{vap,T_2}) + \frac{\Delta H_{vap}}{RT_2}$$

This can be rearranged to

$$\ln(P_{vap,T_1}) - \ln(P_{vap,T_2}) = \frac{\Delta H_{vap}}{R}\left(\frac{1}{T_2} - \frac{1}{T_1}\right)$$

Equation (10.5) is called the *Clausius–Clapeyron equation.*

or

$$\ln\left(\frac{P_{vap,T_1}}{P_{vap,T_2}}\right) = \frac{\Delta H_{vap}}{R}\left(\frac{1}{T_2} - \frac{1}{T_1}\right) \tag{10.5}$$

Interactive Example 10.6

Sign in at http://login.cengagebrain.com to try this Interactive Example in **OWL**.

Calculating Vapor Pressure

The vapor pressure of water at 25°C is 23.8 torr, and the heat of vaporization of water at 25°C is 43.9 kJ/mol. Calculate the vapor pressure of water at 50.°C.

Solution

We will use Equation (10.5):

$$\ln\left(\frac{P_{vap,T_1}}{P_{vap,T_2}}\right) = \frac{\Delta H_{vap}}{R}\left(\frac{1}{T_2} - \frac{1}{T_1}\right)$$

For water we have

$$P_{vap,T_1} = 23.8 \text{ torr}$$
$$T_1 = 25 + 273 = 298 \text{ K}$$
$$T_2 = 50. + 273 = 323 \text{ K}$$
$$\Delta H_{vap} = 43.9 \text{ kJ/mol} = 43{,}900 \text{ J/mol}$$
$$R = 8.3145 \text{ J/K} \cdot \text{mol}$$

In solving this problem, we ignore the fact that ΔH_{vap} is slightly temperature dependent.

Thus

$$\ln\left(\frac{23.8 \text{ torr}}{P_{vap,T_2}\text{(torr)}}\right) = \frac{43{,}900 \text{ J/mol}}{8.3145 \text{ J/K} \cdot \text{mol}}\left(\frac{1}{323 \text{ K}} - \frac{1}{298 \text{ K}}\right)$$

$$\ln\left(\frac{23.8}{P_{vap,T_2}}\right) = -1.37$$

Taking the antilog (see Appendix 1.2) of both sides gives

$$\frac{23.8}{P_{vap,T_2}} = 0.254$$
$$P_{vap,T_2} = 93.7 \text{ torr}$$

See Exercises 10.91 through 10.94

Richard Megna/Fundamental Photographs

Figure 10.41 | Iodine being heated, causing it to sublime, forming crystals of $I_2(s)$ on the bottom of a test tube cooled by ice.

Like liquids, solids have vapor pressures. Figure 10.41 shows iodine vapor forming solid iodine on the bottom of a cooled dish. Under normal conditions iodine **sublimes**; that is, it goes directly from the solid to the gaseous state without passing through the liquid state. **Sublimation** also occurs with dry ice (solid carbon dioxide).

Figure 10.42 | The heating curve (not drawn to scale) for a given quantity of water where energy is added at a constant rate. The plateau at the boiling point is longer than the plateau at the melting point because it takes almost seven times more energy (and thus seven times the heating time) to vaporize liquid water than to melt ice. The slopes of the other lines are different because the different states of water have different molar heat capacities (the energy required to raise the temperature of 1 mole of a substance by 1°C).

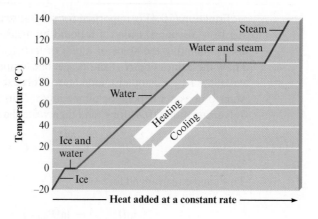

Ionic solids such as NaCl and NaF have very high melting points and enthalpies of fusion because of the strong ionic forces in these solids. At the other extreme is $O_2(s)$, a molecular solid containing nonpolar molecules with weak intermolecular forces. (See Table 10.9.)

The melting and boiling points will be defined more precisely later in this section.

Changes of State

What happens when a solid is heated? Typically, it will melt to form a liquid. If the heating continues, the liquid will at some point boil and form the vapor phase. This process can be represented by a **heating curve**: a plot of temperature versus time for a process where energy is added at a constant rate.

The heating curve for water is given in Fig. 10.42. As energy flows into the ice, the random vibrations of the water molecules increase as the temperature rises. Eventually, the molecules become so energetic that they break loose from their lattice positions, and the change from solid to liquid occurs. This is indicated by a plateau at 0°C on the heating curve. At this temperature, called the *melting point,* all the added energy is used to disrupt the ice structure by breaking the hydrogen bonds, thus increasing the potential energy of the water molecules. The enthalpy change that occurs at the melting point when a solid melts is called the **heat of fusion**, or more accurately, the **enthalpy of fusion**, ΔH_{fus}. The melting points and enthalpies of fusion for several representative solids are listed in Table 10.9.

The temperature remains constant until the solid has completely changed to liquid; then it begins to increase again. At 100°C the liquid water reaches its *boiling point,* and the temperature then remains constant as the added energy is used to vaporize the liquid. When the liquid is completely changed to vapor, the temperature again begins to rise. Note that changes of state are physical changes; although intermolecular forces have been overcome, no chemical bonds have been broken. If the water vapor were heated to much higher temperatures, the water molecules would break down into the individual atoms. This would be a chemical change, since covalent bonds are broken. We no longer have water after this occurs.

Table 10.9 | Melting Points and Enthalpies of Fusion for Several Representative Solids

Compound	Melting Point (°C)	Enthalpy of Fusion (kJ/mol)
O_2	−218	0.45
HCl	−114	1.99
HI	−51	2.87
CCl_4	−23	2.51
$CHCl_3$	−64	9.20
H_2O	0	6.02
NaF	992	29.3
NaCl	801	30.2

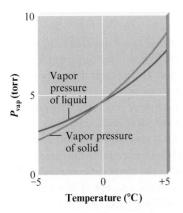

The melting and boiling points for a substance are determined by the vapor pressures of the solid and liquid states. Figure 10.43 shows the vapor pressures of solid and liquid water as functions of temperature near 0°C. Note that below 0°C the vapor pressure of ice is less than the vapor pressure of liquid water. Also note that the vapor pressure of ice has a larger temperature dependence than that of the liquid. That is, the vapor pressure of ice increases more rapidly for a given rise in temperature than does the vapor pressure of water. Thus as the temperature of the solid is increased, a point is eventually reached where the *liquid and solid have identical vapor pressures.* This is the melting point.

These concepts can be demonstrated experimentally using the apparatus illustrated in Fig. 10.44, where ice occupies one compartment and liquid water the other. Consider the following cases.

Case 1

A temperature at which the vapor pressure of the solid is greater than that of the liquid. At this temperature the solid requires a higher pressure than the liquid does to be in equilibrium with the vapor. Thus as vapor is released from the solid to try to achieve equilibrium, the liquid will absorb vapor in an attempt to reduce the vapor pressure to its equilibrium value. The net effect is a conversion from solid to liquid through the vapor phase. In fact, no solid can exist under these conditions. The amount of solid will steadily decrease and the volume of liquid will increase. Finally, there will be only liquid in the right compartment, which will come to equilibrium with the water vapor, and no further changes will occur in the system. This temperature must be above the melting point of ice, since only the liquid state can exist.

Case 2

A temperature at which the vapor pressure of the solid is less than that of the liquid. This is the opposite of the situation in case 1. In this case, the liquid requires a higher pressure than the solid does to be in equilibrium with the vapor, so the liquid will gradually disappear, and the amount of ice will increase. Finally, only the solid will remain, which will achieve equilibrium with the vapor. This temperature must be *below the melting point* of ice, since only the solid state can exist.

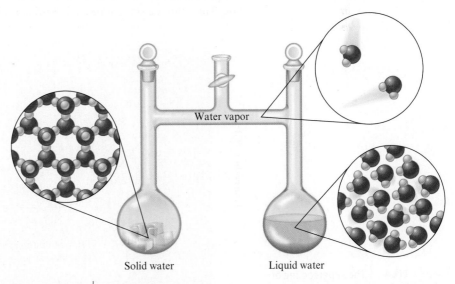

Figure 10.44 | An apparatus that allows solid and liquid water to interact only through the vapor state.

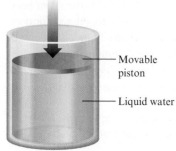

Constant pressure
of 1 atmosphere

Movable
piston

Liquid water

Figure 10.45 | Water in a closed system with a pressure of 1 atm exerted on the piston. No bubbles can form within the liquid as long as the vapor pressure is less than 1 atm.

Case 3

A temperature at which the vapor pressures of the solid and liquid are identical. In this case, the solid and liquid states have the same vapor pressure, so they can coexist in the apparatus at equilibrium simultaneously with the vapor. This temperature represents the *freezing point* where both the solid and liquid states can exist.

We can now describe the melting point of a substance more precisely. The **normal melting point** is defined as *the temperature at which the solid and liquid states have the same vapor pressure under conditions where the total pressure is 1 atmosphere.*

Boiling occurs when the vapor pressure of a liquid becomes equal to the pressure of its environment. The **normal boiling point** of a liquid is *the temperature at which the vapor pressure of the liquid is exactly 1 atmosphere.* This concept is illustrated in Fig. 10.45. At temperatures where the vapor pressure of the liquid is less than 1 atmosphere, no bubbles of vapor can form because the pressure on the surface of the liquid is greater than the pressure in any spaces in the liquid where the bubbles are trying to form. Only when the liquid reaches a temperature at which the pressure of vapor in the spaces in the liquid is 1 atmosphere can bubbles form and boiling occur.

However, changes of state do not always occur exactly at the boiling point or melting point. For example, water can be readily **supercooled**; that is, it can be cooled below 0°C at 1 atm pressure and remain in the liquid state. Supercooling occurs because, as it is cooled, the water may not achieve the degree of organization necessary to form ice at 0°C, and thus it continues to exist as the liquid. At some point the correct ordering occurs and ice rapidly forms, releasing energy in the exothermic process and bringing the temperature back up to the melting point, where the remainder of the water freezes (Fig. 10.46).

A liquid also can be **superheated**, or raised to temperatures above its boiling point, especially if it is heated rapidly. Superheating can occur because bubble formation in the interior of the liquid requires that many high-energy molecules gather in the same vicinity, and this may not happen at the boiling point, especially if the liquid is heated rapidly. If the liquid becomes superheated, the vapor pressure in the liquid is greater than the atmospheric pressure. Once a bubble does form, since its internal pressure is greater than that of the atmosphere, it can burst before rising to the surface, blowing the surrounding liquid out of the container. This is called *bumping* and has ruined many experiments. It can be avoided by adding boiling chips to the flask containing the liquid. Boiling chips are bits of porous ceramic material containing trapped air that escapes on heating, forming tiny bubbles that act as "starters" for vapor bubble formation. This allows a smooth onset of boiling as the boiling point is reached.

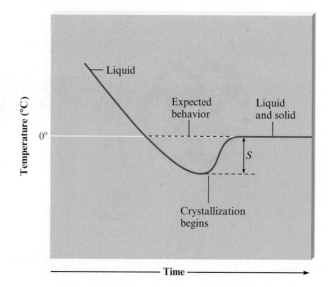

Figure 10.46 | The supercooling of water. The extent of supercooling is given by *S*.

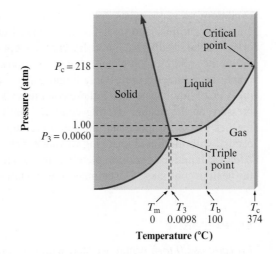

10.9 | Phase Diagrams

A **phase diagram** is a convenient way of representing the phases of a substance as a function of temperature and pressure. For example, the phase diagram for water (Fig. 10.47) shows which state exists at a given temperature and pressure. It is important to recognize that a phase diagram describes conditions and events in a *closed* system of the type represented in Fig. 10.45, where no material can escape into the surroundings and no air is present. Notice that the diagram is not drawn to scale (neither axis is linear). This is done to emphasize certain features of the diagram that will be discussed below.

To show how to interpret the phase diagram for water, we will consider heating experiments at several pressures, shown by the dashed lines in Fig. 10.48.

Experiment 1

Pressure is 1 atm. This experiment begins with the cylinder shown in Fig. 10.45 completely filled with ice at a temperature of $-20°C$ and the piston exerting a pressure of 1 atm directly on the ice (there is no air space). Since at temperatures below $0°C$ the vapor pressure of ice is less than 1 atm—which is the constant external pressure on the piston—no vapor is present in the cylinder. As the cylinder is heated, ice is the only component until the temperature reaches $0°C$, where the ice changes to liquid water as energy is added. This is the normal melting point of water. Note that under these

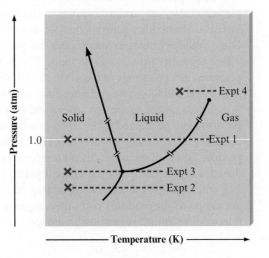

Figure 10.48 | Diagrams of various heating experiments on samples of water in a closed system.

conditions no vapor exists in the system. The vapor pressures of the solid and liquid are equal, but this vapor pressure is less than 1 atm, so no water vapor can exist. This is true on the solid/liquid line everywhere except at the triple point (see Experiment 3). When the solid has completely changed to liquid, the temperature again rises. At this point, the cylinder contains only liquid water. *No vapor is present* because the vapor pressure of liquid water under these conditions is less than 1 atm, the constant external pressure on the piston. Heating continues until the temperature of the liquid water reaches 100°C. At this point, the vapor pressure of liquid water is 1 atm, and boiling occurs, with the liquid changing to vapor. This is the normal boiling point of water. After the liquid has been completely converted to steam, the temperature again rises as the heating continues. The cylinder now contains only water vapor.

Experiment 2

Pressure is 2.0 torr. Again, we start with ice as the only component in the cylinder at −20°C. The pressure exerted by the piston in this case is only 2.0 torr. As heating proceeds, the temperature rises to −10°C, where the ice changes directly to vapor, a process known as *sublimation*. Sublimation occurs when the vapor pressure of ice is equal to the external pressure, which in this case is only 2.0 torr. No liquid water appears under these conditions because the vapor pressure of liquid water is always greater than 2.0 torr, and thus it cannot exist at this pressure. If liquid water were placed in a cylinder under such a low pressure, it would vaporize immediately at temperatures above −10°C or freeze at temperatures below −10°C.

Experiment 3

Pressure is 4.58 torr. Again, we start with ice as the only component in the cylinder at −20°C. In this case the pressure exerted on the ice by the piston is 4.58 torr. As the cylinder is heated, no new phase appears until the temperature reaches 0.01°C (273.16 K). At this point, called the **triple point**, solid and liquid water have identical vapor pressures of 4.58 torr. Thus *at 0.01°C (273.16 K) and 4.58 torr all three states of water are present.* In fact, *only* under these conditions can all three states of water coexist in a closed system.

Experiment 4

Pressure is 225 atm. In this experiment we start with liquid water in the cylinder at 300°C; the pressure exerted by the piston on the water is 225 atm. Liquid water can be present at this temperature because of the high external pressure. As the temperature increases, something happens that we did not see in the first three experiments: The liquid gradually changes into a vapor but goes through an intermediate "fluid" region, which is neither true liquid nor vapor. This is quite unlike the behavior at lower temperatures and pressures, say at 100°C and 1 atm, where the temperature remains constant while a definite phase change from liquid to vapor occurs. This unusual behavior occurs because the conditions are beyond the critical point for water. The **critical temperature** can be defined as the temperature above which the vapor cannot be liquefied no matter what pressure is applied. The **critical pressure** is the pressure required to produce liquefaction *at* the critical temperature. Together, the critical temperature and the critical pressure define the **critical point**. For water the critical point is 374°C and 218 atm. Note that the liquid/vapor line on the phase diagram for water ends at the critical point. Beyond this point the transition from one state to another involves the intermediate "fluid" region just described.

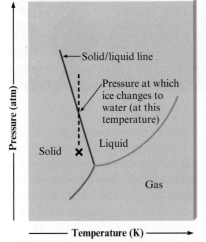

Figure 10.49 | The phase diagram for water. At point X on the phase diagram, water is a solid. However, as the external pressure is increased while the temperature remains constant (indicated by the vertical dotted line), the solid/liquid line is crossed and the ice melts.

Water boils at 89°C in Leadville, Colorado.

Applications of the Phase Diagram for Water

There are several additional interesting features of the phase diagram for water. Note that the solid/liquid boundary line has a negative slope. This means that the melting point of ice *decreases* as the external pressure *increases*. This behavior, which is opposite to that observed for most substances, occurs because the density of ice is *less* than that of liquid water at the melting point. The maximum density of water occurs at 4°C; when liquid water freezes, its volume increases.

We can account for the effect of pressure on the melting point of water using the following reasoning. At the melting point, liquid and solid water coexist—they are in dynamic equilibrium, since the rate at which ice is melting is just balanced by the rate at which the water is freezing. What happens if we apply pressure to this system? When subjected to increased pressure, matter reduces its volume. This behavior is most dramatic for gases but also occurs for condensed states. Since a given mass of ice at 0°C has a larger volume than the same mass of liquid water, the system can reduce its volume in response to the increased pressure by changing to liquid. Thus at 0°C and an external pressure greater than 1 atm, water is liquid. In other words, the freezing point of water is less than 0°C when the pressure is greater than 1 atm.

Figure 10.49 illustrates the effect of pressure on ice. At the point X on the phase diagram, ice is subjected to increased pressure at constant temperature. Note that as the pressure is increased, the solid/liquid line is crossed, indicating that the ice melts. This phenomenon may be important in ice skating. The narrow blade of the skate exerts a large pressure, since the skater's weight is supported by the small area of the blade. Also, the frictional heating due to the moving skate contributes to the melting of the ice.* After the blade passes, the liquid refreezes as normal pressure and temperature return. Without this lubrication effect due to the thawing ice, ice skating would not be the smooth, graceful activity that many people enjoy.

Ice's lower density has other implications. When water freezes in a pipe or an engine block, it will expand and break the container. This is why water pipes are insulated in cold climates and antifreeze is used in water-cooled engines. The lower density of ice also means that ice formed on rivers and lakes will float, providing a layer of insulation that helps prevent bodies of water from freezing solid in the winter. Aquatic life can therefore continue to live through periods of freezing temperatures.

A liquid boils at the temperature where the vapor pressure of the liquid equals the external pressure. Thus the boiling point of a substance, like the melting point, depends on the external pressure. This is why water boils at different temperatures at different elevations (Table 10.10), and any cooking carried out in boiling water will be

---

*The physics of ice skating is quite complex, and there is disagreement about whether the pressure or the frictional heating of the ice skate is most important. See "Letter to the Editor," by R. Silberman, *J. Chem. Ed.* **65** (1988): 186.

Table 10.10 | Boiling Point of Water at Various Locations

Location	Feet Above Sea Level	P_{atm} (torr)	Boiling Point (°C)
Top of Mt. Everest, Tibet	29,028	240	70
Top of Mt. McKinley, Alaska	20,320	340	79
Top of Mt. Whitney, Calif.	14,494	430	85
Leadville, Colo.	10,150	510	89
Top of Mt. Washington, N.H.	6293	590	93
Boulder, Colo.	5430	610	94
Madison, Wis.	900	730	99
New York City, N.Y.	10	760	100
Death Valley, Calif.	−282	770	100.3

Chemical connections

Making Diamonds at Low Pressures: Fooling Mother Nature

In 1955 Robert H. Wentorf, Jr., accomplished something that borders on alchemy—he turned peanut butter into diamonds. He and his coworkers at the General Electric Research and Development Center also changed roofing pitch, wood, coal, and many other carbon-containing materials into diamonds, using a process involving temperatures of $\approx 2000°C$ and pressures of $\approx 10^5$ atm. Although the first diamonds made by this process looked like black sand because of the impurities present, the process has now been developed to a point such that beautiful, clear, gem-quality diamonds can be produced. General Electric now has the capacity to produce 150 million carats (30,000 kg) of diamonds annually (virtually all of which is "diamond grit" used for industrial purposes such as abrasive coatings on cutting tools). The production of large, gem-quality diamonds by this process is still too expensive to compete with the natural sources of these stones. However, this may change as methods are developed for making diamonds at low pressures.

The high temperatures and pressures used in the GE process for making diamonds make sense if one looks at the accompanying phase diagram for carbon. Note that graphite—not diamond—is the most stable form of carbon under ordinary conditions of temperature and pressure. However, diamond becomes more stable than graphite at very high pressures (as one would expect from the greater density of diamond). The high temperature used in the GE process is necessary to disrupt the bonds in graphite so that diamond (the most stable form of carbon at the high pressures used in the process) can form. Once the diamond is produced, the elemental carbon is "trapped" in this form at normal conditions (25°C, 1 atm) because the reaction back to the graphite form is so slow. That is, even though graphite is more stable than diamond at 25°C and 1 atm, diamond can exist almost indefinitely because the conversion to graphite is a very *slow* reaction. As a result, diamonds formed at the high pressures found deep in the earth's crust can be brought to the earth's surface

affected by this variation. For example, it takes longer to hard-boil an egg in Leadville, Colorado (elevation: 10,150 ft), than in San Diego, California (sea level), since water boils at a lower temperature in Leadville.

Critical Thinking

Ice is less dense than liquid water, as evidenced by the fact that ice floats in a glass of water. What if ice was more dense than liquid water? How would this affect the phase diagram for water?

As we mentioned earlier, the phase diagram for water describes a closed system. Therefore, we must be very cautious in using the phase diagram to explain the behavior of water in a natural setting, such as on the earth's surface. For example, in dry climates (low humidity), snow and ice seem to sublime—a minimum amount of slush is produced. Wet clothes put on an outside line at temperatures below 0°C freeze and then dry while frozen. However, the phase diagram (Fig. 10.45) shows that ice should *not* be able to sublime at normal atmospheric pressures. What is happening in these cases? Ice in the natural environment is not in a closed system. The pressure is provided by the atmosphere rather than by a solid piston. This means that the vapor produced over the ice can escape from the immediate region as soon as it is formed. The vapor does not come to equilibrium with the solid, and the ice slowly disappears. Sublimation, which seems forbidden by the phase diagram, does in fact occur under these conditions, although it is not the sublimation under equilibrium conditions described by the phase diagram.

by natural geologic processes and continue to exist for millions of years.*

We have seen that diamond formed in the laboratory at high pressures is "trapped" in this form, but this process is very expensive. Can diamond be formed at low pressures? The phase diagram for carbon says no. However, researchers have found that under the right conditions diamonds can be "grown" at low pressures. The process used is called *chemical vapor deposition* (*CVD*). CVD uses an energy source to release carbon atoms from a compound such as methane into a steady flow of hydrogen gas (some of which is

dissociated to produce hydrogen atoms). The carbon atoms then deposit as a diamond film on a surface maintained at a temperature between 600 and 900°C. Why does diamond form on this surface rather than the favored graphite? Nobody is sure, but it has been suggested that at these relatively high temperatures the diamond structure grows faster than the graphite structure, so diamond is favored under these conditions. It also has been suggested that the hydrogen atoms present react much faster with graphite fragments than with diamond fragments, effectively removing any graphite from the growing film. Once it forms, of course, diamond is trapped. The major advantage of CVD is that there is no need for the extraordinarily high pressures used in the traditional process for synthesizing diamonds.

The first products with diamond films are already on the market. Audiophiles can buy tweeters that have diaphragms coated with a thin diamond film that limits sound

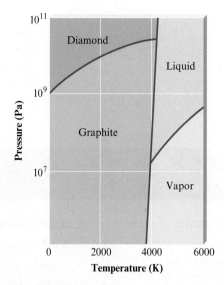

The phase diagram for carbon.

distortion. Watches with diamond-coated crystals are planned, as are diamond-coated windows in infrared scanning devices used in analytical instruments and missile-guidance systems. These applications represent only the beginning for diamond-coated products.

---

*In Morocco, a 50-km-long slab called Beni Bousera contains chunks of graphite that were probably once diamonds formed in the deposit when it was buried 150 km underground. As this slab slowly rose to the surface over millions of years, the very slow reaction changing diamond to graphite had time to occur. On the other hand, in the diamond-rich kimberlite deposits in South Africa, which rise to the surface much faster, the diamonds have not had sufficient time to revert to graphite.

The Phase Diagram for Carbon Dioxide

The phase diagram for carbon dioxide (Fig. 10.50) differs from that for water. The solid/liquid line has a positive slope, since solid carbon dioxide is more dense than liquid carbon dioxide. The triple point for carbon dioxide occurs at 5.1 atm and −56.6°C, and the critical point occurs at 72.8 atm and 31°C. At a pressure of 1 atm, solid carbon dioxide sublimes at −78°C, a property that leads to its common name, *dry ice*. No liquid phase occurs under normal atmospheric conditions, making dry ice a convenient refrigerant.

A carbon dioxide fire extinguisher being used to put out a fire.

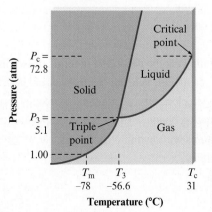

Figure 10.50 | The phase diagram for carbon dioxide. The liquid state does not exist at a pressure of 1 atm. The solid/liquid line has a positive slope, since the density of solid carbon dioxide is greater than that of liquid carbon dioxide.

Carbon dioxide is often used in fire extinguishers, where it exists as a liquid at 25°C under high pressures. Liquid carbon dioxide released from the extinguisher into the environment at 1 atm immediately changes to a vapor. Being heavier than air, this vapor smothers the fire by keeping oxygen away from the flame. The liquid/vapor transition is highly endothermic, so cooling also results, which helps to put out the fire.

For review

Key terms

Section 10.1
condensed states
intermolecular forces
dipole–dipole attraction
hydrogen bonding
London dispersion forces

Section 10.2
surface tension
capillary action
viscosity

Section 10.3
crystalline solid
amorphous solid
lattice
unit cell
X-ray diffraction
ionic solid
molecular solid
atomic solid

Section 10.4
closest packing
hexagonal closest packed (hcp) structure
cubic closest packed (ccp) structure
band model
molecular orbital (MO) model
alloy
substitutional alloy
interstitial alloy

Section 10.5
network solid
silica
silicate
glass
ceramic
semiconductor
n-type semiconductor
p-type semiconductor
p–n junction

Condensed states of matter: liquids and solids

> Held together by forces among the component molecules, atoms, or ions
> Liquids exhibit properties such as surface tension, capillary action, and viscosity that depend on the forces among the components

Dipole–dipole forces

> Attractions among molecules with dipole moments
> Hydrogen bonding is a particularly strong form of dipole–dipole attraction
>> Occurs in molecules containing hydrogen bonded to a highly electronegative element such as nitrogen, oxygen, or fluorine
>> Produces unusually high boiling points

London dispersion forces

> Caused by instantaneous dipoles that form in atoms or nonpolar molecules

Crystalline solids

> Have a regular arrangement of components often represented as a lattice; the smallest repeating unit of the lattice is called the unit cell
> Classified by the types of components:
>> Atomic solids (atoms)
>> Ionic solids (ions)
>> Molecular solids (molecules)
> Arrangement of the components can be determined by X-ray analysis

Metals

> Structure is modeled by assuming atoms to be uniform spheres
>> Closest packing
>>> Hexagonal
>>> Cubic
> Metallic bonding can be described in terms of two models
>> Electron sea model: valence electrons circulate freely among the metal cations
>> Band model: electrons are assumed to occupy molecular orbitals
>>> Conduction bands: closely spaced molecular orbitals with empty electron spaces
> Alloys: mixtures with metallic properties
>> Substitutional
>> Interstitial

Key terms

Network solids

> Contain giant networks of atoms covalently bound together
> Examples are diamond and graphite
> Silicates are network solids containing Si—O—Si bridges that form the basis for many rocks, clays, and ceramics

Semiconductors

> Very pure silicon is "doped" with other elements
>> n-type: doping atoms typically contain five valence electrons (one more than silicon)
>> p-type: doping elements typically contain three valence electrons
> Modern electronics are based on devices with p–n junctions

Molecular solids

> Components are discrete molecules
> Intermolecular forces are typically weak, leading to relatively low boiling and melting points

Ionic solids

> Components are ions
> Interionic forces are relatively strong, leading to solids with high melting and boiling points
> Many structures consist of closest packing of the larger ions with the smaller ions in tetrahedral or octahedral holes

Phase changes

> The change from liquid to gas (vapor) is called vaporization or evaporation
> Condensation is the reverse of vaporization
> Equilibrium vapor pressure: the pressure that occurs over a liquid or solid in a closed system when the rate of evaporation equals the rate of condensation
>> Liquids whose components have high intermolecular forces have relatively low vapor pressures
>> Normal boiling point: the temperature at which the vapor pressure of a liquid equals one atmosphere
>> Normal melting point: the temperature at which a solid and its liquid have the same vapor pressure (at 1 atm external pressure)
> Phase diagram
>> Shows what state exists at a given temperature and pressure in a closed system
>> Triple point: temperature at which all three phases exist simultaneously
>> Critical point: defined by the critical temperature and pressure
>>> Critical temperature: the temperature above which the vapor cannot be liquefied no matter the applied pressure
>>> Critical pressure: the pressure required to produce liquefaction at the critical temperature

Review questions   *Answers to the Review Questions can be found on the Student website (accessible from* **www.cengagebrain.com**).

1. What are intermolecular forces? How do they differ from intramolecular forces? What are dipole–dipole forces? How do typical dipole–dipole forces differ from hydrogen bonding interactions? In what ways are they similar? What are London dispersion forces? How do typical London dispersion forces differ from dipole–dipole forces? In what ways are they similar? Describe the relationship between molecular size and strength of London dispersion forces. Place the major types of intermolecular forces in order of increasing strength. Is there some overlap? That is, can the strongest London dispersion forces be greater than some dipole–dipole forces? Give an example of such an instance.

2. Define the following terms, and describe how each depends on the strength of the intermolecular forces.
   a. surface tension
   b. viscosity
   c. melting point
   d. boiling point
   e. vapor pressure

3. Compare and contrast solids, liquids, and gases.

4. Distinguish between the items in the following pairs.
   a. crystalline solid; amorphous solid
   b. ionic solid; molecular solid
   c. molecular solid; network solid
   d. metallic solid; network solid

5. What is a lattice? What is a unit cell? Describe a simple cubic unit cell. How many net atoms are contained in a simple cubic unit cell? How is the radius of the atom related to the cube edge length for a simple cubic unit cell? Answer the same questions for the body-centered cubic unit cell and for the face-centered unit cell.

6. What is closest packing? What is the difference between hexagonal closest packing and cubic closest packing? What is the unit cell for each closest packing?

7. Use the band model to describe differences among insulators, conductors, and semiconductors. Also use the band model to explain why each of the following increases the conductivity of a semiconductor.
   a. increasing the temperature
   b. irradiating with light
   c. adding an impurity
   How do conductors and semiconductors differ as to the effect of temperature on electrical conductivity? How can an n-type semiconductor be produced from pure germanium? How can a p-type semiconductor be produced from pure germanium?

8. Describe, in general, the structures of ionic solids. Compare and contrast the structure of sodium chloride and zinc sulfide. How many tetrahedral holes and octahedral holes are there per closest packed anion? In zinc sulfide, why are only one-half of the tetrahedral holes filled with cations?

9. Define each of the following.
   a. evaporation
   b. condensation
   c. sublimation
   d. boiling
   e. melting
   f. enthalpy of vaporization
   g. enthalpy of fusion
   h. heating curve

10. Why is the enthalpy of vaporization for water much greater than its enthalpy of fusion? What does this say about the changes in intermolecular forces in going from solid to liquid to vapor? What do we mean when we say that a liquid is *volatile?* Do volatile liquids have large or small vapor pressures at room temperature? What strengths of intermolecular forces occur in highly volatile liquids?

11. Compare and contrast the phase diagrams of water and carbon dioxide. Why doesn't CO_2 have a normal melting point and a normal boiling point, whereas water does? The slopes of the solid–liquid lines in the phase diagrams of H_2O and CO_2 are different. What do the slopes of the solid–liquid lines indicate in terms of the relative densities of the solid and liquid states for each substance? How do the melting points of H_2O and CO_2 depend on pressure? How do the boiling points of H_2O and CO_2 depend on pressure? Rationalize why the critical temperature for H_2O is greater than that for CO_2.

Active Learning Questions

These questions are designed to be used by groups of students in class.

1. It is possible to balance a paper clip on the surface of water in a beaker. If you add a bit of soap to the water, however, the paper clip sinks. Explain how the paper clip can float and why it sinks when soap is added.

2. Consider a sealed container half-filled with water. Which statement best describes what occurs in the container?
   a. Water evaporates until the air is saturated with water vapor; at this point, no more water evaporates.
   b. Water evaporates until the air is overly saturated (supersaturated) with water, and most of this water recondenses; this cycle continues until a certain amount of water vapor is present, and then the cycle ceases.
   c. Water does not evaporate because the container is sealed.
   d. Water evaporates, and then water evaporates and recondenses simultaneously and continuously.
   e. Water evaporates until it is eventually all in vapor form.
   Explain each choice. Justify your choice, and for choices you did not pick, explain what is wrong with them.

3. Explain the following: You add 100 mL water to a 500-mL round-bottom flask and heat the water until it is boiling. You

remove the heat and stopper the flask, and the boiling stops. You then run cool water over the neck of the flask, and the boiling begins again. It seems as though you are boiling water by cooling it.

4. Is it possible for the dispersion forces in a particular substance to be stronger than the hydrogen bonding forces in another substance? Explain your answer.

5. Does the nature of intermolecular forces change when a substance goes from a solid to a liquid, or from a liquid to a gas? What causes a substance to undergo a phase change?

6. Why do liquids have a vapor pressure? Do all liquids have vapor pressures? Explain. Do solids exhibit vapor pressure? Explain. How does vapor pressure change with changing temperature? Explain.

7. Water in an open beaker evaporates over time. As the water is evaporating, is the vapor pressure increasing, decreasing, or staying the same? Why?

8. What is the vapor pressure of water at 100°C? How do you know?

9. Refer to Fig. 10.42. Why doesn't temperature increase continuously over time? That is, why does the temperature stay constant for periods of time?

10. Which are stronger, intermolecular or intramolecular forces for a given molecule? What observation(s) have you made that support this? Explain.

11. Why does water evaporate?

A blue question or exercise number indicates that the answer to that question or exercise appears at the back of this book and a solution appears in the *Solutions Guide,* as found on PowerLecture.

Questions

12. Rationalize why chalk (calcium carbonate) has a higher melting point than motor oil (large compound made from carbon and hydrogen), which has a higher melting point than water and engages in relatively strong hydrogen bonding interactions.

13. In the diagram below, which lines represent the hydrogen bonding?

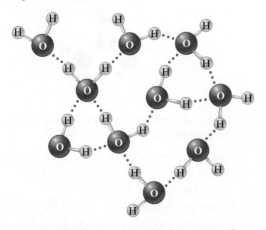

a. the dotted lines between the hydrogen atoms of one water molecule and the oxygen atoms of a different water molecule

b. the solid lines between a hydrogen atom and oxygen atom in the same water molecule

c. Both the solid lines and dotted lines represent hydrogen bonding.

d. There are no hydrogen bonds represented in the diagram.

14. Hydrogen bonding is a special case of very strong dipole–dipole interactions possible among only certain atoms. What atoms in addition to hydrogen are necessary for hydrogen bonding? How does the small size of the hydrogen atom contribute to the unusual strength of the dipole–dipole forces involved in hydrogen bonding?

15. Atoms are assumed to touch in closest packed structures, yet every closest packed unit cell contains a significant amount of empty space. Why?

16. Define *critical temperature* and *critical pressure.* In terms of the kinetic molecular theory, why is it impossible for a substance to exist as a liquid above its critical temperature?

17. Use the kinetic molecular theory to explain why a liquid gets cooler as it evaporates from an insulated container.

18. Will a crystalline solid or an amorphous solid give a simpler X-ray diffraction pattern? Why?

19. What is an alloy? Explain the differences in structure between substitutional and interstitial alloys. Give an example of each type.

20. Describe what is meant by a dynamic equilibrium in terms of the vapor pressure of a liquid.

21. How does each of the following affect the rate of evaporation of a liquid in an open dish?
   a. intermolecular forces
   b. temperature
   c. surface area

22. A common response to hearing that the temperature in New Mexico is 105°F is, "It's not that bad; it's a dry heat," whereas at the same time the summers in Atlanta, Georgia, are characterized as "dreadful," even though the air temperature is typically lower. What role does humidity play in how our bodies regulate temperature?

23. When a person has a severe fever, one therapy used to reduce the fever is an "alcohol rub." Explain how the evaporation of alcohol from a person's skin removes heat energy from the body.

24. Why is a burn from steam typically much more severe than a burn from boiling water?

25. When wet laundry is hung on a clothesline on a cold winter day, it will freeze but eventually dry. Explain.

26. Cake mixes and other packaged foods that require cooking often contain special directions for use at high elevations. Typically these directions indicate that the food should be cooked longer above 5000 ft. Explain why it takes longer to cook something at higher elevations.

27. You have three covalent compounds with three very different boiling points. All of the compounds have similar molar mass and relative shape. Explain how these three compounds could have very different boiling points.

28. Compare and contrast the structures of the following solids.
   a. diamond versus graphite
   b. silica versus silicates versus glass

29. Compare and contrast the structures of the following solids.

    a. $CO_2(s)$ versus $H_2O(s)$

    b. $NaCl(s)$ versus $CsCl(s)$; see Exercise 69 for the structures.

30. Silicon carbide (SiC) is an extremely hard substance that acts as an electrical insulator. Propose a structure for SiC.

31. How could you tell experimentally if TiO_2 is an ionic solid or a network solid?

32. A common prank on college campuses is to switch the salt and sugar on dining hall tables, which is usually easy because the substances look so much alike. Yet, despite the similarity in their appearance, these two substances differ greatly in their properties, since one is a molecular solid and the other is an ionic solid. How do the properties differ and why?

33. A plot of ln (P_{vap}) versus $1/T$ (K) is linear with a negative slope. Why is this the case?

34. Iodine, like most substances, exhibits only three phases: solid, liquid, and vapor. The triple point of iodine is at 90 torr and 115°C. Which of the following statements concerning liquid I_2 must be true? Explain your answer.

    a. $I_2(l)$ is more dense than $I_2(g)$.

    b. $I_2(l)$ cannot exist above 115°C.

    c. $I_2(l)$ cannot exist at 1 atmosphere pressure.

    d. $I_2(l)$ cannot have a vapor pressure greater than 90 torr.

    e. $I_2(l)$ cannot exist at a pressure of 10 torr.

Exercises

In this section similar exercises are paired.

Intermolecular Forces and Physical Properties

35. Identify the most important types of interparticle forces present in the solids of each of the following substances.

    a. Ar   e. CH_4

    b. HCl   f. CO

    c. HF   g. $NaNO_3$

    d. $CaCl_2$

36. Identify the most important types of interparticle forces present in the solids of each of the following substances.

    a. $BaSO_4$ e. CsI

    b. H_2S f. P_4

    c. Xe   g. NH_3

    d. C_2H_6

37. Predict which substance in each of the following pairs would have the greater intermolecular forces.

    a. CO_2 or OCS

    b. SeO_2 or SO_2

    c. $CH_3CH_2CH_2NH_2$ or $H_2NCH_2CH_2NH_2$

    d. CH_3CH_3 or H_2CO

    e. CH_3OH or H_2CO

38. Consider the compounds Cl_2, HCl, F_2, NaF, and HF. Which compound has a boiling point closest to that of argon? Explain.

39. Rationalize the difference in boiling points for each of the following pairs of substances:

    a. *n*-pentane $CH_3CH_2CH_2CH_2CH_3$ 36.2°C

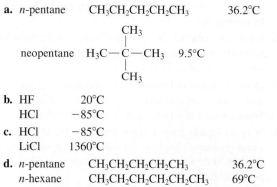

    b. HF   20°C
       HCl   −85°C

    c. HCl   −85°C
       LiCl   1360°C

    d. *n*-pentane $CH_3CH_2CH_2CH_2CH_3$ 36.2°C
       n-hexane $CH_3CH_2CH_2CH_2CH_2CH_3$ 69°C

40. Consider the following electrostatic potential diagrams:

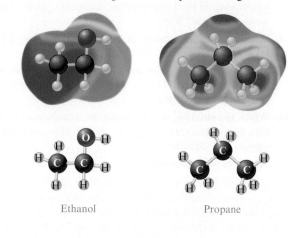

Ethanol   Propane

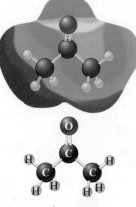

Acetone

Rank the compounds from lowest to highest boiling point and explain your answer.

41. In each of the following groups of substances, pick the one that has the given property. Justify your answer.

    a. highest boiling point: HBr, Kr, or Cl_2

    b. highest freezing point: H_2O, NaCl, or HF

    c. lowest vapor pressure at 25°C: Cl_2, Br_2, or I_2

    d. lowest freezing point: N_2, CO, or CO_2

    e. lowest boiling point: CH_4, CH_3CH_3, or $CH_3CH_2CH_3$

    f. highest boiling point: HF, HCl, or HBr

g. lowest vapor pressure at 25°C: $CH_3CH_2CH_3$, $CH_3\overset{\overset{\textstyle O}{\|}}{C}CH_3$, or $CH_3CH_2CH_2OH$

42. In each of the following groups of substances, pick the one that has the given property. Justify each answer.

   a. highest boiling point: CCl_4, CF_4, CBr_4

   b. lowest freezing point: LiF, F_2, HCl

   c. smallest vapor pressure at 25°C: CH_3OCH_3, CH_3CH_2OH, $CH_3CH_2CH_3$

   d. greatest viscosity: H_2S, HF, H_2O_2

   e. greatest heat of vaporization: H_2CO, CH_3CH_3, CH_4

   f. smallest enthalpy of fusion: I_2, $CsBr$, CaO

Properties of Liquids

43. The shape of the meniscus of water in a glass tube is different from that of mercury in a glass tube. Why?

H$_2$O in glass   Hg in glass

44. Explain why water forms into beads on a waxed car finish.

45. Hydrogen peroxide (H_2O_2) is a syrupy liquid with a relatively low vapor pressure and a normal boiling point of 152.2°C. Rationalize the differences of these physical properties from those of water.

46. Carbon diselenide (CSe_2) is a liquid at room temperature. The normal boiling point is 125°C, and the melting point is −45.5°C. Carbon disulfide (CS_2) is also a liquid at room temperature with normal boiling and melting points of 46.5°C and −111.6°C, respectively. How do the strengths of the intermolecular forces vary from CO_2 to CS_2 to CSe_2? Explain.

Structures and Properties of Solids

47. X rays from a copper X-ray tube ($\lambda = 154$ pm) were diffracted at an angle of 14.22 degrees by a crystal of silicon. Assuming first-order diffraction ($n = 1$ in the Bragg equation), what is the interplanar spacing in silicon?

48. The second-order diffraction ($n = 2$) for a gold crystal is at an angle of 22.20° for X rays of 154 pm. What is the spacing between these crystal planes?

49. A topaz crystal has an interplanar spacing (d) of 1.36 Å ($1 \text{ Å} = 1 \times 10^{-10}$ m). Calculate the wavelength of the X ray that should be used if $\theta = 15.0°$ (assume $n = 1$).

50. X rays of wavelength 2.63 Å were used to analyze a crystal. The angle of first-order diffraction ($n = 1$ in the Bragg equation) was 15.55 degrees. What is the spacing between crystal planes, and what would be the angle for second-order diffraction ($n = 2$)?

51. Calcium has a cubic closest packed structure as a solid. Assuming that calcium has an atomic radius of 197 pm, calculate the density of solid calcium.

52. Nickel has a face-centered cubic unit cell. The density of nickel is 6.84 g/cm³. Calculate a value for the atomic radius of nickel.

53. A certain form of lead has a cubic closest packed structure with an edge length of 492 pm. Calculate the value of the atomic radius and the density of lead.

54. Iridium (Ir) has a face-centered cubic unit cell with an edge length of 383.3 pm. Calculate the density of solid iridium.

55. You are given a small bar of an unknown metal X. You find the density of the metal to be 10.5 g/cm³. An X-ray diffraction experiment measures the edge of the face-centered cubic unit cell as 4.09 Å ($1 \text{ Å} = 10^{-10}$ m). Identify X.

56. A metallic solid with atoms in a face-centered cubic unit cell with an edge length of 392 pm has a density of 21.45 g/cm³. Calculate the atomic mass and the atomic radius of the metal. Identify the metal.

57. Titanium metal has a body-centered cubic unit cell. The density of titanium is 4.50 g/cm³. Calculate the edge length of the unit cell and a value for the atomic radius of titanium. (*Hint:* In a body-centered arrangement of spheres, the spheres touch across the body diagonal.)

58. Barium has a body-centered cubic structure. If the atomic radius of barium is 222 pm, calculate the density of solid barium.

59. The radius of gold is 144 pm, and the density is 19.32 g/cm³. Does elemental gold have a face-centered cubic structure or a body-centered cubic structure?

60. The radius of tungsten is 137 pm and the density is 19.3 g/cm³. Does elemental tungsten have a face-centered cubic structure or a body-centered cubic structure?

61. What fraction of the total volume of a cubic closest packed structure is occupied by atoms? (*Hint:* $V_{sphere} = \frac{4}{3}\pi r^3$.) What fraction of the total volume of a simple cubic structure is occupied by atoms? Compare the answers.

62. Iron has a density of 7.86 g/cm³ and crystallizes in a body-centered cubic lattice. Show that only 68% of a body-centered lattice is actually occupied by atoms, and determine the atomic radius of iron.

63. Explain how doping silicon with either phosphorus or gallium increases the electrical conductivity over that of pure silicon.

64. Explain how a p–n junction makes an excellent rectifier.

65. Selenium is a semiconductor used in photocopying machines. What type of semiconductor would be formed if a small amount of indium impurity is added to pure selenium?

66. The Group 3A/Group 5A semiconductors are composed of equal amounts of atoms from Group 3A and Group 5A—for example, InP and GaAs. These types of semiconductors are used in light-emitting diodes and solid-state lasers. What would you add to make a p-type semiconductor from pure GaAs? How would you dope pure GaAs to make an n-type semiconductor?

67. The band gap in aluminum phosphide (AlP) is 2.5 electron-volts ($1 \text{ eV} = 1.6 \times 10^{-19}$ J). What wavelength of light is emitted by an AlP diode?

68. An aluminum antimonide solid-state laser emits light with a wavelength of 730. nm. Calculate the band gap in joules.

69. The structures of some common crystalline substances are shown below. Show that the net composition of each unit cell corresponds to the correct formula of each substance.

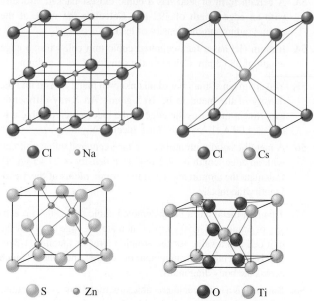

○ Cl   • Na   ○ Cl   ○ Cs

○ S   • Zn   ● O   ○ Ti

70. The unit cell for nickel arsenide is shown below. What is the formula of this compound?

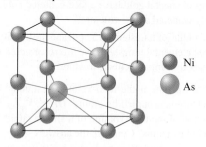

○ Ni
○ As

71. Cobalt fluoride crystallizes in a closest packed array of fluoride ions with the cobalt ions filling one-half of the octahedral holes. What is the formula of this compound?

72. The compounds Na_2O, CdS, and ZrI_4 all can be described as cubic closest packed anions with the cations in tetrahedral holes. What fraction of the tetrahedral holes is occupied for each case?

73. What is the formula for the compound that crystallizes with a cubic closest packed array of sulfur ions, and that contains zinc ions in $\frac{1}{8}$ of the tetrahedral holes and aluminum ions in $\frac{1}{2}$ of the octahedral holes?

74. Assume the two-dimensional structure of an ionic compound, M_xA_y, is

What is the empirical formula of this ionic compound?

75. A certain metal fluoride crystallizes in such a way that the fluoride ions occupy simple cubic lattice sites, while the metal ions occupy the body centers of *half* the cubes. What is the formula of the metal fluoride?

76. The structure of manganese fluoride can be described as a simple cubic array of manganese ions with fluoride ions at the center of each edge of the cubic unit cell. What is the charge of the manganese ions in this compound?

77. The unit cell of MgO is shown below.

Does MgO have a structure like that of NaCl or ZnS? If the density of MgO is 3.58 g/cm³, estimate the radius (in centimeters) of the O^{2-} anions and the Mg^{2+} cations.

78. In solid KCl the smallest distance between the centers of a potassium ion and a chloride ion is 314 pm. Calculate the length of the edge of the unit cell and the density of KCl, assuming it has the same structure as sodium chloride.

79. The CsCl structure is a simple cubic array of chloride ions with a cesium ion at the center of each cubic array (see Exercise 69). Given that the density of cesium chloride is 3.97 g/cm³, and assuming that the chloride and cesium ions touch along the body diagonal of the cubic unit cell, calculate the distance between the centers of adjacent Cs^+ and Cl^- ions in the solid. Compare this value with the expected distance based on the sizes of the ions. The ionic radius of Cs^+ is 169 pm, and the ionic radius of Cl^- is 181 pm.

80. MnO has either the NaCl type structure or the CsCl type structure (see Exercise 69). The edge length of the MnO unit cell is 4.47×10^{-8} cm and the density of MnO is 5.28 g/cm³.

   a. Does MnO crystallize in the NaCl or the CsCl type structure?

   b. Assuming that the ionic radius of oxygen is 140. pm, estimate the ionic radius of manganese.

81. What type of solid will each of the following substances form?

   a. CO_2 **g.** KBr
   b. SiO_2 **h.** H_2O
   c. Si   **i.** NaOH
   d. CH_4 **j.** U
   e. Ru   **k.** $CaCO_3$
   f. I_2 **l.** PH_3

82. What type of solid will each of the following substances form?

   a. diamond   **g.** NH_4NO_3

   b. PH_3 **h.** SF_2

   c. H_2 **i.** Ar

   d. Mg   **j.** Cu

   e. KCl   **k.** $C_6H_{12}O_6$

   f. quartz

83. The memory metal, nitinol, is an alloy of nickel and titanium. It is called a *memory metal* because after being deformed, a piece of nitinol wire will return to its original shape. The structure of nitinol consists of a simple cubic array of Ni atoms and an inner penetrating simple cubic array of Ti atoms. In the extended lattice, a Ti atom is found at the center of a cube of Ni atoms; the reverse is also true.

   a. Describe the unit cell for nitinol.

   b. What is the empirical formula of nitinol?

   c. What are the coordination numbers (number of nearest neighbors) of Ni and Ti in nitinol?

84. Superalloys have been made of nickel and aluminum. The alloy owes its strength to the formation of an ordered phase, called the *gamma-prime phase,* in which Al atoms are at the corners of a cubic unit cell and Ni atoms are at the face centers. What is the composition (relative numbers of atoms) for this phase of the nickel–aluminum superalloy?

85. Perovskite is a mineral containing calcium, titanium, and oxygen. Two different representations of the unit cell are shown below. Show that both these representations give the same formula and the same number of oxygen atoms around each titanium atom.

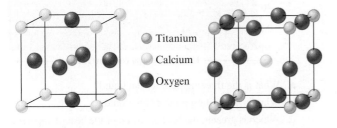

   ○ Titanium

   ○ Calcium

   ● Oxygen

86. A mineral crystallizes in a cubic closest packed array of oxygen ions with aluminum ions in some of the octahedral holes and magnesium ions in some of the tetrahedral holes. Deduce the formula of this mineral and predict the fraction of octahedral holes and tetrahedral holes that are filled by the various cations.

87. Materials containing the elements Y, Ba, Cu, and O that are superconductors (electrical resistance equals zero) at temperatures above that of liquid nitrogen were recently discovered. The structures of these materials are based on the perovskite structure. Were they to have the ideal perovskite structure, the superconductor would have the structure shown in part (a) of the following figure.

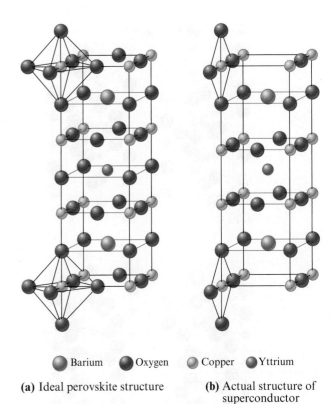

   ○ Barium   ● Oxygen   ○ Copper   ● Yttrium

(a) Ideal perovskite structure   **(b)** Actual structure of superconductor

   a. What is the formula of this ideal perovskite material?

   b. How is this structure related to the perovskite structure shown in Exercise 85?

   These materials, however, do not act as superconductors unless they are deficient in oxygen. The structure of the actual superconducting phase appears to be that shown in part (b) of the figure.

   c. What is the formula of this material?

88. The structures of another class of ceramic, high-temperature superconductors are shown in the following figure on the next page.

   a. Determine the formula of each of these four superconductors.

   b. One of the structural features that appears to be essential for high-temperature superconductivity is the presence of planar sheets of copper and oxygen atoms. As the number of sheets in each unit cell increases, the temperature for the onset of superconductivity increases. Order the four structures from lowest to the highest superconducting temperature.

   c. Assign oxidation states to Cu in each structure assuming Tl exists as Tl^{3+}. The oxidation states of Ca, Ba, and O are assumed to be $+2$, $+2$, and -2, respectively.

   d. It also appears that copper must display a mixture of oxidation states for a material to exhibit superconductivity. Explain how this occurs in these materials as well as in the superconductor in Exercise 87.

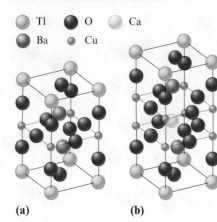

Tl ⬤ O ⬤ Ca
Ba ⬤ Cu

(a)   (b)

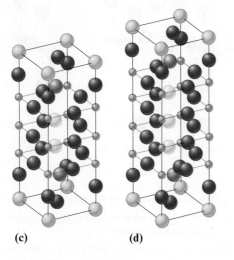

(c)   (d)

Phase Changes and Phase Diagrams

89. Plot the following data and determine ΔH_{vap} for magnesium and lithium. In which metal is the bonding stronger?

Vapor Pressure (mm Hg)	Temperature (°C)	
	Li	Mg
1.	750.	620.
10.	890.	740.
100.	1080.	900.
400.	1240.	1040.
760.	1310.	1110.

90. From the following data for liquid nitric acid, determine its heat of vaporization and normal boiling point.

Temperature (°C)	Vapor Pressure (mm Hg)
0.	14.4
10.	26.6
20.	47.9
30.	81.3
40.	133
50.	208
80.	670.

91. In Breckenridge, Colorado, the typical atmospheric pressure is 520. torr. What is the boiling point of water ($\Delta H_{vap} = 40.7$ kJ/mol) in Breckenridge?

92. The temperature inside a pressure cooker is 115°C. Calculate the vapor pressure of water inside the pressure cooker. What would be the temperature inside the pressure cooker if the vapor pressure of water was 3.50 atm?

93. Carbon tetrachloride, CCl_4, has a vapor pressure of 213 torr at 40.°C and 836 torr at 80.°C. What is the normal boiling point of CCl_4?

94. Diethyl ether ($CH_3CH_2OCH_2CH_3$) was one of the first chemicals used as an anesthetic. At 34.6°C, diethyl ether has a vapor pressure of 760. torr, and at 17.9°C, it has a vapor pressure of 400. torr. What is the ΔH of vaporization for diethyl ether?

95. A substance, X, has the following properties:

		Specific Heat Capacities	
ΔH_{vap}	20. kJ/mol	$C(s)$	3.0 J/g · °C
ΔH_{fus}	5.0 kJ/mol	$C(l)$	2.5 J/g · °C
bp	75°C	$C(g)$	1.0 J/g · °C
mp	−15°C		

Sketch a heating curve for substance X starting at −50.°C.

96. Use the heating–cooling curve below to answer the following questions.

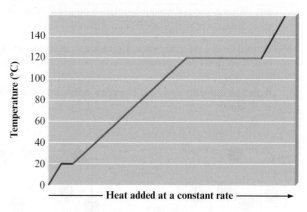

a. What is the freezing point of the liquid?

b. What is the boiling point of the liquid?

c. Which is greater, the heat of fusion or the heat of vaporization? Explain each term and explain how the heating–cooling curve above helps you to answer the question.

97. The molar heat of fusion of sodium metal is 2.60 kJ/mol, whereas its heat of vaporization is 97.0 kJ/mol.

a. Why is the heat of vaporization so much larger than the heat of fusion?

b. What quantity of heat would be needed to *melt* 1.00 g sodium at its normal melting point?

c. What quantity of heat would be needed to *vaporize* 1.00 g sodium at its normal boiling point?

d. What quantity of heat would be evolved if 1.00 g sodium vapor condensed at its normal boiling point?

98. The molar heat of fusion of benzene (C_6H_6) is 9.92 kJ/mol. Its molar heat of vaporization is 30.7 kJ/mol. Calculate the heat required to melt 8.25 g benzene at its normal melting point. Calculate the heat required to vaporize 8.25 g benzene at its normal boiling point. Why is the heat of vaporization more than three times the heat of fusion?

99. What quantity of energy does it take to convert 0.500 kg ice at $-20.°C$ to steam at $250.°C$? Specific heat capacities: ice, 2.03 J/g · °C; liquid, 4.2 J/g · °C; steam, 2.0 J/g · °C; $\Delta H_{vap} = 40.7$ kJ/mol; $\Delta H_{fus} = 6.02$ kJ/mol.

100. Consider a 75.0-g sample of $H_2O(g)$ at $125°C$. What phase or phases are present when 215 kJ of energy is removed from this sample? (See Exercise 99.)

101. An ice cube tray contains enough water at $22.0°C$ to make 18 ice cubes that each has a mass of 30.0 g. The tray is placed in a freezer that uses CF_2Cl_2 as a refrigerant. The heat of vaporization of CF_2Cl_2 is 158 J/g. What mass of CF_2Cl_2 must be vaporized in the refrigeration cycle to convert all the water at $22.0°C$ to ice at $-5.0°C$? The heat capacities for $H_2O(s)$ and $H_2O(l)$ are 2.03 J/g · °C and 4.18 J/g · °C, respectively, and the enthalpy of fusion for ice is 6.02 kJ/mol.

102. A 0.250-g chunk of sodium metal is cautiously dropped into a mixture of 50.0 g water and 50.0 g ice, both at 0°C. The reaction is

$$2Na(s) + 2H_2O(l) \longrightarrow 2NaOH(aq) + H_2(g) \quad \Delta H = -368 \text{ kJ}$$

Assuming no heat loss to the surroundings, will the ice melt? Assuming the final mixture has a specific heat capacity of 4.18 J/g · °C, calculate the final temperature. The enthalpy of fusion for ice is 6.02 kJ/mol.

103. Consider the phase diagram given below. What phases are present at points A through H? Identify the triple point, normal boiling point, normal freezing point, and critical point. Which phase is denser, solid or liquid?

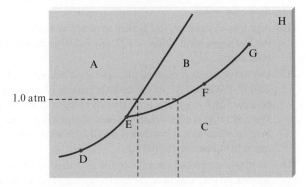

104. Sulfur exhibits two solid phases, rhombic and monoclinic. Use the accompanying phase diagram for sulfur to answer the following questions. (The phase diagram is not to scale.)

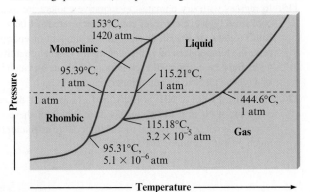

a. How many triple points are in the phase diagram?
b. What phases are in equilibrium at each of the triple points?
c. What is the stable phase at 1 atm and $100.°C$?
d. What are the normal melting point and the normal boiling point of sulfur?
e. Which is the densest phase?
f. At a pressure of 1.0×10^{-5} atm, can rhombic sulfur sublime?
g. What phase changes occur when the pressure on a sample of sulfur at $100.°C$ is increased from 1.0×10^{-8} atm to 1500 atm?

105. Use the accompanying phase diagram for carbon to answer the following questions.

a. How many triple points are in the phase diagram?
b. What phases can coexist at each triple point?
c. What happens if graphite is subjected to very high pressures at room temperature?
d. If we assume that the density increases with an increase in pressure, which is more dense, graphite or diamond?

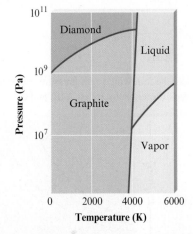

106. Like most substances, bromine exists in one of the three typical phases. Br_2 has a normal melting point of $-7.2°C$ and a normal boiling point of $59°C$. The triple point for Br_2 is $-7.3°C$ and 40 torr, and the critical point is $320°C$ and 100 atm. Using this information, sketch a phase diagram for bromine indicating the points described above. Based on your phase diagram, order the three phases from least dense to most dense. What is the stable phase of Br_2 at room temperature and 1 atm? Under what temperature conditions can liquid bromine never exist? What phase changes occur as the temperature of a sample of bromine at 0.10 atm is increased from $-50°C$ to $200°C$?

107. The melting point of a fictional substance X is $225°C$ at 10.0 atm. If the density of the solid phase of X is 2.67 g/cm^3 and the density of the liquid phase is 2.78 g/cm^3 at 10.0 atm, predict whether the normal melting point of X will be less than, equal to, or greater than $225°C$. Explain.

108. Consider the following data for xenon:

Triple point:	$-121°C$, 280 torr
Normal melting point:	$-112°C$
Normal boiling point:	$-107°C$

Which is more dense, Xe(s) or Xe(l)? How do the melting point and boiling point of xenon depend on pressure?

Additional Exercises

109. Some of the physical properties of H_2O and D_2O are as follows:

Property	H_2O	D_2O
Density at 20°C (g/mL)	0.997	1.108
Boiling point (°C)	100.00	101.41
Melting point (°C)	0.00	3.79
$\Delta H°_{vap}$ (kJ/mol)	40.7	41.61
$\Delta H°_{fus}$ (kJ/mol)	6.02	6.3

Account for the differences. (*Note:* D is a symbol often used for ^2H, the deuterium isotope of hydrogen.)

110. Rationalize the following boiling points:

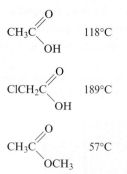

CH_3C 118°C

$ClCH_2C$ 189°C

CH_3C 57°C

111. Consider the following vapor pressure versus temperature plot for three different substances: A, B, and C.

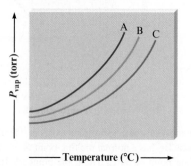

If the three substances are CH_4, SiH_4, and NH_3, match each curve to the correct substance.

112. Consider the following enthalpy changes:

$$F^- + HF \longrightarrow FHF^- \qquad \Delta H = -155 \text{ kJ/mol}$$
$$(CH_3)_2C{=}O + HF \longrightarrow (CH_3)_2C{=}O\text{---}HF$$
$$\Delta H = -46 \text{ kJ/mol}$$
$$H_2O(g) + HOH(g) \longrightarrow H_2O\text{---}HOH \text{ (in ice)}$$
$$\Delta H = -21 \text{ kJ/mol}$$

How do the strengths of hydrogen bonds vary with the electronegativity of the element to which hydrogen is bonded? Where in the preceding series would you expect hydrogen bonds of the following type to fall?

$$\overset{|}{N}\text{---}HO\text{---} \quad \text{and} \quad \overset{|}{N}\text{---}H\text{---}N$$

113. The unit cell for a pure xenon fluoride compound is shown below. What is the formula of the compound?

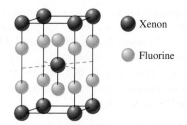

● Xenon

○ Fluorine

114. Boron nitride (BN) exists in two forms. The first is a slippery solid formed from the reaction of BCl_3 with NH_3, followed by heating in an ammonia atmosphere at 750°C. Subjecting the first form of BN to a pressure of 85,000 atm at 1800°C produces a second form that is the second hardest substance known. Both forms of BN remain solids to 3000°C. Suggest structures for the two forms of BN.

115. Consider the following data concerning four different substances.

Compound	Conducts Electricity as a Solid	Other Properties
B_2H_6	No	Gas at 25°C
SiO_2	No	High mp
CsI	No	Aqueous solution Conducts electricity
W	Yes	High mp

Label the four substances as either ionic, network, metallic, or molecular solids.

116. Argon has a cubic closest packed structure as a solid. Assuming that argon has a radius of 190. pm, calculate the density of solid argon.

117. Dry nitrogen gas is bubbled through liquid benzene (C_6H_6) at 20.0°C. From 100.0 L of the gaseous mixture of nitrogen and benzene, 24.7 g benzene is condensed by passing the mixture through a trap at a temperature where nitrogen is gaseous and the vapor pressure of benzene is negligible. What is the vapor pressure of benzene at 20.0°C?

118. A 20.0-g sample of ice at −10.0°C is mixed with 100.0 g water at 80.0°C. Calculate the final temperature of the mixture assuming no heat loss to the surroundings. The heat capacities of $H_2O(s)$ and $H_2O(l)$ are 2.03 and 4.18 J/g · °C, respectively, and the enthalpy of fusion for ice is 6.02 kJ/mol.

119. In regions with dry climates, evaporative coolers are used to cool air. A typical electric air conditioner is rated at 1.00×10^4 Btu/h (1 Btu, or British thermal unit = amount of energy to raise the temperature of 1 lb water by 1°F). What quantity of water must be evaporated each hour to dissipate as much heat as a typical electric air conditioner?

120. The critical point of NH_3 is 132°C and 111 atm, and the critical point of N_2 is −147°C and 34 atm. Which of these substances cannot be liquefied at room temperature no matter how much pressure is applied? Explain.

ChemWork Problems

These multiconcept problems (and additional ones) are found interactively online with the same type of assistance a student would get from an instructor.

121. Which of the following compound(s) exhibit only London dispersion intermolecular forces? Which compound(s) exhibit hydrogen-bonding forces? Considering only the compounds without hydrogen-bonding interactions, which compounds have dipole–dipole intermolecular forces?

    a. SF_4 **d.** HF

    b. CO_2 **e.** ICl_5

    c. CH_3CH_2OH **f.** XeF_4

122. Which of the following statements about intermolecular forces is(are) *true*?

    a. London dispersion forces are the only type of intermolecular force that nonpolar molecules exhibit.

    b. Molecules that have only London dispersion forces will always be gases at room temperature (25°C).

    c. The hydrogen-bonding forces in NH_3 are stronger than those in H_2O.

    d. The molecules in $SO_2(g)$ exhibit dipole–dipole intermolecular interactions.

    e. $CH_3CH_2CH_3$ has stronger London dispersion forces than does CH_4.

123. Which of the following statements is(are) *true*?

    a. LiF will have a higher vapor pressure at 25°C than H_2S.

    b. HF will have a lower vapor pressure at −50°C than HBr.

    c. Cl_2 will have a higher boiling point than Ar.

    d. HCl is more soluble in water than in CCl_4.

    e. MgO will have a higher vapor pressure at 25°C than CH_3CH_2OH.

124. Aluminum has an atomic radius of 143 pm and forms a solid with a cubic closest packed structure. Calculate the density of solid aluminum in g/cm^3.

125. Pyrolusite is a mineral containing manganese ions and oxide ions. Its structure can best be described as a body-centered cubic array of manganese ions with two oxide ions inside the unit cell and two oxide ions each on two faces of the cubic unit cell. What is the charge on the manganese ions in pyrolusite?

126. The structure of the compound K_2O is best described as a cubic closest packed array of oxide ions with the potassium ions in tetrahedral holes. What percent of the tetrahedral holes are occupied in this solid?

127. What type of solid (network, metallic, Group 8A, ionic, or molecular) will each of the following substances form?

    a. Kr   **d.** SiO_2

    b. SO_2 **e.** NH_3

    c. Ni   **f.** Pt

128. Some ice cubes at 0°C with a total mass of 403 g are placed in a microwave oven and subjected to 750. W (750. J/s) of energy for 5.00 minutes. What is the final temperature of the water? Assume all the energy of the microwave is absorbed by the water, and assume no heat loss by the water.

129. The enthalpy of vaporization for acetone is 32.0 kJ/mol. The normal boiling point for acetone is 56.5°C. What is the vapor pressure of acetone at 23.5°C?

130. Choose the statements that correctly describe the following phase diagram.

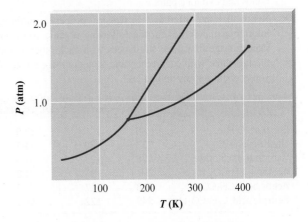

    a. If the temperature is raised from 50 K to 400 K at a pressure of 1 atm, the substance boils at approximately 185 K.

    b. The liquid phase of this substance cannot exist under conditions of 2 atm at any temperature.

    c. The triple point occurs at approximately 165 K.

    d. At a pressure of 1.5 atm, the melting point of the substance is approximately 370 K.

    e. The critical point occurs at approximately 1.7 atm and 410 K.

Challenge Problems

131. When 1 mole of benzene is vaporized at a constant pressure of 1.00 atm and at its boiling point of 353.0 K, 30.79 kJ of energy (heat) is absorbed and the volume change is +28.90 L. What are ΔE and ΔH for this process?

132. You and a friend each synthesize a compound with the formula $XeCl_2F_2$. Your compound is a liquid and your friend's compound is a gas (at the same conditions of temperature and pressure). Explain how the two compounds with the same formulas can exist in different phases at the same conditions of pressure and temperature.

133. Using the heats of fusion and vaporization for water given in Exercise 99, calculate the change in enthalpy for the sublimation of water:

$$H_2O(s) \longrightarrow H_2O(g)$$

Using the ΔH value given in Exercise 112 and the number of hydrogen bonds formed with each water molecule, estimate what portion of the intermolecular forces in ice can be accounted for by hydrogen bonding.

134. Consider a perfectly insulated and sealed container. Determine the minimum volume of a container such that a gallon of water at 25°C will evaporate completely. If the container is a cube, determine the dimensions in feet. Assume the density of water is $0.998\ g/cm^3$.

135. Consider two different organic compounds, each with the formula C_2H_6O. One of these compounds is a liquid at room conditions and the other is a gas. Write Lewis structures consistent with this observation, and explain your answer. (*Hint:* The oxygen atom in both structures satisfies the octet rule with two bonds and two lone pairs.)

136. Rationalize the differences in physical properties in terms of intermolecular forces for the following organic compounds. Compare the first three substances with each other, compare the last three with each other, and then compare all six. Can you account for any anomalies?

	bp (°C)	mp (°C)	ΔH_{vap} (kJ/mol)
Benzene, C_6H_6	80	6	33.9
Naphthalene, $C_{10}H_8$	218	80	51.5
Carbon tetrachloride	76	−23	31.8
Acetone, CH_3COCH_3	56	−95	31.8
Acetic acid, CH_3CO_2H	118	17	39.7
Benzoic acid, $C_6H_5CO_2H$	249	122	68.2

137. Consider the following melting point data:

Compound	NaCl	$MgCl_2$	$AlCl_3$	$SiCl_4$	PCl_3	SCl_2	Cl_2
mp (°C)	801	708	190	−70	−91	−78	−101
Compound	NaF	MgF_2	AlF_3	SiF_4	PF_5	SF_6	F_2
mp (°C)	997	1396	1040	−90	−94	−56	−220

Account for the trends in melting points in terms of interparticle forces.

138. Some ionic compounds contain a mixture of different charged cations. For example, wüstite is an oxide that contains both Fe^{2+} and Fe^{3+} cations and has a formula of $Fe_{0.950}O_{1.00}$. Calculate the fraction of iron ions present as Fe^{3+}. What fraction of the sites normally occupied by Fe^{2+} must be vacant in this solid?

139. Some ionic compounds contain a mixture of different charged cations. For example, some titanium oxides contain a mixture of Ti^{2+} and Ti^{3+} ions. Consider a certain oxide of titanium that is 28.31% oxygen by mass and contains a mixture of Ti^{2+} and Ti^{3+} ions. Determine the formula of the compound and the relative numbers of Ti^{2+} and Ti^{3+} ions.

140. Spinel is a mineral that contains 37.9% aluminum, 17.1% magnesium, and 45.0% oxygen, by mass, and has a density of 3.57 g/cm³. The edge of the cubic unit cell measures 809 pm. How many of each type of ion are present in the unit cell?

141. Mn crystallizes in the same type of cubic unit cell as Cu. Assuming that the radius of Mn is 5.6% larger than the radius of Cu and the density of copper is 8.96 g/cm³, calculate the density of Mn.

142. You are asked to help set up a historical display in the park by stacking some cannonballs next to a Revolutionary War cannon. You are told to stack them by starting with a triangle in which each side is composed of four touching cannonballs. You are to continue stacking them until you have a single ball on the top centered over the middle of the triangular base.

a. How many cannonballs do you need?

b. What type of closest packing is displayed by the cannonballs?

c. The four corners of the pyramid of cannonballs form the corners of what type of regular geometric solid?

143. Some water is placed in a sealed glass container connected to a vacuum pump (a device used to pump gases from a container), and the pump is turned on. The water appears to boil and then freezes. Explain these changes using the phase diagram for water. What would happen to the ice if the vacuum pump was left on indefinitely?

144. The molar enthalpy of vaporization of water at 373 K and 1.00 atm is 40.7 kJ/mol. What fraction of this energy is used to change the internal energy of the water, and what fraction is used to do work against the atmosphere? (*Hint:* Assume that water vapor is an ideal gas.)

145. For a simple cubic array, solve for the volume of an interior sphere (cubic hole) in terms of the radius of a sphere in the array.

146. Rubidium chloride has the sodium chloride structure at normal pressures but assumes the cesium chloride structure at high pressures. (See Exercise 69.) What ratio of densities is expected for these two forms? Does this change in structure make sense on the basis of simple models? The ionic radius is 148 pm for Rb^+ and 181 pm for Cl^-.

Integrative Problems

These problems require the integration of multiple concepts to find the solutions.

147. A 0.132-mole sample of an unknown semiconducting material with the formula XY has a mass of 19.0 g. The element X has an electron configuration of $[Kr]5s^24d^{10}$. What is this semiconducting material? A small amount of the Y atoms in the semiconductor is replaced with an equivalent amount of atoms with an electron configuration of $[Ar]4s^23d^{10}4p^5$. Does this correspond to n-type or p-type doping?

148. A metal burns in air at 600°C under high pressure to form an oxide with formula MO_2. This compound is 23.72% oxygen by mass. The distance between touching atoms in a cubic closest packed crystal of this metal is 269.0 pm. What is this metal? What is its density?

149. One method of preparing elemental mercury involves roasting cinnabar (HgS) in quicklime (CaO) at 600.°C followed by condensation of the mercury vapor. Given the heat of vaporization of mercury (296 J/g) and the vapor pressure of mercury at 25.0°C (2.56×10^{-3} torr), what is the vapor pressure of the condensed mercury at 300.°C? How many atoms of mercury are present in the mercury vapor at 300.°C if the reaction is conducted in a closed 15.0-L container?

Marathon Problem

This problem is designed to incorporate several concepts and techniques into one situation.

150. General Zod has sold Lex Luthor what Zod claims to be a new copper-colored form of kryptonite, the only substance that can harm Superman. Lex, not believing in honor among thieves, decided to carry out some tests on the supposed kryptonite. From previous tests, Lex knew that kryptonite is a metal having a specific heat capacity of 0.082 J/g · °C and a density of 9.2 g/cm³.

Lex Luthor's first experiment was an attempt to find the specific heat capacity of kryptonite. He dropped a 10 g ± 3 g sample of the metal into a boiling water bath at a temperature of 100.0°C ± 0.2°C. He waited until the metal had reached the bath temperature and then quickly transferred it to 100 g ± 3 g of water that was contained in a calorimeter at an initial temperature of 25.0°C ± 0.2°C. The final temperature of the metal and water was 25.2°C. Based on these results, is it possible to distinguish between copper and kryptonite? Explain.

When Lex found that his results from the first experiment were inconclusive, he decided to determine the density of the sample. He managed to steal a better balance and determined the mass of another portion of the purported kryptonite to be 4 g ± 1 g. He dropped this sample into water contained in a 25-mL graduated cylinder and found that it displaced a volume of 0.42 mL ± 0.02 mL. Is the metal copper or kryptonite? Explain.

Lex was finally forced to determine the crystal structure of the metal General Zod had given him. He found that the cubic unit cell contained four atoms and had an edge length of 600. pm. Explain how this information enabled Lex to identify the metal as copper or kryptonite.

Will Lex be going after Superman with the kryptonite or seeking revenge on General Zod? What improvements could he have made in his experimental techniques to avoid performing the crystal structure determination?

Appendixes

| Mathematical Procedures

A1.1 | Exponential Notation

The numbers characteristic of scientific measurements are often very large or very small; thus it is convenient to express them using powers of 10. For example, the number 1,300,000 can be expressed as 1.3×10^6, which means multiply 1.3 by 10 six times, or

$$1.3 \times 10^6 \times 1.3 \times \underline{10 \times 10 \times 10 \times 10 \times 10 \times 10}$$
$$10^6 = 1 \text{ million}$$

Note that each multiplication by 10 moves the decimal point one place to the right:

$$1.3 \times 10 = 13.$$
$$13 \times 10 = 130.$$
$$130 \times 10 = 1300.$$
$$\vdots$$

Thus the easiest way to interpret the notation 1.3×10^6 is that it means move the decimal point in 1.3 to the right six times:

$$1.3 \times 10^6 = 1\,3\,0\,0\,0\,0\,0 = 1,300,000$$
$$1\,2\,3\,4\,5\,6$$

Using this notation, the number 1985 can be expressed as 1.985×10^3. Note that the usual convention is to write the number that appears before the power of 10 as a number between 1 and 10. To end up with the number 1.985, which is between 1 and 10, we had to move the decimal point three places to the left. To compensate for that, we must multiply by 10^3, which says that to get the intended number we start with 1.985 and move the decimal point three places to the right; that is:

$$1.985 \times 10^3 = 1\,9\,8\,5.$$
$$1\,2\,3$$

Some other examples are given below.

Number	Exponential Notation
5.6	5.6×10^0 or 5.6×1
39	3.9×10^1
943	9.43×10^2
1126	1.126×10^3

So far we have considered numbers greater than 1. How do we represent a number such as 0.0034 in exponential notation? We start with a number between 1 and 10 and *divide* by the appropriate power of 10:

$$0.0034 = \frac{3.4}{10 \times 10 \times 10} = \frac{3.4}{10^3} = 3.4 \times 10^{-3}$$

Division by 10 moves the decimal point one place to the *left*. Thus the number

$$0.\ 0\ 0\ 0\ 0\ 0\ 1\ 4$$
$$7\ 6\ 5\ 4\ 3\ 2\ 1$$

can be written as 1.4×10^7.

To summarize, we can write any number in the form

$$N \times 10^{\pm n}$$

where N is between 1 and 10 and the exponent n is an integer. If the sign preceding n is positive, it means the decimal point in N should be moved n places to the right. If a negative sign precedes n, the decimal point in N should be moved n places to the left.

Multiplication and Division

When two numbers expressed in exponential notation are multiplied, the initial numbers are multiplied and the exponents of 10 are added:

$$(M \times 10^m)(N \times 10^n) = (MN) \times 10^{m+n}$$

For example (to two significant figures, as required),

$$(3.2 \times 10^4)(2.8 \times 10^3) = 9.0 \times 10^7$$

When the numbers are multiplied, if a result greater than 10 is obtained for the initial number, the decimal point is moved one place to the left and the exponent of 10 is increased by 1:

$$(5.8 \times 10^2)(4.3 \times 10^8) = 24.9 \times 10^{10}$$
$$= 2.49 \times 10^{11}$$
$$= 2.5\ \times 10^{11} \quad \text{(two significant figures)}$$

Division of two numbers expressed in exponential notation involves normal division of the initial numbers and *subtraction* of the exponent of the divisor from that of the dividend. For example,

$$\frac{4.8 \times 10^8}{\underbrace{2.1 \times 10^3}_{\text{Divisor}}} = \frac{4.8}{2.1} \times 10^{(8-3)} = 2.3 \times 10^5$$

If the initial number resulting from the division is less than 1, the decimal point is moved one place to the right and the exponent of 10 is decreased by 1. For example,

$$\frac{6.4 \times 10^3}{8.3 \times 10^5} = \frac{6.4}{8.3} \times 10^{(3-5)} = 0.77 \times 10^{-2}$$
$$= 7.7 \times 10^{-3}$$

Addition and Subtraction

To add or subtract numbers expressed in exponential notation, *the exponents of the numbers must be the same.* For example, to add 1.31×10^5 and 4.2×10^4, we must rewrite one number so that the exponents of both are the same. The number 1.31×10^5 can be written 13.1×10^4, since moving the decimal point one place to the right can be compensated for by decreasing the exponent by 1. Now we can add the numbers:

$$13.1 \times 10^4$$
$$+\ 4.2 \times 10^4$$
$$\overline{17.3 \times 10^4}$$

In correct exponential notation the result is expressed as 1.73×10^5.

To perform addition or subtraction with numbers expressed in exponential notation, only the initial numbers are added or subtracted. The exponent of the result is the same as those of the numbers being added or subtracted. To subtract 1.8×10^2 from 8.99×10^3, we write

$$
\begin{array}{r}
8.99 \times 10^3 \\
-0.18 \times 10^3 \\
\hline
8.81 \times 10^3
\end{array}
$$

Powers and Roots

When a number expressed in exponential notation is taken to some power, the initial number is taken to the appropriate power and the exponent of 10 is multiplied by that power:

$$(N \times 10^n)^m = N^m \times 10^{m \cdot n}$$

For example,*

$$
\begin{aligned}
(7.5 \times 10^2)^3 &= 7.5^3 \times 10^{3 \cdot 2} \\
&= 422 \times 10^6 \\
&= 4.22 \times 10^8 \\
&= 4.2 \ \times 10^8 \quad \text{(two significant figures)}
\end{aligned}
$$

When a root is taken of a number expressed in exponential notation, the root of the initial number is taken and the exponent of 10 is divided by the number representing the root:

$$\sqrt{N \times 10^n} = (n \times 10^n)^{1/2} = \sqrt{N} \times 10^{n/2}$$

For example,

$$
\begin{aligned}
(2.9 \times 10^6)^{1/2} &= \sqrt{2.9} \times 10^{6/2} \\
&= 1.7 \times 10^3
\end{aligned}
$$

Because the exponent of the result must be an integer, we may sometimes have to change the form of the number so that the power divided by the root equals an integer. For example,

$$
\begin{aligned}
\sqrt{1.9 \times 10^3} = (1.9 \times 10^3)^{1/2} &= (0.19 \times 10^4)^{1/2} \\
&= \sqrt{0.19} \times 10^2 \\
&= 0.44 \times 10^2 \\
&= 4.4 \times 10^1
\end{aligned}
$$

In this case, we moved the decimal point one place to the left and increased the exponent from 3 to 4 to make $n/2$ an integer.

The same procedure is followed for roots other than square roots. For example,

$$
\begin{aligned}
\sqrt[3]{6.9 \times 10^5} = (6.9 \times 10^5)^{1/3} &= (0.69 \times 10^6)^{1/3} \\
&= \sqrt[3]{0.69} \times 10^2 \\
&= 0.88 \times 10^2 \\
&= 8.8 \times 10^1
\end{aligned}
$$

and

$$
\begin{aligned}
\sqrt[3]{4.6 \times 10^{10}} = (4.6 \times 10^{10})^{1/3} &= (46 \times 10^9)^{1/3} \\
&= \sqrt[3]{46} \times 10^3 \\
&= 3.6 \times 10^3
\end{aligned}
$$

*Refer to the instruction booklet for your calculator for directions concerning how to take roots and powers of numbers.

A1.2 | Logarithms

A logarithm is an exponent. Any number N can be expressed as follows:

$$N = 10^x$$

For example,

$$1000 = 10^3$$
$$100 = 10^2$$
$$10 = 10^1$$
$$1 = 10^0$$

The common, or base 10, logarithm of a number is the power to which 10 must be taken to yield the number. Thus, since $1000 = 10^3$,

$$\log 1000 = 3$$

Similarly,

$$\log 100 = 2$$
$$\log 10 = 1$$
$$\log 1 = 0$$

For a number between 10 and 100, the required exponent of 10 will be between 1 and 2. For example, $65 = 10^{1.8129}$; that is, $\log 65 = 1.8129$. For a number between 100 and 1000, the exponent of 10 will be between 2 and 3. For example, $650 = 10^{2.8129}$ and $\log 650 = 2.8129$.

A number N greater than 0 and less than 1 can be expressed as follows:

$$N = 10^{-x} = \frac{1}{10^x}$$

For example,

$$0.001 = \frac{1}{1000} = \frac{1}{10^3} = 10^{-3}$$

$$0.01 = \frac{1}{100} = \frac{1}{10^2} = 10^{-2}$$

$$0.1 = \frac{1}{10} = \frac{1}{10^1} = 10^{-1}$$

Thus

$$\log 0.001 = -3$$
$$\log 0.01 = -2$$
$$\log 0.1 = -1$$

Although common logs are often tabulated, the most convenient method for obtaining such logs is to use an electronic calculator. On most calculators the number is first entered and then the log key is punched. The log of the number then appears in the display.* Some examples are given below. You should reproduce these results on your calculator to be sure that you can find common logs correctly.

Number	Common Log
36	1.56
1849	3.2669
0.156	−0.807
1.68×10^{-5}	−4.775

*Refer to the instruction booklet for your calculator for the exact sequence to obtain logarithms.

Note that the number of digits after the decimal point in a common log is equal to the number of significant figures in the original number.

Since logs are simply exponents, they are manipulated according to the rules for exponents. For example, if $A = 10^x$ and $B = 10^y$, then their product is

$$A \cdot B = 10^x \cdot 10^y = 10^{x+y}$$

and

$$\log AB = x + y = \log A + \log B$$

For division, we have

$$\frac{A}{B} = \frac{10^x}{10^y} = 10^{x-y}$$

and

$$\log \frac{A}{B} = x - y = \log A - \log B$$

For a number raised to a power, we have

$$A^n = (10^x)^n = 10^{nx}$$

and

$$\log A^n = nx = n \log A$$

It follows that

$$\log \frac{1}{A^n} = \log A^{-n} = -n \log A$$

or, for $n = 1$,

$$\log \frac{1}{A} = -\log A$$

When a common log is given, to find the number it represents, we must carry out the process of exponentiation. For example, if the log is 2.673, then $N = 10^{2.673}$. The process of exponentiation is also called taking the antilog, or the inverse logarithm. This operation is usually carried out on calculators in one of two ways. The majority of calculators require that the log be entered first and then the keys (INV) and (LOG) pressed in succession. For example, to find $N = 10^{2.673}$ we enter 2.673 and then press (INV) and (LOG). The number 471 will be displayed; that is, $N = 471$. Some calculators have a (10^x) key. In that case, the log is entered first and then the (10^x) key is pressed. Again, the number 471 will be displayed.

Natural logarithms, another type of logarithm, are based on the number 2.7183, which is referred to as e. In this case, a number is represented as $N = e^x = 2.7183^x$. For example,

$$N = 7.15 = e^x$$
$$\ln 7.15 = x = 1.967$$

To find the natural log of a number using a calculator, the number is entered and then the (ln) key is pressed. Use the following examples to check your technique for finding natural logs with your calculator:

Number (e^x)	Natural Log (x)
784	6.664
1.61×10^3	7.384
1.00×10^{-7}	−16.118
1.00	0

If a natural logarithm is given, to find the number it represents, exponentiation to the base e (2.7183) must be carried out. With many calculators this is done using a key marked $[\,e^x\,]$ (the natural log is entered, with the correct sign, and then the $[\,e^x\,]$ key is pressed). The other common method for exponentiation to base e is to enter the natural log and then press the $[\,\text{INV}\,]$ and $[\,\ln\,]$ keys in succession. The following examples will help you check your technique:

ln $N(x)$	$N(e^x)$
3.256	25.9
−5.169	5.69×10^{-3}
13.112	4.95×10^5

Since natural logarithms are simply exponents, they are also manipulated according to the mathematical rules for exponents given earlier for common logs.

A1.3 | Graphing Functions

In interpreting the results of a scientific experiment, it is often useful to make a graph. If possible, the function to be graphed should be in a form that gives a straight line. The equation for a straight line (a *linear equation*) can be represented by the general form

$$y = mx + b$$

where y is the *dependent variable*, x is the *independent variable*, m is the *slope*, and b is the *intercept* with the y axis.

To illustrate the characteristics of a linear equation, the function $y = 3x + 4$ is plotted in Fig. A.1. For this equation $m = 3$ and $b = 4$. Note that the y intercept occurs when $x = 0$. In this case the intercept is 4, as can be seen from the equation ($b = 4$).

The slope of a straight line is defined as the ratio of the rate of change in y to that in x:

$$m = \text{slope} = \frac{\Delta y}{\Delta x}$$

For the equation $y = 3x + 4$, y changes three times as fast as x (since x has a coefficient of 3). Thus the slope in this case is 3. This can be verified from the graph. For the triangle shown in Fig. A.1,

$$\Delta y = 34 - 10 = 24 \quad \text{and} \quad \Delta x = 10 - 2 = 8$$

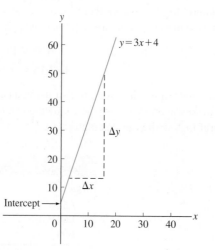

Figure A.1 | Graph of the linear equation $y = 3x + 4$.

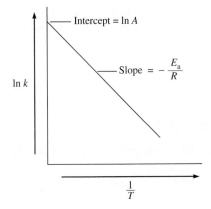

Figure A.2 | Graph of ln k versus $1/T$.

Table A.1 | Some Useful Linear Equations in Standard Form

Equation ($y = mx + b$)	What Is Plotted (y vs. x)	Slope (m)	Intercept (b)	Section in Text
$[A] = -kt + [A]_0$	$[A]$ vs. t	$-k$	$[A]_0$	12.4
$\ln[A] = -kt + \ln[A]_0$	$\ln[A]$ vs. t	$-k$	$\ln[A]_0$	12.4
$\dfrac{1}{[A]} = kt + \dfrac{1}{[A]_0}$	$\dfrac{1}{[A]}$ vs. t	k	$\dfrac{1}{[A]_0}$	12.4
$\ln P_{vap} = -\dfrac{\Delta H_{vap}}{R}\left(\dfrac{1}{T}\right) + C$	$\ln P_{vap}$ vs. $\dfrac{1}{T}$	$\dfrac{-\Delta H_{vap}}{R}$	C	10.8

Thus

$$\text{Slope} = \frac{\Delta y}{\Delta x} = \frac{24}{8} = 3$$

The preceding example illustrates a general method for obtaining the slope of a line from the graph of that line. Simply draw a triangle with one side parallel to the y axis and the other parallel to the x axis as shown in Fig. A.1. Then determine the lengths of the sides to give y and x, respectively, and compute the ratio $\Delta y/\Delta x$.

Sometimes an equation that is not in standard form can be changed to the form $y = mx + b$ by rearrangement or mathematical manipulation. An example is the equation $k = Ae^{-E_a/RT}$ described in Section 12.7, where A, E_a, and R are constants; k is the dependent variable; and $1/T$ is the independent variable. This equation can be changed to standard form by taking the natural logarithm of both sides,

$$\ln k = \ln Ae^{-E_a/RT} = \ln A + \ln e^{-E_a/RT} = \ln A - \frac{E_a}{RT}$$

noting that the log of a product is equal to the sum of the logs of the individual terms and that the natural log of $e^{-E_a/RT}$ is simply the exponent $-E_a/RT$. Thus, in standard form, the equation $k = Ae^{-E_a/RT}$ is written

A plot of $\ln k$ versus $1/T$ (Fig. A.2) gives a straight line with slope $-E_a/R$ and intercept $\ln A$.

Other linear equations that are useful in the study of chemistry are listed in standard form in Table A.1.

A1.4 | Solving Quadratic Equations

A *quadratic equation,* a polynomial in which the highest power of x is 2, can be written as

$$ax^2 + bx + c = 0$$

One method for finding the two values of x that satisfy a quadratic equation is to use the *quadratic formula:*

$$x = \frac{-b \pm \sqrt{b^2 - 4ac}}{2a}$$

where a, b, and c are the coefficients of x^2, x, and the constant, respectively. For example, in determining $[H^+]$ in a solution of $1.0 \times 10^{-4}\ M$ acetic acid the following expression arises:

$$1.8 \times 10^{-5} = \frac{x^2}{1.0 \times 10^{-4} - x}$$

which yields

$$x^2 + (1.8 \times 10^{-5})x - 1.8 \times 10^{-9} = 0$$

where $a = 1$, $b = 1.8 \times 10^{-5}$, and $c = -1.8 \times 10^{-9}$. Using the quadratic formula, we have

$$x = \frac{-b \pm \sqrt{b^2 - 4ac}}{2a}$$

$$= \frac{-1.8 \times 10^{-5} \pm \sqrt{3.24 \times 10^{-10} - (4)(1)(-1.8 \times 10^{-9})}}{2(1)}$$

$$= \frac{-1.8 \times 10^{-5} \pm \sqrt{3.24 \times 10^{-10} + 7.2 \times 10^{-9}}}{2}$$

$$= \frac{-1.8 \times 10^{-5} \pm \sqrt{7.5 \times 10^{-9}}}{2}$$

$$= \frac{-1.8 \times 10^{-5} \pm 8.7 \times 10^{-5}}{2}$$

Thus

$$x = \frac{6.9 \times 10^{-5}}{2} = 3.5 \times 10^{-5}$$

and

$$x = \frac{-10.5 \times 10^{-5}}{2} = -5.2 \times 10^{-5}$$

Note that there are two roots, as there always will be, for a polynomial in x^2. In this case x represents a concentration of H^+ (see Section 14.3). Thus the positive root is the one that solves the problem, since a concentration cannot be a negative number.

A second method for solving quadratic equations is by *successive approximations,* a systematic method of trial and error. A value of x is guessed and substituted into the equation everywhere x (or x^2) appears, except for one place. For example, for the equation

$$x^2 + (1.8 \times 10^{-5})x - 1.8 \times 10^{-9} = 0$$

we might guess $x = 2 \times 10^{-5}$. Substituting that value into the equation gives

$$x^2 + (1.8 \times 10^{-5})(2 \times 10^{-5}) - 1.8 \times 10^{-9} = 0$$

or

$$x^2 = 1.8 \times 10^{-9} - 3.6 \times 10^{-10} = 1.4 \times 10^{-9}$$

Thus

$$x = 3.7 \times 10^{-5}$$

Note that the guessed value of x (2×10^{5}) is not the same as the value of x that is calculated (3.7×10^{-5}) after inserting the estimated value. This means that $x = 2 \times 10^{-5}$ is not the correct solution, and we must try another guess. We take the calculated value (3.7×10^{-5}) as our next guess:

$$x^2 + (1.8 \times 10^{-5})(3.7 \times 10^{-5}) - 1.8 \times 10^{-9} = 0$$
$$x^2 = 1.8 \times 10^{-9} - 6.7 \times 10^{-10} = 1.1 \times 10^{-9}$$

Thus

$$x = 3.3 \times 10^{-5}$$

Now we compare the two values of x again:

Guessed: $x = 3.7 \times 10^{-5}$

Calculated: $x = 3.3 \times 10^{-5}$

These values are closer but not close enough. Next we try 3.3×10^{-5} as our guess:

$$x^2 + (1.8 \times 10^{-5})(3.3 \times 10^{-5}) - 1.8 \times 10^{-9} = 0$$
$$x^2 = 1.8 \times 10^{-9} - 5.9 \times 10^{-10} = 1.2 \times 10^{-9}$$

Thus

$$x = 3.5 \times 10^{-5}$$

Again we compare:

Guessed: $x = 3.3 \times 10^{-5}$

Calculated: $x = 3.5 \times 10^{-5}$

Next we guess $x = 3.5 \times 10^{-5}$ to give

$$x^2 + (1.8 \times 10^{-5})(3.5 \times 10^{-5}) - 1.8 \times 10^{-9} = 0$$
$$x^2 = 1.8 \times 10^{-9} - 6.3 \times 10^{-10} = 1.2 \times 10^{-9}$$

Thus

$$x = 3.5 \times 10^{-5}$$

Now the guessed value and the calculated value are the same; we have found the correct solution. Note that this agrees with one of the roots found with the quadratic formula in the first method.

To further illustrate the method of successive approximations, we will solve Example 14.17 using this procedure. In solving for $[H^+]$ for $0.010\ M$ H_2SO_4, we obtain the following expression:

$$1.2 \times 10^{-2} = \frac{x(0.010 + x)}{0.010 - x}$$

which can be rearranged to give

$$x = (1.2 \times 10^{-2})\left(\frac{0.010 - x}{0.010 + x}\right)$$

We will guess a value for x, substitute it into the right side of the equation, and then calculate a value for x. In guessing a value for x, we know it must be less than 0.010, since a larger value will make the calculated value for x negative and the guessed and calculated values will never match. We start by guessing $x = 0.005$.

The results of the successive approximations are shown in the following table:

Trial	Guessed Value for x	Calculated Value for x
1	0.0050	0.0040
2	0.0040	0.0051
3	0.00450	0.00455
4	0.00452	0.00453

Note that the first guess was close to the actual value and that there was oscillation between 0.004 and 0.005 for the guessed and calculated values. For trial 3, an average of these values was used as the guess, and this led rapidly to the correct value (0.0045 to the correct number of significant figures). Also, note that it is useful to carry extra digits until the correct value is obtained. That value can then be rounded off to the correct number of significant figures.

The method of successive approximations is especially useful for solving polynomials containing x to a power of 3 or higher. The procedure is the same as for quadratic equations: Substitute a guessed value for x into the equation for every x term but one, and then solve for x. Continue this process until the guessed and calculated values agree.

A1.5 | Uncertainties in Measurements

Like all the physical sciences, chemistry is based on the results of measurements. Every measurement has an inherent uncertainty, so if we are to use the results of measurements to reach conclusions, we must be able to estimate the sizes of these uncertainties.

For example, the specification for a commercial 500-mg acetaminophen (the active painkiller in Tylenol) tablet is that each batch of tablets must contain 450 to 550 mg of acetaminophen per tablet. Suppose that chemical analysis gave the following results for a batch of acetaminophen tablets: 428 mg, 479 mg, 442 mg, and 435 mg. How can we use these results to decide if the batch of tablets meets the specification? Although the details of how to draw such conclusions from measured data are beyond the scope of this text, we will consider some aspects of how this is done. We will focus here on the types of experimental uncertainty, the expression of experimental results, and a simplified method for estimating experimental uncertainty when several types of measurement contribute to the final result.

Types of Experimental Error

There are two types of experimental uncertainty (error). A variety of names are applied to these types of errors:

$$\text{Precision} \longleftrightarrow \text{random error} \equiv \text{indeterminate error}$$
$$\text{Accuracy} \longleftrightarrow \text{systematic error} \equiv \text{determinate error}$$

The difference between the two types of error is well illustrated by the attempts to hit a target shown in Fig. 1.7 in Chapter 1.

Random error is associated with every measurement. To obtain the last significant figure for any measurement, we must always make an estimate. For example, we interpolate between the marks on a meter stick, a buret, or a balance. The precision of replicate measurements (repeated measurements of the same type) reflects the size of the random errors. Precision refers to the reproducibility of replicate measurements.

The accuracy of a measurement refers to how close it is to the true value. An inaccurate result occurs as a result of some flaw (systematic error) in the measurement: the presence of an interfering substance, incorrect calibration of an instrument, operator error, and so on. The goal of chemical analysis is to eliminate systematic error, but random errors can only be minimized. In practice, an experiment is almost always done to find an unknown value (the true value is not known—someone is trying to obtain that value by doing the experiment). In this case the precision of several replicate determinations is used to assess the accuracy of the result. The results of the replicate experiments are expressed as an average (which we assume is close to the true value) with an error limit that gives some indication of how close the average value may be to the true value. The error limit represents the uncertainty of the experimental result.

Expression of Experimental Results

If we perform several measurements, such as for the analysis for acetaminophen in painkiller tablets, the results should express two things: the average of the measurements and the size of the uncertainty.

There are two common ways of expressing an average: the mean and the median. The mean $(\bar{x})$ is the arithmetic average of the results, or

$$\text{Mean} = \bar{x} = \sum_{i=1}^{n} \frac{x_i}{n} = \frac{x_1 + x_2 + \cdots + x_n}{n}$$

where Σ means take the sum of the values. The mean is equal to the sum of all the measurements divided by the number of measurements. For the acetaminophen results given previously, the mean is

$$\bar{x} = \frac{428 + 479 + 442 + 435}{4} = 446 \text{ mg}$$

The median is the value that lies in the middle among the results. Half the measurements are above the median and half are below the median. For results of 465 mg, 485 mg, and 492 mg, the median is 485 mg. When there is an even number of results, the median is the average of the two middle results. For the acetaminophen results, the median is

$$\frac{442 + 435}{2} = 438 \text{ mg}$$

There are several advantages to using the median. If a small number of measurements is made, one value can greatly affect the mean. Consider the results for the analysis of acetaminophen: 428 mg, 479 mg, 442 mg, and 435 mg. The mean is 446 mg, which is larger than three of the four weights. The median is 438 mg, which lies near the three values that are relatively close to one another.

In addition to expressing an average value for a series of results, we must express the uncertainty. This usually means expressing either the precision of the measurements or the observed range of the measurements. The range of a series of measurements is defined by the smallest value and the largest value. For the analytical results on the acetaminophen tablets, the range is from 428 mg to 479 mg. Using this range, we can express the results by saying that the true value lies between 428 mg and 479 mg. That is, we can express the amount of acetaminophen in a typical tablet as 446 ± 33 mg, where the error limit is chosen to give the observed range (approximately).

The most common way to specify precision is by the standard deviation, s, which for a small number of measurements is given by the formula

$$s = \left[\frac{\sum_{i=1}^{n} (x_i - \bar{x})^2}{n - 1} \right]^{1/2}$$

where x_i is an individual result, $\bar{x}$ is the average (either mean or median), and n is the total number of measurements. For the acetaminophen example, we have

$$s = \left[\frac{(428 - 446)^2 + (479 - 446)^2 + (442 - 446)^2 + (435 - 446)^2}{4 - 1} \right]^{1/2} = 23$$

Thus we can say the amount of acetaminophen in the typical tablet in the batch of tablets is 446 mg with a sample standard deviation of 23 mg. Statistically this means that any additional measurement has a 68% probability (68 chances out of 100) of being between 423 mg ($446 - 23$) and 469 mg ($446 + 23$). Thus the standard deviation is a measure of the precision of a given type of determination.

The standard deviation gives us a means of describing the precision of a given type of determination using a series of replicate results. However, it is also useful to be able to estimate the precision of a procedure that involves several measurements by combining the precisions of the individual steps. That is, we want to answer the following question: How do the uncertainties propagate when we combine the results of several different types of measurements? There are many ways to deal with the propagation of uncertainty. We will discuss only one simple method here.

A Simplified Method for Estimating Experimental Uncertainty

To illustrate this method, we will consider the determination of the density of an irregularly shaped solid. In this determination we make three measurements. First, we measure the mass of the object on a balance. Next, we must obtain the volume of the solid. The easiest method for doing this is to partially fill a graduated cylinder with a liquid and record the volume. Then we add the solid and record the volume again. The difference in the measured volumes is the volume of the solid. We can then calculate the density of the solid from the equation

$$D = \frac{M}{V_2 - V_1}$$

where M is the mass of the solid, V_1 is the initial volume of liquid in the graduated cylinder, and V_2 is the volume of liquid plus solid. Suppose we get the following results:

$$M = 23.06 \text{ g}$$
$$V_1 = 10.4 \text{ mL}$$
$$V_2 = 13.5 \text{ mL}$$

The calculated density is

$$\frac{23.06 \text{ g}}{13.5 \text{ mL} - 10.4 \text{ mL}} = 7.44 \text{ g/mL}$$

Now suppose that the precision of the balance used is ± 0.02 g and that the volume measurements are precise to ± 0.05 mL. How do we estimate the uncertainty of the density? We can do this by assuming a worst case. That is, we assume the largest uncertainties in all measurements, and see what combinations of measurements will give the largest and smallest possible results (the greatest range). Since the density is the mass divided by the volume, the largest value of the density will be that obtained using the largest possible mass and the smallest possible volume:

Largest possible mass = 23.06 + .02

$$D_{max} = \frac{23.08}{13.45 - 10.45} = 7.69 \text{ g/mL}$$

Smallest possible V_2 Largest possible V_1

The smallest value of the density is

Smallest possible mass

$$D_{min} = \frac{23.04}{13.35 - 10.35} = 7.20 \text{ g/mL}$$

Largest possible V_2 Smallest possible V_1

Thus the calculated range is from 7.20 to 7.69 and the average of these values is 7.44. The error limit is the number that gives the high and low range values when added and subtracted from the average. Therefore, we can express the density as 7.44 ± 0.25 g/mL, which is the average value plus or minus the quantity that gives the range calculated by assuming the largest uncertainties.

Analysis of the propagation of uncertainties is useful in drawing qualitative conclusions from the analysis of measurements. For example, suppose that we obtained the preceding results for the density of an unknown alloy and we want to know if it is one of the following alloys:

Alloy A: $D = 7.58$ g/mL

Alloy B: $D = 7.42$ g/mL

Alloy C: $D = 8.56$ g/mL

We can safely conclude that the alloy is not C. But the values of the densities for alloys A and B are both within the inherent uncertainty of our method. To distinguish between A and B, we need to improve the precision of our determination: The obvious choice is to improve the precision of the volume measurement.

The worst-case method is very useful in estimating uncertainties when the results of several measurements are combined to calculate a result. We assume the maximum uncertainty in each measurement and calculate the minimum and maximum possible result. These extreme values describe the range and thus the error limit.

APPENDIX 2 | The Quantitative Kinetic Molecular Model

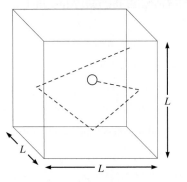

Figure A.3 | An ideal gas particle in a cube whose sides are of length L. The particle collides elastically with the walls in a random, straight-line motion.

We have seen that the kinetic molecular model successfully accounts for the properties of an ideal gas. This appendix will show in some detail how the postulates of the kinetic molecular model lead to an equation corresponding to the experimentally obtained ideal gas equation.

Recall that the particles of an ideal gas are assumed to be volumeless, to have no attraction for each other, and to produce pressure on their container by colliding with the container walls.

Suppose there are n moles of an ideal gas in a cubical container with sides each of length L. Assume each gas particle has a mass m and that it is in rapid, random, straight-line motion colliding with the walls, as shown in Fig. A.3. The collisions will be assumed to be *elastic*—no loss of kinetic energy occurs. We want to compute the force on the walls from the colliding gas particles and then, since pressure is force per unit area, to obtain an expression for the pressure of the gas.

Before we can derive the expression for the pressure of a gas, we must first discuss some characteristics of velocity. Each particle in the gas has a particular velocity u that can be divided into components u_x, u_y, and u_z, as shown in Fig. A.4. First, using u_x and u_y and the Pythagorean theorem, we can obtain u_{xy} as shown in Fig. A.4(c):

$$u_{xy}^2 = u_x^2 + u_y^2$$

↗   ↖ ↖
Hypotenuse of   Sides of
right triangle   right triangle

Then, constructing another triangle as shown in Fig. A.4(c), we find

$$u^2 = u_{xy}^2 + u_z^2$$

or

$$u^2 = u_x^2 + u_y^2 + u_z^2$$

Now let's consider how an "average" gas particle moves. For example, how often does this particle strike the two walls of the box that are perpendicular to the x axis? It is important to realize that only the x component of the velocity affects the particle's impacts on these two walls, as shown in Fig. A.5(a). The larger the x component of the velocity, the faster the particle travels between these two walls, and the more impacts per unit of time it will make on these walls. Remember, the pressure of the gas is due to these collisions with the walls.

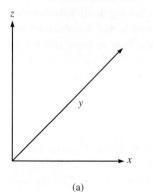

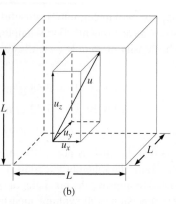

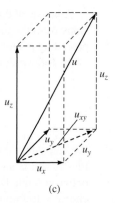

(a)   (b)   (c)

Figure A.4 | (a) The Cartesian coordinate axes.

(b) The velocity u of any gas particle can be broken down into three mutually perpendicular components: u_x, u_y, and u_z. This can be represented as a rectangular solid with sides u_x, u_y, and u_z and body diagonal u.

(c) In the xy plane,

$$u_x^2 + u_y^2 = u_{xy}^2$$

by the Pythagorean theorem. Since u_{xy} and u_x are also perpendicular,

$$u^2 = u_{xy}^2 + u_z^2 = u_x^2 + u_y^2 + u_z^2$$

The collision frequency (collisions per unit of time) with the two walls that are perpendicular to the x axis is given by

$$(\text{Collision frequency})_x = \frac{\text{velocity in the } x \text{ direction}}{\text{distance between the walls}}$$

$$= \frac{u_x}{L}$$

Next, what is the force of a collision? Force is defined as mass times acceleration (change in velocity per unit of time):

$$F = ma = m\left(\frac{\Delta u}{\Delta t}\right)$$

where F represents force, a represents acceleration, Δu represents a change in velocity, and Δt represents a given length of time.

Since we assume that the particle has constant mass, we can write

$$F = \frac{m\Delta u}{\Delta t} = \frac{\Delta(mu)}{\Delta t}$$

The quantity mu is the momentum of the particle (momentum is the product of mass and velocity), and the expression $F = \Delta(mu)/\Delta t$ implies that force is the change in momentum per unit of time. When a particle hits a wall perpendicular to the x axis, as shown in Fig. A.5(b), an elastic collision results in an *exact reversal* of the x component of velocity. That is, the *sign,* or direction, of u_x reverses when the particle collides with one of the walls perpendicular to the x axis. Thus the final momentum is the *negative,* or opposite, of the initial momentum. Remember that an elastic collision means that there is no change in the *magnitude* of the velocity. The change in momentum in the x direction is then

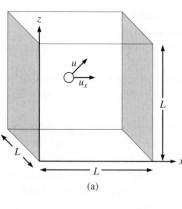

(a)

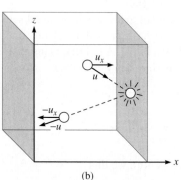

(b)

Figure A.5 | (a) Only the x component of the gas particle's velocity affects the frequency of impacts on the shaded walls, the walls that are perpendicular to the x axis.
(b) For an elastic collision, there is an exact reversal of the x component of the velocity and of the total velocity. The change in momentum (final − initial) is then

$$-mu_x - mu_x = -2mu_x$$

Change in momentum $= \Delta(mu_x) =$ final momentum $-$ initial momentum

$$= -mu_x - mu_x$$

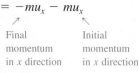

$$= -2mu_x$$

But we are interested in the force the gas particle exerts on the walls of the box. Since we know that every action produces an equal but opposite reaction, the change in momentum with respect to the wall on impact is $-(-2mu_x)$, or $2mu_x$.

Recall that since force is the change in momentum per unit of time,

$$\text{Force}_x = \frac{\Delta(mu_x)}{\Delta t}$$

for the walls perpendicular to the x axis.

This expression can be obtained by multiplying the change in momentum per impact by the number of impacts per unit of time:

$$\text{Force}_x = (2mu_x)\left(\frac{u_x}{L}\right) = \text{change in momentum per unit of time}$$

Change in momentum per impact

Impacts per unit of time

That is,

$$\text{Force}_x = \frac{2mu_x^2}{L}$$

So far we have considered only the two walls of the box perpendicular to the x axis. We can assume that the force on the two walls perpendicular to the y axis is given by

$$\text{Force}_y = \frac{2mu_y^2}{L}$$

and that on the two walls perpendicular to the z axis by

$$\text{Force}_z = \frac{2mu_z^2}{L}$$

Since we have shown that

$$u^2 = u_x^2 + u_y^2 + u_z^2$$

the total force on the box is

$$\text{Force}_{\text{TOTAL}} = \text{force}_x + \text{force}_y + \text{force}_z$$

$$= \frac{2mu_x^2}{L} + \frac{2mu_y^2}{L} + \frac{2mu_z^2}{L}$$

$$= \frac{2m}{L}(u_x^2 + u_y^2 + u_z^2) = \frac{2m}{L}(u^2)$$

Now since we want the average force, we use the average of the square of the velocity $(\overline{u^2})$ to obtain

$$\overline{\text{Force}}_{\text{TOTAL}} = \frac{2m}{L}(\overline{u^2})$$

Next, we need to compute the pressure (force per unit of area)

$$\text{Pressure due to "average" particle} = \frac{\overline{\text{force}}_{\text{TOTAL}}}{\text{area}_{\text{TOTAL}}}$$

$$= \frac{\dfrac{2m\overline{u^2}}{L}}{6L^2} = \frac{m\overline{u^2}}{3L^3}$$

The 6 sides of the cube

Area of each side

Since the volume V of the cube is equal to L^3, we can write

$$\text{Pressure} = P = \frac{m\overline{u^2}}{3V}$$

So far we have considered the pressure on the walls due to a single, "average" particle. Of course, we want the pressure due to the entire gas sample. The number of particles in a given gas sample can be expressed as follows:

$$\text{Number of gas particles} = nN_A$$

where n is the number of moles and N_A is Avogadro's number.

The total pressure on the box due to n moles of a gas is therefore

$$P = nN_A\frac{m\overline{u^2}}{3V}$$

Next we want to express the pressure in terms of the kinetic energy of the gas molecules. Kinetic energy (the energy due to motion) is given by $\frac{1}{2}mu^2$, where m is the mass and u is the velocity. Since we are using the average of the velocity squared $(\overline{u^2})$, and since $m\overline{u^2} = 2(\frac{1}{2}m\overline{u^2})$, we have

$$P = \left(\frac{2}{3}\right)\frac{nN_A(\frac{1}{2}m\overline{u^2})}{V}$$

or

$$\frac{PV}{n} = \left(\frac{2}{3}\right)N_A(\tfrac{1}{2}m\overline{u^2})$$

Thus, based on the postulates of the kinetic molecular model, we have been able to derive an equation that has the same form as the ideal gas equation,

$$\frac{PV}{n} = RT$$

This agreement between experiment and theory supports the validity of the assumptions made in the kinetic molecular model about the behavior of gas particles, at least for the limiting case of an ideal gas.

APPENDIX 3 | Spectral Analysis

Although volumetric and gravimetric analyses are still very commonly used, spectroscopy is the technique most often used for modern chemical analysis. *Spectroscopy* is the study of electromagnetic radiation emitted or absorbed by a given chemical species. Since the quantity of radiation absorbed or emitted can be related to the quantity of the absorbing or emitting species present, this technique can be used for quantitative analysis. There are many spectroscopic techniques, as electromagnetic radiation spans a wide range of energies to include X rays; ultraviolet, infrared, and visible light; and microwaves, to name a few of its familiar forms. We will consider here only one procedure, which is based on the absorption of visible light.

If a liquid is colored, it is because some component of the liquid absorbs visible light. In a solution the greater the concentration of the light-absorbing substance, the more light absorbed, and the more intense the color of the solution.

The quantity of light absorbed by a substance can be measured by a *spectrophotometer,* shown schematically in Fig. A.6. This instrument consists of a source that emits all wavelengths of light in the visible region (wavelengths of ~400 to 700 nm); a monochromator, which selects a given wavelength of light; a sample holder for the solution being measured; and a detector, which compares the intensity of incident light

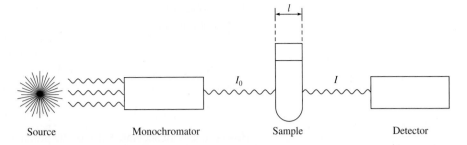

Figure A.6 | A schematic diagram of a simple spectrophotometer. The source emits all wavelengths of visible light, which are dispersed using a prism or grating and then focused, one wavelength at a time, onto the sample. The detector compares the intensity of the incident light (I_0) to the intensity of the light after it has passed through the sample (I).

I_0 to the intensity of light after it has passed through the sample I. The ratio I/I_0, called the *transmittance,* is a measure of the fraction of light that passes through the sample. The amount of light absorbed is given by the *absorbance A,* where

$$A = -\log \frac{I}{I_0}$$

The absorbance can be expressed by the *Beer–Lambert law:*

$$A = \epsilon l c$$

where ϵ is the molar absorptivity or the molar extinction coefficient (in L/mol · cm), l is the distance the light travels through the solution (in cm), and c is the concentration of the absorbing species (in mol/L). The Beer–Lambert law is the basis for using spectroscopy in quantitative analysis. If ϵ and l are known, measuring A for a solution allows us to calculate the concentration of the absorbing species in the solution.

Suppose we have a pink solution containing an unknown concentration of $Co^2(aq)$ ions. A sample of this solution is placed in a spectrophotometer, and the absorbance is measured at a wavelength where ϵ for $Co^{2+}(aq)$ is known to be 12 L/mol · cm. The absorbance A is found to be 0.60. The width of the sample tube is 1.0 cm. We want to determine the concentration of $Co^{2+}(aq)$ in the solution. This problem can be solved by a straightforward application of the Beer–Lambert law,

$$A = \epsilon l c$$

where

$$A = 0.60$$

$$\epsilon = \frac{12 \text{ L}}{\text{mol} \cdot \text{cm}}$$

$$l = \text{light path} = 1.0 \text{ cm}$$

Solving for the concentration gives

$$c = \frac{A}{\epsilon l} = \frac{0.06}{\left(12\dfrac{\text{L}}{\text{mol} \cdot \text{cm}}\right)(1.0 \text{ cm})} = 5.0 \times 10^{-2} \text{ mol/L}$$

To obtain the unknown concentration of an absorbing species from the measured absorbance, we must know the product ϵl, since

$$c = \frac{A}{\epsilon l}$$

We can obtain the product ϵl by measuring the absorbance of a solution of *known* concentration, since

Measured using a
↙ spectrophotometer

$$\epsilon l = \frac{A}{c}$$

↖ Known from making up
the solution

However, a more accurate value of the product ϵl can be obtained by plotting A versus c for a series of solutions. Note that the equation $A = \epsilon l c$ gives a straight line with slope ϵl when A is plotted against c.

For example, consider the following typical spectroscopic analysis. A sample of steel from a bicycle frame is to be analyzed to determine its manganese content. The procedure involves weighing out a sample of the steel, dissolving it in strong acid, treating the resulting solution with a very strong oxidizing agent to convert all the manganese to permanganate ion (MnO_4^-), and then using spectroscopy to determine the concentration of the intensely purple MnO_4^- ions in the solution. To do this, however, the value of ϵl for MnO_4^- must be determined at an appropriate wavelength. The absorbance values for four solutions with known MnO_4^- concentrations were measured to give the following data:

Solution	Concentration of MnO_4^- (mol/L)	Absorbance
1	7.00×10^{-5}	0.175
2	1.00×10^{-4}	0.250
3	2.00×10^{-4}	0.500
4	3.50×10^{-4}	0.875

A plot of absorbance versus concentration for the solutions of known concentration is shown in Fig. A.7. The slope of this line (change in A/change in c) is 2.48×10^3 L/mol. This quantity represents the product ϵl.

A sample of the steel weighing 0.1523 g was dissolved and the unknown amount of manganese was converted to MnO_4^- ions. Water was then added to give a solution with a final volume of 100.0 mL. A portion of this solution was placed in a spectro-

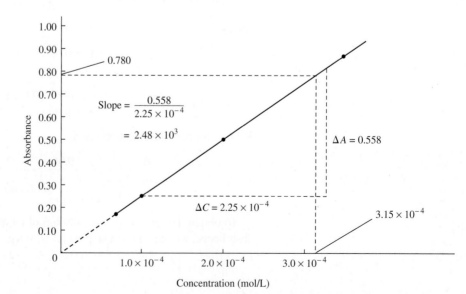

Figure A.7 | A plot of absorbance versus concentration of MnO_4^- in a series of solutions of known concentration.

photometer, and its absorbance was found to be 0.780. Using these data, we want to calculate the percent manganese in the steel. The MnO_4^- ions from the manganese in the dissolved steel sample show an absorbance of 0.780. Using the Beer–Lambert law, we calculate the concentration of MnO_4^- in this solution:

$$c = \frac{A}{\epsilon l} = \frac{0.780}{2.48 \times 10^3 \text{ L/mol}} \times 3.15 \times 10^{-4} \text{ mol/L}$$

There is a more direct way for finding c. Using a graph such as that in Fig. A.7 (often called a *Beer's law plot*), we can read the concentration that corresponds to $A = 0.780$. This interpolation is shown by dashed lines on the graph. By this method, $c = 3.15 \times 10^{-4}$ mol/L, which agrees with the value obtained above.

Recall that the original 0.1523-g steel sample was dissolved, the manganese was converted to permanganate, and the volume was adjusted to 100.0 mL. We now know that $[MnO_4]$ in that solution is $3.15 \times 10^{-4}\ M$. Using this concentration, we can calculate the total number of moles of MnO_4^- in that solution:

$$\text{mol } MnO_4^- = 100.0 \text{ mL} \times \frac{1 \text{ L}}{1000 \text{ mL}} \times 3.15 \times 10^{-4} \frac{\text{mol}}{\text{L}}$$

$$= 3.15 \times 10^{-5} \text{ mol}$$

Since each mole of manganese in the original steel sample yields a mole of MnO_4, that is,

$$1 \text{ mol Mn} \xrightarrow{\text{Oxidation}} 1 \text{ mol } MnO_4^-$$

the original steel sample must have contained 3.15×10^{-5} mole of manganese. The mass of manganese present in the sample is

$$3.15 \times 10^{-5} \text{ mol Mn} \times \frac{54.938 \text{ g of Mn}}{1 \text{ mol Mn}} = 1.73 \times 10^{-3} \text{ g of Mn}$$

Since the steel sample weighed 0.1523 g, the present manganese in the steel is

$$\frac{1.73 \times 10^{-3} \text{ g of Mn}}{1.523 \times 10^{-1} \text{ g of sample}} \times 100 = 1.14\%$$

This example illustrates a typical use of spectroscopy in quantitative analysis. The steps commonly involved are as follows:

1. Preparation of a calibration plot (a Beer's law plot) by measuring the absorbance values of a series of solutions with known concentrations

2. Measurement of the absorbance of the solution of unknown concentration

3. Use of the calibration plot to determine the unknown concentration

---

| APPENDIX 4 | Selected Thermodynamic Data |

Note: All values are assumed precise to at least ± 1.

Substance and State	ΔH_f° (kJ/mol)	ΔG_f° (kJ/mol)	S° (J/K · mol)	Substance and State	ΔH_f° (kJ/mol)	ΔG_f° (kJ/mol)	S° (J/K · mol)
Aluminum				Barium			
Al(s)	0	0	28	Ba(s)	0	0	67
Al$_2$O$_3$(s)	−1676	−1582	51	BaCO$_3$(s)	−1219	−1139	112
Al(OH)$_3$(s)	−1277			BaO(s)	−582	−552	70
AlCl$_3$(s)	−704	−629	111	Ba(OH)$_2$(s)	−946		

(continued)

Appendix Four (continued)

Substance and State	ΔH_f° (kJ/mol)	ΔG_f° (kJ/mol)	S° (J/K · mol)
Barium, *continued*			
$BaSO_4(s)$	−1465	−1353	132
Beryllium			
$Be(s)$	0	0	10
$BeO(s)$	−599	−569	14
$Be(OH)_2(s)$	−904	−815	47
Bromine			
$Br_2(l)$	0	0	152
$Br_2(g)$	31	3	245
$Br_2(aq)$	−3	4	130
$Br^-(aq)$	−121	−104	82
$HBr(g)$	−36	−53	199
Cadmium			
$Cd(s)$	0	0	52
$CdO(s)$	−258	−228	55
$Cd(OH)_2(s)$	−561	−474	96
$CdS(s)$	−162	−156	65
$CdSO_4(s)$	−935	−823	123
Calcium			
$Ca(s)$	0	0	41
$CaC_2(s)$	−63	−68	70
$CaCO_3(s)$	−1207	−1129	93
$CaO(s)$	−635	−604	40
$Ca(OH)_2(s)$	−987	−899	83
$Ca_3(PO_4)_2(s)$	−4126	−3890	241
$CaSO_4(s)$	−1433	−1320	107
$CaSiO_3(s)$	−1630	−1550	84
Carbon			
$C(s)$ (graphite)	0	0	6
$C(s)$ (diamond)	2	3	2
$CO(g)$	−110.5	−137	198
$CO_2(g)$	−393.5	−394	214
$CH_4(g)$	−75	−51	186
$CH_3OH(g)$	−201	−163	240
$CH_3OH(l)$	−239	−166	127
$H_2CO(g)$	−116	−110	219
$HCOOH(g)$	−363	−351	249
$HCN(g)$	135.1	125	202
$C_2H_2(g)$	227	209	201
$C_2H_4(g)$	52	68	219
$CH_3CHO(g)$	−166	−129	250
$C_2H_5OH(l)$	−278	−175	161
$C_2H_6(g)$	−84.7	−32.9	229.5
$C_3H_6(g)$	20.9	62.7	266.9
$C_3H_8(g)$	−104	−24	270
$C_2H_4O(g)$ (ethylene oxide)	−53	−13	242
$CH_2{=}CHCN(g)$	185.0	195.4	274
$CH_3COOH(l)$	−484	−389	160
$C_6H_{12}O_6(s)$	−1275	−911	212
CCl_4	−135	−65	216
Chlorine			
$Cl_2(g)$	0	0	223
$Cl_2(aq)$	−23	7	121

Substance and State	ΔH_f° (kJ/mol)	ΔG_f° (kJ/mol)	S° (J/K · mol)
Chlorine, *continued*			
$Cl^-(aq)$	−167	−131	57
$HCl(g)$	−92	−95	187
Chromium			
$Cr(s)$	0	0	24
$Cr_2O_3(s)$	−1128	−1047	81
$CrO_3(s)$	−579	−502	72
Copper			
$Cu(s)$	0	0	33
$CuCO_3(s)$	−595	−518	88
$Cu_2O(s)$	−170	−148	93
$CuO(s)$	−156	−128	43
$Cu(OH)_2(s)$	−450	−372	108
$CuS(s)$	−49	−49	67
Fluorine			
$F_2(g)$	0	0	203
$F2(aq)$	−333	−279	−14
$HF(g)$	−271	−273	174
Hydrogen			
$H_2(g)$	0	0	131
$H(g)$	217	203	115
$H^+(aq)$	0	0	0
$OH^-(aq)$	−230	−157	−11
$H_2O(l)$	−286	−237	70
$H_2O(g)$	−242	−229	189
Iodine			
$I_2(s)$	0	0	116
$I_2(g)$	62	19	261
$I_2(aq)$	23	16	137
$I^-(aq)$	−55	−52	106
Iron			
$Fe(s)$	0	0	27
$Fe_3C(s)$	21	15	108
$Fe_{0.95}O(s)$ (wustite)	−264	−240	59
FeO	−272	−255	61
$Fe_3O_4(s)$ (magnetite)	−1117	−1013	146
$Fe_2O_3(s)$ (hematite)	−826	−740	90
$FeS(s)$	−95	−97	67
$FeS_2(s)$	−178	−166	53
$FeSO_4(s)$	−929	−825	121
Lead			
$Pb(s)$	0	0	65
$PbO_2(s)$	−277	−217	69
$PbS(s)$	−100	−99	91
$PbSO_4(s)$	−920	−813	149
Magnesium			
$Mg(s)$	0	0	33
$MgCO_3(s)$	−1113	−1029	66
$MgO(s)$	−602	−569	27
$Mg(OH)_2(s)$	−925	−834	64
Manganese			
$Mn(s)$	0	0	32

Substance and State	ΔH_f° (kJ/mol)	ΔG_f° (kJ/mol)	S° (J/K · mol)
Manganese, *continued*			
MnO(s)	−385	−363	60
Mn₃O₄(s)	−1387	−1280	149
Mn₂O₃(s)	−971	−893	110
MnO₂(s)	−521	−466	53
MnO₄⁻(aq)	−543	−449	190
Mercury			
Hg(l)	0	0	76
Hg₂Cl₂(s)	−265	−211	196
HgCl₂(s)	−230	−184	144
HgO(s)	−90	−59	70
HgS(s)	−58	−49	78
Nickel			
Ni(s)	0	0	30
NiCl₂(s)	−316	−272	107
NiO(s)	−241	−213	38
Ni(OH)₂(s)	−538	−453	79
NiS(s)	−93	−90	53
Nitrogen			
N₂(g)	0	0	192
NH₃(g)	−46	−17	193
NH₃(aq)	−80	−27	111
NH₄⁺(aq)	−132	−79	113
NO(g)	90	87	211
NO₂(g)	34	52	240
N₂O(g)	82	104	220
N₂O₄(g)	10	98	304
N₂O₄(l)	−20	97	209
N₂O₅(s)	−42	134	178
N₂H₄(l)	51	149	121
N₂H₃CH₃(l)	54	180	166
HNO₃(aq)	−207	−111	146
HNO₃(l)	−174	−81	156
NH₄ClO₄(s)	−295	−89	186
NH₄Cl(s)	−314	−203	96
Oxygen			
O₂(g)	0	0	205
O(g)	249	232	161
O₃(g)	143	163	239
Phosphorus			
P(s) (white)	0	0	41
P(s) (red)	−18	−12	23
P(s) (black)	−39	−33	23
P₄(g)	59	24	280
PF₅(g)	−1578	−1509	296
PH₃(g)	5	13	210
H₃PO₄(s)	−1279	21119	110
H₃PO₄(l)	−1267	—	—
H₃PO₄(aq)	−1288	−1143	158
P₄O₁₀(s)	−2984	2−2698	229
Potassium			
K(s)	0	0	64
KCl(s)	−436	−408	83

Substance and State	ΔH_f° (kJ/mol)	ΔG_f° (kJ/mol)	S° (J/K · mol)
Potassium, *continued*			
KClO₃(s)	−391	−290	143
KClO₄(s)	−433	−304	151
K₂O(s)	−361	−322	98
K₂O₂(s)	−496	−430	113
KO₂(s)	−283	−238	117
KOH(s)	−425	−379	79
KOH(aq)	−481	2−440	9.20
Silicon			
SiO₂(s) (quartz)	−911	−856	42
SiCl₄(l)	−687	−620	240
Silver			
Ag(s)	0	0	43
Ag⁺(aq)	105	77	73
AgBr(s)	−100	−97	107
AgCN(s)	146	164	84
AgCl(s)	−127	−110	96
Ag₂CrO₄(s)	−712	−622	217
AgI(s)	−62	−66	115
Ag₂O(s)	−31	−11	122
Ag₂S(s)	−32	−40	146
Sodium			
Na(s)	0	0	51
Na⁺(aq)	−240	−262	59
NaBr(s)	−360	−347	84
Na₂CO₃(s)	−1131	−1048	136
NaHCO₃(s)	−948	−852	102
NaCl(s)	−411	−384	72
NaH(s)	−56	−33	40
NaI(s)	−288	−282	91
NaNO₂(s)	−359		
NaNO₃(s)	−467	−366	116
Na₂O(s)	−416	−377	73
Na₂O₂(s)	−515	−451	95
NaOH(s)	−426	−381	64
NaOH(aq)	−470	−419	50
Sulfur			
S(s) (rhombic)	0	0	32
S(s) (monoclinic)	0.3	0.1	33
S²⁻(aq)	33	86	215
S₈(g)	102	50	431
SF₆(g)	−1209	−1105	292
H₂S(g)	−21	−34	206
SO₂(g)	−297	−300	248
SO₃(g)	−396	−371	257
SO₄²⁻(aq)	−909	−745	20
H₂SO₄(l)	−814	−690	157
H₂SO₄(aq)	−909	−745	20
Tin			
Sn(s) (white)	0	0	52
Sn(s) (gray)	22	0.1	44
SnO(s)	−285	−257	56
SnO₂(s)	−581	−520	52

(continued)

Appendix Four (continued)

Substance and State	ΔH_f° (kJ/mol)	ΔG_f° (kJ/mol)	S° (J/K · mol)
Tin, *continued*			
$Sn(OH)_2(s)$	-561	-492	155
Titanium			
$TiCl_4(g)$	-763	-727	355
$TiO_2(s)$	-945	-890	50
Uranium			
$U(s)$	0	0	50
$UF_6(s)$	-2137	-2008	228
$UF_6(g)$	-2113	-2029	380
$UO_2(s)$	-1084	-1029	78
$U_3O_8(s)$	-3575	-3393	282
$UO_3(s)$	-1230	-1150	99

Substance and State	ΔH_f° (kJ/mol)	ΔG_f° (kJ/mol)	S° (J/K · mol)
Xenon			
$Xe(g)$	0	0	170
$XeF_2(g)$	-108	-48	254
$XeF_4(s)$	-251	-121	146
$XeF_6(g)$	-294		
$XeO_3(s)$	402		
Zinc			
$Zn(s)$	0	0	42
$ZnO(s)$	-348	-318	44
$Zn(OH)_2(s)$	-642		
$ZnS(s)$ (wurtzite)	-193		
$ZnS(s)$ (zinc blende)	-206	-201	58
$ZnSO_4(s)$	-983	-874	120

APPENDIX 5 Equilibrium Constants and Reduction Potentials

A5.1 | Values of K_a for Some Common Monoprotic Acids

Name	Formula	Value of K_a
Hydrogen sulfate ion	HSO_4^-	1.2×10^{-2}
Chlorous acid	$HClO_2$	1.2×10^{-2}
Monochloracetic acid	$HC_2H_2ClO_2$	1.35×10^{-3}
Hydrofluoric acid	HF	7.2×10^{-4}
Nitrous acid	HNO_2	4.0×10^{-4}
Formic acid	HCO_2H	1.8×10^{-4}
Lactic acid	$HC_3H_5O_3$	1.38×10^{-4}
Benzoic acid	$HC_7H_5O_2$	6.4×10^{-5}
Acetic acid	$HC_2H_3O_2$	1.8×10^{-5}
Hydrated aluminum(III) ion	$[Al(H_2O)_6]^{3+}$	1.4×10^{-5}
Propanoic acid	$HC_3H_5O_2$	1.3×10^{-5}
Hypochlorous acid	$HOCl$	3.5×10^{-8}
Hypobromous acid	$HOBr$	2×10^{-9}
Hydrocyanic acid	HCN	6.2×10^{-10}
Boric acid	H_3BO_3	5.8×10^{-10}
Ammonium ion	NH_4^+	5.6×10^{-10}
Phenol	HOC_6H_5	1.6×10^{-10}
Hypoiodous acid	HOI	2×10^{-11}

A5.2 | Stepwise Dissociation Constants for Several Common Polyprotic Acids

Name	Formula	K_{a_1}	K_{a_2}	K_{a_3}
Phosphoric acid	H_3PO_4	7.5×10^{-3}	6.2×10^{-8}	4.8×10^{-13}
Arsenic acid	H_3AsO_4	5.5×10^{-3}	1.7×10^{-7}	5.1×10^{-12}
Carbonic acid	H_2CO_3	4.3×10^{-7}	5.6×10^{-11}	
Sulfuric acid	H_2SO_4	Large	1.2×10^{-2}	
Sulfurous acid	H_2SO_3	1.5×10^{-2}	1.0×10^{-7}	
Hydrosulfuric acid	H_2S	1.0×10^{-7}	$\sim 10^{-19}$	
Oxalic acid	$H_2C_2O_4$	6.5×10^{-2}	6.1×10^{-5}	
Ascorbic acid (vitamin C)	$H_2C_6H_6O_6$	7.9×10^{-5}	1.6×10^{-12}	
Citric acid	$H_3C_6H_5O_7$	8.4×10^{-4}	1.8×10^{-5}	4.0×10^{-6}

A5.3 | Values of K_b for Some Common Weak Bases

Name	Conjugate Formula	Acid	K_b
Ammonia	NH_3	NH_4^+	1.8×10^{-5}
Methylamine	CH_3NH_2	$CH_3NH_3^+$	4.38×10^{-4}
Ethylamine	$C_2H_5NH_2$	$C_2H_5NH_3^+$	5.6×10^{-4}
Diethylamine	$(C_2H_5)_2NH$	$(C_2H_5)_2NH_2^+$	1.3×10^{-3}
Triethylamine	$(C_2H_5)_3N$	$(C_2H_5)_3NH^+$	4.0×10^{-4}
Hydroxylamine	$HONH_2$	$HONH_3^+$	1.1×10^{-8}
Hydrazine	H_2NNH_2	$H_2NNH_3^+$	3.0×10^{-6}
Aniline	$C_6H_5NH_2$	$C_6H_5NH_3^+$	3.8×10^{-10}
Pyridine	C_5H_5N	$C_5H_5NH^+$	1.7×10^{-9}

A5.4 | K_{sp} Values at 25°C for Common Ionic Solids

Ionic Solid	K_{sp} (at 25°C)	Ionic Solid	K_{sp} (at 25°C)	Ionic Solid	K_{sp} (at 25°C)
Fluorides		Hg_2CrO_4*	2×10^{-9}	$Ni(OH)_2$	1.6×10^{-16}
BaF_2	2.4×10^{-5}	$BaCrO_4$	8.5×10^{-11}	$Zn(OH)_2$	4.5×10^{-17}
MgF_2	6.4×10^{-9}	Ag_2CrO_4	9.0×10^{-12}	$Cu(OH)_2$	1.6×10^{-19}
PbF_2	4×10^{-8}	$PbCrO_4$	2×10^{-16}	$Hg(OH)_2$	3×10^{-26}
SrF_2	7.9×10^{-10}			$Sn(OH)_2$	3×10^{-27}
CaF_2	4.0×10^{-11}	**Carbonates**		$Cr(OH)_3$	6.7×10^{-31}
		$NiCO_3$	1.4×10^{-7}	$Al(OH)_3$	2×10^{-32}
Chlorides		$CaCO_3$	8.7×10^{-9}	$Fe(OH)_3$	4×10^{-38}
$PbCl_2$	1.6×10^{-5}	$BaCO_3$	1.6×10^{-9}	$Co(OH)_3$	2.5×10^{-43}
$AgCl$	1.6×10^{-10}	$SrCO_3$	7×10^{-10}		
Hg_2Cl_2*	1.1×10^{-18}	$CuCO_3$	2.5×10^{-10}	**Sulfides**	
		$ZnCO_3$	2×10^{-10}	MnS	2.3×10^{-13}
Bromides		$MnCO_3$	8.8×10^{-11}	FeS	3.7×10^{-19}
$PbBr_2$	4.6×10^{-6}	$FeCO_3$	2.1×10^{-11}	NiS	3×10^{-21}
$AgBr$	5.0×10^{-13}	Ag_2CO_3	8.1×10^{-12}	CoS	5×10^{-22}
Hg_2Br_2*	1.3×10^{-22}	$CdCO_3$	5.2×10^{-12}	ZnS	2.5×10^{-22}
		$PbCO_3$	1.5×10^{-15}	SnS	1×10^{-26}
Iodides		$MgCO_3$	1×10^{-5}	CdS	1.0×10^{-28}
PbI_2	1.4×10^{-8}	Hg_2CO_3*	9.0×10^{-15}	PbS	7×10^{-29}
AgI	1.5×10^{-16}			CuS	8.5×10^{-45}
Hg_2I_2*	4.5×10^{-29}	**Hydroxides**		Ag_2S	1.6×10^{-49}
		$Ba(OH)_2$	5.0×10^{-3}	HgS	1.6×10^{-54}
Sulfates		$Sr(OH)_2$	3.2×10^{-4}		
$CaSO_4$	6.1×10^{-5}	$Ca(OH)_2$	1.3×10^{-6}	**Phosphates**	
Ag_2SO_4	1.2×10^{-5}	$AgOH$	2.0×10^{-8}	Ag_3PO_4	1.8×10^{-18}
$SrSO_4$	3.2×10^{-7}	$Mg(OH)_2$	8.9×10^{-12}	$Sr_3(PO_4)_2$	1×10^{-31}
$PbSO_4$	1.3×10^{-8}	$Mn(OH)_2$	2×10^{-13}	$Ca_3(PO_4)_2$	1.3×10^{-32}
$BaSO_4$	1.5×10^{-9}	$Cd(OH)_2$	5.9×10^{-15}	$Ba_3(PO_4)_2$	6×10^{-39}
		$Pb(OH)_2$	1.2×10^{-15}	$Pb_3(PO_4)_2$	1×10^{-54}
Chromates		$Fe(OH)_2$	1.8×10^{-15}		
$SrCrO_4$	3.6×10^{-5}	$Co(OH)_2$	2.5×10^{-16}		

*Contains Hg_2^{2+} ions. $K_{sp} = [Hg_2^{2+}][X^-]^2$ for Hg_2X_2 salts.

A5.5 | Standard Reduction Potentials at 25°C (298 K) for Many Common Half-Reactions

Half-Reaction	$\mathscr{E}°$ (V)	Half-Reaction	$\mathscr{E}°$ (V)
$F_2 + 2e^- \rightarrow 2F^-$	2.87	$O_2 + 2H_2O + 4e^- \rightarrow 4OH^-$	0.40
$Ag^{2+} + e^- \rightarrow Ag^+$	1.99	$Cu^{2+} + 2e^- \rightarrow Cu$	0.34
$Co^{3+} + e^- \rightarrow Co^{2+}$	1.82	$Hg_2Cl_2 + 2e^- \rightarrow 2Hg + 2Cl^-$	0.34
$H_2O_2 + 2H^+ + 2e^- \rightarrow 2H_2O$	1.78	$AgCl + e^- \rightarrow Ag + Cl^-$	0.22
$Ce^{4+} + e^- \rightarrow Ce^{3+}$	1.70	$SO_4^{2-} + 4H^+ + 2e^- \rightarrow H_2SO_3 + H_2O$	0.20
$PbO_2 + 4H^+ + SO_4^{2-} + 2e^- \rightarrow PbSO_4 + 2H_2O$	1.69	$Cu^{2+} + e^- \rightarrow Cu^+$	0.16
$MnO_4^- + 4H^+ + 3e^- \rightarrow MnO_2 + 2H_2O$	1.68	$2H^+ + 2e^- \rightarrow H_2$	0.00
$2e^- + 2H^+ + IO_4^- \rightarrow IO_3^- + H_2O$	1.60	$Fe^{3+} + 3e^- \rightarrow Fe$	-0.036
$MnO_4^- + 8H^+ + 5e^- \rightarrow Mn^{2+} + 4H_2O$	1.51	$Pb^{2+} + 2e^- \rightarrow Pb$	-0.13
$Au^{3+} + 3e^- \rightarrow Au$	1.50	$Sn^{2+} + 2e^- \rightarrow Sn$	-0.14
$PbO_2 + 4H^+ + 2e^- \rightarrow Pb^{2+} + 2H_2O$	1.46	$Ni^{2+} + 2e^- \rightarrow Ni$	-0.23
$Cl_2 + 2e^- \rightarrow 2Cl^-$	1.36	$PbSO_4 + 2e^- \rightarrow Pb + SO_4^{2-}$	-0.35
$Cr_2O_7^{2-} + 14H^+ + 6e^- \rightarrow 2Cr^{3+} + 7H_2O$	1.33	$Cd^{2+} + 2e^- \rightarrow Cd$	-0.40
$O_2 + 4H^+ + 4e^- \rightarrow 2H_2O$	1.23	$Fe^{2+} + 2e^- \rightarrow Fe$	-0.44
$MnO_2 + 4H^+ + 2e^- \rightarrow Mn^{2+} + 2H_2O$	1.21	$Cr^{3+} + e^- \rightarrow Cr^{2+}$	-0.50
$IO_3^- + 6H^+ + 5e^- \rightarrow \frac{1}{2}I_2 + 3H_2O$	1.20	$Cr^{3+} + 3e^- \rightarrow Cr$	-0.73
$Br_2 + 2e^- \rightarrow 2Br^-$	1.09	$Zn^{2+} + 2e^- \rightarrow Zn$	-0.76
$VO_2^+ + 2H^+ + e^- \rightarrow VO^{2+} + H_2O$	1.00	$2H_2O + 2e^- \rightarrow H_2 + 2OH^-$	-0.83
$AuCl_4^- + 3e^- \rightarrow Au + 4Cl^-$	0.99	$Mn^{2+} + 2e^- \rightarrow Mn$	-1.18
$NO_3^- + 4H^+ + 3e^- \rightarrow NO + 2H_2O$	0.96	$Al^{3+} + 3e^- \rightarrow Al$	-1.66
$ClO_2 + e^- \rightarrow ClO_2^-$	0.954	$H_2 + 2e^- \rightarrow 2H^-$	-2.23
$2Hg^{2+} + 2e^- \rightarrow Hg_2^{2+}$	0.91	$Mg^{2+} + 2e^- \rightarrow Mg$	-2.37
$Ag^+ + e^- \rightarrow Ag$	0.80	$La^{3+} + 3e^- \rightarrow La$	-2.37
$Hg_2^{2+} + 2e^- \rightarrow 2Hg$	0.80	$Na^+ + e^- \rightarrow Na$	-2.71
$Fe^{3+} + e^- \rightarrow Fe^{2+}$	0.77	$Ca^{2+} + 2e^- \rightarrow Ca$	-2.76
$O_2 + 2H^+ + 2e^- \rightarrow H_2O_2$	0.68	$Ba^{2+} + 2e^- \rightarrow Ba$	-2.90
$MnO_4^- + e^- \rightarrow MnO_4^{2-}$	0.56	$K^+ + e^- \rightarrow K$	-2.92
$I_2 + 2e^- \rightarrow 2I^-$	0.54	$Li^+ + e^- \rightarrow Li$	-3.05
$Cu^+ + e^- \rightarrow Cu$	0.52		

Length

SI unit: meter (m)
1 meter = 1.0936 yards
1 centimeter = 0.39370 inch
1 inch = 2.54 centimeters (exactly)
1 kilometer = 0.62137 mile
1 mile = 5280 feet
= 1.6093 kilometers
1 angstrom = 10^{-10} meter
= 100 picometers

Mass

SI unit: kilogram (kg)
1 kilogram = 1000 grams
= 2.2046 pounds
1 pound = 453.59 grams
= 0.45359 kilogram
= 16 ounces
1 ton = 2000 pounds
= 907.185 kilograms
1 metric ton = 1000 kilograms
= 2204.6 pounds
1 atomic mass unit = 1.66056×10^{-27} kilograms

Volume

SI unit: cubic meter (m^3)
1 liter = 10^{-3} m^3
= 1 dm^3
= 1.0567 quarts
1 gallon = 4 quarts
= 8 pints
= 3.7854 liters
1 quart = 32 fluid ounces
= 0.94633 liter

Temperature

SI unit: kelvin (K)
0 K = $-273.15°C$
= $-459.67°F$
K = $°C + 273.15$
$°C = \frac{5}{9}°F - 32$
$°F = \frac{9}{5}°C + 32$

Energy

SI unit: joule (J)
1 joule = 1 kg · m^2/s^2
= 0.23901 calorie
= 9.4781×10^{-4} btu
(British thermal unit)
1 calorie = 4.184 joules
= 3.965×10^{-3} btu
1 btu = 1055.06 joules
= 252.2 calories

Pressure

SI unit: pascal (Pa)
1 pascal = 1 N/m^2
= 1 kg/m · s^2
1 atmosphere = 101.325 kilopascals
= 760 torr (mm Hg)
= 14.70 pounds per
square inch
1 bar = 10^5 pascals

Glossary

Accuracy the agreement of a particular value with the true value. (1.4)

Acid a substance that produces hydrogen ions in solution; a proton donor. (2.8; 4.2; 4.8)

Acid–base indicator a substance that marks the end point of an acid–base titration by changing color. (15.5)

Acid dissociation constant (K_a) the equilibrium constant for a reaction in which a proton is removed from an acid by H_2O to form the conjugate base and H_3O^+. (14.1)

Acid rain a result of air pollution by sulfur dioxide. (5.10)

Acidic oxide a covalent oxide that dissolves in water to give an acidic solution. (14.10)

Actinide series a group of 14 elements following actinium in the periodic table, in which the $5f$ orbitals are being filled. (7.11; 20.1)

Activated complex (transition state) the arrangement of atoms found at the top of the potential energy barrier as a reaction proceeds from reactants to products. (12.6)

Activation energy the threshold energy that must be overcome to produce a chemical reaction. (12.6)

Addition polymerization a type of polymerization in which the monomers simply add together to form the polymer, with no other products. (22.5)

Addition reaction a reaction in which atoms add to a carbon–carbon multiple bond. (22.2)

Adsorption the collection of one substance on the surface of another. (12.7)

Air pollution contamination of the atmosphere, mainly by the gaseous products of transportation and production of electricity. (5.10)

Alcohol an organic compound in which the hydroxyl group is a substituent on a hydrocarbon. (22.4)

Aldehyde an organic compound containing the carbonyl group bonded to at least one hydrogen atom. (22.4)

Alkali metal a Group 1A metal. (2.7; 20.2)

Alkaline earth metal a Group 2A metal. (2.7; 20.4)

Alkane a saturated hydrocarbon with the general formula C_nH_{2n+2}. (22.1)

Alkene an unsaturated hydrocarbon containing a carbon–carbon double bond. The general formula is C_nH_{2n}. (22.2)

Alkyne an unsaturated hydrocarbon containing a triple carbon–carbon bond. The general formula is C_nH_{2n-2}. (22.2)

Alloy a substance that contains a mixture of elements and has metallic properties. (10.4)

Alloy steel a form of steel containing carbon plus other metals such as chromium, cobalt, manganese, and molybdenum. (21.8)

Alpha (α) particle a helium nucleus. (19.1)

Alpha-particle production a common mode of decay for radioactive nuclides in which the mass number changes. (19.1)

Amine an organic base derived from ammonia in which one or more of the hydrogen atoms are replaced by organic groups. (14.6; 22.4)

α-Amino acid an organic acid in which an amino group and an R group are attached to the carbon atom next to the carboxyl group. (22.6)

Amorphous solid a solid with considerable disorder in its structure. (10.3)

Ampere the unit of electric current equal to one coulomb of charge per second. (18.8)

Amphoteric substance a substance that can behave either as an acid or as a base. (14.2)

Angular momentum quantum number (ℓ) the quantum number relating to the shape of an atomic orbital, which can assume any integral value from 0 to $n-1$ for each value of n. (7.6)

Anion a negative ion. (2.6)

Anode the electrode in a galvanic cell at which oxidation occurs. (18.2)

Antibonding molecular orbital an orbital higher in energy than the atomic orbitals of which it is composed. (9.2)

Aqueous solution a solution in which water is the dissolving medium or solvent. (4)

Aromatic hydrocarbon one of a special class of cyclic unsaturated hydrocarbons, the simplest of which is benzene. (22.3)

Arrhenius concept a concept postulating that acids produce hydrogen ions in aqueous solution, while bases produce hydroxide ions. (14.1)

Arrhenius equation the equation representing the rate constant as $k = Ae^{-E_a/RT}$, where A represents the product of the collision frequency and the steric factor, and $e^{-E_a/RT}$ is the fraction of collisions with sufficient energy to produce a reaction. (12.6)

Atactic chain a polymer chain in which the substituent groups such as CH_3 are randomly distributed along the chain. (22.5)

Atmosphere the mixture of gases that surrounds the earth's surface. (5.10)

Atomic number the number of protons in the nucleus of an atom. (2.5; 18)

Atomic radius half the distance between the nuclei in a molecule consisting of identical atoms. (7.12)

Atomic solid a solid that contains atoms at the lattice points. (10.3)

Atomic weight the weighted average mass of the atoms in a naturally occurring element. (2.3)

Aufbau principle the principle stating that as protons are added one by one to the nucleus to build up the elements, electrons are similarly added to hydrogen-like orbitals. (7.11)

Autoionization the transfer of a proton from one molecule to another of the same substance. (14.2)

Avogadro's law equal volumes of gases at the same temperature and pressure contain the same number of particles. (5.2)

Avogadro's number the number of atoms in exactly 12 grams of pure ^{12}C, equal to 6.022×10^{23}. (3.3)

Ball-and-stick model a molecular model that distorts the sizes of atoms but shows bond relationships clearly. (2.6)

Band model a molecular model for metals in which the electrons are assumed to travel around the metal crystal in molecular orbitals formed from the valence atomic orbitals of the metal atoms. (10.4)

Barometer a device for measuring atmospheric pressure. (5.1)

Base a substance that produces hydroxide ions in aqueous solution; a proton acceptor. (4.8)

Basic oxide an ionic oxide that dissolves in water to produce a basic solution. (14.10)

Basic oxygen process a process for producing steel by oxidizing and removing the impurities in iron using a high-pressure blast of oxygen. (21.8)

Battery a group of galvanic cells connected in series. (18.6)

Beta (β) particle an electron produced in radioactive decay. (19.1)

Beta-particle production a decay process for radioactive nuclides in which the mass number remains constant and the atomic number changes. The net effect is to change a neutron to a proton. (19.1)

Bidentate ligand a ligand that can form two bonds to a metal ion. (21.3)

Bimolecular step a reaction involving the collision of two molecules. (12.5)

Binary compound a two-element compound. (2.8)

Binding energy (nuclear) the energy required to decompose a nucleus into its component nucleons. (19.5)

Biomolecule a molecule responsible for maintaining and/or reproducing life. (22)

Blast furnace a furnace in which iron oxide is reduced to iron metal by using a very strong blast of hot air to produce carbon monoxide from coke, and then using this gas as a reducing agent for the iron. (21.8)

Bond energy the energy required to break a given chemical bond. (8.1)

Bond length the distance between the nuclei of the two atoms connected by a bond; the distance where the total energy of a diatomic molecule is minimal. (8.1)

Bond order the difference between the number of bonding electrons and the number of antibonding electrons, divided by two. It is an index of bond strength. (9.2)

Bonding molecular orbital an orbital lower in energy than the atomic orbitals of which it is composed. (9.2)

Bonding pair an electron pair found in the space between two atoms. (8.9)

Borane a covalent hydride of boron. (20.5)

Boyle's law the volume of a given sample of gas at constant temperature varies inversely with the pressure. (5.2)

Breeder reactor a nuclear reactor in which fissionable fuel is produced while the reactor runs. (19.6)

Brønsted–Lowry model a model proposing that an acid is a proton donor, and a base is a proton acceptor. (14.1)

Buffered solution a solution that resists a change in its pH when either hydroxide ions or protons are added. (15.2)

Buffering capacity the ability of a buffered solution to absorb protons or hydroxide ions without a significant change in pH;

determined by the magnitudes of [HA] and [A$^-$] in the solution. (15.3)

Calorimetry the science of measuring heat flow. (6.2)

Capillary action the spontaneous rising of a liquid in a narrow tube. (10.2)

Carbohydrate a polyhydroxyl ketone, a polyhydroxyl aldehyde, or a polymer composed of these. (22.6)

Carbon steel an alloy of iron containing up to about 1.5% carbon. (21.8)

Carboxyhemoglobin a stable complex of hemoglobin and carbon monoxide that prevents normal oxygen uptake in the blood. (21.7)

Carboxyl group the —COOH group in an organic acid. (14.2; 22.4)

Carboxylic acid an organic compound containing the carboxyl group; an acid with the general formula RCOOH. (22.4)

Catalyst a substance that speeds up a reaction without being consumed. (12.7)

Cathode the electrode in a galvanic cell at which reduction occurs. (18.2)

Cathode rays the "rays" emanating from the negative electrode (cathode) in a partially evacuated tube; a stream of electrons. (2.4)

Cathodic protection a method in which an active metal, such as magnesium, is connected to steel to protect it from corrosion. (18.6)

Cation a positive ion. (2.6)

Cell potential (electromotive force) the driving force in a galvanic cell that pulls electrons from the reducing agent in one compartment to the oxidizing agent in the other. (18.2)

Ceramic a nonmetallic material made from clay and hardened by firing at high temperature; it contains minute silicate crystals suspended in a glassy cement. (10.5)

Chain reaction (nuclear) a self-sustaining fission process caused by the production of neutrons that proceed to split other nuclei. (19.6)

Charles's law the volume of a given sample of gas at constant pressure is directly proportional to the temperature in kelvins. (5.2)

Chelating ligand (chelate) a ligand having more than one atom with a lone pair that can be used to bond to a metal ion. (21.3)

Chemical bond the force or, more accurately, the energy that holds two atoms together in a compound. (2.6)

Chemical change the change of substances into other substances through a reorganization of the atoms; a chemical reaction. (1.9)

Chemical equation a representation of a chemical reaction showing the relative numbers of reactant and product molecules. (3.8)

Chemical equilibrium a dynamic reaction system in which the concentrations of all reactants and products remain constant as a function of time. (13)

Chemical formula the representation of a molecule in which the symbols for the elements are used to indicate the types of atoms present and subscripts are used to show the relative numbers of atoms. (2.6)

Chemical kinetics the area of chemistry that concerns reaction rates. (12)

Chemical stoichiometry the calculation of the quantities of material consumed and produced in chemical reactions. (3)

Chirality the quality of having nonsuperimposable mirror images. (21.4)

Chlor–alkali process the process for producing chlorine and sodium hydroxide by electrolyzing brine in a mercury cell. (18.9)

Chromatography the general name for a series of methods for separating mixtures by using a system with a mobile phase and a stationary phase. (1.9)

Coagulation the destruction of a colloid by causing particles to aggregate and settle out. (11.8)

Codons organic bases in sets of three that form the genetic code. (22.6)

Colligative properties properties of a solution that depend only on the number, and not on the identity, of the solute particles. (11.5)

Collision model a model based on the idea that molecules must collide to react; used to account for the observed characteristics of reaction rates. (12.7)

Colloid (colloidal dispersion) a suspension of particles in a dispersing medium. (11.8)

Combustion reaction the vigorous and exothermic reaction that takes place between certain substances, particularly organic compounds, and oxygen. (22.1)

Common ion effect the shift in an equilibrium position caused by the addition or presence of an ion involved in the equilibrium reaction. (15.1)

Complete ionic equation an equation that shows all substances that are strong electrolytes as ions. (4.6)

Complex ion a charged species consisting of a metal ion surrounded by ligands. (16.3; 21.1)

Compound a substance with constant composition that can be broken down into elements by chemical processes. (1.9)

Concentration cell a galvanic cell in which both compartments contain the same components, but at different concentrations. (18.5)

Condensation the process by which vapor molecules re-form a liquid. (10.8)

Condensation polymerization a type of polymerization in which the formation of a small molecule, such as water, accompanies the extension of the polymer chain. (22.5)

Condensation reaction a reaction in which two molecules are joined, accompanied by the elimination of a water molecule. (20.3)

Condensed states of matter liquids and solids. (10.1)

Conjugate acid the species formed when a proton is added to a base. (14.1)

Conjugate acid–base pair two species related to each other by the donating and accepting of a single proton. (14.1)

Conjugate base what remains of an acid molecule after a proton is lost. (14.1)

Continuous spectrum a spectrum that exhibits all the wavelengths of visible light. (7.3)

Control rods rods in a nuclear reactor composed of substances that absorb neutrons. These rods regulate the power level of the reactor. (19.6)

Coordinate covalent bond a metal–ligand bond resulting from the interaction of a Lewis base (the ligand) and a Lewis acid (the metal ion). (21.3)

Coordination compound a compound composed of a complex ion and counter ions sufficient to give no net charge. (21.3)

Coordination isomerism isomerism in a coordination compound in which the composition of the coordination sphere of the metal ion varies. (21.4)

Coordination number the number of bonds formed between the metal ion and the ligands in a complex ion. (21.3)

Copolymer a polymer formed from the polymerization of more than one type of monomer. (22.5)

Core electron an inner electron in an atom; one not in the outermost (valence) principal quantum level. (7.11)

Corrosion the process by which metals are oxidized in the atmosphere. (18.7)

Coulomb's law $E = 2.31 \times 10^{-19} \left(\dfrac{Q_1 Q_2}{r} \right)$, where E is the energy of interaction between a pair of ions, expressed in joules; r is the distance between the ion centers in nm; and Q_1 and Q_2 are the numerical ion charges. (8.1)

Counterions anions or cations that balance the charge on the complex ion in a coordination compound. (21.3)

Covalent bonding a type of bonding in which electrons are shared by atoms. (2.6; 8.1)

Critical mass the mass of fissionable material required to produce a self-sustaining chain reaction. (19.6)

Critical point the point on a phase diagram at which the temperature and pressure have their critical values; the endpoint of the liquid–vapor line. (10.9)

Critical pressure the minimum pressure required to produce liquefaction of a substance at the critical temperature. (10.9)

Critical reaction (nuclear) a reaction in which exactly one neutron from each fission event causes another fission event, thus sustaining the chain reaction. (19.6)

Critical temperature the temperature above which vapor cannot be liquefied no matter what pressure is applied. (10.9)

Crosslinking the existence of bonds between adjacent chains in a polymer, thus adding strength to the material. (22.5)

Crystal field model a model used to explain the magnetism and colors of coordination complexes through the splitting of the d orbital energies. (21.6)

Crystalline solid a solid with a regular arrangement of its components. (10.3)

Cubic closest packed (ccp) structure a solid modeled by the closest packing of spheres with an *abcabc* arrangement of layers; the unit cell is face-centered cubic. (10.4)

Cyanidation a process in which crushed gold ore is treated with an aqueous cyanide solution in the presence of air to dissolve the gold. Pure gold is recovered by reduction of the ion to the metal. (21.8)

Cyclotron a type of particle accelerator in which an ion introduced at the center is accelerated in an expanding spiral path by the use of alternating electrical fields in the presence of a magnetic field. (19.3)

Cytochromes a series of iron-containing species composed of heme and a protein. Cytochromes are the principal electron-transfer molecules in the respiratory chain. (21.7)

Dalton's law of partial pressures for a mixture of gases in a container, the total pressure exerted is the sum of the pressures that each gas would exert if it were alone. (5.5)

Degenerate orbitals a group of orbitals with the same energy. (7.7)

# # # # # #

Dehydrogenation reaction a reaction in which two hydrogen atoms are removed from adjacent carbons of a saturated hydrocarbon, giving an unsaturated hydrocarbon. (22.1)

Denaturation the breaking down of the three-dimensional structure of a protein resulting in the loss of its function. (22.6)

Denitrification the return of nitrogen from decomposed matter to the atmosphere by bacteria that change nitrates to nitrogen gas. (20.2)

Density a property of matter representing the mass per unit volume. (1.8)

Deoxyribonucleic acid (DNA) a huge nucleotide polymer having a double-helical structure with complementary bases on the two strands. Its major functions are protein synthesis and the storage and transport of genetic information. (22.6)

Desalination the removal of dissolved salts from an aqueous solution. (11.6)

Dialysis a phenomenon in which a semipermeable membrane allows transfer of both solvent molecules and small solute molecules and ions. (11.6)

Diamagnetism a type of magnetism, associated with paired electrons, that causes a substance to be repelled from the inducing magnetic field. (9.3)

Differential rate law an expression that gives the rate of a reaction as a function of concentrations; often called the rate law. (12.2)

Diffraction the scattering of light from a regular array of points or lines, producing constructive and destructive interference. (7.2)

Diffusion the mixing of gases. (5.7)

Dilution the process of adding solvent to lower the concentration of solute in a solution. (4.3)

Dimer a molecule formed by the joining of two identical monomers. (22.5)

Dipole–dipole attraction the attractive force resulting when polar molecules line up so that the positive and negative ends are close to each other. (10.1)

Dipole moment a property of a molecule whose charge distribution can be represented by a center of positive charge and a center of negative charge. (8.3)

Direct reduction furnace a furnace in which iron oxide is reduced to iron metal using milder reaction conditions than in a blast furnace. (21.8)

Disaccharide a sugar formed from two monosaccharides joined by a glycoside linkage. (22.6)

Disproportionation reaction a reaction in which a given element is both oxidized and reduced. (20.7)

Distillation a method for separating the components of a liquid mixture that depends on differences in the ease of vaporization of the components. (1.9)

Disulfide linkage an S—S bond that stabilizes the tertiary structure of many proteins. (22.6)

Double bond a bond in which two pairs of electrons are shared by two atoms. (8.8)

Downs cell a cell used for electrolyzing molten sodium chloride. (18.9)

Dry cell battery a common battery used in calculators, watches, radios, and portable audio players. (18.6)

Dual nature of light the statement that light exhibits both wave and particulate properties. (7.2)

Effusion the passage of a gas through a tiny orifice into an evacuated chamber. (5.7)

Electrical conductivity the ability to conduct an electric current. (4.2)

Electrochemistry the study of the interchange of chemical and electrical energy. (18)

Electrolysis a process that involves forcing a current through a cell to cause a nonspontaneous chemical reaction to occur. (18.8)

Electrolyte a material that dissolves in water to give a solution that conducts an electric current. (4.2)

Electrolytic cell a cell that uses electrical energy to produce a chemical change that would otherwise not occur spontaneously. (18.8)

Electromagnetic radiation radiant energy that exhibits wavelike behavior and travels through space at the speed of light in a vacuum. (7.1)

Electron a negatively charged particle that moves around the nucleus of an atom. (2.4)

Electron affinity the energy change associated with the addition of an electron to a gaseous atom. (7.12)

Electron capture a process in which one of the inner-orbital electrons in an atom is captured by the nucleus. (19.1)

Electron spin quantum number a quantum number representing one of the two possible values for the electron spin; either $+\frac{1}{2}$ or $-\frac{1}{2}$. (7.8)

Electronegativity the tendency of an atom in a molecule to attract shared electrons to itself. (8.2)

Element a substance that cannot be decomposed into simpler substances by chemical or physical means. (1.9)

Elementary step a reaction whose rate law can be written from its molecularity. (12.5)

$E = mc^2$ Einstein's equation proposing that energy has mass; E is energy, m is mass, and c is the speed of light. (7.2)

Empirical formula the simplest whole-number ratio of atoms in a compound. (3.6)

Enantiomers isomers that are nonsuperimposable mirror images of each other. (21.4)

End point the point in a titration at which the indicator changes color. (4.8)

Endothermic refers to a reaction where energy (as heat) flows into the system. (6.1)

Energy the capacity to do work or to cause heat flow. (6.1)

Enthalpy a property of a system equal to $E + PV$, where E is the internal energy of the system, P is the pressure of the system, and V is the volume of the system. At constant pressure the change in enthalpy equals the energy flow as heat. (6.2)

Enthalpy (heat) of fusion the enthalpy change that occurs to melt a solid at its melting point. (10.8)

Entropy a thermodynamic function that measures randomness or disorder. (17.1)

Enzyme a large molecule, usually a protein, that catalyzes biological reactions. (12.7)

Equilibrium constant the value obtained when equilibrium concentrations of the chemical species are substituted in the equilibrium expression. (13.2)

Equilibrium expression the expression (from the law of mass action) obtained by multiplying the product concentrations and dividing by the multiplied reactant concentrations, with each concentration raised to a power represented by the coefficient in the balanced equation. (13.2)

Equilibrium point (thermodynamic definition) the position where the free energy of a reaction system has its lowest possible value. (17.8)

Equilibrium position a particular set of equilibrium concentrations. (13.2)

Equivalence point (stoichiometric point) the point in a titration when enough titrant has been added to react exactly with the substance in solution being titrated. (4.9; 15.4)

Ester an organic compound produced by the reaction between a carboxylic acid and an alcohol. (22.4)

Exothermic refers to a reaction where energy (as heat) flows out of the system. (6.1)

Exponential notation expresses a number as $N \times 10^M$, a convenient method for representing a very large or very small number and for easily indicating the number of significant figures. (1.5)

Faraday a constant representing the charge on one mole of electrons; 96,485 coulombs. (18.4)

Filtration a method for separating the components of a mixture containing a solid and a liquid. (1.9)

First law of thermodynamics the energy of the universe is constant; same as the law of conservation of energy. (6.1)

Fission the process of using a neutron to split a heavy nucleus into two nuclei with smaller mass numbers. (19.6)

Flotation process a method of separating the mineral particles in an ore from the gangue that depends on the greater wettability of the mineral pieces. (21.8)

Formal charge the charge assigned to an atom in a molecule or polyatomic ion derived from a specific set of rules. (8.12)

Formation constant (stability constant) the equilibrium constant for each step of the formation of a complex ion by the addition of an individual ligand to a metal ion or complex ion in aqueous solution. (16.3)

Formula equation an equation representing a reaction in solution showing the reactants and products in undissociated form, whether they are strong or weak electrolytes. (4.6)

Fossil fuel coal, petroleum, or natural gas; consists of carbon-based molecules derived from decomposition of once-living organisms. (6.5)

Frasch process the recovery of sulfur from underground deposits by melting it with hot water and forcing it to the surface by air pressure. (20.6)

Free energy a thermodynamic function equal to the enthalpy (H) minus the product of the entropy (S) and the Kelvin temperature (T); $G = H - TS$. Under certain conditions the change in free energy for a process is equal to the maximum useful work. (17.4)

Free radical a species with an unpaired electron. (22.5)

Frequency the number of waves (cycles) per second that pass a given point in space. (7.1)

Fuel cell a galvanic cell for which the reactants are continuously supplied. (18.6)

Functional group an atom or group of atoms in hydrocarbon derivatives that contains elements in addition to carbon and hydrogen. (22.4)

Fusion the process of combining two light nuclei to form a heavier, more stable nucleus. (19.6)

Galvanic cell a device in which chemical energy from a spontaneous redox reaction is changed to electrical energy that can be used to do work. (18.2)

Galvanizing a process in which steel is coated with zinc to prevent corrosion. (18.7)

Gamma (γ) ray a high-energy photon. (19.1)

Gangue the impurities (such as clay or sand) in an ore. (21.8)

Geiger–Müller counter (Geiger counter) an instrument that measures the rate of radioactive decay based on the ions and electrons produced as a radioactive particle passes through a gas-filled chamber. (19.4)

Gene a given segment of the DNA molecule that contains the code for a specific protein. (22.6)

Geometrical (*cis–trans*) isomerism isomerism in which atoms or groups of atoms can assume different positions around a rigid ring or bond. (21.4; 22.2)

Glass an amorphous solid obtained when silica is mixed with other compounds, heated above its melting point, and then cooled rapidly. (10.5)

Glass electrode an electrode for measuring pH from the potential difference that develops when it is dipped into an aqueous solution containing H^+ ions. (18.5)

Glycoside linkage a C—O—C bond formed between the rings of two cyclic monosaccharides by the elimination of water. (22.6)

Graham's law of effusion the rate of effusion of a gas is inversely proportional to the square root of the mass of its particles. (5.7)

Greenhouse effect a warming effect exerted by the earth's atmosphere (particularly CO_2 and H_2O) due to thermal energy retained by absorption of infrared radiation. (6.5)

Ground state the lowest possible energy state of an atom or molecule. (7.4)

Group (of the periodic table) a vertical column of elements having the same valence electron configuration and showing similar properties. (2.7)

Haber process the manufacture of ammonia from nitrogen and hydrogen, carried out at high pressure and high temperature with the aid of a catalyst. (3.10; 20.2)

Half-life (of a radioactive sample) the time required for the number of nuclides in a radioactive sample to reach half of the original value. (19.2)

Half-life (of a reactant) the time required for a reactant to reach half of its original concentration. (12.4)

Half-reactions the two parts of an oxidation–reduction reaction, one representing oxidation, the other reduction. (4.10; 17.1)

Halogen a Group 7A element. (2.7; 20.7)

Halogenation the addition of halogen atoms to unsaturated hydrocarbons. (22.2)

Hard water water from natural sources that contains relatively large concentrations of calcium and magnesium ions. (20.4)

Heat energy transferred between two objects due to a temperature difference between them. (6.1)

Heat capacity the amount of energy required to raise the temperature of an object by one degree Celsius. (6.2)

Heat of fusion the enthalpy change that occurs to melt a solid at its melting point. (10.8)

Heat of hydration the enthalpy change associated with placing gaseous molecules or ions in water; the sum of the energy needed to expand the solvent and the energy released from the solvent–solute interactions. (11.2)

Heat of solution the enthalpy change associated with dissolving a solute in a solvent; the sum of the energies needed to expand both solvent and solute in a solution and the energy released from the solvent–solute interactions. (11.2)

Heat of vaporization the energy required to vaporize one mole of a liquid at a pressure of one atmosphere. (10.8)

Heating curve a plot of temperature versus time for a substance where energy is added at a constant rate. (10.8)

Heisenberg uncertainty principle a principle stating that there is a fundamental limitation to how precisely both the position and momentum of a particle can be known at a given time. (7.5)

Heme an iron complex. (21.7)

Hemoglobin a biomolecule composed of four myoglobin-like units (proteins plus heme) that can bind and transport four oxygen molecules in the blood. (21.7)

Henderson–Hasselbalch equation an equation giving the relationship between the pH of an acid–base system and the concentrations of base and acid:
$$pH = pK_a + \log\left(\frac{[base]}{[acid]}\right). \quad (15.2)$$

Henry's law the amount of a gas dissolved in a solution is directly proportional to the pressure of the gas above the solution. (11.3)

Hess's law in going from a particular set of reactants to a particular set of products, the enthalpy change is the same whether the reaction takes place in one step or in a series of steps; in summary, enthalpy is a state function. (6.3)

Heterogeneous equilibrium an equilibrium involving reactants and/or products in more than one phase. (13.4)

Hexagonal closest packed (hcp) structure a structure composed of closest packed spheres with an *ababab* arrangement of layers; the unit cell is hexagonal. (10.4)

Homogeneous equilibrium an equilibrium system where all reactants and products are in the same phase. (13.4)

Homopolymer a polymer formed from the polymerization of only one type of monomer. (22.5)

Hund's rule the lowest energy configuration for an atom is the one having the maximum number of unpaired electrons allowed by the Pauli exclusion principle in a particular set of degenerate orbitals, with all unpaired electrons having parallel spins. (7.11)

Hybrid orbitals a set of atomic orbitals adopted by an atom in a molecule different from those of the atom in the free state. (9.1)

Hybridization a mixing of the native orbitals on a given atom to form special atomic orbitals for bonding. (9.1)

Hydration the interaction between solute particles and water molecules. (4.1)

Hydride a binary compound containing hydrogen. The hydride ion, H^-, exists in ionic hydrides. The three classes of hydrides are covalent, interstitial, and ionic. (20.3)

Hydrocarbon a compound composed of carbon and hydrogen. (22.1)

Hydrocarbon derivative an organic molecule that contains one or more elements in addition to carbon and hydrogen. (22.4)

Hydrogen bonding unusually strong dipole–dipole attractions that occur among molecules in which hydrogen is bonded to a highly electronegative atom. (10.1)

Hydrogenation reaction a reaction in which hydrogen is added, with a catalyst present, to a carbon–carbon multiple bond. (22.2)

Hydrohalic acid an aqueous solution of a hydrogen halide. (20.7)

Hydrometallurgy a process for extracting metals from ores by use of aqueous chemical solutions. Two steps are involved: selective leaching and selective precipitation. (21.8)

Hydronium ion the H_3O^+ ion; a hydrated proton. (14.1)

Hypothesis one or more assumptions put forth to explain the observed behavior of nature. (1.2)

Ideal gas law an equation of state for a gas, where the state of the gas is its condition at a given time; expressed by $PV = nRT$, where P = pressure, V = volume, n = moles of the gas, R = the universal gas constant, and T = absolute temperature. This equation expresses behavior approached by real gases at high T and low P. (5.3)

Ideal solution a solution whose vapor pressure is directly proportional to the mole fraction of solvent present. (11.4)

Indicator a chemical that changes color and is used to mark the end point of a titration. (4.8; 15.5)

Integrated rate law an expression that shows the concentration of a reactant as a function of time. (12.2)

Interhalogen compound a compound formed by the reaction of one halogen with another. (20.7)

Intermediate a species that is neither a reactant nor a product but that is formed and consumed in the reaction sequence. (12.5)

Intermolecular forces relatively weak interactions that occur between molecules. (10.1)

Internal energy a property of a system that can be changed by a flow of work, heat, or both; $\Delta E = q + w$, where ΔE is the change in the internal energy of the system, q is heat, and w is work. (6.1)

Ion an atom or a group of atoms that has a net positive or negative charge. (2.6)

Ion exchange (water softening) the process in which an ion-exchange resin removes unwanted ions (for example, Ca^{2+} and Mg^{2+}) and replaces them with Na^+ ions, which do not interfere with soap and detergent action. (20.4)

Ion pairing a phenomenon occurring in solution when oppositely charged ions aggregate and behave as a single particle. (11.7)

Ion-product (dissociation) constant (K_w) the equilibrium constant for the auto-ionization of water; $K_w = [H^+][OH^-]$. At 25°C, $K_w = 1.0 \times 10^{-14}$. (14.2)

Ion-selective electrode an electrode sensitive to the concentration of a particular ion in solution. (18.5)

Ionic bonding the electrostatic attraction between oppositely charged ions. (2.6; 8.1)

Ionic compound (binary) a compound that results when a metal reacts with a nonmetal to form a cation and an anion. (8.1)

Ionic solid a solid containing cations and anions that dissolves in water to give a solution containing the separated ions, which are mobile and thus free to conduct an electric current. (2.6; 10.3)

Irreversible process any real process. When a system undergoes the changes State 1 → State 2 → State 1 by any real pathway, the universe is different than before the cyclic process took place in the system. (17.9)

Isoelectronic ions ions containing the same number of electrons. (8.4)

Isomers species with the same formula but different properties. (21.4)

Isotactic chain a polymer chain in which the substituent groups such as CH_3 are all arranged on the same side of the chain. (22.5)

Isotonic solutions solutions having identical osmotic pressures. (11.6)

Isotopes atoms of the same element (the same number of protons) with different numbers of neutrons. They have identical atomic numbers but different mass numbers. (2.5; 18)

Ketone an organic compound containing the carbonyl group

bonded to two carbon atoms. (22.4)

Kinetic energy ($\frac{1}{2}mv^2$) energy due to the motion of an object; dependent on the mass of the object and the square of its velocity. (6.1)

Kinetic molecular theory (KMT) a model that assumes that an ideal gas is composed of tiny particles (molecules) in constant motion. (5.6)

Lanthanide contraction the decrease in the atomic radii of the lanthanide series elements, going from left to right in the periodic table. (21.1)

Lanthanide series a group of 14 elements following lanthanum in the periodic table, in which the $4f$ orbitals are being filled. (7.11; 20.1; 21.1)

Lattice a three-dimensional system of points designating the positions of the centers of the components of a solid (atoms, ions, or molecules). (10.3)

Lattice energy the energy change occurring when separated gaseous ions are packed together to form an ionic solid. (8.5)

Law of conservation of energy energy can be converted from one form to another but can be neither created nor destroyed. (6.1)

Law of conservation of mass mass is neither created nor destroyed. (1.2; 2.2)

Law of definite proportion a given compound always contains exactly the same proportion of elements by mass. (2.2)

Law of mass action a general description of the equilibrium condition; it defines the equilibrium constant expression. (13.2)

Law of multiple proportions a law stating that when two elements form a series of compounds, the ratios of the masses of the second element that combine with one gram of the first element can always be reduced to small whole numbers. (2.2)

Leaching the extraction of metals from ores using aqueous chemical solutions. (21.8)

Lead storage battery a battery (used in cars) in which the anode is lead, the cathode is lead coated with lead dioxide, and the electrolyte is a sulfuric acid solution. (18.6)

Le Châtelier's principle if a change is imposed on a system at equilibrium, the position of the equilibrium will shift in a direction that tends to reduce the effect of that change. (13.7)

Lewis acid an electron-pair acceptor. (14.11)

Lewis base an electron-pair donor. (14.11)

Lewis structure a diagram of a molecule showing how the valence electrons are arranged among the atoms in the molecule. (8.10)

Ligand a neutral molecule or ion having a lone pair of electrons that can be used to form a bond to a metal ion; a Lewis base. (21.3)

Lime–soda process a water-softening method in which lime and soda ash are added to water to remove calcium and magnesium ions by precipitation. (14.6)

Limiting reactant (limiting reagent) the reactant that is completely consumed when a reaction is run to completion. (3.11)

Line spectrum a spectrum showing only certain discrete wavelengths. (7.3)

Linear accelerator a type of particle accelerator in which a changing electrical field is used to accelerate a positive ion along a linear path. (19.3)

Linkage isomerism isomerism involving a complex ion where the ligands are all the same but the point of attachment of at least one of the ligands differs. (21.4)

Liquefaction the transformation of a gas into a liquid. (20.1)

Localized electron (LE) model a model that assumes that a molecule is composed of atoms that are bound together by sharing pairs of electrons using the atomic orbitals of the bound atoms. (8.9)

London dispersion forces the forces, existing among noble gas atoms and nonpolar molecules, that involve an accidental dipole that induces a momentary dipole in a neighbor. (10.1)

Lone pair an electron pair that is localized on a given atom; an electron pair not involved in bonding. (8.9)

Magnetic quantum number (m_ℓ) the quantum number relating to the orientation of an orbital in space relative to the other orbitals with the same ℓ quantum number. It can have integral values between ℓ and $-\ell$, including zero. (7.6)

Main-group (representative) elements elements in the groups labeled 1A, 2A, 3A, 4A, 5A, 6A, 7A, and 8A in the periodic table. The group number gives the sum of the valence s and p electrons. (7.11; 18.1)

Major species the components present in relatively large amounts in a solution. (14.3)

Manometer a device for measuring the pressure of a gas in a container. (5.1)

Mass the quantity of matter in an object. (1.3)

Mass defect the change in mass occurring when a nucleus is formed from its component nucleons. (19.5)

Mass number the total number of protons and neutrons in the atomic nucleus of an atom. (2.5; 18)

Mass percent the percent by mass of a component of a mixture (11.1) or of a given element in a compound. (3.5)

Mass spectrometer an instrument used to determine the relative masses of atoms by the deflection of their ions on a magnetic field. (3.2)

Matter the material of the universe. (1.9)

Messenger RNA (mRNA) a special RNA molecule built in the cell nucleus that migrates into the cytoplasm and participates in protein synthesis. (22.6)

Metal an element that gives up electrons relatively easily and is lustrous, malleable, and a good conductor of heat and electricity. (2.7)

Metalloids (semimetals) elements along the division line in the periodic table between metals and nonmetals. These elements exhibit both metallic and nonmetallic properties. (7.13; 20.1)

Metallurgy the process of separating a metal from its ore and preparing it for use. (20.1; 21.8)

Millimeters of mercury (mm Hg) a unit of pressure, also called a torr, 760 mm Hg = 760 torr = 101,325 Pa = 1 standard atmosphere. (5.1)

Mineral a relatively pure compound as found in nature. (21.8)

Model (theory) a set of assumptions put forth to explain the observed behavior of matter. The models of chemistry usually involve assumptions about the behavior of individual atoms or molecules. (1.2)

Moderator a substance used in a nuclear reactor to slow down the neutrons. (19.6)

Molal boiling-point elevation constant a constant characteristic of a particular solvent that gives the change in boiling point as a function of solution molality; used in molecular weight determinations. (11.5)

Molal freezing-point depression constant a constant characteristic of a particular solvent that gives the change in freezing point as a function of the solution molality; used in molecular weight determinations (11.5)

Molality the number of moles of solute per kilogram of solvent in a solution. (11.1)

Molar heat capacity the energy required to raise the temperature of one mole of a substance by one degree Celsius. (6.2)

Molar mass the mass in grams of one mole of molecules or formula units of a substance; also called *molecular weight*. (3.4)

Molar volume the volume of one mole of an ideal gas; equal to 22.42 liters at STP. (5.4)

Molarity moles of solute per volume of solution in liters. (4.3; 11.1)

Mole (mol) the number equal to the number of carbon atoms in exactly 12 grams of pure ^{12}C: Avogadro's number. One mole represents 6.022×10^{23} units. (3.3)

Mole fraction the ratio of the number of moles of a given component in a mixture to the total number of moles in the mixture. (5.5; 11.1)

Mole ratio (stoichiometry) the ratio of moles of one substance to moles of another substance in a balanced chemical equation. (3.9)

Molecular formula the exact formula of a molecule, giving the types of atoms and the number of each type. (3.6)

Molecular orbital (MO) model a model that regards a molecule as a collection of nuclei and electrons, where the electrons are assumed to occupy orbitals much as they do in atoms, but having the orbitals extend over the entire molecule. In this model the electrons are assumed to be delocalized rather than always located between a given pair of atoms. (9.2; 10.4)

Molecular orientations (kinetics) orientations of molecules during collisions, some of which can lead to reaction while others cannot. (12.6)

Molecular solid a solid composed of neutral molecules at the lattice points. (10.3)

Molecular structure the three-dimensional arrangement of atoms in a molecule. (8.13)

Molecularity the number of species that must collide to produce the reaction represented by an elementary step in a reaction mechanism. (12.5)

Molecule a bonded collection of two or more atoms of the same or different elements. (2.6)

Monodentate (unidentate) ligand a ligand that can form one bond to a metal ion. (21.3)

Monoprotic acid an acid with one acidic proton. (14.2)

Monosaccharide (simple sugar) a polyhydroxy ketone or aldehyde containing from three to nine carbon atoms. (22.6)

Myoglobin an oxygen-storing biomolecule consisting of a heme complex and a proton. (21.7)

Natural law a statement that expresses generally observed behavior. (1.2)

Nernst equation an equation relating the potential of an electrochemical cell to the concentrations of the cell components:

$$\mathscr{E} = \mathscr{E}° - \frac{0.0591}{n} \log(Q) \text{ at } 25°C \text{ (18.5)}$$

Net ionic equation an equation for a reaction in solution, where strong electrolytes are written as ions, showing only those components that are directly involved in the chemical change. (4.6)

Network solid an atomic solid containing strong directional covalent bonds. (10.5)

Neutralization reaction an acid–base reaction. (4.8)

Neutron a particle in the atomic nucleus with mass virtually equal to the proton's but with no charge. (2.5; 19)

Nitrogen cycle the conversion of N_2 to nitrogen-containing compounds, followed by the return of nitrogen gas to the atmosphere by natural decay processes. (20.2)

Nitrogen fixation the process of transforming N_2 to nitrogen-containing compounds useful to plants. (20.2)

Nitrogen-fixing bacteria bacteria in the root nodules of plants that can convert atmospheric nitrogen to ammonia and other nitrogen-containing compounds useful to plants. (20.2)

Noble gas a Group 8A element. (2.7; 20.8)

Node an area of an orbital having zero electron probability. (7.7)

Nonelectrolyte a substance that, when dissolved in water, gives a nonconducting solution. (4.2)

Nonmetal an element not exhibiting metallic characteristics. Chemically, a typical nonmetal accepts electrons from a metal. (2.7)

Normal boiling point the temperature at which the vapor pressure of a liquid is exactly one atmosphere. (10.8)

Normal melting point the temperature at which the solid and liquid states have the same vapor pressure under conditions where the total pressure on the system is one atmosphere. (10.8)

Normality the number of equivalents of a substance dissolved in a liter of solution. (11.1)

Nuclear atom an atom having a dense center of positive charge (the nucleus) with electrons moving around the outside. (2.4)

Nuclear transformation the change of one element into another. (19.3)

Nucleon a particle in an atomic nucleus, either a neutron or a proton. (19)

Nucleotide a monomer of the nucleic acids composed of a five-carbon sugar, a nitrogen-containing base, and phosphoric acid. (22.6)

Nucleus the small, dense center of positive charge in an atom. (2.4)

Nuclide the general term applied to each unique atom; represented by $^A_Z X$, where X is the symbol for a particular element. (19)

Octet rule the observation that atoms of nonmetals tend to form the most stable molecules when they are surrounded by eight electrons (to fill their valence orbitals). (8.10)

Open hearth process a process for producing steel by oxidizing and removing the impurities in molten iron using external heat and a blast of air or oxygen. (21.8)

Optical isomerism isomerism in which the isomers have opposite effects on plane-polarized light. (21.4)

Orbital a specific wave function for an electron in an atom. The square of this function gives the probability distribution for the electron. (7.5) *d*-**Orbital splitting** a splitting of the *d* orbitals of the metal ion in a complex such that the orbitals pointing at the ligands have higher energies than those pointing between the ligands. (21.6)

Order (of reactant) the positive or negative exponent, determined by experiment, of the reactant concentration in a rate law. (12.2)

Organic acid an acid with a carbon-atom backbone; often contains the carboxyl group. (14.2)

Organic chemistry the study of carbon-containing compounds (typically chains of carbon atoms) and their properties. (22)

Osmosis the flow of solvent into a solution through a semipermeable membrane. (11.6)

Osmotic pressure (π) the pressure that must be applied to a solution to stop osmosis; $\pi = MRT$. (11.6)

Ostwald process a commercial process for producing nitric acid by the oxidation of ammonia. (20.2)

Oxidation an increase in oxidation state (a loss of electrons). (4.9; 17.1)

Oxidation–reduction (redox) reaction a reaction in which one or more electrons are transferred. (4.9; 17.1)

Oxidation states a concept that provides a way to keep track of electrons in oxidation–reduction reactions according to certain rules. (4.9; 21.3)

Oxidizing agent (electron acceptor) a reactant that accepts electrons from another reactant. (4.9; 17.1)

Oxyacid an acid in which the acidic proton is attached to an oxygen atom. (14.2)

Ozone O_3, the form of elemental oxygen in addition to the much more common O_2. (20.5)

Paramagnetism a type of induced magnetism, associated with unpaired electrons, that causes a substance to be attracted into the inducing magnetic field. (9.3)

Partial pressures the independent pressures exerted by different gases in a mixture. (5.5)

Particle accelerator a device used to accelerate nuclear particles to very high speeds. (19.3)

Pascal the SI unit of pressure; equal to newtons per meter squared. (5.1)

Pauli exclusion principle in a given atom no two electrons can have the same set of four quantum numbers. (7.8)

Peptide linkage the bond resulting from the condensation reaction between amino acids; represented by:

$$\begin{matrix} & O & & H \\ & \| & & | \\ - & C & - & N & - \end{matrix}$$

(22.6)

Percent dissociation the ratio of the amount of a substance that is dissociated at equilibrium to the initial concentration of the substance in a solution, multiplied by 100. (14.5)

Percent yield the actual yield of a product as a percentage of the theoretical yield. (3.11)

Periodic table a chart showing all the elements arranged in columns with similar chemical properties. (2.7)

pH curve (titration curve) a plot showing the pH of a solution being analyzed as a function of the amount of titrant added. (15.4)

pH scale a log scale based on 10 and equal to $-\log[H^+]$; a convenient way to represent solution acidity. (14.3)

Phase diagram a convenient way of representing the phases of a substance in a closed system as a function of temperature and pressure. (10.9)

Phenyl group the benzene molecule minus one hydrogen atom. (22.3)

Photochemical smog air pollution produced by the action of light on oxygen, nitrogen oxides, and unburned fuel from auto exhaust to form ozone and other pollutants. (5.10)

Photon a quantum of electromagnetic radiation. (7.2)

Physical change a change in the form of a substance, but not in its chemical composition; chemical bonds are not broken in a physical change. (1.9)

Pi (π) bond a covalent bond in which parallel *p* orbitals share an electron pair occupying the space above and below the line joining the atoms. (9.1)

Planck's constant the constant relating the change in energy for a system to the frequency of the electromagnetic radiation absorbed or emitted; equal to 6.626×10^{-34} J · s. (7.2)

Polar covalent bond a covalent bond in which the electrons are not shared equally because one atom attracts them more strongly than the other. (8.1)

Polar molecule a molecule that has a permanent dipole moment. (4.1)

Polyatomic ion an ion containing a number of atoms. (2.6)

Polyelectronic atom an atom with more than one electron. (7.9)

Polymer a large, usually chainlike molecule built from many small molecules (monomers). (22.5)

Polymerization a process in which many small molecules (monomers) are joined together to form a large molecule. (22.2)

Polypeptide a polymer formed from amino acids joined together by peptide linkages. (22.6)

Polyprotic acid an acid with more than one acidic proton. It dissociates in a stepwise manner, one proton at a time. (14.7)

Porous disk a disk in a tube connecting two different solutions in a galvanic cell that allows ion flow without extensive mixing of the solutions. (18.2)

Porphyrin a planar ligand with a central ring structure and various substituent groups at the edges of the ring. (21.7)

Positional probability a type of probability that depends on the number of arrangements in space that yield a particular state. (17.1)

Positron production a mode of nuclear decay in which a particle is formed having the same mass as an electron but opposite charge. The net effect is to change a proton to a neutron. (19.1)

Potential energy energy due to position or composition. (6.1)

Precipitation reaction a reaction in which an insoluble substance forms and separates from the solution. (4.5)

Precision the degree of agreement among several measurements of the same quantity; the reproducibility of a measurement. (1.4)

Primary structure (of a protein) the order (sequence) of amino acids in the protein chain. (22.6)

Principal quantum number (n) the quantum number relating to the size and energy of an orbital; it can have any positive integer value. (7.6)

Probability distribution the square of the wave function indicating the probability of finding an electron at a particular point in space. (7.5)

Product a substance resulting from a chemical reaction. It is shown to the right of the arrow in a chemical equation. (3.8)

Protein a natural high-molecular-weight polymer formed by condensation reactions between amino acids. (22.6)

Proton a positively charged particle in an atomic nucleus. (2.5; 19)

Pure substance a substance with constant composition. (1.9)

Pyrometallurgy recovery of a metal from its ore by treatment at high temperatures. (21.8)

Quantization the concept that energy can occur only in discrete units called *quanta*. (7.2)

Rad a unit of radiation dosage corresponding to 10^{-2} J of energy deposited per kilogram of tissue (from *r*adiation *a*bsorbed *d*ose). (19.7)

Radioactive decay (radioactivity) the spontaneous decomposition of a nucleus to form a different nucleus. (2.4; 19.1)

Radiocarbon dating (carbon-14 dating) a method for dating ancient wood or cloth based on the rate of radioactive decay of the nuclide $^{14}_6C$. (19.4)

Radiotracer a radioactive nuclide, introduced into an organism for diagnostic purposes, whose pathway can be traced by monitoring its radioactivity. (19.4)

Random error an error that has an equal probability of being high or low. (1.4)

Raoult's law the vapor pressure of a solution is directly proportional to the mole fraction of solvent present. (11.4)

Rate constant the proportionality constant in the relationship between reaction rate and reactant concentrations. (12.2)

Rate of decay the change in the number of radioactive nuclides in a sample per unit time. (19.2)

Rate-determining step the slowest step in a reaction mechanism, the one determining the overall rate. (12.5)

Rate law (differential rate law) an expression that shows how the rate of reaction depends on the concentration of reactants. (12.2)

Reactant a starting substance in a chemical reaction. It appears to the left of the arrow in a chemical equation. (3.8)

Reaction mechanism the series of elementary steps involved in a chemical reaction. (12.5)

Reaction quotient, Q a quotient obtained by applying the law of mass action to initial concentrations rather than to equilibrium concentrations. (13.5)

Reaction rate the change in concentration of a reactant or product per unit time. (12.1)

Reactor core the part of a nuclear reactor where the fission reaction takes place. (19.6)

Reducing agent (electron donor) a reactant that donates electrons to another substance to reduce the oxidation state of one of its atoms. (4.9; 17.1)

Reduction a decrease in oxidation state (a gain of electrons). (4.9; 17.1)

Rem a unit of radiation dosage that accounts for both the energy of the dose and its effectiveness in causing biological damage (from *r*oentgen *e*quivalent for *m*an). (19.7)

Resonance a condition occurring when more than one valid Lewis structure can be written for a particular molecule. The actual electronic structure is not represented by any one of the Lewis structures but by the average of all of them. (8.12)

Reverse osmosis the process occurring when the external pressure on a solution causes a net flow of solvent through a semipermeable membrane from the solution to the solvent. (11.6)

Reversible process a cyclic process carried out by a hypothetical pathway, which leaves the universe exactly the same as it was before the process. No real process is reversible. (17.9)

Ribonucleic acid (RNA) a nucleotide polymer that transmits the genetic information stored in DNA to the ribosomes for protein synthesis. (22.6)

Roasting a process of converting sulfide minerals to oxides by heating in air at temperatures below their melting points. (21.8)

Root mean square velocity the square root of the average of the squares of the individual velocities of gas particles. (5.6)

Salt an ionic compound. (14.8)

Salt bridge a U-tube containing an electrolyte that connects the two compartments of a galvanic cell, allowing ion flow without extensive mixing of the different solutions. (18.1)

Scientific method the process of studying natural phenomena, involving observations, forming laws and theories, and testing of theories by experimentation. (1.2)

Scintillation counter an instrument that measures radioactive decay by sensing the flashes of light produced in a substance by the radiation. (19.4)

Second law of thermodynamics in any spontaneous process, there is always an increase in the entropy of the universe. (17.2)

Secondary structure (of a protein) the three-dimensional structure of the protein chain (for example, α-helix, random coil, or pleated sheet). (22.6)

Selective precipitation a method of separating metal ions from an aqueous mixture by using a reagent whose anion forms a precipitate with only one or a few of the ions in the mixture. (4.7; 16.2)

Semiconductor a substance conducting only a slight electric current at room temperature, but showing increased conductivity at higher temperatures. (10.5)

Semipermeable membrane a membrane that allows solvent but not solute molecules to pass through. (11.6)

SI system International System of units based on the metric system and units derived from the metric system. (1.3)

Side chain (of amino acid) the hydrocarbon group on an amino acid represented by H, CH_3, or a more complex substituent. (22.6)

Sigma (σ) bond a covalent bond in which the electron pair is shared in an area centered on a line running between the atoms. (9.1)

Significant figures the certain digits and the first uncertain digit of a measurement. (1.4)

Silica the fundamental silicon–oxygen compound, which has the empirical formula SiO_2, and forms the basis of quartz and certain types of sand. (10.5)

Silicates salts that contain metal cations and polyatomic silicon–oxygen anions that are usually polymeric. (10.5)

Single bond a bond in which one pair of electrons is shared by two atoms. (8.8)

Smelting a metallurgical process that involves reducing metal ions to the free metal. (21.8)

Solubility the amount of a substance that dissolves in a given volume of solvent at a given temperature. (4.2)

Solubility product constant the constant for the equilibrium expression representing the dissolving of an ionic solid in water. (16.1)

Solute a substance dissolved in a liquid to form a solution. (4.2; 11.1)

Solution a homogeneous mixture. (1.9)

Solvent the dissolving medium in a solution. (4.2)

Somatic damage radioactive damage to an organism resulting in its sickness or death. (19.7)

Space-filling model a model of a molecule showing the relative sizes of the atoms and their relative orientations. (2.6)

Specific heat capacity the energy required to raise the temperature of one gram of a substance by one degree Celsius. (6.2)

Spectator ions ions present in solution that do not participate directly in a reaction. (4.6)

Spectrochemical series a listing of ligands in order based on their ability to produce d-orbital splitting. (21.6)

Spontaneous fission the spontaneous splitting of a heavy nuclide into two lighter nuclides. (19.1)

Spontaneous process a process that occurs without outside intervention. (17.1)

Standard atmosphere a unit of pressure equal to 760 mm Hg. (5.1)

Standard enthalpy of formation the enthalpy change that accompanies the formation of one mole of a compound at 25°C from its elements, with all substances in their standard states at that temperature. (6.4)

Standard free energy change the change in free energy that will occur for one unit of reaction if the reactants in their standard states are converted to products in their standard states. (17.6)

Standard free energy of formation the change in free energy that accompanies the formation of one mole of a substance from its constituent elements with all reactants and products in their standard states. (17.6)

Standard hydrogen electrode a platinum conductor in contact with 1 M H$^+$ ions and bathed by hydrogen gas at one atmosphere. (18.3)

Standard reduction potential the potential of a half-reaction under standard state conditions, as measured against the potential of the standard hydrogen electrode. (18.3)

Standard solution a solution whose concentration is accurately known. (4.3)

Standard state a reference state for a specific substance defined according to a set of conventional definitions. (6.4)

Standard temperature and pressure (STP) the condition 0°C and 1 atmosphere of pressure. (5.4)

Standing wave a stationary wave as on a string of a musical instrument; in the wave mechanical model, the electron in the hydrogen atom is considered to be a standing wave. (7.5)

State function (property) a property that is independent of the pathway. (6.1)

States (of matter) the three different forms in which matter can exist; solid, liquid, and gas. (1.9)

Stereoisomerism isomerism in which all the bonds in the isomers are the same but the spatial arrangements of the atoms are different. (21.4)

Steric factor the factor (always less than 1) that reflects the fraction of collisions with orientations that can produce a chemical reaction. (12.6)

Stoichiometric quantities quantities of reactants mixed in exactly the correct amounts so that all are used up at the same time. (3.10)

Strong acid an acid that completely dissociates to produce an H$^+$ ion and the conjugate base. (4.2; 14.2)

Strong base a metal hydroxide salt that completely dissociates into its ions in water. (4.2; 14.6)

Strong electrolyte a material that, when dissolved in water, gives a solution that conducts an electric current very efficiently. (4.2)

Structural formula the representation of a molecule in which the relative positions of the atoms are shown and the bonds are indicated by lines. (2.6)

Structural isomerism isomerism in which the isomers contain the same atoms but one or more bonds differ. (21.4; 22.1)

Subcritical reaction (nuclear) a reaction in which less than one neutron causes another fission event and the process dies out. (19.6)

Sublimation the process by which a substance goes directly from the solid to the gaseous state without passing through the liquid state. (10.8)

Subshell a set of orbitals with a given azimuthal quantum number. (7.6)

Substitution reaction (hydrocarbons) a reaction in which an atom, usually a halogen, replaces a hydrogen atom in a hydrocarbon. (22.1)

Supercooling the process of cooling a liquid below its freezing point without its changing to a solid. (10.8)

Supercritical reaction (nuclear) a reaction in which more than one neutron from each fission event causes another fission event. The process rapidly escalates to a violent explosion. (19.6)

Superheating the process of heating a liquid above its boiling point without its boiling. (10.8)

Superoxide a compound containing the O$_2^-$ anion. (19.2)

Surface tension the resistance of a liquid to an increase in its surface area. (10.2)

Surroundings everything in the universe surrounding a thermodynamic system. (6.1)

Syndiotactic chain a polymer chain in which the substituent groups such as CH$_3$ are arranged on alternate sides of the chain. (22.5)

Syngas synthetic gas, a mixture of carbon monoxide and hydrogen, obtained by coal gasification. (6.6)

System (thermodynamic) that part of the universe on which attention is to be focused. (6.1)

Systematic error an error that always occurs in the same direction. (1.4)

Tempering a process in steel production that fine-tunes the proportions of carbon crystals and cementite by heating to intermediate temperatures followed by rapid cooling. (21.8)

Termolecular step a reaction involving the simultaneous collision of three molecules. (12.5)

Tertiary structure (of a protein) the overall shape of a protein, long and narrow or globular, maintained by different types of intramolecular interactions. (22.6)

Theoretical yield the maximum amount of a given product that can be formed when the limiting reactant is completely consumed. (3.11)

Theory a set of assumptions put forth to explain some aspect of the observed behavior of matter. (1.2)

Thermal pollution the oxygen-depleting effect on lakes and rivers of using water for industrial cooling and returning it to its natural source at a higher temperature. (11.3)

Thermodynamic stability (nuclear) the potential energy of a particular nucleus as compared to the sum of the potential energies of its component protons and neutrons. (19.1)

Thermodynamics the study of energy and its interconversions. (6.1)

Thermoplastic polymer a substance that when molded to a certain shape under appropriate conditions can later be remelted. (22.5)

Thermoset polymer a substance that when molded to a certain shape under pressure and high temperatures cannot be softened again or dissolved. (22.5)

Third law of thermodynamics the entropy of a perfect crystal at 0 K is zero. (17.5)

Titration a technique in which one solution is used to analyze another. (4.8)

Torr another name for millimeter of mercury (mm Hg). (5.1)

Transfer RNA (tRNA) a small RNA fragment that finds specific amino acids and attaches them to the protein chain as dictated by the codons in mRNA. (22.6)

Transition metals several series of elements in which inner orbitals (*d* or *f* orbitals) are being filled. (7.11; 20.1)

Transuranium elements the elements beyond uranium that are made artificially by particle bombardment. (19.3)

Triple bond a bond in which three pairs of electrons are shared by two atoms. (8.8)

Triple point the point on a phase diagram at which all three states of a substance are present. (10.9)

Tyndall effect the scattering of light by particles in a suspension. (11.8)

Uncertainty (in measurement) the characteristic that any measurement involves estimates and cannot be exactly reproduced. (1.4)

Unimolecular step a reaction step involving only one molecule. (12.5)

Unit cell the smallest repeating unit of a lattice. (10.3)

Unit factor method an equivalence statement between units used for converting from one unit to another. (1.6)

Universal gas constant the combined proportionality constant in the ideal gas law; $0.08206 \text{ L} \cdot \text{atm/K} \cdot \text{mol}$ or $8.3145 \text{ J/K} \cdot \text{mol}$. (5.3)

Valence electrons the electrons in the outermost principal quantum level of an atom. (7.11)

Valence shell electron-pair repulsion (VSEPR) model a model whose main postulate is that the structure around a given atom in a molecule is determined principally by minimizing electron-pair repulsions. (8.13)

Van der Waals equation a mathematical expression for describing the behavior of real gases. (5.8)

van't Hoff factor the ratio of moles of particles in solution to moles of solute dissolved. (11.7)

Vapor pressure the pressure of the vapor over a liquid at equilibrium. (10.8)

Vaporization (evaporization) the change in state that occurs when a liquid evaporates to form a gas. (10.8)

Viscosity the resistance of a liquid to flow. (10.2)

Volt the unit of electrical potential defined as one joule of work per coulomb of charge transferred. (18.2)

Voltmeter an instrument that measures cell potential by drawing electric current through a known resistance. (18.2)

Volumetric analysis a process involving titration of one solution with another. (4.8)

Vulcanization a process in which sulfur is added to rubber and the mixture is heated, causing crosslinking of the polymer chains and thus adding strength to the rubber. (22.5)

Wave function a function of the coordinates of an electron's position in three-dimensional space that describes the properties of the electron. (7.5)

Wave mechanical model a model for the hydrogen atom in which the electron is assumed to behave as a standing wave. (7.5)

Wavelength the distance between two consecutive peaks or troughs in a wave. (7.1)

Weak acid an acid that dissociates only slightly in aqueous solution. (4.2; 14.2)

Weak base a base that reacts with water to produce hydroxide ions to only a slight extent in aqueous solution. (4.2; 14.6)

Weak electrolyte a material that, when dissolved in water, gives a solution that conducts only a small electric current. (4.2)

Weight the force exerted on an object by gravity. (1.3)

Work force acting over a distance. (6.1)

X-ray diffraction a technique for establishing the structure of crystalline solids by directing X rays of a single wavelength at a crystal and obtaining a diffraction pattern from which interatomic spaces can be determined. (10.3)

Zone of nuclear stability the area encompassing the stable nuclides on a plot of their positions as a function of the number of protons and the number of neutrons in the nucleus. (19.1)

Zone refining a metallurgical process for obtaining a highly pure metal that depends on continuously melting the impure material and recrystallizing the pure metal. (21.8)

Answers to Selected Exercises

The answers listed here are from the *Complete Solutions Guide,* in which rounding is carried out at each intermediate step in a calculation in order to show the correct number of significant figures for that step. Therefore, an answer given here may differ in the last digit from the result obtained by carrying extra digits throughout the entire calculation and rounding at the end (the procedure you should follow).

Chapter 1

17. A law summarizes what happens, e.g., law of conservation of mass in a chemical reaction or the ideal gas law, $PV = nRT$. A theory (model) is an attempt to explain why something happens. Dalton's atomic theory explains why mass is conserved in a chemical reaction. The kinetic molecular theory explains why pressure and volume are inversely related at constant temperature and moles of gas. **19.** The fundamental steps are (1) making observations; (2) formulating hypotheses; (3) performing experiments to test the hypotheses. The key to the scientific method is performing experiments to test hypotheses. If after the test of time, the hypotheses seem to account satisfactorily for some aspect of natural behavior, then the set of tested hypotheses turns into a theory (model). However, scientists continue to perform experiments to refine or replace existing theories. **21.** A qualitative observation expresses what makes something what it is; it does not involve a number; e.g., the air we breathe is a mixture of gases, ice is less dense than water, rotten milk stinks. The SI units are mass in grams, length in meters, and volume in the derived units of m^3. The assumed uncertainty in a number is ± 1 in the last significant figure of the number. The precision of an instrument is related to the number of significant figures associated with an experimental reading on that instrument. Different instruments for measuring mass, length, or volume have varying degrees of precision. Some instruments only give a few significant figures for a measurement while others will give more significant figures. **23.** Significant figures are the digits we associate with a number. They contain all of the certain digits and the first uncertain digit (the first estimated digit). What follows is one thousand indicated to varying numbers of significant figures: 1000 or 1×10^3 (1 S.F.); 1.0×10^3 (2 S.F.); 1.00×10^3 (3 S.F.); 1000. or 1.000×10^3 (4 S.F.). To perform the calculation, the addition/subtraction significant rule is applied to $1.5 - 1.0$. The result of this is the one significant figure answer of 0.5. Next, the multiplication/division rule is applied to 0.5/0.50. A one significant number divided by a two significant number yields an answer with one significant figure (answer = 1). **25.** The slope of the T_F vs. T_C plot is 1.8 (= 9/5) and the y-intercept is 32°F. The slope of T_C vs. T_K plot is 1 and the y-intercept is -273°C. **27. a.** exact; **b.** inexact; **c.** exact; **d.** inexact **29. a.** 3; **b.** 4; **c.** 4; **d.** 1; **e.** 7; **f.** 1; **g.** 3; **h.** 3 **31. a.** 3.42×10^{-4}; **b.** 1.034×10^4; **c.** 1.7992×10^1; **d.** 3.37×10^5 **33.** $2.85 + 0.280 = 3.13$ mL; the graduated cylinder on the left limits the precision of the total volume; it is the least precise measuring device between the two graduated cylinders. **35. a.** 641.0; **b.** 1.327; **c.** 77.34; **d.** 3215; **e.** 0.420 **37. a.** 188.1; **b.** 12; **c.** 4×10^{-7}; **d.** 6.3×10^{-26}; **e.** 4.9; Uncertainty appears in the first decimal place. The average of several numbers can be only as precise as the least precise number. Averages can be exceptions to the significant figure rules. **f.** 0.22 **39. a.** 84.3 mm; **b.** 2.41 m; **c.** 2.945×10^{-5} cm; **d.** 14.45 km; **e.** 2.353×10^5 mm; **f.** 0.9033 μm **41. a.** 8 lb and 9.9 oz; $20\frac{1}{4}$ in; **b.** 4.0×10^4 km, 4.0×10^7 m; **c.** 1.2×10^{-2} m³, 12 L, 730 in³, 0.42 ft³ **43. a.** 4.00×10^2 rods; 10.0 furlongs; 2.01×10^3 m; 2.01 km; **b.** 8390.0 rods; 209.75 furlongs; 42,195 m; 42.195 km **45. a.** 0.373 kg, 0.822 lb; **b.** 31.1 g, 156 carats; **c.** 19.3 cm³ **47.** 24 capsules **49.** 2.91×10^9 knots; 3.36×10^9 mi/h **51.** To the proper number of significant figures, the car is traveling at 40. mi/h, which would not violate the speed limit. **53.** $1.47/lb **55.** 7×10^5 kg mercury **57. a.** -273°C, 0 K;

b. $-40.$°C, 233 K; **c.** 20.°C, 293 K; **d.** 4×10^7.°C, 4×10^7 K **59. a.** 312.4 K; 102.6°F; **b.** 248 K; -13°F; **c.** 0 K; -459°F; **d.** 1074 K; 1470°F **61.** 160.°C $= 320.$°F **63. a.** $T_X = (T_C + 10°C)\dfrac{5°X}{14°C}$; **b.** 11.4°X; **c.** 152°C; 425 K; 306°F **65.** It will float (density $= 0.80$ g/cm³). **67.** 1×10^6 g/cm³ **69. a.** 0.28 cm³; **b.** 49 carats **71.** 3.8 g/cm³ **73. a.** Both are the same mass; **b.** 1.0 mL mercury; **c.** Both are the same mass; **d.** 1.0 L benzene **75. a.** 1.0 kg feather; **b.** 100 g water; **c.** same **77.** 2.77 cm **79. a.** Picture iv represents a gaseous compound. Pictures ii and iii also contain a gaseous compound but have a gaseous element present. **b.** Picture vi represents a mixture of two gaseous elements. **c.** Picture v represents a solid element. **d.** Pictures ii and iii both represent a mixture of a gaseous element and a gaseous compound. **81. a.** heterogeneous; **b.** homogeneous (hopefully); **c.** homogeneous; **d.** homogeneous (hopefully); **e.** heterogeneous; **f.** heterogeneous **83. a.** pure; **b.** mixture; **c.** mixture; **d.** pure; **e.** mixture (copper and zinc); **f.** pure; **g.** mixture; **h.** mixture; **i.** mixture. Iron and uranium are elements. Water is a compound. Table salt is usually a homogeneous mixture composed mostly of sodium chloride (NaCl), but it usually will contain other substances that help absorb water vapor (an anticaking agent). **85.** Compound **87. a.** physical; **b.** chemical; **c.** physical; **d.** chemical **89.** 0.010 kg **91.** 1.0×10^5 bags **93.** 3.0×10^{17} m **95. a.** 1.1 g; **b.** 4.9 g **97.** 56.56°C **99. a.** Volume $\times$ density $=$ mass; the orange block is more dense. Since mass (orange) $>$ mass (blue) and volume (orange) $<$ volume (blue), the density of the orange block must be greater to account for the larger mass of the orange block. **b.** Which block is more dense cannot be determined. Since mass (orange) $>$ mass (blue) and volume (orange) $>$ volume (blue), the density of the orange block may or may not be larger than the blue block. If the blue block is more dense, then its density cannot be so large that the mass of the smaller blue block becomes larger than the orange block mass. **c.** The blue block is more dense. Since mass (blue) $=$ mass (orange) and volume (blue) $<$ volume (orange), the density of the blue block must be larger to equate the masses. **d.** The blue block is more dense. Since mass (blue) $>$ mass (orange) and the volumes are equal, the density of the blue block must be larger to give the blue block the larger mass. **101.** 8.5 ± 0.5 g/cm³ **109.** In a subtraction, the result gets smaller, but the uncertainties add. If the two numbers are very close together, the uncertainty may be larger than the result. For example, let us assume we want to take the difference of the following two measured quantities, $999{,}999 \pm 2$ and $999{,}996 \pm 2$. The difference is 3 ± 4. Because of the larger uncertainty, subtracting two similar numbers is bad practice. **111. a.** 2%; **b.** 2.2%; **c.** 0.2% **113.** $d_{old} = 8.8$ g/cm³, $d_{new} = 7.17$ g/cm³; $d_{new}/d_{old} = $ mass$_{new}$/mass$_{old} = 0.81$; The difference in mass is accounted for by the difference in the alloy used (if the assumptions are correct). **115.** 7.0% **117. a.** One possibility is that rope B is not attached to anything and rope A and rope C are connected via a pair of pulleys and/or gears; **b.** Try to pull rope B out of the box. Measure the distance moved by C for a given movement of A. Hold either A or C firmly while pulling on the other rope. **119.** $160./person; 3.20×10^3 nickels/person; 85.6 £/person **121.** 200.0°F $= 93.33$°C; -100.0°F $= -73.33$°C; 93.33°C $= 366.48$ K; -73.33°C $= 199.82$ K; difference of temperatures in °C $= 166.66$; difference of temperatures in K $= 166.66$; No, there is not a difference of 300 degrees in °C or K.

Chapter 2

17. A compound will always contain the same numbers (and types) of atoms. A given amount of hydrogen will react only with a specific amount of oxygen. Any excess oxygen will remain unreacted. **19.** Law of conservation of mass: mass is neither created nor destroyed. The mass before a chemical reaction always equals the mass after a chemical reaction. Law of definite proportion: a given compound always contains exactly the same proportion of elements by mass. Water is always 1 g H for every 8 g oxygen. Law of multiple proportions: When two elements form a series of compounds, the ratios of the mass of the second element that combine with one gram of the first element can always be reduced to small whole numbers. For CO_2 and CO discussed in Section 2.2, the mass ratios of oxygen that react with 1 g of carbon in each compound are in a 2:1 ratio.   **21.** J. J. Thomson's study of cathode-ray tubes led him to postulate the existence of negatively charged particles, which we now call electrons. Ernest Rutherford and his alpha bombardment of metal foil experiments led him to postulate the nuclear atom—an atom with a tiny dense center of positive charge (the nucleus) with electrons moving about the nucleus at relatively large distances away; the distance is so large that an atom is mostly empty space.   **23.** The number and arrangement of electrons in an atom determines how the atom will react with other atoms. The electrons determine the chemical properties of an atom. The number of neutrons present determine the isotope identity and the mass number.   **25.** 2; the ratio increases   **27.** Metallic character increases as one goes down a family and decreases from left to right across the periodic table.   **29. a.** This represents ionic bonding, which is the electrostatic attraction between anions and cations. **b.** This represents covalent bonding, where electrons are shared between two atoms. This could be the space-filling model for H_2O or SF_2 or NO_2, etc.   **31.** Statements a and b are true. Counting over in the periodic table, element 118 will be the next noble gas (a nonmetal). For statement c, hydrogen has mostly nonmetallic properties. For statement d, a family of elements is also known as a group of elements. For statement e, two items are incorrect. When a metal reacts with a nonmetal, an ionic compound is produced and the formula of the compound would be AX_2 (since alkaline earth metals for $+2$ ions and halogens form -1 ions in ionic compounds). The correct statement would be: When alkaline earth metal, A, reacts with a halogen, X, the formula of the ionic compound formed should be AX_2.   **33. a.** The composition of a substance depends on the number of atoms of each element making up the compound (depends on the formula of the compound) and not on the composition of the mixture from which it was formed. **b.** Avogadro's hypothesis implies that volume ratios are equal to molecule ratios at constant temperature and pressure. $H_2(g) + Cl_2(g) \rightarrow 2HCl(g)$; from the balanced equation, the volume of HCl produced will be twice the volume of H_2 (or Cl_2) reacted.   **35.** 299 g   **37.** All the masses of hydrogen in these three compounds can be expressed as simple whole-number ratios. The g H/g N in hydrazine, ammonia, and hydrogen azide are in the ratios 6:9:1.   **39.** For CO and CO_2, it is easiest to concentrate on the mass of oxygen that combines with 1 g of carbon. From the formulas (two oxygen atoms per carbon atom in CO_2 vs. one oxygen atom per carbon atom in CO), CO_2 will have twice the mass of oxygen that combines per gram of carbon as compared to CO. For CO_2 and C_3O_2, it is easiest to concentrate on the mass of carbon that combines with 1 g of oxygen. From the formulas (three carbon atoms per two oxygen atoms in C_3O_2 vs. one carbon atom per two oxygen atoms in CO_2), C_3O_2 will have three times the mass of carbon that combines per gram of oxygen as compared to CO_2. As expected, the mass ratios are whole numbers as predicted by the law of multiple proportions.   **41.** Mass is conserved in a chemical reaction because atoms are conserved. Chemical reactions involve the reorganization of atoms, so formulas change in a chemical reaction, but the number and types of atoms do not change. Because the atoms do not change in a chemical reaction, mass must not change. In this equation we have two oxygen atoms and four hydrogen atoms both before and after the reaction occurs.   **43.** O, 7.94; Na, 22.8; Mg, 11.9; O and Mg are incorrect by a factor of ≈ 2; correct formulas are H_2O, Na_2O, and MgO.   **45.** Using $r = 5 \times 10^{-14}$ cm, $d_{nucleus} = 3 \times 10^{15}$ g/cm^3; using $r = 1 \times 10^{-8}$ cm, $d_{atom} = 0.4$ g/cm^3 **47.** 37   **49.** sodium, Na; radium, Ra; iron, Fe; gold, Au; manganese, Mn; lead, Pb   **51.** Sn, tin; Pt, platinum; Hg, mercury; Mg, magnesium; K, potassium; Ag, silver   **53. a.** Metals: Mg, Ti, Au, Bi, Ge, Eu, Am; nonmetals: Si, B, At, Rn, Br; **b.** metalloids: Si, Ge, B, At. The elements at the boundary between the metals and the nonmetals are B, Si, Ge, As, Sb, Te, Po, and At. These elements are all considered metalloids. Aluminum has mostly properties of metals.   **55. a.** transition metals; **b.** alkaline earth metals; **c.** alkali metals; **d.** noble gases; **e.** halogens   **57. a.** $^{17}_{8}O$; **b.** $^{37}_{17}Cl$; **c.** $^{60}_{27}Co$; **d.** $^{57}_{26}Fe$; **e.** $^{131}_{53}I$; **f.** $^{7}_{3}Li$ **59. a.** $^{23}_{11}Na$ **b.** $^{19}_{9}F$ **c.** $^{16}_{8}O$ **61. a.** 35 p, 44 n, 35 e; **b.** 35 p, 46 n, 35 e; **c.** 94 p, 145 n, 94 e; **d.** 55 p, 78 n, 55 e; **e.** 1 p, 2 n, 1 e;

f. 26 p, 30 n, 26 e   **63. a.** 56 p; 54 e; **b.** 30 p; 28 e; **c.** 7 p; 10 e; **d.** 37 p; 36 e; **e.** 27 p; 24 e; **f.** 52 p; 54 e; **g.** 35 p; 36 e   **65.** $^{151}_{63}Eu^{3+}$; $^{118}_{50}Sn^{2+}$ **67.** $^{238}_{92}U$, 92 p, 146 n, 92 e, 0; $^{40}_{20}Ca^{2+}$, 20 p, 20 n, 18 e, 2+; $^{51}_{23}V^{3+}$, 23 p, 28 n, 20 e, 3+; $^{89}_{39}Y$, 39 p, 50 n, 39 e, 0; $^{79}_{35}Br^{-}$, 35 p, 44 n, 36 e, 1−; $^{31}_{15}P^{3-}$, 15 p, 16 n, 18 e, 3−   **69. a.** lose 2 e^- to form Ra^{2+}; **b.** lose 3 e^- to form In^{3+}; **c.** gain 3 e^- to form P^{3-}; **d.** gain 2 e^- to form Te^{2-}; **e.** gain 1 e^- to form Br^-; **f.** lose 1 e^- to form Rb^+ **71. a.** sodium bromide; **b.** rubidium oxide; **c.** calcium sulfide; **d.** aluminum iodide; **e.** SrF_2; **f.** Al_2Se_3; **g.** K_3N; **h.** Mg_3P_2 **73. a.** cesium fluoride; **b.** lithium nitride; **c.** silver sulfide (Silver forms only +1 ions so no Roman numerals are needed); **d.** manganese(IV) oxide; **e.** titanium(IV) oxide; **f.** strontium phosphide   **75. a.** barium sulfite; **b.** sodium nitrite; **c.** potassium permanganate; **d.** potassium dichromate   **77. a.** dinitrogen tetroxide; **b.** iodine trichloride; **c.** sulfur dioxide; **d.** diphosphorus pentasulfide   **79. a.** copper(I) iodide; **b.** copper(II) iodide; **c.** cobalt(II) iodide; **d.** sodium carbonate; **e.** sodium hydrogen carbonate or sodium bicarbonate; **f.** tetrasulfur tetranitride; **g.** selenium tetrachloride; **h.** sodium hypochlorite; **i.** barium chromate; **j.** ammonium nitrate   **81.** selenate; selenite; tellurate; tellurite   **83. a.** SF_2; **b.** SF_6; **c.** NaH_2PO_4; **d.** Li_3N; **e.** $Cr_2(CO_3)_3$; **f.** SnF_2; **g.** $NH_4C_2H_3O_2$; **h.** NH_4HSO_4; **i.** $Co(NO_3)_3$; **j.** Hg_2Cl_2; Mercury(I) exists as Hg_2^{2+} ions. **k.** $KClO_3$; **l.** NaH   **85. a.** Na_2O; **b.** Na_2O_2; **c.** KCN; **d.** $Cu(NO_3)_2$; **e.** $SeBr_4$; **f.** HIO_2; **g.** PbS_2; **h.** CuCl; **i.** GaAs (from the positions in the periodic table, Ga^{3+} and As^{3-} are the predicted ions); **j.** CdSe; **k.** ZnS; **l.** HNO_2; **m.** P_2O_5 **87. a.** nitric acid, HNO_3; **b.** perchloric acid, $HClO_4$; **c.** acetic acid, $HC_2H_3O_2$; **d.** sulfuric acid, H_2SO_4; **e.** phosphoric acid, H_3PO_4 **89.** Yes, 1.0 g H would react with 37.0 g ^{37}Cl and 1.0 g H would react with 35.0 g ^{35}Cl. No, the mass ratio of H/Cl would always be 1 g H/37 g Cl for ^{37}Cl and 1 g H/35 g Cl for ^{35}Cl. As long as we had pure ^{35}Cl or pure ^{37}Cl, these ratios will always hold. If we have a mixture (such as the natural abundance of chlorine), the ratio will also be constant as long as the composition of the mixture of the two isotopes does not change.   **91.** 26 p, 27 n, 24 e   **93.** Only statement a is true. For statement b, X has 34 protons. For statement c, X has 45 neutrons. For statement d, X is selenium.   **95. a.** lead(II) acetate; **b.** copper(II) sulfate; **c.** calcium oxide; **d.** magnesium sulfate; **e.** magnesium hydroxide; **f.** calcium sulfate; **g.** dinitrogen monoxide or nitrous oxide (common)   **97.** X = Ra, 142 neutrons   **99. a.** Ca_3N_2; calcium nitride; **b.** K_2O; potassium oxide; **c.** RbF; rubidium fluoride; **d.** MgS; magnesium sulfide; **e.** BaI_2; barium iodide; **f.** Al_2Se_3; aluminum selenide; **g.** Cs_3P; cesium phosphide; **h.** $InBr_3$; indium(III) bromide (In forms compounds with +1 and +3 ions. You would predict a +3 ion from the position of In in the periodic table.)   **101. a.** P; 15 p; 16 n; **b.** I; 53 p; 74 n; **c.** K; 19 p; 20 n **d.** Yb; 70 p; 103 n   **109.** Cu, Ag, and Au   **111.** C:H = 8:18 or 4:9   **113.** The alchemists were incorrect. The solid residue must have come from the flask.   **115. a.** The compounds are isomers of each other. Isomers are compounds with the same formula but the atoms are attached differently, resulting in different properties. **b.** When wood burns, most of the solid material is converted to gases, which escape. **c.** Atoms are not an indivisible particle. Atoms are composed of electrons, neutrons, and protons. **d.** The two hydride samples contain different isotopes of either hydrogen and/or lithium. Isotopes may have different masses but have similar chemical properties.   **117.** CO_2 **119.** tantalum(V) oxide; the formula would have the same subscripts, Ta_2S_5; 40 protons   **121.** Ge^{4+}; ^{99}Tc

Chapter 3

23. From the relative abundances, there would be 9889 atoms of ^{12}C and 111 atoms of ^{13}C in the 10,000-atom sample. The average mass of carbon is independent of the sample size; it will always be 12.01 amu. The total mass would be 1.201×10^5 amu. For 1 mole of carbon (6.022×10^{23} atoms C), the average mass would still be 12.01 amu. There would be 5.955×10^{23} atoms of ^{12}C and 6.68×10^{21} atoms of ^{13}C. The total mass would be 7.233×10^{24} amu. The total mass in grams is 12.01 g/mol.   **25.** Each person would have 90 trillion dollars.   **27.** Only in b are the empirical formulas the same for both compounds illustrated. In b, general formulas of X_2Y_4 and XY_2 are illustrated, and both have XY_2 for an empirical formula.   **29.** The mass percent of a compound is a constant no matter what amount of substance is present. Compounds always have constant composition.   **31.** c   **33.** The theoretical yield is the stoichiometric amount of product that should form if the limiting reactant is completely consumed and the reaction has 100% yield.   **35.** The infor-

mation needed is mostly the coefficients in the balanced equation and the molar masses of the reactants and products. For percent yield, we would need the actual yield of the reaction and the amounts of reactants used.
a. mass of CB produced = 1.00×10^4 molecules $A_2B_2 \times$

$$\frac{1 \text{ mol } A_2B_2}{6.022 \times 10^{23} \text{ molecules } A_2B_2} \times \frac{2 \text{ mol CB}}{1 \text{ mol } A_2B_2} \times \frac{\text{molar mass of CB}}{\text{mol CB}}$$

b. atoms of A produced = 1.00×10^4 molecules $A_2B_2 \times \frac{2 \text{ atoms A}}{1 \text{ molecule } A_2B_2}$

c. mol of C reacted = 1.00×10^4 molecules $A_2B_2 \times$

$$\frac{1 \text{ mol } A_2B_2}{6.022 \times 10^{23} \text{ molecules } A_2B_2} \times \frac{2 \text{ mol C}}{1 \text{ mol } A_2B_2}$$

d. % yield = $\frac{\text{actual mass}}{\text{theoretical mass}} \times 100$; The theoretical mass of CB produced was calculated in part a. If the actual mass of CB produced is given, then the percent yield can be determined for the reaction using the percent yield equation. **37.** 207.2 u, Pb **39.** 185 u **41.** 48% ^{151}Eu and 52% ^{153}Eu **43.** There are three peaks in the mass spectrum, each 2 mass units apart. This is consistent with two isotopes, differing in mass by two mass units. **45.** 4.64×10^{-20} g Fe **47.** 1.00×10^{22} atoms C **49.** Al_2O_3, 101.96 g/mol; Na_3AlF_6, 209.95 g/mol **51. a.** 17.03 g/mol; **b.** 32.05 g/mol; **c.** 252.08 g/mol **53. a.** 0.0587 mol NH_3; **b.** 0.0312 mol N_2H_4; **c.** 3.97×10^{-3} mol $(NH_4)_2Cr_2O_7$ **55. a.** 85.2 g NH_3; **b.** 160. g N_2H_4; **c.** 1260 g $(NH_4)_2Cr_2O_7$ **57. a.** 70.1 g N; **b.** 140. g N; **c.** 140. g N **59. a.** 3.54×10^{22} molecules NH_3; **b.** 1.88×10^{22} molecules N_2H_4; **c.** 2.39×10^{21} formula units $(NH_4)_2Cr_2O_7$ **61. a.** 3.54×10^{22} atoms N; **b.** 3.76×10^{22} atoms N; **c.** 4.78×10^{21} atoms N **63.** 2.77×10^{19} molecules; 3.26 mg Cl **65. a.** 0.9393 mol; **b.** 2.17×10^{-4} mol; **c.** 2.5×10^{-8} mol **67. a.** 4.01×10^{22} atoms N; **b.** 5.97×10^{22} atoms N; **c.** 3.67×10^{22} atoms N; **d.** 6.54×10^{22} atoms N **69.** 176.12 g/mol; 2.839×10^{-3} mol; 1.710×10^{21} molecules **71. a.** 165.39 g/mol; **b.** 3.023 mol; **c.** 3.3 g; **d.** 5.5×10^{22} atoms; **e.** 1.6 g; **f.** 1.373×10^{-19} g **73. a.** 50.00% C, 5.595% H, 44.41% O; **b.** 55.80% C, 7.025% H, 37.18% O; **c.** 67.90% C, 5.699% H, 26.40% N **75.** NO **77.** 6.54×10^4 g/mol **79. a.** 39.99% C, 6.713% H, 53.30% O; **b.** 40.00% C, 6.714% H, 53.29% O; **c.** 40.00% C, 6.714% H, 53.29% O (all the same except for rounding differences) **81. a.** NO_2; **b.** CH_2; **c.** P_2O_5; **d.** CH_2O **83.** $C_3H_6O_2$ **85.** compound I: HgO; compound II: Hg_2O **87.** SN; S_4N_4 **89.** C_3H_5Cl; $C_6H_{10}Cl_2$ **91.** CH_2O **93.** C_3H_4, C_9H_{12} **95. a.** $C_6H_{12}O_6(s) + 6O_2(g) \rightarrow 6CO_2(g) + 6H_2O(g)$; **b.** $Fe_2S_3(s) + 6HCl(g) \rightarrow 2FeCl_3(s) + 3H_2S(g)$; **c.** $CS_2(l) + 2NH_3(g) \rightarrow H_2S(g) + NH_4SCN(s)$ **97.** $2H_2O_2(aq) \xrightarrow{MnO_2} 2H_2O(l) + O_2(g)$ **99. a.** $3Ca(OH)_2(aq) + 2H_3PO_4(aq) \rightarrow 6H_2O(l) + Ca_3(PO_4)_2(s)$; **b.** $Al(OH)_3(s) + 3HCl(aq) \rightarrow AlCl_3(aq) + 3H_2O(l)$; **c.** $2AgNO_3(aq) + H_2SO_4(aq) \rightarrow Ag_2SO_4(s) + 2HNO_3(aq)$ **101. a.** $2C_6H_6(l) + 15O_2(g) \rightarrow 12CO_2(g) + 6H_2O(g)$; **b.** $2C_4H_{10}(g) + 13O_2(g) \rightarrow 8CO_2(g) + 10H_2O(g)$; **c.** $C_{12}H_{22}O_{11}(s) + 12O_2(g) \rightarrow 12CO_2(g) + 11H_2O(g)$; **d.** $4Fe(s) + 3O_2(g) \rightarrow 2Fe_2O_3(s)$; **e.** $4FeO(s) + O_2(g) \rightarrow 2Fe_2O_3(s)$ **103. a.** $SiO_2(s) + 2C(s) \rightarrow Si(s) + 2CO(g)$; **b.** $SiCl_4(l) + 2Mg(s) \rightarrow Si(s) + 2MgCl_2(s)$; **c.** $Na_2SiF_6(s) + 4Na(s) \rightarrow Si(s) + 6NaF(s)$ **105.** 7.26 g Al; 21.5 g Fe_2O_3; 13.7 g Al_2O_3 **107.** 4.355 kg **109. a.** 0.076 g; **b.** 0.052 g **111.** 32 kg **113.** 7.2×10^4 g **115.** $2NO(g) + O_2(g) \rightarrow 2NO_2(g)$; NO is limiting. **117. a.** 1.22×10^3 g; 284 g H_2 unreacted **119.** 0.301 g; 3.6×10^{-2} g HCl unreacted **121.** 2.81×10^6 g HCN; 5.63×10^6 g H_2O **123.** 82.9% **125.** 1.20×10^3 kg = 1.20 metric tons **127.** $^{12}C^1H_2^{16}O$ **129.** 6 **131.** 71.40% C, 8.689% H, 5.648% F, 14.26% O **133.** $C_8H_{11}O_3N$ **135.** 1360 g/mol **137.** 9.25×10^{22} H atoms **139.** 4.30×10^{-2} mol; 2.50 g **141.** 5 **143.** 42.8% **145.** 81.1 g **147.** 86.2% **149.** $C_{20}H_{30}O$ **151.** 14.3% X, 85.7% Z **161.** I: NH_3; II: N_2H_4; III: HN_3. If we set the atomic mass of H equal to 1.008, the atomic mass for nitrogen is 14.01. **163.** $C_7H_5N_3O_6$ **165.** 5.7 g FeO and 22.4 g Fe_2O_3 **167.** 40.08% **169.** N_4H_6 **171.** 1.8×10^6 g $Cu(NH_3)_4Cl_2$; 5.9×10^5 g NH_3 **173.** 207 u, Pb **175.** Al_2Se_3 **177.** 0.48 mol **179. a.** 113 Fe atoms; **b.** mass = 9.071×10^{-20} g; **c.** 540 Ru atoms **181.** M = Y, X = Cl, yttrium(III) chloride; 1.84 g

Chapter 4

13. a. Polarity is a term applied to covalent compounds. Polar covalent compounds have an unequal sharing of electrons in bonds that results in an

unequal charge distribution in the overall molecule. Polar molecules have a partial negative end and a partial positive end. These are not full charges as in ionic compounds, but are charges much less in magnitude. Water is a polar molecule and dissolves other polar solutes readily. The oxygen end of water (the partial negative end of the polar water molecule) aligns with the partial positive end of the polar solute while the hydrogens of water (the partial positive end of the polar water molecule) align with the partial negative end of the solute. These opposite charged attractions stabilize polar solutes in water. This process is called hydration. Nonpolar solutes do not have permanent partial negative and partial positive ends; nonpolar solutes are not stabilized in water and do not dissolve. **b.** KF is a soluble ionic compound so it is a strong electrolyte. KF(*aq*) actually exists as separate hydrated K^+ ions and hydrated F^- ions in solution: $C_6H_{12}O_6$ is a polar covalent molecule that is a nonelectrolyte. $C_6H_{12}O_6$ is hydrated as described in part a. **c.** RbCl is a soluble ionic compound so it exists as separate hydrated Rb^+ ions and hydrated Cl^- ions in solution. AgCl is an insoluble ionic compound so the ions stay together in solution and fall to the bottom of the container as a precipitate. **d.** HNO_3 is a strong acid and exists as separate hydrated H^+ ions and hydrated NO_3^- ions in solution. CO is a polar covalent molecule and is hydrated as explained in part a. **15.** Only statement b is true. A concentrated solution can also contain a nonelectrolyte dissolved in water, e.g., concentrated sugar water. Acids are either strong or weak electrolytes. Some ionic compounds are not soluble in water, so they are not labeled as a specific type of electrolyte. **17.** Bromides: NaBr, KBr, and NH_4Br (and others) would be soluble and AgBr, $PbBr_2$, and Hg_2Br_2 would be insoluble. Sulfates: Na_2SO_4, K_2SO_4, and $(NH_4)_2SO_4$ (and others) would be soluble and $BaSO_4$, $CaSO_4$, and $PbSO_4$ (or Hg_2SO_4) would be insoluble. Hydroxides: NaOH, KOH, and $Ca(OH)_2$ (and others) would be soluble and $Al(OH)_3$, $Fe(OH)_3$, and $Cu(OH)_2$ (and others) would be insoluble. Phosphates: Na_3PO_4, K_3PO_4, and $(NH_4)_3PO_4$ (and others) would be soluble and Ag_3PO_4, $Ca_3(PO_4)_2$, and $FePO_4$ (and others) would be insoluble. Lead: $PbCl_2$, $PbBr_2$, PbI_2, $Pb(OH)_2$, $PbSO_4$, and PbS (and others) would be insoluble. $Pb(NO_3)_2$ would be a soluble Pb^{2+} salt. **19.** The Brønsted–Lowry definitions are best for our purposes. An acid is a proton donor and a base is a proton acceptor. A proton is an H^+ ion. Neutral hydrogen has 1 electron and 1 proton, so an H^+ ion is just a proton. An acid–base reaction is the transfer of an H^+ ion (a proton) from an acid to a base. **21. a.** The species reduced is the element that gains electrons. The reducing agent causes reduction to occur by itself being oxidized. The reducing agent is generally listed as the entire formula of the compound/ion that contains the element oxidized. **b.** The species oxidized is the element that loses electrons. The oxidizing agent causes oxidation to occur by itself being reduced. The oxidizing agent is generally listed as the entire formula of the compound/ion that contains the element reduced. **c.** For simple binary ionic compounds, the actual charges on the ions are the oxidation states. For covalent compounds, nonzero oxidation states are imaginary charges the elements would have if they were held together by ionic bonds (assuming the bond is between two different nonmetals). Nonzero oxidation states for elements in covalent compounds are not actual charges. Oxidation states for covalent compounds are a bookkeeping method to keep track of electrons in a reaction.

23.

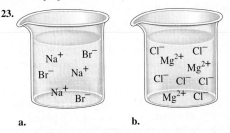

a.   **b.**

c. For answers c–i, we will describe what should be in each solution. For c, the drawing should have three times as many NO_3^- anions as Al^{3+} cations. **d.** The drawing should have twice as many NH_4^+ cations as SO_4^{2-} anions. **e.** The drawing should have equal numbers of Na^+ cations and OH^- anions. **f.** The drawing should have equal numbers of Fe^{2+} cations and SO_4^{2-} anions. **g.** The drawing should have equal numbers of K^+ cations and MnO_4^- anions. **h.** The drawing should have equal numbers of H^+ cations and ClO_4^- anions. **i.** The drawing should have equal numbers of NH_4^+ cations and $C_2H_3O_2^-$

anions. **25.** $CaCl_2(s) \rightarrow Ca^{2+}(aq) + 2Cl^-(aq)$ **27. a.** 0.2677 M; **b.** 1.255 $\times 10^{-3}$ M; **c.** 8.065 $\times 10^{-3}$ M **29. a.** $M_{Ca^{2+}} = 1.00$ M, $M_{NO_3^-} = 2.00$ M; **b.** $M_{Na^+} = 4.0$ M, $M_{SO_4^{2-}} = 2.0$ M; **c.** $M_{NH_4^+} = M_{Cl^-} = 0.187$ M; **d.** $M_{K^+} = 0.0564$ M, $M_{PO_4^{3-}} = 0.0188$ M **31.** 100.0 mL of 0.30 M $AlCl_3$ **33.** 4.00 g **35. a.** Place 20.0 g NaOH in a 2-L volumetric flask; add water to dissolve the NaOH and fill to the mark. **b.** Add 500. mL of the 1.00 M NaOH stock solution to a 2-L volumetric flask; fill to the mark with water. **c.** As in a, instead using 38.8 g K_2CrO_4. **d.** As in b, instead using 114 mL of 1.75 M K_2CrO_4 stock solution. **37.** $M_{NH_4^+} = 0.272$ M, $M_{SO_4^{2-}} = 0.136$ M **39.** 4.5 M **41.** 5.94 $\times 10^{-8}$ M **43.** Aluminum nitrate, magnesium chloride, and rubidium sulfate are soluble. **45. a.** No precipitate forms; **b.** $Al(OH)_3(s)$; **c.** $CaSO_4(s)$; **d.** $NiS(s)$ **47. a.** No reaction occurs because all possible products are soluble salts. **b.** $2Al(NO_3)_3(aq) + 3Ba(OH)_2(aq) \rightarrow 2Al(OH)_3(s) + 3Ba(NO_3)_2(aq)$; $2Al^{3+}(aq) + 6NO_3^-(aq) + 3Ba^{2+}(aq) + 6OH^-(aq) \rightarrow 2Al(OH)_3(s) + 3Ba^{2+}(aq) + 6NO_3^-(aq)$; $Al^{3+}(aq) + 3OH^-(aq) \rightarrow Al(OH)_3(s)$; **c.** $CaCl_2(aq) + Na_2SO_4(aq) \rightarrow CaSO_4(s) + 2NaCl(aq)$; $Ca^{2+}(aq) + 2Cl^-(aq) + 2Na^+(aq) + SO_4^{2-}(aq) \rightarrow CaSO_4(s) + 2Na^+(aq) + 2Cl^-(aq)$; $Ca^{2+}(aq) + SO_4^{2-}(aq) \rightarrow CaSO_4(s)$; **d.** $K_2S(aq) + Ni(NO_3)_2(aq) \rightarrow 2KNO_3(aq) + NiS(s)$; $2K^+(aq) + S^{2-}(aq) + Ni^{2+}(aq) + 2NO_3^-(aq) \rightarrow 2K^+(aq) + 2NO_3^-(aq) + NiS(s)$; $Ni^{2+}(aq) + S^{2-}(aq) \rightarrow NiS(s)$ **49. a.** $CuSO_4(aq) + Na_2S(aq) \rightarrow CuS(s) + Na_2SO_4(aq)$; $Cu^{2+}(aq) + S^{2-}(aq) \rightarrow CuS(s)$; the gray spheres are the Na^+ spectator ions and the blue-green spheres are the SO_4^{2-} spectator ions; **b.** $CoCl_2(aq) + 2NaOH(aq) \rightarrow Co(OH)_2(s) + 2NaCl(aq)$; $Co^{2+}(aq) + 2OH^-(aq) \rightarrow Co(OH)_2(s)$; the gray spheres are the Na^+ spectator ions and the green spheres are the Cl^- spectator ions; **c.** $AgNO_3(aq) + KI(aq) \rightarrow AgI(s) + KNO_3(aq)$; $Ag^+(aq) + I^-(aq) \rightarrow AgI(s)$; the red spheres are the K^+ spectator ions and the blue spheres are the NO_3^- spectator ions **51. a.** $Ba^{2+}(aq) + SO_4^{2-}(aq) \rightarrow BaSO_4(s)$; **b.** $Pb^{2+}(aq) + 2Cl^-(aq) \rightarrow PbCl_2(s)$; **c.** no reaction; **d.** no reaction; **e.** $Cu^{2+}(aq) + 2OH^-(aq) \rightarrow Cu(OH)_2(s)$ **53.** Ca^{2+}, Sr^{2+}, or Ba^{2+} could all be present. **55.** 0.607 g **57.** 0.520 g $Al(OH)_3$ **59.** 2.82 g **61. a.** $2KOH(aq) + Mg(NO_3)_2(aq) \rightarrow Mg(OH)_2(aq) + 2KNO_3(aq)$; **b.** magnesium hydroxide; **c.** 0.583 g; **d.** 0 M OH^-, 5.00×10^{-2} M Mg^{2+}, 0.100 M K^+, 0.200 M NO_3^- **63.** 23 u; Na **65. a.** $2HClO_4(aq) + Mg(OH)_2(s) \rightarrow Mg(ClO_4)_2(aq) + 2H_2O(l)$; $2H^+(aq) + 2ClO_4^-(aq) + Mg(OH)_2(s) \rightarrow Mg^{2+}(aq) + 2ClO_4^-(aq) + 2H_2O(l)$; $2H^+(aq) + Mg(OH)_2(s) \rightarrow Mg^{2+}(aq) + 2H_2O(l)$; **b.** $HCN(aq) + NaOH(aq) \rightarrow NaCN(aq) + H_2O(l)$; $HCN(aq) + Na^+(aq) + OH^-(aq) \rightarrow Na^+(aq) + CN^-(aq) + H_2O(l)$; $HCN(aq) + OH^-(aq) \rightarrow H_2O(l) + CN^-(aq)$; **c.** $HCl(aq) + NaOH(aq) \rightarrow NaCl(aq) + H_2O(l)$; $H^+(aq) + Cl^-(aq) + Na^+(aq) + OH^-(aq) \rightarrow Na^+(aq) + Cl^-(aq) + H_2O(l)$; $H^+(aq) + OH^-(aq) \rightarrow H_2O(l)$ **67. a.** $KOH(aq) + HNO_3(aq) \rightarrow H_2O(l) + KNO_3(aq)$; **b.** $Ba(OH)_2(aq) + 2HCl(aq) \rightarrow 2H_2O(l) + BaCl_2(aq)$; **c.** $3HClO_4(aq) + Fe(OH)_3(s) \rightarrow 3H_2O(l) + Fe(ClO_4)_3(aq)$; **d.** $AgOH(s) + HBr(aq) \rightarrow AgBr(s) + H_2O(l)$; **e.** $Sr(OH)_2(aq) + 2HI(aq) \rightarrow 2H_2O(l) + SrI_2(aq)$ **69. a.** 100. mL; **b.** 66.7 mL; **c.** 50.0 mL **71.** 2.0×10^{-2} M excess OH^- **73.** 0.102 M **75.** 43.8 mL **77.** 0.4178 g **79. a.** K, +1; O, −2; Mn, +7; **b.** Ni, +4; O, −2; **c.** Na, +1; Fe, +2; O, −2; H, +1 **d.** H, +1; O, −2; N, −3; P, +5; **e.** P, +3; O, −2; **f.** O, −2; Fe, $+\frac{8}{3}$; **g.** O, −2; F, −1; Xe, +6; **h.** S, +4; F, −1; **i.** C, +2; O, −2; **j.** C, 0; H, +1; O, −2 **81. a.** −3; **b.** −3; **c.** −2; **d.** +2; **e.** +1; **f.** +4; **g.** +3; **h.** +5; **i.** 0

83.

Redox?	Oxidizing Agent	Reducing Agent	Substance Oxidized	Substance Reduced
a. Yes	Ag^+	Cu	Cu	Ag^+
b. No	—	—	—	—
c. No	—	—	—	—
d. Yes	$SiCl_4$	Mg	Mg	$SiCl_4$ (Si)
e. No	—	—	—	—

In b, c, and e, no oxidation numbers change from reactants to products. **85.** Each sodium atom will lose one electron, and each fluorine atom will gain one electron. Two sodium atoms are required to react with one fluorine molecule in order to equalize the electrons lost with the electrons gained. $2Na(s) + F_2(g) \rightarrow 2NaF(s)$ **87. a.** $2C_2H_6(g) + 7O_2(g) \rightarrow 4CO_2(g) + 6H_2O(g)$; **b.** $Mg(s) + 2HCl(aq) \rightarrow Mg^{2+}(aq) + 2Cl^-(aq) + H_2(g)$; **c.** $2Co^{3+}(aq) + Ni(s) \rightarrow 2Co^{2+}(aq) + Ni^{2+}(aq)$; **d.** $Zn(s) + H_2SO_4(aq) \rightarrow ZnSO_4(aq) + H_2(g)$ **89.** We should weigh out between 4.24 and 4.32 g of KIO_3. We should

weigh it to the nearest milligram or 0.1 mg. Dissolve the KIO_3 in water, and dilute to the mark in a 1-L volumetric flask. This will produce a solution whose concentration is within the limits and is known to at least the fourth decimal place. **91.** increase by 0.019 M **93. a.** $AgNO_3$, $Pb(NO_3)_2$, and $Hg_2(NO_3)_2$ would form precipitates with the Cl^- ion; $Ag^+(aq) + Cl^-(aq) \rightarrow AgCl(s)$; $Pb^{2+}(aq) + 2Cl^-(aq) \rightarrow PbCl_2(s)$; $Hg_2^{2+}(aq) + 2Cl^-(aq) \rightarrow Hg_2Cl_2(s)$; **b.** Na_2SO_4, Na_2CO_3, and Na_3PO_4 would form precipitates with the Ca^{2+} ion; $Ca^{2+}(aq) + SO_4^{2-}(aq) \rightarrow CaSO_4(s)$; $Ca^{2+}(aq) + CO_3^{2-}(aq) \rightarrow CaCO_3(s)$; $3Ca^{2+}(aq) + 2PO_4^{3-}(aq) \rightarrow Ca_3(PO_4)_2(s)$; **c.** NaOH, Na_2S, and Na_2CO_3 would form precipitates with the Fe^{3+} ion; $Fe^{3+}(aq) + 3OH^-(aq) \rightarrow Fe(OH)_3(s)$; $2Fe^{3+}(aq) + 3S^{2-}(aq) \rightarrow Fe_2S_3(s)$; $2Fe^{3+}(aq) + 3CO_3^{2-}(aq) \rightarrow Fe_2(CO_3)_3(s)$; **d.** $BaCl_2$, $Pb(NO_3)_2$, and $Ca(NO_3)_2$ would form precipitates with the SO_4^{2-} ion; $Ba^{2+}(aq) + SO_4^{2-}(aq) \rightarrow BaSO_4(s)$; $Pb^{2+}(aq) + SO_4^{2-}(aq) \rightarrow PbSO_4(s)$; $Ca^{2+}(aq) + SO_4^{2-}(aq) \rightarrow CaSO_4(s)$; **e.** Na_2SO_4, NaCl, and NaI would form precipitates with the Hg_2^{2+} ion; $Hg_2^{2+}(aq) + SO_4^{2-}(aq) \rightarrow Hg_2SO_4(s)$; $Hg_2^{2+}(aq) + 2Cl^-(aq) \rightarrow Hg_2Cl_2(s)$; $Hg_2^{2+}(aq) + 2I^-(aq) \rightarrow Hg_2I_2(s)$; **f.** NaBr, Na_2CrO_4, and Na_3PO_4 would form precipitates with the Ag^+ ion; $Ag^+(aq) + Br^-(aq) \rightarrow AgBr(s)$; $2Ag^+(aq) + CrO_4^{2-}(aq) \rightarrow Ag_2CrO_4(s)$; $3Ag^+(aq) + PO_4^{3-}(aq) \rightarrow Ag_3PO_4(s)$ **95.** Ba **97.** 1.465% **99.** 2.00 M **101.** 180.2 g/mol **103.** $C_6H_8O_6$ **105.** 4.05% **107.** 63.74% **109.** Chromium is reduced and carbon is oxidized. Twelve (12) electrons are transferred in the balanced reaction. **121. a.** 2.5×10^{-8} M; **b.** 8.4×10^{-9} M; **c.** 1.33×10^{-4} M; **d.** 2.8×10^{-7} M **123. a.** 24.8% Co, 29.7% Cl, 5.09% H, 40.4% O; **b.** $CoCl_2 \cdot 6H_2O$; **c.** $CoCl_2 \cdot 6H_2O(aq) + 2AgNO_3(aq) \rightarrow 2AgCl(s) + Co(NO_3)_2(aq) + 6H_2O(l)$, $CoCl_2 \cdot 6H_2O(aq) + 2NaOH(aq) \rightarrow Co(OH)_2(s) + 2NaCl(aq) + 6H_2O(l)$, $4Co(OH)_2(s) + O_2(g) \rightarrow 2Co_2O_3(s) + 4H_2O(l)$ **125.** 14.6 g Zn and 14.4 g Ag **127.** 0.123 g SO_4^{2-}, 60.0% SO_4^{2-}; 61% K_2SO_4 and 39% Na_2SO_4 **129.** 4.90 M **131.** Y, 2.06 mL/min; Z, 4.20 mL/min **133.** 57.6 mL **135.** 4.7×10^{-2} M **137.** Citric acid has three acidic hydrogens per citric acid molecule. **139.** 0.07849 ± 0.00016 M or 0.0785 ± 0.0002 M **141.** $3(NH_4)_2CrO_4(aq) + 2Cr(NO_2)_3(aq) \rightarrow 6NH_4NO_2(aq) + Cr_2(CrO_4)_3(s)$; 7.34 g **143.** X = Se; H_2Se is hydroselenic acid; 0.252 g

Chapter 5

21. higher than; 13.6 times taller; When the pressure of the column of liquid standing on the surface of the liquid is equal to the pressure exerted by air on the rest of the surface of the liquid, then the height of the column of liquid is a measure of atmospheric pressure. Because water is 13.6 times less dense than mercury, the column of water must be 13.6 times longer than that of mercury to match the force exerted by the columns of liquid standing on the surface. **23.** The P versus $1/V$ plot is incorrect. The plot should be linear with positive slope and a y-intercept of zero. $PV = k$ so $P = k (1/V)$, which is in the form of the straight-line equation $y = mx + b$. **25.** $d = $ (molar mass)P/RT; density is directly proportional to the molar mass of the gas. Helium, which has the smallest molar mass of all the noble gases, will have the smallest density. **27.** At STP ($T = 273.2$ K and $P = 1.000$ atm), the volume of 1.000 mole of gas is:

$$V = \frac{nRT}{P} = \frac{1.000\,\text{mol} \times \dfrac{0.08206\,\text{L atm}}{\text{K mol}} \times 273.2\,\text{K}}{1.000\,\text{atm}} = 22.42\,\text{L}$$

The volume of 1.000 mole of any gas is 22.42 L, assuming the gas behaves ideally. Therefore, the molar volume of $He(g)$ and $N_2(g)$ at STP both equal 22.42 L/mol. If the temperature increases to 25.0°C (298.2 K), the molar volume of a gas will be larger than 22.42 L/mol because volume and temperature are directly related at constant pressure. If 1.000 mole of a gas is collected over water at a total pressure of 1.000 atm, the partial pressure of the collected gas will be less than 1.000 atm because water vapor is present ($P_{total} = P_{gas} + P_{H_2O}$). At some partial pressure below 1.000 atm, the molar volume of a gas will be larger the 22.42 L/mol because volume and pressure are inversely related at constant temperature. **29.** No; At any nonzero Kelvin temperature, there is a distribution of kinetic energies. Similarly, there is a distribution of velocities at any nonzero Kelvin temperature. **31.** $2NH_3(g) \rightarrow N_2(g) + 3H_2(g)$; At constant P and T, volume is directly proportional to the moles of gas present. In the reaction, the moles of gas double as reactants are converted to products, so the volume of the container should double. At constant V and T, P is directly proportional to the moles of gas present. As the moles of gas double, the pressure will double. The partial pressure of N_2 will be 1/2 the initial pressure of NH_3 and the partial pressure of H_2 will be 3/2 the initial pressure of NH_3. The partial pressure of H_2 will be three times the partial pressure of N_2.   **33.** CO_2 **35.** From the plot, Ne appears to behave most ideally.   **37. a.** 3.6×10^3 mm Hg; **b.** 3.6×10^3 torr; **c.** 4.9×10^5 Pa; **d.** 71 psi   **39.** 65 torr, 8.7×10^3 Pa, 8.6×10^{-2} atm   **41. a.** 642 torr, 0.845 atm; 8.56×10^4 Pa; **b.** 975 torr; 1.28 atm; 1.30×10^5 Pa; **c.** 517 torr; 850. torr   **43.** The balloon will burst.   **45.** 0.89 mol   **47. a.** 14.0 L; **b.** 4.72×10^{-2} mol; **c.** 678 K; **d.** 133 atm   **49.** 88.9 g He; 44.8 g H_2 **51.** 0.0449 mol   **53. a.** 69.6 K; **b.** 32.3 atm   **55.** 1.27 mol   **57.** $P_B = 2P_A$ **59. a.** 12.8 atm; **b.** 161 K; **c.** 620. K   **61.** 5.1×10^4 torr   **63.** The volume of the balloon increases from 1.00 L to 2.82 L, so the change in volume is 1.82 L.   **65.** 3.21 g Al   **67.** 135 g NaN_3 **69.** 1.5×10^7 g Fe, 2.6×10^7 g 98% H_2SO_4 **71.** 33.2 g   **73.** 2.47 mol H_2O **75. a.** $2CH_4(g) + 2NH_3(g) + 3O_2(g) \rightarrow 2HCN(g) + 6H_2O(g)$; **b.** 13.3 L   **77.** Cl_2 **79.** 12.6 g/L   **81.** $P_{He} = 0.50$ atm, $P_{Ne} = 0.30$ atm, $P_{Ar} = 0.20$ atm   **83.** 1.1 atm, $P_{CO_2} = 1.1$ atm, $P_{TOTAL} = 2.1$ atm   **85.** $P_{H_2} = 317$ torr, $P_{N_2} = 50.7$ torr, $P_{TOTAL} = 368$ torr   **87.** $P_{He} = 50.0$ torr; $P_{Ne} = 76.0$ torr; $P_{Ar} = 90.0$ torr; $P_{TOTAL} = 216.0$ torr   **89. a.** $\chi_{CH_4} = 0.412$, $\chi_{O_2} = 0.588$; **b.** 0.161 mol; **c.** 1.06 g CH_4, 3.03 g O_2 **91.** 0.990 atm; 0.625 g   **93.** 18.0%   **95.** 46.5%   **97.** $P_{TOTAL} = 6.0$ atm; $P_{H_2} = 1.5$ atm; $P_{N_2} = 4.5$ atm   **99.** $P_{N_2} = 0.98$ atm; $P_{TOTAL} = 2.9$ atm   **101.** Both $CH_4(g)$ and $N_2(g)$ have the same average kinetic energy at the various temperatures. 273 K, 5.65×10^{-21} J/molecule; 546 K, 1.13×10^{-20} J/molecule   **103.** CH_4: 652 m/s (273 K); 921 m/s (546 K); N_2: 493 m/s (273 K); 697 m/s (546 K)   **105.** The number of gas particles is constant, so at constant moles of gas, either a temperature change or a pressure change results in the smaller volume. If the temperature is constant, an increase in the external pressure would cause the volume to decrease. Gases are mostly empty space, so gases are easily compressible. If the pressure is constant, a decrease in temperature would cause the volume to decrease. As the temperature is lowered, the gas particles move with a slower average velocity and don't collide with the container walls as frequently and as forcefully. As a result, the internal pressure decreases. In order to keep the pressure constant, the volume of the container must decrease in order to increase the gas particle collisions per unit area.

107.

	Average KE	Average Velocity	Wall-Collision Frequency
a.	Increase	Increase	Increase
b.	Decrease	Decrease	Decrease
c.	Same	Same	Increase
d.	Same	Same	Increase

109. a. All the same; **b.** Flask C   **111.** CF_2Cl_2 **113.** The relative rates of effusion of $^{12}C^{16}O$, $^{12}C^{17}O$, and $^{12}C^{18}O$ are 1.04, 1.02, and 1.00. Advantage: CO_2 isn't as toxic as CO. Disadvantage: Can get a mixture of oxygen isotopes in CO_2, so some species would effuse at about the same rate.   **115. a.** 12.24 atm; **b.** 12.13 atm; **c.** The ideal gas law is high by 0.91%.   **117.** 5.2×10^{-6} atm; 1.3×10^{14} atoms He/cm^3 **119.** $S(s) + O_2(g) \rightarrow SO_2(g)$; $2SO_2(g) + O_2(g) \rightarrow 2SO_3(g)$; $SO_3(g) + H_2O(l) \rightarrow H_2SO_4(aq)$ **121. a.** 0.19 torr; **b.** 6.6×10^{21} molecules CO/m^3; **c.** 6.6×10^{15} molecules CO/cm^3

123. a.   **b.**

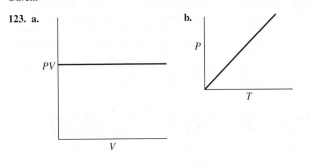

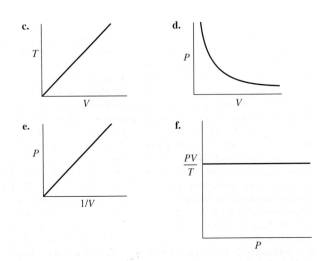

125. 0.772 atm $\cdot$ L; in Example 5.3, 1.0 mole of gas was present at 0°C. The moles of gas and/or the temperature must have been different for Boyle's data.   **127.** $P_{He} = 582$ torr, $P_{Xe} = 18$ torr   **129.** 13.3%   **131.** 1490   **133.** 24 torr   **135.** 4.1×10^6 L air; 7.42×10^5 L H_2 **137.** C_2H_3N; C_2H_3N **139.** $C_{12}H_{21}NO$; $C_{24}H_{42}N_2O_2$ **149.** 13.4% CaO, 86.6% BaO   **151.** C_2H_6 **153. a.** 8.7×10^3 L air/min; **b.** $\chi_{CO} = 0.0017$, $\chi_{CO_2} = 0.032$, $\chi_{O_2} = 0.13$, $\chi_{N_2} = 0.77$, $\chi_{H_2O} = 0.067$ **155. a.** 1.01×10^4 g; **b.** 6.65×10^4 g; **c.** 8.7×10^3 g   **157. a.** Due to air's larger average molar mass, a given volume of air at a given set of conditions has a higher density than helium. We need to heat the air to greater than 25°C to lower the air density (by driving air out of the hot-air balloon) until the density is the same as that for helium (at 25°C and 1.00 atm). **b.** 2150 K   **159.** C_3H_8 is the most likely identity.   **161.** 0.023 mol   **163.** 4.8 g/L; UF_3 will effuse 1.02 times faster.

Chapter 6

11. Path-dependent functions for a trip from Chicago to Denver are those quantities that depend on the route taken. One can fly directly from Chicago to Denver or one could fly from Chicago to Atlanta to Los Angeles and then to Denver. Some path-dependent quantities are miles traveled, fuel consumption of the airplane, time traveling, airplane snacks eaten, etc. State functions are path-independent; they only depend on the initial and final states. Some state functions for an airplane trip from Chicago to Denver would be longitude change, latitude change, elevation change, and overall time zone change.   **13.** Both q and w are negative.   **15. a.** 446 kJ released; **b.** 446 kJ released

17.

$$H_2O(l) + \tfrac{1}{2}CO_2(g) \rightarrow \tfrac{1}{2}CH_4(g) + O_2(g) \qquad \Delta H_1 = -\tfrac{1}{2}(-891 \text{ kJ})$$
$$\tfrac{1}{2}CH_4(g) + O_2(g) \rightarrow \tfrac{1}{2}CO_2(g) + H_2O(g) \qquad \Delta H_2 = \tfrac{1}{2}(-803 \text{ kJ})$$
$$\overline{H_2O(l) \rightarrow H_2O(g) \qquad \Delta H = \Delta H_1 + \Delta H_2 = 44 \text{ kJ}}$$

19. The zero points for ΔH_f° values are elements in their standard state. All substances are measured in relationship to this zero point.   **21.** No matter how insulated your thermos bottle, some heat will always escape into the surroundings. If the temperature of the thermos bottle (the surroundings) is high, less heat initially will escape from the coffee (the system); as a result your coffee will stay hotter for a longer period of time.   **23.** Fossil fuels contain carbon; the incomplete combustion of fossil fuels produces $CO(g)$ instead of $CO_2(g)$. This occurs when the amount of oxygen reacting is not sufficient to convert all of the carbon in fossil fuels to CO_2. Carbon monoxide is a poisonous gas to humans.   **25.** 150 J   **27. a.** Potential energy is energy due to position. Initially, ball A has a higher potential energy than ball B because the position of ball A is higher than that of ball B. In the final position, ball B has the higher position, so ball B has the higher potential energy. **b.** As ball A rolled down the hill, some of the potential energy lost by A was converted to random motion of the components of the hill (frictional heating). The remainder of the lost potential energy was added to B to initially increase its kinetic energy and then to increase its potential energy.   **29.** 16 kJ   **31. a.** 41 kJ; **b.** 35 kJ; **c.** 47 kJ; **d.** part a only   **33.** 375 J heat transferred to the system   **35.** -13.2 kJ   **37.** 11.0 L   **39.** $q = 30.9$ kJ,

$w = -12.4$ kJ, $\Delta E = 18.5$ kJ   **41.** This is an endothermic reaction, so heat must be absorbed to convert reactants into products. The high-temperature environment of internal combustion engines provides the heat. **43. a.** endothermic; **b.** exothermic; **c.** exothermic; **d.** endothermic   **45. a.** 1650 kJ released; **b.** 826 kJ released; **c.** 7.39 kJ released; **d.** 34.4 kJ released   **47.** 4400 g C_3H_8 **49.** When a liquid is converted into a gas, there is an increase in volume. The 2.5 kJ/mol quantity can be considered as the work done by the vaporization process in pushing back the atmosphere.   **51.** $H_2O(l)$; 2.30×10^3 J; Hg(l); 140°C   **53.** Al(s) 55. 311 K   **57.** 23.7°C   **59.** 0.25 J/°C · g   **61.** -66 kJ/mol   **63.** 170 J/g; 20. kJ/mol   **65.** 39.2°C   **67. a.** 31.5 kJ/°C; **b.** -1.10×10^3 kJ/mol   **69.** -220.8 kJ   **71.** 1268 kJ; No, because this reaction is very endothermic, it would not be a practical way of making ammonia due to the high energy costs. **73.** -233 kJ   **75.** -713 kJ   **77.** The enthalpy change for the formation of one mole of a compound from its elements, with all substances in their standard states. Na(s) + $\frac{1}{2}Cl_2(g) \rightarrow$ NaCl(s); $H_2(g) + \frac{1}{2}O_2(g) \rightarrow H_2O(l)$; $6C_{graphite}(s) + 6H_2(g) + 3O_2(g) \rightarrow C_6H_{12}O_6(s)$; Pb($s$) + $S_{rhombic}(s)$ + $2O_2(g) \rightarrow PbSO_4(s)$ **79. a.** $-940.$ kJ; **b.** -265 kJ; **c.** -176 kJ   **81. a.** -908 kJ, -112 kJ, $-140.$ kJ; **b.** $12NH_3(g) + 21O_2(g) \rightarrow 8HNO_3(aq) + 4NO(g) + 14H_2O(g)$, exothermic **83.** -2677 kJ   **85.** -169 kJ/mol   **87.** 132 kJ   **89.** -29.67 kJ/g   **91.** For $C_3H_8(g)$, -50.37 kJ/g vs. -47.7 kJ/g for octane. Because of the low boiling point of propane, there are extra safety hazards associated with storing the propane in high-pressure compressed gas tanks.   **93.** 1.05×10^5 L   **95.** 4900 g   **97. a.** $2SO_2(g) + O_2(g) \rightarrow 2SO_3(g)$; $w > 0$; **b.** $COCl_2(g) \rightarrow CO(g) + Cl_2(g)$; $w < 0$; **c.** $N_2(g) + O_2(g) \rightarrow 2NO(g)$; $w = 0$; Compare the sum of the coefficients of all the product gases in the balanced equation to the sum of the coefficients of all the reactant gases. When a balanced reaction has more moles of product gases than moles of reactant gases, then the reaction will expand in volume (ΔV is positive) and the system does work on the surroundings ($w < 0$). When a balanced reaction has a decrease in the moles of gas from reactants to products, then the reaction will contract in volume (ΔV is negative) and the surroundings does compression work on the system ($w > 0$). When there is no change in the moles of gas from reactants to products, then $\Delta V = 0$ and $w = 0$.   **99. a.** $C_{12}H_{22}O_{11}(s) + 12\ O_2(g) \rightarrow 12\ CO_2(g) + 11\ H_2O(l)$; **b.** -5630 kJ/mol; **c.** -5630 kJ/mol   **101.** 25 J   **103.** 4.2 kJ heat released   **105.** The calculated ΔH value will be less positive (smaller) than it should be.   **107.** 25.91°C   **109.** $\frac{1}{2}C + F \rightarrow A + B + D$; 47.0 kJ   **111. a.** 632 kJ; **b.** $C_2H_2(g)$ **113.** 28 g   **115. a.** -361 kJ; **b.** -199 kJ; **c.** -227 kJ; **d.** -112 kJ   **125.** $w = -2940$ J; $\Delta E = 27.8$ kJ   **127.** 25 m² **129.** 1×10^4 steps   **131.** 56.9 kJ   **133.** 1.74 kJ **135.** 3.3 cm

Chapter 7

19. The equations relating the terms are $v\lambda = c$, $E = hv$, and $E = hc/\lambda$. From the equations, wavelength and frequency are inversely related, photon energy and frequency are directly related, and photon energy and wavelength are inversely related. The unit of 1 joule (J) = 1 kg m²/s². This is why you must change mass units to kg when using the de Broglie equation.   **21.** The *photoelectric effect* refers to the phenomenon in which electrons are emitted from the surface of a metal when light strikes it. The light must have a certain minimum frequency (energy) in order to remove electrons from the surface of a metal. Light having a frequency below the minimum results in no electrons being emitted, whereas light at or higher than the minimum frequency does cause electrons to be emitted. For light having a frequency higher than the minimum frequency, the excess energy is transferred into kinetic energy for the emitted electron. Albert Einstein explained the photoelectric effect by applying quantum theory.   **23.** Example 7.3 calculates the de Broglie wavelength of a ball and of an electron. The ball has a wave length on the order of 10^{-34} m. This is incredibly short and, as far as the wave-particle duality is concerned, the wave properties of large objects are insignificant. The electron, with its tiny mass, also has a short wavelength: on the order of 10^{-10} m. However, this wavelength is significant as it is on the same order as the spacing between atoms in a typical crystal. For very tiny objects like electrons, the wave properties are important. The wave properties must be considered, along with the particle properties, when hypothesizing about the electron motion in an atom. **25.** The Bohr model was an important step in the development of the current quantum mechanical model of the atom. The idea that electrons can occupy only certain, allowed energy levels is illustrated nicely (and relatively easily).

We talk about the Bohr model to present the idea of quantized energy levels. **27.** When the p and d orbital functions are evaluated at various points in space, the results sometimes have positive values and sometimes have negative values. The term *phase* is often associated with the $+$ and $-$ signs. For example, a sine wave has alternating positive and negative phases. This is analogous to the positive and negative values (phases) in the p and d orbitals.   **29.** If one more electron is added to a half-filled subshell, electron–electron repulsions will increase because two electrons must now occupy the same atomic orbital. This may slightly decrease the stability of the atom.   **31.** The valence electrons are strongly attracted to the nucleus for elements with large ionization energies. One would expect these species to readily accept another electron and have very exothermic electron affinities. The noble gases are an exception; they have a large ionization energy but an endothermic electron affinity. Noble gases have a filled valence shell. The added electron must go into a higher n value atomic orbital, which would have a significantly higher energy. This is unfavorable.   **33.** For hydrogen and hydrogen-like (one electron) ions, all orbitals with the same value of n have the same energy. For polyatomic atoms/ions, the energy of the orbitals also depends on ℓ. Because there are more nondegenerate energy levels for polyatomic atoms/ions as compared with hydrogen, there are many more possible electronic transitions resulting in more complicated line spectra.   **35.** Yes, the maximum number of unpaired electrons in any configuration corresponds to a minimum in electron–electron repulsions.   **37.** Ionization energy applies to the removal of the electron from the atom in the gas phase. The work function applies to the removal of an electron from the surface of a solid element.   **39.** 3.84×10^{14} s⁻¹   **41.** 3.0×10^{10} s⁻¹, 2.0×10^{-23} J/photon, 12 J/mol   **43.** 9.4×10^{14} s⁻¹ to 1.1×10^{15} s⁻¹   **45.** Wave a has the longer wavelength (4.0×10^{-4} m). Wave b has the higher frequency (1.5×10^{12} s⁻¹) and larger photon energy (9.9×10^{-22} J). Since both of these waves represent electromagnetic radiation, they both should travel at the same speed, c, the speed of light. Both waves represent infrared radiation.   **47.** 1.50×10^{23} atoms   **49.** 427.7 nm   **51.** 276 nm   **53. a.** 2.4×10^{-11} m; **b.** 3.4×10^{-34} m   **55.** 1.6×10^{-27} kg   **57. a.** 656.7 nm (visible); **b.** 486.4 nm (visible); **c.** 121.6 nm (ultraviolet)

59.

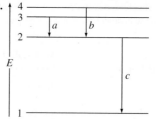

61. 7462 nm; 93.79 nm   **63.** $n = 1 \rightarrow n = 5$, $\lambda = 95.00$ nm; $n = 2 \rightarrow n = 6$, $\lambda = 410.4$ nm; visible light has sufficient energy for the $n = 2 \rightarrow n = 6$ transition but does not have sufficient energy for the $n = 1 \rightarrow n = 5$ transition.   **65.** $n = 1$, 91.20 nm; $n = 2$, 364.8 nm   **67.** $n = 2$ **69. a.** 5.79×10^{-4} m; **b.** 3.64×10^{-33} m; **c.** The diameter of an H atom is roughly 1.0×10^{-8} cm. The uncertainty in position is much larger than the size of the atom. **d.** The uncertainty is insignificant compared to the size of a baseball. **71.** $n = 1, 2, 3, \ldots$; $\ell = 0, 1, 2, \ldots (n - 1)$; $m_\ell = -\ell, \ldots, -2, -1, 0, 1, 2, \ldots, +\ell$.   **73. b.** For $\ell = 3$, m_ℓ can range from -3 to $+3$; thus $+4$ is not allowed. **c.** n cannot equal zero. **d.** ℓ cannot be a negative number.   **75.** ψ^2 gives the probability of finding the electron at that point.   **77.** He: $1s^2$; Ne: $1s^2 2s^2 2p^6$; Ar: $1s^2 2s^2 2p^6 3s^2 3p^6$; each peak in the diagram corresponds to a subshell with different values of n. Corresponding subshells are closer to the nucleus for heavier elements because of the increased nuclear charge. **79.** 3; 1; 5; 25; 16   **81. a.** 32; **b.** 8; **c.** 25; **d.** 10; **e.** 6

83. a.

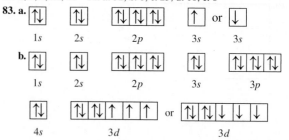

c.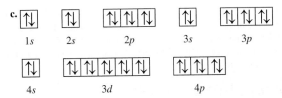

1s   2s   2p   3s   3p

4s   3d   4p

85. Si: $1s^22s^22p^63s^23p^2$ or [Ne]$3s^23p^2$; Ga: $1s^22s^22p^63s^23p^64s^23d^{10}4p^1$ or [Ar]$4s^23d^{10}4p^1$; As: [Ar]$4s^23d^{10}4p^3$; Ge: [Ar]$4s^23d^{10}4p^2$; Al: [Ne]$3s^23p^1$; Cd: [Kr]$5s^24d^{10}$; S: [Ne]$3s^23p^4$; Se: [Ar]$4s^23d^{10}4p^4$ **87.** Sc: $1s^22s^22p^63s^23p^6$ $4s^23d^1$; Fe: $1s^22s^22p^63s^23p^63d^6$; P:$1s^22s^22p^63s^23p^3$; Cs: $1s^22s^22p^63s^23p^6$ $4s^23d^{10}4p^65s^24d^{10}5p^66s^1$; Eu: $1s^22s^22p^63s^23p^64s^23d^{10}4p^65s^24d^{10}5p^66s^24f^65d^1$ (Actual: [Xe]$6s^24f^7$); Pt: $1s^22s^22p^63s^23p^64s^23d^{10}4p^65s^24d^{10}5p^66s^24f^{14}5d^8$ (Actual: [Xe]$6s^14f^{14}5d^9$); Xe: $1s^22s^22p^63s^23p^64s^23d^{10}4p^65s^24d^{10}5p^6$; Br: $1s^22s^2$ $2p^63s^23p^64s^23d^{10}4p^5$ **89. a.** I: [Kr]$5s^24d^{10}5p^5$; **b.** element 120: [Rn]$7s^25f^{14}$ $6d^{10}7p^68s^2$; **c.** Rn: [Xe]$6s^24f^{14}5d^{10}6p^6$; **d.** Cr: [Ar]$4s^13d^5$ **91. a.** S; **b.** N; **c.** Se$^-$ **93. a.** 18; **b.** 30; **c.** 8; **d.** 40

95. B: $1s^22s^22p^1$ N: $1s^22s^22p^3$

	n	ℓ	m_ℓ	m_s		n	ℓ	m_ℓ	m_s
1s	1	0	0	$+\frac{1}{2}$	1s	1	0	0	$+\frac{1}{2}$
1s	1	0	0	$-\frac{1}{2}$	1s	1	0	0	$-\frac{1}{2}$
2s	2	0	0	$+\frac{1}{2}$	2s	2	0	0	$+\frac{1}{2}$
2s	2	0	0	$-\frac{1}{2}$	2s	2	0	0	$-\frac{1}{2}$
2p	2	1	-1	$+\frac{1}{2}$	2p	2	1	-1	$+\frac{1}{2}$
					2p	2	1	0	$+\frac{1}{2}$
					2p	2	1	$+1$	$+\frac{1}{2}$

For boron, there are six possibilities for the $2p$ electron. For nitrogen, all the $2p$ electrons could have $m_s = -\frac{1}{2}$. **97.** 1A: 1, ns^1, Li, $2s^1$; 2A: 2, ns^2, Ra, $7s^2$; 3A: 3, ns^2np^1, Ga, $4s^24p^1$; 4A: 4, ns^2np^2, Si, $3s^23p^2$; 5A: 5, ns^2np^3, Sb, $5s^25p^3$; 6A: 6, ns^2np^4, Po, $6s^26p^4$; 7A: 7, ns^2np^5, 117, $7s^27p^5$; 8A: 8, ns^2np^6, Ne, $2s^22p^6$ **99.** none; an excited state; energy released **101.** C, O, Si, S, Ti, Ni, Ge, Se **103.** Li (1 unpaired electron), N (3 unpaired electrons), Ni (2 unpaired electrons), and Te (2 unpaired electrons) are all expected to be paramagnetic because they have unpaired electrons. **105. a.** S < Se < Te; **b.** Br < Ni < K; **c.** F < Si < Ba **107. a.** Te < Se < S; **b.** K < Ni < Br; **c.** Ba < Si < F **109. a.** He; **b.** Cl; **c.** element 116; **d.** Si; **e.** Na$^+$ **111. a.** [Rn]$7s^25f^{14}6d^4$; **b.** W; **c.** SgO_3 and SgO_4^{2-} probably would form (similar to Cr). **113.** Se is an exception to the general ionization trend. There are extra electron–electron repulsions in Se because two electrons are in the same $4p$ orbital, resulting in a lower ionization energy than expected. **115. a.** As we remove succeeding electrons, the electron being removed is closer to the nucleus, and there are fewer electrons left repelling it. The remaining electrons are more strongly attracted to the nucleus, and it takes more energy to remove these electrons. **b.** Al: $1s^22s^22p^63s^23p^1$. For I_4, we begin removing an electron with $n = 2$. For I_3, we remove an electron with $n = 3$ (the last valence electron). In going from $n = 3$ to $n = 2$, there is a big jump in ionization energy because the $n = 2$ electrons are closer to the nucleus on average than the $n = 3$ electrons. Since the $n = 2$ electrons are closer, on average, to the nucleus, they are held more tightly and require a much larger amount of energy to remove compared to the $n = 3$ electrons. In general, valence electrons are much easier to remove than inner-core electrons. **117. a.** C, Br; **b.** N, Ar; **c.** C, Br **119.** Al (-44), Si (-120), P (-74), S (-200.4), Cl (-348.7); Based on the increasing nuclear charge, we would expect the electron affinity (EA) values to become more exothermic as we go from left to right in the period. Phosphorus is out of line. The reaction for the EA of P is

$$P(g) + e^- \rightarrow P^-(g)$$

$$[\text{Ne}]3s^23p^3 \qquad [\text{Ne}]3s^23p^4$$

The additional electron in P$^-$ will have to go into an orbital that already has one electron. There will be greater repulsions between the paired electrons in P$^-$, causing the EA of P to be less favorable than predicted based solely on attractions to the nucleus. **121. a.** Se < S; **b.** I < Br < F < Cl **123.** Electron–

electron repulsions are much greater in O$^-$ than in S$^-$ because the electron goes into a smaller $2p$ orbital versus the larger $3p$ orbital in sulfur. This results in a more favorable (more exothermic) EA for sulfur. **125. a.** Se$^{3+}(g)$ $\rightarrow$ Se$^{4+}(g)$ + e$^-$; **b.** S$^-(g)$ + e$^-$ $\rightarrow$ S$^{2-}(g)$; **c.** Fe$^{3+}(g)$ + e$^-$ $\rightarrow$ Fe$^{2+}(g)$; **d.** Mg(g) $\rightarrow$ Mg$^+(g)$ + e$^-$ **127.** potassium peroxide, K$_2$O$_2$; K^{2+} unstable **129.** 6.582×10^{14} s^{-1}; 4.361×10^{-19} J **131.** Yes; the ionization energy general trend decreases down a group and the atomic radius trend increases down a group. The data in Table 7.8 confirm both of these general trends. **133. a.** $6\text{Li}(s) + \text{N}_2(g) \rightarrow 2\text{Li}_3\text{N}(s)$; **b.** $2\text{Rb}(s) + \text{S}(s) \rightarrow \text{Rb}_2\text{S}(s)$ **135.** No; lithium metal is very reactive. It will react somewhat violently with water, making it completely unsuitable for human consumption. Lithium has a small first ionization energy, so it is more likely that the lithium prescribed will be in the form of a soluble lithium salt (a soluble ionic compound with Li$^+$ as the cation). **137.** 386 nm **139.** 200 s **141.** $\lambda = 4.104 \times 10^{-5}$ cm so violet light is emitted. **143. a.** true for H only; **b.** true for all atoms; **c.** true for all atoms **145.** $1p$: $n = 1$, $\ell = 1$ is not possible; $3f$: $n = 3$, $\ell = 3$ is not possible; $2d$: $n = 2$, $\ell = 2$ is not possible. In all three incorrect cases, $n = \ell$. The maximum value ℓ can have is $n - 1$, not n. **147.** smallest size: O; largest size: K; smallest 1st ionization energy: K; largest 1st ionization energy: N **149.** Valence electrons are easier to remove than inner-core electrons. The large difference in energy between I_2 and I_3 indicates that this element has two valence electrons. The element is most likely an alkaline earth metal because alkaline earth metals have two valence electrons. **151. a.** 146 kJ; **b.** 407 kJ; **c.** 1117 kJ; **d.** 1524 kJ **163.** S **165. a.** line A, $n = 6$ to $n = 3$; line B, $n = 5$ to $n = 3$; **b.** 121.6 nm **167.** X = C, $m = 5$; C^{5+} is the ion. **169.** For $r = a_0$ and $\theta = 0°$, $\psi^2 = 2.46 \times 10^{28}$. For $r = a_0$ and $\theta = 90°$, $\psi^2 = 0$. As expected, the xy plane is a node for the $2p_z$ atomic orbital.

171. a.

1						2
3						4

5	6	7	8	9	10	11	12
13	14	15	16	17	18	19	20

b. 2, 4, 12, and 20; **c.** There are many possibilities. One example of each formula is XY = 1 + 11, XY$_2$ = 6 + 11, X$_2$Y = 1 + 10, XY$_3$ = 7 + 11, and X$_2$Y$_3$ = 7 + 10; **d.** 6; **e.** 0; **f.** 18 **173.** The ratios for Mg, Si, P, Cl, and Ar are about the same. However, the ratios for Na, Al, and S are higher. For Na, the second IE is extremely high because the electron is taken from $n = 2$ (the first electron is taken from $n = 3$). For Al, the first electron requires a bit less energy than expected by the trend due to the fact it is a $3p$ electron. For S, the first electron requires a bit less energy than expected by the trend due to electrons being paired in one of the p orbitals. **175. a.** $Z_{\text{eff}} = 43.33$; **b.** Silver is element 47, so Z = 47 for silver. Our calculated Z_{eff} value is less than 47. Electrons in other orbitals can penetrate the $1s$ orbital. Thus a $1s$ electron can be slightly shielded from the nucleus, giving a Z_{eff} close to but less than Z. **177.** Solving for the molar mass of the element gives 40.2 g/mol; this is calcium. **179. a.** Fr = [Rn]$7s^1$, Fr$^+$ = [Rn] = [Xe]$6s^24f^{14}5d^{10}6p^6$; **b.** 7.7×10^{22} Fr atoms; **c.** 2.27790×10^{-22} g

Chapter 8

15. a. This diagram represents a polar covalent bond as in HCl. In a polar covalent bond, there is an electron-rich region (indicated by the red color) and an electron-poor region (indicated by the blue color). In HCl, the more electronegative Cl atom has a slightly greater ability to attract the bonding electrons, which in turn produces a dipole moment. **b.** This diagram represents an ionic bond as in NaCl. Here, the electronegativity differences between the Na and Cl are so great that the valence electron of sodium is transferred to the chlorine atom. This results in the formation of a cation, an anion, and an ionic bond. **c.** This diagram represents a pure covalent bond as in H$_2$. Both atoms attract the bonding electrons equally, so there is no bond dipole formed. This is illustrated in the electrostatic potential diagram as there is no one region that is red nor one region that is blue (there is no specific partial negative end and no specific partial positive end). **17.** (NH$_4$)$_2$SO$_4$ and Ca$_3$(PO$_4$)$_2$ are compounds with both ionic and covalent bonds. **19.** Electronegativity increases left to right across the periodic table and decreases from top to bottom. Hydrogen has an electronegativity value between B and C in the second row, and identical to P in the third row. Going further down the periodic table, H has an electronegativity value between As and Se (row 4) and identical to Te (row 5). It is important to know where hydrogen fits into the electronegativity trend, especially for rows 2 and 3. If you know where H fits into the trend, then you can predict bond dipole directions for nonmetals bonded to hydrogen. **21.** For ions, concentrate on the number of protons and the number of electrons present. The species whose nucleus holds the electrons most tightly will be smallest. For example, compare the size of an anion to the neutral atom. The anion has more electrons held by the same number of protons in the nucleus. These electrons will not be held as tightly, resulting in a bigger size for the anion as compared to the neutral atom. For isoelectronic ions, the same number of electrons are held by different numbers of protons in the various ions. The ion with the most protons holds the electrons tightest and has smallest size. **23.** Fossil fuels contain a lot of carbon and hydrogen atoms. Combustion of fossil fuels (reaction with O_2) produces CO_2 and H_2O. Both these compounds have very strong bonds. Because much stronger product bonds are formed than reactant bonds broken, combustion reactions are very exothermic.

25. $\quad 0 \quad 0 \quad 0 \qquad -1 \quad 0 \; +1 \qquad\quad +1 \quad 0 \; -1$

$$\ddot{O}=C=\ddot{O} \longleftrightarrow \; :\ddot{O}-C\equiv O: \longleftrightarrow \; :O\equiv C-\ddot{O}:$$

The formal charges are shown above the atoms in the three Lewis structures. The best Lewis structure for CO_2, from a formal charge standpoint, is the first structure with each oxygen double bonded to carbon. This structure has a formal charge of zero on all atoms (which is preferred). The other two resonance structures have nonzero formal charges on the oxygens, making them less reasonable. For CO_2, we usually ignore the last two resonance structures and think of the first structure as the true Lewis structure for CO_2. **27. a.** C < N < O; **b.** Se < S < Cl; **c.** Sn < Ge < Si; **d.** Tl < Ge < S **29. a.** Ge—F; **b.** P—Cl; **c.** S—F; **d.** Ti—Cl **31.** Order of electronegativity from Fig. 8.3: **a.** C (2.5) < N (3.0) < O (3.5), same; **b.** Se (2.4) < S (2.5) < Cl (3.0), same; **c.** Si = Ge = Sn (1.8), different; **d.** Tl (1.8) = Ge (1.8) < S (2.5), different. Most polar bonds using actual electronegativity values: **a.** Si—F and Ge—F equal polarity (Ge—F predicted); **b.** P—Cl (same as predicted); **c.** S—F (same as predicted); **d.** Ti—Cl (same as predicted) **33.** Incorrect: b, d, e; **b.** $^{\delta-}$Cl—I$^{\delta+}$; **d.** nonpolar bond so no dipole moment; **e.** $^{\delta-}$O—P$^{\delta+}$ **35. a.** ionic; **b.** covalent; **c.** polar covalent; **d.** ionic; **e.** polar covalent; **f.** covalent **37.** F—H > O—H > N—H > C—H > P—H **39. a.** Has a permanent dipole; **b.** has no permanent dipole; **c.** has no permanent dipole; **d.** has a permanent dipole; **e.** has no permanent dipole; **f.** has no permanent dipole. **41.** Al^{3+}: $[He]2s^22p^6$; Ba^{2+}: $[Kr]5s^24d^{10}5p^6$; Se^{2-}: $[Ar]4s^23d^{10}4p^6$; I^-: $[Kr]5s^24d^{10}5p^6$ **43. a.** Li^+ and N^{3-} are the expected ions. The formula of the compound would be Li_3N (lithium nitride). **b.** Ga^{3+} and O^{2-}; Ga_2O_3, gallium(III) oxide or gallium oxide; **c.** Rb^+ and Cl^-; RbCl, rubidium chloride; **d.** Ba^{2+} and S^{2-}; BaS, barium sulfide **45. a.** Mg^{2+} and Al^{3+}: $1s^22s^22p^6$; K^+: $1s^22s^22p^63s^23p^6$; **b.** N^{3-}, O^{2-}, and F^-: $1s^22s^22p^6$; Te^{2-}: $[Kr]5s^24d^{10}5p^6$ **47. a.** Sc^{3+}; **b.** Te^{2-}; **c.** Ce^{4+}, Ti^{4+}; **d.** Ba^{2+} **49. a.** F^-, O^{2-}, or N^{3-}; **b.** Cl^-, S^{2-}, or P^{3-}; **c.** F^-, O^{2-}, or N^{3-}; **d.** Br^-, Se^{2-}, or As^{3-} **51.** Mg^{2+}, Na^+, F^-, O^{2-}, and N^{3-} (plus others) are all isoelectronic with neon. $Mg^{2+} < Na^+ < F^- < O^{2-} < N^{3-}$ **53. a.** Cu > Cu^+ > Cu^{2+}; **b.** Pt^{2+} > Pd^{2+} > Ni^{2+}; **c.** O^{2-} > O^- > O; **d.** La^{3+} > Eu^{3+} > Gd^{3+} > Yb^{3+}; **e.** Te^{2-} > I^- > Cs^+ > Ba^{2+} > La^{3+} **55. a.** NaCl, Na^+ smaller than K^+; **b.** LiF, F^- smaller than Cl^-; **c.** MgO, O^{2-} greater charge than OH^-; **d.** $Fe(OH)_3$, Fe^{3+} greater charge than Fe^{2+}; **e.** Na_2O, O^{2-} greater charge than Cl^-; **f.** MgO, Mg^{2+} smaller than Ba^{2+}, and O^{2-} smaller than S^{2-}. **57.** -411 kJ/mol **59.** The lattice energy for $Mg^{2+}O^{2-}$ will be much more exothermic than for Mg^+O^-. **61.** 161 kJ/mol **63.** Ca^{2+} has greater charge than Na^+, and Se^{2-} is smaller than Te^{2-}. Charge differences affect lattice energy values more than size differences, and we expect the trend from most exothermic to least exothermic to be:

CaSe	>	CaTe	>	Na_2Se	>	Na_2Te
(-2862)		(-2721)		(-2130)		(-2095)

65. a. -183 kJ; **b.** -109 kJ **67.** -42 kJ **69.** -1282 kJ **71.** -1228 kJ **73.** 485 kJ/mol **75. a.** Using standard enthalpies of formation, $\Delta H° = -184$ kJ vs. -183 kJ from bond energies; **b.** Using standard enthalpies of formation, $\Delta H = -92$ kJ vs. -109 kJ from bond energies. Bond energies give a reasonably good estimate for ΔH, especially when all reactants and products are gases. **77. a.** Using SF_4 data: $D_{SF} = 342.5$ kJ/mol. Using SF_6

data: $D_{SF} = 327.0$ kJ/mol. **b.** The S—F bond energy in the table is 327 kJ/mol. The value in the table was based on the S—F bond in SF_6. **c.** $S(g)$ and $F(g)$ are not the most stable forms of the elements at 25°C. The most stable forms are $S_8(s)$ and $F_2(g)$; $\Delta H_f° = 0$ for these two species. **79.** 158.4 kJ/mol vs. 160. kJ/mol in Table 8.4

81. a. $:\ddot{F}-\ddot{F}:$ **b.** $\ddot{O}=\ddot{O}$ **c.** $:C\equiv O:$

d. $H-\underset{\underset{H}{|}}{\overset{\overset{H}{|}}{C}}-H$ **e.** $H-\underset{\underset{H}{|}}{\overset{}{N}}-H$ **f.** $H-\ddot{O}-H$ **g.** $H-\ddot{F}:$

83. a. $:\ddot{Cl}-\underset{\underset{:\ddot{Cl}:}{|}}{\overset{\overset{:\ddot{Cl}:}{|}}{C}}-\ddot{Cl}:$ **b.** $:\ddot{Cl}-\underset{\underset{:\ddot{Cl}:}{|}}{N}-\ddot{Cl}:$

c. $:\ddot{Cl}-\ddot{Se}-\ddot{Cl}:$ **d.** $:\ddot{I}-\ddot{Cl}:$

85. H—Be—H $\quad H-\underset{\underset{H\quad H}{}}{B}$

87. PF_5 (Lewis structure of PF_5 with 5 F around P)

SF_4 (Lewis structure of SF_4)

ClF_3 (Lewis structure of ClF_3)

Br_3^- $\left[:\ddot{Br}-\ddot{Br}-\ddot{Br}:\right]^-$

Row 3 and heavier nonmetals can have more than 8 electrons around them when they have to. Row 3 and heavier elements have empty d orbitals that are close in energy to valence s and p orbitals. These empty d orbitals can accept extra electrons.

89. a. NO_2^- $\left[\ddot{O}=N-\ddot{O}:\right]^- \longleftrightarrow \left[:\ddot{O}-N=\ddot{O}\right]^-$

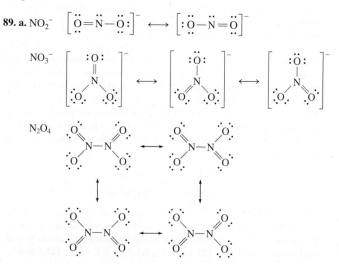

b.

OCN^- $\left[:\ddot{O}-C\equiv N:\right]^- \longleftrightarrow \left[\ddot{O}=C=\ddot{N}:\right]^- \longleftrightarrow \left[:O\equiv C-\ddot{N}:\right]^-$

SCN^- $\left[:\ddot{S}-C\equiv N:\right]^- \longleftrightarrow \left[\ddot{S}=C=\ddot{N}:\right]^- \longleftrightarrow \left[:S\equiv C-\ddot{N}:\right]^-$

N_3^- $\left[:\ddot{N}-N\equiv N:\right]^- \longleftrightarrow \left[\ddot{N}=N=\ddot{N}:\right]^- \longleftrightarrow \left[:N\equiv N-\ddot{N}:\right]^-$

91.

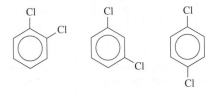

93. With resonance all carbon–carbon bonds are equivalent (we indicate this with a circle in the ring), giving three different structures:

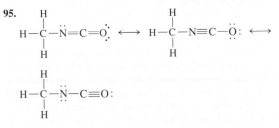

Localized double bonds would give four unique structures.

95.

H—C(H)(H)—$\ddot{N}$=C=$\ddot{O}$: $\longleftrightarrow$ H—C(H)(H)—$\ddot{N}$=C=$\ddot{O}$: (second with triple bond) $\longleftrightarrow$

H—C(H)(H)—$\ddot{N}$—C≡O:

97. Longest → shortest C—O bond: $CH_3OH > CO_3^{2-} > CO_2 > CO$; weakest → strongest C—O bond: $CH_3OH < CO_3^{2-} < CO_2 < CO$

99.

$\overset{+1}{F}$ double bonded to $\overset{-1}{B}$ with two $\overset{0}{:F:}$ single bonded

$\overset{0}{:\ddot{F}:}$ bonded to $\overset{0}{B}$ with two $\overset{0}{:\ddot{F}:}$

The first Lewis structure obeys the octet rule but has a $+1$ formal charge on the most electronegative element there is, fluorine, and a negative formal charge on a much less electronegative element, boron. This is just the opposite of what we expect: negative formal charge on F and positive formal charge on B. The other Lewis structure does not obey the octet rule for B but has a zero formal charge on each element in BF_3. Since structures generally want to minimize formal charge, BF_3 with only single bonds is best from a formal charge point of view.   **101.** a–f and h all have similar Lewis structures:

:Y:
|
:Y—X—Y:
|
:Y:

g. ClO_3^- $\left[:\ddot{O}-\overset{}{Cl}-\ddot{O}:\right]^-$ with $:\ddot{O}:$ below Cl

Formal charges: **a.** $+1$; **b.** $+2$; **c.** $+3$; **d.** $+1$; **e.** $+2$; **f.** $+4$; **g.** $+2$; **h.** $+1$

103. $:\ddot{F}-\ddot{O}-\ddot{O}-\ddot{F}:$

Formal charge: 0   0   0   0
Oxidation number: -1 $+1$ $+1$ -1

Oxidation numbers are more useful. We are forced to assign $+1$ as the oxidation number for oxygen. Oxygen is very electronegative and $+1$ is not a stable oxidation state for this element.

105. $:\ddot{Cl}-\ddot{S}-\ddot{S}-\ddot{Cl}:$

107.

H   C≡N: H
| | |
H—C=C—C—$\ddot{O}$—C—H
              ‖ |
            :O: H

109. [83] **a.** tetrahedral, $109.5°$; **b.** trigonal pyramid, $<109.5°$; **c.** V-shaped, $<109.5°$; **d.** linear, no bond angle in diatomic molecules. [89] **a.** NO_2^-: V-shaped, $\sim120°$; NO_3^-: trigonal planar, $120°$; N_2O_4: trigonal planar about both N atoms, $120°$; **b.** all are linear, $180°$ **111.** Br_3^-: linear; ClF_3: T-shaped; SF_4: see-saw   **113. a.** trigonal planar; $120°$; **b.** V-shaped; $\sim120°$ **115. a.** linear; $180°$; **b.** T-shaped; $\sim90°$; **c.** see-saw; $\sim90°$ and $\sim120°$; **d.** trigonal bipyramid; $90°$ and $120°$ **117.** SeO_2 (bond dipoles do not cancel each other out in SeO_2)   **119.** ICl_3 and TeF_4 (bond dipoles do not cancel each other out in ICl_3 and TeF_4)

121. a. OCl_2 [structure with O bonded to two Cl]   V-shaped, polar

KrF_2 $[:\ddot{F}-\ddot{Kr}-\ddot{F}:]$ Linear, nonpolar

BeH_2 H—Be—H   Linear, nonpolar

SO_2 [S double/single bonded to two O]   V-shaped, polar

(one other resonance structure possible)

b. SO_3 [S with one double bond O on top and two O's]   Trigonal planar, nonpolar

(two other resonance structures possible)

NF_3 [N bonded to three F with lone pair]   Trigonal pyramid, polar

IF_3 [I bonded to three F]   T-shaped, polar

c. CF_4 [C bonded to four F]   Tetrahedral, nonpolar

SeF_4 [Se bonded to four F]   See-saw, polar

KrF_4 [Kr bonded to four F]   Square planar, nonpolar

d. IF_5 [I bonded to five F]   Square pyramid, polar

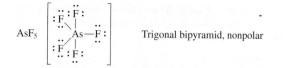

AsF₅   Trigonal bipyramid, nonpolar

123. Element E is a halogen (F, Cl, Br, or I); trigonal pyramid; <109.5° **125.** The polar bonds are symmetrically arranged about the central atoms, and all the individual bond dipoles cancel to give no net dipole moment for each molecule, i.e., the molecules are nonpolar. **127. a.** radius: $N^+ < N <$ N^-; I.E.: $N^- < N < N^+$; **b.** radius: $Cl^+ < Cl < Se < Se^-$; I.E.: $Se^- <$ $Se < Cl < Cl^+$; **c.** radius: $Sr^{2+} < Rb^+ < Br^-$; I.E.: $Br^- < Rb^+ < Sr^{2+}$; **129. a.** 1549 kJ; **b.** 1390. kJ **c.** 1312 kJ; **d.** 1599 kJ **131. a.** NaBr: In NaBr₂, the sodium ion would have a +2 charge assuming each bromine has a −1 charge. Sodium doesn't form stable Na^{2+} compounds. **b.** ClO_4^-: ClO_4 has 31 valence electrons so it is impossible to satisfy the octet rule for all atoms in ClO_4. The extra electron from the −1 charge in ClO_4^- allows for complete octets for all atoms. **c.** XeO_4: We can't draw a Lewis structure that obeys the octet rule for SO_4 (30 electrons), unlike XeO_4 (32 electrons). **d.** SeF_4: Both compounds require the central atom to expand its octet. O is too small and doesn't have low-energy d orbitals to expand its octet (which is true for all row 2 elements). **133. a.** Both have one or more 180° bond angles; both are made up entirely of Xe and Cl; both have the individual bond dipoles arranged so they cancel each other (both are nonpolar); both have lone pairs on the central Xe atom; both have a central Xe atom that has more than 8 electrons around it. **b.** All have lone pairs on the central atom; all have a net dipole moment (all are polar). **135.** Yes, each structure has the same number of effective pairs around the central atom. (We count a multiple bond as a single group of electrons.)

137.

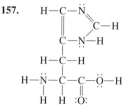

The lone pair of electrons around Te exerts a stronger repulsion than the bonding pairs. The stronger repulsion pushes the four square planar F atoms away from the lone pair, reducing the bond angles between the axial F atom and the square planar F atoms. **147. a.** -5×10^{-18} J; **b.** -6×10^{-18} J **149.** 17 kJ/mol **151.** See Fig. 8.11 to see the data supporting MgO as an ionic compound. Note that the lattice energy is large enough to overcome all of the other processes (removing 2 electrons from Mg, and so on). The bond energy for O_2 (247 kJ/mol) and electron affinity (737 kJ/mol) are the same when making CO. However, the energy needed to ionize carbon to form a C^{2+} ion must be too large. Figure 7.32 shows that the first ionization energy for carbon is about 350 kJ/mol greater than the first IE for magnesium. If all other numbers were equal, the overall energy change would be 250 kJ/mol (see Fig. 8.11). It is not unreasonable to assume that the second ionization energy for carbon is more than 250 kJ/mol greater than the second ionization energy for magnesium. This would result in a positive ΔH value for the formation of CO as an ionic compound. One wouldn't expect CO to be ionic if the energetics were unfavorable. **153.** As the halogen atoms get larger, it becomes more difficult to fit three halogen atoms around the small nitrogen atom, and the NX_3 molecule becomes less stable. **155.** reaction i: −2636 kJ; reaction ii: −3471 kJ; reaction iii: −3543 kJ; Reaction iii yields the most energy per kg (−8085 kJ/kg)

157.

H—C—N
        ‖   \
        | C—H
        C—N—H
        |
    H—C—H
        |
H—N—C—C—O—H
    | | ‖
    H   H :O:

We would expect 120° bond angles about the carbon atom labeled 1 and ~109.5° bond angles about the nitrogen atom labeled 2. The nitrogen bond angles should be slightly smaller than 109.5° due to the lone pair of electrons on nitrogen.

159. a. Two possible structures exist; each has a T-shaped molecular structure:

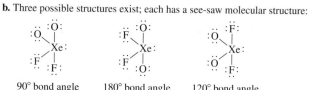

90° bond angle between I atoms   180° bond angle between I atoms

b. Three possible structures exist; each has a see-saw molecular structure:

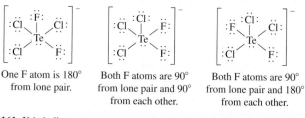

90° bond angle between O atoms   180° bond angle between O atoms   120° bond angle between O atoms

c. Three possible structures exist; each has a square pyramid molecular structure:

One F atom is 180° from lone pair.   Both F atoms are 90° from lone pair and 90° from each other.   Both F atoms are 90° from lone pair and 180° from each other.

161. X is iodine; square pyramid **163.** $[Ar]4s^13d^5 = Cr$; $[Ne]3s^23p^3 = P$; $[Ar]4s^23d^{10}4p^3 = As$; $[Ne]3s^23p^5 = Cl$; Cr < As < P < Cl

Chapter 9

9. In hybrid orbital theory, some or all of the valence atomic orbitals of the central atom in a molecule are mixed together to form hybrid orbitals; these hybrid orbitals point to where the bonded atoms and lone pairs are oriented. The σ bonds are formed from the hybrid orbitals overlapping head to head with an appropriate orbital on the bonded atom. The π bonds in hybrid orbital theory are formed from unhybridized p atomic orbitals. The p orbitals overlap side to side to form the π bond where the π electrons occupy the space above and below a line joining the atoms π the internuclear axis). Assuming the z-axis is the internuclear axis, then the p_z atomic orbital will always be hybridized whether the hybridization is sp, sp^2, sp^3, dsp^3 or d^2sp^3. For sp hybridization, the p_x and p_y atomic orbitals are unhybridized; they are used to form two π bonds to the bonded atom(s). For sp^2 hybridization, either the p_x or p_y atomic orbital is hybridized (along with the s and p_z orbitals); the other p orbital is used to form a π bond to a bonded atom. For sp^3 hybridization, the s and all of the p orbitals are hybridized; no unhybridized p atomic orbitals are present, so typical π bonds do not form with sp^3 hybridization. For dsp^3 and d^2sp^3 hybridization, we just mix in one or two d orbitals into the hybridization process. Which specific d orbitals are used is not important to our discussion. **11.** We use d orbitals when we have to; i.e., we use d orbitals when the central atom on a molecule has more than eight electrons around it. The d orbitals are necessary to accommodate the electrons over eight. Row 2 elements never have more than eight electrons around them so they never hybridize d orbitals. We rationalize this by saying there are no d orbitals close in energy to the valence 2s and 2p orbitals (2d orbitals are forbidden energy levels). However, for row 3 and heavier elements, there are 3d, 4d, 5d, etc. orbitals that will be close in energy to the valence s and p orbitals. It is row 3 and heavier nonmetals that hybridize d orbitals when they have to. For sulfur, the valence electrons are in 3s and 3p orbitals. Therefore, 3d orbitals are closest in energy and are available for hybridization. Arsenic would hybridize 4d orbitals to go with the valence 4s and 4p orbitals while iodine would hybridize 5d orbitals since the valence electrons are in $n = 5$. **13.** Bonding and antibonding molecular orbitals are both solutions to the quantum mechanical treatment of the molecule. Bonding orbitals form when in-phase orbitals combine to give constructive interference. This results in enhanced electron probability located between the two nuclei. The end result is that a bonding MO is lower in energy than the atomic orbitals of which it is composed. Antibonding orbitals form when out-of-phase orbitals combine. The mismatched phases produce destructive interference leading to a node in the electron probability between the two nuclei. With electron distribution pushed to the outside, the energy of an antibonding orbital is higher than the energy of the atomic orbitals of which it is composed. **15.** The localized electron model does not deal effectively with molecules containing unpaired electrons. We can draw all of the possible resonance structures for NO, but still not have a good feel for whether the bond in NO is weaker or stronger than the bond in NO⁻. MO theory can handle odd electron species without any modifications. In addition, hybrid orbital theory does not predict that NO⁻ is paramagnetic. The MO theory correctly makes this prediction.

17. H₂O: :O:
             H   H

H₂O has a tetrahedral arrangement of the electron pairs about the O atom that requires sp^3 hybridization. Two of the sp^3 hybrid orbitals are used to form bonds to the two hydrogen atoms, and the other two sp^3 hybrid orbitals hold the two lone pairs on oxygen.

19. H₂CO: :O:
                 ‖
                 C
               H   H

The central carbon atom has a trigonal planar arrangement of the electron pairs that requires sp^2 hybridization. Two of the sp^2 hybrid orbitals are used to form the two bonds to hydrogen. The other sp^2 hybrid orbital forms the σ bond to oxygen. The unchanged (unhybridized) p orbital on carbon is used to form the π bond between carbon and oxygen.

21. Ethane: H   H
                  | |
              H—C—C—H
                  | |
                  H   H

The carbon atoms are sp^3 hybridized. The six C—H bonds are formed from the sp^3 hybrid orbitals on C with the $1s$ atomic orbitals from the hydrogen atoms. The carbon–carbon bond is formed from an sp^3 hybrid orbital on each C atom.

Ethanol: H   H
           | | ..
       H—C—C—O—H
           | |
           H   H

The two C atoms and the O atom are all sp^3 hybridized. All bonds are formed from these sp^3 hybrid orbitals. The C—H and O—H bonds form from sp^3 hybrid orbitals and the $1s$ atomic orbitals from the hydrogen atom. The C—C and C—O bonds are formed from sp^3 hybrid orbitals on each atom. **23.** [83] All are sp^3 hybridized. [89] **a.** NO₂⁻, sp^2; NO₃⁻, sp^2; N₂O₄, both N atoms are sp^2 hybridized; **b.** All are sp hybridized. **25.** All exhibit dsp^3 hybridization. **27.** The molecules in Exercise 113 all exhibit sp^2 hybridization about the central atom; the molecules in Exercise 114 all exhibit sp^3 hybridization about the central atom. **29. a.** tetrahedral, 109.5°, sp^3, non-polar

                    :F:
                     |
              :F—C—F:
                    |
                   :F:

b. trigonal pyramid, <109.5°, sp^3, polar

                :F—N:
                :F: F:

c. V-shaped, <109.5°, sp^3, polar

                  :O:
                :F   F:

d. trigonal planar, 120°, sp^2, nonpolar

                   :F:
                    |
                    B
                 :F   F:

e. linear, 180°, sp, nonpolar

              H—Be—H

f.    a ≈120°, see-saw, dsp^3, polar
                               b ≈90°

g. [AsF₅ structure]   a = 90°, trigonal bipyramid, dsp^3, nonpolar
                           b = 120°

h. :F—Kr—F: linear, 180°, dsp^3, nonpolar

i.   :F: square planar, 90°, d^2sp^3, nonpolar
          :F—Kr—F:
               :F:

j.   :F: :F: octahedral, 90°, d^2sp^3, nonpolar
          :F—Se—F:
            :F: :F:

k.   :F: :F: **l.** :F:
          F—I—F |
        :F: :F: :I—F:
                                    |
                                   :F:

square pyramid, ≈90°,   T-shaped, ≈90°,
d^2sp^3, polar dsp^3, polar

31. The π bond forces all six atoms into the same plane. **33. a.** There are 33 σ and 9 π bonds. **b.** All C atoms are sp^2 hybridized since all have a trigonal planar arrangement of the electrons. **35.** Biacetyl:

All CCO angles are 120°. The six atoms are not forced to lie in the same plane. 11 σ and 2 π bonds; Acetoin:

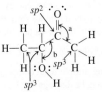

angle a = 120°, angle b = 109.5°, 13 σ bonds and 1 π bond

37. a.

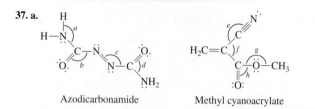

Azodicarbonamide   Methyl cyanoacrylate

Note: NH$_2$, CH$_2$, and CH$_3$ are shorthand for carbon atoms singly bonded to hydrogen atoms. **b.** In azodicarbonamide, the two carbon atoms are sp^2 hybridized. The two nitrogen atoms with hydrogens attached are sp^3 hybridized and the other two nitrogens are sp^2 hybridized. In methyl cyanoacrylate, the CH$_3$ carbon is sp^3 hybridized, the carbon with the triple bond is sp hybridized, and the other three carbons are sp^2 hybridized. **c.** Azodicarbonamide contains 3 π bonds and methyl cyanoacrylate contains 4 π bonds. **d.** a: $\approx$109.5°, b: 120°, c: $\approx$120°, d: 120°, e: 180°, f: 120°, g: $\approx$109.5°, h: 120°   **39.** To complete the Lewis structure, add lone pairs to complete octets for each atom. **a.** 6; **b.** 4; **c.** The center N in —N=N=N group; **d.** 33 σ; **e.** 5 π; **f.** 180°; **g.** $\approx$109.5°; **h.** sp **41. a.** The bonding molecular orbital is on the right and the antibonding molecular orbital is on the left. The bonding MO has the greatest electron probability between the nuclei, while the antibonding MO has greatest electron probability on either side of the nuclei. **b.** The bonding MO is lower in energy. Because the electrons in the bonding MO have the greatest probability of lying between the two nuclei, these electrons are attracted to two different nuclei, resulting in a lower energy.   **43. a.** H$_2^+$, H$_2$, H$_2^-$; **b.** He$_2^{2+}$ and He$_2^+$ **45. a.** $(\sigma_{2s})^2$; B.O. = 1; diamagnetic (0 unpaired electrons); **b.** $(\sigma_{2s})^2(\sigma_{2s}^*)^2(\pi_{2p})^4$; B.O. = 2; diamagnetic (0 unpaired electrons); **c.** $(\sigma_{3s})^2(\sigma_{3s}^*)^2(\sigma_{3p})^2(\pi_{3p})^4(\pi_{3p}^*)^2$; B.O. = 2; paramagnetic (2 unpaired electrons)   **47.** When O$_2$ loses an electron, it comes from an antibonding orbital, which strengthens the bond from a bond order of 2 to a bond order of 2.5. When N$_2$ loses an electron, it comes from a bonding orbital, which changes the bond order from 3 to 2.5 (the bond weakens).   **49.** CO: $(\sigma_{2s})^2(\sigma_{2s}^*)^2(\pi_{2p})^4(\sigma_{2p})^2$ B.O. = (8 − 2)/2 = 3; 0 unpaired electrons; O$_2$: $(\sigma_{2s})2(\sigma_{2s}^*)^2(\sigma_{2p})^2(\pi_{2p})^4(\pi_{2p}^*)^2$ B.O. = (8 − 4)/2 = 2; 2 unpaired electrons; The most obvious differences are that CO has a larger bond order than O$_2$ (3 vs. 2) and that CO is diamagnetic, whereas O$_2$ is paramagnetic.   **51. a.** $(\sigma_{2s})^2(\sigma_{2s}^*)^2(\pi_{2p})^4(\sigma_{2p})^2$; B.O. = 3; diamagnetic; **b.** $(\sigma_{2s})^2(\sigma_{2s}^*)^2(\pi_{2p})^4(\sigma_{2p})^1$; B.O. = 2.5; paramagnetic; **c.** $(\sigma_{2s})^2(\sigma_{2s}^*)^2(\pi_{2p})^4$; B.O. = 2; diamagnetic; bond length: CO < CO$^+$ < CO^{2+}; bond energy: CO^{2+} < CO$^+$ < CO   **53.** H$_2$; B$_2$; C$_2^{2-}$ **55.**

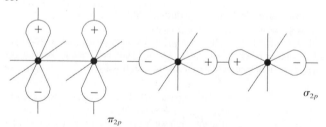

57. a. The electrons would be closer to F on the average. The F atom is more electronegative than the H atom, and the $2p$ orbital of F is lower in energy than the $1s$ orbital of H; **b.** The bonding MO would have more fluorine $2p$ character because it is closer in energy to the fluorine $2p$ orbital; **c.** The antibonding MO would place more electron density closer to H and would have a greater contribution from the higher-energy hydrogen $1s$ atomic orbital.   **59.** [:C≡C:]$^{2-}$ sp hybrid orbitals form the σ bond and the two unhybridized p atomic orbitals from each carbon form the two π bonds.

MO: $(\sigma_{2s})^2(\sigma_{2s}^*)^2(\pi_{2p})^4(\sigma_{2p})^2$, B.O. = (8 − 2)/2 = 3

Both give the same picture, a triple bond composed of a σ bond and two π bonds. Both predict the ion to be diamagnetic. Lewis structures deal well with diamagnetic (all electrons paired) species. The Lewis model cannot really predict magnetic properties.   **61.** O$_3$ and NO$_2^-$ have identical Lewis structures, so we need to discuss only one of them. The Lewis structures for O$_3$ are

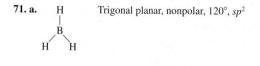

Localized electron model: The central oxygen atom is sp^2 hybridized, which is used to form the two σ bonds and hold the lone pair of electrons. An unchanged (unhybridized) p atomic orbital forms the π bond with the neighboring oxygen atoms. The actual structure of O$_3$ is an average of the two resonance structures. Molecular orbital model: There are two localized σ bonds and a π bond that is delocalized over the entire surface of the molecule. The delocalized π bond results from overlap of a p atomic oribtal on each oxygen atom in O$_3$.
63. a. Trigonal pyramid; sp^3

b. Tetrahedral; sp^3

c. Square pyramid; d^2sp^3

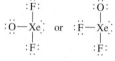

d. T-shaped; dsp^3

e. Trigonal bipyramid; dsp^3

65. a. No, some atoms are attached differently; **b.** Structure 1: All N atoms are sp^3 hybridized, all C atoms are sp^2 hybridized; structure 2: All C and N atoms are sp^2 hybridized; **c.** The first structure with the carbon–oxygen double bonds is slightly more stable.
67.

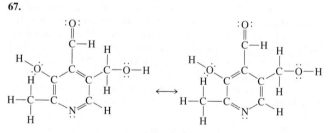

a. 21 σ bonds; 4 π bonds (The electrons in the 3 π bonds in the ring are delocalized.) **b.** angles a, c, and g: $\approx$109.5°; angles b, d, e, and f: $\approx$120°; **c.** 6 sp^2 carbons; **d.** 4 sp^3 atoms; **e.** Yes, the π electrons in the ring are delocalized. The atoms in the ring are all sp^2 hybridized. This leaves a p orbital perpendicular to the plane of the ring from each atom. Overlap of all six of these p orbitals results in a π molecular orbital system where the electrons are delocalized above and below the plane of the ring (similar to benzene in Fig. 9.47 of the text).   **69.** 267 kJ/mol; this amount of energy must be supplied to break the π bond.

71. a.

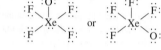

Trigonal planar, nonpolar, 120°, sp^2

b. N_2F_2

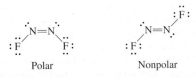

Polar   Nonpolar

V-shaped about both N atoms, ≈120°, sp^2

These are distinctly different molecules.

c.

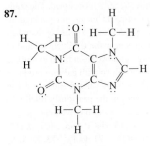

Trigonal planar about all C atoms, nonpolar, 120°, sp^2

73. a. The NNO structure is correct. From the Lewis structures we would predict both NNO and NON to be linear, but NON would be nonpolar. NNO is polar.

b. $\ddot{\text{N}}=\text{N}=\ddot{\text{O}} \longleftrightarrow :\text{N}\equiv\text{N}-\ddot{\ddot{\text{O}}}: \longleftrightarrow :\ddot{\ddot{\text{N}}}-\text{N}\equiv\text{O}:$

        −1 +1 0   0 +1 −1   −2 +1 +1 Formal
                                                    charges

The central N is sp hybridized. We can probably ignore the third resonance structure on the basis of formal charge. **c.** sp hybrid orbitals on the center N overlap with atomic orbitals (or hybrid orbitals) on the other two atoms to form two σ bonds. The remaining p orbitals on the center N overlap with p orbitals on the other N to form two π bonds.   **75.** F_2: $(\sigma_{2s})^2(\sigma_{2s}*)^2(\sigma_{2p})^2$ $(\pi_{2p})^4(\pi_{2p}*)^4$; F_2 should have a lower ionization energy than F. The electron removed from F_2 is in a $\pi_{2p}*$ antibonding molecular orbital, which is higher in energy than the $2p$ atomic orbitals from which the electron in atomic fluorine is removed. Since the electron removed from F_2 is higher in energy than the electron removed from F, it should be easier to remove an electron from F_2 than from F.   **77.** π molecular orbital

87.

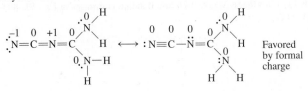

6 C and N atoms are sp^2 hybridized;
6 C and N atoms are sp^3 hybridized;
0 C and N atoms are sp hybridized;
25 σ bonds and 4 π bonds

89. a. NCN^{2-}:

$$\left[:\ddot{\text{N}}=\text{C}=\ddot{\text{N}}:\right]^{2-} \longleftrightarrow \left[:\text{N}\equiv\text{C}-\ddot{\ddot{\text{N}}}:\right]^{2-} \longleftrightarrow \left[:\ddot{\ddot{\text{N}}}-\text{C}\equiv\text{N}:\right]^{2-}$$

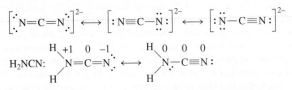

H_2NCN:

Favored by formal charge

Dicyandiamide:

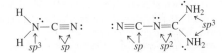

Favored by formal charge

Melamine:

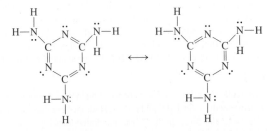

b. NCN^{2-}: C is sp hybridized. Each resonance structure predicts a different hybridization for the N atoms. For the remaining compounds, we will predict hybrids for the favored resonance structures only.

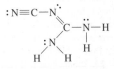

Melamine: N in NH_2 groups are all sp^3 hybridized. Atoms in ring are all sp^2 hybridized; **c.** NCN^{2-}: 2 σ and 2 π bonds; H_2NCN: 4 σ and 2 π bonds; dicyandiamide: 9 σ and 3 π bonds; melamine: 15 σ and 3 π bonds; **d.** The π system forces the ring to be planar just as the benzene ring is planar. **e.** The structure

$:\text{N}\equiv\text{C}-\ddot{\text{N}}$ ... (structure)

best agrees with experiment because it has three different CN bonds. This structure is also favored on the basis of formal charge.   **91. a.** 25-nm light has sufficient energy to ionize N and N_2 and to break the triple bond in N_2. Thus, N_2, N_2^+, N, and N^+ will all be present, assuming excess N_2. **b.** 85.33 nm $< \lambda \le 127$ nm; **c.** The ionization energy of a substance is the energy it takes to completely remove an electron. N_2: $(\sigma_{2s})^2(\sigma_{2s}*)^2(\pi_{2p})^4(\sigma_{2p})^2$; the electron removed from N_2 is in the σ_{2p} molecular orbital, which is lower in energy than the $2p$ atomic orbital from which the electron in atomic nitrogen is removed. Since the electron removed from N_2 is lower in energy than the electron removed in N, then the ionization energy of N_2 is greater than the ionization energy of N. **93.** Both reactions apparently involve only the breaking of the N—Cl bond. However, in the reaction ONCl → NO + Cl some energy is released in forming the stronger NO bond, lowering the value of ΔH. Therefore, the apparent N—Cl bond energy is artificially low for this reaction. The first reaction involves only the breaking of the N—Cl bond.   **95.** The ground state MO electron configuration for He_2 is $(\sigma_{1s})^2(\sigma_{1s}*)^2$ giving a bond order of 0. Therefore, He_2 molecules are not predicted to be stable (and are not stable) in the lowest-energy ground state. However, in a high-energy environment, the electron(s) from the antibonding orbitals in He_2 can be promoted into higher-energy bonding orbitals, thus giving a nonzero bond order and a "reason" to form. For example, a possible excited-state MO electron configuration for He_2 would be $(\sigma_{1s})^2(\sigma_{1s}*)^1(\sigma_{2s})^1$, giving a bond order of $(3 − 1)/2 = 1$. Thus excited He_2 molecules can form, but they spontaneously break apart as the electron(s) fall back to the ground state, where the bond order equals zero. **97.** The species with the smallest ionization energy has the highest energy electron. O_2, N_2^{2-}, N_2^-, and O_2^+ all have at least one electron in the high-energy $\pi_{2p}*$ orbitals. Because N_2^{2-} has the highest ratio of electrons to protons, the $\pi_{2p}*$ electrons are least attracted to the nuclei and easiest to remove, translating into the smallest ionization energy.   **99. a.** The CO bond is polar, with the negative end at the more electronegative oxygen atom. We would expect metal cations to be attracted to and to bond to the oxygen end of CO on the basis of electronegativity. **b.** The formal charge on C is −1, and the formal charge on O is +1. From formal charge, we would expect metal cations to bond to the carbon (with the negative formal charge). **c.** In molecular orbital theory, only orbitals with proper symmetry overlap to form bonding orbitals. The metals that form bonds to CO are usually transition metals, all of which have outer electrons in the d orbitals. The only molecular orbitals of CO that have proper symmetry to overlap with d orbitals are the $\pi_{2p}*$ orbitals, whose shape is similar to that of the d orbitals. Since the antibonding molecular orbitals have more carbon character, one would expect the bond to form through carbon.   **101. a.** Li_2 bond order = 1; B_2 bond order = 1; **b.** 4 electrons must be removed; **c.** 4.5×10^5 kJ   **103.** T-shaped and dsp^3 hybridized

Chapter 10

13. Intermolecular forces are the forces between molecules that hold the substances together in the solid and liquid phases. Hydrogen bonding is a specific type of intermolecular force. In this figure, the dotted lines represent the hydrogen bonding interactions that hold individual H_2O molecules together in the solid and liquid phases (answer a is correct). The solid lines represent the O—H covalent bonds.   **15.** Atoms have an approximately spherical shape. It is impossible to pack spheres together without some empty space between the spheres.   **17.** Evaporation takes place when some molecules at the surface of a liquid have enough energy to break the intermolecular forces holding them in the liquid phase. When a liquid evaporates, the molecules that escape have high kinetic energies. The average kinetic energy of the remaining molecules is lower; thus the temperature of the liquid is lower.   **19.** An alloy is a substance that contains a mixture of elements and has metallic properties. In a substitutional alloy, some of the host metal atoms are replaced by other metal atoms of similar size (e.g., in brass, pewter, plumber's solder). An interstitial alloy is formed when some of the interstices (holes) in the closest packed metal structure are occupied by smaller atoms (e.g., in carbon steels).   **21. a.** As intermolecular forces increase, the rate of evaporation decreases. **b.** increase T: increase rate; **c.** increase surface area: increase rate   **23.** $C_2H_5OH(l) \rightarrow C_2H_5OH(g)$ is an endothermic process. Heat is absorbed when liquid ethanol vaporizes; the internal heat from the body provides this heat, which results in the cooling of the body.   **25.** Sublimation will occur, allowing water to escape as $H_2O(g)$.   **27.** The strength of intermolecular forces determines relative boiling points. The types of intermolecular forces for covalent compounds are London dispersion forces, dipole forces, and hydrogen bonding. Because the three compounds are assumed to have similar molar mass and shape, the strength of the London dispersion forces will be about equal among the three compounds. One of the compounds will be nonpolar so it only has London dispersion forces. The other two compounds will be polar so they have additional dipole forces and will boil at a higher temperature than the nonpolar compound. One of the polar compounds will have an H covalently bonded to either N, O, or F. This gives rise to the strongest type of covalent intermolecular force, hydrogen bonding. This compound exhibiting hydrogen bonding will have the highest boiling point while the polar compound with no hydrogen bonding will boil at an intermediate temperature.   **29. a.** Both CO_2 and H_2O are molecular solids. Both have an ordered array of the individual molecules, with the molecular units occupying the lattice points. A difference within each solid lattice is the strength of the intermolecular forces. CO_2 is nonpolar and only exhibits London dispersion forces. H_2O exhibits the relatively strong hydrogen bonding interactions. The difference in strength is evidenced by the solid phase change that occurs at 1 atm. CO_2 sublimes at a relatively low temperature of $-78°C$. In sublimation, all of the intermolecular forces are broken. However, H_2O doesn't have a solid phase change until 0°C, and in this phase change from ice to water, only a fraction of the intermolecular forces are broken. The higher temperature and the fact that only a portion of the intermolecular forces are broken are attributed to the strength of the intermolecular forces in H_2O as compared to CO_2. Related to the intermolecular forces are the relative densities of the solid and liquid phases for these two compounds. $CO_2(s)$ is denser than $CO_2(l)$ while $H_2O(s)$ is less dense than $H_2O(l)$. For $CO_2(s)$ and for most solids, the molecules pack together as close as possible, which is why solids are usually more dense than the liquid phase. Water is an exception to this. Water molecules are particularly well suited to interact with each other because each molecule has two polar O—H bonds and two lone pairs on each oxygen. This can lead to the association of four

hydrogen atoms with each oxygen: two by covalent bonds and two by dipoles. To keep this symmetric arrangement (which maximizes the hydrogen bonding interactions), the $H_2O(s)$ molecules occupy positions that create empty space in the lattice. This translates into a smaller density for $H_2O(s)$ (less mass per unit volume). **b.** Both NaCl and CsCl are ionic compounds with the anions at the lattice points of the unit cell and the cations occupying the empty spaces created by the anions (called holes). In NaCl, the Cl^- anions occupy the lattice points of a face-centered unit cell with the Na^+ cations occupying the octahedral holes. Octahedral holes are the empty spaces created by six Cl^- ions. CsCl has the Cl^- ions at the lattice points of a simple cubic unit cell with the Cs^+ cations occupying the middle of the cube.   **31.** If TiO_2 conducts electricity as a liquid, then it would be ionic.   **33.** In the $\ln(P_{vap})$ versus $1/T$ plot, the slope of the straight line is equal to $-\Delta H_{vap}/R$. Because ΔH_{vap} is always positive, the slope of the line will always be negative. **35. a.** LD (London dispersion); **b.** dipole, LD; **c.** hydrogen bonding, LD; **d.** ionic; **e.** LD; **f.** dipole, LD; **g.** ionic   **37. a.** OCS; **b.** SeO_2; **c.** $H_2NCH_2CH_2NH_2$; **d.** H_2CO; **e.** CH_3OH **39. a.** Neopentane is more compact than n-pentane. There is less surface-area contact among neopentane molecules. This leads to weaker London dispersion forces and a lower boiling point. **b.** HF is capable of hydrogen bonding; HCl is not. **c.** LiCl is ionic, and HCl is a molecular solid with only dipole forces and London dispersion forces. Ionic forces are much stronger than the forces for molecular solids. **d.** n-Hexane is a larger molecule, so it has stronger London dispersion forces.   **41. a.** HBr has dipole forces in addition to LD forces; **b.** NaCl, stronger ionic forces; **c.** I_2, larger molecule so stronger LD forces; **d.** N_2, smallest nonpolar compound present, has weakest LD forces; **e.** CH_4, smallest nonpolar compound present, has weakest LD forces; **f.** HF, can form relatively strong hydrogen bonding interactions, unlike the other compounds; **g.** $CH_3CH_2CH_2OH$, unlike others, has relatively strong hydrogen bonding.   **43.** H_2O is attracted to glass while Hg is not.   **45.** The structure of H_2O_2 produces greater hydrogen bonding than water.   **47.** 313 pm   **49.** 0.704 Å   **51.** 1.54 g/cm^3 **53.** 174 pm; 11.6 g/cm^3 **55.** Ag   **57.** edge, 328 pm; radius, 142 pm   **59.** face-centered cubic unit cell   **61.** For a cubic closest packed structure, 74.06% of the volume of each unit cell is occupied by atoms; in a simple cubic unit cell structure, 52.36% is occupied. The cubic closest packed structure provides the more efficient means for packing atoms.   **63.** Doping silicon with phosphorus produces an n-type semiconductor. The phosphorus adds electrons at energies near the conduction band of silicon. Electrons do not need as much energy to move from filled to unfilled energy levels so conduction increases. Doping silicon with gallium produces a p-type semiconductor. Because gallium has fewer valence electrons than silicon, holes (unfilled energy levels) at energies in the previously filled molecular orbitals are created, which induces greater electron movement (greater conductivity).   **65.** p-type   **67.** 5.0×10^2 nm   **69.** NaCl: $4Na^+, 4Cl^-$; CsCl: $1Cs^+, 1Cl^-$; ZnS: $4Zn^{2+}, 4S^{2-}$; TiO_2: $2Ti^{4+}, 4O^{2-}$ **71.** CoF_2 **73.** $ZnAl_2S_4$ **75.** MF_2 **77.** NaCl structure; $r_{O^{2-}} = 1.49 \times 10^{-8}$ cm; $r_{Mg^{2+}} = 6.15 \times 10^{-9}$ cm   **79.** The calculated distance between ion centers is 358 pm. Ionic radii give a distance of 350. pm. The distance calculated from the density is 8 pm (2.3%) greater than that calculated from the table of ionic radii.   **81. a.** CO_2: molecular; **b.** SiO_2: network; **c.** Si: atomic, network; **d.** CH_4: molecular; **e.** Ru: atomic, metallic; **f.** I_2: molecular; **g.** KBr: ionic; **h.** H_2O: molecular; **i.** NaOH: ionic; **j.** U: atomic, metallic; **k.** $CaCO_3$: ionic; **l.** PH_3: molecular   **83. a.** The unit cell consists of Ni at the cube corners and Ti at the body center or Ti at the cube corners and Ni at the body center. **b.** NiTi; **c.** Both have a coordination number of 8.   **85.** $CaTiO_3$; each structure has six oxygens around each Ti.   **87. a.** $YBa_2Cu_3O_9$; **b.** The structure of this superconductor material is based on the second perovskite structure. The $YBa_2Cu_3O_9$ structure is three of these cubic perovskite unit cells stacked on top of each other. The oxygens are in the same places, Cu takes the place of Ti, two Ca are replaced by two Ba, and one Ca is replaced by Y. **c.** $YBa_2Cu_3O_7$ **89.** Li, 158 kJ/mol; Mg, 139 kJ/mol. Bonding is stronger in Li.   **91.** 89°C   **93.** 77°C

95.

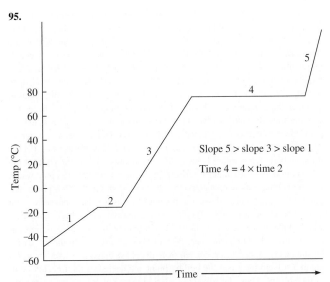

Slope 5 > slope 3 > slope 1

Time 4 = 4 × time 2

97. a. Much more energy is required to break the intermolecular forces when going from a liquid to a gas than is required to go from a solid to a liquid. Hence ΔH_{vap} is much larger than ΔH_{fus}; **b.** 113 J; **c.** 4220 J; **d.** 4220 J released **99.** 1680 kJ **101.** 1490 g **103.** A: solid; B: liquid; C: vapor; D: solid + vapor; E: solid + liquid + vapor (triple point); F: liquid + vapor; G: liquid + vapor (critical point); H: vapor; the first dashed line (at the lower temperature) is the normal freezing point, and the second dashed line is the normal boiling point. The solid phase is denser. **105. a.** two; **b.** higher pressure triple point: graphite, diamond, and liquid; lower pressure triple point: graphite, liquid and vapor; **c.** It is converted to diamond (the more dense solid form); **d.** Diamond is more dense, which is why graphite can be converted to diamond by applying pressure. **107.** Because the density of the liquid phase is greater than the density of the solid phase, the slope of the solid/liquid boundary line is negative (as in H_2O). With a negative slope, the melting points increase with a decrease in pressure so the normal melting point of X should be greater than 225°C. **109.** As the physical properties indicate, the intermolecular forces are slightly stronger in D_2O than in H_2O. **111.** A: CH_4; B: SiH_4; C: NH_3 **113.** XeF_2 **115.** B_2H_6, molecular; SiO_2, network; CsI, ionic; W, metallic **117.** 57.8 torr **119.** 4.65 kg/h **131.** $\Delta E = 27.86$ kJ/mol; $\Delta H = 30.79$ kJ/mol **133.** 46.7 kJ/mol; 90.%

135.

```
   H  H                      H        H
   |  | ..                   |    ..  |
H—C——C——O—H     and    H—C——O——C—H
   |  |  ¨¨                  |    ¨¨  |
   H  H                      H        H
```

   Liquid (H-bonding)   Gas (no H-bonding)

The first structure with the —OH bond is capable of forming hydrogen bonding; the other structure is not. Therefore, the liquid (which has the stronger intermolecular forces) is the first structure, and the gas is the second structure. **137.** The solids with high melting points (NaCl, $MgCl_2$, NaF, MgF_2, AlF_3) are all ionic solids. $SiCl_4$, SiF_4, Cl_2, F_2, PF_5, and SF_6 are nonpolar covalent molecules with LD forces. PCl_3 and SCl_2 are polar molecules with LD and dipole forces. In these 8 molecular substances the intermolecular forces are weak and the melting points low. $AlCl_3$ is intermediate. The melting point indicates there are stronger forces present than in the nonmetal halides, but not as strong as for an ionic solid. $AlCl_3$ illustrates a gradual transition from ionic to covalent bonding; from an ionic solid to discrete molecules. **139.** $TiO_{1.182}$ or $Ti_{0.8462}O$; 63.7% Ti^{2+}, 36.3% Ti^{3+} (a 1.75:1 ion ratio) **141.** 6.58 g/cm³

143.

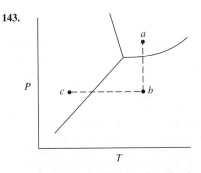

As P is lowered, we go from a to b on the phase diagram. The water boils. The boiling of water is endothermic and the water is cooled $(b \rightarrow c)$, forming some ice. If the pump is left on, the ice will sublime until none is left. This is the basis of freeze drying.

145. The volume of the hole is $\frac{4}{3}\pi[(\sqrt{3} - 1)r]^3$ **147.** CdS; n-type

149. 253 torr; 6.38×10^{22} atoms

Chapter 11

11. 9.74 M **13.** 160 mL **15.** 4.5 M **17.** As the temperature increases, the gas molecules will have a greater average kinetic energy. A greater fraction of the gas molecules in solution will have kinetic energy greater than the attractive forces between the gas molecules and the solvent molecules. More gas molecules will escape to the vapor phase, and the solubility of the gas will decrease. **19.** The levels of the liquids in each beaker will become constant when the concentration of solute is the same in both beakers. Because the solute is less volatile, the beaker on the right will have a larger volume when the concentrations become equal. Water will initially condense in this beaker in a larger amount than solute is evaporating, while the net change occurring initially in the other beaker is for water to evaporate in a larger amount than solute is condensing. Eventually the rate that solute and H_2O leave and return to each beaker will become equal when the concentrations become equal. **21.** No. For an ideal solution, $\Delta H_{soln} = 0$. **23.** Normality is the number of equivalents per liter of solution. For an acid or a base, an equivalent is the mass of acid or base that can furnish 1 mole of protons (if an acid) or accept 1 mole of protons (if a base). A proton is an H^+ ion. Molarity is defined as the moles of solute per liter of solution. When the number of equivalents equals the number of moles of solute, then normality = molarity. This is true for acids that only have one acidic proton in them and for bases that accept only one proton per formula unit. Examples of acids where equivalents = moles solute are HCl, HNO_3, HF, and $HC_2H_3O_2$. Examples of bases where equivalents = moles solute are NaOH, KOH, and NH_3. When equivalents ≠ moles solute, then normality ≠ molarity. This is true for acids that donate more than one proton (H_2SO_4, H_3PO_4, H_2CO_3, etc.) and for bases that react with more than one proton per formula unit [$Ca(OH)_2$, $Ba(OH)_2$, $Sr(OH)_2$, etc.]. **25.** Only statement b is true. A substance freezes when the vapor pressures of the liquid and solid phases are the same. When a solute is added to water, the vapor pressure of the solution at 0°C is less than the vapor pressure of the solid; the net result is for any ice present to convert to liquid in order to try to equalize the vapor pressures (which never can occur at 0°C). A lower temperature is needed to equalize the vapor pressures of water and ice, hence the freezing point is depressed. For statement a, the vapor pressure of a solution is directly related to the mole fraction of solvent (not solute) by Raoult's law. For statement c, colligative properties depend on the number of solute particles present and not on the identity of the solute. For statement d, the boiling point of water is increased because the sugar solute decreases the vapor pressure of the water; a higher temperature is required for the vapor pressure of the solution to equal the external pressure so boiling can occur. **27.** Isotonic solutions are those that have identical osmotic pressures. Crenation and hemolysis refer to phenomena that occur when red blood cells are bathed in solutions having a mismatch in osmotic pressure between the inside and the outside of the cell. When red blood cells are in a solution having a higher osmotic pressure than that of the cells, the cells shrivel as there is a net transfer of water out of the cells. This is called crenation. Hemolysis occurs when the red blood cells are bathed in a solution having lower osmotic pressure than that inside the cell. Here, the cells rupture as there is a net transfer of water to inside the red blood cells. **29.** 1.06 g/mL; 0.0180 mole fraction H_3PO_4, 0.9820 mole fraction H_2O; 0.981 mol/L; 1.02 mol/kg **31.** HCl: 12 M, 17 m, 0.23; HNO_3: 16 M, 37 m, 0.39; H_2SO_4: 18 M, 200 m, 0.76; $HC_2H_3O_2$: 17 M, 2000 m, 0.96; NH_3: 15 M, 23 m, 0.29 **33.** 35%; 0.39; 7.3 m; 3.1 M **35.** 10.1% by mass; 2.45 mol/kg **37.** 23.9%; 1.6 m, 0.028, 4.11 N **39.** $NaI(s) \rightarrow Na^+(aq) + I^-(aq)$ $\Delta H_{soln} = -8$ kJ/mol **41.** The attraction of water molecules for Al^{3+} and OH^- cannot overcome the larger lattice energy of $Al(OH)_3$. **43. a.** CCl_4; **b.** H_2O; **c.** H_2O; **d.** CCl_4; **e.** H_2O; **f.** H_2O; **g.** CCl_4 **45. a.** NH_3; **b.** CH_3CN; **c.** CH_3COOH **47.** As the length of the hydrocarbon chain increases, the solubility decreases because the nonpolar hydrocarbon chain interacts poorly with the polar water molecules. **49.** 1.04×10^{-3} mol/L · atm; 1.14×10^{-3} mol/L **51.** 50.0 torr **53.** 0.918 **55.** 23.2 torr at 25°C; 70.0 torr at 45°C **57. a.** 290 torr; **b.** 0.69 **59.** $\chi_{methanol} = \chi_{propanol} = 0.500$ **61.** solution c **63. a.** These two molecules are named acetone (CH_3COCH_3) and water. As discussed in Section 11.4 on nonideal solutions, acetone–water solutions exhibit negative deviations from Raoult's law. Acetone and water have the ability to hydrogen bond with each other, which gives the solution stronger intermolecular forces compared to the pure states of both solute and solvent. In the pure state, acetone cannot hydrogen bond with itself, so the middle diagram illustrating negative deviations from Raoult's law is the correct choice for acetone–water solutions. **b.** These two molecules are named ethanol (CH_3CH_2OH) and water. Ethanol–water solutions show positive deviations from Raoult's law. Both substances can hydrogen bond in the pure state, and they can continue this in solution. However, the solute–solvent interactions are weaker for ethanol–water solutions due to the significant nonpolar part of ethanol (CH_3—CH_2 is the nonpolar part of ethanol). This nonpolar part of ethanol weakens the intermolecular forces in solution, so the first diagram illustrating positive deviations from Raoult's law is the correct choice for ethanol–water solutions. **c.** These two molecules are named heptane (C_7H_{16}) and hexane (C_6H_{14}). Heptane and hexane are very similar nonpolar substances; both are composed entirely of nonpolar C—C bonds and relatively nonpolar C—H bonds, and both have a similar size and shape. Solutions of heptane and hexane should be ideal, so the third diagram illustrating no deviation from Raoult's law is the correct choice for heptane–hexane solutions. **d.** These two molecules are named heptane (C_7H_{16}) and water. The interactions between the nonpolar heptane molecules and the polar water molecules will certainly be weaker in solution compared to the pure solvent and pure solute interactions. This results in positive deviations from Raoult's law (the first diagram). **65.** 101.5°C **67.** 14.8 g $C_3H_8O_3$ **69.** $T_f = -29.9$°C, $T_b = 108.2$°C **71.** 6.6×10^{-2} mol/kg; 590 g/mol (610 g/mol if no rounding of numbers) **73. a.** $\Delta T = 2.0 \times 10^{-5}$°C, $\pi = 0.20$ torr; **b.** Osmotic pressure is better for determining the molar mass of large molecules. A temperature change of 10^{-5}°C is very difficult to measure. A change in height of a column of mercury by 0.2 mm is not as hard to measure precisely. **75.** 2.51×10^5 g/mol **77.** Dissolve 210 g sucrose in some water and dilute to 1.0 L in a volumetric flask. To get 0.62 ± 0.01 mol/L, we need 212 ± 3 g sucrose. **79. a.** 0.010 m Na_3PO_4 and 0.020 m KCl; **b.** 0.020 m HF; **c.** 0.020 m $CaBr_2$ **81. a.** $T_f = -13$°C; $T_b = 103.5$°C; **b.** $T_f = -4.7$°C; $T_b = 101.3$°C **83.** 1.67 **85. a.** $T_f = -0.28$°C; $T_b = 100.077$°C; **b.** $T_f = -0.37$°C; $T_b = 100.10$°C **87.** 2.63 (0.0225 m), 2.60 (0.0910 m), 2.57 (0.278 m); $i_{average} = 2.60$ **89. a.** yes; **b.** no **91.** Benzoic acid is capable of hydrogen bonding, but a significant part of benzoic acid is the nonpolar benzene ring. In benzene, a hydrogen bonded dimer forms.

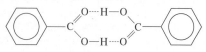

The dimer is relatively nonpolar and thus more soluble in benzene than in water. Benzoic acid would be more soluble in a basic solution because of the reaction $C_6H_5CO_2H + OH^- \rightarrow C_6H_5CO_2^- + H_2O$. By removing the proton from benzoic acid, an anion forms, and like all anions, the species becomes more soluble in water. **93. a.** 26.6 kJ/mol; **b.** -657 kJ/mol **95. a.** Water boils when the vapor pressure equals the pressure above the water. In an open pan, $P_{atm} \approx 1.0$ atm. In a pressure cooker, $P_{inside} > 1.0$ atm and water boils at a higher temperature. The higher the cooking temperature, the faster the cooking time. **b.** Salt dissolves in water, forming a solution with a melting point lower than that of pure water ($\Delta T_f = K_f m$). This happens in water on the surface of ice. If it is not too cold, the ice melts. This process won't occur if the ambient temperature is lower than the depressed freezing point of the salt solution. **c.** When water freezes from a solution, it freezes as pure water, leaving behind a more concentrated salt solution. **d.** On the CO_2 phase diagram, the triple point is above 1 atm and $CO_2(g)$ is the stable phase at 1 atm and room temperature. $CO_2(l)$ can't exist at normal atmospheric pressures, which explains why dry ice sublimes rather than boils. In a fire extinguisher, $P > 1$ atm and $CO_2(l)$ can exist. When CO_2 is released from the fire extinguisher, $CO_2(g)$ forms as predicted from the phase diagram. **e.** Adding a solute to a solvent increases the boiling point and decreases the freezing point of the solvent. Thus, the solvent is a liquid over a wider range of temperatures when a solute is dissolved. **97.** 0.600 **99.** $P_{ideal} = 188.6$ torr; $\chi_{acetone} = 0.512, \chi_{methanol} = 0.488$; the actual vapor pressure of the solution is smaller than the ideal vapor pressure, so this solution exhibits a negative deviation from Raoult's law. This occurs when solute–solvent attractions are stronger than for the pure substances. **101.** 776 g/mol **103.** $C_2H_4O_3$; 151 g/mol (exp.); 152.10 g/mol (calc.); $C_4H_8O_6$ **105.** 1.97% NaCl **107. a.** 100.77°C; **b.** 23.1 mm Hg; **c.** Assume an ideal solution; assume no ions form ($i = 1$); assume the solute is nonvolatile. **109.** 639 g/mol; 33.7 torr

119. 30.% A: $\chi_A = \dfrac{0.30y}{0.70x + 0.30y}$, $\chi_B = 1 - \chi_A$;

50.% A: $\chi_A = \dfrac{y}{x + y}$, $\chi_B = 1 - \dfrac{y}{x + y}$;

80.% A: $\chi_A = \dfrac{0.80y}{0.20x + 0.80y}$, $\chi_B = 1 - \chi_A$;

30.% A: $\chi_A^V = \dfrac{0.30x}{0.30x + 0.70y}$, $\chi_B^V = 1 - \dfrac{0.30x}{0.30x + 0.70y}$;

50.% A: $\chi_A^V = \dfrac{x}{x + y}$, $\chi_B^V = 1 - \chi_A^V$;

80.% A: $\chi_A^V = \dfrac{0.80x}{0.80x + 0.20y}$, $\chi_B^V = 1 - \chi_A^V$

121. 72.5% sucrose and 27.5% NaCl by mass; 0.313 **123.** 0.050 **125.** 44% naphthalene, 56% anthracene **127.** -0.20°C, 100.056°C **129. a.** 46 L; **b.** No; a reverse osmosis system that applies 8.0 atm can only purify water with solute concentrations less than 0.32 mol/L. Salt water has a solute concentration of $2(0.60\ M) = 1.2\ M$ ions. The solute concentration of salt water is much too high for this reverse osmosis unit to work. **131.** $i = 3.00$; $CdCl_2$

Chapter 12

11. In a unimolecular reaction, a single reactant molecule decomposes to products. In a bimolecular reaction, two molecules collide to give products. The probability of the simultaneous collision of three molecules with enough energy and orientation is very small, making termolecular steps very unlikely. **13.** All of these choices would affect the rate of the reaction, but only b and c affect the rate by affecting the value of the rate constant k. The value of the rate constant is dependent on temperature. It also depends on the activation energy. A catalyst will change the value of k because the activation energy changes. Increasing the concentration (partial pressure) of either O_2 or NO does not affect the value of k, but it does increase the rate of the reaction because both concentrations appear in the rate law. **15.** The average rate decreases with time because the reverse reaction occurs more frequently as the concentration of products increases. Initially, with no products present, the rate of the forward reaction is at its fastest; but as time goes on, the rate gets slower and slower since products are converting back into reactants. The instantaneous rate will also decrease with time. The only rate that is constant is the initial rate. This is the instantaneous rate taken at $t \approx 0$. At this time, the amount of products is insignificant and the rate of the reaction only depends on the rate of the forward reaction.   **17.** When the rate doubles as the concentration quadruples, the order is $1/2$. For a reactant that has an order of -1, the rate will decrease by a factor of $1/2$ when the concentrations are doubled.   **19.** Two reasons are: **1)** the collision must involve enough energy to produce the reaction; i.e., the collision energy must equal or exceed the activation energy. **2)** the relative orientation of the reactants must allow formation of any new bonds necessary to produce products.   **21.** Enzymes are very efficient catalysts. As is true for all catalysts, enzymes speed up a reaction by providing an alternative pathway for reactants to convert to products. This alternative pathway has a smaller activation energy and, hence, a faster rate. Also true is that catalysts are not used up in the overall chemical reaction. Once an enzyme comes in contact with the correct reagent, the chemical reaction quickly occurs, and the enzyme is then free to catalyze another reaction. Because of the efficiency of the reaction step, only a relatively small amount of enzyme is needed to catalyze a specific reaction, no matter how complex the reaction.   **23.** P_4: 6.0×10^{-4} mol/L $\cdot$ s; H_2: 3.6×10^{-3} mol/L $\cdot$ s   **25. a.** average rate of decomposition of $H_2O_2 = 2.31 \times 10^{-5}$ mol/L $\cdot$ s, rate of production of $O_2 = 1.16 \times 10^{-5}$ mol/L $\cdot$ s; **b.** average rate of decomposition of $H_2O_2 = 1.16 \times 10^{-5}$ mol/L $\cdot$ s, rate of production of $O_2 = 5.80 \times 10^{-6}$ mol/L $\cdot$ s   **27. a.** mol/L $\cdot$ s; **b.** mol/L $\cdot$ s; **c.** s^{-1}; **d.** L/mol $\cdot$ s; **e.** L^2/mol^2 $\cdot$ s   **29. a.** rate $= k[NO]^2[Cl_2]$; **b.** 1.8×10^2 L^2/mol^2 $\cdot$ min   **31. a.** rate $= k[NOCl]^2$; **b.** 6.6×10^{-29} cm^3/molecules $\cdot$ s; **c.** 4.0×10^{-8} L/mol $\cdot$ s   **33. a.** rate $= k[I^-][OCl^-]$; **b.** 3.7 L/mol $\cdot$ s; **c.** 0.083 mol/L $\cdot$ s   **35. a.** first order in Hb and first order in CO; **b.** rate $= k[Hb][CO]$; **c.** 0.280 L/μmol $\cdot$ s; **d.** 2.26 μmol/L $\cdot$ s   **37.** rate $= k[H_2O_2]$; $\ln[H_2O_2] = -kt + \ln[H_2O_2]_0$; $k = 8.3 \times 10^{-4}$ s^{-1}; 0.037 M **39.** rate $= k[NO_2]^2$; $\dfrac{1}{[NO_2]} = kt + \dfrac{1}{[NO_2]_0}$; $k = 2.08 \times 10^{-4}$ L/mol $\cdot$ s; 0.131 M **41. a.** rate $= k$; $[C_2H_5OH] = -kt + [C_2H_5OH]_0$; because slope $= -k$, $k = 4.00 \times 10^{-5}$ mol/L $\cdot$ s; **b.** 156 s; **c.** 313 s   **43.** rate $= k[C_4H_6]^2$; $\dfrac{1}{[C_4H_6]} = kt + \dfrac{1}{[C_4H_6]_0}$; $k = 1.4 \times 10^{-2}$ L/mol $\cdot$ s   **45.** second order; 0.1 M **47. a.** $[A] = -kt + [A]_0$; **b.** 1.0×10^{-2} s; **c.** 2.5×10^{-4} M **49.** 9.2×10^{-3} s^{-1}; 75 s   **51.** 150. s   **53.** 12.5 s   **55. a.** $1.1 \times 10^{-2} M$; **b.** 0.025 M **57.** 1.6 L/mol $\cdot$ s   **59. a.** rate $= k[CH_3NC]$; **b.** rate $= k[O_3][NO]$; **c.** rate $= k[O_3]$; **d.** rate $= k[O_3][O]$ **61.** Rate $= k[C_4H_9Br]$; $C_4H_9Br + 2H_2O \rightarrow C_4H_9OH + Br^- + H_3O^+$; the intermediates are $C_4H_9^+$ and $C_4H_9OH_2^+$.

63.

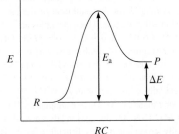

65. 341 kJ/mol   **67.** The graph of $\ln(k)$ versus $1/T$ is linear with slope $= -E_a/R = -1.2 \times 10^4$ K; $E_a = 1.0 \times 10^2$ kJ/mol   **69.** 9.5×10^{-5} L/mol $\cdot$ s   **71.** $51°C$ **73.** $H_3O^+(aq) + OH^-(aq) \rightarrow 2H_2O(l)$ should have the faster rate. H_3O^+ and OH^- will be electrostatically attracted to each other; Ce^{4+} and Hg_2^{2+} will repel each other (so E_a is much larger).   **75. a.** NO; **b.** NO_2; **c.** 2.3   **77.** $CH_2D—CH_2D$ should be the product. If the mechanism is possible, then the reaction must be $C_2H_4 + D_2 \rightarrow CH_2DCH_2D$. If we got this product, then we could conclude that this is a possible mechanism. If we got some other product, e.g., CH_3CHD_2, then we would conclude that the mechanism is wrong. Even though this mechanism correctly predicts the products of the reaction, we cannot say conclusively that this is the correct mechanism; we might be able to conceive of other mechanisms that would give the same product as our proposed one.   **79.** The rate depends on the number of reactant molecules adsorbed on the surface of the catalyst. This quantity is proportional to the concentration of reactant. However, when all the catalyst surface sites are occupied, the rate becomes independent of the concentration of reactant.   **81.** $215°C$ **83. a.** 20 min; **b.** 30 min; **c.** 15 min   **85.** 427 s

87. 1.0×10^2 kJ/mol   **89.** 6.58×10^{-6} mol/L $\cdot$ s   **91. a.** 25 kJ/mol; **b.** 12 s;

c.

T	Interval	$54 - 2$(Intervals)
$21.0°C$	16.3 s	$21°C$
$27.8°C$	13.0 s	$28°C$
$30.0°C$	12 s	$30.°C$

This rule of thumb gives excellent agreement to two significant figures.   **93. a.** 115 L^3/mol^3 $\cdot$ s; **b.** 87.0 s; **c.** $[A] = 1.27 \times 10^{-5} M$, $[B] = 1.00$ M **103.** rate $= \dfrac{k[I^-][OCl^-]}{[OH^-]}$; $k = 6.0 \times 10^1$ s^{-1} **105. a.** first order with respect to both reactants; **b.** rate $= k[NO][O_3]$; **c.** $k' = 1.8$ s^{-1}; $k'' = 3.6$ s^{-1}; **d.** $k = 1.8 \times 10^{-14}$ cm^3/molecules $\cdot$ s   **107. a.** For a three-step reaction with the first step limiting, the energy-level diagram could be

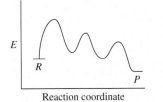

Reaction coordinate

Note that the heights of the second and third humps must be lower than the first-step activation energy. However, the height of the third hump could be higher than the second hump. One cannot determine this absolutely from the information in the problem.

b.

$F_2 \rightarrow 2F$	slow
$F + H_2 \rightarrow HF + H$	fast
$H + F \rightarrow HF$	fast
$F_2 + H_2 \rightarrow 2HF$	

c. F_2 was the limiting reactant.   **109. a.** $[B] \gg [A]$ so that $[B]$ can be considered constant over the experiments. (This gives us a pseudo-order rate law equation.) **b.** Rate $= k[A]^2[B]$, $k = 0.050$ L^2/mol^2 $\cdot$ s   **111.** Rate $= k[A][B]^2$, $k = 1.4 \times 10^{-2}$ L^2/mol^2 $\cdot$ s   **113.** 2.20×10^{-5} s^{-1}; 5.99×10^{21} molecules   **115.** 1.3×10^{-5} s^{-1}; 112 torr

Chapter 13

11. No, equilibrium is a dynamic process. Both the forward and reverse reactions are occurring at equilibrium, just at equal rates. Thus the forward and reverse reactions will distribute ^{14}C atoms between CO and CO_2.   **13.** A large value for K ($K \gg 1$) indicates there are relatively large concentrations of product gases and/or solutes as compared with the concentrations of reactant gases and/or solutes at equilibrium. A reaction with a large K value is a good source of products.   **15.** 4 molecules H_2O, 2 molecules CO, 4 molecules H_2, and 4 molecules CO_2 are present at equilibrium.   **17.** K and K_p are equilibrium constants as determined by the law of mass action. For K, the units used for concentrations are mol/L, and for K_p, partial pressures in units of atm are used (generally). Q is called the reaction quotient. Q has the exact same form as K or K_p, but instead of equilibrium concentrations, initial concentrations are used to calculate the Q value. Q is of use when it is compared to the K value. When $Q = K$ (or when $Q_p = K_p$), the reaction is at equilibrium. When $Q \neq K$, the reaction is not at equilibrium and one can determine what has to be the net change for the system to get to equilibrium.   **19.** We always try to make good assumptions that simplify the math. In some problems, we can set up the problem so that the net change, x, that must occur to reach equilibrium is a small number. This comes in handy when you have expressions like $0.12 - x$ or $0.727 + 2x$. When x is small, we assume that it makes little difference when subtracted from or added to some relatively big number. When this is true, $0.12 - x \approx 0.12$ and $0.727 + 2x \approx 0.727$. If the assumption holds by the 5% rule, then the assumption is assumed valid. The 5% rule refers to x (or $2x$ or $3x$, etc.) that was assumed small compared to some number. If x (or $2x$ or $3x$, etc.) is less than 5% of the number the assumption was made against, then the assumption will be assumed valid. If the 5% rule fails to work, one can generally use a math procedure called the method of successive approximations to solve the quadratic or cubic equation.

21. a. $K = \dfrac{[NO]^2}{[N_2][O_2]}$; **b.** $K = \dfrac{[NO_2]^2}{[N_2O_4]}$; **c.** $K = \dfrac{[SiCl_4][H_2]^2}{[SiH_4][Cl_2]^2}$;

d. $K = \dfrac{[PCl_3]^2[Br_2]^3}{[PBr_3]^2[Cl_2]^3}$ **23. a.** 0.11; **b.** 77; **c.** 8.8; **d.** 1.7×10^{-4} **25.** 4.0×10^6

27. 1.7×10^{-5} **29.** 6.3×10^{-13} **31.** 4.6×10^3

33. a. $K = \dfrac{[H_2O]}{[NH_3]^2[CO_2]}$, $K_p = \dfrac{P_{H_2O}}{P_{NH_3}^2 \times P_{CO_2}}$; **b.** $K = [N_2][Br_2]^3$,

$K_p = P_{N_2} \times P_{Br_2}^3$; **c.** $K = [O_2]^3$, $K_p = P_{O_2}^3$; **d.** $K = \dfrac{[H_2O]}{[H_2]}$, $K_p = \dfrac{P_{H_2O}}{P_{H_2}}$

35. only reaction d   **37.** 8.0×10^9 **39. a.** not at equilibrium; $Q > K$, shift left; **b.** at equilibrium; **c.** not at equilibrium; $Q > K$, shift left   **41. a.** decrease; **b.** no change; **c.** no change; **d.** increase   **43.** 0.16 mol   **45.** 3.4   **47.** 0.056   **49.** $[N_2]_0 = 10.0$ M, $[H_2]_0 = 11.0$ M **51.** $[SO_3] = [NO] = 1.06 M$; $[SO_2] = [NO_2] = 0.54 M$ **53.** 7.8×10^{-2} atm   **55.** $P_{SO_2} = 0.38$ atm; $P_{O_2} = 0.44$ atm; $P_{SO_3} = 0.12$ atm   **57. a.** $[NO] = 0.032$ M, $[Cl_2] = 0.016$ M, $[NOCl] = 1.0$ M; **b.** $[NO] = [NOCl] = 1.0$ M, $[Cl_2] = 1.6 \times 10^{-5} M$; **c.** $[NO] = 8.0 \times 10^{-3} M$, $[Cl_2] = 1.0 M$, $[NOCl] = 2.0 M$ **59.** $[CO_2] = 0.39$ M, $[CO] = 8.6 \times 10^{-3}$ M, $[O_2] = 4.3 \times 10^{-3}$ M **61.** 0.27 atm   **63. a.** no effect; **b.** shifts left; **c.** shifts right   **65. a.** right; **b.** right; **c.** no effect; **d.** left; **e.** no effect   **67. a.** left; **b.** right; **c.** left; **d.** no effect; **e.** no effect; **f.** right   **69.** increase   **71.** 2.6×10^{81} **73.** 6.74×10^{-6} **75. a.** 0.379 atm; **b.** 0.786   **77.** $[Fe^{3+}] = 2 \times 10^{-4}$ M, $[SCN^-] = 0.08$ M, $[FeSCN^{2+}] = 0.020$ M **79.** 1.43×10^{-2} atm   **81.** pink   **83.** Added OH^- reacts with H^+ to produce H_2O. As H^+ is removed, the reaction shifts right to produce more H^+ and CrO_4^{2-}. Because more CrO_4^{2-} is produced, the solution turns yellow.   **85.** 9.0×10^{-3} M **87.** 0.50   **89.** 3.0×10^{-6} M **99.** $P_{CO} = 0.58$ atm, $P_{CO_2} = 1.65$ atm   **101.** $[NOCl] = 2.0 M$, $[NO] = 0.050 M$, $[Cl_2] = 0.025 M$ **103.** 2.1×10^{-3} atm   **105.** $P_{NO_2} = 0.704$ atm, $P_{N_2O_4} = 0.12$ atm   **107. a.** Assuming mol/L units for concentrations, $K = 8.16 \times 10^{47}$; **b.** 33.9 L   **109.** 0.63   **111.** 0.240 atm   **113. a.** 2.33×10^{-4}; **b.** The argon gas will increase the volume of the container. This is because the container is a constant-pressure system, and if the number of moles increases at constant T and P, the volume must increase. An increase in volume will dilute the concentrations of all gaseous reactants and gaseous products. Because there are more moles of product gases than reactant gases (3 moles vs. 2 moles), the dilution will decrease the numerator of K more than the denominator will decrease. This causes $Q < K$ and the reaction shifts right to get back to equilibrium. Because temperature was unchanged, the value of K will not change. K is a constant as long as temperature is constant.   **115.** 192 g; 1.3 atm   **117.** 33

Chapter 14

19. as an acid: $HCO_3^-(aq) + H_2O(l) \rightleftharpoons CO_3^{2-}(aq) + H_3O^+(aq)$; as a base: $HCO_3^-(aq) + H_2O(l) \rightleftharpoons H_2CO_3(aq) + OH^-(aq)$; as an acid: $H_2PO_4^-(aq) + H_2O(l) \rightleftharpoons HPO_4^{2-}(aq) + H_3O^+(aq)$; as a base: $H_2PO_4^-(aq) + H_2O(l) \rightleftharpoons H_3PO_4(aq) + OH^-(aq)$.   **21.** b, c, and d   **23.** 10.78 (4 significant figures); 6.78 (3 significant figures); 0.78 (2 significant figures); a pH value is a logarithm. The numbers to the left of the decimal place identify the power of 10 to which $[H^+]$ is expressed in scientific notation—for example, 10^{-11}, 10^{-7}, 10^{-1}. The number of decimal places in a pH value identifies the number of significant figures in $[H^+]$. In all three pH values, the $[H^+]$ should be expressed only to two significant figures since these pH values have only two decimal places.

25.

NH_3	$+$	NH_3	$\rightleftharpoons$	NH_2^-	$+$	NH_4^+
Acid		Base		Conjugate Base		Conjugate Acid

One of the NH_3 molecules acts as a base and accepts a proton to form NH_4^+. The other NH_3 molecule acts as an acid and donates a proton to form NH_2^-. NH_4^+ is the conjugate acid of the NH_3 base. In the reverse reaction, NH_4^+ donates a proton. NH_2^- is the conjugate base of the NH_3 acid. In the reverse reaction, NH_2^- accepts a proton. Conjugate acid–base pairs differ only by an H^+ in the formula.   **27. a.** These would be 0.10 M solutions of strong acids like HCl, HBr, HI, HNO_3, H_2SO_4, or $HClO_4$. **b.** These are salts of the conjugate acids of the bases in Table 14.3. These conjugate acids are all weak acids. Three examples would be 0.10 M solutions of NH_4Cl, $CH_3NH_3NO_3$, and $C_2H_5NH_3Br$. Note that the anions used to form these salts are conjugate bases of strong acids; this is because they have no acidic or basic properties in

water (with the exception of HSO_4^-, which has weak acid properties). **c.** These would be 0.10 M solutions of strong bases like LiOH, NaOH, KOH, RbOH, CsOH, $Ca(OH)_2$, $Sr(OH)_2$, and $Ba(OH)_2$. **d.** These are salts of the conjugate bases of the neutrally charged weak acids in Table 14.2. The conjugate bases of weak acids are weak bases themselves. Three examples would be 0.10 M solutions of $NaClO_2$, $KC_2H_3O_2$, and CaF_2. The cations used to form these salts are Li^+, Na^+, K^+, Rb^+, Cs^+, Ca^{2+}, Sr^{2+}, and Ba^{2+} since these cations have no acidic or basic properties in water. Notice that these are the cations of the strong bases that you should memorize. **e.** There are two ways to make a neutral salt. The easiest way is to combine a conjugate base of a strong acid (except for HSO_4^-) with one of the cations from the strong bases. These ions have no acidic/basic properties in water so salts of these ions are neutral. Three examples would be 0.10 M solutions of NaCl, KNO_3, and SrI_2. Another type of strong electrolyte that can produce neutral solutions are salts that contain an ion with weak acid properties combined with an ion of opposite charge having weak base properties. If the K_a for the weak acid ion is equal to the K_b for the weak base ion, then the salt will produce a neutral solution. The most common example of this type of salt is ammonium acetate, $NH_4C_2H_3O_2$. For this salt, K_a for $NH_4^+ = K_b$ for $C_2H_3O_2^- = 5.6 \times 10^{-10}$. This salt, at any concentration, produces a neutral solution.

29. a. $H_2O(l) + H_2O(l) \rightleftharpoons H_3O^+(aq) + OH^-(aq)$ or

$\qquad H_2O(l) \rightleftharpoons H^+(aq) + OH^-(aq)$ $\quad K = K_w = [H^+][OH^-]$

b. $HF(aq) + H_2O(l) \rightleftharpoons F^-(aq) + H_3O^+(aq)$ or

$\quad HF(aq) \rightleftharpoons H^+(aq) + F^-(aq)$ $\quad K = K_a = \dfrac{[H^+][F^-]}{[HF]}$

c. $C_5H_5N(aq) + H_2O(l) \rightleftharpoons C_5H_5NH^+(aq) + OH^-(aq)$

$$K = K_b = \dfrac{[C_5H_5NH^+][OH^-]}{[C_5H_5N]}$$

31. a. This expression holds true for solutions of strong acids having a concentration greater than 1.0×10^{-6} M. For example, 0.10 M HCl, 7.8 M HNO_3, and 3.6×10^{-4} M $HClO_4$ are solutions where this expression holds true. **b.** This expression holds true for solutions of weak acids where the two normal assumptions hold. The two assumptions are that the contribution of H^+ from water is negligible and that the acid is less than 5% dissociated in water (from the assumption that x is small compared to some number). This expression will generally hold true for solutions of weak acids having a K_a value less than 1×10^{-4}, as long as there is a significant amount of weak acid present. Three example solutions are 1.5 M $HC_2H_3O_2$, 0.10 M HOCl, and 0.72 M HCN. **c.** This expression holds true for strong bases that donate 2 OH^- ions per formula unit. As long as the concentration of the base is above 5×10^{-7} M, this expression will hold true. Three examples are 5.0×10^{-3} M $Ca(OH)_2$, 2.1×10^{-4} M $Sr(OH)_2$, and 9.1×10^{-5} M $Ba(OH)_2$. **d.** This expression holds true for solutions of weak bases where the two normal assumptions hold. The assumptions are that the OH^- contribution from water is negligible and that the base is less than 5% ionized in water. For the 5% rule to hold, you generally need bases with $K_b < 1 \times 10^{-4}$ and concentrations of weak base greater than 0.10 M. Three examples are 0.10 M NH_3, 0.54 M $C_6H_5NH_2$, and 1.1 M C_5H_5N. **33.** One reason HF is a weak acid is that the H—F bond is unusually strong and thus, is difficult to break. This contributes to the reluctance of the HF molecules to dissociate in water. **35. a.** $HClO_4(aq) + H_2O(l) \rightarrow H_3O^+(aq) + ClO_4^-(aq)$ or $HClO_4(aq) \rightarrow H^+(aq) + ClO_4^-(aq)$; water is commonly omitted from K_a reactions. **b.** $CH_3CH_2CO_2H(aq) \rightleftharpoons H^+(aq) + CH_3CH_2CO_2^-(aq)$; **c.** $NH_4^+(aq) \rightleftharpoons H^+(aq) + NH_3(aq)$ **37. a.** H_2O, base; H_2CO_3, acid; H_3O^+, conjugate acid; HCO_3^-, conjugate base; **b.** $C_5H_5NH^+$, acid; H_2O, base; C_5H_5N, conjugate base; H_3O^+, conjugate acid; **c.** HCO_3^-, base; $C_5H_5NH^+$, acid; H_2CO_3, conjugate acid; C_5H_5N, conjugate base   **39. a.** $HClO_4$, strong acid; **b.** HOCl, weak acid; **c.** H_2SO_4, strong acid; **d.** H_2SO_3, weak acid   **41.** $HClO_4 > HClO_2 > NH_4^+ > H_2O$ **43. a.** HCl; **b.** HNO_2; **c.** HCN since it has a larger K_a value.   **45. a.** 1.0×10^{-7} M, neutral; **b.** 12 M, basic; **c.** 8.3×10^{-16} M, acidic; **d.** 1.9×10^{-10} M, acidic   **47. a.** endothermic; **b.** $[H^+] = [OH^-] = 2.34 \times 10^{-7}$ M **49.** [45] **a.** pH = pOH = 7.00; **b.** pH = 15.08, pOH = −1.08; **c.** pH = −1.08, pOH = 15.08; **d.** pH = 4.27, pOH = 9.73 [46] **a.** pH = 14.18, pOH = −0.18; **b.** pH = −0.44, pOH = 14.44; **c.** pH = pOH = 7.00; **d.** pH = 10.86, pOH = 3.14   **51. a.** pH = 6.88, pOH = 7.12, $[H^+] = 1.3 \times 10^{-7} M$, $[OH^-] = 7.6 \times 10^{-8} M$, acidic; **b.** pH = 0.92, pOH = 13.08, $[H^+] = 0.12 M$,

$[OH^-] = 8.4 \times 10^{-14}$ M, acidic; **c.** pH = 10.89, pOH = 3.11, $[H^+] = 1.3 \times 10^{-11}$ M, $[OH^-] = 7.8 \times 10^{-4}$ M, basic; **d.** pH = pOH = 7.00, $[H^+] = [OH^-] = 1.0 \times 10^{-7}$ M, neutral   **53.** pOH = 11.9, $[H^+] = 8 \times 10^{-3}$ M, $[OH^-] = 1 \times 10^{-12}$ M, acidic   **55. a.** H^+, ClO_4^-, H_2O; 0.602; **b.** H^+, NO_3^-, H_2O; 0.602   **57. a.** 1.00; **b.** −0.70; **c.** 7.00   **59.** 3.2×10^{-3} M **61.** Add 4.2 mL of 12 M HCl to water with mixing; add enough water to bring the solution volume to 1600 mL.   **63. a.** HNO_2 and H_2O, 2.00; **b.** $HC_2H_3O_2$ and H_2O, 2.68   **65.** $[H^+] = [F^-] = 3.5 \times 10^{-3}$ M, $[OH^-] = 2.9 \times 10^{-12}$ M, $[HF] = 0.017$ M, 2.46   **67.** $[HC_3H_5O_2] = 0.099$ M, $[C_3H_5O_2^-] = [H^+] = 1.1 \times 10^{-3}$ M, $[OH^-] = 9.1 \times 10^{-12}$, pH = 2.96, 1.1% dissociated.   **69.** 1.96   **71.** 1.57; 5.9×10^{-9} M **73. a.** 0.60%; **b.** 1.9%; **c.** 5.8%; **d.** Dilution shifts equilibrium to the side with the greater number of particles (% dissociation increases). **e.** $[H^+]$ also depends on initial concentration of weak acid.   **75.** 1.4×10^{-4} **77.** 0.16   **79.** 0.024 M **81.** 6×10^{-3}

83. a. $NH_3(aq) + H_2O(l) \rightleftharpoons NH_4^+(aq) + OH^-(aq)$ $K_b = \dfrac{[NH_4^+][OH^-]}{[NH_3]}$;

b. $C_5H_5N(aq) + H_2O(l) \rightleftharpoons C_5H_5NH^+(aq) + OH^-(aq)$
$$K_b = \dfrac{[C_5H_5NH^+][OH^-]}{[C_5H_5N]}$$

85. $NH_3 > C_5H_5N > H_2O > NO_3^-$ **87. a.** $C_6H_5NH_2$; **b.** $C_6H_5NH_2$; **c.** OH^-; **d.** CH_3NH_2 **89. a.** 13.00; **b.** 7.00; **c.** 14.30   **91. a.** K^+, OH^-, and H_2O, 0.015 M, 12.18; **b.** Ba^{2+}, OH^-, and H_2O, 0.030 M, 0.16 g   **95.** NH_3 and H_2O, 1.6×10^{-3} M, 11.20   **97. a.** $[OH^-] = 8.9 \times 10^{-3}$ M, $[H^+] = 1.1 \times 10^{-12}$ M, 11.96; **b.** $[OH^-] = 4.7 \times 10^{-5}$ M, $[H^+] = 2.1 \times 10^{-10}$ M, 9.68   **99.** 12.00   **101. a.** 1.3%; **b.** 4.2%; **c.** 6.4%   **103.** 1.0×10^{-9}

105. $H_2SO_3(aq) \rightleftharpoons HSO_3^-(aq) + H^+(aq)$ K_{a_1} reaction
$HSO_3^-(aq) \rightleftharpoons SO_3^{2-}(aq) + H^+(aq)$ K_{a_2} reaction

107. 3.00   **109.** 4.00; 1.0×10^{-19} M **111.** −0.30   **113.** HCl > NH_4Cl > KNO_3 > KCN > KOH   **115.** OCl^- **117. a.** $[OH^-] = [H^+] = 1.0 \times 10^{-7}$ M, pH = 7.00; **b.** $[OH^-] = 2.4 \times 10^{-5}$ M, $[H^+] = 4.2 \times 10^{-10}$ M, pH = 9.38   **119. a.** 5.82; **b.** 10.95   **121.** $[HN_3] = [OH^-] = 2.3 \times 10^{-6}$ M, $[Na^+] = 0.010$ M, $[N_3^-] = 0.010$ M, $[H^+] = 4.3 \times 10^{-9}$ M **123.** NaF   **125.** 3.66   **127.** 3.08   **129. a.** neutral; **b.** basic; $NO_2^- + H_2O \rightleftharpoons HNO_2 + OH^-$; **c.** acidic; $C_5H_5NH^+ \rightleftharpoons C_5H_5N + H^+$; **d.** acidic because NH_4^+ is a stronger acid than NO_2^- is a base; $NH_4^+ \rightleftharpoons NH_3 + H^+$; $NO_2^- + H_2O \rightleftharpoons HNO_2 + OH^-$; **e.** basic; $OCl^- + H_2O \rightleftharpoons HOCl + OH^-$; **f.** basic because OCl^- is a stronger base than NH_4^+ is an acid; $OCl^- + H_2O \rightleftharpoons HOCl + OH^-$, $NH_4^+ \rightleftharpoons NH_3 + H^+$ **131. a.** $HIO_3 < HBrO_3$; as the electronegativity of the central atom increases, acid strength increases. **b.** $HNO_2 < HNO_3$; as the number of oxygen atoms attached to the central atom increases, acid strength increases. **c.** HOI < HOCl; same reasoning as in part a. **d.** $H_3PO_3 < H_3PO_4$; same reasoning as in part b. **133. a.** $H_2O < H_2S < H_2Se$; acid strength increases as bond energy decreases. **b.** $CH_3CO_2H < FCH_2CO_2H < F_2CHCO_2H < F_3CCO_2H$; as the electronegativity of the neighboring atoms increases, acid strength increases. **c.** $NH_4^+ < HONH_3^+$; same reasoning as in part b. **d.** $NH_4^+ < PH_4^+$; same reasoning as in part a. **135. a.** basic; $CaO(s) + H_2O(l) \rightarrow Ca(OH)_2(aq)$; **b.** acidic; $SO_2(g) + H_2O(l) \rightarrow H_2SO_3(aq)$; **c.** acidic; $Cl_2O(g) + H_2O(l) \rightarrow 2HOCl(aq)$ **137. a.** $B(OH)_3$, acid; H_2O, base; **b.** Ag^+, acid; NH_3, base; **c.** BF_3, acid; F^-, base **139.** $Al(OH)_3(s) + 3H^+(aq) \rightarrow Al^{3+}(aq) + 3H_2O(l)$; $Al(OH)_3(s) + OH^-(aq) \rightarrow Al(OH)_4^-(aq)$ **141.** Fe^{3+}; because it is smaller with a greater positive charge, Fe^{3+} will be more strongly attracted to a lone pair of electrons from a Lewis base.   **143.** 990 mL H_2O **145.** NH_4Cl **147. a.** $Hb(O_2)_4$ in lungs, HbH_4^{4+} in cells; **b.** Decreasing $[CO_2]$ will decrease $[H^+]$, favoring $Hb(O_2)_4$ formation. Breathing into a bag raises $[CO_2]$. **c.** $NaHCO_3$ lowers the acidity from accumulated CO_2.   **149.** 4.2×10^{-2} M **151. a.** 2.62; **b.** 2.4%; **c.** 8.48   **153. a.** HA is a weak acid. Most of the acid is present as HA molecules; only one set of H^+ and A^- ions is present. In a strong acid, all of the acid would be dissociated into H^+ and A^- ions. **b.** 2.2×10^{-3} **155.** 9.2×10^{-7} **157. a.** 1.66; **b.** Fe^{2+} ions will produce a less acidic solution (higher pH) due to the lower charge on Fe^{2+} as compared with Fe^{3+}. As the charge on a metal ion increases, acid strength of the hydrated ion increases.   **159.** acidic; $HSO_4^- \rightleftharpoons SO_4^{2-} + H^+$; 1.54   **161. a.** H_2SO_3; **b.** $HClO_3$; **c.** H_3PO_3; NaOH and KOH are ionic compounds composed of either Na^+ or K^+ cations and OH^- anions. When soluble ionic compounds dissolve in

water, they form the ions from which they are formed. The acids in this problem are all covalent compounds. When these acids dissolve in water, the covalent bond between oxygen and hydrogen breaks to form H^+ ions.   **171.** 7.20.   **173.** 4540 mL   **175.** 4.17   **177.** 0.022 M **179.** 917 mL of water evaporated   **181.** PO_4^{3-}, $K_b = 0.021$; HPO_4^{2-}, $K_b = 1.6 \times 10^{-7}$; $H_2PO_4^-$, $K_b = 1.3 \times 10^{-12}$; from the K_b values, PO_4^{3-} is the strongest base.   **183. a.** basic; **b.** acidic; **c.** basic; **d.** acidic; **e.** acidic   **185.** 1.0×10^{-3} **187.** 5.4×10^{-4} **189.** 2.5×10^{-3}

Chapter 15

9. When an acid dissociates, ions are produced. A common ion is when one of the product ions in a particular equilibrium is added from an outside source. For a weak acid dissociating to its conjugate base and H^+, the common ion would be the conjugate base; this would be added by dissolving a soluble salt of the conjugate base into the acid solution. The presence of the conjugate base from an outside source shifts the equilibrium to the left so less acid dissociates.   **11.** The more weak acid and conjugate base present, the more H^+ and/or OH^- that can be absorbed by the buffer without significant pH change. When the concentrations of weak acid and conjugate base are equal (so that pH = pK_a), the buffer system is equally efficient at absorbing either H^+ or OH^-. If the buffer is overloaded with weak acid or with conjugate base, then the buffer is not equally efficient at absorbing either H^+ or OH^-.   **13. a.** Let's call the acid HA, which is a weak acid. When HA is present in the beakers, it exists in the undissociated form, making it a weak acid. A strong acid would exist as separate H^+ and A^- ions. **b.** beaker c → beaker a → beaker e → beaker b → beaker d **c.** pH = pK_a when a buffer solution is present that has equal concentrations of the weak acid and conjugate base. This is beaker e. **d.** The equivalence point is when just enough OH^- has been added to exactly react with all of the acid present initially. This is beaker b. **e.** Past the equivalence point, the pH is dictated by the concentration of excess OH^- added from the strong base. We can ignore the amount of hydroxide added by the weak conjugate base that is also present. This is beaker d.   **15.** The three key points to emphasize in your sketch are the initial pH, the pH at the halfway point to equivalence, and the pH at the equivalence point. For the two weak bases titrated, pH = pK_a at the halfway point to equivalence (50.0 mL HCl added) because [weak base] = [conjugate acid] at this point. For the initial pH, the strong base has the highest pH (most basic), while the weakest base has the lowest pH (least basic). At the equivalence point, the strong base titration has pH = 7.0. The weak bases titrated have acidic pHs at the equivalence point because the conjugate acids of the weak bases titrated are the major species present. The weakest base has the strongest conjugate acid so its pH will be lowest (most acidic) at the equivalence point.

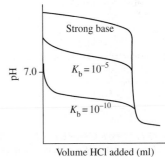

17. Only the third beaker represents a buffer solution. A weak acid and its conjugate base both must be present in large quantities in order to have a buffer solution. This is only the case in the third beaker. The first beaker represents a beaker full of strong acid that is 100% dissociated. The second beaker represents a weak acid solution. In a weak acid solution, only a small fraction of the acid is dissociated. In this representation, 1/10 of the weak acid has dissociated. The only B^- present in this beaker is from the dissociation of the weak acid. A buffer solution has B^- added from another source.   **19.** When strong acid or strong base is added to a sodium bicarbonate/sodium carbonate buffer mixture, the strong acid/base is neutralized. The reaction goes to completion resulting in the strong acid/base being replaced with a weak acid/base. This results in a new buffer solution. The reactions are $H^+(aq) + CO_3{}^{2-}(aq) \rightarrow HCO_3{}^-(aq)$; $OH^-(aq) + HCO_3{}^-(aq) \rightarrow CO_3{}^{2-}(aq) + H_2O(l)$　**21. a.** 2.96; **b.** 8.94; **c.** 7.00; **d.** 4.89　**23.** 1.1% vs. 1.3×10^{-2}% dissociated; the presence of $C_3H_5O_2{}^-$ in solution 21d greatly inhibits the dissociation of $HC_3H_5O_2$. This is called the *common ion effect.*　**25. a.** 1.70; **b.** 5.49; **c.** 1.70; **d.** 4.71　**27. a.** 4.29; **b.** 12.30; **c.** 12.30; **d.** 5.07　**29.** solution d; solution d is a buffer solution that resists pH changes.　**31.** 3.40　**33.** 3.48; 3.22　**35. a.** 5.14; **b.** 4.34; **c.** 5.14; **d.** 4.34　**37.** 4.37　**39. a.** 7.97; **b.** 8.73; both solutions have an initial pH = 8.77. The two solutions differ in their buffer capacities. Solution b with the larger concentrations has the greater capacity to resist pH change.　**41.** 15 g　**43. a.** 0.19; **b.** 0.59; **c.** 1.0; **d.** 1.9　**45.** $1.3 \times 10^{-2}\ M$　**47.** HOCl; there are many possibilities. One possibility is a solution with [HOCl] = 1.0 M and [NaOCl] = 0.35 M.　**49.** 8.18; add 0.20 mole NaOH　**51.** solution d　**53. a.** 1.0 mole; **b.** 0.30 mole; **c.** 1.3 moles　**55. a.** ~22 mL base added; **b.** buffer region is from ~1 mL to ~21 mL base added. The maximum buffering region would be from ~5 mL to ~17 mL base added with the halfway point to equivalence (~11 mL) as the best buffer point. **c.** ~11 mL base added; **d.** 0 mL base added; **e.** ~22 mL base added (the stoichiometric point); **f.** any point after the stoichiometric point (volume base added > ~22 mL)　**57. a.** 0.699; **b.** 0.854; **c.** 1.301; **d.** 7.00; **e.** 12.15　**59. a.** 2.72; **b.** 4.26; **c.** 4.74; **d.** 5.22; **e.** 8.79; **f.** 12.15

61.

Volume (mL)	pH
0.0	2.43
4.0	3.14
8.0	3.53
12.5	3.86
20.0	4.46
24.0	5.24
24.5	5.6
24.9	6.3
25.0	8.28
25.1	10.3
26.0	11.30
28.0	11.75
30.0	11.96

See *Solutions Guide* for pH plot.

63.

Volume (mL)	pH
0.0	11.11
4.0	9.97
8.0	9.58
12.5	9.25
20.0	8.65
24.0	7.87
24.5	7.6
24.9	6.9
25.0	5.28
25.1	3.7
26.0	2.71
28.0	2.24
30.0	2.04

See *Solutions Guide* for pH plot.

65. a. 4.19, 8.45; **b.** 10.74; 5.96; **c.** 0.89, 7.00　**67.** 2.1×10^{-6}　**69. a.** yellow; **b.** 8.0; **c.** blue　**71.** phenolphthalein　**73.** Phenol red is one possible indicator for the titration in Exercise 57. Phenolphthalein is one possible indicator for the titration in Exercise 59.　**75.** Phenolphthalein is one possible indicator for Exercise 61. Bromcresol green is one possible indicator for Exercise 63.　**77.** The pH is between 5 and 8.　**79. a.** yellow; **b.** green; **c.** yellow; **d.** blue　**81.** $pOH = pK_b + \log\dfrac{[\text{acid}]}{[\text{base}]}$　**83. a.** The optimal buffer pH is about 8.1; **b.** 0.083 at pH = 7.00; 8.3 at pH = 9.00; **c.** 8.08; 7.95　**85. a.** potassium fluoride + HCl; **b.** benzoic acid + NaOH; **c.** acetic acid + sodium acetate; **d.** HOCl + NaOH; **e.** ammonium chloride + NaOH　**87. a.** $1.1 \approx 1$; **b.** A best buffer has approximately equal concentrations of weak acid and conjugate base, so pH $\approx$ pK_a for a best buffer. The pK_a

value for an $H_3PO_4/H_2PO_4{}^-$ buffer is $-\log(7.5 \times 10^{-3}) = 2.12$. A pH of 7.15 is too high for an $H_3PO_4/H_2PO_4{}^-$ buffer to be effective. At this high of a pH, there would be so little H_3PO_4 present that we could hardly consider it a buffer; this solution would not be effective in resisting pH changes, especially when a strong base is added.　**89. a.** 1.8×10^9; **b.** 5.6×10^4; **c.** 1.0×10^{14}　**91.** 4.4 L　**93.** 65 mL　**95.** 180. g/mol; 3.3×10^{-4}; assumed acetylsalicylic acid is a weak monoprotic acid.　**97.** 0.210 M　**99.** 1.74×10^{-8}　**109.** 49 mL　**111.** 3.9 L　**113. a.** 200.0 mL; **b.** i. H_2A, H_2O; ii. H_2A, HA^-, H_2O, Na^+; iii. HA^-, H_2O, Na^+; iv. HA^-, A^{2-}, H_2O, Na^+; v. A^{2-}, H_2O, Na^+; vi. A^{2-}, H_2O, Na^+, OH^-; **c.** $K_{a_1} = 1 \times 10^{-4}$; $K_{a_2} = 1 \times 10^{-8}$　**115. a.** The major species present at the various points after H^+ reacts completely are A: $CO_3{}^{2-}$, H_2O, Na^+; B: $CO_3{}^{2-}$, $HCO_3{}^-$, H_2O, Cl^-, Na^+; C: $HCO_3{}^-$, H_2O, Cl^-, Na^+; D: $HCO_3{}^-$, CO_2 (H_2CO_3), H_2O, Cl^-, Na^+; E: CO_2 (H_2CO_3), H_2O, Cl^-, Na^+; F: H^+, CO_2 (H_2CO_3), H_2O, Cl^-, Na^+; **b.** B: pH = 10.25; D: pH = 6.37　**117.** pH $\approx$ 5.0; $K_a \approx 1 \times 10^{-10}$　**119.** 3.00　**121.** 2.78

Chapter 16

9. K_{sp} values can only be compared to determine relative solubilities when the salts produce the same number of ions. Here, Ag_2S and CuS do not produce the same number of ions when they dissolve, so each has a different mathematical relationship between the K_{sp} value and the molar solubility. To determine which salt has the larger molar solubility, you must do the actual calculations and compare the two molar solubility values.　**11. i.** This is the result when you have a salt that breaks up into two ions. Examples of these salts would be AgCl, $SrSO_4$, $BaCrO_4$, and $ZnCO_3$. **ii.** This is the result when you have a salt that breaks up into three ions, either two cations and one anion or one cation and two anions. Some examples are SrF_2, Hg_2I_2, and Ag_2SO_4. **iii.** This is the result when you have a salt that breaks up into four ions, either three cations and one anion (Ag_3PO_4) or one cation and three anions (ignoring the hydroxides, there are no examples of this type of salt in Table 16.1). **iv.** This is the result when you have a salt that breaks up into five ions, either three cations and two anions [$Sr_3(PO_4)_2$] or two cations and three anions (no examples of this type of salt are in Table 16.1).　**13.** For the K_{sp} reaction of a salt dissolving into its respective ions, the common ion would be if one of the ions in the salt was added from an outside source. When a common ion is present, the K_{sp} equilibrium shifts to the left resulting in less of the salt dissolving into its ions.　**15.** Some people would automatically think that an increase in temperature would increase the solubility of a salt. This is not always the case because some salts show a decrease in solubility as temperature increases. The two major methods used to increase solubility of a salt both involve removing one of the ions in the salt by reaction. If the salt has an ion with basic properties, adding H^+ will increase the solubility of the salt because the added H^+ will react with the basic ion, thus removing it from solution. More salt dissolves in order to make up for the lost ion. Some examples of salts with basic ions are AgF, $CaCO_3$, and $Al(OH)_3$. The other way to remove an ion is to form a complex ion. For example, the Ag^+ ion in silver salts forms the complex ion $Ag(NH_3)_2{}^+$ as ammonia is added. Silver salts increase their solubility as NH_3 is added because the Ag^+ ion is removed through complex ion formation.　**17.** In 2.0 M NH_3, the soluble complex ion $Ag(NH_3)_2{}^+$ forms, which increases the solubility of AgCl(s). The reaction is $AgCl(s) + 2NH_3 \rightleftharpoons Ag(NH_3)_2{}^+ + Cl^-$. In 2.0 M NH_4NO_3, NH_3 is only formed by the dissociation of the weak acid $NH_4{}^+$. There is not enough NH_3 produced by this reaction to dissolve AgCl(s) by the formation of the complex ion.　**19. a.** $AgC_2H_3O_2(s) \rightleftharpoons Ag^+ (aq) + C_2H_3O_2{}^- (aq)$; $K_{sp} = [Ag^+][C_2H_3O_2{}^-]$; **b.** $Al(OH)_3(s) \rightleftharpoons Al^{3+} (aq) + 3OH^- (aq)$; $K_{sp} = [Al^{3+}][OH^-]^3$; **c.** $Ca_3(PO_4)_2(s) \rightleftharpoons 3Ca^{2+}(aq) + 2PO_4{}^{3-}(aq)$; $K_{sp} = [Ca^{2+}]^3[PO_4{}^{3-}]^2$　**21. a.** 2.3×10^{-9}; **b.** 8.20×10^{-19}　**23.** 1.4×10^{-8}　**25.** 3.92×10^{-5}　**27. a.** 1.6×10^{-5} mol/L; **b.** 9.3×10^{-5} mol/L; **c.** 6.5×10^{-7} mol/L　**29.** 0.89 g　**31.** 1.3×10^{-4} mol/L　**33.** 2×10^{-11} mol/L　**35. a.** CaF_2; **b.** $FePO_4$　**37. a.** 4×10^{-17} mol/L; **b.** 4×10^{-11} mol/L; **c.** 4×10^{-29} mol/L　**39. a.** 1.4×10^{-2} mol/L; **b.** 1.2×10^{-3} mol/L; **c.** 3.9×10^{-3} mol/L　**41.** 2.3×10^{-11} mol/L　**43.** 3.5×10^{-10}　**45.** If the anion in the salt can act as a base in water, then the solubility of the salt will increase as the solution becomes more acidic. Added H^+ will react with the base, forming the conjugate acid. As the basic anion is removed, more of the salt will dissolve to replenish the basic anion. The salts with basic anions are Ag_3PO_4, $CaCO_3$, $CdCO_3$, and $Sr_3(PO_4)_2$. Hg_2Cl_2 and PbI_2 do not have any pH dependence because Cl^- and I^- are terrible bases (the conjugate bases of strong acids).

$$Ag_3PO_4(s) + H^+(aq) \longrightarrow 3Ag^+(aq) + HPO_4^{2-}(aq) \xrightarrow{\text{excess } H^+}$$
$$3Ag^+(aq) + H_3PO_4(aq)$$

$$CaCO_3(s) + H^+(aq) \longrightarrow Ca^{2+}(aq) + HCO_3^-(aq) \xrightarrow{\text{excess } H^+}$$
$$Ca^{2+}(aq) + H_2CO_3(aq) \, [H_2O(l) + CO_2(g)]$$

$$CdCO_3(s) + H^+(aq) \longrightarrow Cd^{2+}(aq) + HCO_3^-(aq) \xrightarrow{\text{excess } H^+}$$
$$Cd^{2+}(aq) + H_2CO_3(aq) \, [H_2O(l) + CO_2(g)]$$

$$Sr_3(PO_4)_2(s) + 2H^+(aq) \longrightarrow 3Sr^{2+}(aq) + 2HPO_4^{2-}(aq) \xrightarrow{\text{excess } H^+}$$
$$3Sr^{2+}(aq) + 2H_3PO_4(aq)$$

47. 1.5×10^{-19} g **49.** No precipitate forms. **51.** $PbF_2(s)$ will not form.
53. $[K^+] = 0.160\ M$, $[C_2O_4^{2-}] = 3.3 \times 10^{-7}\ M$, $[Ba^{2+}] = 0.0700\ M$, $[Br^-] = 0.300\ M$ **55.** 7.5×10^{-6} mol/L **57.** $[AgNO_3] > 5.6 \times 10^{-5}\ M$
59. $PbS(s)$ will form first, followed by $Pb_3(PO_4)_2(s)$, and $PbF_2(s)$ will form last.
61. a.
$$Ni^{2+} + CN^- \rightleftharpoons NiCN^+$$
$$NiCN^+ + CN^- \rightleftharpoons Ni(CN)_2$$
$$Ni(CN)_2 + CN^- \rightleftharpoons Ni(CN)_3^-$$
$$\underline{Ni(CN)_3^- + CN^- \rightleftharpoons Ni(CN)_4^{2-}}$$
$$Ni^{2+} + 4CN^- \rightleftharpoons Ni(CN)_4^{2-}$$
b.
$$V^{3+} + C_2O_4^{2-} \rightleftharpoons VC_2O_4^+$$
$$VC_2O_4^+ + C_2O_4^{2-} \rightleftharpoons V(C_2O_4)_2^-$$
$$\underline{V(C_2O_4)_2^- + C_2O_4^{2-} \rightleftharpoons V(C_2O_4)_3^{3-}}$$
$$V^{3+} + 3C_2O_4^{2-} \rightleftharpoons V(C_2O_4)_3^{3-}$$

63. 1.0×10^{42} **65.** $Hg^{2+}(aq) + 2I^-(aq) \rightleftharpoons HgI_2(s)$ (orange precipitate); $HgI_2(s) + 2I^-(aq) \rightleftharpoons HgI_4^{2-}(aq)$ (soluble complex ion) **67.** $3.3 \times 10^{-32}\ M$
69. a. $1.0 \times 10^{-3}\ M$; **b.** $2.0 \times 10^{-7}\ M$; **c.** $8.0 \times 10^{-15}\ M$ **71. a.** 1.2×10^{-8} mol/L; **b.** 1.5×10^{-4} mol/L; **c.** The presence of NH_3 increases the solubility of AgI. Added NH_3 removes Ag^+ from solution by forming the complex ion $Ag(NH_3)_2^+$. As Ag^+ is removed, more AgI will dissolve to replenish the Ag^+ concentration. **73.** 4.7×10^{-2} mol/L **75.** Test tube 1: added Cl^- reacts with Ag^+ to form the silver chloride precipitate. The net ionic equation is $Ag^+(aq) + Cl^-(aq) \rightarrow AgCl(s)$. Test tube 2: added NH_3 reacts with Ag^+ ions to form the soluble complex ion $Ag(NH_3)_2^+$. As this complex ion forms, Ag^+ is removed from solution, which causes AgCl(s) to dissolve. When enough NH_3 is added, then all of the silver chloride precipitate will dissolve. The equation is $AgCl(s) + 2NH_3(aq) \rightarrow Ag(NH_3)_2^+(aq) + Cl^-(aq)$. Test tube 3: added H^+ reacts with the weak base NH_3 to form NH_4^+. As NH_3 is removed, Ag^+ ions are released to solution, which can then react with Cl^- to re-form AgCl(s). The equations are $Ag(NH_3)_2^+(aq) + 2H^+(aq) \rightarrow Ag^+(aq) + 2NH_4^+(aq)$ and $Ag^+(aq) + Cl^-(aq) \rightarrow AgCl(s)$. **77.** 1.7 g AgCl; 5×10^{-9} mol/L **79.** 2.7×10^{-5} mol/L; the solubility of hydroxyapatite will increase as a solution gets more acidic, since both phosphate and hydroxide can react with H^+. 6×10^{-8} mol/L; the hydroxyapatite in the tooth enamel is converted to the less soluble fluorapatite by fluoride-treated water. The less soluble fluorapatite will then be more difficult to dissolve, making teeth less susceptible to decay. **81.** Addition of more than 9.0×10^{-6} g $Ca(NO_3)_2$ should start precipitation of $CaF_2(s)$. **83.** 7.0×10^{-8} **85.** 6.2×10^5
87. a. 6.7×10^{-6} mol/L; **b.** 1.2×10^{-13} mol/L; **c.** $Pb(OH)_2(s)$ will not form since $Q < K_{sp}$. **89. a.** 1.6×10^{-6}; **b.** 0.056 mol/L **91.** $Ba(OH)_2$, pH = 13.34; $Sr(OH)_2$, pH = 12.93; $Ca(OH)_2$, pH = 12.15 **99. a.** 0.33 mol/L;
b. 0.33 M; **c.** $4.8 \times 10^{-3}\ M$ **101. a.** 7.1×10^{-7} mol/L; **b.** 8.7×10^{-3} mol/L;
c. The presence of NH_3 increases the solubility of AgBr. Added NH_3 removes Ag^+ from solution by forming the complex ion, $Ag(NH_3)_2^+$. As Ag^+ is removed, more AgBr(s) will dissolve to replenish the Ag^+ concentration.
d. 0.41 g AgBr; **e.** Added HNO_3 will have no effect on the AgBr(s) solubility in pure water. Neither H^+ nor NO_3^- react with Ag^+ or Br^- ions. Br^- is the conjugate base of the strong acid HBr, so it is a terrible base. However, added

HNO_3 will reduce the solubility of AgBr(s) in the ammonia solution. NH_3 is a weak base ($K_b = 1.8 \times 10^{-5}$). Added H^+ will react with NH_3 to form NH_4^+. As NH_3 is removed, a smaller amount of the $Ag(NH_3)_2^+$ complex ion will form, resulting in a smaller amount of AgBr(s) that will dissolve.
103. 5.7×10^{-2} mol/L **105.** 3 M **107.** $4.8 \times 10^{-11}\ M$ **109. a.** 5.8×10^{-4} mol/L; **b.** greater; F^- is a weak base ($K_b = 1.4 \times 10^{-11}$), so some of the F^- is removed by reaction with water. As F^- is removed, more SrF_2 will dissolve; **c.** 3.5×10^{-3} mol/L **111.** pH = 0.70; $0.20\ M\ Ba^{2+}$; 23 g $BaSO_4(s)$

Chapter 17
11. Living organisms need an external source of energy to carry out these processes. For all processes combined, ΔS_{univ} must be greater than zero (the 2nd law). **13.** As any process occurs, ΔS_{univ} will increase; ΔS_{univ} cannot decrease. Time also goes in one direction, just as ΔS_{univ} goes in one direction. **15.** This reaction is kinetically slow but thermodynamically favorable ($\Delta G < 0$). Thermodynamics only tells us if a reaction can occur. To answer the question will the reaction occur, one also needs to consider the kinetics (speed of reaction). The ultraviolet light provides the activation energy for this slow reaction to occur. **17.** $\Delta S_{surr} = -\Delta H/T$; heat flow ($\Delta H$) into or out of the system dictates ΔS_{surr}. If heat flows into the surroundings, the random motions of the surroundings increase, and the entropy of the surroundings increases. The opposite is true when heat flows from the surroundings into the system (an endothermic reaction). Although the driving force described here really results from the change in entropy of the surroundings, it is often described in terms of energy. Nature tends to seek the lowest possible energy. **19.** Note that these substances are not in the solid state but are in the aqueous state; water molecules are also present. There is an apparent increase in ordering when these ions are placed in water. The hydrating water molecules must be in a highly ordered state when surrounding these anions. **21.** One can determine $\Delta S°$ and $\Delta H°$ for the reaction using the standard entropies and standard enthalpies of formation in Appendix 4, then use the equation $\Delta G° = \Delta H° - T\Delta S°$. One can also use the standard free energies of formation in Appendix 4. And finally, one can use Hess's law to calculate $\Delta G°$. Here, reactions having known $\Delta G°$ values are manipulated to determine $\Delta G°$ for a different reaction. For temperatures other than 25°C, $\Delta G°$ is estimated using the $\Delta G° = \Delta H° - T\Delta S°$ equation. The assumptions made are that the $\Delta H°$ and $\Delta S°$ values determined from Appendix 4 data are temperature independent. We use the same $\Delta H°$ and $\Delta S°$ values as determined when $T = 25°C$, then plug in the new temperature in kelvins into the equation to estimate $\Delta G°$ at the new temperature. **23.** The light source for the first reaction is necessary for kinetic reasons. The first reaction is just too slow to occur unless a light source is available. The kinetics of a reaction are independent of the thermodynamics of a reaction. Even though the first reaction is more favorable thermodynamically (assuming standard conditions), it is unfavorable for kinetic reasons. The second reaction has a negative $\Delta G°$ value and is a fast reaction, so the second reaction occurs very quickly. When considering if a reaction will occur, thermodynamics and kinetics must both be considered. **25.** a, b, c
27. Possible arrangements for one molecule:

1 way   1 way

Both are equally probable.
Possible arrangements for two molecules:

1 way   2 ways   1 way
                Most probable

Possible arrangement for three molecules:

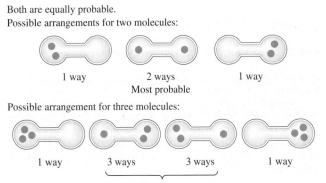

1 way   3 ways   3 ways   1 way

Equally most probable

29. We draw all of the possible arrangements of the two particles in the three levels.

2 kJ	—	—	x	—	x	xx
1 kJ	—	x	—	xx	x	—
0 kJ	xx	x	x	—	—	—
Total $E =$	0 kJ	1 kJ	2 kJ	2 kJ	3 kJ	4 kJ

The most likely total energy is 2 kJ.   **31. a.** H_2 at 100°C and 0.5 atm; **b.** N_2 at STP; **c.** $H_2O(l)$ **33. a.** negative; **b.** positive   **35.** Spontaneous ($\Delta G < 0$) for b, c, d   **37.** 89.3 J/K · mol   **39. a.** yes ($\Delta G < 0$); **b.** 196 K   **41. a.** negative; **b.** negative; **c.** negative; **d.** positive   **43. a.** $C_{graphite}(s)$; **b.** $C_2H_5OH(g)$; **c.** $CO_2(g)$ **45. a.** negative, -186 J/K; **b.** positive, 187 J/K; **c.** hard to predict since $\Delta n = 0$; 138 J/K   **47.** 262 J/K · mol   **49. a.** ΔH and ΔS are both positive; **b.** $S_{rhombic}$ **51. a.** ΔH and ΔS are both negative; **b.** low temperatures   **53. a.** $\Delta H^\circ = -803$ kJ, $\Delta S^\circ = -4$ J/K, $\Delta G^\circ = -802$ kJ; **b.** $\Delta H^\circ = 2802$ kJ, $\Delta S^\circ = -262$ J/K, $\Delta G^\circ = 2880.$ kJ; **c.** $\Delta H^\circ = -416$ kJ, $\Delta S^\circ = -209$ J/K, $\Delta G^\circ = -354$ kJ; **d.** $\Delta H^\circ = -176$ kJ, $\Delta S^\circ = -284$ J/K, $\Delta G^\circ = -91$ kJ   **55.** -5.40 kJ; 328.6 K; ΔG° is negative below 328.6 K.   **57.** -817 kJ   **59.** -731 kJ/mol   **61. a.** 53 kJ; **b.** No, the reaction is not spontaneous at standard concentrations and 298 K. **c.** $T > 630$ K   **63.** $CH_4(g) + CO_2(g) \rightarrow CH_3CO_2H(l)$, $\Delta H^\circ = -16$ kJ, $\Delta S^\circ = -240.$ J/K, $\Delta G^\circ = 56$ kJ; $CH_3OH(g) + CO(g) \rightarrow CH_3CO_2H(l)$, $\Delta H^\circ = -173$ kJ, $\Delta S^\circ = -278$ J/K, $\Delta G^\circ = -90.$ kJ; the second reaction is preferred. It should be run at temperatures below 622 K.   **65.** -188 kJ   **67. a.** shifts right; **b.** no shift since the reaction is at equilibrium; **c.** shifts left   **69.** -198 kJ; 5.07×10^{34} **71.** 8.72; 0.0789   **73.** 140 kJ   **75. a.** 2.22×10^5; **b.** 94.3   **77.** -71 kJ/mol   **79.** $\Delta H^\circ = 1.1 \times 10^5$ J/mol; $\Delta S^\circ = 330$ J/K · mol; the major difference in the plot is the slope of the line. An endothermic process has a negative slope for the $\ln(K)$ versus $1/T$ plot, whereas an exothermic process has a positive slope.   **81.** 447 J/K · mol   **83.** decreases; ΔS will be negative since 2 moles of gaseous reactants form 1 mole of gaseous product. For ΔG to be negative, ΔH must be negative (exothermic). For exothermic reactions, K decreases as T increases, so the ratio of the partial pressure of PCl_5 to the partial pressure of PCl_3 will decrease.   **85.** 43.7 K   **87.** 7.0×10^{-4} **89.** 60   **91.** ΔS is more favorable (less negative) for reaction 2 than for reaction 1, resulting in $K_2 > K_1$. In reaction 1, seven particles in solution form one particle. In reaction 2, four particles form one, which results in a smaller decrease in positional probability than for reaction 1.   **93.** 725 K   **103. a.** Vessel 1: At 0°C, this system is at equilibrium, so $\Delta S_{univ} = 0$ and $\Delta S = \Delta S_{surr}$. Because the vessel is perfectly insulated, $q = 0$ so $\Delta S_{surr} = 0 = \Delta S_{sys}$. **b.** Vessel 2: The presence of salt in water lowers the freezing point of water to a temperature below 0°C. In vessel 2 the conversion of ice into water will be spontaneous at 0°C, so $\Delta S_{univ} > 0$. Because the vessel is perfectly insulated, $\Delta S_{surr} = 0$. Therefore, ΔS_{sys} must be positive ($\Delta S > 0$) in order for ΔS_{univ} to be positive.   **105.** $\Delta H^\circ = 286$ kJ; $\Delta G^\circ = 326$ kJ; $K = 7.22 \times 10^{-58}$; $P_{O_3} = 3.3 \times 10^{-41}$ atm; this partial pressure represents one molecule of ozone per 9.5×10^{17} L of air. Equilibrium is probably not maintained under the conditions because the concentration of ozone is not large enough to maintain equilibrium   **107. a.**

$$k_f = A \exp\left(\frac{-E_a}{RT}\right) \text{ and } k_r = A \exp\left(\frac{-(E_a - \Delta G^\circ)}{RT}\right),$$

$$\frac{k_f}{k_r} = \exp\left(\frac{-E_a}{RT} + \frac{(E_a - \Delta G^\circ)}{RT}\right) = \exp\left(\frac{-\Delta G^\circ}{RT}\right)$$

From $\Delta G^\circ = -RT \ln K$, K also equals the same expression: $K = \exp\left(\frac{-\Delta G^\circ}{RT}\right)$, so $K = \frac{k_f}{k_r}$. **b.** A catalyst increases the value of the rate constant (increases rate) by lowering the activation energy. For the equilibrium constant K to remain constant, both k_f and k_r must increase by the same factor. Therefore, a catalyst must increase the rate of both the forward and the reverse reactions.   **109. a.** 0.333; **b.** $P_A = 1.50$ atm; $P_B = 0.50$ atm; **c.** $\Delta G = \Delta G^\circ + RT \ln(P_B/P_A) = 2722 \text{ J} - 2722 \text{ J} = 0$ **111.** greater than 7.5 torr   **113.** 6 M   **115.** 61 kJ/mol   **117.** -4.1 kJ/mol

Chapter 18

15. Oxidation: increase in oxidation number, loss of electrons; reduction: decrease in oxidation number, gain of electrons   **17.** Reactions a, b, and c are oxidation–reduction reactions.

Oxidizing Agent	Reducing Agent	Substance Oxidized	Substance Reduced
a. H_2O	CH_4	$CH_4(C)$	$H_2O(H)$
b. $AgNO_3$	Cu	Cu	$AgNO_3(Ag)$
c. HCl	Zn	Zn	HCl(H)

19. Electrochemistry is the study of the interchange of chemical and electrical energy. A redox (oxidation–reduction) reaction is a reaction in which one or more electrons are transferred. In a galvanic cell, a spontaneous redox reaction occurs which produces an electric current. In an electrolytic cell, electricity is used to force a nonspontaneous redox reaction to occur.   **21.** Magnesium is an alkaline earth metal; Mg will oxidize to Mg^{2+}. The oxidation state of hydrogen in HCl is $+1$. To be reduced, the oxidation state of H must decrease. The obvious choice for the hydrogen product is $H_2(g)$ where hydrogen has a zero oxidation state. The balanced reaction is: $Mg(s) + 2HCl(aq) \rightarrow MgCl_2(aq) + H_2(g)$. Mg goes from the 0 to the $+2$ oxidation state by losing two electrons. Each H atom goes from the $+1$ to the 0 oxidation state by gaining one electron. Since there are two H atoms in the balanced equation, then a total of two electrons are gained by the H atoms. Hence, two electrons are transferred in the balanced reaction. When the electrons are transferred directly from Mg to H^+, no work is obtained. In order to harness this reaction to do useful work, we must control the flow of electrons through a wire. This is accomplished by making a galvanic cell, which separates the reduction reaction from the oxidation reaction in order to control the flow of electrons through a wire to produce a voltage.   **23.** An extensive property is one that depends on the amount of substance. The free energy change for a reaction depends on whether 1 mole of product is produced or 2 moles of product is produced or 1 million moles of product is produced. This is not the case for cell potentials, which do not depend on the amount of substance. The equation that relates ΔG to E is $\Delta G = -nFE$. It is the n term that converts the intensive property E into the extensive property ΔG. n is the number of moles of electrons transferred in the balanced reaction that ΔG is associated with.   **25.** A potential hazard when jump-starting a car is that the electrolysis of $H_2O(l)$ can occur. When $H_2O(l)$ is electrolyzed, the products are the explosive gas mixture of $H_2(g)$ and $O_2(g)$. A spark produced during jump-starting a car could ignite any $H_2(g)$ and $O_2(g)$ produced. Grounding the jumper cable far from the battery minimizes the risk of a spark nearby the battery where $H_2(g)$ and $O_2(g)$ could be collecting.   **27.** You need to know the identity of the metal so you know which molar mass to use. You need to know the oxidation state of metal ion in the salt so the moles of electrons transferred can be determined. And finally, you need to know the amount of current and the time the current was passed through the electrolytic cell. If you know these four quantities, then the mass of metal plated out can be calculated.   **29. a.** $3I^-(aq) + 2H^+(aq) + ClO^-(aq) \rightarrow I_3^-(aq) + Cl^-(aq) + H_2O(l)$; **b.** $7H_2O(l) + 4H^+(aq) + 3As_2O_3(s) + 4NO_3^-(aq) \rightarrow 4NO(g) + 6H_3AsO_4(aq)$; **c.** $16H^+(aq) + 2MnO_4^-(aq) + 10Br^-(aq) \rightarrow 5Br_2(l) + 2Mn^{2+}(aq) + 8H_2O(l)$; **d.** $8H^+(aq) + 3CH_3OH(aq) + Cr_2O_7^{2-}(aq) \rightarrow 2Cr^{3+}(aq) + 3CH_2O(aq) + 7H_2O(l)$ **31. a.** $2H_2O(l) + Al(s) + MnO_4^-(aq) \rightarrow Al(OH)_4^-(aq) + MnO_2(s)$; **b.** $2OH^-(aq) + Cl_2(g) \rightarrow Cl^-(aq) + OCl^-(aq) + H_2O(l)$; **c.** $OH^-(aq) + H_2O(l) + NO_2^-(aq) + 2Al(s) \rightarrow NH_3(g) + 2AlO_2^-(aq)$ **33.** $4NaCl(aq) + 2H_2SO_4(aq) + MnO_2(s) \rightarrow 2Na_2SO_4(aq) + MnCl_2(aq) + Cl_2(g) + 2H_2O(l)$ **35.** The reducing agent causes reduction to occur; it does this by containing the species that is oxidized. Oxidation occurs at the anode, so the reducing agent will be in the anode compartment. The oxidizing agent causes oxidation to occur; it does this by containing the species that is reduced. Reduction occurs at the cathode, so the oxidizing agent will be in the cathode compartment. Electron flow is always from the anode compartment to the cathode compartment.   **37.** See Fig. 18.3 of the text for a typical galvanic cell. The anode compartment contains the oxidation half-reaction compounds/ions, and the cathode compartment contains the reduction half-reaction compounds/ions. The electrons flow from the anode to the cathode. For each of the following answers, all solutes are 1.0 M and all gases are at 1.0 atm. **a.** $7H_2O(l) + 2Cr^{3+}(aq) + 3Cl_2(g) \rightarrow Cr_2O_7^{2-}(aq) + 6Cl^-(aq) + 14H^+(aq)$; cathode: Pt electrode; Cl_2 bubbled into solution, Cl^- in solution; anode: Pt electrode; Cr^{3+}, H^+, and $Cr_2O_7^{2-}$ in solution; **b.** $Cu^{2+}(aq) + Mg(s) \rightarrow Cu(s) + Mg^{2+}(aq)$; cathode: Cu electrode; Cu^{2+} in solution; anode: Mg electrode; Mg^{2+} in solution **39. a.** 0.03 V; **b.** 2.71 V **41.** See Exercise 37 for a description of a galvanic cell. For each of the following answers, all solutes are 1.0 M and all gases are at 1.0 atm. In the salt bridge, cations flow to the cathode and anions flow to the anode. **a.** $Cl_2(g) + 2Br^-(aq) \rightarrow Br_2(aq) + 2Cl^-(aq)$, $\mathscr{E}^\circ = 0.27$ V; cathode: Pt electrode; $Cl_2(g)$ bubbled in, Cl^- in solution; anode: Pt electrode; Br_2 and Br^- in solution; **b.** $3H_2O(l) + 5IO_4^-(aq) + 2Mn^{2+}(aq) \rightarrow 5IO_3^-(aq) + 2MnO_4^-(aq) + 6H^+(aq)$, $\mathscr{E}^\circ = 0.09$ V; cathode: Pt electrode; IO_4^-, IO_3^-, and H_2SO_4 (as a source of H^+) in solution; anode: Pt electrode; Mn^{2+}, MnO_4^-, and H_2SO_4 in solution **43.** 37a. $Pt|Cr^{3+}$ (1.0 M), H^+ (1.0 M), $Cr_2O_7^{2-}$ (1.0 M)$||Cl_2$ (1.0 atm)$|Cl^-$ (1.0 M)$|$Pt; 37b. $Mg|Mg^{2+}$ (1.0 M)$||Cu^{2+}$ (1.0 M)$|$Cu; 41a. $Pt|Br^-$ (1.0 M), Br_2 (1.0 M)$||Cl_2$ (1.0 atm)$|Cl^-$ (1.0 M)$|$Pt; 41b. $Pt|Mn^{2+}$ (1.0 M), MnO_4^- (1.0 M), H^+ (1.0 M)$||IO_4^-$ (1.0 M), H^+ (1.0 M), IO_3^- (1.0 M) $|$Pt **45. a.** $Au^{3+}(aq) + 3Cu^+(aq) \rightarrow 3Cu^{2+}(aq) + Au(s)$, $\mathscr{E}^\circ = 1.34$ V; **b.** $2VO_2^+(aq) + 4H^+(aq) + Cd(s) \rightarrow Cd^{2+}(aq) + 2VO^{2+}(aq) + 2H_2O(l)$, $\mathscr{E}^\circ = 1.40$ V **47. a.** $16H^+ + 2MnO_4^- + 10I^- \rightarrow 5I_2 + 2Mn^{2+} + 8H_2O$, $\mathscr{E}^\circ_{cell} = 0.97$ V, spontaneous; **b.** $16H^+ + 2MnO_4^- + 10F^- \rightarrow 5F_2 + 2Mn^{2+} + 8H_2O$, $\mathscr{E}^\circ_{cell} = -1.36$ V, not spontaneous **49.** $\mathscr{E}^\circ = 0.41$ V, $\Delta G^\circ = -79$ kJ **51.** 45a. -388 kJ; 45b. $-270.$ kJ **53.** -0.829 V; the two values agree to two significant figures. **55.** 1.24 V **57.** $K^+ < H_2O < Cd^{2+} < I_2 < AuCl_4^- < IO_3^-$ **59. a.** no; **b.** yes; **c.** yes; **61. a.** Of the species available, Ag^+ would be the best oxidizing agent because it has the largest $\mathscr{E}^\circ$ value. **b.** Of the species available, Zn would be the best reducing agent because it has the largest $-\mathscr{E}^\circ$ value. **c.** $SO_4^{2-}(aq)$ can oxidize Pb(s) and Zn(s) at standard conditions. When $SO_4^{2-}(aq)$ is coupled with these reagents, $\mathscr{E}^\circ_{cell}$ is positive. **d.** Al(s) can reduce $Ag^+(aq)$ and $Zn^{2+}(aq)$ at standard conditions because $\mathscr{E}^\circ_{cell} > 0$. **63. a.** $Cr_2O_7^{2-}$, O_2, MnO_2, IO_3^-; **b.** $PbSO_4$, Cd^{2+}, Fe^{2+}, Cr^{3+}, Zn^{2+}, H_2O **65.** $ClO^-(aq) + 2NH_3(aq) \rightarrow Cl^-(aq) + N_2H_4(aq) + H_2O(l)$, $\mathscr{E}^\circ_{cell} = 1.00$ V; because $\mathscr{E}^\circ_{cell}$ is positive for this reaction, at standard conditions ClO^- can spontaneously oxidize NH_3 to the somewhat toxic N_2H_4. **67. a.** larger; **b.** smaller **69.** Electron flow is always from the anode to the cathode. For the cells with a nonzero cell potential, we will identify the cathode, which means the other compartment is the anode. **a.** 0; **b.** 0.018 V; compartment with $[Ag^+] = 2.0$ M is cathode; **c.** 0.059 V; compartment with $[Ag^+] = 1.0$ M is cathode; **d.** 0.26 V; compartment with $[Ag^+] = 1.0$ M is cathode; **e.** 0 **71.** 2.12 V **73.** 1.09 V **75.** [37]. **a.** $\Delta G^\circ = -20$ kJ; 1×10^3; **b.** $\Delta G^\circ = -523$ kJ; 5.12×10^{91}; [41]. **a.** $\Delta G^\circ = -52$ kJ; 1.4×10^9; **b.** $\Delta G^\circ = -90$ kJ; 2×10^{15} **77. a.** $Fe^{2+}(aq) + Zn(s) \rightarrow Zn^{2+}(aq) + Fe(s)$ $\mathscr{E}^\circ_{cell} = 0.32$ V; **b.** -62 kJ; 6.8×10^{10}; **c.** 0.20 V **79. a.** 0.23 V; **b.** 1.2×10^{-5} M **81.** 0.16 V, copper is oxidized. **83.** 1.7×10^{-30} **85. a.** no reaction; **b.** $Cl_2(g) + 2I^-(aq) \rightarrow I_2(s) + 2Cl^-(aq)$, $\mathscr{E}^\circ_{cell} = 0.82$ V; $\Delta G^\circ = -160$ kJ; $K = 5.6 \times 10^{27}$; **c.** no reaction; **d.** $4Fe^{2+}(aq) + 4H^+(aq) + O_2(g) \rightarrow 4Fe^{3+}(aq) + 2H_2O(l)$, $\mathscr{E}^\circ_{cell} = 0.46$ V; $\Delta G^\circ = -180$ kJ; $K = 1.3 \times 10^{31}$; **87.** 0.151 V; -29.1 kJ **89.** 5.1×10^{-20} **91.** -0.14 V **93. a.** 30. hours; **b.** 33 s; **c.** 1.3 hours **95. a.** 16 g; **b.** 25 g; **c.** 71 g; **d.** 4.9 g **97.** Bi **99.** 9.12 L F_2 (anode), 29.2 g K (cathode) **101.** 1×10^5 A **103.** 1.14×10^{-2} M **105.** Au followed by Ag followed by Ni followed by Cd **107.** Cathode: $2H_2O + 2e^- \rightarrow H_2(g) + 2OH^-$; anode: $2H_2O \rightarrow O_2(g) + 4H^+ + 4e^-$ **109. a.** cathode: $Ni^{2+} + 2e^- \rightarrow Ni$; anode: $2Br^- \rightarrow Br_2 + 2e^-$; **b.** cathode: $Al^{3+} + 3e^- \rightarrow Al$; anode: $2F^- \rightarrow F_2 + 2e^-$; **c.** cathode: $Mn^{2+} + 2e^- \rightarrow Mn$; anode: $2I^- \rightarrow I_2 + 2e^-$ **111. a.** cathode: $Ni^{2+} + 2e^- \rightarrow Ni$; anode: $2Br^- \rightarrow Br_2 + 2e^-$; **b.** cathode: $2H_2O + 2e^- \rightarrow H_2 + 2OH^-$; anode: $2H_2O \rightarrow O_2 + 4H^+ + 4e^-$; **c.** cathode: $2H_2O + 2e^- \rightarrow H_2 + 2OH^-$; anode: $2I^- \rightarrow I_2 + 2e^-$ **113.** 0.250 mol **115. a.** 0.10 V, SCE is anode; **b.** 0.53 V, SCE is anode; **c.** 0.02 V, SCE is cathode; **d.** 1.90 V, SCE is cathode; **e.** 0.47 V, SCE is cathode **117. a.** decrease; **b.** increase; **c.** decrease; **d.** decrease; **e.** same **119. a.** $\Delta G^\circ = -582$ kJ; $K = 3.45 \times 10^{102}$; $\mathscr{E}^\circ = 1.01$ V; **b.** -0.65 V **121.** Aluminum has the ability to form a durable oxide coating over its surface. Once the HCl dissolves this oxide coating, Al is exposed to H^+ and is easily oxidized to Al^{3+}. Thus, the Al foil disappears after the oxide coating is dissolved. **123.** 1.14 V **125.** $w_{max} = -13{,}200$ kJ; the work done can be no larger than the free energy change. If the process were reversible all of the free energy released would go into work, but this does not occur in any real process. Fuel cells are more efficient in converting chemical energy to electrical energy; they are also less massive. Major disadvantage: They are expensive. **127.** 0.98 V **129.** 7.44×10^4 A **131.** To produce 1.0 kg Al by the Hall-Heroult process requires 5.4×10^4 kJ of energy. To melt 1.0 kg Al requires 4.0×10^2 kJ of energy. It is feasible to recycle Al by melting the metal because, in theory, it takes less than 1% of the energy required to produce the same amount of Al by the Hall-Heroult process. **133.** +3 **141.** $\mathscr{E}^\circ = \frac{T\Delta S^\circ}{nF} - \frac{\Delta H^\circ}{nF}$; if we graph $\mathscr{E}^\circ$ versus T, we should get a straight line ($y = mx + b$). The slope of the line is equal to $\Delta S^\circ/nF$ and the y-intercept is equal to $-\Delta H^\circ/nF$. $\mathscr{E}^\circ$ will have little temperature dependence for cell reactions with ΔS° close to zero. **143.** 9.8×10^{-6} **145.** 2.39×10^{-7} **147. a.** ±0.02 pH units; $\pm6 \times 10^{-6}$ M H^+; **b.** ±0.001 V **149. a.** 0.16 V; **b.** 8.6 mol **151.** $[Ag^+] = 4.6 \times 10^{-18}$ M; $[Ni^{2+}] = 1.5$ M **153.** 0.64 V **155. a.** 5.77×10^{10}; **b.** -12.2 kJ/mol **157.** Osmium(IV) nitrate; $[Ar]4s^13d^{10}$

Chapter 19

1. The characteristic frequencies of energies emitted in a nuclear reaction suggest that discrete energy levels exist in the nucleus. The extra stability of certain numbers of nucleons and the predominance of nuclei with even numbers of nucleons suggest that the nuclear structure might be described by using quantum numbers. **3.** β-particle production has the net effect of turning a neutron into a proton. Radioactive nuclei having too many neutrons typically undergo β-particle decay. Positron production has the net effect of turning a proton into a neutron. Nuclei having too many protons typically undergo positron decay. **5.** The transuranium elements are the elements having more protons than uranium. They are synthesized by bombarding heavier nuclei with neutrons and positive ions in a particle accelerator. **7.** $\Delta E = \Delta mc^2$; The key difference is the mass change when going from reactants to products. In chemical reactions, the mass change is indiscernible. In nuclear processes, the mass change is discernible. It is the conversion of this discernible mass change into energy that results in the huge energies associated with nuclear processes. **9.** The temperatures of fusion reactions are so high that all physical containers would be destroyed. At these high temperatures, most of the electrons are stripped from the atoms. A plasma of gaseous ions is formed that can be controlled by magnetic fields. **11. a.** $^{3}_{1}H \rightarrow ^{3}_{2}He + ^{0}_{-1}e$; **b.** $^{8}_{3}Li \rightarrow ^{8}_{4}Be + ^{0}_{-1}e$, $^{8}_{4}Be \rightarrow 2\,^{4}_{2}He$; overall reaction: $^{8}_{3}Li \rightarrow 2\,^{4}_{2}He + ^{0}_{-1}e$; **c.** $^{7}_{4}Be + ^{0}_{-1}e \rightarrow ^{7}_{3}Li$; **d.** $^{8}_{5}B \rightarrow ^{8}_{4}Be + ^{0}_{+1}e$ **13. a.** $^{234}_{90}Th$; this is α-particle production. **b.** $^{0}_{-1}e$; this is β-particle production. **15. a.** $^{68}_{31}Ga + ^{0}_{-1}e \rightarrow ^{68}_{30}Zn$; **b.** $^{62}_{29}Cu \rightarrow ^{0}_{+1}e + ^{62}_{28}Ni$; **c.** $^{212}_{87}Fr \rightarrow ^{4}_{2}He + ^{208}_{85}At$; **d.** $^{129}_{51}Sb \rightarrow ^{0}_{-1}e + ^{129}_{52}Te$ **17.** 7 α particles; 4 β particles **19. a.** $^{241}_{95}Am \rightarrow ^{4}_{2}He + ^{237}_{93}Np$; **b.** $^{209}_{83}Bi$; **c.** The intermediate radionuclides are $^{237}_{93}Np$, $^{233}_{91}Pa$, $^{233}_{92}U$, $^{229}_{90}Th$, $^{225}_{88}Ra$, $^{225}_{89}Ac$, $^{221}_{87}Fr$, $^{217}_{85}At$, $^{213}_{83}Bi$, $^{213}_{84}Po$, and $^{209}_{82}Pb$. **21.** $^{53}_{26}Fe$ has too many protons. It will undergo positron production, electron capture, and/or alpha-particle production. $^{59}_{26}Fe$ has too many neutrons and will undergo beta-particle production. (See Table 19.2 of the text.) **23. a.** $^{249}_{98}Cf + ^{18}_{8}O \rightarrow ^{263}_{106}Sg + 4\,^{1}_{0}n$; **b.** $^{259}_{104}Rf$ **25.** 690 hours **27.** ^{81}Kr is most stable since it has the longest half-life. ^{73}Kr is "hottest" since it decays very rapidly due to its very short half-life. ^{73}Kr, 81s; ^{74}Kr, 34.5 min; ^{76}Kr, 44.4 h; ^{81}Kr, 6.3×10^5 yr **29.** 15 μg $^{47}CaCO_3$ should be ordered at a minimum. **31.** 19.6% **33.** The fraction that remains is 0.0041, or 0.41%. **35.** 26 g **37.** 2.3 counts per minute per gram of C. No; for a 10.-mg C sample, it would take roughly 40 min to see a single disintegration. This is too long to wait, and the background radiation would probably be much greater than the ^{14}C activity. Thus ^{14}C dating is not practical for very small samples. **39.** 3.8×10^9 yr **41.** 4.3×10^6 kg/s **43.** ^{232}Pu, -1.715×10^{14} J/mol; ^{231}Pa, -1.714×10^{14} J/mol **45.** ^{12}C: 1.230×10^{-12} J/nucleon; ^{235}U: 1.2154×10^{-12} J/nucleon; since ^{56}Fe is the most stable known nucleus, the binding energy per nucleon for ^{56}Fe would be larger than that of ^{12}C or ^{235}U. (See Fig. 19.9 of the text.) **47.** 6.01513 u **49.** -2.0×10^{10} J/g of hydrogen nuclei **51.** The Geiger–Müller tube has a certain response time. After the gas in the tube ionizes to produce a "count," some time must elapse for the gas to return to an electrically neutral state. The response of the tube levels off because, at high activities, radioactive particles are entering the tube faster than the tube can respond to them.   **53.** Water is produced in this reaction by removing an OH group from one substance and an H from the other substance. There are two ways to do this:

i.

$$CH_3C\overset{O}{\overline{}}\!\!\!\overset{\parallel}{}\!\overline{(OH + H)}\!-\!\overset{18}{O}CH_3 \longrightarrow CH_3\overset{O}{\overset{\parallel}{C}}\!-\!\overset{18}{O}CH_3 + HO\!-\!H$$

ii.

$$CH_3C\overset{O}{\overset{\parallel}{O}}\!-\!\overline{(H + H\,\overset{18}{O})}\!-\!CH_3 \longrightarrow CH_3\overset{O}{\overset{\parallel}{C}}O\!-\!CH_3 + H\!-\!\overset{18}{O}H$$

Because the water produced is not radioactive, methyl acetate forms by the first reaction, where all of the oxygen-18 ends up in methyl acetate.   **55.** 2 neutrons; 4 β particles   **57.** Strontium. Xe is chemically unreactive and not readily incorporated into the body. Sr can be easily oxidized to Sr^{2+}. Strontium is in the same family as calcium and could be absorbed and concentrated in the body in a fashion similar to Ca^{2+}. The chemical properties determine where radioactive material may be concentrated in the body or how easily it may be excreted.   **59. a.** unstable; beta production; **b.** stable; **c.** unstable; positron production or electron capture; **d.** unstable, positron production, electron capture, or alpha production.   **61.** 3800 decays/s   **63.** The third-life will be the time required for the number of nuclides to reach one-third of the original value ($N_0/3$). The third-life of this nuclide is 49.8 years.   **65.** 1975   **67.** 900 g ^{235}U **69.** 7 × 10^5 m/s; 8 × 10^{-16} J/nuclei;   **71.** All evolved $O_2(g)$ comes from water.   **79.** 77% ^{238}U and 23% ^{235}U **81.** Assuming that (1) the radionuclide is long lived enough that no significant decay occurs during the time of the experiment, and (2) the total activity is uniformly distributed only in the rat's blood, $V = 10.$ mL.   **83. a.** $^{12}_{6}C$; **b.** ^{13}N, ^{13}C, ^{14}N, ^{15}O, and ^{15}N; **c.** $-5.950 × 10^{11}$ J/mol ^{1}H **85.** 4.3 × 10^{-29}   **87.** $^{249}_{97}Bk + ^{22}_{10}Ne \rightarrow ^{267}_{107}Bh + 4\ ^{1}_{0}n$; 62.7 s; $[Rn]7s^25f^{14}6d^5$

Chapter 20

1. The gravity of the earth cannot keep the light H_2 molecules in the atmosphere.   **3.** The acidity decreases. Solutions of Be^{2+} are acidic, while solutions of the other M^{2+} ions are neutral.   **5.** For Groups 1A–3A, the small size of H (as compared to Li), Be (as compared to Mg), and B (as compared to Al) seems to be the reason why these elements have nonmetallic properties, while others in Groups 1A–3A are strictly metallic. The small size of H, Be, and B also causes these species to polarize the electron cloud in nonmetals, thus forcing a sharing of electrons when bonding occurs. For Groups 4A–6A, a major difference between the first and second members of a group is the ability to form π bonds. The smaller elements form stable π bonds, while the larger elements do not exhibit good overlap between parallel p orbitals and, in turn, do not form strong π bonds. For Group 7A, the small size of F as compared to Cl is used to explain the low electron affinity of F and the weakness of the F—F bond.   **7.** In order to maximize hydrogen bonding interactions in the solid phase, ice is forced into an open structure. This open structure is why $H_2O(s)$ is less dense than $H_2O(l)$.   **9.** Group 1A and 2A metals are all easily oxidized. They must be produced in the absence of materials (H_2O, O_2) that are capable of oxidizing them.   **11. a.** $\Delta H° = 207$ kJ, $\Delta S° = 216$ J/K; **b.** $T > 958$ K   **13.** $4Li(s) + O_2(g) \rightarrow 2Li_2O(s)$; $2Li(s) + S(s) \rightarrow Li_2S(s)$; $2Li(s) + Cl_2(g) \rightarrow 2LiCl(s)$; $12Li(s) + P_4(s) \rightarrow 4Li_3P(s)$; $2Li(s) + H_2(g) \rightarrow 2LiH(s)$; $2Li(s) + 2H_2O(l) \rightarrow 2LiOH(aq) + H_2(g)$; $2Li(s) + 2HCl(aq) \rightarrow 2LiCl(aq) + H_2(g)$ **15.** When lithium reacts with excess oxygen, Li_2O forms, which is composed of Li^+ and O_2^- ions. This is called an oxide salt. When sodium reacts with oxygen, Na_2O_2 forms, which is composed of Na^+ and O_2^{2-} ions. This is called a peroxide salt. When potassium (or rubidium or cesium) reacts with oxygen, KO_2 forms, which is composed of K^+ and O_2^- ions. For your information, this is called a superoxide salt. So the three types of alkali metal oxides that can form differ in the oxygen anion part of the formula (O^{2-} vs. O_2^{2-} vs. O_2^-). Each of these anions have unique bonding arrangements and oxidation states.   **17.** The small size of the Li^+ cation results in a much greater attraction to water. The attraction to water is not so great for the other alkali metal ions. Thus lithium salts tend to absorb water.   **19.** $CaCO_3(s) + H_2SO_4(aq) \rightarrow CaSO_4(aq) + H_2O(l) + CO_2(g)$ **21.** 3.84 × 10^6 g Ba   **23.** Beryllium has a small size and a large electronegativity as compared with the other alkaline earth metals. The electronegativity of Be is so high that it does not readily give up electrons to nonmetals,

as is the case for the other alkaline earth metals. Instead, Be has significant covalent character in its bonds; it prefers to share valence electrons rather than give them up to form ionic bonds.   **25.** element 113: $[Rn]7s^25f^{14}6d^{10}7p^1$; element 113 would fall below Tl in the periodic table. We would expect element 113, like Tl, to form +1 and +3 oxidation states in its compounds.   **27.** $B_2H_6(g) + 3O_2(g) \rightarrow 2B(OH)_3(s)$ **29.** $2Ga(s) + 3F_2(g) \rightarrow 2GaF_3(s)$; $4Ga(s) + 3O_2(g) \rightarrow 2Ga_2O_3(s)$; $2Ga(s) + 3S(s) \rightarrow Ga_2S_3(s)$; $2Ga(s) + 6HCl(aq) \rightarrow 2GaCl_3(aq) + 3H_2(g)$ **31.** An amphoteric substance is one that can behave as either an acid or a base. Al_2O_3 dissolves in both acidic and basic solutions. The reactions are $Al_2O_3(s) + 6H^+(aq) \rightarrow 2Al^{3+}(aq) + 3H_2O(l)$ and $Al_2O_3(s) + 2OH^-(aq) + 3H_2O(l) \rightarrow 2Al(OH)_4^-(aq)$.   **33.** Compounds containing Si—Si single and multiple bonds are rare, unlike compounds of carbon. The bond strengths of the Si—Si and C—C single bonds are similar. The difference in bonding properties must be for other reasons. One reason is that silicon does not form strong pi bonds, unlike carbon. Another reason is that silicon forms particularly strong sigma bonds to oxygen, resulting in compounds with Si—O bonds instead of Si—Si bonds.   **35.** O=C=O The darker green orbitals about carbon are sp hybrid orbitals. The lighter green orbitals about each oxygen are sp^2 hybrid orbitals, and the gold orbitals about all of the atoms are unhybridized p atomic orbitals. In each double bond in CO_2, one σ bond and one π bond exist. The two carbon–oxygen σ bonds are formed from overlap of sp hybrid orbitals from carbon with an sp^2 hybrid orbital from each oxygen. The two carbon–oxygen π bonds are formed from side-to-side overlap of the unhybridized p atomic orbitals from carbon with an unhybridized p atomic orbital from each oxygen, as illustrated in the figure.   **37. a.** $SiO_2(s) + 2C(s) \rightarrow Si(s) + 2CO(g)$; **b.** $SiCl_4(l) + 2Mg(s) \rightarrow Si(s) + 2MgCl_2(s)$; **c.** $Na_2SiF_6(s) + 4Na(s) \rightarrow Si(s) + 6NaF(s)$ **39.** 2:1   **41.** Nitrogen's small size does not provide room for all four oxygen atoms, making NO_4^{3-} unstable. Phosphorus is larger so PO_4^{3-} is more stable. To form NO_3^-, a pi bond must form. Phosphorus doesn't form strong pi bonds as readily as nitrogen.   **43.** $2Bi_2S_3(s) + 9O_2(g) \rightarrow 2Bi_2O_3(s) + 6SO_2(g)$; $2Bi_2O_3(s) + 3C(s) \rightarrow 4Bi(s) + 3CO_2(g)$; $2Sb_2S_3(s) + 9O_2(g) \rightarrow 2Sb_2O_3(s) + 6SO_2(g)$; $2Sb_2O_3(s) + 3C(s) \rightarrow 4Sb(s) + 3CO_2(g)$

45.

Trigonal bipyramid; dsp^3

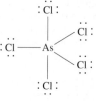
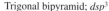

Trigonal bipyramid; dsp^3

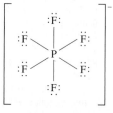

Octahedral; d^2sp^3

Nitrogen does not have low-energy d orbitals it can use to expand its octet. Both NF_5 and NCl_6^- would require nitrogen to have more than 8 valence electrons around it; this never happens.   **47.** $\frac{1}{2}N_2(g) + \frac{1}{2}O_2(g) \rightarrow NO(g)$; $\Delta G° = \Delta G°_{f(NO)}$; NO (and some other oxides of nitrogen) have weaker bond's as compared with the triple bond of N_2 and the double bond of O_2. Because of this, NO (and some other oxides of nitrogen) has a higher (positive) standard free energy of formation as compared to the relatively stable N_2 and O_2 molecules.   **49.** The pollution provides nitrogen and phosphorus nutrients so the algae can grow. The algae consume dissolved oxygen, causing fish to die.   **51.** The acidic hydrogens in the oxyacids of phosphorus all are bonded to oxygen. The hydrogens bonded directly to phosphorus are not acidic. H_3PO_4 has three oxygen-bonded hydrogens, and it is a triprotic acid. H_3PO_3 has only two of the hydrogens bonded to oxygen, and it is a diprotic acid. The third oxyacid of phosphorus, H_3PO_2, has only one of the hydrogens bonded to an oxygen; it is a monoprotic acid. **53.** 821 nm **55.** $H_2SeO_4(aq) + 3SO_2(g) \rightarrow Se(s) + 3SO_3(g) + H_2O(l)$ **57.** In the upper atmosphere, O_3 acts as a filter for ultraviolet (UV) radiation:

$$O_3 \xrightarrow{hv} O_2 + O$$

O_3 is also a powerful oxidizing agent. It irritates the lungs and eyes, and at high concentration, it is toxic. The smell of a "spring thunderstorm" is O_3 formed during lightning discharges. Toxic materials don't necessarily smell bad. For example, HCN smells like almonds. **59.** The M.O. electron configuration of O_2 has two unpaired electrons in the degenerate pi antibonding (π_{2p}^*) orbitals. A substance with unpaired electrons is paramagnetic. **61.** From the following Lewis structure, each oxygen atom has a tetrahedral arrangement of electron pairs. Therefore, bond angles are $\approx 109.5°$, and each O is sp^3 hybridized.

$$:\ddot{F}—\ddot{O}—\ddot{O}—\ddot{F}:$$

Formal charge: 0   0   0   0
Oxidation state: −1   +1   +1   −1

Oxidation states are more useful. We are forced to assign +1 as the oxidation state for oxygen. Oxygen is very electronegative, and +1 is not a stable oxidation state for this element.

63.

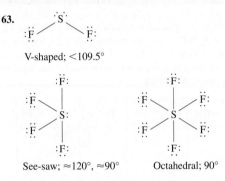

V-shaped; $<109.5°$

See-saw; $\approx 120°$, $\approx 90°$ Octahedral; $90°$

OF_4 would not form. This compound would require oxygen to have more than 8 valence electrons around it. This never occurs for oxygen; oxygen does not have low-energy d orbitals it can use to expand its octet. **65.** The oxyacid strength increases as the number of oxygens in the formula increases. Therefore, the order of the oxyacids from weakest to strongest acid is $HOCl < HClO_2 < HClO_3 < HClO_4$. **67. a.** IO_4^-; **b.** IO_3^-; **c.** IF_2^-; **d.** IF_4^-; **e.** IF_6^- **69.** Helium is unreactive and doesn't combine with any other elements. It is a very light gas and would easily escape the earth's gravitational pull as the planet was formed. **71.** One would expect RnF_2, RnF_4, and maybe RnF_6 to form in fashion similar to XeF_2, XeF_4, and XeF_6. The chemistry of radon is difficult to study because radon isotopes are all radioactive. The hazards of dealing with radioactive materials are immense. **73.** $N_2H_4(l) + O_2(g) \rightarrow N_2(g) + 2H_2O(g)$; $\Delta H = -590.$ kJ **75.** If the compound contained Ga(II), it would be paramagnetic and if the compound contained Ga(I) and Ga(III), it would be diamagnetic. Paramagnetic compounds have an apparent greater mass in a magnetic field. **77.** 9.3×10^{-5} mol/L **79. a.** $AgCl(s) \xrightarrow{hv} Ag(s) + Cl$; the reactive chlorine atom is trapped in the crystal. When light is removed, Cl reacts with silver atoms to re-form AgCl; i.e., the reverse reaction occurs. In pure AgCl, the Cl atoms escape, making the reverse reaction impossible. **b.** Over time, chlorine is lost and the dark silver metal is permanent. **81.** 6.5×10^{27} atoms **83.** Strontium and calcium are both alkaline earth metals, so both have similar chemical properties. Since milk is a good source of calcium, strontium could replace some calcium in milk without much difficulty. **85.** +6 oxidation state: SO_4^{2-}, SO_3, SF_6; +4 oxidation state: SO_3^{2-}, SO_2, SF_4; +2 oxidation state: SCl_2; 0 oxidation state: S_8 and all other elemental forms of sulfur; −2 oxidation state: H_2S, Na_2S **97.** Ca; 12.698 **99.** I **101.** For the reaction

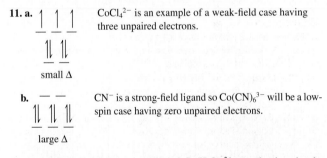

$$N—N\!=\!\ddot{O}: \longrightarrow NO_2 + NO$$

the activation energy must in some way involve the breaking of a nitrogen–nitrogen single bond. For the reaction

$$N—N\!=\!\ddot{O}: \longrightarrow O_2 + N_2O$$

at some point nitrogen–oxygen bonds must be broken. N—N single bonds (160 kJ/mol) are weaker than N—O single bonds (201 kJ/mol). In addition, resonance structures indicate that there is more double-bond character in the N—O bonds than in the N—N bond. Thus NO_2 and NO are preferred by kinetics because of the lower activation energy. **103.** 5.89 **105.** 20. g **107. a.** 7.1 g; **b.** 979 nm; this electromagnetic radiation is not visible to humans; it is in the infrared region of the electromagnetic radiation spectrum; **c.** n-type **109. a.** +6; **b.** 4.42

Chapter 21

5. The lanthanide elements are located just before the $5d$ transition metals. The lanthanide contraction is the steady decrease in the atomic radii of the lanthanide elements when going from left to right across the periodic table. As a result of the lanthanide contraction, the sizes of the $4d$ and $5d$ elements are very similar. This leads to a greater similarity in the chemistry of the $4d$ and $5d$ elements in a given vertical group. **7.** No; both the *trans* and the *cis* forms of $Co(NH_3)_4Cl_2^+$ have mirror images that are superimposable. For the *cis* form, the mirror image only needs a 90° rotation to produce the original structure. Hence, neither the *trans* nor the *cis* form is optically active. **9.** $Fe_2O_3(s) + 6H_2C_2O_4(aq) \rightarrow 2Fe(C_2O_4)_3^{3-}(aq) + 3H_2O(l) + 6H^+(aq)$; the oxalate anion forms a soluble complex ion with iron in rust (Fe_2O_3), which allows rust stains to be removed.

11. a.
↑ ↑ ↑
↑↓ ↑↓
small Δ

$CoCl_4^{2-}$ is an example of a weak-field case having three unpaired electrons.

b.
— —
↑↓ ↑↓ ↑↓
large Δ

CN^- is a strong-field ligand so $Co(CN)_6^{3-}$ will be a low-spin case having zero unpaired electrons.

13. From Table 21.16, the red octahedral $Co(H_2O)_6^{2+}$ complex ion absorbs blue-green light ($\lambda \approx 490$ nm), whereas the blue tetrahedral $CoCl_4^{2-}$ complex ion absorbs orange light ($\lambda \approx 600$ nm). Because tetrahedral complexes have a d-orbital splitting much less than octahedral complexes, one would expect the tetrahedral complex to have a smaller energy difference between split d orbitals. This translates into longer-wavelength light absorbed ($E = hc/\lambda$) for tetrahedral complex ions compared to octahedral complex ions. Information from Table 21.16 confirms this. **15.** SCN^-, NO_2^-, and OCN^- can form linkage isomers; all are able to bond to the metal ion in two different ways. **17.** Sc^{3+} has no electrons in d orbitals. Ti^{3+} and V^{3+} have d electrons present. The color of transition metal complexes results from electron transfer between split d orbitals. If no d electrons are present, no electron transfer can occur, and the compounds are not colored. **19.** At high altitudes, the oxygen content of air is lower, so less oxyhemoglobin is formed which diminishes the transport of oxygen in the blood. A serious illness called high-altitude sickness can result from the decrease of O_2 in the blood. High-altitude acclimatization is the phenomenon that occurs in the human body in response to the lower amounts of oxyhemoglobin in the blood. This response is to produce more hemoglobin, and, hence, increase the oxyhemoglobin in the blood. High-altitude acclimatization takes several weeks to take hold for people moving from lower altitudes to higher altitudes. **21. a.** Ni: $[Ar]4s^23d^8$; **b.** Cd: $[Kr]5s^24d^{10}$; **c.** Zr: $[Kr]5s^24d^2$; **d.** Os: $[Xe]6s^24f^{14}5d^6$ **23. a.** Ti: $[Ar]4s^23d^2$; Ti^{2+}: $[Ar]3d^2$; Ti^{4+}: $[Ne]3s^23p^6$ or $[Ar]$; **b.** Re: $[Xe]6s^24f^{14}5d^5$; Re^{2+}: $[Xe]4f^{14}5d^5$; Re^{3+}: $[Xe]4f^{14}5d^4$; **c.** Ir: $[Xe]6s^24f^{14}5d^7$; Ir^{2+}: $[Xe]4f^{14}5d^7$; Ir^{3+}: $[Xe]4f^{14}5d^6$ **25. a.** Fe^{3+}: $[Ar]3d^5$; **b.** Ag^+: $[Kr]4d^{10}$; **c.** Ni^{2+}: $[Ar]3d^8$; **d.** Cr^{3+}: $[Ar]3d^3$ **27. a.** molybdenum(IV)

sulfide, molybdenum(VI) oxide; **b.** MoS_2, +4; MoO_3, +6; $(NH_4)_2Mo_2O_7$, +6; $(NH_4)_6Mo_7O_{24} \cdot 4H_2O$, +6   **29.** NH_3 is a weak base that produces OH^- ions in solution. The white precipitate is $Cu(OH)_2(s)$. $Cu^{2+}(aq) + 2OH^-(aq) \rightarrow Cu(OH)_2(s)$; with excess NH_3 present, Cu^{2+} forms a soluble complex ion, $Cu(NH_3)_4{}^{2+}$. $Cu(OH)_2(s) + 4NH_3(aq) \rightarrow Cu(NH_3)_4{}^{2+}(aq) + 2OH^-(aq)$ **31.** $[Co(NH_3)_6]I_3$: 3 moles of AgI; $[Pt(NH_3)_4I_2]I_2$: 2 moles of AgI; $Na_2[PtI_6]$: 0 moles of AgI; $[Cr(NH_3)_4I_2]I$: 1 mole of AgI   **33. a.** hexacyanomanganate(II) ion; **b.** *cis*-tetraamminedichlorocobalt(III) ion; **c.** pentaamminechlorocobalt(II) ion   **35. a.** hexaamminecobalt(II) chloride; **b.** hexaaquacobalt(III) iodide; **c.** potassium tetrachloroplatinate(II); **d.** potassium hexachloroplatinate(II); **e.** pentaamminechlorocobalt(III) chloride; **f.** triamminetrinitrocobalt(III)   **37. a.** $K_2[CoCl_4]$; **b.** $[Pt(H_2O)(CO)_3]Br_2$; **c.** $Na_3[Fe(CN)_2(C_2O_4)_2]$; **d.** $[Cr(NH_3)_3Cl(H_2NCH_2CH_2NH_2)]I_2$
39. a.

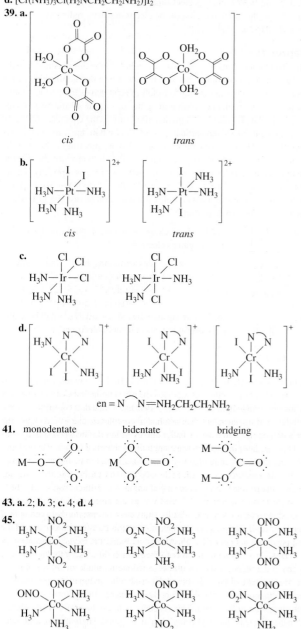

41. monodentate   bidentate   bridging

43. a. 2; **b.** 3; **c.** 4; **d.** 4
45.

47. $Cr(acac)_3$ and *cis*-$Cr(acac)_2(H_2O)_2$ are optically active.   **49.** With five electrons each in a different orbital, diagram (a) is for the weak-field $Fe(H_2O)_6{}^{3+}$ complex ion. With three electrons, diagram (b) is for the $Cr(NH_3)_5Cl^{2+}$ complex ion. With six electrons all paired up, diagram (c) is for the strong-field $Co(NH_3)_4Br_2{}^+$ complex ion.

51. a. Fe^{2+}

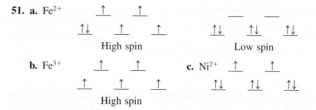

b. Fe^{3+} **c.** Ni^{2+}

53. weak field   **55. a.** 0; **b.** 2; **c.** 2   **57.** $Co(CN)_6{}^{3-} < Co(en)_3{}^{3+} < Co(H_2O)_6{}^{3+} < CoI_6{}^{3-}$ **59.** The violet complex ion absorbs yellow-green light ($\lambda \approx 570$ nm), the yellow complex ion absorbs blue light ($\lambda \approx 450$ nm), and the green complex ion absorbs red light ($\lambda \approx 650$ nm). The violet complex ion is $Cr(H_2O)_6{}^{3+}$, the yellow complex ion is $Cr(NH_3)_6{}^{3+}$, and the green complex ion is $Cr(H_2O)_4Cl_2{}^+$.   **61.** $CoBr_4{}^{2-}$ is a tetrahedral complex ion, while $CoBr_6{}^{4-}$ is an octahedral complex ion. Since tetrahedral *d*-orbital splitting is less than one-half the octahedral *d*-orbital splitting, the octahedral complex ion ($CoBr_6{}^{4-}$) will absorb higher-energy light, which will have a shorter wavelength than 3.4×10^{-6} m ($E = hc/\lambda$).   **63.** 5   **65. a.** -11 kJ; **b.** $\Delta H° = 172.5$ kJ; $\Delta S° = 176$ J/K; $T > 980$. K   **67.** Fe_2O_3: iron has a $+3$ oxidation state; Fe_3O_4: iron has a $+8/3$ oxidation state. The three iron ions in Fe_3O_4 must have a total charge of $+8$. The only combination that works is to have two Fe^{3+} ions and one Fe^{2+} ion per formula unit. This makes sense from the other formula for magnetite, $FeO \cdot Fe_2O_3$. FeO has an Fe^{2+} ion and Fe_2O_3 has two Fe^{3+} ions.   **69.** $8CN^-(aq) + 4Ag(s) + O_2(g) + 2H_2O(l) \rightarrow 4Ag(CN)_2{}^-(aq) + 4OH^-(aq)$ **71.** The molecular formula is $EuC_{15}H_{21}O_6$. Because each $acac^-$ is $C_5H_7O_2{}^-$, an abbreviated molecular formula is $Eu(acac)_3$.   **73.** 0.66 V; -130 kJ; 2.2×10^{22} **75.** There are four geometrical isomers (labeled i–iv). Isomers iii and iv are optically active, and the nonsuperimposable mirror images are shown.

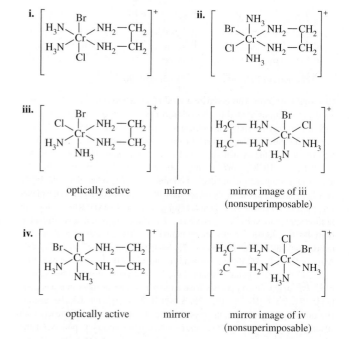

77. Octahedral Cr^{2+} complexes should be used. Cr^{2+}: $[Ar]3d^4$; high-spin (weak-field) Cr^{2+} complexes have four unpaired electrons and low-spin (strong-field) Cr^{2+} complexes have two unpaired electrons. Ni^{2+}: $[Ar]3d^8$; octahedral Ni^{2+} complexes will always have two unpaired electrons, whether high or low spin. Therefore, Ni^{2+} complexes cannot be used to distinguish weak- from strong-field ligands by examining magnetic properties. Alternatively, the ligand field strengths can be measured using visible spectra. Either Cr^{2+} or Ni^{2+} complexes can be used for this method.   **79. a.** $[Co(C_5H_5N)_6]Cl_3$; **b.** $[Cr(NH_3)_5I]I_2$; **c.** $[Ni(NH_2CH_2CH_2NH_2)_3]Br_2$; **d.** $K_2[Ni(CN)_4]$; **e.** $[Pt(NH_3)_4Cl_2][PtCl_4]$ **81.** 60   **89.** $Ni^{2+} = d^8$; if ligands

A and B produced very similar crystal fields, the cis-$[NiA_2B_4]^{2+}$ complex ion would give the following octahedral crystal field diagram for a d^8 ion:

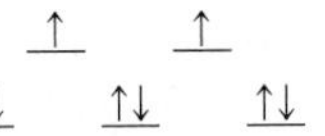

This is paramagnetic.

Because it is given that the complex ion is diamagnetic, the A and B ligands must produce different crystal fields, giving a unique d-orbital splitting diagram that would result in a diamagnetic species.　**91. a.** -0.26 V; **b.** From standard reduction potentials, Co^{3+} ($\mathscr{E}° = 1.82$ V) is a much stronger oxidizing agent than $Co(en)_3^{3+}$ ($\mathscr{E}° = -0.26$ V); **c.** In aqueous solution, Co^{3+} forms the hydrated transition metal complex, $Co(H_2O)_6^{3+}$. In both complexes, $Co(H_2O)_6^{3+}$ and $Co(en)_3^{3+}$, cobalt exists as Co^{3+}, which has 6 d electrons. If we assume a strong-field case for each complex ion, then the d-orbital splitting diagram for each has the six electrons paired in the lower-energy t_{2g} orbitals. When each complex ion gains an electron, the electron enters the higher-energy e_g orbitals. Since en is a stronger-field ligand than H_2O, then the d-orbital splitting is larger for $Co(en)_3^{3+}$, and it takes more energy to add an electron to $Co(en)_3^{3+}$ than to $Co(H_2O)_6^{3+}$. Therefore, it is more favorable for $Co(H_2O)_6^{3+}$ to gain an electron than for $Co(en)_3^{3+}$ to gain an electron.　**93.** No, in all three cases six bonds are formed between Ni^{2+} and nitrogen. So ΔH values should be similar. $\Delta S°$ for formation of the complex ion is most negative for 6 NH_3 molecules reacting with a metal ion (7 independent species become 1). For penten reacting with a metal ion, 2 independent species become 1, so $\Delta S°$ is the least negative. Thus the chelate effect occurs because the more bonds a chelating agent can form to the metal, the less unfavorable $\Delta S°$ becomes for the formation of the complex ion, and the larger the formation constant.

95.

97. The coordinate system for the complex ion is shown below. From the coordinate system, the CN^- ligands are in a square planar arrangement. Since CN^- produces a much stronger crystal field, the diagram will most resemble that of a square planar complex:

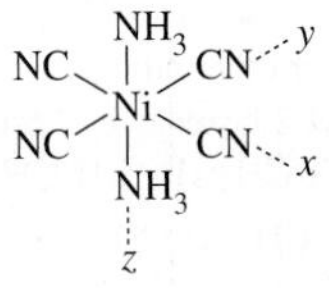

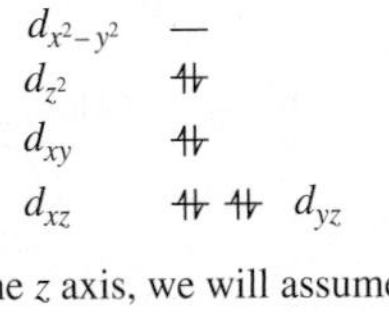

With the NH_3 ligands on the z axis, we will assume that the d_{z^2} orbital is destabilized more than the d_{xy} orbital. This may or may not be the case.

99. $[Cr(NH_3)_5I]I_2$; octahedral

Chapter 22

1. a. 1-sec-butylpropane

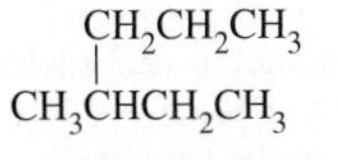

3-methylhexane is correct.

c. 2-ethylpentane

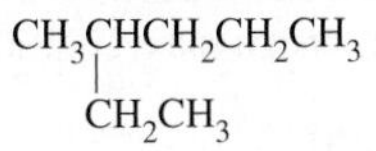

3-methylhexane is correct.

b. 4-methylhexane

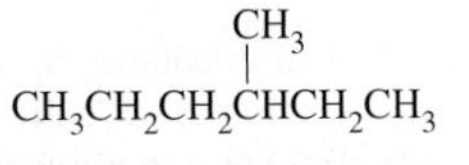

3-methylhexane is correct.

d. 1-ethyl-1-methylbutane

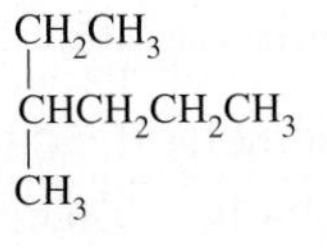

3-methylhexane is correct.

e. 3-methylhexane

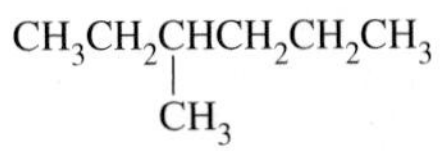

f. 4-ethylpentane

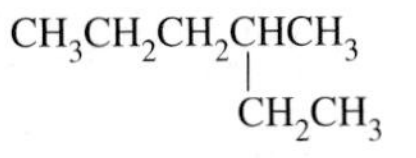

3-methylhexane is correct.

All six of these are the same compound. They only differ from each other by rotations about one or more carbon–carbon single bonds. Only one isomer of C_7H_{16} is present in all of these names: 3-methylhexane.

3. a.

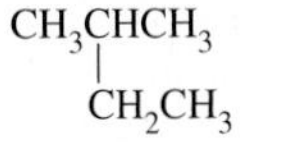

The longest chain is 4 carbons long. The correct name is 2-methylbutane.

b.

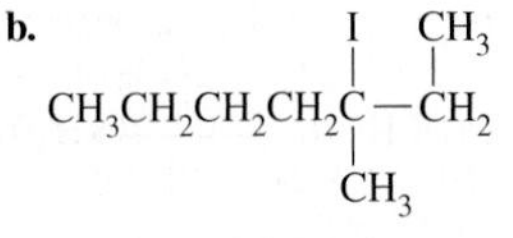

The longest chain is 7 carbons long and we would start the numbering system at the other end for lowest possible numbers. The correct name is 3-iodo-3-methylheptane.

c.

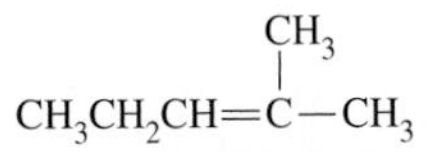

This compound cannot exhibit $cis-$ $trans$ isomerism since one of the double-bonded carbons has the same two groups (CH_3) attached. The numbering system should also start at the other end to give the double bond the lowest possible number. 2-methyl-2-pentene is correct.

d.

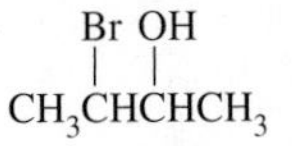

The OH functional group gets the lowest number. 3-bromo-2-butanol is correct.

5. Hydrocarbons are nonpolar substances exhibiting only London dispersion forces. Size and shape are the two most important structural features relating to the strength of London dispersion forces. For size, the bigger the molecule (the larger the molar mass), the stronger the London dispersion forces and the higher the boiling point. For shape, the more branching present in a compound, the weaker the London dispersion forces and the lower the boiling point.　**7.** The amide functional group is:

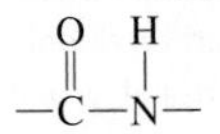

When the amine end of one amino acid reacts with the carboxylic acid end of another amino acid, the two amino acids link together by forming an amide functional group. A polypeptide has many amino acids linked together, with each linkage made by the formation of an amide functional group. Because all linkages result in the presence of the amide functional group, the resulting polymer is called a polyamide. For nylon, the monomers also link together by forming the amide functional group (the amine end of one monomer reacts with a carboxylic acid end of another monomer to give the amide functional group linkage). Hence nylon is also a polyamide. The correct order of strength is polyhydrocarbon < polyester < polyamide. The difference in strength is related to the types of intermolecular forces present. All of these polymers have London dispersion forces. However, polyhydrocarbons only have London dispersion forces. The polar ester group in polyesters and the polar amide group in polyamides give rise to additional dipole forces. The polyamide has the ability to form relatively strong hydrogen bonding interactions, hence why it would form the strongest fibers.

9. a.

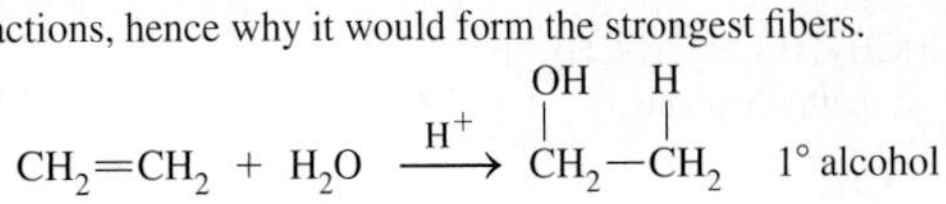

b.

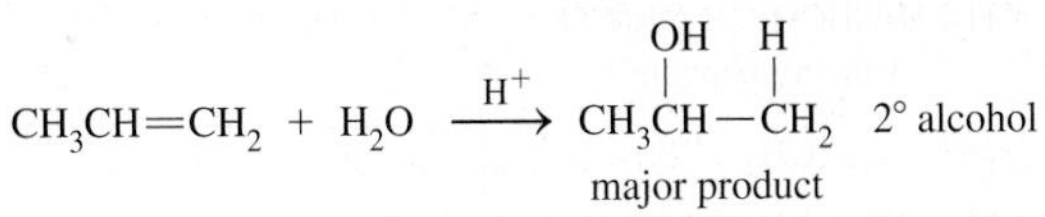

c.

$$CH_3C=CH_2 + H_2O \xrightarrow{H^+} CH_3\overset{OH}{\underset{CH_3}{C}}-\overset{H}{CH_2} \quad 3° \text{ alcohol}$$

$$\underset{CH_3}{}$$

major product

d.

$$CH_3CH_2OH \xrightarrow{\text{oxidation}} CH_3\overset{O}{\overset{\|}{CH}} \quad \text{aldehyde}$$

e.

$$CH_3\overset{OH}{\underset{}{CHCH_3}} \xrightarrow{\text{oxidation}} CH_3\overset{O}{\overset{\|}{C}}CH_3 \quad \text{ketone}$$

f.

$$CH_3CH_2CH_2OH \xrightarrow{\text{oxidation}} CH_3CH_2\overset{O}{\overset{\|}{C}}-OH \quad \text{carboxylic acid}$$

or

$$CH_3CH_2\overset{O}{\overset{\|}{CH}} \xrightarrow{\text{oxidation}} CH_3CH_2\overset{O}{\overset{\|}{C}}-OH$$

g.

$$CH_3OH + HO\overset{O}{\overset{\|}{C}}CH_3 \longrightarrow CH_3-O-\overset{O}{\overset{\|}{C}}CH_3 + H_2O \quad \text{ester}$$

11. a. A polyester forms when an alcohol functional group reacts with a carboxylic acid functional group. The monomer for a homopolymer polyester must have an alcohol functional group and a carboxylic acid functional group present in the structure. **b.** A polyamide forms when an amine functional group reacts with a carboxylic acid functional group. For a copolymer polyamide, one monomer would have at least two amine functional groups present and the other monomer would have at least two carboxylic acid functional groups present. For polymerization to occur, each monomer must have two reactive functional groups present. **c.** To form an addition polymer, a carbon–carbon double bond must be present. To form a polyester, the monomer would need the alcohol and carboxylic acid functional groups present. To form a polyamide, the monomer would need the amine and carboxylic acid functional groups present. The two possibilities are for the monomer to have a carbon–carbon double bond, an alcohol functional group, and a carboxylic acid functional group all present, or to have a carbon–carbon double bond, an amine functional group, and a carboxylic acid functional group present.

13. $CH_3-CH_2-CH_2-CH_2-CH_2-CH_3$

$CH_3-\overset{CH_3}{\underset{}{CH}}-CH_2-CH_2-CH_3$

$CH_3-CH_2-\overset{CH_3}{\underset{}{CH}}-CH_2-CH_3$

$CH_3-\overset{CH_3}{\underset{CH_3}{C}}-CH_2-CH_3$

$CH_3-\overset{CH_3}{\underset{}{CH}}-\overset{CH_3}{\underset{}{CH}}-CH_3$

15. a.

$CH_3\overset{CH_3}{\underset{}{CHCH_2CH_2CH_2CH_3}}$
2-methylheptane

$CH_3CH_2\overset{CH_3}{\underset{}{CHCH_2CH_2CH_3}}$
3-methylheptane

$CH_3CH_2CH_2\overset{CH_3}{\underset{}{CHCH_2CH_2CH_3}}$
4-methylheptane

b.

$CH_3-\overset{CH_3}{\underset{CH_3}{C}}-\overset{CH_3}{\underset{CH_3}{C}}-CH_3$
2,2,3,3-tetramethylbutane

17. a. $CH_3\overset{CH_3}{\underset{}{CHCH_3}}$ **b.** $CH_3\overset{CH_3}{\underset{}{CHCH_2CH_3}}$

c. $CH_3\overset{CH_3}{\underset{}{CHCH_2CH_2CH_3}}$ **d.** $CH_3\overset{CH_3}{\underset{}{CHCH_2CH_2CH_2CH_3}}$

19. a. $CH_3CH_2-\overset{}{\underset{CH_2-CH-CH_3}{CH}}-CH_2CH_2CH_3$
                $\underset{CH_3}{}$

b.

$CH_3-\overset{CH_3}{\underset{CH_3}{C}}-CH_2-\overset{}{\underset{CH_3}{CH}}-CH_3$

c. $CH_3-\overset{}{\underset{CH_3-C-CH_3}{CH}}-CH_2CH_2CH_3$
           $\underset{CH_3}{}$

d. 4-ethyl-2-methylheptane; 2,2,3-trimethylhexane

21. a. 2,2,4-trimethylhexane; **b.** 5-methylnonane; **c.** 2,2,4,4-tetramethylpentane; **d.** 3-ethyl-3-methyloctane

23. $CH_3-CH_2-CH_2-CH_3$;

$$\begin{array}{c} \text{H} \quad \text{H} \\ | \quad\quad | \\ \text{H}-\text{C}-\text{C}-\text{H} \\ | \quad\quad | \\ \text{H}-\text{C}-\text{C}-\text{H} \\ | \quad\quad | \\ \text{H} \quad \text{H} \end{array}$$

Each carbon is bonded to four other atoms.

25. a. 1-butene; **b.** 4-methyl-2-hexene; **c.** 2,5-dimethyl-3-heptene

27. a. $CH_3CH_2CH=CHCH_2CH_3$; **b.** $CH_3CH=CHCH=CHCH_2CH_3$;

c. $CH_3\overset{}{\underset{CH_3}{CHCH}}=CHCH_2CH_2CH_2CH_3$

29. a.

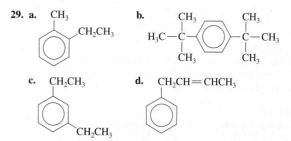

31. a. 1,3-dichlorobutane; **b.** 1,1,1-trichlorobutane; **c.** 2,3-dichloro-2, 4-dimethylhexane; **d.** 1,2-difluoroethane   **33.** CH_2Cl-CH_2Cl, 1,2-dichloroethane: There is free rotation about the C—C single bond that doesn't lead to different compounds. $CHCl=CHCl$, 1,2-dichloroethene: There is no rotation about the C=C double bond. This creates the *cis* and *trans* isomers, which are different compounds.   **35.** [25], compounds b and c; [27], all compounds

37. $CH_2=CHCH_2CH_2CH_3$ $CH_3CH=CHCH_2CH_3$

     $CH_2=C\overset{}{\underset{CH_3}{CH_2CH_3}}$ $CH_3C\overset{}{\underset{CH_3}{=}}CHCH_3$

     $CH_3\overset{}{\underset{CH_3}{CHCH}}=CH_2$

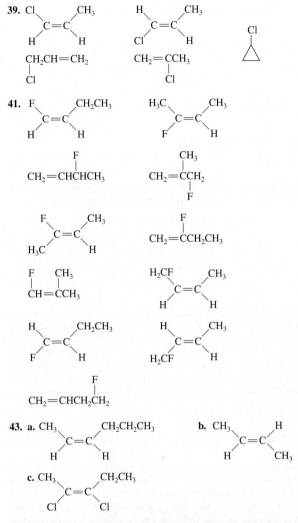

39.

41.

43. a. CH₃ C=C CH₂CH₂CH₃ ... **b.** CH₃ C=C H ... **c.** CH₃ C=C CH₂CH₃

45. a. 3 monochloro isomers of *n*-pentane; **b.** 4 monochloro isomers of 2-methylbutane; **c.** 3 monochloro isomers of 2,4-dimethylpentane; **d.** 4 monochloro isomers of methylcyclobutane **47. a.** ketone; **b.** aldehyde; **c.** carboxylic acid; **d.** amine

49. a.

b. 5 carbons in ring and the carbon in —CO₂H: sp^2; the other two carbons: sp^3; **c.** 24 sigma bonds, 4 pi bonds **51. a.** 3-chloro-1-butanol, primary; **b.** 3-methyl-3-hexanol, tertiary; **c.** 2-methylcyclopentanol, secondary **53.** 1-pentanol; 2-pentanol; 3-pentanol; 2-methyl-1-butanol; 2-methyl-2-butanol; 3-methyl-2-butanol; 3-methyl-1-butanol; 2,2-dimethyl-1-propanol; 6 ethers **55. a.** 4,5-dichloro-3-hexanone; **b.** 2,3-dimethylpentanal; **c.** 3-methylbenzaldehyde or *m*-methylbenzaldehyde **57. a.** 4-chlorobenzoic acid or *p*-chlorobenzoic acid; **b.** 3-ethyl-2-methylhexanoic acid; **c.** methanoic acid (common name = formic acid) **59.** Only statement d is false.

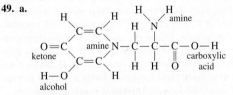

2-butenal: HCCH=CHCH₃ .

The formula of 2-butenal is C_4H_6O, while the alcohol has a formula of C_4H_8O.

61. a. CH₃CH₂CH₂CH₃

b. Cl Cl Cl Cl | | | | CH₂CHCHCH | | CH₃ CH₃ **c.** + HCl

d. $C_4H_8(g) + 6O_2(g) \longrightarrow 4CO_2(g) + 4H_2O(g)$

63. For the iron-catalyzed reaction, one of the *ortho* or *para* hydrogens in benzene is replaced by chlorine. When an iron catalyst is not present, the benzene hydrogens are unreactive. To substitute for an alkane hydrogen, light must be present. For toluene, the light-catalyzed reaction substitutes a chlorine for a hydrogen in the methyl group attached to the benzene ring.

65. a.

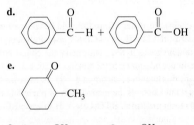

b.

c. No reaction

d.

e.

f.

67. a. CH₃CH=CH₂ + Br₂ → CH₃CHBrCH₂Br;

b. OH | CH₃—CH—CH₃ →(oxidation) CH₃—C—CH₃ (with =O);

c. CH₃ | CH₂=C—CH₃ + H₂O →(H⁺) CH₃ | CH₂—C—CH₃ | | H OH ;

d. CH₃CH₂CH₂OH →(KMnO₄) CH₃CH₂C—OH (with =O)

69. a.

CH₃C—OH + HOCH₂(CH₂)₆CH₃ ⟶

CH₃C—O—CH₂(CH₂)₆CH₃ + H₂O;

b.

CH₃CH₂C—OH + HOCH₂(CH₂)₄CH₃ ⟶

CH₃CH₂C—O—CH₂(CH₂)₄CH₃ + H₂O

71. CFCl=CF₂

73.

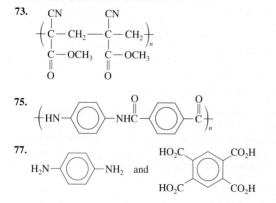

75.

77.

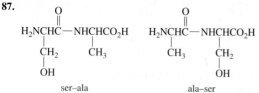

79. Divinylbenzene has two reactive double bonds that are used during formation of the polymer. The key is for the double bonds to insert themselves into two different polymer chains. When this occurs, the two chains are bonded together (are crosslinked). The chains cannot move past each other because of the crosslinks making the polymer more rigid. **81. a.** The polymer from 1,2-diaminoethane and terephthalic acid is stronger because of the possibility of hydrogen bonding between chains. **b.** The polymer of

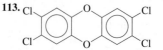

is more rigid because the chains are stiffer due to the rigid benzene rings in the chains. **c.** Polyacetylene is $nHC\equiv CH \rightarrow -(CH=CH-)_n$. Polyacetylene is more rigid because the double bonds in the chains make the chains stiffer. **83. a.** serine; tyrosine; threonine; **b.** aspartic acid; glutamic acid; **c.** histidine; lysine; arginine; tryptophan; **d.** glutamine; asparagine **85. a.** aspartic acid and phenylalanine; **b.** Aspartame contains the methyl ester of phenylalanine. This ester can hydrolyze to form methanol, $RCO_2CH_3 + H_2O \rightleftharpoons RCO_2H + CH_3OH$.

87.

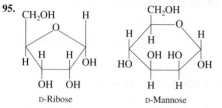

ser–ala   ala–ser

89. a. Six tetrapeptides are possible. From NH_2 to CO_2H end: phe–phe–gly–gly, gly–gly–phe–phe, gly–phe–phe–gly, phe–gly–gly–phe, phe–gly–phe–gly, gly–phe–gly–phe; **b.** Twelve tetrapeptides are possible. From NH_2 to CO_2H end: phe–phe–gly–ala, phe–phe–ala–gly, phe–gly–phe–ala, phe–gly–ala–phe, phe–ala–gly–phe, phe–ala–phe–gly, gly–phe–phe–ala, gly–phe–ala–phe, gly–ala–phe–phe, ala–phe–phe–gly, ala–phe–gly–phe, ala–gly–phe–phe **91.** Ionic: his, lys, or arg with asp or glu; hydrogen bonding: ser, glu, tyr, his, arg, asn, thr, asp, gln, or lys with any amino acid; covalent: cys with cys; London dispersion: all amino acids with nonpolar R groups (gly, ala, pro, phe, ile, trp, met, leu, val); dipole–dipole: tyr, thr, and ser with each other **93.** Glutamic acid has a polar R group and valine has a nonpolar R group. The change in polarity of the R groups could affect the tertiary structure of hemoglobin and affect the ability of hemoglobin to bond to oxygen.

95.

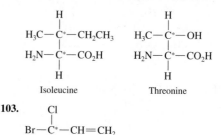

D-Ribose   D-Mannose

97. aldohexose: glucose, mannose, galactose; aldopentose: ribose, arabinose; ketohexose: fructose; ketopentose: ribulose **99.** They differ in the orientation of a hydroxy group on a particular carbon. Starch is composed from α-D-glucose, and cellulose is composed from β-D-glucose.

101. The chiral carbons are marked with asterisks.

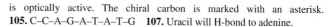

Isoleucine   Threonine

103.

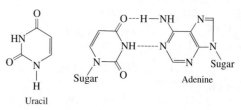

is optically active. The chiral carbon is marked with an asterisk.
105. C–C–A–G–A–T–A–T–G **107.** Uracil will H-bond to adenine.

Uracil   Adenine

109. a. glu: CTT, CTC; val: CAA, CAG, CAT, CAC; met: TAC; trp: ACC; phe: AAA, AAG; asp: CTA, CTG; **b.** ACC–CTT–AAA–TAC or ACC–CTC–AAA–TAC or ACC–CTT–AAG–TAC or ACC–CTC–AAG– TAC; **c.** four (see answer to part b); **d.** met–asp–phe; **e.** TAC–CTA–AAG; TAC–CTA–AAA; TAC–CTG–AAA **111. a.** 2,3,5,6-tetramethyloctane; **b.** 2,2,3,5-tetramethylheptane; **c.** 2,3,4-trimethylhexane; **d.** 3-methyl-1-pentyne

113.

There are many possibilities for isomers. Any structure with four chlorines replacing four hydrogens in any of the numbered positions would be an isomer; i.e., 1,2,3,4-tetrachloro-dibenzo-*p*-dioxin is a possible isomer. **115.** $-23°C$: CH_3-O-CH_3; 78.5°C: CH_3-CH_2-OH **117.** Alcohols consist of two parts, the polar OH group and the nonpolar hydrocarbon chain attached to the OH group. As the length of the nonpolar hydrocarbon chain increases, the solubility of the alcohol decreases. In methyl alcohol (methanol), the polar OH group can override the effect of the nonpolar CH_3 group, and methyl alcohol is soluble in water. In stearyl alcohol, the molecule consists mostly of the long nonpolar hydrocarbon chain, so it is insoluble in water. **119.** *n*-hexane, 69°C; pentanal, 103°C; 1-pentanol, 137°C; butanoic acid, 164°C. **121.** 2-propanone; propanoic acid **123.** In nylon, hydrogen-bonding interactions occur due to the presence of N—H bonds in the polymer. For a given polymer chain length, there are more N—H groups in nylon-46 as compared to nylon-6. Hence, nylon-46 forms a stronger polymer compared to nylon-6 due to the increased hydrogen-bonding interactions.

125. a.

b. Repeating unit:

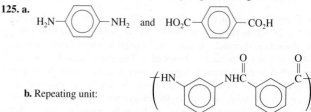

The two polymers differ in the substitution pattern on the benzene rings. The Kevlar chain is straighter, and there is more efficient hydrogen bonding between Kevlar chains than between Nomex chains. **127. a.** The bond angles in the ring are about 60°. VSEPR predicts bond angles close to 109°. The bonding electrons are much closer together than they prefer, resulting in strong electron–electron repulsions. Thus ethylene oxide is unstable (reactive). **b.** The ring opens up during polymerization and the monomers link together through the formation of O—C bonds.

$$+O-CH_2CH_2-O-CH_2CH_2-O-CH_2CH_2+_n$$

129. H$_2$N—CH—CO$_2$H   or   H$_2$N—CH—CO$_2^-$Na$^+$
                 | |
        CH$_2$CH$_2$CO$_2^-$Na$^+$ CH$_2$CH$_2$CO$_2$H

The first structure is MSG, which is impossible for you to predict.   **131.** $\Delta G = \Delta H - T\Delta S$; for the reaction, we break a P–O and O–H bond and form a P–O and O–H bond, so $\Delta H \approx 0$ based on bond dissociation energies. ΔS for this process is negative (unfavorable) because positional probability decreases. Thus, $\Delta G > 0$ due to the unfavorable ΔS term, and the reaction is not expected to be spontaneous.

133.

                        CH$_3$ O
                          | ‖
    1.0 M H$^+$:   H$_3$N$^+$—CH—C—OH;

                        CH$_3$ O
                          | ‖
    1.0 M OH$^-$:   H$_2$N—CH—C—O$^-$

135. Both ΔH and ΔS are positive values.   **145.** 6.07   **147. a.** No; the mirror image is superimposable.

b.

        OH   OH   OH   OH
         | | | |
        .C————C.   .C————C.
     H′ ⋮   ⋮ ′CO$_2$H   HO$_2$C′ ⋮   ⋮ ′H
        CO$_2$H H   H   CO$_2$H

149.

                                                        O
                                                        ‖
HC≡C−C≡C−HC=C=CH−HC=CH−HC=CH−CH$_2$−C−OH
13 12 11 10   9 8 7   6 5   4 3   2   1

151. a. CH$_3$CHCH$_2$CH$_3$
                 |
                CH$_3$

b. CH$_2$=CCH$_3$
              |
              CH$_3$

c.

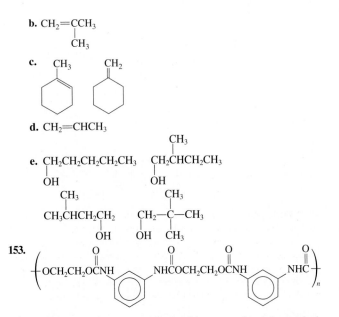

d. CH$_2$=CHCH$_3$

e. CH$_2$CH$_2$CH$_2$CH$_2$CH$_3$ CH$_2$CHCH$_2$CH$_3$
         | | |
         OH   OH   CH$_3$

                CH$_3$ CH$_3$
                  | |
        CH$_3$CHCH$_2$CH$_2$ CH$_2$−C−CH$_3$
                  | | |
                  OH   OH   CH$_3$

153.

$$\left(OCH_2CH_2OCNH-\bigcirc-NHCOCH_2CH_2OCNH-\bigcirc-NHC\right)_n$$

155. a. The temperature of the rubber band increases when it is stretched; **b.** exothermic (heat is released); **c.** As the chains are stretched, they line up more closely resulting in stronger London dispersion forces between the chains. Heat is released as the strength of the intermolecular forces increases. **d.** ΔG is positive and ΔS is negative; **e.** The structure of the stretched polymer chains is more ordered (has a smaller positional probability) than in unstretched rubber. Therefore, entropy decreases as the rubber band is stretched.   **157.** 0.11%   **159. a.** 37.50%; **b.** The hybridization changes from sp^2 to sp^3; **c.** 3,4-dimethyl-3-hexanol

Index